THE WILEY BICENTENNIAL—KNOWLEDGE FOR GENERATIONS

Each generation has its unique needs and aspirations. When Charles Wiley first opened his small printing shop in lower Manhattan in 1807, it was a generation of boundless potential searching for an identity. And we were there, helping to define a new American literary tradition. Over half a century later, in the midst of the Second Industrial Revolution, it was a generation focused on building the future. Once again, we were there, supplying the critical scientific, technical, and engineering knowledge that helped frame the world. Throughout the 20th Century, and into the new millennium, nations began to reach out beyond their own borders and a new international community was born. Wiley was there, expanding its operations around the world to enable a global exchange of ideas, opinions, and know-how.

For 200 years, Wiley has been an integral part of each generation's journey, enabling the flow of information and understanding necessary to meet their needs and fulfill their aspirations. Today, bold new technologies are changing the way we live and learn. Wiley will be there, providing you the must-have knowledge you need to imagine new worlds, new possibilities, and new opportunities.

Generations come and go, but you can always count on Wiley to provide you the knowledge you need, when and where you need it!

WILLIAM J. PESCE
PRESIDENT AND CHIEF EXECUTIVE OFFICER

PETER BOOTH WILEY
CHAIRMAN OF THE BOARD

INSTRUCTOR'S SOLUTIONS MANUAL, VOLUME 1

Sen-Ben Liao
Lawrence Livermore National Laboratory

to accompany

FUNDAMENTALS OF PHYSICS

EIGHTH EDITION

David Halliday
University of Pittsburgh

Robert Resnick
Rensselaer Polytechnic Institute

Jearl Walker
Cleveland State University

Copyright § 2008 John Wiley & Sons, Inc. All rights reserved.

Excerpts from this work may be reproduced by instructors for distribution on a not-for-profit basis for testing or instructional purposes only to students enrolled in courses for which the textbook has been adopted. *Any other reproduction or translation of this work beyond that permitted by Sections 107 or 108 of the 1976 United States Copyright Act without the permission of the copyright owner is unlawful. Requests for permission or further information should be addressed to the Permissions Department, John Wiley & Sons, Inc., 111 River Street, Hoboken, NJ 07030-5774.*

To order books or for customer service, please call 1-800-CALL-WILEY (225-5945).

ISBN-10 978-0-470-11404-9

Printed in the United States of America.

10 9 8 7 6 5 4 3 2 1

Printed and bound by BindRite Graphics

Preface

This book includes the solutions to the PROBLEMS sections of the 8th edition of **Fundamentals of Physics** by Halliday, Resnick and Walker. We have not included solutions or discussions that pertain to the Questions sections. These solutions have been typed using Microsoft Word and MathType™ equation editor. The solution files are available on the Instructor's Companion Website (www.wiley.com/college/halliday). Additional information regarding MathType™ can be found at www.mathtype.com

The author has put great time and effort into writing high quality solutions. He welcomes comments and suggestions from readers, please also report any errors that you may find. The author's email address is: senben_liao@yahoo.com

Note to adopters regarding the Instructor's Solutions Manual:

Thank you for adopting **Fundamentals of Physics** 8e by Halliday, Resnick and Walker! We are pleased to provide you with a variety of support material to help you in the teaching of your course.

Please note that all of this material is copyrighted by John Wiley & Sons, Inc. and is explicitly intended for use only at your institution.

Our providing this material does not carry with it permission to distribute it beyond your institution. Before putting any of this material on a website, we ask that you request formal permission from us to do so. Please write to: Permissions Department, John Wiley & Sons, Inc. 111 River St. Hoboken, NJ 07030, or fill out a simple on-line form on our website www.wiley.com/about/permissions. In most cases we will grant such permission PROVIDED THAT THE WEBSITE IS PASSWORD PROTECTED.

Our goal is to prevent students from other campuses from being able to access your materials. We trust that you can understand how that might undermine the efforts of your colleagues at other institutions.

Table of Contents

Chapter 1	1
Chapter 2	21
Chapter 3	77
Chapter 4	111
Chapter 5	175
Chapter 6	221
Chapter 7	277
Chapter 8	307
Chapter 9	375
Chapter 10	437
Chapter 11	481
Chapter 12	531
Chapter 13	573
Chapter 14	617
Chapter 15	645
Chapter 16	691
Chapter 17	733
Chapter 18	771
Chapter 19	805
Chapter 20	843

Chapter 1

1. The metric prefixes (micro, pico, nano, ...) are given for ready reference on the inside front cover of the textbook (see also Table 1–2).

(a) Since 1 km = 1×10^3 m and 1 m = 1×10^6 μm,

$$1\,\text{km} = 10^3\,\text{m} = (10^3\,\text{m})(10^6\,\mu\text{m/m}) = 10^9\,\mu\text{m}.$$

The given measurement is 1.0 km (two significant figures), which implies our result should be written as 1.0×10^9 μm.

(b) We calculate the number of microns in 1 centimeter. Since 1 cm = 10^{-2} m,

$$1\,\text{cm} = 10^{-2}\,\text{m} = (10^{-2}\,\text{m})(10^6\,\mu\text{m/m}) = 10^4\,\mu\text{m}.$$

We conclude that the fraction of one centimeter equal to 1.0 μm is 1.0×10^{-4}.

(c) Since 1 yd = (3 ft)(0.3048 m/ft) = 0.9144 m,

$$1.0\,\text{yd} = (0.91\,\text{m})(10^6\,\mu\text{m/m}) = 9.1 \times 10^5\,\mu\text{m}.$$

2. (a) Using the conversion factors 1 inch = 2.54 cm exactly and 6 picas = 1 inch, we obtain

$$0.80\,\text{cm} = (0.80\,\text{cm})\left(\frac{1\,\text{inch}}{2.54\,\text{cm}}\right)\left(\frac{6\,\text{picas}}{1\,\text{inch}}\right) \approx 1.9\,\text{picas}.$$

(b) With 12 points = 1 pica, we have

$$0.80\,\text{cm} = (0.80\,\text{cm})\left(\frac{1\,\text{inch}}{2.54\,\text{cm}}\right)\left(\frac{6\,\text{picas}}{1\,\text{inch}}\right)\left(\frac{12\,\text{points}}{1\,\text{pica}}\right) \approx 23\,\text{points}.$$

3. Using the given conversion factors, we find

(a) the distance d in *rods* to be

$$d = 4.0\,\text{furlongs} = \frac{(4.0\,\text{furlongs})(201.168\,\text{m/furlong})}{5.0292\,\text{m/rod}} = 160\,\text{rods},$$

(b) and that distance *in chains* to be

$$d = \frac{(4.0 \text{ furlongs})(201.168 \text{ m/furlong})}{20.117 \text{ m/chain}} = 40 \text{ chains.}$$

4. The conversion factors 1 gry =1/10 line, 1 line=1/12 inch and 1 point = 1/72 inch imply that

$$1 \text{ gry} = (1/10)(1/12)(72 \text{ points}) = 0.60 \text{ point.}$$

Thus, $1 \text{ gry}^2 = (0.60 \text{ point})^2 = 0.36 \text{ point}^2$, which means that $0.50 \text{ gry}^2 = 0.18 \text{ point}^2$.

5. Various geometric formulas are given in Appendix E.

(a) Expressing the radius of the Earth as

$$R = (6.37 \times 10^6 \text{ m})(10^{-3} \text{ km/m}) = 6.37 \times 10^3 \text{ km,}$$

its circumference is $s = 2\pi R = 2\pi(6.37 \times 10^3 \text{ km}) = 4.00 \times 10^4$ km.

(b) The surface area of Earth is $A = 4\pi R^2 = 4\pi (6.37 \times 10^3 \text{ km})^2 = 5.10 \times 10^8 \text{ km}^2$.

(c) The volume of Earth is $V = \frac{4\pi}{3} R^3 = \frac{4\pi}{3} (6.37 \times 10^3 \text{ km})^3 = 1.08 \times 10^{12} \text{ km}^3$.

6. From Figure 1.6, we see that 212 S is equivalent to 258 W and 212 – 32 = 180 S is equivalent to 216 – 60 = 156 Z. The information allows us to convert S to W or Z.

(a) In units of W, we have

$$50.0 \text{ S} = (50.0 \text{ S})\left(\frac{258 \text{ W}}{212 \text{ S}}\right) = 60.8 \text{ W}$$

(b) In units of Z, we have

$$50.0 \text{ S} = (50.0 \text{ S})\left(\frac{156 \text{ Z}}{180 \text{ S}}\right) = 43.3 \text{ Z}$$

7. The volume of ice is given by the product of the semicircular surface area and the thickness. The area of the semicircle is $A = \pi r^2/2$, where r is the radius. Therefore, the volume is

$$V = \frac{\pi}{2} r^2 z$$

where z is the ice thickness. Since there are 10^3 m in 1 km and 10^2 cm in 1 m, we have

$$r = (2000 \, \text{km}) \left(\frac{10^3 \, \text{m}}{1 \, \text{km}} \right) \left(\frac{10^2 \, \text{cm}}{1 \, \text{m}} \right) = 2000 \times 10^5 \, \text{cm}.$$

In these units, the thickness becomes

$$z = 3000 \, \text{m} = (3000 \, \text{m}) \left(\frac{10^2 \, \text{cm}}{1 \, \text{m}} \right) = 3000 \times 10^2 \, \text{cm}$$

which yields $V = \frac{\pi}{2} \left(2000 \times 10^5 \, \text{cm} \right)^2 \left(3000 \times 10^2 \, \text{cm} \right) = 1.9 \times 10^{22} \, \text{cm}^3$.

8. We make use of Table 1-6.

(a) We look at the first ("cahiz") column: 1 fanega is equivalent to what amount of cahiz? We note from the already completed part of the table that 1 cahiz equals a dozen fanega. Thus, 1 fanega = $\frac{1}{12}$ cahiz, or 8.33×10^{-2} cahiz. Similarly, "1 cahiz = 48 cuartilla" (in the already completed part) implies that 1 cuartilla = $\frac{1}{48}$ cahiz, or 2.08×10^{-2} cahiz. Continuing in this way, the remaining entries in the first column are 6.94×10^{-3} and 3.47×10^{-3}.

(b) In the second ("fanega") column, we similarly find 0.250, 8.33×10^{-2}, and 4.17×10^{-2} for the last three entries.

(c) In the third ("cuartilla") column, we obtain 0.333 and 0.167 for the last two entries.

(d) Finally, in the fourth ("almude") column, we get $\frac{1}{2} = 0.500$ for the last entry.

(e) Since the conversion table indicates that 1 almude is equivalent to 2 medios, our amount of 7.00 almudes must be equal to 14.0 medios.

(f) Using the value (1 almude = 6.94×10^{-3} cahiz) found in part (a), we conclude that 7.00 almudes is equivalent to 4.86×10^{-2} cahiz.

(g) Since each decimeter is 0.1 meter, then 55.501 cubic decimeters is equal to 0.055501 m^3 or 55501 cm^3. Thus, 7.00 almudes = $\frac{7.00}{12}$ fanega = $\frac{7.00}{12}$ (55501 cm^3) = 3.24×10^4 cm^3.

9. We use the conversion factors found in Appendix D.

$$1 \text{ acre} \cdot \text{ft} = (43{,}560 \text{ ft}^2) \cdot \text{ft} = 43{,}560 \text{ ft}^3$$

Since 2 in. = (1/6) ft, the volume of water that fell during the storm is

$$V = (26 \text{ km}^2)(1/6 \text{ ft}) = (26 \text{ km}^2)(3281 \text{ft/km})^2(1/6 \text{ ft}) = 4.66 \times 10^7 \text{ ft}^3.$$

Thus,

$$V = \frac{4.66 \times 10^7 \text{ ft}^3}{4.3560 \times 10^4 \text{ ft}^3/\text{acre} \cdot \text{ft}} = 1.1 \times 10^3 \text{ acre} \cdot \text{ft}.$$

10. A day is equivalent to 86400 seconds and a meter is equivalent to a million micrometers, so

$$\frac{(3.7 \text{ m})(10^6 \,\mu\text{m/m})}{(14 \text{ day})(86400 \text{ s/day})} = 3.1 \,\mu\text{m/s}.$$

11. A week is 7 days, each of which has 24 hours, and an hour is equivalent to 3600 seconds. Thus, two weeks (a fortnight) is 1209600 s. By definition of the micro prefix, this is roughly $1.21 \times 10^{12} \,\mu\text{s}$.

12. The metric prefixes (micro (μ), pico, nano, ...) are given for ready reference on the inside front cover of the textbook (also, Table 1–2).

(a) $1 \,\mu\text{century} = (10^{-6} \text{ century})\left(\dfrac{100 \text{ y}}{1 \text{ century}}\right)\left(\dfrac{365 \text{ day}}{1 \text{ y}}\right)\left(\dfrac{24 \text{ h}}{1 \text{ day}}\right)\left(\dfrac{60 \text{ min}}{1 \text{ h}}\right) = 52.6 \text{ min}.$

(b) The percent difference is therefore

$$\frac{52.6 \text{ min} - 50 \text{ min}}{52.6 \text{ min}} = 4.9\%.$$

13. (a) Presuming that a French decimal day is equivalent to a regular day, then the ratio of weeks is simply 10/7 or (to 3 significant figures) 1.43.

(b) In a regular day, there are 86400 seconds, but in the French system described in the problem, there would be 10^5 seconds. The ratio is therefore 0.864.

14. We denote the pulsar rotation rate f (for frequency).

$$f = \frac{1 \text{ rotation}}{1.55780644887275 \times 10^{-3} \text{ s}}$$

(a) Multiplying f by the time-interval $t = 7.00$ days (which is equivalent to 604800 s, if we ignore *significant figure* considerations for a moment), we obtain the number of rotations:

$$N = \left(\frac{1 \text{ rotation}}{1.55780644887275 \times 10^{-3} \text{ s}}\right)(604800 \text{ s}) = 388238218.4$$

which should now be rounded to 3.88×10^8 rotations since the time-interval was specified in the problem to three significant figures.

(b) We note that the problem specifies the *exact* number of pulsar revolutions (one million). In this case, our unknown is t, and an equation similar to the one we set up in part (a) takes the form $N = ft$, or

$$1 \times 10^6 = \left(\frac{1 \text{ rotation}}{1.55780644887275 \times 10^{-3} \text{ s}}\right)t$$

which yields the result $t = 1557.80644887275$ s (though students who do this calculation on their calculator might not obtain those last several digits).

(c) Careful reading of the problem shows that the time-uncertainty *per revolution* is $\pm 3 \times 10^{-17}$ s. We therefore expect that as a result of one million revolutions, the uncertainty should be $(\pm 3 \times 10^{-17})(1 \times 10^6) = \pm 3 \times 10^{-11}$ s.

15. The time on any of these clocks is a straight-line function of that on another, with slopes $\neq 1$ and y-intercepts $\neq 0$. From the data in the figure we deduce

$$t_C = \frac{2}{7}t_B + \frac{594}{7}, \quad t_B = \frac{33}{40}t_A - \frac{662}{5}.$$

These are used in obtaining the following results.

(a) We find

$$t'_B - t_B = \frac{33}{40}(t'_A - t_A) = 495 \text{ s}$$

when $t'_A - t_A = 600$ s.

(b) We obtain $t'_C - t_C = \frac{2}{7}(t'_B - t_B) = \frac{2}{7}(495) = 141$ s.

(c) Clock B reads $t_B = (33/40)(400) - (662/5) \approx 198$ s when clock A reads $t_A = 400$ s.

(d) From $t_C = 15 = (2/7)t_B + (594/7)$, we get $t_B \approx -245$ s.

16. Since a change of longitude equal to 360° corresponds to a 24 hour change, then one expects to change longitude by $360°/24 = 15°$ before resetting one's watch by 1.0 h.

17. None of the clocks advance by exactly 24 h in a 24-h period but this is not the most important criterion for judging their quality for measuring time intervals. What is important is that the clock advance by the same amount in each 24-h period. The clock reading can then easily be adjusted to give the correct interval. If the clock reading jumps around from one 24-h period to another, it cannot be corrected since it would impossible to tell what the correction should be. The following gives the corrections (in seconds) that must be applied to the reading on each clock for each 24-h period. The entries were determined by subtracting the clock reading at the end of the interval from the clock reading at the beginning.

CLOCK	Sun.-Mon.	Mon.-Tues.	Tues.-Wed.	Wed.-Thurs.	Thurs.-Fri.	Fri.-Sat.
A	−16	−16	−15	−17	−15	−15
B	−3	+5	−10	+5	+6	−7
C	−58	−58	−58	−58	−58	−58
D	+67	+67	+67	+67	+67	+67
E	+70	+55	+2	+20	+10	+10

Clocks C and D are both good timekeepers in the sense that each is consistent in its daily drift (relative to WWF time); thus, C and D are easily made "perfect" with simple and predictable corrections. The correction for clock C is less than the correction for clock D, so we judge clock C to be the best and clock D to be the next best. The correction that must be applied to clock A is in the range from 15 s to 17s. For clock B it is the range from -5 s to +10 s, for clock E it is in the range from -70 s to -2 s. After C and D, A has the smallest range of correction, B has the next smallest range, and E has the greatest range. From best to worst, the ranking of the clocks is C, D, A, B, E.

18. The last day of the 20 centuries is longer than the first day by

$$(20 \text{ century})(0.001 \text{ s/century}) = 0.02 \text{ s}.$$

The average day during the 20 centuries is $(0 + 0.02)/2 = 0.01$ s longer than the first day. Since the increase occurs uniformly, the cumulative effect T is

$$T = (\text{average increase in length of a day})(\text{number of days})$$
$$= \left(\frac{0.01 \text{ s}}{\text{day}}\right)\left(\frac{365.25 \text{ day}}{\text{y}}\right)(2000 \text{ y})$$
$$= 7305 \text{ s}$$

or roughly two hours.

19. When the Sun first disappears while lying down, your line of sight to the top of the Sun is tangent to the Earth's surface at point A shown in the figure. As you stand, elevating your eyes by a height h, the line of sight to the Sun is tangent to the Earth's surface at point B.

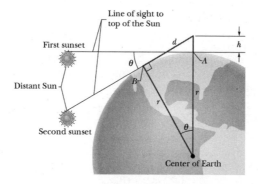

Let d be the distance from point B to your eyes. From Pythagorean theorem, we have

$$d^2 + r^2 = (r+h)^2 = r^2 + 2rh + h^2$$

or $d^2 = 2rh + h^2$, where r is the radius of the Earth. Since $r \gg h$, the second term can be dropped, leading to $d^2 \approx 2rh$. Now the angle between the two radii to the two tangent points A and B is θ, which is also the angle through which the Sun moves about Earth during the time interval $t = 11.1$ s. The value of θ can be obtained by using

$$\frac{\theta}{360°} = \frac{t}{24 \text{ h}}.$$

This yields

$$\theta = \frac{(360°)(11.1 \text{ s})}{(24 \text{ h})(60 \text{ min/h})(60 \text{ s/min})} = 0.04625°.$$

Using $d = r \tan \theta$, we have $d^2 = r^2 \tan^2 \theta = 2rh$, or

$$r = \frac{2h}{\tan^2 \theta}.$$

Using the above value for θ and $h = 1.7$ m, we have $r = 5.2 \times 10^6$ m.

20. The density of gold is

$$\rho = \frac{m}{V} = \frac{19.32 \text{ g}}{1 \text{ cm}^3} = 19.32 \text{ g/cm}^3.$$

(a) We take the volume of the leaf to be its area A multiplied by its thickness z. With density $\rho = 19.32$ g/cm^3 and mass $m = 27.63$ g, the volume of the leaf is found to be

$$V = \frac{m}{\rho} = 1.430 \text{ cm}^3.$$

We convert the volume to SI units:

$$V = \left(1.430 \text{ cm}^3\right) \left(\frac{1 \text{ m}}{100 \text{ cm}}\right)^3 = 1.430 \times 10^{-6} \text{ m}^3.$$

Since $V = Az$ with $z = 1 \times 10^{-6}$ m (metric prefixes can be found in Table 1–2), we obtain

$$A = \frac{1.430 \times 10^{-6} \text{ m}^3}{1 \times 10^{-6} \text{ m}} = 1.430 \text{ m}^2.$$

(b) The volume of a cylinder of length ℓ is $V = A\ell$ where the cross-section area is that of a circle: $A = \pi r^2$. Therefore, with $r = 2.500 \times 10^{-6}$ m and $V = 1.430 \times 10^{-6}$ m^3, we obtain

$$\ell = \frac{V}{\pi r^2} = 7.284 \times 10^4 \text{ m} = 72.84 \text{ km}.$$

21. We introduce the notion of density:

$$\rho = \frac{m}{V}$$

and convert to SI units: 1 g = 1×10^{-3} kg.

(a) For volume conversion, we find 1 cm^3 = $(1 \times 10^{-2}$m$)^3$ = 1×10^{-6}m^3. Thus, the density in kg/m^3 is

$$1 \text{ g/cm}^3 = \left(\frac{1 \text{ g}}{\text{cm}^3}\right)\left(\frac{10^{-3} \text{ kg}}{\text{g}}\right)\left(\frac{\text{cm}^3}{10^{-6} \text{ m}^3}\right) = 1 \times 10^3 \text{ kg/m}^3.$$

Thus, the mass of a cubic meter of water is 1000 kg.

(b) We divide the mass of the water by the time taken to drain it. The mass is found from $M = \rho V$ (the product of the volume of water and its density):

$$M = \left(5700 \text{ m}^3\right)\left(1 \times 10^3 \text{ kg/m}^3\right) = 5.70 \times 10^6 \text{ kg}.$$

The time is $t = (10\text{h})(3600 \text{ s/h}) = 3.6 \times 10^4$ s, so the *mass flow rate R* is

$$R = \frac{M}{t} = \frac{5.70 \times 10^6 \text{ kg}}{3.6 \times 10^4 \text{ s}} = 158 \text{ kg/s}.$$

22. (a) We find the volume in cubic centimeters

$$193 \text{ gal} = (193 \text{ gal})\left(\frac{231 \text{ in}^3}{1 \text{ gal}}\right)\left(\frac{2.54 \text{ cm}}{1 \text{ in}}\right)^3 = 7.31 \times 10^5 \text{ cm}^3$$

and subtract this from 1×10^6 cm^3 to obtain 2.69×10^5 cm^3. The conversion gal $\rightarrow$ in^3 is given in Appendix D (immediately below the table of Volume conversions).

(b) The volume found in part (a) is converted (by dividing by $(100 \text{ cm/m})^3$) to 0.731 m^3, which corresponds to a mass of

$$(1000 \text{ kg/m}^3)(0.731 \text{ m}^2) = 731 \text{ kg}$$

using the density given in the problem statement. At a rate of 0.0018 kg/min, this can be filled in

$$\frac{731 \text{ kg}}{0.0018 \text{ kg/min}} = 4.06 \times 10^5 \text{ min} = 0.77 \text{ y}$$

after dividing by the number of minutes in a year (365 days)(24 h/day) (60 min/h).

23. If M_E is the mass of Earth, m is the average mass of an atom in Earth, and N is the number of atoms, then $M_E = Nm$ or $N = M_E/m$. We convert mass m to kilograms using Appendix D (1 u = 1.661×10^{-27} kg). Thus,

$$N = \frac{M_E}{m} = \frac{5.98 \times 10^{24} \text{ kg}}{(40 \text{ u})(1.661 \times 10^{-27} \text{ kg/u})} = 9.0 \times 10^{49}.$$

24. (a) The volume of the cloud is $(3000 \text{ m})\pi(1000 \text{ m})^2 = 9.4 \times 10^9$ m^3. Since each cubic meter of the cloud contains from 50×10^6 to 500×10^6 water drops, then we conclude that the entire cloud contains from 4.7×10^{18} to 4.7×10^{19} drops. Since the volume of each drop is $\frac{4}{3}\pi(10 \times 10^{-6} \text{ m})^3 = 4.2 \times 10^{-15}$ m^3, then the total volume of water in a cloud is from 2×10^3 to 2×10^4 m^3.

(b) Using the fact that $1 \text{ L} = 1 \times 10^3 \text{ cm}^3 = 1 \times 10^{-3} \text{ m}^3$, the amount of water estimated in part (a) would fill from 2×10^6 to 2×10^7 bottles.

(c) At 1000 kg for every cubic meter, the mass of water is from two million to twenty million kilograms. The coincidence in numbers between the results of parts (b) and (c) of this problem is due to the fact that each liter has a mass of one kilogram when water is at its normal density (under standard conditions).

25. We introduce the notion of density, $\rho = m/V$, and convert to SI units: 1000 g = 1 kg, and 100 cm = 1 m.

(a) The density ρ of a sample of iron is

$$\rho = \left(7.87 \text{ g/cm}^3\right)\left(\frac{1 \text{ kg}}{1000 \text{ g}}\right)\left(\frac{100 \text{ cm}}{1 \text{ m}}\right)^3 = 7870 \text{ kg/m}^3.$$

If we ignore the empty spaces between the close-packed spheres, then the density of an individual iron atom will be the same as the density of any iron sample. That is, if M is the mass and V is the volume of an atom, then

$$V = \frac{M}{\rho} = \frac{9.27 \times 10^{-26} \text{ kg}}{7.87 \times 10^3 \text{ kg/m}^3} = 1.18 \times 10^{-29} \text{ m}^3.$$

(b) We set $V = 4\pi R^3/3$, where R is the radius of an atom (Appendix E contains several geometry formulas). Solving for R, we find

$$R = \left(\frac{3V}{4\pi}\right)^{1/3} = \left(\frac{3(1.18 \times 10^{-29} \text{ m}^3)}{4\pi}\right)^{1/3} = 1.41 \times 10^{-10} \text{ m}.$$

The center-to-center distance between atoms is twice the radius, or 2.82×10^{-10} m.

26. If we estimate the "typical" large domestic cat mass as 10 kg, and the "typical" atom (in the cat) as 10 u ≈ 2×10^{-26} kg, then there are roughly (10 kg)/(2×10^{-26} kg) ≈ 5×10^{26} atoms. This is close to being a factor of a thousand greater than Avogradro's number. Thus this is roughly a kilomole of atoms.

27. According to Appendix D, a nautical mile is 1.852 km, so 24.5 nautical miles would be 45.374 km. Also, according to Appendix D, a mile is 1.609 km, so 24.5 miles is 39.4205 km. The difference is 5.95 km.

28. The metric prefixes (micro (μ), pico, nano, ...) are given for ready reference on the inside front cover of the textbook (see also Table 1–2). The surface area A of each grain of sand of radius $r = 50$ μm $= 50 \times 10^{-6}$ m is given by $A = 4\pi(50 \times 10^{-6})^2 = 3.14 \times 10^{-8}$ m^2 (Appendix E contains a variety of geometry formulas). We introduce the notion of density, $\rho = m/V$, so that the mass can be found from $m = \rho V$, where $\rho = 2600$ kg/m^3. Thus, using $V = 4\pi r^3/3$, the mass of each grain is

$$m = \rho V = \rho\left(\frac{4\pi r^3}{3}\right) = \left(2600\ \frac{\text{kg}}{\text{m}^3}\right)\frac{4\pi\left(50\times 10^{-6}\ \text{m}\right)^3}{3} = 1.36\times 10^{-9}\ \text{kg}.$$

We observe that (because a cube has six equal faces) the indicated surface area is 6 m². The number of spheres (the grains of sand) N that have a total surface area of 6 m² is given by

$$N = \frac{6\ \text{m}^2}{3.14\times 10^{-8}\ \text{m}^2} = 1.91\times 10^8.$$

Therefore, the total mass M is $M = Nm = (1.91\times 10^8)(1.36\times 10^{-9}\ \text{kg}) = 0.260$ kg.

29. The volume of the section is $(2500\ \text{m})(800\ \text{m})(2.0\ \text{m}) = 4.0\times 10^6\ \text{m}^3$. Letting "$d$" stand for the thickness of the mud after it has (uniformly) distributed in the valley, then its volume there would be $(400\ \text{m})(400\ \text{m})d$. Requiring these two volumes to be equal, we can solve for d. Thus, $d = 25$ m. The volume of a small part of the mud over a patch of area of 4.0 m² is $(4.0)d = 100\ \text{m}^3$. Since each cubic meter corresponds to a mass of 1900 kg (stated in the problem), then the mass of that small part of the mud is 1.9×10^5 kg.

30. To solve the problem, we note that the first derivative of the function with respect to time gives the rate. Setting the rate to zero gives the time at which an extreme value of the variable mass occurs; here that extreme value is a maximum.

(a) Differentiating $m(t) = 5.00t^{0.8} - 3.00t + 20.00$ with respect to t gives

$$\frac{dm}{dt} = 4.00t^{-0.2} - 3.00.$$

The water mass is the greatest when $dm/dt = 0$, or at $t = (4.00/3.00)^{1/0.2} = 4.21$ s.

(b) At $t = 4.21$ s, the water mass is

$$m(t = 4.21\ \text{s}) = 5.00(4.21)^{0.8} - 3.00(4.21) + 20.00 = 23.2\ \text{g}.$$

(c) The rate of mass change at $t = 2.00$ s is

$$\left.\frac{dm}{dt}\right|_{t=2.00\ \text{s}} = \left[4.00(2.00)^{-0.2} - 3.00\right]\text{g/s} = 0.48\ \text{g/s} = 0.48\frac{\text{g}}{\text{s}}\cdot\frac{1\ \text{kg}}{1000\ \text{g}}\cdot\frac{60\ \text{s}}{1\ \text{min}}$$

$$= 2.89\times 10^{-2}\ \text{kg/min}.$$

(d) Similarly, the rate of mass change at $t = 5.00$ s is

$$\left.\frac{dm}{dt}\right|_{t=2.00\,s} = \left[4.00(5.00)^{-0.2} - 3.00\right] \text{g/s} = -0.101 \text{ g/s} = -0.101 \frac{\text{g}}{\text{s}} \cdot \frac{1 \text{ kg}}{1000 \text{ g}} \cdot \frac{60 \text{ s}}{1 \text{ min}}$$
$$= -6.05 \times 10^{-3} \text{ kg/min}.$$

31. The mass density of the candy is

$$\rho = \frac{m}{V} = \frac{0.0200 \text{ g}}{50.0 \text{ mm}^3} = 4.00 \times 10^{-4} \text{ g/mm}^3 = 4.00 \times 10^{-4} \text{ kg/cm}^3.$$

If we neglect the volume of the empty spaces between the candies, then the total mass of the candies in the container when filled to height h is $M = \rho A h$, where $A = (14.0 \text{ cm})(17.0 \text{ cm}) = 238 \text{ cm}^2$ is the base area of the container that remains unchanged. Thus, the rate of mass change is given by

$$\frac{dM}{dt} = \frac{d(\rho A h)}{dt} = \rho A \frac{dh}{dt} = (4.00 \times 10^{-4} \text{ kg/cm}^3)(238 \text{ cm}^2)(0.250 \text{ cm/s})$$
$$= 0.0238 \text{ kg/s} = 1.43 \text{ kg/min}.$$

32. Table 7 can be completed as follows:

(a) It should be clear that the first column (under "wey") is the reciprocal of the first row – so that $\frac{9}{10} = 0.900$, $\frac{3}{40} = 7.50 \times 10^{-2}$, and so forth. Thus, 1 pottle = 1.56×10^{-3} wey and 1 gill = 8.32×10^{-6} wey are the last two entries in the first column.

(b) In the second column (under "chaldron"), clearly we have 1 chaldron = 1 caldron (that is, the entries along the "diagonal" in the table must be 1's). To find out how many chaldron are equal to one bag, we note that 1 wey = 10/9 chaldron = 40/3 bag so that $\frac{1}{12}$ chaldron = 1 bag. Thus, the next entry in that second column is $\frac{1}{12} = 8.33 \times 10^{-2}$. Similarly, 1 pottle = 1.74×10^{-3} chaldron and 1 gill = 9.24×10^{-6} chaldron.

(c) In the third column (under "bag"), we have 1 chaldron = 12.0 bag, 1 bag = 1 bag, 1 pottle = 2.08×10^{-2} bag, and 1 gill = 1.11×10^{-4} bag.

(d) In the fourth column (under "pottle"), we find 1 chaldron = 576 pottle, 1 bag = 48 pottle, 1 pottle = 1 pottle, and 1 gill = 5.32×10^{-3} pottle.

(e) In the last column (under "gill"), we obtain 1 chaldron = 1.08×10^5 gill, 1 bag = 9.02×10^3 gill, 1 pottle = 188 gill, and, of course, 1 gill = 1 gill.

(f) Using the information from part (c), 1.5 chaldron = (1.5)(12.0) = 18.0 bag. And since each bag is 0.1091 m^3 we conclude 1.5 chaldron = (18.0)(0.1091) = 1.96 m^3.

33. The first two conversions are easy enough that a *formal* conversion is not especially called for, but in the interest of *practice makes perfect* we go ahead and proceed formally:

(a) $11 \text{ tuffets} = (11 \text{ tuffets})\left(\dfrac{2 \text{ peck}}{1 \text{ tuffet}}\right) = 22 \text{ pecks}.$

(b) $11 \text{ tuffets} = (11 \text{ tuffets})\left(\dfrac{0.50 \text{ Imperial bushel}}{1 \text{ tuffet}}\right) = 5.5 \text{ Imperial bushels}.$

(c) $11 \text{ tuffets} = (5.5 \text{ Imperial bushel})\left(\dfrac{36.3687 \text{ L}}{1 \text{ Imperial bushel}}\right) \approx 200 \text{ L}.$

34. (a) Using the fact that the area A of a rectangle is (width) × (length), we find

$$A_{\text{total}} = (3.00 \text{ acre}) + (25.0 \text{ perch})(4.00 \text{ perch})$$
$$= (3.00 \text{ acre})\left(\dfrac{(40 \text{ perch})(4 \text{ perch})}{1 \text{ acre}}\right) + 100 \text{ perch}^2$$
$$= 580 \text{ perch}^2.$$

We multiply this by the perch^2 → rood conversion factor (1 rood/40 perch^2) to obtain the answer: $A_{\text{total}} = 14.5$ roods.

(b) We convert our intermediate result in part (a):

$$A_{\text{total}} = (580 \text{ perch}^2)\left(\dfrac{16.5 \text{ ft}}{1 \text{ perch}}\right)^2 = 1.58 \times 10^5 \text{ ft}^2.$$

Now, we use the feet → meters conversion given in Appendix D to obtain

$$A_{\text{total}} = (1.58 \times 10^5 \text{ ft}^2)\left(\dfrac{1 \text{ m}}{3.281 \text{ ft}}\right)^2 = 1.47 \times 10^4 \text{ m}^2.$$

35. (a) Dividing 750 miles by the expected "40 miles per gallon" leads the tourist to believe that the car should need 18.8 gallons (in the U.S.) for the trip.

(b) Dividing the two numbers given (to high precision) in the problem (and rounding off) gives the conversion between U.K. and U.S. gallons. The U.K. gallon is larger than the U.S gallon by a factor of 1.2. Applying this to the result of part (a), we find the answer for part (b) is 22.5 gallons.

36. The customer expects a volume $V_1 = 20 \times 7056$ in^3 and receives $V_2 = 20 \times 5826$ in^3, the difference being $\Delta V = V_1 - V_2 = 24600$ in^3, or

$$\Delta V = (24600 \text{ in}^3)\left(\frac{2.54 \text{ cm}}{1 \text{ inch}}\right)^3 \left(\frac{1 \text{ L}}{1000 \text{ cm}^3}\right) = 403 \text{ L}$$

where Appendix D has been used.

37. (a) Using Appendix D, we have 1 ft = 0.3048 m, 1 gal = 231 in.^{3}, and 1 in.^{3} = 1.639 × 10^{-2} L. From the latter two items, we find that 1 gal = 3.79 L. Thus, the quantity 460 ft^2/gal becomes

$$460 \text{ ft}^2/\text{gal} = \left(\frac{460 \text{ ft}^2}{\text{gal}}\right)\left(\frac{1 \text{ m}}{3.28 \text{ ft}}\right)^2 \left(\frac{1 \text{ gal}}{3.79 \text{ L}}\right) = 11.3 \text{ m}^2/\text{L}.$$

(b) Also, since 1 m^3 is equivalent to 1000 L, our result from part (a) becomes

$$11.3 \text{ m}^2/\text{L} = \left(\frac{11.3 \text{ m}^2}{\text{L}}\right)\left(\frac{1000 \text{ L}}{1 \text{ m}^3}\right) = 1.13 \times 10^4 \text{ m}^{-1}.$$

(c) The inverse of the original quantity is $(460 \text{ ft}^2/\text{gal})^{-1} = 2.17 \times 10^{-3}$ gal/ft^2.

(d) The answer in (c) represents the volume of the paint (in gallons) needed to cover a square foot of area. From this, we could also figure the paint thickness [it turns out to be about a tenth of a millimeter, as one sees by taking the reciprocal of the answer in part (b)].

38. The total volume V of the real house is that of a triangular prism (of height $h = 3.0$ m and base area $A = 20 \times 12 = 240$ m^2) in addition to a rectangular box (height $h' = 6.0$ m and same base). Therefore,

$$V = \frac{1}{2}hA + h'A = \left(\frac{h}{2} + h'\right)A = 1800 \text{ m}^3.$$

(a) Each dimension is reduced by a factor of 1/12, and we find

$$V_{\text{doll}} = (1800 \text{ m}^3)\left(\frac{1}{12}\right)^3 \approx 1.0 \text{ m}^3.$$

(b) In this case, each dimension (relative to the real house) is reduced by a factor of 1/144. Therefore,

$$V_{\text{miniature}} = (1800 \text{ m}^3)\left(\frac{1}{144}\right)^3 \approx 6.0 \times 10^{-4} \text{ m}^3.$$

39. Using the (exact) conversion 2.54 cm = 1 in. we find that 1 ft = (12)(2.54)/100 = 0.3048 m (which also can be found in Appendix D). The volume of a cord of wood is 8 × 4 × 4 = 128 ft^3, which we convert (multiplying by 0.3048^3) to 3.6 m^3. Therefore, one cubic meter of wood corresponds to 1/3.6 ≈ 0.3 cord.

40. (a) In atomic mass units, the mass of one molecule is (16 + 1 + 1)u = 18 u. Using Eq. 1–9, we find

$$18u = (18u)\left(\frac{1.6605402 \times 10^{-27} \text{ kg}}{1u}\right) = 3.0 \times 10^{-26} \text{ kg}.$$

(b) We divide the total mass by the mass of each molecule and obtain the (approximate) number of water molecules:

$$N \approx \frac{1.4 \times 10^{21}}{3.0 \times 10^{-26}} \approx 5 \times 10^{46}.$$

41. (a) The difference between the total amounts in "freight" and "displacement" tons, (8 − 7)(73) = 73 barrels bulk, represents the extra M&M's that are shipped. Using the conversions in the problem, this is equivalent to (73)(0.1415)(28.378) = 293 U.S. bushels.

(b) The difference between the total amounts in "register" and "displacement" tons, (20 − 7)(73) = 949 barrels bulk, represents the extra M&M's are shipped. Using the conversions in the problem, this is equivalent to (949)(0.1415)(28.378) = 3.81 × 10^3 U.S. bushels.

42. (a) The receptacle is a volume of (40 cm)(40 cm)(30 cm) = 48000 cm^3 = 48 L = (48)(16)/11.356 = 67.63 standard bottles, which is a little more than 3 nebuchadnezzars (the largest bottle indicated). The remainder, 7.63 standard bottles, is just a little less than 1 methuselah. Thus, the answer to part (a) is 3 nebuchadnezzars and 1 methuselah.

(b) Since 1 methuselah.= 8 standard bottles, then the extra amount is 8 − 7.63 = 0.37 standard bottle.

(c) Using the conversion factor 16 standard bottles = 11.356 L, we have

$$0.37 \text{ standard bottle} = (0.37 \text{ standard bottle})\left(\frac{11.356 \text{ L}}{16 \text{ standard bottles}}\right) = 0.26 \text{ L}.$$

43. The volume of one unit is 1 cm^3 = 1 × 10^{-6} m^3, so the volume of a mole of them is 6.02 × 10^{23} cm^3 = 6.02 × 10^{17} m^3. The cube root of this number gives the edge length: 8.4×10^5 m^3. This is equivalent to roughly 8 × 10^2 kilometers.

44. Equation 1-9 gives (to very high precision!) the conversion from atomic mass units to kilograms. Since this problem deals with the ratio of total mass (1.0 kg) divided by the mass of one atom (1.0 u, but converted to kilograms), then the computation reduces to simply taking the reciprocal of the number given in Eq. 1-9 and rounding off appropriately. Thus, the answer is 6.0×10^{26}.

45. We convert meters to astronomical units, and seconds to minutes, using

$$1000 \text{ m} = 1 \text{ km}$$
$$1 \text{ AU} = 1.50 \times 10^8 \text{ km}$$
$$60 \text{ s} = 1 \text{ min}.$$

Thus, 3.0×10^8 m/s becomes

$$\left(\frac{3.0 \times 10^8 \text{ m}}{\text{s}}\right)\left(\frac{1 \text{ km}}{1000 \text{ m}}\right)\left(\frac{\text{AU}}{1.50 \times 10^8 \text{ km}}\right)\left(\frac{60 \text{ s}}{\text{min}}\right) = 0.12 \text{ AU/min}.$$

46. The volume of the water that fell is

$$V = (26 \text{ km}^2)(2.0 \text{ in.}) = (26 \text{ km}^2)\left(\frac{1000 \text{ m}}{1 \text{ km}}\right)^2 (2.0 \text{ in.})\left(\frac{0.0254 \text{ m}}{1 \text{ in.}}\right)$$
$$= (26 \times 10^6 \text{ m}^2)(0.0508 \text{ m})$$
$$= 1.3 \times 10^6 \text{ m}^3.$$

We write the mass-per-unit-volume (density) of the water as:

$$\rho = \frac{m}{V} = 1 \times 10^3 \text{ kg/m}^3.$$

The mass of the water that fell is therefore given by $m = \rho V$:

$$m = (1 \times 10^3 \text{ kg/m}^3)(1.3 \times 10^6 \text{ m}^3) = 1.3 \times 10^9 \text{ kg}.$$

47. A million milligrams comprise a kilogram, so 2.3 kg/week is 2.3×10^6 mg/week. Figuring 7 days a week, 24 hours per day, 3600 second per hour, we find 604800 seconds are equivalent to one week. Thus, $(2.3 \times 10^6$ mg/week$)/(604800$ s/week$) = 3.8$ mg/s.

48. The mass of the pig is 3.108 slugs, or $(3.108)(14.59) = 45.346$ kg. Referring now to the corn, a U.S. bushel is 35.238 liters. Thus, a value of 1 for the *corn-hog ratio* would be equivalent to $35.238/45.346 = 0.7766$ in the indicated metric units. Therefore, a value of 5.7 for the *ratio* corresponds to $5.7(0.777) \approx 4.4$ in the indicated metric units.

49. Two jalapeño peppers have spiciness = 8000 SHU, and this amount multiplied by 400 (the number of people) is 3.2×10^6 SHU, which is roughly ten times the SHU value for a single habanero pepper. More precisely, 10.7 habanero peppers will provide that total required SHU value.

50. The volume removed in one year is

$$V = (75 \times 10^4 \text{ m}^2)(26 \text{ m}) \approx 2 \times 10^7 \text{ m}^3$$

which we convert to cubic kilometers: $V = (2 \times 10^7 \text{ m}^3)\left(\dfrac{1 \text{ km}}{1000 \text{ m}}\right)^3 = 0.020 \text{ km}^3$.

51. The number of seconds in a year is 3.156×10^7. This is listed in Appendix D and results from the product

$$(365.25 \text{ day/y})(24 \text{ h/day})(60 \text{ min/h})(60 \text{ s/min}).$$

(a) The number of shakes in a second is 10^8; therefore, there are indeed more shakes per second than there are seconds per year.

(b) Denoting the age of the universe as 1 u-day (or 86400 u-sec), then the time during which humans have existed is given by

$$\dfrac{10^6}{10^{10}} = 10^{-4} \text{ u-day},$$

which may also be expressed as $(10^{-4} \text{ u-day})\left(\dfrac{86400 \text{ u-sec}}{1 \text{ u-day}}\right) = 8.6 \text{ u-sec}$.

52. Abbreviating wapentake as "wp" and assuming a hide to be 110 acres, we set up the ratio 25 wp/11 barn along with appropriate conversion factors:

$$\dfrac{(25 \text{ wp})\left(\dfrac{100 \text{ hide}}{1 \text{ wp}}\right)\left(\dfrac{110 \text{ acre}}{1 \text{ hide}}\right)\left(\dfrac{4047 \text{ m}^2}{1 \text{ acre}}\right)}{(11 \text{ barn})\left(\dfrac{1 \times 10^{-28} \text{ m}^2}{1 \text{ barn}}\right)} \approx 1 \times 10^{36}.$$

53. (a) Squaring the relation 1 ken = 1.97 m, and setting up the ratio, we obtain

$$\dfrac{1 \text{ ken}^2}{1 \text{ m}^2} = \dfrac{1.97^2 \text{ m}^2}{1 \text{ m}^2} = 3.88.$$

(b) Similarly, we find

$$\frac{1 \text{ ken}^3}{1 \text{ m}^3} = \frac{1.97^3 \text{ m}^3}{1 \text{ m}^3} = 7.65.$$

(c) The volume of a cylinder is the circular area of its base multiplied by its height. Thus,

$$\pi r^2 h = \pi (3.00)^2 (5.50) = 156 \text{ ken}^3.$$

(d) If we multiply this by the result of part (b), we determine the volume in cubic meters: $(155.5)(7.65) = 1.19 \times 10^3 \text{ m}^3$.

54. The mass in kilograms is

$$(28.9 \text{ piculs}) \left(\frac{100 \text{ gin}}{1 \text{ picul}} \right) \left(\frac{16 \text{ tahil}}{1 \text{ gin}} \right) \left(\frac{10 \text{ chee}}{1 \text{ tahil}} \right) \left(\frac{10 \text{ hoon}}{1 \text{ chee}} \right) \left(\frac{0.3779 \text{ g}}{1 \text{ hoon}} \right)$$

which yields 1.747×10^6 g or roughly 1.75×10^3 kg.

55. In the simplest approach, we set up a ratio for the total increase in *horizontal depth x* (where $\Delta x = 0.05$ m is the increase in horizontal depth per step)

$$x = N_{\text{steps}} \Delta x = \left(\frac{4.57}{0.19} \right) (0.05 \text{ m}) = 1.2 \text{ m}.$$

However, we can approach this more carefully by noting that if there are $N = 4.57/.19 \approx 24$ rises then under normal circumstances we would expect $N - 1 = 23$ runs (horizontal pieces) in that staircase. This would yield $(23)(0.05 \text{ m}) = 1.15$ m, which - to two significant figures - agrees with our first result.

56. Since one atomic mass unit is $1 \text{ u} = 1.66 \times 10^{-24}$ g (see Appendix D), the mass of one mole of atoms is about $m = (1.66 \times 10^{-24} \text{ g})(6.02 \times 10^{23}) = 1 \text{ g}$. On the other hand, the mass of one mole of atoms in the common Eastern mole is

$$m' = \frac{75 \text{ g}}{7.5} = 10 \text{ g}$$

Therefore, in atomic mass units, the average mass of one atom in the common Eastern mole is

$$\frac{m'}{N_A} = \frac{10 \text{ g}}{6.02 \times 10^{23}} = 1.66 \times 10^{-23} \text{ g} = 10 \text{ u}.$$

57. (a) When θ is measured in radians, it is equal to the arc length s divided by the radius R. For a very large radius circle and small value of θ, such as we deal with in Fig. 1–9,

the arc may be approximated as the straight line-segment of length 1 AU. First, we convert $\theta = 1$ arcsecond to radians:

$$(1 \text{ arcsecond})\left(\frac{1 \text{ arcminute}}{60 \text{ arcsecond}}\right)\left(\frac{1°}{60 \text{ arcminute}}\right)\left(\frac{2\pi \text{ radian}}{360°}\right)$$

which yields $\theta = 4.85 \times 10^{-6}$ rad. Therefore, one parsec is

$$R_o = \frac{s}{\theta} = \frac{1 \text{ AU}}{4.85 \times 10^{-6}} = 2.06 \times 10^5 \text{ AU}.$$

Now we use this to convert $R = 1$ AU to parsecs:

$$R = (1 \text{ AU})\left(\frac{1 \text{ pc}}{2.06 \times 10^5 \text{ AU}}\right) = 4.9 \times 10^{-6} \text{ pc}.$$

(b) Also, since it is straightforward to figure the number of seconds in a year (about 3.16×10^7 s), and (for constant speeds) distance = speed × time, we have

$$1 \text{ ly} = (186,000 \text{ mi/s})(3.16 \times 10^7 \text{ s}) \; 5.9 \times 10^{12} \text{ mi}$$

which we convert to AU by dividing by 92.6×10^6 (given in the problem statement), obtaining 6.3×10^4 AU. Inverting, the result is $1 \text{ AU} = 1/6.3 \times 10^4 = 1.6 \times 10^{-5}$ ly.

58. The volume of the filled container is 24000 cm^3 = 24 liters, which (using the conversion given in the problem) is equivalent to 50.7 pints (U.S). The expected number is therefore in the range from 1317 to 1927 Atlantic oysters. Instead, the number received is in the range from 406 to 609 Pacific oysters. This represents a shortage in the range of roughly 700 to 1500 oysters (the answer to the problem). Note that the minimum value in our answer corresponds to the minimum Atlantic minus the maximum Pacific, and the maximum value corresponds to the maximum Atlantic minus the minimum Pacific.

59. (a) For the minimum (43 cm) case, 9 cubit converts as follows:

$$9 \text{ cubit} = (9 \text{ cubit})\left(\frac{0.43 \text{ m}}{1 \text{ cubit}}\right) = 3.9 \text{ m}.$$

And for the maximum (43 cm) case we obtain

$$9 \text{ cubit} = (9 \text{ cubit})\left(\frac{0.53 \text{ m}}{1 \text{ cubit}}\right) = 4.8 \text{ m}.$$

(b) Similarly, with 0.43 m → 430 mm and 0.53 m → 530 mm, we find 3.9×10^3 mm and 4.8×10^3 mm, respectively.

(c) We can convert length and diameter first and then compute the volume, or first compute the volume and then convert. We proceed using the latter approach (where d is diameter and ℓ is length).

$$V_{\text{cylinder, min}} = \frac{\pi}{4}\ell d^2 = 28 \text{ cubit}^3 = \left(28 \text{ cubit}^3\right)\left(\frac{0.43 \text{ m}}{1 \text{ cubit}}\right)^3 = 2.2 \text{ m}^3.$$

Similarly, with 0.43 m replaced by 0.53 m, we obtain $V_{\text{cylinder, max}} = 4.2 \text{ m}^3$.

60. (a) We reduce the stock amount to British teaspoons:

$$\begin{aligned}
1 \text{ breakfastcup} &= 2 \times 8 \times 2 \times 2 = 64 \text{ teaspoons} \\
1 \text{ teacup} &= 8 \times 2 \times 2 = 32 \text{ teaspoons} \\
6 \text{ tablespoons} &= 6 \times 2 \times 2 = 24 \text{ teaspoons} \\
1 \text{ dessertspoon} &= 2 \text{ teaspoons}
\end{aligned}$$

which totals to 122 British teaspoons, or 122 U.S. teaspoons since liquid measure is being used. Now with one U.S cup equal to 48 teaspoons, upon dividing $122/48 \approx 2.54$, we find this amount corresponds to 2.5 U.S. cups plus a remainder of precisely 2 teaspoons. In other words,

122 U.S. teaspoons = 2.5 U.S. cups + 2 U.S. teaspoons.

(b) For the nettle tops, one-half quart is still one-half quart.

(c) For the rice, one British tablespoon is 4 British teaspoons which (since dry-goods measure is being used) corresponds to 2 U.S. teaspoons.

(d) A British saltspoon is $\frac{1}{2}$ British teaspoon which corresponds (since dry-goods measure is again being used) to 1 U.S. teaspoon.

Chapter 2

1. We use Eq. 2-2 and Eq. 2-3. During a time t_c when the velocity remains a positive constant, speed is equivalent to velocity, and distance is equivalent to displacement, with $\Delta x = v\, t_c$.

(a) During the first part of the motion, the displacement is $\Delta x_1 = 40$ km and the time interval is

$$t_1 = \frac{(40\text{ km})}{(30\text{ km}/\text{h})} = 1.33\text{ h}.$$

During the second part the displacement is $\Delta x_2 = 40$ km and the time interval is

$$t_2 = \frac{(40\text{ km})}{(60\text{ km}/\text{h})} = 0.67\text{ h}.$$

Both displacements are in the same direction, so the total displacement is

$$\Delta x = \Delta x_1 + \Delta x_2 = 40\text{ km} + 40\text{ km} = 80\text{ km}.$$

The total time for the trip is $t = t_1 + t_2 = 2.00$ h. Consequently, the average velocity is

$$v_{avg} = \frac{(80\text{ km})}{(2.0\text{ h})} = 40\text{ km}/\text{h}.$$

(b) In this example, the numerical result for the average speed is the same as the average velocity 40 km/h.

(c) As shown below, the graph consists of two contiguous line segments, the first having a slope of 30 km/h and connecting the origin to $(t_1, x_1) = (1.33\text{ h}, 40\text{ km})$ and the second having a slope of 60 km/h and connecting (t_1, x_1) to $(t, x) = (2.00\text{ h}, 80\text{ km})$. From the graphical point of view, the slope of the dashed line drawn from the origin to (t, x) represents the average velocity.

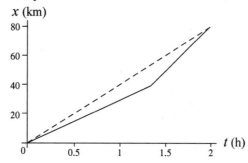

2. Average speed, as opposed to average velocity, relates to the total distance, as opposed to the net displacement. The distance D up the hill is, of course, the same as the distance down the hill, and since the speed is constant (during each stage of the motion) we have speed = D/t. Thus, the average speed is

$$\frac{D_{up} + D_{down}}{t_{up} + t_{down}} = \frac{2D}{\dfrac{D}{v_{up}} + \dfrac{D}{v_{down}}}$$

which, after canceling D and plugging in v_{up} = 40 km/h and v_{down} = 60 km/h, yields 48 km/h for the average speed.

3. The speed (assumed constant) is v = (90 km/h)(1000 m/km)/(3600 s/h) = 25 m/s. Thus, in 0.50 s, the car travels (0.50 s)(25 m/s) ≈ 13 m.

4. Huber's speed is

$$v_0 = (200 \text{ m})/(6.509 \text{ s}) = 30.72 \text{ m/s} = 110.6 \text{ km/h},$$

where we have used the conversion factor 1 m/s = 3.6 km/h. Since Whittingham beat Huber by 19.0 km/h, his speed is v_1=(110.6 km/h + 19.0 km/h)=129.6 km/h, or 36 m/s (1 km/h = 0.2778 m/s). Thus, the time through a distance of 200 m for Whittingham is

$$\Delta t = \frac{\Delta x}{v_1} = \frac{200 \text{ m}}{36 \text{ m/s}} = 5.554 \text{ s}.$$

5. Using $x = 3t - 4t^2 + t^3$ with SI units understood is efficient (and is the approach we will use), but if we wished to make the units explicit we would write

$$x = (3 \text{ m/s})t - (4 \text{ m/s}^2)t^2 + (1 \text{ m/s}^3)t^3.$$

We will quote our answers to one or two significant figures, and not try to follow the significant figure rules rigorously.

(a) Plugging in $t = 1$ s yields $x = 3 - 4 + 1 = 0$.

(b) With $t = 2$ s we get $x = 3(2) - 4(2)^2 + (2)^3 = -2$ m.

(c) With $t = 3$ s we have $x = 0$ m.

(d) Plugging in $t = 4$ s gives $x = 12$ m.

For later reference, we also note that the position at $t = 0$ is $x = 0$.

(e) The position at $t = 0$ is subtracted from the position at $t = 4$ s to find the displacement $\Delta x = 12$ m.

(f) The position at $t = 2$ s is subtracted from the position at $t = 4$ s to give the displacement $\Delta x = 14$ m. Eq. 2-2, then, leads to

$$v_{avg} = \frac{\Delta x}{\Delta t} = \frac{14 \text{ m}}{2 \text{ s}} = 7 \text{ m/s}.$$

(g) The horizontal axis is $0 \le t \le 4$ with SI units understood.

Not shown is a straight line drawn from the point at $(t, x) = (2, -2)$ to the highest point shown (at $t = 4$ s) which would represent the answer for part (f).

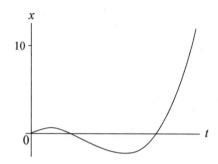

6. (a) Using the fact that time = distance/velocity while the velocity is constant, we find

$$v_{avg} = \frac{73.2 \text{ m} + 73.2 \text{ m}}{\frac{73.2 \text{ m}}{1.22 \text{ m/s}} + \frac{73.2 \text{ m}}{3.05 \text{ m}}} = 1.74 \text{ m/s}.$$

(b) Using the fact that distance = vt while the velocity v is constant, we find

$$v_{avg} = \frac{(1.22 \text{ m/s})(60 \text{ s}) + (3.05 \text{ m/s})(60 \text{ s})}{120 \text{ s}} = 2.14 \text{ m/s}.$$

(c) The graphs are shown below (with meters and seconds understood). The first consists of two (solid) line segments, the first having a slope of 1.22 and the second having a slope of 3.05. The slope of the dashed line represents the average velocity (in both graphs). The second graph also consists of two (solid) line segments, having the same slopes as before — the main difference (compared to the first graph) being that the stage involving higher-speed motion lasts much longer.

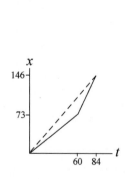 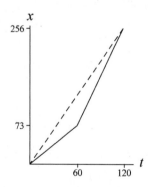

7. Converting to seconds, the running times are $t_1 = 147.95$ s and $t_2 = 148.15$ s, respectively. If the runners were equally fast, then

$$S_{avg1} = S_{avg2} \Rightarrow \frac{L_1}{t_1} = \frac{L_2}{t_2}.$$

From this we obtain

$$L_2 - L_1 = \left(\frac{t_2}{t_1} - 1\right) L_1 = \left(\frac{148.15}{147.95} - 1\right) L_1 = 0.00135 L_1 \approx 1.4 \text{ m}$$

where we set $L_1 \approx 1000$ m in the last step. Thus, if L_1 and L_2 are no different than about 1.4 m, then runner 1 is indeed faster than runner 2. However, if L_1 is shorter than L_2 by more than 1.4 m, then runner 2 would actually be faster.

8. Let v_w be the speed of the wind and v_c be the speed of the car.

(a) Suppose during time interval t_1, the car moves in the same direction as the wind. Then its effective speed is $v_{eff,1} = v_c + v_w$, and the distance traveled is $d = v_{eff,1} t_1 = (v_c + v_w) t_1$. On the other hand, for the return trip during time interval t_2, the car moves in the opposite direction of the wind and the effective speed would be $v_{eff,2} = v_c - v_w$. The distance traveled is $d = v_{eff,2} t_2 = (v_c - v_w) t_2$. The two expressions can be rewritten as

$$v_c + v_w = \frac{d}{t_1} \quad \text{and} \quad v_c - v_w = \frac{d}{t_2}$$

Adding the two equations and dividing by two, we obtain $v_c = \frac{1}{2}\left(\frac{d}{t_1} + \frac{d}{t_2}\right)$. Thus, method 1 gives the car's speed v_c in windless situation.

(b) If method 2 is used, the result would be

$$v'_c = \frac{d}{(t_1 + t_2)/2} = \frac{2d}{t_1 + t_2} = \frac{2d}{\dfrac{d}{v_c + v_w} + \dfrac{d}{v_c - v_w}} = \frac{v_c^2 - v_w^2}{v_c} = v_c \left[1 - \left(\frac{v_w}{v_c}\right)^2\right].$$

The fractional difference would be

$$\frac{v_c - v'_c}{v_c} = \left(\frac{v_w}{v_c}\right)^2 = (0.0240)^2 = 5.76 \times 10^{-4}.$$

9. The values used in the problem statement make it easy to see that the first part of the trip (at 100 km/h) takes 1 hour, and the second part (at 40 km/h) also takes 1 hour. Expressed in decimal form, the time left is 1.25 hour, and the distance that remains is 160 km. Thus, a speed $v = (160 \text{ km})/(1.25 \text{ h}) = 128$ km/h is needed.

10. The amount of time it takes for each person to move a distance L with speed v_s is $\Delta t = L/v_s$. With each additional person, the depth increases by one body depth d

(a) The rate of increase of the layer of people is

$$R = \frac{d}{\Delta t} = \frac{d}{L/v_s} = \frac{dv_s}{L} = \frac{(0.25 \text{ m})(3.50 \text{ m/s})}{1.75 \text{ m}} = 0.50 \text{ m/s}$$

(b) The amount of time required to reach a depth of $D = 5.0$ m is

$$t = \frac{D}{R} = \frac{5.0 \text{ m}}{0.50 \text{ m/s}} = 10 \text{ s}$$

11. Recognizing that the gap between the trains is closing at a constant rate of 60 km/h, the total time which elapses before they crash is t = (60 km)/(60 km/h) = 1.0 h. During this time, the bird travels a distance of $x = vt$ = (60 km/h)(1.0 h) = 60 km.

12. (a) Let the fast and the slow cars be separated by a distance d at $t = 0$. If during the time interval $t = L/v_s = (12.0 \text{ m})/(5.0 \text{ m/s}) = 2.40$ s in which the slow car has moved a distance of $L = 12.0$ m, the fast car moves a distance of $vt = d + L$ to join the line of slow cars, then the shock wave would remain stationary. The condition implies a separation of

$$d = vt - L = (25 \text{ m/s})(2.4 \text{ s}) - 12.0 \text{ m} = 48.0 \text{ m}.$$

(b) Let the initial separation at $t = 0$ be $d = 96.0$ m. At a later time t, the slow and the fast cars have traveled $x = v_s t$ and the fast car joins the line by moving a distance $d + x$. From

$$t = \frac{x}{v_s} = \frac{d+x}{v},$$

we get

$$x = \frac{v_s}{v - v_s} d = \frac{5.00 \text{ m/s}}{25.0 \text{ m/s} - 5.00 \text{ m/s}} (96.0 \text{ m}) = 24.0 \text{ m},$$

which in turn gives $t = (24.0 \text{ m})/(5.00 \text{ m/s}) = 4.80$ s. Since the rear of the slow-car pack has moved a distance of $\Delta x = x - L = 24.0 \text{ m} - 12.0 \text{ m} = 12.0$ m downstream, the speed of the rear of the slow-car pack, or equivalently, the speed of the shock wave, is

$$v_{shock} = \frac{\Delta x}{t} = \frac{12.0 \text{ m}}{4.80 \text{ s}} = 2.50 \text{ m/s}.$$

(c) Since $x > L$, the direction of the shock wave is downstream.

13. (a) Denoting the travel time and distance from San Antonio to Houston as T and D, respectively, the average speed is

$$S_{avg1} = \frac{D}{T} = \frac{(55 \text{ km/h})(T/2) + (90 \text{ km/h})(T/2)}{T} = 72.5 \text{ km/h}$$

which should be rounded to 73 km/h.

(b) Using the fact that time = distance/speed while the speed is constant, we find

$$S_{avg2} = \frac{D}{T} = \frac{D}{\frac{D/2}{55 \text{ km/h}} + \frac{D/2}{90 \text{ km/h}}} = 68.3 \text{ km/h}$$

which should be rounded to 68 km/h.

(c) The total distance traveled (2D) must not be confused with the net displacement (zero). We obtain for the two-way trip

$$S_{avg} = \frac{2D}{\frac{D}{72.5 \text{ km/h}} + \frac{D}{68.3 \text{ km/h}}} = 70 \text{ km/h}.$$

(d) Since the net displacement vanishes, the average velocity for the trip in its entirety is zero.

(e) In asking for a *sketch*, the problem is allowing the student to arbitrarily set the distance D (the intent is *not* to make the student go to an Atlas to look it up); the student can just as easily arbitrarily set T instead of D, as will be clear in the following discussion. We briefly describe the graph (with kilometers-per-hour understood for the slopes): two contiguous line segments, the first having a slope of 55 and connecting the origin to $(t_1, x_1) = (T/2, 55T/2)$ and the second having a slope of 90 and connecting (t_1, x_1) to (T, D) where $D = (55 + 90)T/2$. The average velocity, from the graphical point of view, is the slope of a line drawn from the origin to (T, D). The graph (not drawn to scale) is depicted below:

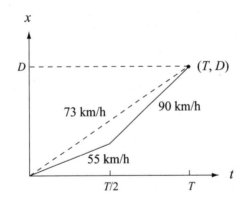

14. We use the functional notation $x(t)$, $v(t)$ and $a(t)$ in this solution, where the latter two quantities are obtained by differentiation:

$$v(t) = \frac{dx(t)}{dt} = -12t \quad \text{and} \quad a(t) = \frac{dv(t)}{dt} = -12$$

with SI units understood.

(a) From $v(t) = 0$ we find it is (momentarily) at rest at $t = 0$.

(b) We obtain $x(0) = 4.0$ m

(c) and (d) Requiring $x(t) = 0$ in the expression $x(t) = 4.0 - 6.0t^2$ leads to $t = \pm 0.82$ s for the times when the particle can be found passing through the origin.

(e) We show both the asked-for graph (on the left) as well as the "shifted" graph which is relevant to part (f). In both cases, the time axis is given by $-3 \le t \le 3$ (SI units understood).

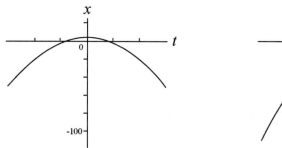
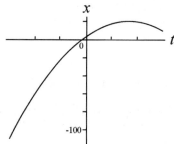

(f) We arrived at the graph on the right (shown above) by adding $20t$ to the $x(t)$ expression.

(g) Examining where the slopes of the graphs become zero, it is clear that the shift causes the $v = 0$ point to correspond to a larger value of x (the top of the second curve shown in part (e) is higher than that of the first).

15. We use Eq. 2-4. to solve the problem.

(a) The velocity of the particle is

$$v = \frac{dx}{dt} = \frac{d}{dt}(4 - 12t + 3t^2) = -12 + 6t.$$

Thus, at $t = 1$ s, the velocity is $v = (-12 + (6)(1)) = -6$ m/s.

(b) Since $v < 0$, it is moving in the negative x direction at $t = 1$ s.

(c) At $t = 1$ s, the *speed* is $|v| = 6$ m/s.

(d) For $0 < t < 2$ s, $|v|$ decreases until it vanishes. For $2 < t < 3$ s, $|v|$ increases from zero to the value it had in part (c). Then, $|v|$ is larger than that value for $t > 3$ s.

(e) Yes, since v smoothly changes from negative values (consider the $t = 1$ result) to positive (note that as $t \to +\infty$, we have $v \to +\infty$). One can check that $v = 0$ when $t = 2$ s.

(f) No. In fact, from $v = -12 + 6t$, we know that $v > 0$ for $t > 2$ s.

16. Using the general property $\frac{d}{dx}\exp(bx) = b\exp(bx)$, we write

$$v = \frac{dx}{dt} = \left(\frac{d(19t)}{dt}\right) \cdot e^{-t} + (19t) \cdot \left(\frac{de^{-t}}{dt}\right).$$

If a concern develops about the appearance of an argument of the exponential ($-t$) apparently having units, then an explicit factor of $1/T$ where $T = 1$ second can be inserted and carried through the computation (which does not change our answer). The result of this differentiation is

$$v = 16(1 - t)e^{-t}$$

with t and v in SI units (s and m/s, respectively). We see that this function is zero when $t = 1$ s. Now that we know *when* it stops, we find out *where* it stops by plugging our result $t = 1$ into the given function $x = 16te^{-t}$ with x in meters. Therefore, we find $x = 5.9$ m.

17. We use Eq. 2-2 for average velocity and Eq. 2-4 for instantaneous velocity, and work with distances in centimeters and times in seconds.

(a) We plug into the given equation for x for $t = 2.00$ s and $t = 3.00$ s and obtain $x_2 = 21.75$ cm and $x_3 = 50.25$ cm, respectively. The average velocity during the time interval $2.00 \leq t \leq 3.00$ s is

$$v_{avg} = \frac{\Delta x}{\Delta t} = \frac{50.25 \text{ cm} - 21.75 \text{ cm}}{3.00 \text{ s} - 2.00 \text{ s}}$$

which yields $v_{avg} = 28.5$ cm/s.

(b) The instantaneous velocity is $v = \frac{dx}{dt} = 4.5t^2$, which, at time $t = 2.00$ s, yields $v = (4.5)(2.00)^2 = 18.0$ cm/s.

(c) At $t = 3.00$ s, the instantaneous velocity is $v = (4.5)(3.00)^2 = 40.5$ cm/s.

(d) At $t = 2.50$ s, the instantaneous velocity is $v = (4.5)(2.50)^2 = 28.1$ cm/s.

(e) Let t_m stand for the moment when the particle is midway between x_2 and x_3 (that is, when the particle is at $x_m = (x_2 + x_3)/2 = 36$ cm). Therefore,

$$x_m = 9.75 + 1.5t_m^3 \Rightarrow t_m = 2.596$$

in seconds. Thus, the instantaneous speed at this time is $v = 4.5(2.596)^2 = 30.3$ cm/s.

(f) The answer to part (a) is given by the slope of the straight line between $t = 2$ and $t = 3$ in this x-vs-t plot. The answers to parts (b), (c), (d) and (e) correspond to the slopes of tangent lines (not shown but easily imagined) to the curve at the appropriate points.

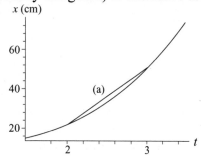

18. We use the functional notation $x(t)$, $v(t)$ and $a(t)$ and find the latter two quantities by differentiating:

$$v(t) = \frac{dx(t)}{t} = -15t^2 + 20 \quad \text{and} \quad a(t) = \frac{dv(t)}{dt} = -30t$$

with SI units understood. These expressions are used in the parts that follow.

(a) From $0 = -15t^2 + 20$, we see that the only positive value of t for which the particle is (momentarily) stopped is $t = \sqrt{20/15} = 1.2$ s.

(b) From $0 = -30t$, we find $a(0) = 0$ (that is, it vanishes at $t = 0$).

(c) It is clear that $a(t) = -30t$ is negative for $t > 0$

(d) The acceleration $a(t) = -30t$ is positive for $t < 0$.

(e) The graphs are shown below. SI units are understood.

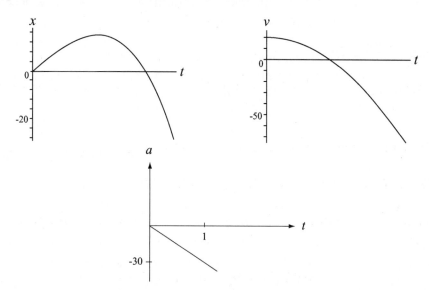

19. We represent its initial direction of motion as the $+x$ direction, so that $v_0 = +18$ m/s and $v = -30$ m/s (when $t = 2.4$ s). Using Eq. 2-7 (or Eq. 2-11, suitably interpreted) we find

$$a_{avg} = \frac{(-30 \text{ m/s}) - (+1 \text{ m/s})}{2.4 \text{ s}} = -20 \text{ m/s}^2$$

which indicates that the average acceleration has magnitude 20 m/s² and is in the opposite direction to the particle's initial velocity.

20. (a) Taking derivatives of $x(t) = 12t^2 - 2t^3$ we obtain the velocity and the acceleration functions:

$$v(t) = 24t - 6t^2 \quad \text{and} \quad a(t) = 24 - 12t$$

with length in meters and time in seconds. Plugging in the value $t = 3$ yields $x(3) = 54$ m.

(b) Similarly, plugging in the value $t = 3$ yields $v(3) = 18$ m/s.

(c) For $t = 3$, $a(3) = -12$ m/s².

(d) At the maximum x, we must have $v = 0$; eliminating the $t = 0$ root, the velocity equation reveals $t = 24/6 = 4$ s for the time of maximum x. Plugging $t = 4$ into the equation for x leads to $x = 64$ m for the largest x value reached by the particle.

(e) From (d), we see that the x reaches its maximum at $t = 4.0$ s.

(f) A maximum v requires $a = 0$, which occurs when $t = 24/12 = 2.0$ s. This, inserted into the velocity equation, gives $v_{max} = 24$ m/s.

(g) From (f), we see that the maximum of v occurs at $t = 24/12 = 2.0$ s.

(h) In part (e), the particle was (momentarily) motionless at $t = 4$ s. The acceleration at that time is readily found to be $24 - 12(4) = -24$ m/s².

(i) The *average velocity* is defined by Eq. 2-2, so we see that the values of x at $t = 0$ and $t = 3$ s are needed; these are, respectively, $x = 0$ and $x = 54$ m (found in part (a)). Thus,

$$v_{avg} = \frac{54-0}{3-0} = 18 \text{ m/s} \quad .$$

21. In this solution, we make use of the notation $x(t)$ for the value of x at a particular t. The notations $v(t)$ and $a(t)$ have similar meanings.

(a) Since the unit of ct^2 is that of length, the unit of c must be that of length/time², or m/s² in the SI system.

(b) Since bt^3 has a unit of length, b must have a unit of length/time³, or m/s³.

(c) When the particle reaches its maximum (or its minimum) coordinate its velocity is zero. Since the velocity is given by $v = dx/dt = 2ct - 3bt^2$, $v = 0$ occurs for $t = 0$ and for

$$t = \frac{2c}{3b} = \frac{2(3.0 \text{ m/s}^2)}{3(2.0 \text{ m/s}^3)} = 1.0 \text{ s}.$$

For $t = 0$, $x = x_0 = 0$ and for $t = 1.0$ s, $x = 1.0$ m $> x_0$. Since we seek the maximum, we reject the first root ($t = 0$) and accept the second ($t = 1$s).

(d) In the first 4 s the particle moves from the origin to $x = 1.0$ m, turns around, and goes back to

$$x(4 \text{ s}) = (3.0 \text{ m/s}^2)(4.0 \text{ s})^2 - (2.0 \text{ m/s}^3)(4.0 \text{ s})^3 = -80 \text{ m}.$$

The total path length it travels is 1.0 m + 1.0 m + 80 m = 82 m.

(e) Its displacement is $\Delta x = x_2 - x_1$, where $x_1 = 0$ and $x_2 = -80$ m. Thus, $\Delta x = -80$ m.

The velocity is given by $v = 2ct - 3bt^2 = (6.0 \text{ m/s}^2)t - (6.0 \text{ m/s}^3)t^2$.

(f) Plugging in $t = 1$ s, we obtain

$$v(1 \text{ s}) = (6.0 \text{ m/s}^2)(1.0 \text{ s}) - (6.0 \text{ m/s}^3)(1.0 \text{ s})^2 = 0.$$

(g) Similarly, $v(2 \text{ s}) = (6.0 \text{ m/s}^2)(2.0 \text{ s}) - (6.0 \text{ m/s}^3)(2.0 \text{ s})^2 = -12 \text{ m/s}$.

(h) $v(3 \text{ s}) = (6.0 \text{ m/s}^2)(3.0 \text{ s}) - (6.0 \text{ m/s}^3)(3.0 \text{ s})^2 = -36 \text{ m/s}$.

(i) $v(4 \text{ s}) = (6.0 \text{ m/s}^2)(4.0 \text{ s}) - (6.0 \text{ m/s}^3)(4.0 \text{ s})^2 = -72 \text{ m/s}$.

The acceleration is given by $a = dv/dt = 2c - 6bt = 6.0 \text{ m/s}^2 - (12.0 \text{ m/s}^3)t$.

(j) Plugging in $t = 1$ s, we obtain

$$a(1 \text{ s}) = 6.0 \text{ m/s}^2 - (12.0 \text{ m/s}^3)(1.0 \text{ s}) = -6.0 \text{ m/s}^2.$$

(k) $a(2 \text{ s}) = 6.0 \text{ m/s}^2 - (12.0 \text{ m/s}^3)(2.0 \text{ s}) = -18 \text{ m/s}^2$.

(l) $a(3 \text{ s}) = 6.0 \text{ m/s}^2 - (12.0 \text{ m/s}^3)(3.0 \text{ s}) = -30 \text{ m/s}^2$.

(m) $a(4 \text{ s}) = 6.0 \text{ m/s}^2 - (12.0 \text{ m/s}^3)(4.0 \text{ s}) = -42 \text{ m/s}^2$.

22. We use Eq. 2-2 (average velocity) and Eq. 2-7 (average acceleration). Regarding our coordinate choices, the initial position of the man is taken as the origin and his direction of motion during 5 min $\leq t \leq$ 10 min is taken to be the positive x direction. We also use the fact that $\Delta x = v\Delta t'$ when the velocity is constant during a time interval $\Delta t'$.

(a) The entire interval considered is $\Delta t = 8 - 2 = 6$ min which is equivalent to 360 s, whereas the sub-interval in which he is *moving* is only $\Delta t' = 8 - 5 = 3$ min $= 180$ s. His position at $t = 2$ min is $x = 0$ and his position at $t = 8$ min is $x = v\Delta t' = (2.2)(180) = 396$ m. Therefore,

$$v_{avg} = \frac{396 \text{ m} - 0}{360 \text{ s}} = 1.10 \text{ m/s}.$$

(b) The man is at rest at $t = 2$ min and has velocity $v = +2.2$ m/s at $t = 8$ min. Thus, keeping the answer to 3 significant figures,

$$a_{avg} = \frac{2.2 \text{ m/s} - 0}{360 \text{ s}} = 0.00611 \text{ m/s}^2.$$

(c) Now, the entire interval considered is $\Delta t = 9 - 3 = 6$ min (360 s again), whereas the sub-interval in which he is moving is $\Delta t' = 9 - 5 = 4$ min $= 240$ s). His position at $t = 3$ min is $x = 0$ and his position at $t = 9$ min is $x = v\Delta t' = (2.2)(240) = 528$ m. Therefore,

$$v_{avg} = \frac{528 \text{ m} - 0}{360 \text{ s}} = 1.47 \text{ m/s}.$$

(d) The man is at rest at $t = 3$ min and has velocity $v = +2.2$ m/s at $t = 9$ min. Consequently, $a_{avg} = 2.2/360 = 0.00611$ m/s^2 just as in part (b).

(e) The horizontal line near the bottom of this x-vs-t graph represents the man standing at $x = 0$ for $0 \leq t < 300$ s and the linearly rising line for $300 \leq t \leq 600$ s represents his constant-velocity motion. The dotted lines represent the answers to part (a) and (c) in the sense that their slopes yield those results.

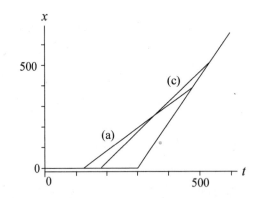

The graph of v-vs-t is not shown here, but would consist of two horizontal "steps" (one at $v = 0$ for $0 \leq t < 300$ s and the next at $v = 2.2$ m/s for $300 \leq t \leq 600$ s). The indications of the average accelerations found in parts (b) and (d) would be dotted lines connecting the "steps" at the appropriate t values (the slopes of the dotted lines representing the values of a_{avg}).

23. We use $v = v_0 + at$, with $t = 0$ as the instant when the velocity equals +9.6 m/s.

(a) Since we wish to calculate the velocity for a time *before* $t = 0$, we set $t = -2.5$ s. Thus, Eq. 2-11 gives

$$v = (9.6 \text{ m/s}) + (3.2 \text{ m/s}^2)(-2.5 \text{ s}) = 1.6 \text{ m/s}.$$

(b) Now, $t = +2.5$ s and we find

$$v = (9.6 \text{ m/s}) + (3.2 \text{ m/s}^2)(2.5 \text{ s}) = 18 \text{ m/s}.$$

24. The constant-acceleration condition permits the use of Table 2-1.

(a) Setting $v = 0$ and $x_0 = 0$ in $v^2 = v_0^2 + 2a(x - x_0)$, we find

$$x = -\frac{1}{2}\frac{v_0^2}{a} = -\frac{1}{2}\frac{(5.00 \times 10^6)^2}{-1.25 \times 10^{14}} = 0.100 \text{ m}.$$

Since the muon is slowing, the initial velocity and the acceleration must have opposite signs.

(b) Below are the time-plots of the position x and velocity v of the muon from the moment it enters the field to the time it stops. The computation in part (a) made no reference to t, so that other equations from Table 2-1 (such as $v = v_0 + at$ and $x = v_0 t + \frac{1}{2}at^2$) are used in making these plots.

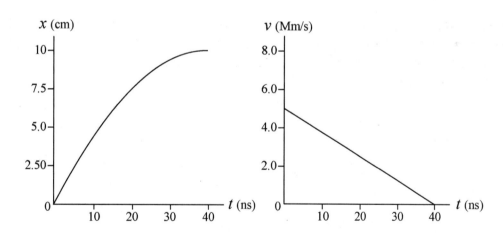

25. The constant acceleration stated in the problem permits the use of the equations in Table 2-1.

(a) We solve $v = v_0 + at$ for the time:

$$t = \frac{v - v_0}{a} = \frac{\frac{1}{10}(3.0 \times 10^8 \text{ m/s})}{9.8 \text{ m/s}^2} = 3.1 \times 10^6 \text{ s}$$

which is equivalent to 1.2 months.

(b) We evaluate $x = x_0 + v_0 t + \frac{1}{2}at^2$, with $x_0 = 0$. The result is

$$x = \frac{1}{2}\left(9.8 \text{ m/s}^2\right)(3.1\times 10^6 \text{s})^2 = 4.6\times 10^{13} \text{ m}.$$

26. We take $+x$ in the direction of motion, so $v_0 = +24.6$ m/s and $a = -4.92$ m/s^2. We also take $x_0 = 0$.

(a) The time to come to a halt is found using Eq. 2-11:

$$0 = v_0 + at \Rightarrow t = \frac{24.6 \text{ m/s}}{-4.92 \text{ m/s}^2} = 5.00 \text{ s}.$$

(b) Although several of the equations in Table 2-1 will yield the result, we choose Eq. 2-16 (since it does not depend on our answer to part (a)).

$$0 = v_0^2 + 2ax \Rightarrow x = -\frac{(24.6 \text{ m/s})^2}{2(-4.92 \text{ m/s}^2)} = 61.5 \text{ m}.$$

(c) Using these results, we plot $v_0 t + \frac{1}{2}at^2$ (the x graph, shown next, on the left) and $v_0 + at$ (the v graph, on the right) over $0 \le t \le 5$ s, with SI units understood.

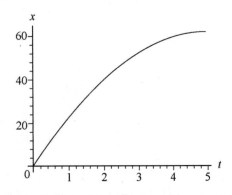
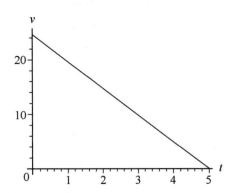

27. Assuming constant acceleration permits the use of the equations in Table 2-1. We solve $v^2 = v_0^2 + 2a(x - x_0)$ with $x_0 = 0$ and $x = 0.010$ m. Thus,

$$a = \frac{v^2 - v_0^2}{2x} = \frac{(5.7\times 10^5 \text{ m/s})^2 - (1.5\times 10^5 \text{ m/s})^2}{2(0.010 \text{ m})} = 1.62\times 10^{15} \text{ m/s}^2.$$

28. In this problem we are given the initial and final speeds, and the displacement, and asked to find the acceleration. We use the constant-acceleration equation given in Eq. 2-16, $v^2 = v_0^2 + 2a(x - x_0)$.

(a) With $v_0 = 0$, $v = 1.6$ m/s and $\Delta x = 5.0 \mu$m, the acceleration of the spores during the launch is

$$a = \frac{v^2 - v_0^2}{2x} = \frac{(1.6 \text{ m/s})^2}{2(5.0 \times 10^{-6} \text{ m})} = 2.56 \times 10^5 \text{ m/s}^2 = 2.6 \times 10^4 g$$

(b) During the speed-reduction stage, the acceleration is

$$a = \frac{v^2 - v_0^2}{2x} = \frac{0 - (1.6 \text{ m/s})^2}{2(1.0 \times 10^{-3} \text{ m})} = -1.28 \times 10^3 \text{ m/s}^2 = -1.3 \times 10^2 g$$

The negative sign means that the spores are decelerating.

29. We separate the motion into two parts, and take the direction of motion to be positive. In part 1, the vehicle accelerates from rest to its highest speed; we are given $v_0 = 0$; $v = 20$ m/s and $a = 2.0$ m/s^2. In part 2, the vehicle decelerates from its highest speed to a halt; we are given $v_0 = 20$ m/s; $v = 0$ and $a = -1.0$ m/s^2 (negative because the acceleration vector points opposite to the direction of motion).

(a) From Table 2-1, we find t_1 (the duration of part 1) from $v = v_0 + at$. In this way, $20 = 0 + 2.0t_1$ yields $t_1 = 10$ s. We obtain the duration t_2 of part 2 from the same equation. Thus, $0 = 20 + (-1.0)t_2$ leads to $t_2 = 20$ s, and the total is $t = t_1 + t_2 = 30$ s.

(b) For part 1, taking $x_0 = 0$, we use the equation $v^2 = v_0^2 + 2a(x - x_0)$ from Table 2-1 and find

$$x = \frac{v^2 - v_0^2}{2a} = \frac{(20 \text{ m/s})^2 - (0)^2}{2(2.0 \text{ m/s}^2)} = 100 \text{ m}.$$

This position is then the *initial* position for part 2, so that when the same equation is used in part 2 we obtain

$$x - 100 \text{ m} = \frac{v^2 - v_0^2}{2a} = \frac{(0)^2 - (20 \text{ m/s})^2}{2(-1.0 \text{ m/s}^2)}.$$

Thus, the final position is $x = 300$ m. That this is also the total distance traveled should be evident (the vehicle did not "backtrack" or reverse its direction of motion).

30. The acceleration is found from Eq. 2-11 (or, suitably interpreted, Eq. 2-7).

$$a = \frac{\Delta v}{\Delta t} = \frac{(1020 \text{ km/h})\left(\frac{1000 \text{ m/km}}{3600 \text{ s/h}}\right)}{1.4 \text{ s}} = 202.4 \text{ m/s}^2.$$

In terms of the gravitational acceleration g, this is expressed as a multiple of 9.8 m/s^2 as follows:

$$a = \left(\frac{202.4 \text{ m/s}^2}{9.8 \text{ m/s}^2}\right)g = 21g.$$

36 CHAPTER 2

31. We assume the periods of acceleration (duration t_1) and deceleration (duration t_2) are periods of constant a so that Table 2-1 can be used. Taking the direction of motion to be $+x$ then $a_1 = +1.22$ m/s^2 and $a_2 = -1.22$ m/s^2. We use SI units so the velocity at $t = t_1$ is $v = 305/60 = 5.08$ m/s.

(a) We denote Δx as the distance moved during t_1, and use Eq. 2-16:

$$v^2 = v_0^2 + 2a_1 \Delta x \Rightarrow \Delta x = \frac{(5.08 \text{ m/s})^2}{2(1.22 \text{ m/s}^2)} = 10.59 \text{ m} \approx 10.6 \text{ m}.$$

(b) Using Eq. 2-11, we have

$$t_1 = \frac{v - v_0}{a_1} = \frac{5.08 \text{ m/s}}{1.22 \text{ m/s}^2} = 4.17 \text{ s}.$$

The deceleration time t_2 turns out to be the same so that $t_1 + t_2 = 8.33$ s. The distances traveled during t_1 and t_2 are the same so that they total to $2(10.59 \text{ m}) = 21.18$ m. This implies that for a distance of 190 m $-$ 21.18 m $=$ 168.82 m, the elevator is traveling at constant velocity. This time of constant velocity motion is

$$t_3 = \frac{168.82 \text{ m}}{5.08 \text{ m/s}} = 33.21 \text{ s}.$$

Therefore, the total time is 8.33 s + 33.21 s $\approx$ 41.5 s.

32. We choose the positive direction to be that of the initial velocity of the car (implying that $a < 0$ since it is slowing down). We assume the acceleration is constant and use Table 2-1.

(a) Substituting $v_0 = 137$ km/h $= 38.1$ m/s, $v = 90$ km/h $= 25$ m/s, and $a = -5.2$ m/s^2 into $v = v_0 + at$, we obtain

$$t = \frac{25 \text{ m/s} - 38 \text{ m/s}}{-5.2 \text{ m/s}^2} = 2.5 \text{ s}.$$

(b) We take the car to be at $x = 0$ when the brakes are applied (at time $t = 0$). Thus, the coordinate of the car as a function of time is given by

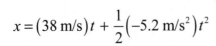

in SI units. This function is plotted from $t = 0$ to $t = 2.5$ s on the graph below. We have not shown the v-vs-t graph here; it is a descending straight line from v_0 to v.

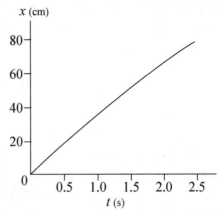

33. The problem statement (see part (a)) indicates that $a =$ constant, which allows us to use Table 2-1.

(a) We take $x_0 = 0$, and solve $x = v_0 t + \frac{1}{2} a t^2$ (Eq. 2-15) for the acceleration: $a = 2(x - v_0 t)/t^2$. Substituting $x = 24.0$ m, $v_0 = 56.0$ km/h $= 15.55$ m/s and $t = 2.00$ s, we find

$$a = \frac{2(24.0\text{m} - (15.55\text{m/s})(2.00\text{s}))}{(2.00\text{s})^2} = -3.56 \text{m/s}^2,$$

or $|a| = 3.56$ m/s^2. The negative sign indicates that the acceleration is opposite to the direction of motion of the car. The car is slowing down.

(b) We evaluate $v = v_0 + at$ as follows:

$$v = 15.55 \text{ m/s} - (3.56 \text{ m/s}^2)(2.00 \text{ s}) = 8.43 \text{ m/s}$$

which can also be converted to 30.3 km/h.

34. (a) Eq. 2-15 is used for part 1 of the trip and Eq. 2-18 is used for part 2:

$$\Delta x_1 = v_{o1} t_1 + \frac{1}{2} a_1 t_1^2 \qquad \text{where } a_1 = 2.25 \text{ m/s}^2 \text{ and } \Delta x_1 = \frac{900}{4} \text{ m}$$

$$\Delta x_2 = v_2 t_2 - \frac{1}{2} a_2 t_2^2 \qquad \text{where } a_2 = -0.75 \text{ m/s}^2 \text{ and } \Delta x_2 = \frac{3(900)}{4} \text{ m}$$

In addition, $v_{o1} = v_2 = 0$. Solving these equations for the times and adding the results gives $t = t_1 + t_2 = 56.6$ s.

(b) Eq. 2-16 is used for part 1 of the trip:

$$v^2 = (v_{o1})^2 + 2a_1 \Delta x_1 = 0 + 2(2.25)(\tfrac{900}{4}) = 1013 \text{ m}^2/\text{s}^2$$

which leads to $v = 31.8$ m/s for the maximum speed.

35. (a) From the figure, we see that $x_0 = -2.0$ m. From Table 2-1, we can apply $x - x_0 = v_0 t + \frac{1}{2} a t^2$ with $t = 1.0$ s, and then again with $t = 2.0$ s. This yields two equations for the two unknowns, v_0 and a:

$$0.0 - (-2.0 \text{ m}) = v_0 (1.0 \text{ s}) + \frac{1}{2} a (1.0 \text{ s})^2$$

$$6.0 \text{ m} - (-2.0 \text{ m}) = v_0 (2.0 \text{ s}) + \frac{1}{2} a (2.0 \text{ s})^2.$$

Solving these simultaneous equations yields the results $v_0 = 0$ and $a = 4.0$ m/s^2.

(b) The fact that the answer is positive tells us that the acceleration vector points in the $+x$ direction.

36. We assume the train accelerates from rest ($v_0 = 0$ and $x_0 = 0$) at $a_1 = +1.34 \text{ m/s}^2$ until the midway point and then decelerates at $a_2 = -1.34 \text{ m/s}^2$ until it comes to a stop ($v_2 = 0$) at the next station. The velocity at the midpoint is v_1 which occurs at $x_1 = 806/2 = 403$ m.

(a) Eq. 2-16 leads to

$$v_1^2 = v_0^2 + 2a_1 x_1 \Rightarrow v_1 = \sqrt{2(1.34 \text{ m/s}^2)(403 \text{ m})} = 32.9 \text{ m/s}.$$

(b) The time t_1 for the accelerating stage is (using Eq. 2-15)

$$x_1 = v_0 t_1 + \frac{1}{2} a_1 t_1^2 \Rightarrow t_1 = \sqrt{\frac{2(403 \text{ m})}{1.34 \text{ m/s}^2}} = 24.53 \text{ s}.$$

Since the time interval for the decelerating stage turns out to be the same, we double this result and obtain $t = 49.1$ s for the travel time between stations.

(c) With a "dead time" of 20 s, we have $T = t + 20 = 69.1$ s for the total time between start-ups. Thus, Eq. 2-2 gives

$$v_{avg} = \frac{806 \text{ m}}{69.1 \text{ s}} = 11.7 \text{ m/s}.$$

(d) The graphs for x, v and a as a function of t are shown below. SI units are understood. The third graph, $a(t)$, consists of three horizontal "steps" — one at 1.34 during $0 < t < 24.53$ and the next at -1.34 during $24.53 < t < 49.1$ and the last at zero during the "dead time" $49.1 < t < 69.1$).

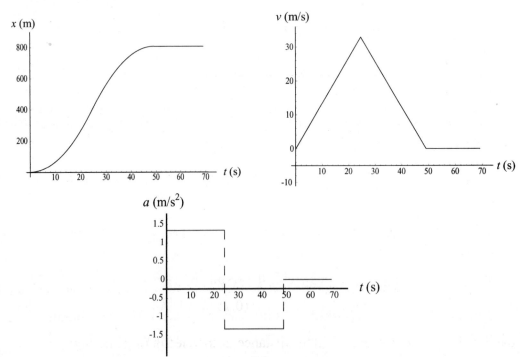

37. (a) We note that $v_A = 12/6 = 2$ m/s (with two significant figures understood). Therefore, with an initial x value of 20 m, car A will be at $x = 28$ m when $t = 4$ s. This must be the value of x for car B at that time; we use Eq. 2-15:

$$28 \text{ m} = (12 \text{ m/s})t + \tfrac{1}{2} a_B t^2 \quad \text{where } t = 4.0 \text{ s}.$$

This yields $a_B = -2.5$ m/s^2.

(b) The question is: using the value obtained for a_B in part (a), are there other values of t (besides $t = 4$ s) such that $x_A = x_B$? The requirement is

$$20 + 2t = 12t + \tfrac{1}{2} a_B t^2$$

where $a_B = -5/2$. There are two distinct roots unless the discriminant $\sqrt{10^2 - 2(-20)(a_B)}$ is zero. In our case, it is zero – which means there is only one root. The cars are side by side only once at $t = 4$ s.

(c) A sketch is shown below. It consists of a straight line (x_A) tangent to a parabola (x_B) at $t = 4$.

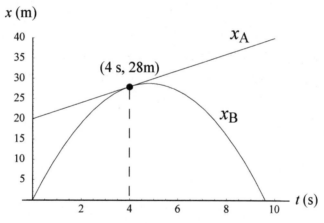

(d) We only care about real roots, which means $10^2 - 2(-20)(a_B) \geq 0$. If $|a_B| > 5/2$ then there are no (real) solutions to the equation; the cars are never side by side.

(e) Here we have $10^2 - 2(-20)(a_B) > 0 \Rightarrow$ two real roots. The cars are side by side at two different times.

38. We take the direction of motion as $+x$, so $a = -5.18$ m/s^2, and we use SI units, so $v_0 = 55(1000/3600) = 15.28$ m/s.

(a) The velocity is constant during the reaction time T, so the distance traveled during it is
$$d_r = v_0 T - (15.28 \text{ m/s})(0.75 \text{ s}) = 11.46 \text{ m}.$$

We use Eq. 2-16 (with $v = 0$) to find the distance d_b traveled during braking:

$$v^2 = v_0^2 + 2ad_b \quad \Rightarrow \quad d_b = -\frac{(15.28 \text{ m/s})^2}{2(-5.18 \text{ m/s}^2)}$$

which yields d_b = 22.53 m. Thus, the total distance is $d_r + d_b$ = 34.0 m, which means that the driver *is* able to stop in time. And if the driver were to continue at v_0, the car would enter the intersection in t = (40 m)/(15.28 m/s) = 2.6 s which is (barely) enough time to enter the intersection before the light turns, which many people would consider an acceptable situation.

(b) In this case, the total distance to stop (found in part (a) to be 34 m) is greater than the distance to the intersection, so the driver cannot stop without the front end of the car being a couple of meters into the intersection. And the time to reach it at constant speed is 32/15.28 = 2.1 s, which is too long (the light turns in 1.8 s). The driver is caught between a rock and a hard place.

39. The displacement (Δx) for each train is the "area" in the graph (since the displacement is the integral of the velocity). Each area is triangular, and the area of a triangle is 1/2(base) × (height). Thus, the (absolute value of the) displacement for one train (1/2)(40 m/s)(5 s) = 100 m, and that of the other train is (1/2)(30 m/s)(4 s) = 60 m. The initial "gap" between the trains was 200 m, and according to our displacement computations, the gap has narrowed by 160 m. Thus, the answer is 200 – 160 = 40 m.

40. Let d be the 220 m distance between the cars at t = 0, and v_1 be the 20 km/h = 50/9 m/s speed (corresponding to a passing point of x_1 = 44.5 m) and v_2 be the 40 km/h =100/9 m/s speed (corresponding to passing point of x_2 = 76.6 m) of the red car. We have two equations (based on Eq. 2-17):

$$d - x_1 = v_0 t_1 + \tfrac{1}{2} a\, t_1^2 \qquad \text{where } t_1 = x_1/v_1$$

$$d - x_2 = v_0 t_2 + \tfrac{1}{2} a\, t_2^2 \qquad \text{where } t_2 = x_2/v_2$$

We simultaneously solve these equations and obtain the following results:

(a) v_0 = – 13.9 m/s. or roughly – 50 km/h (the negative sign means that it's along the –x direction).

(b) a = – 2.0 m/s^2 (the negative sign means that it's along the –x direction).

41. The positions of the cars as a function of time are given by

$$x_r(t) = x_{r0} + \tfrac{1}{2} a_r t^2 = (-35.0 \text{ m}) + \tfrac{1}{2} a_r t^2$$
$$x_g(t) = x_{g0} + v_g t = (270 \text{ m}) - (20 \text{ m/s})t$$

where we have substituted the velocity and not the speed for the green car. The two cars pass each other at $t = 12.0$ s when the graphed lines cross. This implies that

$$(270 \text{ m}) - (20 \text{ m/s})(12.0 \text{ s}) = 30 \text{ m} = (-35.0 \text{ m}) + \frac{1}{2} a_r (12.0 \text{ s})^2$$

which can be solved to give $a_r = 0.90 \text{ m/s}^2$.

42. In this solution we elect to wait until the last step to convert to SI units. Constant acceleration is indicated, so use of Table 2-1 is permitted. We start with Eq. 2-17 and denote the train's initial velocity as v_t and the locomotive's velocity as v_ℓ (which is also the final velocity of the train, if the rear-end collision is barely avoided). We note that the distance Δx consists of the original gap between them D as well as the forward distance traveled during this time by the locomotive $v_\ell t$. Therefore,

$$\frac{v_t + v_\ell}{2} = \frac{\Delta x}{t} = \frac{D + v_\ell t}{t} = \frac{D}{t} + v_\ell.$$

We now use Eq. 2-11 to eliminate time from the equation. Thus,

$$\frac{v_t + v_\ell}{2} = \frac{D}{(v_\ell - v_t)/a} + v_\ell$$

which leads to

$$a = \left(\frac{v_t + v_\ell}{2} - v_\ell\right)\left(\frac{v_\ell - v_t}{D}\right) = -\frac{1}{2D}(v_\ell - v_t)^2.$$

Hence,

$$a = -\frac{1}{2(0.676 \text{ km})}\left(29 \frac{\text{km}}{\text{h}} - 161 \frac{\text{km}}{\text{h}}\right)^2 = -12888 \text{ km/h}^2$$

which we convert as follows:

$$a = (-12888 \text{ km/h}^2)\left(\frac{1000 \text{ m}}{1 \text{ km}}\right)\left(\frac{1 \text{ h}}{3600 \text{ s}}\right)^2 = -0.994 \text{ m/s}^2$$

so that its *magnitude* is $|a| = 0.994 \text{ m/s}^2$. A graph is shown below for the case where a collision is just avoided (x along the vertical axis is in meters and t along the horizontal axis is in seconds). The top (straight) line shows the motion of the locomotive and the bottom curve shows the motion of the passenger train.

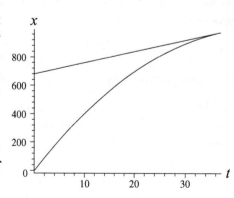

The other case (where the collision is not quite avoided) would be similar except that the slope of the bottom curve would be greater than that of the top line at the point where they meet.

43. (a) Note that 110 km/h is equivalent to 30.56 m/s. During a two second interval, you travel 61.11 m. The decelerating police car travels (using Eq. 2-15) 51.11 m. In light of the fact that the initial "gap" between cars was 25 m, this means the gap has narrowed by 10.0 m – that is, to a distance of 15.0 m between cars.

(b) First, we add 0.4 s to the considerations of part (a). During a 2.4 s interval, you travel 73.33 m. The decelerating police car travels (using Eq. 2-15) 58.93 m during that time. The initial distance between cars of 25 m has therefore narrowed by 14.4 m. Thus, at the start of your braking (call it t_o) the gap between the cars is 10.6 m. The speed of the police car at t_o is $30.56 - 5(2.4) = 18.56$ m/s. Collision occurs at time t when $x_{you} = x_{police}$ (we choose coordinates such that your position is $x = 0$ and the police car's position is $x = 10.6$ m at t_o). Eq. 2-15 becomes, for each car:

$$x_{police} - 10.6 = 18.56(t - t_o) - \tfrac{1}{2}(5)(t - t_o)^2$$
$$x_{you} = 30.56(t - t_o) - \tfrac{1}{2}(5)(t - t_o)^2 \ .$$

Subtracting equations, we find

$$10.6 = (30.56 - 18.56)(t - t_o) \implies 0.883 \text{ s} = t - t_o.$$

At that time your speed is $30.56 + a(t - t_o) = 30.56 - 5(0.883) \approx 26$ m/s (or 94 km/h).

44. Neglect of air resistance justifies setting $a = -g = -9.8$ m/s^2 (where *down* is our $-y$ direction) for the duration of the fall. This is constant acceleration motion, and we may use Table 2-1 (with Δy replacing Δx).

(a) Using Eq. 2-16 and taking the negative root (since the final velocity is downward), we have

$$v = -\sqrt{v_0^2 - 2g\Delta y} = -\sqrt{0 - 2(9.8 \text{ m/s}^2)(-1700 \text{ m})} = -183 \text{ m/s}.$$

Its magnitude is therefore 183 m/s.

(b) No, but it is hard to make a convincing case without more analysis. We estimate the mass of a raindrop to be about a gram or less, so that its mass and speed (from part (a)) would be less than that of a typical bullet, which is good news. But the fact that one is dealing with *many* raindrops leads us to suspect that this scenario poses an unhealthy situation. If we factor in air resistance, the final speed is smaller, of course, and we return to the relatively healthy situation with which we are familiar.

45. We neglect air resistance, which justifies setting $a = -g = -9.8$ m/s^2 (taking *down* as the $-y$ direction) for the duration of the fall. This is constant acceleration motion, which justifies the use of Table 2-1 (with Δy replacing Δx).

(a) Starting the clock at the moment the wrench is dropped ($v_0 = 0$), then $v^2 = v_0^2 - 2g\Delta y$ leads to

$$\Delta y = -\frac{(-24 \text{ m/s})^2}{2(9.8 \text{ m/s}^2)} = -29.4 \text{ m}$$

so that it fell through a height of 29.4 m.

(b) Solving $v = v_0 - gt$ for time, we find:

$$t = \frac{v_0 - v}{g} = \frac{0 - (-24 \text{ m/s})}{9.8 \text{ m/s}^2} = 2.45 \text{ s}.$$

(c) SI units are used in the graphs, and the initial position is taken as the coordinate origin. In the interest of saving space, we do not show the acceleration graph, which is a horizontal line at -9.8 m/s^2.

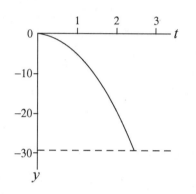
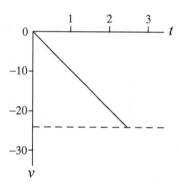

46. We neglect air resistance, which justifies setting $a = -g = -9.8$ m/s^2 (taking *down* as the $-y$ direction) for the duration of the fall. This is constant acceleration motion, which justifies the use of Table 2-1 (with Δy replacing Δx).

(a) Noting that $\Delta y = y - y_0 = -30$ m, we apply Eq. 2-15 and the quadratic formula (Appendix E) to compute t:

$$\Delta y = v_0 t - \frac{1}{2} g t^2 \Rightarrow t = \frac{v_0 \pm \sqrt{v_0^2 - 2g\Delta y}}{g}$$

which (with $v_0 = -12$ m/s since it is downward) leads, upon choosing the positive root (so that $t > 0$), to the result:

$$t = \frac{-12 \text{ m/s} + \sqrt{(-12 \text{ m/s})^2 - 2(9.8 \text{ m/s}^2)(-30 \text{ m})}}{9.8 \text{ m/s}^2} = 1.54 \text{ s}.$$

(b) Enough information is now known that any of the equations in Table 2-1 can be used to obtain v; however, the one equation that does *not* use our result from part (a) is Eq. 2-16:

$$v = \sqrt{v_0^2 - 2g\Delta y} = 27.1 \text{ m/s}$$

where the positive root has been chosen in order to give *speed* (which is the magnitude of the velocity vector).

47. We neglect air resistance for the duration of the motion (between "launching" and "landing"), so $a = -g = -9.8$ m/s^2 (we take downward to be the $-y$ direction). We use the equations in Table 2-1 (with Δy replacing Δx) because this is $a =$ constant motion.

(a) At the highest point the velocity of the ball vanishes. Taking $y_0 = 0$, we set $v = 0$ in $v^2 = v_0^2 - 2gy$, and solve for the initial velocity: $v_0 = \sqrt{2gy}$. Since $y = 50$ m we find $v_0 = 31$ m/s.

(b) It will be in the air from the time it leaves the ground until the time it returns to the ground ($y = 0$). Applying Eq. 2-15 to the entire motion (the rise and the fall, of total time $t > 0$) we have

$$y = v_0 t - \frac{1}{2}gt^2 \implies t = \frac{2v_0}{g}$$

which (using our result from part (a)) produces $t = 6.4$ s. It is possible to obtain this without using part (a)'s result; one can find the time just for the rise (from ground to highest point) from Eq. 2-16 and then double it.

(c) SI units are understood in the x and v graphs shown. In the interest of saving space, we do not show the graph of a, which is a horizontal line at -9.8 m/s^2.

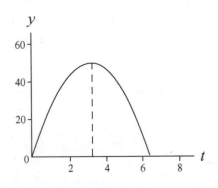
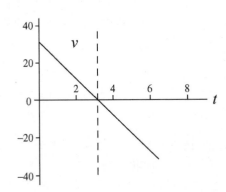

48. We neglect air resistance, which justifies setting $a = -g = -9.8$ m/s^2 (taking *down* as the $-y$ direction) for the duration of the motion. We are allowed to use Table 2-1 (with Δy replacing Δx) because this is constant acceleration motion. The ground level is taken to correspond to the origin of the y axis.

(a) Using $y = v_0 t - \frac{1}{2}gt^2$, with $y = 0.544$ m and $t = 0.200$ s, we find

$$v_0 = \frac{y + gt^2/2}{t} = \frac{0.544 \text{ m} + (9.8 \text{ m/s}^2)(0.200 \text{ s})^2/2}{0.200 \text{ s}} = 3.70 \text{ m/s}.$$

(b) The velocity at $y = 0.544$ m is

$$v = v_0 - gt = 3.70 \text{ m/s} - (9.8 \text{ m/s}^2)(0.200 \text{ s}) = 1.74 \text{ m/s}.$$

(c) Using $v^2 = v_0^2 - 2gy$ (with different values for y and v than before), we solve for the value of y corresponding to maximum height (where $v = 0$).

$$y = \frac{v_0^2}{2g} = \frac{(3.7 \text{ m/s})^2}{2(9.8 \text{ m/s}^2)} = 0.698 \text{ m}.$$

Thus, the armadillo goes $0.698 - 0.544 = 0.154$ m higher.

49. We neglect air resistance, which justifies setting $a = -g = -9.8$ m/s^2 (taking *down* as the $-y$ direction) for the duration of the motion. We are allowed to use Table 2-1 (with Δy replacing Δx) because this is constant acceleration motion. We are placing the coordinate origin on the ground. We note that the initial velocity of the package is the same as the velocity of the balloon, $v_0 = +12$ m/s and that its initial coordinate is $y_0 = +80$ m.

(a) We solve $y = y_0 + v_0 t - \frac{1}{2} g t^2$ for time, with $y = 0$, using the quadratic formula (choosing the positive root to yield a positive value for t).

$$t = \frac{v_0 + \sqrt{v_0^2 + 2 g y_0}}{g} = \frac{12 + \sqrt{12^2 + 2(9.8)(80)}}{9.8} = 5.4 \text{ s}$$

(b) If we wish to avoid using the result from part (a), we could use Eq. 2-16, but if that is not a concern, then a variety of formulas from Table 2-1 can be used. For instance, Eq. 2-11 leads to

$$v = v_0 - gt = 12 - (9.8)(5.4) = -41 \text{ m/s}.$$

Its final *speed* is 41 m/s.

50. The full extent of the bolt's fall is given by $y - y_0 = -\frac{1}{2} g t^2$ where $y - y_0 = -90$ m (if upwards is chosen as the positive y direction). Thus the time for the full fall is found to be $t = 4.29$ s. The first 80% of its free fall distance is given by $-72 = -g \tau^2/2$, which requires time $\tau = 3.83$ s.

(a) Thus, the final 20% of its fall takes $t - \tau = 0.45$ s.

(b) We can find that speed using $v = -g\tau$. Therefore, $|v| = 38$ m/s, approximately.

(c) Similarly, $v_{final} = -gt \Rightarrow |v_{final}| = 42$ m/s.

51. The speed of the boat is constant, given by $v_b = d/t$. Here, d is the distance of the boat from the bridge when the key is dropped (12 m) and t is the time the key takes in falling. To calculate t, we put the origin of the coordinate system at the point where the key is dropped and take the y axis to be positive in the *downward* direction. Taking the time to be zero at the instant the key is dropped, we compute the time t when $y = 45$ m. Since the initial velocity of the key is zero, the coordinate of the key is given by $y = \frac{1}{2} g t^2$. Thus,

$$t = \sqrt{\frac{2y}{g}} = \sqrt{\frac{2(45 \text{ m})}{9.8 \text{ m/s}^2}} = 3.03 \text{ s}.$$

Therefore, the speed of the boat is

$$v_b = \frac{12 \text{ m}}{3.03 \text{ s}} = 4.0 \text{ m/s}.$$

52. The y coordinate of Apple 1 obeys $y - y_{o1} = -\frac{1}{2}g t^2$ where $y = 0$ when $t = 2.0$ s. This allows us to solve for y_{o1}, and we find $y_{o1} = 19.6$ m.

The graph for the coordinate of Apple 2 (which is thrown apparently at $t = 1.0$ s with velocity v_2) is

$$y - y_{o2} = v_2(t - 1.0) - \frac{1}{2}g(t - 1.0)^2$$

where $y_{o2} = y_{o1} = 19.6$ m and where $y = 0$ when $t = 2.25$ s. Thus we obtain $|v_2| = 9.6$ m/s, approximately.

53. (a) With upward chosen as the $+y$ direction, we use Eq. 2-11 to find the initial velocity of the package:

$$v = v_o + at \quad \Rightarrow \quad 0 = v_o - (9.8 \text{ m/s}^2)(2.0 \text{ s})$$

which leads to $v_o = 19.6$ m/s. Now we use Eq. 2-15:

$$\Delta y = (19.6 \text{ m/s})(2.0 \text{ s}) + \frac{1}{2}(-9.8 \text{ m/s}^2)(2.0 \text{ s})^2 \approx 20 \text{ m}.$$

We note that the "2.0 s" in this second computation refers to the time interval $2 < t < 4$ in the graph (whereas the "2.0 s" in the first computation referred to the $0 < t < 2$ time interval shown in the graph).

(b) In our computation for part (b), the time interval ("6.0 s") refers to the $2 < t < 8$ portion of the graph:

$$\Delta y = (19.6 \text{ m/s})(6.0 \text{ s}) + \frac{1}{2}(-9.8 \text{ m/s}^2)(6.0 \text{ s})^2 \approx -59 \text{ m},$$

or $|\Delta y| = 59$ m.

54. We use Eq. 2-16, $v_B^2 = v_A^2 + 2a(y_B - y_A)$, with $a = -9.8$ m/s^2, $y_B - y_A = 0.40$ m, and $v_B = \frac{1}{3} v_A$. It is then straightforward to solve: $v_A = 3.0$ m/s, approximately.

55. (a) We first find the velocity of the ball just before it hits the ground. During contact with the ground its average acceleration is given by

$$a_{avg} = \frac{\Delta v}{\Delta t}$$

where Δv is the change in its velocity during contact with the ground and $\Delta t = 20.0 \times 10^{-3}$ s is the duration of contact. Now, to find the velocity just *before* contact, we put the origin at the point where the ball is dropped (and take $+y$ upward) and take $t = 0$ to be when it is dropped. The ball strikes the ground at $y = -15.0$ m. Its velocity there is found from Eq. 2-16: $v^2 = -2gy$. Therefore,

$$v = -\sqrt{-2gy} = -\sqrt{-2(9.8 \text{ m/s}^2)(-15.0 \text{ m})} = -17.1 \text{ m/s}$$

where the negative sign is chosen since the ball is traveling downward at the moment of contact. Consequently, the average acceleration during contact with the ground is

$$a_{avg} = \frac{0 - (-17.1 \text{ m/s})}{20.0 \times 10^{-3} \text{ s}} = 857 \text{ m/s}^2.$$

(b) The fact that the result is positive indicates that this acceleration vector points upward. In a later chapter, this will be directly related to the magnitude and direction of the force exerted by the ground on the ball during the collision.

56. (a) We neglect air resistance, which justifies setting $a = -g = -9.8$ m/s² (taking *down* as the $-y$ direction) for the duration of the motion. We are allowed to use Eq. 2-15 (with Δy replacing Δx) because this is constant acceleration motion. We use primed variables (except t) with the first stone, which has zero initial velocity, and unprimed variables with the second stone (with initial downward velocity $-v_0$, so that v_0 is being used for the initial *speed*). SI units are used throughout.

$$\Delta y' = 0(t) - \frac{1}{2}gt^2$$

$$\Delta y = (-v_0)(t-1) - \frac{1}{2}g(t-1)^2$$

Since the problem indicates $\Delta y' = \Delta y = -43.9$ m, we solve the first equation for t (finding $t = 2.99$ s) and use this result to solve the second equation for the initial speed of the second stone:

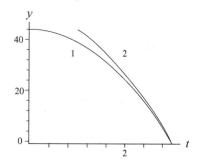

$$-43.9 \text{ m} = (-v_0)(1.99 \text{ s}) - \frac{1}{2}(9.8 \text{ m/s}^2)(1.99 \text{ s})^2$$

which leads to $v_0 = 12.3$ m/s.

(b) The velocity of the stones are given by

$$v'_y = \frac{d(\Delta y')}{dt} = -gt, \qquad v_y = \frac{d(\Delta y)}{dt} = -v_0 - g(t-1)$$

The plot is shown below:

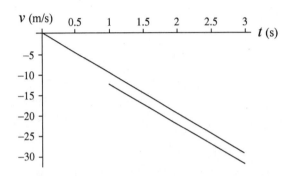

57. The average acceleration during contact with the floor is $a_{avg} = (v_2 - v_1)/\Delta t$, where v_1 is its velocity just before striking the floor, v_2 is its velocity just as it leaves the floor, and Δt is the duration of contact with the floor (12×10^{-3} s).

(a) Taking the y axis to be positively upward and placing the origin at the point where the ball is dropped, we first find the velocity just before striking the floor, using $v_1^2 = v_0^2 - 2gy$. With $v_0 = 0$ and $y = -4.00$ m, the result is

$$v_1 = -\sqrt{-2gy} = -\sqrt{-2(9.8 \text{ m/s}^2)(-4.00 \text{ m})} = -8.85 \text{ m/s}$$

where the negative root is chosen because the ball is traveling downward. To find the velocity just after hitting the floor (as it ascends without air friction to a height of 2.00 m), we use $v^2 = v_2^2 - 2g(y - y_0)$ with $v = 0$, $y = -2.00$ m (it ends up two meters *below* its initial drop height), and $y_0 = -4.00$ m. Therefore,

$$v_2 = \sqrt{2g(y - y_0)} = \sqrt{2(9.8 \text{ m/s}^2)(-2.00 \text{ m} + 4.00 \text{ m})} = 6.26 \text{ m/s}.$$

Consequently, the average acceleration is

$$a_{avg} = \frac{v_2 - v_1}{\Delta t} = \frac{6.26 \text{ m/s} - (-8.85 \text{ m/s})}{12.0 \times 10^{-3} \text{ s}} = 1.26 \times 10^3 \text{ m/s}^2.$$

(b) The positive nature of the result indicates that the acceleration vector points upward. In a later chapter, this will be directly related to the magnitude and direction of the force exerted by the ground on the ball during the collision.

58. To find the "launch" velocity of the rock, we apply Eq. 2-11 to the maximum height (where the speed is momentarily zero)

$$v = v_0 - gt \quad \Rightarrow \quad 0 = v_0 - (9.8 \text{ m/s}^2)(2.5 \text{ s})$$

so that v_0 = 24.5 m/s (with +y up). Now we use Eq. 2-15 to find the height of the tower (taking y_0 = 0 at the ground level)

$$y - y_0 = v_0 t + \frac{1}{2}at^2 \quad \Rightarrow \quad y - 0 = (24.5 \text{ m/s})(1.5 \text{ s}) - \frac{1}{2}(9.8 \text{ m/s}^2)(1.5 \text{ s})^2.$$

Thus, we obtain y = 26 m.

59. We neglect air resistance, which justifies setting $a = -g = -9.8$ m/s² (taking *down* as the $-y$ direction) for the duration of the motion. We are allowed to use Table 2-1 (with Δy replacing Δx) because this is constant acceleration motion. The ground level is taken to correspond to the origin of the y axis.

(a) The time drop 1 leaves the nozzle is taken as $t = 0$ and its time of landing on the floor t_1 can be computed from Eq. 2-15, with $v_0 = 0$ and $y_1 = -2.00$ m.

$$y_1 = -\frac{1}{2}gt_1^2 \quad \Rightarrow \quad t_1 = \sqrt{\frac{-2y}{g}} = \sqrt{\frac{-2(-2.00 \text{ m})}{9.8 \text{ m/s}^2}} = 0.639 \text{ s}.$$

At that moment, the fourth drop begins to fall, and from the regularity of the dripping we conclude that drop 2 leaves the nozzle at $t = 0.639/3 = 0.213$ s and drop 3 leaves the nozzle at $t = 2(0.213 \text{ s}) = 0.426$ s. Therefore, the time in free fall (up to the moment drop 1 lands) for drop 2 is $t_2 = t_1 - 0.213$ s = 0.426 s. Its position at the moment drop 1 strikes the floor is

$$y_2 = -\frac{1}{2}gt_2^2 = -\frac{1}{2}(9.8 \text{ m/s}^2)(0.426 \text{ s})^2 = -0.889 \text{ m},$$

or 89 cm below the nozzle.

(b) The time in free fall (up to the moment drop 1 lands) for drop 3 is $t_3 = t_1 - 0.426$ s = 0.213 s. Its position at the moment drop 1 strikes the floor is

$$y_3 = -\frac{1}{2}gt_3^2 = -\frac{1}{2}(9.8 \text{ m/s}^2)(0.213 \text{ s})^2 = -0.222 \text{ m},$$

or 22 cm below the nozzle.

60. We choose *down* as the +y direction and set the coordinate origin at the point where it was dropped (which is when we start the clock). We denote the 1.00 s duration mentioned in the problem as $t - t'$ where t is the value of time when it lands and t' is one second prior

to that. The corresponding distance is $y - y' = 0.50h$, where y denotes the location of the ground. In these terms, y is the same as h, so we have $h - y' = 0.50h$ or $0.50h = y'$.

(a) We find t' and t from Eq. 2-15 (with $v_0 = 0$):

$$y' = \frac{1}{2}gt'^2 \Rightarrow t' = \sqrt{\frac{2y'}{g}}$$

$$y = \frac{1}{2}gt^2 \Rightarrow t = \sqrt{\frac{2y}{g}}.$$

Plugging in $y = h$ and $y' = 0.50h$, and dividing these two equations, we obtain

$$\frac{t'}{t} = \sqrt{\frac{2(0.50h)/g}{2h/g}} = \sqrt{0.50}.$$

Letting $t' = t - 1.00$ (SI units understood) and cross-multiplying, we find

$$t - 1.00 = t\sqrt{0.50} \Rightarrow t = \frac{1.00}{1 - \sqrt{0.50}}$$

which yields $t = 3.41$ s.

(b) Plugging this result into $y = \frac{1}{2}gt^2$ we find $h = 57$ m.

(c) In our approach, we did not use the quadratic formula, but we did "choose a root" when we assumed (in the last calculation in part (a)) that $\sqrt{0.50} = +0.707$ instead of -0.707. If we had instead let $\sqrt{0.50} = -0.707$ then our answer for t would have been roughly 0.6 s which would imply that $t' = t - 1$ would equal a negative number (indicating a time *before* it was dropped) which certainly does not fit with the physical situation described in the problem.

61. The time t the pot spends passing in front of the window of length $L = 2.0$ m is 0.25 s each way. We use v for its velocity as it passes the top of the window (going up). Then, with $a = -g = -9.8$ m/s² (taking *down* to be the $-y$ direction), Eq. 2-18 yields

$$L = vt - \frac{1}{2}gt^2 \Rightarrow v = \frac{L}{t} - \frac{1}{2}gt.$$

The distance H the pot goes above the top of the window is therefore (using Eq. 2-16 with the *final velocity* being zero to indicate the highest point)

$$H = \frac{v^2}{2g} = \frac{(L/t - gt/2)^2}{2g} = \frac{(2.00 \text{ m}/0.25 \text{ s} - (9.80 \text{ m/s}^2)(0.25 \text{ s})/2)^2}{2(9.80 \text{ m/s}^2)} = 2.34 \text{ m}.$$

62. The graph shows $y = 25$ m to be the highest point (where the speed momentarily vanishes). The neglect of "air friction" (or whatever passes for that on the distant planet) is certainly reasonable due to the symmetry of the graph.

(a) To find the acceleration due to gravity g_p on that planet, we use Eq. 2-15 (with $+y$ up)

$$y - y_0 = vt + \frac{1}{2}g_p t^2 \quad \Rightarrow \quad 25 \text{ m} - 0 = (0)(2.5 \text{ s}) + \frac{1}{2}g_p (2.5 \text{ s})^2$$

so that $g_p = 8.0$ m/s^2.

(b) That same (max) point on the graph can be used to find the initial velocity.

$$y - y_0 = \frac{1}{2}(v_0 + v)t \quad \Rightarrow \quad 25 \text{ m} - 0 = \frac{1}{2}(v_0 + 0)(2.5 \text{ s})$$

Therefore, $v_0 = 20$ m/s.

63. We choose *down* as the $+y$ direction and place the coordinate origin at the top of the building (which has height H). During its fall, the ball passes (with velocity v_1) the top of the window (which is at y_1) at time t_1, and passes the bottom (which is at y_2) at time t_2. We are told $y_2 - y_1 = 1.20$ m and $t_2 - t_1 = 0.125$ s. Using Eq. 2-15 we have

$$y_2 - y_1 = v_1(t_2 - t_1) + \frac{1}{2}g(t_2 - t_1)^2$$

which immediately yields

$$v_1 = \frac{1.20 \text{ m} - \frac{1}{2}(9.8 \text{ m/s}^2)(0.125 \text{ s})^2}{0.125 \text{ s}} = 8.99 \text{ m/s}.$$

From this, Eq. 2-16 (with $v_0 = 0$) reveals the value of y_1:

$$v_1^2 = 2gy_1 \quad \Rightarrow \quad y_1 = \frac{(8.99 \text{ m/s})^2}{2(9.8 \text{ m/s}^2)} = 4.12 \text{ m}.$$

It reaches the ground ($y_3 = H$) at t_3. Because of the symmetry expressed in the problem ("upward flight is a reverse of the fall") we know that $t_3 - t_2 = 2.00/2 = 1.00$ s. And this means $t_3 - t_1 = 1.00$ s $+ 0.125$ s $= 1.125$ s. Now Eq. 2-15 produces

$$y_3 - y_1 = v_1(t_3 - t_1) + \frac{1}{2}g(t_3 - t_1)^2$$

$$y_3 - 4.12 \text{ m} = (8.99 \text{ m/s})(1.125 \text{ s}) + \frac{1}{2}(9.8 \text{ m/s}^2)(1.125 \text{ s})^2$$

which yields $y_3 = H = 20.4$ m.

64. The height reached by the player is $y = 0.76$ m (where we have taken the origin of the y axis at the floor and $+y$ to be upward).

(a) The initial velocity v_0 of the player is

$$v_0 = \sqrt{2gy} = \sqrt{2(9.8 \text{ m/s}^2)(0.76 \text{ m})} = 3.86 \text{ m/s}.$$

This is a consequence of Eq. 2-16 where velocity v vanishes. As the player reaches $y_1 = 0.76 \text{ m} - 0.15 \text{ m} = 0.61 \text{ m}$, his speed v_1 satisfies $v_0^2 - v_1^2 = 2gy_1$, which yields

$$v_1 = \sqrt{v_0^2 - 2gy_1} = \sqrt{(3.86 \text{ m/s})^2 - 2(9.80 \text{ m/s}^2)(0.61 \text{ m})} = 1.71 \text{ m/s}.$$

The time t_1 that the player spends *ascending* in the top $\Delta y_1 = 0.15 \text{ m}$ of the jump can now be found from Eq. 2-17:

$$\Delta y_1 = \frac{1}{2}(v_1 + v)t_1 \Rightarrow t_1 = \frac{2(0.15 \text{ m})}{1.71 \text{ m/s} + 0} = 0.175 \text{ s}$$

which means that the total time spent in that top 15 cm (both ascending and descending) is $2(0.175 \text{ s}) = 0.35 \text{ s} = 350 \text{ ms}$.

(b) The time t_2 when the player reaches a height of 0.15 m is found from Eq. 2-15:

$$0.15 \text{ m} = v_0 t_2 - \frac{1}{2} g t_2^2 = (3.86 \text{ m/s}) t_2 - \frac{1}{2}(9.8 \text{ m/s}^2) t_2^2,$$

which yields (using the quadratic formula, taking the smaller of the two positive roots) $t_2 = 0.041 \text{ s} = 41 \text{ ms}$, which implies that the total time spent in that bottom 15 cm (both ascending and descending) is $2(41 \text{ ms}) = 82 \text{ ms}$.

65. The key idea here is that the speed of the head (and the torso as well) at any given time can be calculated by finding the area on the graph of the head's acceleration versus time, as shown in Eq. 2-26:

$$v_1 - v_0 = \begin{pmatrix} \text{area between the acceleration curve} \\ \text{and the time axis, from } t_0 \text{ to } t_1 \end{pmatrix}$$

(a) From Fig. 2.14a, we see that the head begins to accelerate from rest ($v_0 = 0$) at $t_0 = 110$ ms and reaches a maximum value of 90 m/s^2 at $t_1 = 160$ ms. The area of this region is

$$\text{area} = \frac{1}{2}(160 - 110) \times 10^{-3} \text{s} \cdot (90 \text{ m/s}^2) = 2.25 \text{ m/s}$$

which is equal to v_1, the speed at t_1.

(b) To compute the speed of the torso at $t_1 = 160$ ms, we divide the area into 4 regions: From 0 to 40 ms, region A has zero area. From 40 ms to 100 ms, region B has the shape of a triangle with area

$$\text{area}_B = \frac{1}{2}(0.0600 \text{ s})(50.0 \text{ m/s}^2) = 1.50 \text{ m/s}.$$

From 100 to 120 ms, region C has the shape of a rectangle with area

$$\text{area}_C = (0.0200 \text{ s}) (50.0 \text{ m/s}^2) = 1.00 \text{ m/s}.$$

From 110 to 160 ms, region D has the shape of a trapezoid with area

$$\text{area}_D = \frac{1}{2}(0.0400 \text{ s}) (50.0 + 20.0) \text{ m/s}^2 = 1.40 \text{ m/s}.$$

Substituting these values into Eq. 2-26, with $v_0=0$ then gives

$$v_1 - 0 = 0 + 1.50 \text{ m/s} + 1.00 \text{ m/s} + 1.40 \text{ m/s} = 3.90 \text{ m/s},$$

or $v_1 = 3.90$ m/s.

66. This problem can be solved by noting that velocity can be determined by the graphical integration of acceleration versus time. The speed of the tongue of the salamander is simply equal to the area under the acceleration curve:

$$v = \text{area} = \frac{1}{2}(10^{-2} \text{ s})(100 \text{ m/s}^2) + \frac{1}{2}(10^{-2} \text{ s})(100 \text{ m/s}^2 + 400 \text{ m/s}^2) + \frac{1}{2}(10^{-2} \text{ s})(400 \text{ m/s}^2)$$
$$= 5.0 \text{ m/s}.$$

67. Since $v = dx/dt$ (Eq. 2-4), then $\Delta x = \int v \, dt$, which corresponds to the area under the v vs t graph. Dividing the total area A into rectangular (base × height) and triangular ($\frac{1}{2}$ base × height) areas, we have

$$A = A_{0<t<2} + A_{2<t<10} + A_{10<t<12} + A_{12<t<16}$$
$$= \frac{1}{2}(2)(8) + (8)(8) + \left((2)(4) + \frac{1}{2}(2)(4)\right) + (4)(4)$$

with SI units understood. In this way, we obtain $\Delta x = 100$ m.

68. The key idea here is that the position of an object at any given time can be calculated by finding the area on the graph of the object's velocity versus time, as shown in Eq. 2-25:

$$x_1 - x_0 = \begin{pmatrix} \text{area between the velocity curve} \\ \text{and the time axis, from } t_0 \text{ to } t_1 \end{pmatrix}.$$

(a) To compute the position of the fist at $t = 50$ ms, we divide the area in Fig. 2-37 into two regions. From 0 to 10 ms, region A has the shape of a triangle with area

$$\text{area}_A = \frac{1}{2}(0.010 \text{ s}) (2 \text{ m/s}) = 0.01 \text{ m}.$$

From 10 to 50 ms, region B has the shape of a trapezoid with area

$$\text{area}_B = \frac{1}{2}(0.040 \text{ s})(2+4) \text{ m/s} = 0.12 \text{ m}.$$

Substituting these values into Eq. 2-25, with $x_0=0$ then gives

$$x_1 - 0 = 0 + 0.01 \text{ m} + 0.12 \text{ m} = 0.13 \text{ m},$$

or $x_1 = 0.13$ m.

(b) The speed of the fist reaches a maximum at $t_1 = 120$ ms. From 50 to 90 ms, region C has the shape of a trapezoid with area

$$\text{area}_C = \frac{1}{2}(0.040 \text{ s})(4+5) \text{ m/s} = 0.18 \text{ m}.$$

From 90 to 120 ms, region D has the shape of a trapezoid with area

$$\text{area}_D = \frac{1}{2}(0.030 \text{ s})(5+7.5) \text{ m/s} = 0.19 \text{ m}.$$

Substituting these values into Eq. 2-25, with $x_0=0$ then gives

$$x_1 - 0 = 0 + 0.01 \text{ m} + 0.12 \text{ m} + 0.18 \text{ m} + 0.19 \text{ m} = 0.50 \text{ m},$$

or $x_1 = 0.50$ m.

69. The problem is solved using Eq. 2-26:

$$v_1 - v_0 = \begin{pmatrix} \text{area between the acceleration curve} \\ \text{and the time axis, from } t_0 \text{ to } t_1 \end{pmatrix}$$

To compute the speed of the unhelmeted, bare head at $t_1=7.0$ ms, we divide the area under the a vs. t graph into 4 regions: From 0 to 2 ms, region A has the shape of a triangle with area

$$\text{area}_A = \frac{1}{2}(0.0020 \text{ s})(120 \text{ m/s}^2) = 0.12 \text{ m/s}.$$

From 2 ms to 4 ms, region B has the shape of a trapezoid with area

$$\text{area}_B = \frac{1}{2}(0.0020 \text{ s})(120+140) \text{ m/s}^2 = 0.26 \text{ m/s}.$$

From 4 to 6 ms, region C has the shape of a trapezoid with area

$$\text{area}_C = \frac{1}{2}(0.0020 \text{ s})(140+200) \text{ m/s}^2 = 0.34 \text{ m/s}.$$

From 6 to 7 ms, region D has the shape of a triangle with area

$$\text{area}_D = \frac{1}{2}(0.0010 \text{ s})(200 \text{ m/s}^2) = 0.10 \text{ m/s}.$$

Substituting these values into Eq. 2-26, with $v_0=0$ then gives

$$v_{unhelmeted} = 0.12 \text{ m/s} + 0.26 \text{ m/s} + 0.34 \text{ m/s} + 0.10 \text{ m/s} = 0.82 \text{ m/s}.$$

Carrying out similar calculations for the helmeted head, we have the following results: From 0 to 3 ms, region A has the shape of a triangle with area

$$\text{area}_A = \frac{1}{2}(0.0030 \text{ s})(40 \text{ m/s}^2) = 0.060 \text{ m/s}.$$

From 3 ms to 4 ms, region B has the shape of a rectangle with area

$$\text{area}_B = (0.0010 \text{ s})(40 \text{ m/s}^2) = 0.040 \text{ m/s}.$$

From 4 to 6 ms, region C has the shape of a trapezoid with area

$$\text{area}_C = \frac{1}{2}(0.0020 \text{ s})(40+80) \text{ m/s}^2 = 0.12 \text{ m/s}.$$

From 6 to 7 ms, region D has the shape of a triangle with area

$$\text{area}_D = \frac{1}{2}(0.0010 \text{ s})(80 \text{ m/s}^2) = 0.040 \text{ m/s}.$$

Substituting these values into Eq. 2-26, with $v_0=0$ then gives

$$v_{helmeted} = 0.060 \text{ m/s} + 0.040 \text{ m/s} + 0.12 \text{ m/s} + 0.040 \text{ m/s} = 0.26 \text{ m/s}.$$

Thus, the difference in the speed is

$$\Delta v = v_{unhelmeted} - v_{helmeted} = 0.82 \text{ m/s} - 0.26 \text{ m/s} = 0.56 \text{ m/s}.$$

70. To solve this problem, we note that velocity is equal to the time derivative of a position function, as well as the time integral of an acceleration function, with the integration constant being the initial velocity. Thus, the velocity of particle 1 can be written as

$$v_1 = \frac{dx_1}{dt} = \frac{d}{dt}\left(6.00t^2 + 3.00t + 2.00\right) = 12.0t + 3.00.$$

Similarly, the velocity of particle 2 is

$$v_2 = v_{20} + \int a_2 dt = 20.0 + \int (-8.00t)dt = 20.0 - 4.00t^2.$$

The condition that $v_1 = v_2$ implies

$$12.0t + 3.00 = 20.0 - 4.00t^2 \implies 4.00t^2 + 12.0t - 17.0 = 0$$

which can be solved to give (taking positive root) $t = (-3 + \sqrt{26})/2 = 1.05$ s. Thus, the velocity at this time is $v_1 = v_2 = 12.0(1.05) + 3.00 = 15.6$ m/s.

71. We denote the required time as *t*, assuming the light turns green when the clock reads zero. By this time, the distances traveled by the two vehicles must be the same.

(a) Denoting the acceleration of the automobile as *a* and the (constant) speed of the truck as *v* then

$$\Delta x = \left(\frac{1}{2}at^2\right)_{car} = (vt)_{truck}$$

which leads to

$$t = \frac{2v}{a} = \frac{2(9.5 \text{ m/s})}{2.2 \text{ m/s}^2} = 8.6 \text{ s}.$$

Therefore,

$$\Delta x = vt = (9.5 \text{ m/s})(8.6 \text{ s}) = 82 \text{ m}.$$

(b) The speed of the car at that moment is

$$v_{car} = at = (2.2 \text{ m/s}^2)(8.6 \text{ s}) = 19 \text{ m/s}.$$

72. (a) A constant velocity is equal to the ratio of displacement to elapsed time. Thus, for the vehicle to be traveling at a constant speed v_p over a distance D_{23}, the time delay should be $t = D_{23}/v_p$.

(b) The time required for the car to accelerate from rest to a cruising speed v_p is $t_0 = v_p/a$. During this time interval, the distance traveled is $\Delta x_0 = at_0^2/2 = v_p^2/2a$. The car then moves at a constant speed v_p over a distance $D_{12} - \Delta x_0 - d$ to reach intersection 2, and the time elapsed is $t_1 = (D_{12} - \Delta x_0 - d)/v_p$. Thus, the time delay at intersection 2 should be set to

$$t_{total} = t_r + t_0 + t_1 = t_r + \frac{v_p}{a} + \frac{D_{12} - \Delta x_0 - d}{v_p} = t_r + \frac{v_p}{a} + \frac{D_{12} - (v_p^2/2a) - d}{v_p}$$

$$= t_r + \frac{1}{2}\frac{v_p}{a} + \frac{D_{12} - d}{v_p}$$

73. (a) The derivative (with respect to time) of the given expression for x yields the "velocity" of the spot:

$$v(t) = 9 - \tfrac{9}{4} t^2$$

with 3 significant figures understood. It is easy to see that $v = 0$ when $t = 2.00$ s.

(b) At $t = 2$ s, $x = 9(2) - \tfrac{3}{4}(2)^3 = 12$. Thus, the location of the spot when $v = 0$ is 12.0 cm from left edge of screen.

(c) The derivative of the velocity is $a = -\tfrac{9}{2} t$ which gives an acceleration (leftward) of magnitude 9.00 m/s^2 when the spot is 12 cm from left edge of screen.

(d) Since $v > 0$ for times less than $t = 2$ s, then the spot had been moving rightwards.

(e) As implied by our answer to part (c), it moves leftward for times immediately after $t = 2$ s. In fact, the expression found in part (a) guarantees that for all $t > 2$, $v < 0$ (that is, until the clock is "reset" by reaching an edge).

(f) As the discussion in part (e) shows, the edge that it reaches at some $t > 2$ s cannot be the right edge; it is the left edge ($x = 0$). Solving the expression given in the problem statement (with $x = 0$) for positive t yields the answer: the spot reaches the left edge at $t = \sqrt{12}$ s ≈ 3.46 s.

74. (a) Let the height of the diving board be h. We choose *down* as the $+y$ direction and set the coordinate origin at the point where it was dropped (which is when we start the clock). Thus, $y = h$ designates the location where the ball strikes the water. Let the depth of the lake be D, and the total time for the ball to descend be T. The speed of the ball as it reaches the surface of the lake is then $v = \sqrt{2gh}$ (from Eq. 2-16), and the time for the ball to fall from the board to the lake surface is $t_1 = \sqrt{2h/g}$ (from Eq. 2-15). Now, the time it spends descending in the lake (at constant velocity v) is

$$t_2 = \frac{D}{v} = \frac{D}{\sqrt{2gh}}.$$

Thus, $T = t_1 + t_2 = \sqrt{\frac{2h}{g}} + \frac{D}{\sqrt{2gh}}$, which gives

$$D = T\sqrt{2gh} - 2h = (4.80 \text{ s})\sqrt{(2)(9.80 \text{ m/s}^2)(5.20 \text{ m})} - 2(5.20 \text{ m}) = 38.1 \text{ m}.$$

(b) Using Eq. 2-2, the magnitude of the average velocity is

$$v_{avg} = \frac{D+h}{T} = \frac{38.1 \text{ m} + 5.20 \text{ m}}{4.80 \text{ s}} = 9.02 \text{ m/s}$$

(c) In our coordinate choices, a positive sign for v_{avg} means that the ball is going downward. If, however, upwards had been chosen as the positive direction, then this answer in (b) would turn out negative-valued.

(d) We find v_0 from $\Delta y = v_0 t + \frac{1}{2} g t^2$ with $t = T$ and $\Delta y = h + D$. Thus,

$$v_0 = \frac{h+D}{T} - \frac{gT}{2} = \frac{5.20 \text{ m} + 38.1 \text{ m}}{4.80 \text{ s}} - \frac{(9.8 \text{ m/s}^2)(4.80 \text{ s})}{2} = 14.5 \text{ m/s}$$

(e) Here in our coordinate choices the negative sign means that the ball is being thrown upward.

75. We choose *down* as the $+y$ direction and use the equations of Table 2-1 (replacing x with y) with $a = +g$, $v_0 = 0$ and $y_0 = 0$. We use subscript 2 for the elevator reaching the ground and 1 for the halfway point.

(a) Eq. 2-16, $v_2^2 = v_0^2 + 2a(y_2 - y_0)$, leads to

$$v_2 = \sqrt{2gy_2} = \sqrt{2(9.8 \text{ m/s}^2)(120 \text{ m})} = 48.5 \text{ m/s}.$$

(b) The time at which it strikes the ground is (using Eq. 2-15)

$$t_2 = \sqrt{\frac{2y_2}{g}} = \sqrt{\frac{2(120 \text{ m})}{9.8 \text{ m/s}^2}} = 4.95 \text{ s}.$$

(c) Now Eq. 2-16, in the form $v_1^2 = v_0^2 + 2a(y_1 - y_0)$, leads to

$$v_1 = \sqrt{2gy_1} = \sqrt{2(9.8 \text{ m/s}^2)(60 \text{ m})} = 34.3 \text{ m/s}.$$

(d) The time at which it reaches the halfway point is (using Eq. 2-15)

$$t_1 = \sqrt{\frac{2y_1}{g}} = \sqrt{\frac{2(60 \text{ m})}{9.8 \text{ m/s}^2}} = 3.50 \text{ s}.$$

76. Taking $+y$ to be upward and placing the origin at the point from which the objects are dropped, then the location of diamond 1 is given by $y_1 = -\frac{1}{2} g t^2$ and the location of diamond 2 is given by $y_2 = -\frac{1}{2} g (t-1)^2$. We are starting the clock when the first object is dropped. We want the time for which $y_2 - y_1 = 10$ m. Therefore,

$$-\frac{1}{2}g(t-1)^2 + \frac{1}{2}gt^2 = 10 \quad \Rightarrow \quad t = (10/g) + 0.5 = 1.5 \text{ s}.$$

77. Assuming the horizontal velocity of the ball is constant, the horizontal displacement is

$$\Delta x = v\Delta t$$

where Δx is the horizontal distance traveled, Δt is the time, and v is the (horizontal) velocity. Converting v to meters per second, we have 160 km/h = 44.4 m/s. Thus

$$\Delta t = \frac{\Delta x}{v} = \frac{18.4 \text{ m}}{44.4 \text{ m/s}} = 0.414 \text{ s}.$$

The velocity-unit conversion implemented above can be figured "from basics" (1000 m = 1 km, 3600 s = 1 h) or found in Appendix D.

78. In this solution, we make use of the notation $x(t)$ for the value of x at a particular t. Thus, $x(t) = 50t + 10t^2$ with SI units (meters and seconds) understood.

(a) The average velocity during the first 3 s is given by

$$v_{avg} = \frac{x(3) - x(0)}{\Delta t} = \frac{(50)(3) + (10)(3)^2 - 0}{3} = 80 \text{ m/s}.$$

(b) The instantaneous velocity at time t is given by $v = dx/dt = 50 + 20t$, in SI units. At $t = 3.0$ s, $v = 50 + (20)(3.0) = 110$ m/s.

(c) The instantaneous acceleration at time t is given by $a = dv/dt = 20$ m/s^2. It is constant, so the acceleration at any time is 20 m/s^2.

(d) and (e) The graphs that follow show the coordinate x and velocity v as functions of time, with SI units understood. The dashed line marked (a) in the first graph runs from $t = 0$, $x = 0$ to $t = 3.0$s, $x = 240$ m. Its slope is the average velocity during the first 3s of motion. The dashed line marked (b) is tangent to the x curve at $t = 3.0$ s. Its slope is the instantaneous velocity at $t = 3.0$ s.

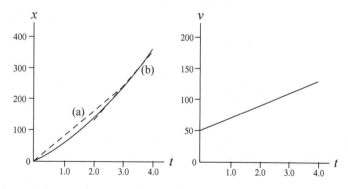

79. We take $+x$ in the direction of motion, so $v_0 = +30$ m/s, $v_1 = +15$ m/s and $a < 0$. The acceleration is found from Eq. 2-11: $a = (v_1 - v_0)/t_1$ where $t_1 = 3.0$ s. This gives $a = -5.0$

m/s². The displacement (which in this situation is the same as the distance traveled) to the point it stops ($v_2 = 0$) is, using Eq. 2-16,

$$v_2^2 = v_0^2 + 2a\Delta x \quad \Rightarrow \quad \Delta x = -\frac{(30 \text{ m/s})^2}{2(-5 \text{ m/s}^2)} = 90 \text{ m}.$$

80. If the plane (with velocity v) maintains its present course, and if the terrain continues its upward slope of 4.3°, then the plane will strike the ground after traveling

$$\Delta x = \frac{h}{\tan \theta} = \frac{35 \text{ m}}{\tan 4.3°} = 465.5 \text{ m} \approx 0.465 \text{ km}.$$

This corresponds to a time of flight found from Eq. 2-2 (with $v = v_{avg}$ since it is constant)

$$t = \frac{\Delta x}{v} = \frac{0.465 \text{ km}}{1300 \text{ km/h}} = 0.000358 \text{ h} \approx 1.3 \text{ s}.$$

This, then, estimates the time available to the pilot to make his correction.

81. The problem consists of two constant-acceleration parts: part 1 with $v_0 = 0$, $v = 6.0$ m/s, $x = 1.8$ m, and $x_0 = 0$ (if we take its original position to be the coordinate origin); and, part 2 with $v_0 = 6.0$ m/s, $v = 0$, and $a_2 = -2.5$ m/s² (negative because we are taking the positive direction to be the direction of motion).

(a) We can use Eq. 2-17 to find the time for the first part

$$x - x_0 = \frac{1}{2}(v_0 + v)t_1 \quad \Rightarrow \quad 1.8 \text{ m} - 0 = \frac{1}{2}(0 + 6.0 \text{ m/s})t_1$$

so that $t_1 = 0.6$ s. And Eq. 2-11 is used to obtain the time for the second part

$$v = v_0 + a_2 t_2 \quad \Rightarrow \quad 0 = 6.0 \text{ m/s} + (-2.5 \text{ m/s}^2)t_2$$

from which $t_2 = 2.4$ s is computed. Thus, the total time is $t_1 + t_2 = 3.0$ s.

(b) We already know the distance for part 1. We could find the distance for part 2 from several of the equations, but the one that makes no use of our part (a) results is Eq. 2-16

$$v^2 = v_0^2 + 2a_2 \Delta x_2 \quad \Rightarrow \quad 0 = (6.0 \text{ m/s})^2 + 2(-2.5 \text{ m/s}^2)\Delta x_2$$

which leads to $\Delta x_2 = 7.2$ m. Therefore, the total distance traveled by the shuffleboard disk is $(1.8 + 7.2)$ m $= 9.0$ m.

82. The time required is found from Eq. 2-11 (or, suitably interpreted, Eq. 2-7). First, we convert the velocity change to SI units:

$$\Delta v = (100 \text{ km/h})\left(\frac{1000 \text{ m/km}}{3600 \text{ s/h}}\right) = 27.8 \text{ m/s}.$$

Thus, $\Delta t = \Delta v/a = 27.8/50 = 0.556$ s.

83. From Table 2-1, $v^2 - v_0^2 = 2a\Delta x$ is used to solve for a. Its minimum value is

$$a_{min} = \frac{v_2^2 - v_0^2}{2\Delta x_{max}} = \frac{(360 \text{ km/h})^2}{2(1.80 \text{ km})} = 36000 \text{ km/h}^2$$

which converts to 2.78 m/s².

84. (a) For the automobile $\Delta v = 55 - 25 = 30$ km/h, which we convert to SI units:

$$a = \frac{\Delta v}{\Delta t} = \frac{(30 \text{ km/h})(\frac{1000 \text{ m/km}}{3600 \text{ s/h}})}{(0.50 \text{ min})(60 \text{ s/min})} = 0.28 \text{ m/s}^2.$$

(b) The change of velocity for the bicycle, for the same time, is identical to that of the car, so its acceleration is also 0.28 m/s².

85. We denote t_r as the reaction time and t_b as the braking time. The motion during t_r is of the constant-velocity (call it v_0) type. Then the position of the car is given by

$$x = v_0 t_r + v_0 t_b + \frac{1}{2}at_b^2$$

where v_0 is the initial velocity and a is the acceleration (which we expect to be negative-valued since we are taking the velocity in the positive direction and we know the car is decelerating). *After* the brakes are applied the velocity of the car is given by $v = v_0 + at_b$. Using this equation, with $v = 0$, we eliminate t_b from the first equation and obtain

$$x = v_0 t_r - \frac{v_0^2}{a} + \frac{1}{2}\frac{v_0^2}{a} = v_0 t_r - \frac{1}{2}\frac{v_0^2}{a}.$$

We write this equation for each of the initial velocities:

$$x_1 = v_{01} t_r - \frac{1}{2}\frac{v_{01}^2}{a}$$

and

$$x_2 = v_{02} t_r - \frac{1}{2}\frac{v_{02}^2}{a}.$$

Solving these equations simultaneously for t_r and a we get

$$t_r = \frac{v_{02}^2 x_1 - v_{01}^2 x_2}{v_{01} v_{02}(v_{02} - v_{01})}$$

and

$$a = -\frac{1}{2}\frac{v_{02}v_{01}^2 - v_{01}v_{02}^2}{v_{02}x_1 - v_{01}x_2}.$$

(a) Substituting $x_1 = 56.7$ m, $v_{01} = 80.5$ km/h $= 22.4$ m/s, $x_2 = 24.4$ m and $v_{02} = 48.3$ km/h $= 13.4$ m/s, we find

$$t_r = \frac{v_{02}^2 x_1 - v_{01}^2 x_2}{v_{01}v_{02}(v_{02}-v_{01})} = \frac{(13.4 \text{ m/s})^2(56.7 \text{ m}) - (22.4 \text{ m/s})^2(24.4 \text{ m})}{(22.4 \text{ m/s})(13.4 \text{ m/s})(13.4 \text{ m/s} - 22.4 \text{ m/s})} = 0.74 \text{ s}.$$

(b) In a similar manner, substituting $x_1 = 56.7$ m, $v_{01} = 80.5$ km/h $= 22.4$ m/s, $x_2 = 24.4$ m and $v_{02} = 48.3$ km/h $= 13.4$ m/s gives

$$a = -\frac{1}{2}\frac{v_{02}v_{01}^2 - v_{01}v_{02}^2}{v_{02}x_1 - v_{01}x_2} = -\frac{1}{2}\frac{(13.4 \text{ m/s})(22.4 \text{ m/s})^2 - (22.4 \text{ m/s})(13.4 \text{ m/s})^2}{(13.4 \text{ m/s})(56.7 \text{ m}) - (22.4 \text{ m/s})(24.4 \text{ m})} = -6.2 \text{ m/s}^2.$$

The *magnitude* of the deceleration is therefore 6.2 m/s². Although rounded off values are displayed in the above substitutions, what we have input into our calculators are the "exact" values (such as $v_{02} = \frac{161}{12}$ m/s).

86. We take the moment of applying brakes to be $t = 0$. The deceleration is constant so that Table 2-1 can be used. Our primed variables (such as $v'_0 = 72$ km/h $= 20$ m/s) refer to one train (moving in the +x direction and located at the origin when $t = 0$) and unprimed variables refer to the other (moving in the –x direction and located at $x_0 = +950$ m when $t = 0$). We note that the acceleration vector of the unprimed train points in the *positive* direction, even though the train is slowing down; its initial velocity is $v_0 = -144$ km/h $= -40$ m/s. Since the primed train has the lower initial speed, it should stop sooner than the other train would (were it not for the collision). Using Eq 2-16, it should stop (meaning $v' = 0$) at

$$x' = \frac{(v')^2 - (v'_0)^2}{2a'} = \frac{0 - (20 \text{ m/s})^2}{-2 \text{ m/s}^2} = 200 \text{ m}.$$

The speed of the other train, when it reaches that location, is

$$v = \sqrt{v_0^2 + 2a\Delta x} = \sqrt{(-40 \text{ m/s})^2 + 2(1.0 \text{ m/s}^2)(200 \text{ m} - 950 \text{ m})} = 10 \text{ m/s}$$

using Eq 2-16 again. Specifically, its velocity at that moment would be –10 m/s since it is still traveling in the –x direction when it crashes. If the computation of v had failed (meaning that a negative number would have been inside the square root) then we would have looked at the possibility that there was no collision and examined how far apart they finally were. A concern that can be brought up is whether the primed train collides before it comes to rest; this can be studied by computing the time it stops (Eq. 2-11 yields $t = 20$

s) and seeing where the unprimed train is at that moment (Eq. 2-18 yields $x = 350$ m, still a good distance away from contact).

87. The y coordinate of Piton 1 obeys $y - y_{o1} = -\frac{1}{2} g t^2$ where $y = 0$ when $t = 3.0$ s. This allows us to solve for y_{o1}, and we find $y_{o1} = 44.1$ m. The graph for the coordinate of Piton 2 (which is thrown apparently at $t = 1.0$ s with velocity v_1) is

$$y - y_{o2} = v_1(t-1.0) - \frac{1}{2} g (t-1.0)^2$$

where $y_{o2} = y_{o1} + 10 = 54.1$ m and where (again) $y = 0$ when $t = 3.0$ s. Thus we obtain $|v_1| = 17$ m/s, approximately.

88. We adopt the convention frequently used in the text: that "up" is the positive y direction.

(a) At the highest point in the trajectory $v = 0$. Thus, with $t = 1.60$ s, the equation $v = v_0 - gt$ yields $v_0 = 15.7$ m/s.

(b) One equation that is not dependent on our result from part (a) is $y - y_0 = vt + \frac{1}{2}gt^2$; this readily gives $y_{max} - y_0 = 12.5$ m for the highest ("max") point measured relative to where it started (the top of the building).

(c) Now we use our result from part (a) and plug into $y - y_0 = v_0 t + \frac{1}{2}gt^2$ with $t = 6.00$ s and $y = 0$ (the ground level). Thus, we have

$$0 - y_0 = (15.68 \text{ m/s})(6.00 \text{ s}) - \frac{1}{2} (9.8 \text{ m/s}^2)(6.00 \text{ s})^2 \ .$$

Therefore, y_0 (the height of the building) is equal to 82.3 m.

89. Integrating (from $t = 2$ s to variable $t = 4$ s) the acceleration to get the velocity and using the values given in the problem, leads to

$$v = v_0 + \int_{t_0}^{t} a\, dt = v_0 + \int_{t_0}^{t} (5.0t) dt = v_0 + \frac{1}{2}(5.0)(t^2 - t_0^2) = 17 + \frac{1}{2}(5.0)(4^2 - 2^2) = 47 \text{ m/s}.$$

90. We take $+x$ in the direction of motion. We use subscripts 1 and 2 for the data. Thus, $v_1 = +30$ m/s, $v_2 = +50$ m/s and $x_2 - x_1 = +160$ m.

(a) Using these subscripts, Eq. 2-16 leads to

$$a = \frac{v_2^2 - v_1^2}{2(x_2 - x_1)} = \frac{(50 \text{ m/s})^2 - (30 \text{ m/s})^2}{2(160 \text{ m})} = 5.0 \text{ m/s}^2 \ .$$

(b) We find the time interval corresponding to the displacement $x_2 - x_1$ using Eq. 2-17:

$$t_2 - t_1 = \frac{2(x_2 - x_1)}{v_1 + v_2} = \frac{2(160 \text{ m})}{30 \text{ m/s} + 50 \text{ m/s}} = 4.0 \text{ s}.$$

(c) Since the train is at rest ($v_0 = 0$) when the clock starts, we find the value of t_1 from Eq. 2-11:

$$v_1 = v_0 + at_1 \Rightarrow t_1 = \frac{30 \text{ m/s}}{5.0 \text{ m/s}^2} = 6.0 \text{ s}.$$

(d) The coordinate origin is taken to be the location at which the train was initially at rest (so $x_0 = 0$). Thus, we are asked to find the value of x_1. Although any of several equations could be used, we choose Eq. 2-17:

$$x_1 = \frac{1}{2}(v_0 + v_1)t_1 = \frac{1}{2}(30 \text{ m/s})(6.0 \text{ s}) = 90 \text{ m}.$$

(e) The graphs are shown below, with SI units assumed.

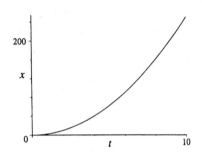
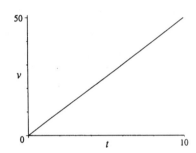

91. We take $+x$ in the direction of motion, so

$$v = (60 \text{ km/h})\left(\frac{1000 \text{ m/km}}{3600 \text{ s/h}}\right) = +16.7 \text{ m/s}$$

and $a > 0$. The location where it starts from rest ($v_0 = 0$) is taken to be $x_0 = 0$.

(a) Eq. 2-7 gives $a_{avg} = (v - v_0)/t$ where $t = 5.4$ s and the velocities are given above. Thus, $a_{avg} = 3.1 \text{ m/s}^2$.

(b) The assumption that $a =$ constant permits the use of Table 2-1. From that list, we choose Eq. 2-17:

$$x = \frac{1}{2}(v_0 + v)\,t = \frac{1}{2}(16.7 \text{ m/s})(5.4 \text{ s}) = 45 \text{ m}.$$

(c) We use Eq. 2-15, now with $x = 250$ m:

$$x = \frac{1}{2}at^2 \Rightarrow t = \sqrt{\frac{2x}{a}} = \sqrt{\frac{2(250 \text{ m})}{3.1 \text{ m/s}^2}}$$

which yields $t = 13$ s.

92. We take the direction of motion as $+x$, take $x_0 = 0$ and use SI units, so $v = 1600(1000/3600) = 444$ m/s.

(a) Eq. 2-11 gives $444 = a(1.8)$ or $a = 247$ m/s^2. We express this as a multiple of g by setting up a ratio:

$$a = \left(\frac{247 \text{ m/s}^2}{9.8 \text{ m/s}^2}\right) g = 25g.$$

(b) Eq. 2-17 readily yields

$$x = \frac{1}{2}(v_0 + v) t = \frac{1}{2}(444 \text{ m/s})(1.8 \text{ s}) = 400 \text{ m}.$$

93. The object, once it is dropped ($v_0 = 0$) is in free-fall ($a = -g = -9.8$ m/s^2 if we take *down* as the $-y$ direction), and we use Eq. 2-15 repeatedly.

(a) The (positive) distance D from the lower dot to the mark corresponding to a certain reaction time t is given by $\Delta y = -D = -\frac{1}{2}gt^2$, or $D = gt^2/2$. Thus, for $t_1 = 50.0$ ms,

$$D_1 = \frac{(9.8 \text{ m/s}^2)(50.0 \times 10^{-3} \text{ s})^2}{2} = 0.0123 \text{ m} = 1.23 \text{ cm}.$$

(b) For $t_2 = 100$ ms, $D_2 = \dfrac{(9.8 \text{ m/s}^2)(100 \times 10^{-3} \text{ s})^2}{2} = 0.049 \text{ m} = 4D_1$.

(c) For $t_3 = 150$ ms, $D_3 = \dfrac{(9.8 \text{ m/s}^2)(150 \times 10^{-3} \text{ s})^2}{2} = 0.11 \text{ m} = 9D_1$.

(d) For $t_4 = 200$ ms, $D_4 = \dfrac{(9.8 \text{ m/s}^2)(200 \times 10^{-3} \text{ s})^2}{2} = 0.196 \text{ m} = 16D_1$.

(e) For $t_4 = 250$ ms, $D_5 = \dfrac{(9.8 \text{ m/s}^2)(250 \times 10^{-3} \text{ s})^2}{2} = 0.306 \text{ m} = 25D_1$.

The velocity v at $t = 6$ (SI units and two significant figures understood) is $v_{\text{given}} + \int_{-2}^{6} a \, dt$. A quick way to implement this is to recall the area of a triangle ($\frac{1}{2}$ base × height). The result is $v = 7$ m/s $+ 32$ m/s $= 39$ m/s.

95. Let D be the distance up the hill. Then

$$\text{average speed} = \frac{\text{total distance traveled}}{\text{total time of travel}} = \frac{2D}{\dfrac{D}{20 \text{ km/h}} + \dfrac{D}{35 \text{ km/h}}} \approx 25 \text{ km/h}.$$

96. Converting to SI units, we have $v = 3400(1000/3600) = 944$ m/s (presumed constant) and $\Delta t = 0.10$ s. Thus, $\Delta x = v\Delta t = 94$ m.

97. The (ideal) driving time before the change was $t = \Delta x/v$, and after the change it is $t' = \Delta x/v'$. The time saved by the change is therefore

$$t - t' = \Delta x \left(\frac{1}{v} - \frac{1}{v'}\right) = \Delta x \left(\frac{1}{55} - \frac{1}{65}\right) = \Delta x (0.0028 \text{ h/mi})$$

which becomes, converting $\Delta x = 700/1.61 = 435$ mi (using a conversion found on the inside front cover of the textbook), $t - t' = (435)(0.0028) = 1.2$ h. This is equivalent to 1 h and 13 min.

98. We obtain the velocity by integration of the acceleration:

$$v - v_0 = \int_0^t (6.1 - 1.2t')dt'.$$

Lengths are in meters and times are in seconds. The student is encouraged to look at the discussion in the textbook in §2-7 to better understand the manipulations here.

(a) The result of the above calculation is

$$v = v_0 + 6.1t - 0.6t^2,$$

where the problem states that $v_0 = 2.7$ m/s. The maximum of this function is found by knowing when its derivative (the acceleration) is zero ($a = 0$ when $t = 6.1/1.2 = 5.1$ s) and plugging that value of t into the velocity equation above. Thus, we find $v = 18$ m/s.

(b) We integrate again to find x as a function of t:

$$x - x_0 = \int_0^t v\,dt' = \int_0^t (v_0 + 6.1t' - 0.6t'^2)\,dt' = v_0 t + 3.05 t^2 - 0.2 t^3.$$

With $x_0 = 7.3$ m, we obtain $x = 83$ m for $t = 6$. This is the correct answer, but one has the right to worry that it might not be; after all, the problem asks for the total distance traveled (and $x - x_0$ is just the *displacement*). If the cyclist backtracked, then his total distance would be greater than his displacement. Thus, we might ask, "did he backtrack?" To do so would require that his velocity be (momentarily) zero at some point (as he reversed his direction of motion). We could solve the above quadratic equation for velocity, for a positive value of t where $v = 0$; if we did, we would find that at $t = 10.6$ s, a reversal does indeed happen. However, in the time interval concerned with in our problem ($0 \le t \le 6$ s), there is no reversal and the displacement is the same as the total distance traveled.

67

99. We neglect air resistance, which justifies setting $a = -g = -9.8$ m/s² (taking *down* as the $-y$ direction) for the duration of the motion. We are allowed to use Table 2-1 (with Δy replacing Δx) because this is constant acceleration motion. When something is thrown straight up and is caught at the level it was thrown from (with a trajectory similar to that shown in Fig. 2-31), the time of flight t is half of its time of ascent t_a, which is given by Eq. 2-18 with $\Delta y = H$ and $v = 0$ (indicating the maximum point).

$$H = vt_a + \frac{1}{2}gt_a^2 \quad \Rightarrow \quad t_a = \sqrt{\frac{2H}{g}}$$

Writing these in terms of the total time in the air $t = 2t_a$ we have

$$H = \frac{1}{8}gt^2 \quad \Rightarrow \quad t = 2\sqrt{\frac{2H}{g}}.$$

We consider two throws, one to height H_1 for total time t_1 and another to height H_2 for total time t_2, and we set up a ratio:

$$\frac{H_2}{H_1} = \frac{\frac{1}{8}gt_2^2}{\frac{1}{8}gt_1^2} = \left(\frac{t_2}{t_1}\right)^2$$

from which we conclude that if $t_2 = 2t_1$ (as is required by the problem) then $H_2 = 2^2 H_1 = 4H_1$.

100. The acceleration is constant and we may use the equations in Table 2-1.

(a) Taking the first point as coordinate origin and time to be zero when the car is there, we apply Eq. 2-17:

$$x = \frac{1}{2}(v + v_0)t = \frac{1}{2}(15.0 \text{ m/s} + v_0)(6.00 \text{ s}).$$

With $x = 60.0$ m (which takes the direction of motion as the $+x$ direction) we solve for the initial velocity: $v_0 = 5.00$ m/s.

(b) Substituting $v = 15.0$ m/s, $v_0 = 5.00$ m/s and $t = 6.00$ s into $a = (v - v_0)/t$ (Eq. 2-11), we find $a = 1.67$ m/s².

(c) Substituting $v = 0$ in $v^2 = v_0^2 + 2ax$ and solving for x, we obtain

$$x = -\frac{v_0^2}{2a} = -\frac{(5.00 \text{ m/s})^2}{2(1.67 \text{ m/s}^2)} = -7.50 \text{ m},$$

or $|x| = 7.50$ m.

(d) The graphs require computing the time when $v = 0$, in which case, we use $v = v_0 + at' = 0$. Thus,

$$t' = \frac{-v_0}{a} = \frac{-5.00 \text{ m/s}}{1.67 \text{ m/s}^2} = -3.0 \text{ s}$$

indicates the moment the car was at rest. SI units are assumed.

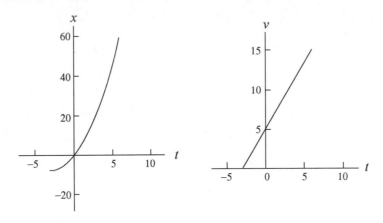

101. Taking the $+y$ direction *downward* and $y_0 = 0$, we have $y = v_0 t + \frac{1}{2} g t^2$ which (with $v_0 = 0$) yields $t = \sqrt{2y/g}$.

(a) For this part of the motion, $y = 50$ m so that

$$t = \sqrt{\frac{2(50 \text{ m})}{9.8 \text{ m/s}^2}} = 3.2 \text{ s}.$$

(b) For this next part of the motion, we note that the total displacement is $y = 100$ m. Therefore, the total time is

$$t = \sqrt{\frac{2(100 \text{ m})}{9.8 \text{ m/s}^2}} = 4.5 \text{ s}.$$

The different between this and the answer to part (a) is the time required to fall through that second 50 m distance: $4.5 - 3.2 = 1.3$ s.

102. Direction of $+x$ is implicit in the problem statement. The initial position (when the clock starts) is $x_0 = 0$ (where $v_0 = 0$), the end of the speeding-up motion occurs at $x_1 = 1100/2 = 550$ m, and the subway comes to a halt ($v_2 = 0$) at $x_2 = 1100$ m.

(a) Using Eq. 2-15, the subway reaches x_1 at

$$t_1 = \sqrt{\frac{2x_1}{a_1}} = \sqrt{\frac{2(550 \text{ m})}{1.2 \text{ m/s}^2}} = 30.3 \text{ s}.$$

The time interval $t_2 - t_1$ turns out to be the same value (most easily seen using Eq. 2-18 so the total time is $t_2 = 2(30.3) = 60.6$ s.

(b) Its maximum speed occurs at t_1 and equals

$$v_1 = v_0 + a_1 t_1 = 36.3 \text{ m/s}.$$

(c) The graphs are shown below:

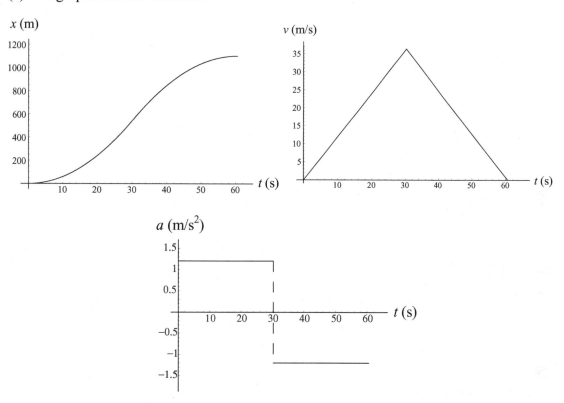

103. This problem consists of two parts: part 1 with constant acceleration (so that the equations in Table 2-1 apply), $v_0 = 0$, $v = 11.0$ m/s, $x = 12.0$ m, and $x_0 = 0$ (adopting the starting line as the coordinate origin); and, part 2 with constant velocity (so that $x - x_0 = vt$ applies) with $v = 11.0$ m/s, $x_0 = 12.0$, and $x = 100.0$ m.

(a) We obtain the time for part 1 from Eq. 2-17

$$x - x_0 = \frac{1}{2}(v_0 + v) t_1 \Rightarrow 12.0 - 0 = \frac{1}{2}(0 + 11.0) t_1$$

so that $t_1 = 2.2$ s, and we find the time for part 2 simply from $88.0 = (11.0)t_2 \rightarrow t_2 = 8.0$ s. Therefore, the total time is $t_1 + t_2 = 10.2$ s.

(b) Here, the total time is required to be 10.0 s, and we are to locate the point x_p where the runner switches from accelerating to proceeding at constant speed. The equations for parts 1 and 2, used above, therefore become

$$x_p - 0 = \frac{1}{2}(0 + 11.0 \text{ m/s})t_1$$

$$100.0 \text{ m} - x_p = (11.0 \text{ m/s})(10.0 \text{ s} - t_1)$$

where in the latter equation, we use the fact that $t_2 = 10.0 - t_1$. Solving the equations for the two unknowns, we find that $t_1 = 1.8$ s and $x_p = 10.0$ m.

104. (a) Using the fact that the area of a triangle is $\frac{1}{2}$ (base) (height) (and the fact that the integral corresponds to the area under the curve) we find, from $t = 0$ through $t = 5$ s, the integral of v with respect to t is 15 m. Since we are told that $x_0 = 0$ then we conclude that $x = 15$ m when $t = 5.0$ s.

(b) We see directly from the graph that $v = 2.0$ m/s when $t = 5.0$ s.

(c) Since $a = dv/dt$ = slope of the graph, we find that the acceleration during the interval $4 < t < 6$ is uniformly equal to -2.0 m/s^2.

(d) Thinking of $x(t)$ in terms of accumulated area (on the graph), we note that $x(1) = 1$ m; using this and the value found in part (a), Eq. 2-2 produces

$$v_{avg} = \frac{x(5) - x(1)}{5 - 1} = \frac{15 \text{ m} - 1 \text{ m}}{4 \text{ s}} = 3.5 \text{ m/s}.$$

(e) From Eq. 2-7 and the values $v(t)$ we read directly from the graph, we find

$$a_{avg} = \frac{v(5) - v(1)}{5 - 1} = \frac{2 \text{ m/s} - 2 \text{ m/s}}{4 \text{ s}} = 0.$$

105. We neglect air resistance, which justifies setting $a = -g = -9.8$ m/s^2 (taking *down* as the $-y$ direction) for the duration of the stone's motion. We are allowed to use Table 2-1 (with Δx replaced by y) because the ball has constant acceleration motion (and we choose $y_0 = 0$).

(a) We apply Eq. 2-16 to both measurements, with SI units understood.

$$v_B^2 = v_0^2 - 2gy_B \Rightarrow \left(\frac{1}{2}v\right)^2 + 2g(y_A + 3) = v_0^2$$

$$v_A^2 = v_0^2 - 2gy_A \Rightarrow v^2 + 2gy_A = v_0^2$$

We equate the two expressions that each equal v_0^2 and obtain

$$\frac{1}{4}v^2 + 2gy_A + 2g(3) = v^2 + 2gy_A \Rightarrow 2g(3) = \frac{3}{4}v^2$$

which yields $v = \sqrt{2g(4)} = 8.85$ m/s.

(b) An object moving upward at A with speed $v = 8.85$ m/s will reach a maximum height $y - y_A = v^2/2g = 4.00$ m above point A (this is again a consequence of Eq. 2-16, now with the "final" velocity set to zero to indicate the highest point). Thus, the top of its motion is 1.00 m above point B.

106. We neglect air resistance, which justifies setting $a = -g = -9.8$ m/s^2 (taking *down* as the $-y$ direction) for the duration of the motion. We are allowed to use Table 2-1 (with Δy replacing Δx) because this is constant acceleration motion. The ground level is taken to correspond to the origin of the y-axis. The total time of fall can be computed from Eq. 2-15 (using the quadratic formula).

$$\Delta y = v_0 t - \frac{1}{2} g t^2 \quad \Rightarrow \quad t = \frac{v_0 + \sqrt{v_0^2 - 2g\Delta y}}{g}$$

with the positive root chosen. With $y = 0$, $v_0 = 0$ and $y_0 = h = 60$ m, we obtain

$$t = \frac{\sqrt{2gh}}{g} = \sqrt{\frac{2h}{g}} = 3.5 \text{ s}.$$

Thus, "1.2 s earlier" means we are examining where the rock is at $t = 2.3$ s:

$$y - h = v_0 (2.3 \text{ s}) - \frac{1}{2} g (2.3 \text{ s})^2 \quad \Rightarrow \quad y = 34 \text{ m}$$

where we again use the fact that $h = 60$ m and $v_0 = 0$.

107. (a) The wording of the problem makes it clear that the equations of Table 2-1 apply, the challenge being that v_0, v, and a are not explicitly given. We can, however, apply $x - x_0 = v_0 t + \frac{1}{2} a t^2$ to a variety of points on the graph and solve for the unknowns from the simultaneous equations. For instance,

$$16 \text{ m} - 0 = v_0 (2.0 \text{ s}) + \frac{1}{2} a (2.0 \text{ s})^2$$

$$27 \text{ m} - 0 = v_0 (3.0 \text{ s}) + \frac{1}{2} a (3.0 \text{ s})^2$$

lead to the values $v_0 = 6.0$ m/s and $a = 2.0$ m/s^2.

(b) From Table 2-1,

$$x - x_0 = v t - \frac{1}{2} a t^2 \quad \Rightarrow \quad 27 \text{ m} - 0 = v(3.0 \text{ s}) - \frac{1}{2} (2.0 \text{ m/s}^2)(3.0 \text{ s})^2$$

which leads to $v = 12$ m/s.

(c) Assuming the wind continues during $3.0 \leq t \leq 6.0$, we apply $x - x_0 = v_0 t + \frac{1}{2} a t^2$ to this interval (where $v_0 = 12.0$ m/s from part (b)) to obtain

$$\Delta x = (12.0 \text{ m/s})(3.0 \text{ s}) + \frac{1}{2}(2.0 \text{ m/s}^2)(3.0 \text{ s})^2 = 45 \text{ m} \quad .$$

108. With $+y$ upward, we have $y_0 = 36.6$ m and $y = 12.2$ m. Therefore, using Eq. 2-18 (the last equation in Table 2-1), we find

$$y - y_0 = vt + \frac{1}{2}gt^2 \implies v = -22.0 \text{ m/s}$$

at $t = 2.00$ s. The term *speed* refers to the magnitude of the velocity vector, so the answer is $|v| = 22.0$ m/s.

109. The bullet starts at rest ($v_0 = 0$) and after traveling the length of the barrel ($\Delta x = 1.2$ m) emerges with the given velocity ($v = 640$ m/s), where the direction of motion is the positive direction. Turning to the constant acceleration equations in Table 2-1, we use

$$\Delta x = \frac{1}{2}(v_0 + v)\,t \,.$$

Thus, we find $t = 0.00375$ s (or about 3.8 ms).

110. During free fall, we ignore the air resistance and set $a = -g = -9.8$ m/s^2 where we are choosing *down* to be the $-y$ direction. The initial velocity is zero so that Eq. 2-15 becomes $\Delta y = -\frac{1}{2}gt^2$ where Δy represents the *negative* of the distance d she has fallen. Thus, we can write the equation as $d = \frac{1}{2}gt^2$ for simplicity.

(a) The time t_1 during which the parachutist is in free fall is (using Eq. 2-15) given by

$$d_1 = 50 \text{ m} = \frac{1}{2}gt_1^2 = \frac{1}{2}\left(9.80 \text{ m/s}^2\right)t_1^2$$

which yields $t_1 = 3.2$ s. The *speed* of the parachutist just before he opens the parachute is given by the positive root $v_1^2 = 2gd_1$, or

$$v_1 = \sqrt{2gh_1} = \sqrt{(2)(9.80 \text{ m/s}^2)(50 \text{ m})} = 31 \text{ m/s}.$$

If the final speed is v_2, then the time interval t_2 between the opening of the parachute and the arrival of the parachutist at the ground level is

$$t_2 = \frac{v_1 - v_2}{a} = \frac{31 \text{ m/s} - 3.0 \text{ m/s}}{2 \text{ m/s}^2} = 14 \text{ s}.$$

This is a result of Eq. 2-11 where *speeds* are used instead of the (negative-valued) velocities (so that final-velocity minus initial-velocity turns out to equal initial-speed minus final-speed); we also note that the acceleration vector for this part of the motion is positive since it points upward (opposite to the direction of motion — which makes it a deceleration). The total time of flight is therefore $t_1 + t_2 = 17$ s.

(b) The distance through which the parachutist falls after the parachute is opened is given by

$$d = \frac{v_1^2 - v_2^2}{2a} = \frac{(31 \text{m/s})^2 - (3.0 \text{ m/s})^2}{(2)(2.0 \text{ m/s}^2)} \approx 240 \text{ m}.$$

In the computation, we have used Eq. 2-16 with both sides multiplied by –1 (which changes the negative-valued Δy into the positive d on the left-hand side, and switches the order of v_1 and v_2 on the right-hand side). Thus the fall begins at a height of $h = 50 + d \approx 290$ m.

111. There is no air resistance, which makes it quite accurate to set $a = -g = -9.8$ m/s² (where downward is the $-y$ direction) for the duration of the fall. We are allowed to use Table 2-1 (with Δy replacing Δx) because this is constant acceleration motion; in fact, when the acceleration changes (during the process of catching the ball) we will again assume constant acceleration conditions; in this case, we have $a_2 = +25g = 245$ m/s².

(a) The time of fall is given by Eq. 2-15 with $v_0 = 0$ and $y = 0$. Thus,

$$t = \sqrt{\frac{2y_0}{g}} = \sqrt{\frac{2(145 \text{ m})}{9.8 \text{ m/s}^2}} = 5.44 \text{ s}.$$

(b) The final velocity for its free-fall (which becomes the initial velocity during the catching process) is found from Eq. 2-16 (other equations can be used but they would use the result from part (a)).

$$v = -\sqrt{v_0^2 - 2g(y - y_0)} = -\sqrt{2gy_0} = -53.3 \text{ m/s}$$

where the negative root is chosen since this is a downward velocity. Thus, the speed is $|v| = 53.3$ m/s.

(c) For the catching process, the answer to part (b) plays the role of an *initial* velocity ($v_0 = -53.3$ m/s) and the final velocity must become zero. Using Eq. 2-16, we find

$$\Delta y_2 = \frac{v^2 - v_0^2}{2a_2} = \frac{-(-53.3 \text{ m/s})^2}{2(245 \text{ m/s}^2)} = -5.80 \text{ m},$$

or $|\Delta y_2| = 5.80$ m. The negative value of Δy_2 signifies that the distance traveled while arresting its motion is downward.

112. We neglect air resistance, which justifies setting $a = -g = -9.8$ m/s² (taking down as the $-y$ direction) for the duration of the motion. We are allowed to use Table 2-1 (with Δy replacing Δx) because this is constant acceleration motion. The ground level is taken to correspond to $y = 0$.

(a) With $y_0 = h$ and v_0 replaced with $-v_0$, Eq. 2-16 leads to

$$v = \sqrt{(-v_0)^2 - 2g(y - y_0)} = \sqrt{v_0^2 + 2gh}\,.$$

The positive root is taken because the problem asks for the speed (the *magnitude* of the velocity).

(b) We use the quadratic formula to solve Eq. 2-15 for t, with v_0 replaced with $-v_0$,

$$\Delta y = -v_0 t - \frac{1}{2}gt^2 \;\Rightarrow\; t = \frac{-v_0 + \sqrt{(-v_0)^2 - 2g\Delta y}}{g}$$

where the positive root is chosen to yield $t > 0$. With $y = 0$ and $y_0 = h$, this becomes

$$t = \frac{\sqrt{v_0^2 + 2gh} - v_0}{g}.$$

(c) If it were thrown upward with that speed from height h then (in the absence of air friction) it would return to height h with that same downward speed and would therefore yield the same final speed (before hitting the ground) as in part (a). An important perspective related to this is treated later in the book (in the context of energy conservation).

(d) Having to travel up before it starts its descent certainly requires more time than in part (b). The calculation is quite similar, however, except for now having $+v_0$ in the equation where we had put in $-v_0$ in part (b). The details follow:

$$\Delta y = v_0 t - \frac{1}{2}gt^2 \;\Rightarrow\; t = \frac{v_0 + \sqrt{v_0^2 - 2g\Delta y}}{g}$$

with the positive root again chosen to yield $t > 0$. With $y = 0$ and $y_0 = h$, we obtain

$$t = \frac{\sqrt{v_0^2 + 2gh} + v_0}{g}.$$

113. During T_r the velocity v_0 is constant (in the direction we choose as $+x$) and obeys $v_0 = D_r/T_r$ where we note that in SI units the velocity is $v_0 = 200(1000/3600) = 55.6$ m/s. During T_b the acceleration is opposite to the direction of v_0 (hence, for us, $a < 0$) until the car is stopped ($v = 0$).

(a) Using Eq. 2-16 (with $\Delta x_b = 170$ m) we find

$$v^2 = v_0^2 + 2a\Delta x_b \quad \Rightarrow \quad a = -\frac{v_0^2}{2\Delta x_b}$$

which yields $|a| = 9.08$ m/s^2.

(b) We express this as a multiple of g by setting up a ratio:

$$a = \left(\frac{9.08 \text{ m/s}^2}{9.8 \text{ m/s}^2}\right)g = 0.926g.$$

(c) We use Eq. 2-17 to obtain the braking time:

$$\Delta x_b = \frac{1}{2}(v_0 + v)T_b \quad \Rightarrow \quad T_b = \frac{2(170 \text{ m})}{55.6 \text{ m/s}} = 6.12 \text{ s}.$$

(d) We express our result for T_b as a multiple of the reaction time T_r by setting up a ratio:

$$T_b = \left(\frac{6.12 \text{ s}}{400 \times 10^{-3} \text{ s}}\right)T_r = 15.3 T_r.$$

(e) Since $T_b > T_r$, most of the full time required to stop is spent in braking.

(f) We are only asked what the *increase* in distance D is, due to $\Delta T_r = 0.100$ s, so we simply have

$$\Delta D = v_0 \Delta T_r = (55.6 \text{ m/s})(0.100 \text{ s}) = 5.56 \text{ m}.$$

114. We assume constant velocity motion and use Eq. 2-2 (with $v_{\text{avg}} = v > 0$). Therefore,

$$\Delta x = v\Delta t = \left(303 \frac{\text{km}}{\text{h}}\left(\frac{1000 \text{ m/km}}{3600 \text{ s/h}}\right)\right)(100 \times 10^{-3} \text{ s}) = 8.4 \text{ m}.$$

Chapter 3

1. A vector $\vec{a}$ can be represented in the *magnitude-angle* notation (a, θ), where

$$a = \sqrt{a_x^2 + a_y^2}$$

is the magnitude and

$$\theta = \tan^{-1}\left(\frac{a_y}{a_x}\right)$$

is the angle $\vec{a}$ makes with the positive x axis.

(a) Given $A_x = -25.0$ m and $A_y = 40.0$ m, $A = \sqrt{(-25.0 \text{ m})^2 + (40.0 \text{ m})^2} = 47.2$ m

(b) Recalling that $\tan \theta = \tan (\theta + 180°)$, $\tan^{-1}[(40.0 \text{ m})/(-25.0 \text{ m})] = -58°$ or $122°$. Noting that the vector is in the third quadrant (by the signs of its x and y components) we see that $122°$ is the correct answer. The graphical calculator "shortcuts" mentioned above are designed to correctly choose the right possibility.

2. The angle described by a full circle is $360° = 2\pi$ rad, which is the basis of our conversion factor.

(a)
$$20.0° = (20.0°)\frac{2\pi \text{ rad}}{360°} = 0.349 \text{ rad}.$$

(b)
$$50.0° = (50.0°)\frac{2\pi \text{ rad}}{360°} = 0.873 \text{ rad}.$$

(c)
$$100° = (100°)\frac{2\pi \text{ rad}}{360°} = 1.75 \text{ rad}.$$

(d)
$$0.330 \text{ rad} = (0.330 \text{ rad})\frac{360°}{2\pi \text{ rad}} = 18.9°.$$

(e)
$$2.10 \text{ rad} = (2.10 \text{ rad})\frac{360°}{2\pi \text{ rad}} = 120°.$$

(f)
$$7.70 \text{ rad} = (7.70 \text{ rad})\frac{360°}{2\pi \text{ rad}} = 441°.$$

3. The x and the y components of a vector $\vec{a}$ lying on the xy plane are given by

$$a_x = a\cos\theta, \quad a_y = a\sin\theta$$

where $a = |\vec{a}|$ is the magnitude and θ is the angle between $\vec{a}$ and the positive x axis.

(a) The x component of $\vec{a}$ is given by $a_x = 7.3 \cos 250° = -2.5$ m.

(b) and the y component is given by $a_y = 7.3 \sin 250° = -6.9$ m.

In considering the variety of ways to compute these, we note that the vector is 70° below the $-x$ axis, so the components could also have been found from $a_x = -7.3 \cos 70°$ and $a_y = -7.3 \sin 70°$. In a similar vein, we note that the vector is 20° to the left from the $-y$ axis, so one could use $a_x = -7.3 \sin 20°$ and $a_y = -7.3 \cos 20°$ to achieve the same results.

4. (a) The height is $h = d \sin\theta$, where $d = 12.5$ m and $\theta = 20.0°$. Therefore, $h = 4.28$ m.

(b) The horizontal distance is $d \cos\theta = 11.7$ m.

5. The vector sum of the displacements $\vec{d}_{storm}$ and $\vec{d}_{new}$ must give the same result as its originally intended displacement $\vec{d}_o = (120 \text{ km})\hat{j}$ where east is $\hat{i}$, north is $\hat{j}$. Thus, we write

$$\vec{d}_{storm} = (100 \text{ km})\hat{i}, \quad \vec{d}_{new} = A\hat{i} + B\hat{j}.$$

(a) The equation $\vec{d}_{storm} + \vec{d}_{new} = \vec{d}_o$ readily yields $A = -100$ km and $B = 120$ km. The magnitude of $\vec{d}_{new}$ is therefore equal to $|\vec{d}_{new}| = \sqrt{A^2 + B^2} = 156$ km.

(b) The direction is $\tan^{-1}(B/A) = -50.2°$ or $180° + (-50.2°) = 129.8°$. We choose the latter value since it indicates a vector pointing in the second quadrant, which is what we expect here. The answer can be phrased several equivalent ways: 129.8° counterclockwise from east, or 39.8° west from north, or 50.2° north from west.

6. (a) With $r = 15$ m and $\theta = 30°$, the x component of $\vec{r}$ is given by

$$r_x = r\cos\theta = (15 \text{ m}) \cos 30° = 13 \text{ m}.$$

(b) Similarly, the y component is given by $r_y = r \sin\theta = (15 \text{ m}) \sin 30° = 7.5$ m.

7. The length unit meter is understood throughout the calculation.

(a) We compute the distance from one corner to the diametrically opposite corner:
$\sqrt{(3.00 \text{ m})^2 + (3.70 \text{ m})^2 + (4.30 \text{ m})^2}$.

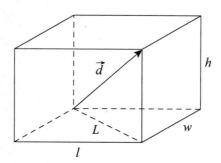

(b) The displacement vector is along the straight line from the beginning to the end point of the trip. Since a straight line is the shortest distance between two points, the length of the path cannot be less than the magnitude of the displacement.

(c) It can be greater, however. The fly might, for example, crawl along the edges of the room. Its displacement would be the same but the path length would be $\ell+w+h=11.0$ m.

(d) The path length is the same as the magnitude of the displacement if the fly flies along the displacement vector.

(e) We take the x axis to be out of the page, the y axis to be to the right, and the z axis to be upward. Then the x component of the displacement is $w = 3.70$ m, the y component of the displacement is 4.30 m, and the z component is 3.00 m. Thus,

$$\vec{d} = (3.70 \text{ m})\hat{i} + (4.30 \text{ m})\hat{j} + (3.00 \text{ m})\hat{k}.$$

An equally correct answer is gotten by interchanging the length, width, and height.

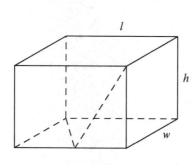 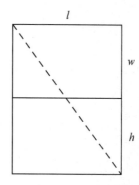

(f) Suppose the path of the fly is as shown by the dotted lines on the upper diagram. Pretend there is a hinge where the front wall of the room joins the floor and lay the wall down as shown on the lower diagram. The shortest walking distance between the lower left back of the room and the upper right front corner is the dotted straight line shown on the diagram. Its length is

$$L_{min} = \sqrt{(w+h)^2 + \ell^2} = \sqrt{(3.70 \text{ m} + 3.00 \text{ m})^2 + (4.30 \text{ m})^2} = 7.96 \text{ m}.$$

8. We label the displacement vectors $\vec{A}$, $\vec{B}$ and $\vec{C}$ (and denote the result of their vector sum as $\vec{r}$). We choose *east* as the $\hat{i}$ direction (+x direction) and *north* as the $\hat{j}$ direction (+y direction). We note that the angle between $\vec{C}$ and the x axis is 60°. Thus,

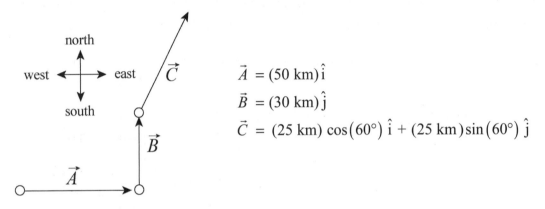

$\vec{A} = (50 \text{ km})\hat{i}$

$\vec{B} = (30 \text{ km})\hat{j}$

$\vec{C} = (25 \text{ km})\cos(60°)\hat{i} + (25 \text{ km})\sin(60°)\hat{j}$

(a) The total displacement of the car from its initial position is represented by

$$\vec{r} = \vec{A} + \vec{B} + \vec{C} = (62.5 \text{ km})\hat{i} + (51.7 \text{ km})\hat{j}$$

which means that its magnitude is

$$|\vec{r}| = \sqrt{(62.5 \text{ km})^2 + (51.7 \text{ km})^2} = 81 \text{ km}.$$

(b) The angle (counterclockwise from +x axis) is $\tan^{-1}(51.7 \text{ km}/62.5 \text{ km}) = 40°$, which is to say that it points 40° *north of east*. Although the resultant $\vec{r}$ is shown in our sketch, it would be a direct line from the "tail" of $\vec{A}$ to the "head" of $\vec{C}$.

9. We write $\vec{r} = \vec{a} + \vec{b}$. When not explicitly displayed, the units here are assumed to be meters.

(a) The x and the y components of $\vec{r}$ are $r_x = a_x + b_x = (4.0 \text{ m}) - (13 \text{ m}) = -9.0 \text{ m}$ and $r_y = a_y + b_y = (3.0 \text{ m}) + (7.0 \text{ m}) = 10 \text{ m}$, respectively. Thus $\vec{r} = (-9.0 \text{ m})\hat{i} + (10 \text{ m})\hat{j}$.

(b) The magnitude of $\vec{r}$ is

$$r = |\vec{r}| = \sqrt{r_x^2 + r_y^2} = \sqrt{(-9.0 \text{ m})^2 + (10 \text{ m})^2} = 13 \text{ m}.$$

(c) The angle between the resultant and the +x axis is given by

$$\theta = \tan^{-1}(r_y/r_x) = \tan^{-1}[(10 \text{ m})/(-9.0 \text{ m})] = -48° \text{ or } 132°.$$

Since the x component of the resultant is negative and the y component is positive, characteristic of the second quadrant, we find the angle is 132° (measured counterclockwise from +x axis).

10. We label the displacement vectors $\vec{A}$, $\vec{B}$ and $\vec{C}$ (and denote the result of their vector sum as $\vec{r}$). We choose *east* as the $\hat{i}$ direction (+x direction) and *north* as the $\hat{j}$ direction (+y direction) All distances are understood to be in kilometers.

(a) The vector diagram representing the motion is shown below:

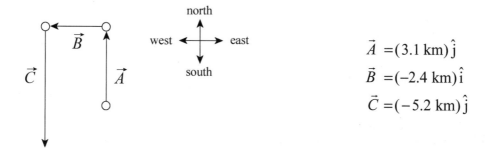

$\vec{A} = (3.1 \text{ km})\hat{j}$

$\vec{B} = (-2.4 \text{ km})\hat{i}$

$\vec{C} = (-5.2 \text{ km})\hat{j}$

(b) The final point is represented by

$$\vec{r} = \vec{A} + \vec{B} + \vec{C} = (-2.4 \text{ km})\hat{i} + (-2.1 \text{ km})\hat{j}$$

whose magnitude is

$$|\vec{r}| = \sqrt{(-2.4 \text{ km})^2 + (-2.1 \text{ km})^2} \approx 3.2 \text{ km}.$$

(c) There are two possibilities for the angle:

$$\theta = \tan^{-1}\left(\frac{-2.1 \text{ km}}{-2.4 \text{ km}}\right) = 41°, \text{ or } 221°.$$

We choose the latter possibility since $\vec{r}$ is in the third quadrant. It should be noted that many graphical calculators have polar ↔ rectangular "shortcuts" that automatically produce the correct answer for angle (measured counterclockwise from the +x axis). We may phrase the angle, then, as 221° counterclockwise from East (a phrasing that sounds peculiar, at best) or as 41° south from west or 49° west from south. The resultant $\vec{r}$ is not shown in our sketch; it would be an arrow directed from the "tail" of $\vec{A}$ to the "head" of $\vec{C}$.

11. We find the components and then add them (as scalars, not vectors). With $d = 3.40$ km and $\theta = 35.0°$ we find $d \cos\theta + d \sin\theta = 4.74$ km.

12. (a) $\vec{a} + \vec{b} = (3.0\hat{i} + 4.0\hat{j})$ m $+ (5.0\hat{i} - 2.0\hat{j})$ m $= (8.0$ m$)\hat{i} + (2.0$ m$)\hat{j}$.

(b) The magnitude of $\vec{a} + \vec{b}$ is

$$|\vec{a} + \vec{b}| = \sqrt{(8.0 \text{ m})^2 + (2.0 \text{ m})^2} = 8.2 \text{ m}.$$

(c) The angle between this vector and the $+x$ axis is $\tan^{-1}[(2.0 \text{ m})/(8.0 \text{ m})] = 14°$.

(d) $\vec{b} - \vec{a} = (5.0\hat{i} - 2.0\hat{j})$ m $- (3.0\hat{i} + 4.0\hat{j})$ m $= (2.0$ m$)\hat{i} - (6.0$ m$)\hat{j}$.

(e) The magnitude of the difference vector $\vec{b} - \vec{a}$ is

$$|\vec{b} - \vec{a}| = \sqrt{(2.0 \text{ m})^2 + (-6.0 \text{ m})^2} = 6.3 \text{ m}.$$

(f) The angle between this vector and the $+x$ axis is $\tan^{-1}[(-6.0 \text{ m})/(2.0 \text{ m})] = -72°$. The vector is 72° *clockwise* from the axis defined by $\hat{i}$.

13. All distances in this solution are understood to be in meters.

(a) $\vec{a} + \vec{b} = [4.0 + (-1.0)]\hat{i} + [(-3.0) + 1.0]\hat{j} + (1.0 + 4.0)\hat{k} = (3.0\hat{i} - 2.0\hat{j} + 5.0\hat{k})$ m.

(b) $\vec{a} - \vec{b} = [4.0 - (-1.0)]\hat{i} + [(-3.0) - 1.0]\hat{j} + (1.0 - 4.0)\hat{k} = (5.0\hat{i} - 4.0\hat{j} - 3.0\hat{k})$ m.

(c) The requirement $\vec{a} - \vec{b} + \vec{c} = 0$ leads to $\vec{c} = \vec{b} - \vec{a}$, which we note is the opposite of what we found in part (b). Thus, $\vec{c} = (-5.0\hat{i} + 4.0\hat{j} + 3.0\hat{k})$ m.

14. The x, y and z components of $\vec{r} = \vec{c} + \vec{d}$ are, respectively,

(a) $r_x = c_x + d_x = 7.4$ m $+ 4.4$ m $= 12$ m,

(b) $r_y = c_y + d_y = -3.8$ m $- 2.0$ m $= -5.8$ m, and

(c) $r_z = c_z + d_z = -6.1$ m $+ 3.3$ m $= -2.8$ m.

15. Reading carefully, we see that the (x, y) specifications for each "dart" are to be interpreted as $(\Delta x, \Delta y)$ descriptions of the corresponding displacement vectors. We combine the different parts of this problem into a single exposition.

(a) Along the x axis, we have (with the centimeter unit understood)

$$30.0 + b_x - 20.0 - 80.0 = -140,$$

which gives $b_x = -70.0$ cm.

(b) Along the y axis we have

$$40.0 - 70.0 + c_y - 70.0 = -20.0$$

which yields $c_y = 80.0$ cm.

(c) The magnitude of the final location (-140, -20.0) is $\sqrt{(-140)^2 + (-20.0)^2} = 141$ cm.

(d) Since the displacement is in the third quadrant, the angle of the overall displacement is given by $\pi + \tan^{-1}[(-20.0)/(-140)]$ or 188° counterclockwise from the $+x$ axis (172° clockwise from the $+x$ axis).

16. If we wish to use Eq. 3-5 in an unmodified fashion, we should note that the angle between $\vec{C}$ and the $+x$ axis is $180° + 20.0° = 200°$.

(a) The x and y components of $\vec{B}$ are given by

$$B_x = C_x - A_x = (15.0 \text{ m}) \cos 200° - (12.0 \text{ m}) \cos 40° = -23.3 \text{ m},$$
$$B_y = C_y - A_y = (15.0 \text{ m}) \sin 200° - (12.0 \text{ m}) \sin 40° = -12.8 \text{ m}.$$

Consequently, its magnitude is $|\vec{B}| = \sqrt{(-23.3 \text{ m})^2 + (-12.8 \text{ m})^2} = 26.6$ m.

(b) The two possibilities presented by a simple calculation for the angle between $\vec{B}$ and the $+x$ axis are $\tan^{-1}[(-12.8 \text{ m})/(-23.3 \text{ m})] = 28.9°$, and $180° + 28.9° = 209°$. We choose the latter possibility as the correct one since it indicates that $\vec{B}$ is in the third quadrant (indicated by the signs of its components). We note, too, that the answer can be equivalently stated as $-151°$.

17. It should be mentioned that an efficient way to work this vector addition problem is with the cosine law for general triangles (and since $\vec{a}, \vec{b}$ and $\vec{r}$ form an isosceles triangle, the angles are easy to figure). However, in the interest of reinforcing the usual systematic approach to vector addition, we note that the angle $\vec{b}$ makes with the $+x$ axis is $30° + 105° = 135°$ and apply Eq. 3-5 and Eq. 3-6 where appropriate.

(a) The x component of $\vec{r}$ is $r_x = (10.0 \text{ m}) \cos 30° + (10.0 \text{ m}) \cos 135° = 1.59$ m.

(b) The y component of $\vec{r}$ is $r_y = (10.0 \text{ m}) \sin 30° + (10.0 \text{ m}) \sin 135° = 12.1$ m.

(c) The magnitude of $\vec{r}$ is $r = |\vec{r}| = \sqrt{(1.59 \text{ m})^2 + (12.1 \text{ m})^2} = 12.2$ m.

(d) The angle between $\vec{r}$ and the +x direction is $\tan^{-1}[(12.1 \text{ m})/(1.59 \text{ m})] = 82.5°$.

18. (a) Summing the x components, we have

$$20 \text{ m} + b_x - 20 \text{ m} - 60 \text{ m} = -140 \text{ m},$$

which gives $b_x = -80$ m.

(b) Summing the y components, we have

$$60 \text{ m} - 70 \text{ m} + c_y - 70 \text{ m} = 30 \text{ m},$$

which implies $c_y = 110$ m.

(c) Using the Pythagorean theorem, the magnitude of the overall displacement is given by $\sqrt{(-140 \text{ m})^2 + (30 \text{ m})^2} \approx 143$ m.

(d) The angle is given by $\tan^{-1}(30/(-140)) = -12°$, (which would be 12° measured clockwise from the –x axis, or 168° measured counterclockwise from the +x axis)

19. Many of the operations are done efficiently on most modern graphical calculators using their built-in vector manipulation and rectangular ↔ polar "shortcuts." In this solution, we employ the "traditional" methods (such as Eq. 3-6). Where the length unit is not displayed, the unit meter should be understood.

(a) Using unit-vector notation,

$$\vec{a} = (50 \text{ m})\cos(30°)\hat{i} + (50 \text{ m})\sin(30°)\hat{j}$$
$$\vec{b} = (50 \text{ m})\cos(195°)\hat{i} + (50 \text{ m})\sin(195°)\hat{j}$$
$$\vec{c} = (50 \text{ m})\cos(315°)\hat{i} + (50 \text{ m})\sin(315°)\hat{j}$$
$$\vec{a} + \vec{b} + \vec{c} = (30.4 \text{ m})\hat{i} - (23.3 \text{ m})\hat{j}.$$

The magnitude of this result is $\sqrt{(30.4 \text{ m})^2 + (-23.3 \text{ m})^2} = 38$ m.

(b) The two possibilities presented by a simple calculation for the angle between the vector described in part (a) and the +x direction are $\tan^{-1}[(-23.2 \text{ m})/(30.4 \text{ m})] = -37.5°$, and $180° + (-37.5°) = 142.5°$. The former possibility is the correct answer since the vector is in the fourth quadrant (indicated by the signs of its components). Thus, the angle is –37.5°, which is to say that it is 37.5° *clockwise* from the +x axis. This is equivalent to 322.5° counterclockwise from +x.

(c) We find

$$\vec{a}-\vec{b}+\vec{c} = [43.3-(-48.3)+35.4]\,\hat{i}-[25-(-12.9)+(-35.4)]\,\hat{j} = (127\,\hat{i}+2.60\,\hat{j})\text{ m}$$

in unit-vector notation. The magnitude of this result is

$$|\vec{a}-\vec{b}+\vec{c}| = \sqrt{(127\text{ m})^2+(2.6\text{ m})^2} \approx 1.30\times 10^2 \text{ m}.$$

(d) The angle between the vector described in part (c) and the +x axis is $\tan^{-1}(2.6\text{ m}/127\text{ m}) \approx 1.2°$.

(e) Using unit-vector notation, $\vec{d}$ is given by $\vec{d} = \vec{a}+\vec{b}-\vec{c} = (-40.4\,\hat{i}+47.4\,\hat{j})$ m, which has a magnitude of $\sqrt{(-40.4\text{ m})^2+(47.4\text{ m})^2} = 62$ m.

(f) The two possibilities presented by a simple calculation for the angle between the vector described in part (e) and the +x axis are $\tan^{-1}(47.4/(-40.4)) = -50.0°$, and $180°+(-50.0°) = 130°$. We choose the latter possibility as the correct one since it indicates that $\vec{d}$ is in the second quadrant (indicated by the signs of its components).

20. Angles are given in 'standard' fashion, so Eq. 3-5 applies directly. We use this to write the vectors in unit-vector notation before adding them. However, a very different-looking approach using the special capabilities of most graphical calculators can be imagined. Wherever the length unit is not displayed in the solution below, the unit meter should be understood.

(a) Allowing for the different angle units used in the problem statement, we arrive at

$$\vec{E} = 3.73\,\hat{i}+4.70\,\hat{j}$$
$$\vec{F} = 1.29\,\hat{i}-4.83\,\hat{j}$$
$$\vec{G} = 1.45\,\hat{i}+3.73\,\hat{j}$$
$$\vec{H} = -5.20\,\hat{i}+3.00\,\hat{j}$$
$$\vec{E}+\vec{F}+\vec{G}+\vec{H} = 1.28\,\hat{i}+6.60\,\hat{j}.$$

(b) The magnitude of the vector sum found in part (a) is $\sqrt{(1.28\text{ m})^2+(6.60\text{ m})^2} = 6.72$ m.

(c) Its angle measured counterclockwise from the +x axis is $\tan^{-1}(6.60/1.28) = 79.0°$.

(d) Using the conversion factor π rad = $180°$, $79.0° = 1.38$ rad.

21. (a) With $\hat{i}$ directed forward and $\hat{j}$ directed leftward, then the resultant is $(5.00\,\hat{i} + 2.00\,\hat{j})$ m. The magnitude is given by the Pythagorean theorem: $\sqrt{(5.00\text{ m})^2 + (2.00\text{ m})^2} = 5.385$ m ≈ 5.39 m.

(b) The angle is $\tan^{-1}(2.00/5.00) \approx 21.8°$ (left of forward).

22. The desired result is the displacement vector, in units of km, $\vec{A}$ = (5.6 km), $90°$ (measured counterclockwise from the $+x$ axis), or $\vec{A} = (5.6\text{ km})\hat{j}$, where $\hat{j}$ is the unit vector along the positive y axis (north). This consists of the sum of two displacements: during the whiteout, $\vec{B} = (7.8$ km), $50°$, or

$$\vec{B} = (7.8\text{ km})(\cos 50°\hat{i} + \sin 50°\,\hat{j}) = (5.01\text{ km})\hat{i} + (5.98\text{ km})\hat{j}$$

and the unknown $\vec{C}$. Thus, $\vec{A} = \vec{B} + \vec{C}$.

(a) The desired displacement is given by $\vec{C} = \vec{A} - \vec{B} = (-5.01\text{ km})\,\hat{i} - (0.38\text{ km})\,\hat{j}$. The magnitude is $\sqrt{(-5.01\text{ km})^2 + (-0.38\text{ km})^2} = 5.0$ km.

(b) The angle is $\tan^{-1}[(-0.38\text{ km})/(-5.01\text{ km})] = 4.3°$, south of due west.

23. The strategy is to find where the camel is ($\vec{C}$) by adding the two consecutive displacements described in the problem, and then finding the difference between that location and the oasis ($\vec{B}$). Using the magnitude-angle notation

$$\vec{C} = (24\,\angle\,-15°) + (8.0\,\angle\,90°) = (23.25\,\angle\,4.41°)$$

so

$$\vec{B} - \vec{C} = (25\,\angle\,0°) - (23.25\,\angle\,4.41°) = (2.5\,\angle\,-45°)$$

which is efficiently implemented using a vector capable calculator in polar mode. The distance is therefore 2.6 km.

24. Let $\vec{A}$ represent the first part of Beetle 1's trip (0.50 m east or $0.5\,\hat{i}$) and $\vec{C}$ represent the first part of Beetle 2's trip intended voyage (1.6 m at $50°$ north of east). For their respective second parts: $\vec{B}$ is 0.80 m at $30°$ north of east and $\vec{D}$ is the unknown. The final position of Beetle 1 is

$$\vec{A} + \vec{B} = (0.5\text{ m})\hat{i} + (0.8\text{ m})(\cos 30°\,\hat{i} + \sin 30°\,\hat{j}) = (1.19\text{ m})\,\hat{i} + (0.40\text{ m})\hat{j}.$$

The equation relating these is $\vec{A}+\vec{B}=\vec{C}+\vec{D}$, where

$$\vec{C}=(1.60 \text{ m})(\cos 50.0°\hat{i}+\sin 50.0°\hat{j})=(1.03 \text{ m})\hat{i}+(1.23 \text{ m})\hat{j}$$

(a) We find $\vec{D}=\vec{A}+\vec{B}-\vec{C}=(0.16 \text{ m})\hat{i}+(-0.83 \text{ m})\hat{j}$, and the magnitude is $D = 0.84$ m.

(b) The angle is $\tan^{-1}(-0.83/0.16)=-79°$ which is interpreted to mean 79° south of east (or 11° east of south).

25. The resultant (along the y axis, with the same magnitude as $\vec{C}$) forms (along with $\vec{C}$) a side of an isosceles triangle (with $\vec{B}$ forming the base). If the angle between $\vec{C}$ and the y axis is $\theta=\tan^{-1}(3/4)=36.87°$, then it should be clear that (referring to the magnitudes of the vectors) $B=2C\sin(\theta/2)$. Thus (since $C=5.0$) we find $B=3.2$.

26. As a vector addition problem, we express the situation (described in the problem statement) as $\vec{A}+\vec{B}=(3A)\hat{j}$, where $\vec{A}=A\hat{i}$ and $B=7.0$ m. Since $\hat{i}\perp\hat{j}$ we may use the Pythagorean theorem to express B in terms of the magnitudes of the other two vectors:

$$B=\sqrt{(3A)^2+A^2} \quad \Rightarrow \quad A=\frac{1}{\sqrt{10}}B=2.2 \text{ m}.$$

27. Let $l_0=2.0$ cm be the length of each segment. The nest is located at the endpoint of segment w.

(a) Using unit-vector notation, the displacement vector for point A is

$$\vec{d}_A=\vec{w}+\vec{v}+\vec{i}+\vec{h}=l_0(\cos 60°\hat{i}+\sin 60°\hat{j})+(l_0\hat{j})+l_0(\cos 120°\hat{i}+\sin 120°\hat{j})+(l_0\hat{j})$$
$$=(2+\sqrt{3})l_0\hat{j}.$$

Therefore, the magnitude of $\vec{d}_A$ is $|\vec{d}_A|=(2+\sqrt{3})(2.0 \text{ cm})=7.5$ cm.

(b) The angle of $\vec{d}_A$ is $\theta=\tan^{-1}(d_{A,y}/d_{A,x})=\tan^{-1}(\infty)=90°$.

(c) Similarly, the displacement for point B is

$$\vec{d}_B=\vec{w}+\vec{v}+\vec{j}+\vec{p}+\vec{o}$$
$$=l_0(\cos 60°\hat{i}+\sin 60°\hat{j})+(l_0\hat{j})+l_0(\cos 60°\hat{i}+\sin 60°\hat{j})+l_0(\cos 30°\hat{i}+\sin 30°\hat{j})+(l_0\hat{i})$$
$$=(2+\sqrt{3}/2)l_0\hat{i}+(3/2+\sqrt{3})l_0\hat{j}.$$

Therefore, the magnitude of $\vec{d}_B$ is

$$|\vec{d}_B| = l_0\sqrt{(2+\sqrt{3}/2)^2 + (3/2+\sqrt{3})^2} = (2.0 \text{ cm})(4.3) = 8.6 \text{ cm}.$$

(d) The direction of $\vec{d}_B$ is

$$\theta_B = \tan^{-1}\left(\frac{d_{B,y}}{d_{B,x}}\right) = \tan^{-1}\left(\frac{3/2+\sqrt{3}}{2+\sqrt{3}/2}\right) = \tan^{-1}(1.13) = 48°.$$

28. Many of the operations are done efficiently on most modern graphical calculators using their built-in vector manipulation and rectangular ↔ polar "shortcuts." In this solution, we employ the "traditional" methods (such as Eq. 3-6).

(a) The magnitude of $\vec{a}$ is $a = \sqrt{(4.0 \text{ m})^2 + (-3.0 \text{ m})^2} = 5.0$ m.

(b) The angle between $\vec{a}$ and the +x axis is $\tan^{-1}[(-3.0 \text{ m})/(4.0 \text{ m})] = -37°$. The vector is 37° *clockwise* from the axis defined by $\hat{i}$.

(c) The magnitude of $\vec{b}$ is $b = \sqrt{(6.0 \text{ m})^2 + (8.0 \text{ m})^2} = 10$ m.

(d) The angle between $\vec{b}$ and the +x axis is $\tan^{-1}[(8.0 \text{ m})/(6.0 \text{ m})] = 53°$.

(e) $\vec{a}+\vec{b} = (4.0 \text{ m}+6.0 \text{ m})\hat{i} + [(-3.0 \text{ m})+8.0 \text{ m}]\hat{j} = (10 \text{ m})\hat{i} + (5.0 \text{ m})\hat{j}$. The magnitude of this vector is $|\vec{a}+\vec{b}| = \sqrt{(10 \text{ m})^2 + (5.0 \text{ m})^2} = 11$ m; we round to two significant figures in our results.

(f) The angle between the vector described in part (e) and the +x axis is $\tan^{-1}[(5.0 \text{ m})/(10 \text{ m})] = 27°$.

(g) $\vec{b}-\vec{a} = (6.0 \text{ m}-4.0 \text{ m})\hat{i} + [8.0 \text{ m}-(-3.0 \text{ m})]\hat{j} = (2.0 \text{ m})\hat{i} + (11 \text{ m})\hat{j}$. The magnitude of this vector is $|\vec{b}-\vec{a}| = \sqrt{(2.0 \text{ m})^2 + (11 \text{ m})^2} = 11$ m, which is, interestingly, the same result as in part (e) (exactly, not just to 2 significant figures) (this curious coincidence is made possible by the fact that $\vec{a} \perp \vec{b}$).

(h) The angle between the vector described in part (g) and the +x axis is $\tan^{-1}[(11 \text{ m})/(2.0 \text{ m})] = 80°$.

(i) $\vec{a}-\vec{b} = (4.0 \text{ m}-6.0 \text{ m})\hat{i} + [(-3.0 \text{ m})-8.0 \text{ m}]\hat{j} = (-2.0 \text{ m})\hat{i} + (-11 \text{ m})\hat{j}$. The magnitude of this vector is $|\vec{a}-\vec{b}| = \sqrt{(-2.0 \text{ m})^2 + (-11 \text{ m})^2} = 11$ m.

(j) The two possibilities presented by a simple calculation for the angle between the vector described in part (i) and the +x direction are $\tan^{-1}[(-11\text{ m})/(-2.0\text{ m})] = 80°$, and $180° + 80° = 260°$. The latter possibility is the correct answer (see part (k) for a further observation related to this result).

(k) Since $\vec{a} - \vec{b} = (-1)(\vec{b} - \vec{a})$, they point in opposite (anti-parallel) directions; the angle between them is 180°.

29. Solving the simultaneous equations yields the answers:

(a) $\vec{d_1} = 4\vec{d_3} = 8\hat{i} + 16\hat{j}$, and

(b) $\vec{d_2} = \vec{d_3} = 2\hat{i} + 4\hat{j}$.

30. The vector equation is $\vec{R} = \vec{A} + \vec{B} + \vec{C} + \vec{D}$. Expressing $\vec{B}$ and $\vec{D}$ in unit-vector notation, we have $(1.69\hat{i} + 3.63\hat{j})$ m and $(-2.87\hat{i} + 4.10\hat{j})$ m, respectively. Where the length unit is not displayed in the solution below, the unit meter should be understood.

(a) Adding corresponding components, we obtain $\vec{R} = (-3.18\text{ m})\hat{i} + (4.72\text{ m})\hat{j}$.

(b) Using Eq. 3-6, the magnitude is

$$|\vec{R}| = \sqrt{(-3.18\text{ m})^2 + (4.72\text{ m})^2} = 5.69\text{ m}.$$

(c) The angle is

$$\theta = \tan^{-1}\left(\frac{4.72\text{ m}}{-3.18\text{ m}}\right) = -56.0° \text{ (with } -x \text{ axis)}.$$

If measured counterclockwise from +x-axis, the angle is then $180° - 56.0° = 124°$. Thus, converting the result to polar coordinates, we obtain

$$(-3.18, 4.72) \rightarrow (5.69 \angle 124°)$$

31. (a) As can be seen from Figure 3-32, the point diametrically opposite the origin (0,0,0) has position vector $a\hat{i} + a\hat{j} + a\hat{k}$ and this is the vector along the "body diagonal."

(b) From the point $(a, 0, 0)$ which corresponds to the position vector $a\hat{i}$, the diametrically opposite point is $(0, a, a)$ with the position vector $a\hat{j} + a\hat{k}$. Thus, the vector along the line is the difference $-a\hat{i} + a\hat{j} + a\hat{k}$.

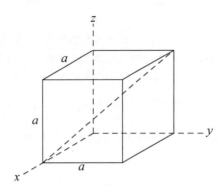

(c) If the starting point is (0, a, 0) with the corresponding position vector $a\hat{j}$, the diametrically opposite point is (a, 0, a) with the position vector $a\hat{i} + a\hat{k}$. Thus, the vector along the line is the difference $a\hat{i} - a\hat{j} + a\hat{k}$.

(d) If the starting point is (a, a, 0) with the corresponding position vector $a\hat{i} + a\hat{j}$, the diametrically opposite point is (0, 0, a) with the position vector $a\hat{k}$. Thus, the vector along the line is the difference $-a\hat{i} - a\hat{j} + a\hat{k}$.

(e) Consider the vector from the back lower left corner to the front upper right corner. It is $a\hat{i} + a\hat{j} + a\hat{k}$. We may think of it as the sum of the vector $a\hat{i}$ parallel to the x axis and the vector $a\hat{j} + a\hat{k}$ perpendicular to the x axis. The tangent of the angle between the vector and the x axis is the perpendicular component divided by the parallel component. Since the magnitude of the perpendicular component is $\sqrt{a^2 + a^2} = a\sqrt{2}$ and the magnitude of the parallel component is a, $\tan\theta = (a\sqrt{2})/a = \sqrt{2}$. Thus θ = 54.7°. The angle between the vector and each of the other two adjacent sides (the y and z axes) is the same as is the angle between any of the other diagonal vectors and any of the cube sides adjacent to them.

(f) The length of any of the diagonals is given by $\sqrt{a^2 + a^2 + a^2} = a\sqrt{3}$.

32. (a) With a = 17.0 m and θ = 56.0° we find $a_x = a\cos\theta$ = 9.51 m.

(b) Similarly, $a_y = a\sin\theta$ = 14.1 m.

(c) The angle relative to the new coordinate system is θ' = (56.0° − 18.0°) = 38.0°. Thus, $a_x' = a\cos\theta'$ = 13.4 m.

(d) Similarly, $a_y' = a\sin\theta'$ = 10.5 m.

33. (a) The scalar (dot) product is (4.50)(7.30)cos(320° − 85.0°) = − 18.8 .

(b) The vector (cross) product is in the $\hat{k}$ direction (by the right-hand rule) with magnitude $|(4.50)(7.30) \sin(320° - 85.0°)| = 26.9$.

34. First, we rewrite the given expression as $4(\vec{d}_{plane} \cdot \vec{d}_{cross})$ where $\vec{d}_{plane} = \vec{d}_1 + \vec{d}_2$ and in the plane of $\vec{d}_1$ and $\vec{d}_2$, and $\vec{d}_{cross} = \vec{d}_1 \times \vec{d}_2$. Noting that $\vec{d}_{cross}$ is perpendicular to the plane of $\vec{d}_1$ and $\vec{d}_2$, we see that the answer must be 0 (the scalar [dot] product of perpendicular vectors is zero).

35. We apply Eq. 3-30 and Eq.3-23. If a vector-capable calculator is used, this makes a good exercise for getting familiar with those features. Here we briefly sketch the method.

(a) We note that $\vec{b} \times \vec{c} = -8.0\hat{i} + 5.0\hat{j} + 6.0\hat{k}$. Thus,

$$\vec{a} \cdot (\vec{b} \times \vec{c}) = (3.0)(-8.0) + (3.0)(5.0) + (-2.0)(6.0) = -21.$$

(b) We note that $\vec{b} + \vec{c} = 1.0\hat{i} - 2.0\hat{j} + 3.0\hat{k}$. Thus,

$$\vec{a} \cdot (\vec{b} + \vec{c}) = (3.0)(1.0) + (3.0)(-2.0) + (-2.0)(3.0) = -9.0.$$

(c) Finally,

$$\vec{a} \times (\vec{b} + \vec{c}) = [(3.0)(3.0) - (-2.0)(-2.0)]\hat{i} + [(-2.0)(1.0) - (3.0)(3.0)]\hat{j}$$
$$+ [(3.0)(-2.0) - (3.0)(1.0)]\hat{k}$$
$$= 5\hat{i} - 11\hat{j} - 9\hat{k}$$

36. We apply Eq. 3-30 and Eq. 3-23.

(a) $\vec{a} \times \vec{b} = (a_x b_y - a_y b_x)\hat{k}$ since all other terms vanish, due to the fact that neither $\vec{a}$ nor $\vec{b}$ have any z components. Consequently, we obtain $[(3.0)(4.0) - (5.0)(2.0)]\hat{k} = 2.0\hat{k}$.

(b) $\vec{a} \cdot \vec{b} = a_x b_x + a_y b_y$ yields $(3.0)(2.0) + (5.0)(4.0) = 26$.

(c) $\vec{a} + \vec{b} = (3.0 + 2.0)\hat{i} + (5.0 + 4.0)\hat{j} \Rightarrow (\vec{a} + \vec{b}) \cdot \vec{b} = (5.0)(2.0) + (9.0)(4.0) = 46$.

(d) Several approaches are available. In this solution, we will construct a $\hat{b}$ unit-vector and "dot" it (take the scalar product of it) with $\vec{a}$. In this case, we make the desired unit-vector by

$$\hat{b} = \frac{\vec{b}}{|\vec{b}|} = \frac{2.0\hat{i} + 4.0\hat{j}}{\sqrt{(2.0)^2 + (4.0)^2}}.$$

We therefore obtain

$$a_b = \vec{a} \cdot \hat{b} = \frac{(3.0)(2.0) + (5.0)(4.0)}{\sqrt{(2.0)^2 + (4.0)^2}} = 5.8.$$

37. Examining the figure, we see that $\vec{a} + \vec{b} + \vec{c} = 0$, where $\vec{a} \perp \vec{b}$.

(a) $|\vec{a} \times \vec{b}| = (3.0)(4.0) = 12$ since the angle between them is 90°.

(b) Using the Right Hand Rule, the vector $\vec{a} \times \vec{b}$ points in the $\hat{i} \times \hat{j} = \hat{k}$, or the +z direction.

(c) $|\vec{a} \times \vec{c}| = |\vec{a} \times (-\vec{a} - \vec{b})| = |-(\vec{a} \times \vec{b})| = 12.$

(d) The vector $-\vec{a} \times \vec{b}$ points in the $-\hat{i} \times \hat{j} = -\hat{k}$, or the $-z$ direction.

(e) $|\vec{b} \times \vec{c}| = |\vec{b} \times (-\vec{a} - \vec{b})| = |-(\vec{b} \times \vec{a})| = |(\vec{a} \times \vec{b})| = 12.$

(f) The vector points in the +z direction, as in part (a).

38. The displacement vectors can be written as (in meters)

$$\vec{d}_1 = (4.50 \text{ m})(\cos 63° \hat{j} + \sin 63° \hat{k}) = (2.04 \text{ m})\hat{j} + (4.01 \text{ m})\hat{k}$$
$$\vec{d}_2 = (1.40 \text{ m})(\cos 30° \hat{i} + \sin 30° \hat{k}) = (1.21 \text{ m})\hat{i} + (0.70 \text{ m})\hat{k}.$$

(a) The dot product of $\vec{d}_1$ and $\vec{d}_2$ is

$$\vec{d}_1 \cdot \vec{d}_2 = (2.04\hat{j} + 4.01\hat{k}) \cdot (1.21\hat{i} + 0.70\hat{k}) = (4.01\hat{k}) \cdot (0.70\hat{k}) = 2.81 \text{ m}^2.$$

(b) The cross product of $\vec{d}_1$ and $\vec{d}_2$ is

$$\begin{aligned}\vec{d}_1 \times \vec{d}_2 &= (2.04\hat{j} + 4.01\hat{k}) \times (1.21\hat{i} + 0.70\hat{k}) \\ &= (2.04)(1.21)(-\hat{k}) + (2.04)(0.70)\hat{i} + (4.01)(1.21)\hat{j} \\ &= (1.43\,\hat{i} + 4.86\hat{j} - 2.48\hat{k}) \text{ m}^2.\end{aligned}$$

(c) The magnitudes of $\vec{d}_1$ and $\vec{d}_2$ are

$$d_1 = \sqrt{(2.04 \text{ m})^2 + (4.01 \text{ m})^2} = 4.50 \text{ m}$$
$$d_2 = \sqrt{(1.21 \text{ m})^2 + (0.70 \text{ m})^2} = 1.40 \text{ m}.$$

Thus, the angle between the two vectors is

$$\theta = \cos^{-1}\left(\frac{\vec{d}_1 \cdot \vec{d}_2}{d_1 d_2}\right) = \cos^{-1}\left(\frac{2.81 \text{ m}^2}{(4.50 \text{ m})(1.40 \text{ m})}\right) = 63.5°.$$

39. Since $ab \cos\phi = a_x b_x + a_y b_y + a_z b_z$,

$$\cos\phi = \frac{a_x b_x + a_y b_y + a_z b_z}{ab}.$$

The magnitudes of the vectors given in the problem are

$$a = |\vec{a}| = \sqrt{(3.00)^2 + (3.00)^2 + (3.00)^2} = 5.20$$

$$b = |\vec{b}| = \sqrt{(2.00)^2 + (1.00)^2 + (3.00)^2} = 3.74.$$

The angle between them is found from

$$\cos\phi = \frac{(3.00)(2.00) + (3.00)(1.00) + (3.00)(3.00)}{(5.20)(3.74)} = 0.926.$$

The angle is $\phi = 22°$.

40. Using the fact that

$$\hat{i} \times \hat{j} = \hat{k}, \quad \hat{j} \times \hat{k} = \hat{i}, \quad \hat{k} \times \hat{i} = \hat{j}$$

we obtain

$$2\vec{A} \times \vec{B} = 2\left(2.00\hat{i} + 3.00\hat{j} - 4.00\hat{k}\right) \times \left(-3.00\hat{i} + 4.00\hat{j} + 2.00\hat{k}\right) = 44.0\hat{i} + 16.0\hat{j} + 34.0\hat{k}.$$

Next, making use of

$$\hat{i} \cdot \hat{i} = \hat{j} \cdot \hat{j} = \hat{k} \cdot \hat{k} = 1$$
$$\hat{i} \cdot \hat{j} = \hat{j} \cdot \hat{k} = \hat{k} \cdot \hat{i} = 0$$

we have

$$3\vec{C} \cdot (2\vec{A} \times \vec{B}) = 3\left(7.00\hat{i} - 8.00\hat{j}\right) \cdot \left(44.0\hat{i} + 16.0\hat{j} + 34.0\hat{k}\right)$$
$$= 3[(7.00)(44.0) + (-8.00)(16.0) + (0)(34.0)] = 540.$$

41. From the definition of the dot product between $\vec{A}$ and $\vec{B}$, $\vec{A}\cdot\vec{B} = AB\cos\theta$, we have

$$\cos\theta = \frac{\vec{A}\cdot\vec{B}}{AB}$$

With $A = 6.00$, $B = 7.00$ and $\vec{A}\cdot\vec{B} = 14.0$, $\cos\theta = 0.333$, or $\theta = 70.5°$.

42. Applying Eq. 3-23, $\vec{F} = q\vec{v}\times\vec{B}$ (where q is a scalar) becomes

$$F_x\hat{i} + F_y\hat{j} + F_z\hat{k} = q\left(v_yB_z - v_zB_y\right)\hat{i} + q\left(v_zB_x - v_xB_z\right)\hat{j} + q\left(v_xB_y - v_yB_x\right)\hat{k}$$

which — plugging in values — leads to three equalities:

$$4.0 = 2\,(4.0B_z - 6.0B_y)$$
$$-20 = 2\,(6.0B_x - 2.0B_z)$$
$$12 = 2\,(2.0B_y - 4.0B_x)$$

Since we are told that $B_x = B_y$, the third equation leads to $B_y = -3.0$. Inserting this value into the first equation, we find $B_z = -4.0$. Thus, our answer is

$$\vec{B} = -3.0\,\hat{i} - 3.0\,\hat{j} - 4.0\,\hat{k}.$$

43. From the figure, we note that $\vec{c}\perp\vec{b}$, which implies that the angle between $\vec{c}$ and the +x axis is 120°. Direct application of Eq. 3-5 yields the answers for this and the next few parts.

(a) $a_x = a\cos 0° = a = 3.00$ m.

(b) $a_y = a\sin 0° = 0$.

(c) $b_x = b\cos 30° = (4.00\text{ m})\cos 30° = 3.46$ m.

(d) $b_y = b\sin 30° = (4.00\text{ m})\sin 30° = 2.00$ m.

(e) $c_x = c\cos 120° = (10.0\text{ m})\cos 120° = -5.00$ m.

(f) $c_y = c\sin 30° = (10.0\text{ m})\sin 120° = 8.66$ m.

(g) In terms of components (first x and then y), we must have

$$-5.00 \text{ m} = p\,(3.00 \text{ m}) + q\,(3.46 \text{ m})$$
$$8.66 \text{ m} = p\,(0) + q\,(2.00 \text{ m}).$$

Solving these equations, we find $p = -6.67$.

(h) Similarly, $q = 4.33$ (note that it's easiest to solve for q first). The numbers p and q have no units.

44. The two vectors are written as, in unit of meters,

$$\vec{d}_1 = 4.0\hat{i} + 5.0\hat{j} = d_{1x}\hat{i} + d_{1y}\hat{j}, \quad \vec{d}_2 = -3.0\hat{i} + 4.0\hat{j} = d_{2x}\hat{i} + d_{2y}\hat{j}$$

(a) The vector (cross) product gives

$$\vec{d}_1 \times \vec{d}_2 = (d_{1x}d_{2y} - d_{1y}d_{2x})\hat{k} = [(4.0)(4.0) - (5.0)(-3.0)]\hat{k} = 31\,\hat{k}$$

(b) The scalar (dot) product gives

$$\vec{d}_1 \cdot \vec{d}_2 = d_{1x}d_{2x} + d_{1y}d_{2y} = (4.0)(-3.0) + (5.0)(4.0) = 8.0.$$

(c)
$$(\vec{d}_1 + \vec{d}_2) \cdot \vec{d}_2 = \vec{d}_1 \cdot \vec{d}_2 + d_2^2 = 8.0 + (-3.0)^2 + (4.0)^2 = 33.$$

(d) Note that the magnitude of the d_1 vector is $\sqrt{16+25} = 6.4$. Now, the dot product is $(6.4)(5.0)\cos\theta = 8$. Dividing both sides by 32 and taking the inverse cosine yields $\theta = 75.5°$. Therefore the component of the d_1 vector along the direction of the d_2 vector is $6.4\cos\theta \approx 1.6$.

45. Although we think of this as a three-dimensional movement, it is rendered effectively two-dimensional by referring measurements to its well-defined plane of the fault.

(a) The magnitude of the net displacement is

$$|\vec{AB}| = \sqrt{|AD|^2 + |AC|^2} = \sqrt{(17.0 \text{ m})^2 + (22.0 \text{ m})^2} = 27.8 \text{m}.$$

(b) The magnitude of the vertical component of $\vec{AB}$ is $|AD| \sin 52.0° = 13.4$ m.

46. Where the length unit is not displayed, the unit meter is understood.

(a) We first note that $a = |\vec{a}| = \sqrt{(3.2)^2 + (1.6)^2} = 3.58$ and $b = |\vec{b}| = \sqrt{(0.50)^2 + (4.5)^2} = 4.53$. Now,

$$\vec{a}\cdot\vec{b} = a_x b_x + a_y b_y = ab\cos\phi$$
$$(3.2)(0.50)+(1.6)(4.5) = (3.58)(4.53)\cos\phi$$

which leads to $\phi = 57°$ (the inverse cosine is double-valued as is the inverse tangent, but we know this is the right solution since both vectors are in the same quadrant).

(b) Since the angle (measured from $+x$) for $\vec{a}$ is $\tan^{-1}(1.6/3.2) = 26.6°$, we know the angle for $\vec{c}$ is $26.6° - 90° = -63.4°$ (the other possibility, $26.6° + 90°$ would lead to a $c_x < 0$). Therefore,
$$c_x = c\cos(-63.4°) = (5.0)(0.45) = 2.2 \text{ m}.$$

(c) Also, $c_y = c\sin(-63.4°) = (5.0)(-0.89) = -4.5$ m.

(d) And we know the angle for $\vec{d}$ to be $26.6° + 90° = 116.6°$, which leads to
$$d_x = d\cos(116.6°) = (5.0)(-0.45) = -2.2 \text{ m}.$$

(e) Finally, $d_y = d\sin 116.6° = (5.0)(0.89) = 4.5$ m.

47. We apply Eq. 3-20 and Eq. 3-27.

(a) The scalar (dot) product of the two vectors is
$$\vec{a}\cdot\vec{b} = ab\cos\phi = (10)(6.0)\cos 60° = 30.$$

(b) The magnitude of the vector (cross) product of the two vectors is
$$|\vec{a}\times\vec{b}| = ab\sin\phi = (10)(6.0)\sin 60° = 52.$$

48. The vectors are shown on the diagram. The x axis runs from west to east and the y axis runs from south to north. Then $a_x = 5.0$ m, $a_y = 0$, $b_x = -(4.0 \text{ m})\sin 35° = -2.29$ m, and $b_y = (4.0 \text{ m})\cos 35° = 3.28$ m.

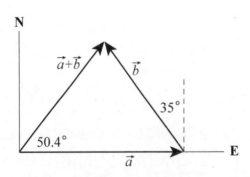

(a) Let $\vec{c} = \vec{a} + \vec{b}$. Then $c_x = a_x + b_x = 5.00 \text{ m} - 2.29 \text{ m} = 2.71$ m and

$c_y = a_y + b_y = 0 + 3.28 \text{ m} = 3.28 \text{ m}$. The magnitude of c is

$$c = \sqrt{c_x^2 + c_y^2} = \sqrt{(2.71\text{m})^2 + (3.28\text{m})^2} = 4.2 \text{ m}.$$

(b) The angle θ that $\vec{c} = \vec{a} + \vec{b}$ makes with the $+x$ axis is

$$\theta = \tan^{-1}\left(\frac{c_y}{c_x}\right) = \tan^{-1}\left(\frac{3.28}{2.71}\right) = 50°.$$

The second possibility ($\theta = 50.4° + 180° = 230.4°$) is rejected because it would point in a direction opposite to $\vec{c}$.

(c) The vector $\vec{b} - \vec{a}$ is found by adding $-\vec{a}$ to $\vec{b}$. The result is shown on the diagram to the right. Let $\vec{c} = \vec{b} - \vec{a}$. The components are $c_x = b_x - a_x = -2.29 \text{ m} - 5.00 \text{ m} = -7.29 \text{ m}$, and $c_y = b_y - a_y = 3.28 \text{ m}$. The magnitude of $\vec{c}$ is $c = \sqrt{c_x^2 + c_y^2} = 8.0 \text{ m}$.

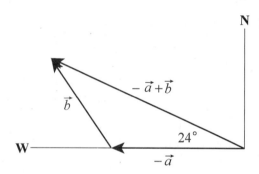

(d) The tangent of the angle θ that $\vec{c}$ makes with the $+x$ axis (east) is

$$\tan\theta = \frac{c_y}{c_x} = \frac{3.28 \text{ m}}{-7.29 \text{ m}} = -4.50.$$

There are two solutions: $-24.2°$ and $155.8°$. As the diagram shows, the second solution is correct. The vector $\vec{c} = -\vec{a} + \vec{b}$ is $24°$ north of west.

49. We choose $+x$ east and $+y$ north and measure all angles in the "standard" way (positive ones are counterclockwise from $+x$). Thus, vector $\vec{d}_1$ has magnitude $d_1 = 4.00$ m (with the unit meter) and direction $\theta_1 = 225°$. Also, $\vec{d}_2$ has magnitude $d_2 = 5.00$ m and direction $\theta_2 = 0°$, and vector $\vec{d}_3$ has magnitude $d_3 = 6.00$ m and direction $\theta_3 = 60°$.

(a) The x-component of $\vec{d}_1$ is $d_{1x} = d_1 \cos\theta_1 = -2.83$ m.

(b) The y-component of $\vec{d}_1$ is $d_{1y} = d_1 \sin\theta_1 = -2.83$ m.

(c) The x-component of $\vec{d}_2$ is $d_{2x} = d_2 \cos\theta_2 = 5.00$ m.

(d) The y-component of $\vec{d}_2$ is $d_{2y} = d_2 \sin\theta_2 = 0$.

(e) The x-component of $\vec{d}_3$ is $d_{3x} = d_3 \cos\theta_3 = 3.00$ m.

(f) The y-component of $\vec{d}_3$ is $d_{3y} = d_3 \sin\theta_3 = 5.20$ m.

(g) The sum of x-components is

$$d_x = d_{1x} + d_{2x} + d_{3x} = -2.83 \text{ m} + 5.00 \text{ m} + 3.00 \text{ m} = 5.17 \text{ m}.$$

(h) The sum of y-components is

$$d_y = d_{1y} + d_{2y} + d_{3y} = -2.83 \text{ m} + 0 + 5.20 \text{ m} = 2.37 \text{ m}.$$

(i) The magnitude of the resultant displacement is

$$d = \sqrt{d_x^2 + d_y^2} = \sqrt{(5.17 \text{ m})^2 + (2.37 \text{ m})^2} = 5.69 \text{ m}.$$

(j) And its angle is $\theta = \tan^{-1}(2.37/5.17) = 24.6°$ which (recalling our coordinate choices) means it points at about 25° north of east.

(k) and (l) This new displacement (the direct line home) when vectorially added to the previous (net) displacement must give zero. Thus, the new displacement is the negative, or opposite, of the previous (net) displacement. That is, it has the same magnitude (5.69 m) but points in the opposite direction (25° south of west).

50. From the figure, it is clear that $\vec{a} + \vec{b} + \vec{c} = 0$, where $\vec{a} \perp \vec{b}$.

(a) $\vec{a} \cdot \vec{b} = 0$ since the angle between them is 90°.

(b) $\vec{a} \cdot \vec{c} = \vec{a} \cdot (-\vec{a} - \vec{b}) = -|\vec{a}|^2 = -16$.

(c) Similarly, $\vec{b} \cdot \vec{c} = -9.0$.

51. Let $\vec{A}$ represent the first part of his actual voyage (50.0 km east) and $\vec{C}$ represent the intended voyage (90.0 km north). We are looking for a vector $\vec{B}$ such that $\vec{A} + \vec{B} = \vec{C}$.

(a) The Pythagorean theorem yields $B = \sqrt{(50.0 \text{ km})^2 + (90.0 \text{ km})^2} = 103 \text{ km}$.

(b) The direction is $\tan^{-1}(50.0 \text{ km}/90.0 \text{ km}) = 29.1°$ west of north (which is equivalent to 60.9° north of due west).

52. If we wish to use Eq. 3-5 directly, we should note that the angles for $\vec{Q}, \vec{R}$ and $\vec{S}$ are 100°, 250° and 310°, respectively, if they are measured counterclockwise from the +x axis.

(a) Using unit-vector notation, with the unit meter understood, we have

$$\vec{P} = 10.0\cos(25.0°)\hat{i} + 10.0\sin(25.0°)\hat{j}$$
$$\vec{Q} = 12.0\cos(100°)\hat{i} + 12.0\sin(100°)\hat{j}$$
$$\vec{R} = 8.00\cos(250°)\hat{i} + 8.00\sin(250°)\hat{j}$$
$$\vec{S} = 9.00\cos(310°)\hat{i} + 9.00\sin(310°)\hat{j}$$
$$\vec{P} + \vec{Q} + \vec{R} + \vec{S} = (10.0 \text{ m})\hat{i} + (1.63 \text{ m})\hat{j}$$

(b) The magnitude of the vector sum is $\sqrt{(10.0 \text{ m})^2 + (1.63 \text{ m})^2} = 10.2 \text{ m}$.

(c) The angle is $\tan^{-1}(1.63 \text{ m}/10.0 \text{ m}) \approx 9.24°$ measured counterclockwise from the +x axis.

53. Noting that the given 130° is measured counterclockwise from the +x axis, the two vectors can be written as

$$\vec{A} = 8.00(\cos 130°\hat{i} + \sin 130°\hat{j}) = -5.14\hat{i} + 6.13\hat{j}$$
$$\vec{B} = B_x\hat{i} + B_y\hat{j} = -7.72\hat{i} - 9.20\hat{j}.$$

(a) The angle between the negative direction of the y axis $(-\hat{j})$ and the direction of $\vec{A}$ is

$$\theta = \cos^{-1}\left(\frac{\vec{A}\cdot(-\hat{j})}{A}\right) = \cos^{-1}\left(\frac{-6.13}{\sqrt{(-5.14)^2 + (6.13)^2}}\right) = \cos^{-1}\left(\frac{-6.13}{8.00}\right) = 140°.$$

Alternatively, one may say that the −y direction corresponds to an angle of 270°, and the answer is simply given by 270°−130° = 140°.

(b) Since the y axis is in the xy plane, and $\vec{A} \times \vec{B}$ is perpendicular to that plane, then the answer is 90.0°.

(c) The vector can be simplified as

$$\vec{A} \times (\vec{B} + 3.00\hat{k}) = (-5.14\hat{i} + 6.13\hat{j}) \times (-7.72\hat{i} - 9.20\hat{j} + 3.00\hat{k})$$
$$= 18.39\hat{i} + 15.42\hat{j} + 94.61\hat{k}$$

Its magnitude is $|\vec{A} \times (\vec{B} + 3.00\hat{k})| = 97.6$. The angle between the negative direction of the y axis ($-\hat{j}$) and the direction of the above vector is

$$\theta = \cos^{-1}\left(\frac{-15.42}{97.6}\right) = 99.1°.$$

54. The three vectors are

$$\vec{d}_1 = 4.0\hat{i} + 5.0\hat{j} - 6.0\hat{k}$$
$$\vec{d}_2 = -1.0\hat{i} + 2.0\hat{j} + 3.0\hat{k}$$
$$\vec{d}_3 = 4.0\hat{i} + 3.0\hat{j} + 2.0\hat{k}$$

(a) $\vec{r} = \vec{d}_1 - \vec{d}_2 + \vec{d}_3 = (9.0 \text{ m})\hat{i} + (6.0 \text{ m})\hat{j} + (-7.0 \text{ m})\hat{k}$.

(b) The magnitude of $\vec{r}$ is $|\vec{r}| = \sqrt{(9.0 \text{ m})^2 + (6.0 \text{ m})^2 + (-7.0 \text{ m})^2} = 12.9 \text{ m}$. The angle between $\vec{r}$ and the z-axis is given by

$$\cos\theta = \frac{\vec{r} \cdot \hat{k}}{|\vec{r}|} = \frac{-7.0 \text{ m}}{12.9 \text{ m}} = -0.543$$

which implies $\theta = 123°$.

(c) The component of $\vec{d}_1$ along the direction of $\vec{d}_2$ is given by $d_\parallel = \vec{d}_1 \cdot \hat{u} = d_1 \cos\varphi$ where φ is the angle between $\vec{d}_1$ and $\vec{d}_2$, and $\hat{u}$ is the unit vector in the direction of $\vec{d}_2$. Using the properties of the scalar (dot) product, we have

$$d_\parallel = d_1\left(\frac{\vec{d}_1 \cdot \vec{d}_2}{d_1 d_2}\right) = \frac{\vec{d}_1 \cdot \vec{d}_2}{d_2} = \frac{(4.0)(-1.0) + (5.0)(2.0) + (-6.0)(3.0)}{\sqrt{(-1.0)^2 + (2.0)^2 + (3.0)^2}} = \frac{-12}{\sqrt{14}} = -3.2 \text{ m}.$$

(d) Now we are looking for $d_\perp$ such that $d_1^2 = (4.0)^2 + (5.0)^2 + (-6.0)^2 = 77 = d_\parallel^2 + d_\perp^2$. From (c), we have

$$d_\perp = \sqrt{77 \text{ m}^2 - (-3.2 \text{ m})^2} = 8.2 \text{ m}.$$

This gives the magnitude of the perpendicular component (and is consistent with what one would get using Eq. 3-27), but if more information (such as the direction, or a full specification in terms of unit vectors) is sought then more computation is needed.

55. The two vectors are given by

$$\vec{A} = 8.00(\cos 130° \hat{i} + \sin 130° \hat{j}) = -5.14\hat{i} + 6.13\hat{j}$$
$$\vec{B} = B_x \hat{i} + B_y \hat{j} = -7.72\hat{i} - 9.20\hat{j}.$$

(a) The dot product of $5\vec{A} \cdot \vec{B}$ is

$$5\vec{A} \cdot \vec{B} = 5(-5.14\hat{i} + 6.13\hat{j}) \cdot (-7.72\hat{i} - 9.20\hat{j}) = 5[(-5.14)(-7.72) + (6.13)(-9.20)]$$
$$= -83.4.$$

(b) In unit vector notation

$$4\vec{A} \times 3\vec{B} = 12\vec{A} \times \vec{B} = 12(-5.14\hat{i} + 6.13\hat{j}) \times (-7.72\hat{i} - 9.20\hat{j}) = 12(94.6\hat{k}) = 1.14 \times 10^3 \hat{k}$$

(c) We note that the azimuthal angle is undefined for a vector along the z axis. Thus, our result is "1.14×10^3, θ not defined, and $\phi = 0°$."

(d) Since $\vec{A}$ is in the xy plane, and $\vec{A} \times \vec{B}$ is perpendicular to that plane, then the answer is 90°.

(e) Clearly, $\vec{A} + 3.00\hat{k} = -5.14\hat{i} + 6.13\hat{j} + 3.00\hat{k}$.

(f) The Pythagorean theorem yields magnitude $A = \sqrt{(5.14)^2 + (6.13)^2 + (3.00)^2} = 8.54$. The azimuthal angle is $\theta = 130°$, just as it was in the problem statement ($\vec{A}$ is the projection onto to the xy plane of the new vector created in part (e)). The angle measured from the +z axis is $\phi = \cos^{-1}(3.00/8.54) = 69.4°$.

56. The two vectors $\vec{d_1}$ and $\vec{d_2}$ are given by $\vec{d_1} = -d_1 \hat{j}$ and $\vec{d_2} = d_2 \hat{i}$.

(a) The vector $\vec{d_2}/4 = (d_2/4)\hat{i}$ points in the +x direction. The ¼ factor does not affect the result.

(b) The vector $\vec{d_1}/(-4) = (d_1/4)\hat{j}$ points in the +y direction. The minus sign (with the "−4") does affect the direction: $-(-y) = +y$.

(c) $\vec{d}_1 \cdot \vec{d}_2 = 0$ since $\hat{i} \cdot \hat{j} = 0$. The two vectors are perpendicular to each other.

(d) $\vec{d}_1 \cdot (\vec{d}_2/4) = (\vec{d}_1 \cdot \vec{d}_2)/4 = 0$, as in part (c).

(e) $\vec{d}_1 \times \vec{d}_2 = -d_1 d_2 (\hat{j} \times \hat{i}) = d_1 d_2 \hat{k}$, in the +z-direction.

(f) $\vec{d}_2 \times \vec{d}_1 = -d_2 d_1 (\hat{i} \times \hat{j}) = -d_1 d_2 \hat{k}$, in the −z-direction.

(g) The magnitude of the vector in (e) is $d_1 d_2$.

(h) The magnitude of the vector in (f) is $d_1 d_2$.

(i) Since $\vec{d}_1 \times (\vec{d}_2/4) = (d_1 d_2/4)\hat{k}$, the magnitude is $d_1 d_2/4$.

(j) The direction of $\vec{d}_1 \times (\vec{d}_2/4) = (d_1 d_2/4)\hat{k}$ is in the +z-direction.

57. The three vectors are
$$\vec{d}_1 = -3.0\hat{i} + 3.0\hat{j} + 2.0\hat{k}$$
$$\vec{d}_2 = -2.0\hat{i} - 4.0\hat{j} + 2.0\hat{k}$$
$$\vec{d}_3 = 2.0\hat{i} + 3.0\hat{j} + 1.0\hat{k}.$$

(a) Since $\vec{d}_2 + \vec{d}_3 = 0\hat{i} - 1.0\hat{j} + 3.0\hat{k}$, we have
$$\vec{d}_1 \cdot (\vec{d}_2 + \vec{d}_3) = (-3.0\hat{i} + 3.0\hat{j} + 2.0\hat{k}) \cdot (0\hat{i} - 1.0\hat{j} + 3.0\hat{k})$$
$$= 0 - 3.0 + 6.0 = 3.0 \text{ m}^2.$$

(b) Using Eq. 3-30, we obtain $\vec{d}_2 \times \vec{d}_3 = -10\hat{i} + 6.0\hat{j} + 2.0\hat{k}$. Thus,
$$\vec{d}_1 \cdot (\vec{d}_2 \times \vec{d}_3) = (-3.0\hat{i} + 3.0\hat{j} + 2.0\hat{k}) \cdot (-10\hat{i} + 6.0\hat{j} + 2.0\hat{k})$$
$$= 30 + 18 + 4.0 = 52 \text{ m}^3.$$

(c) We found $\vec{d}_2 + \vec{d}_3$ in part (a). Use of Eq. 3-30 then leads to
$$\vec{d}_1 \times (\vec{d}_2 + \vec{d}_3) = (-3.0\hat{i} + 3.0\hat{j} + 2.0\hat{k}) \times (0\hat{i} - 1.0\hat{j} + 3.0\hat{k})$$
$$= (11\hat{i} + 9.0\hat{j} + 3.0\hat{k}) \text{ m}^2$$

58. We choose +x east and +y north and measure all angles in the "standard" way (positive ones counterclockwise from +x, negative ones clockwise). Thus, vector $\vec{d}_1$ has magnitude $d_1 = 3.66$ (with the unit meter and three significant figures assumed) and direction $\theta_1 = 90°$. Also, $\vec{d}_2$ has magnitude $d_2 = 1.83$ and direction $\theta_2 = -45°$, and vector $\vec{d}_3$ has magnitude $d_3 = 0.91$ and direction $\theta_3 = -135°$. We add the x and y components, respectively:

$$x:\ d_1 \cos \theta_1 + d_2 \cos \theta_2 + d_3 \cos \theta_3 = 0.65 \text{ m}$$
$$y:\ d_1 \sin \theta_1 + d_2 \sin \theta_2 + d_3 \sin \theta_3 = 1.7 \text{ m}.$$

(a) The magnitude of the direct displacement (the vector sum $\vec{d}_1 + \vec{d}_2 + \vec{d}_3$) is $\sqrt{(0.65 \text{ m})^2 + (1.7 \text{ m})^2} = 1.8$ m.

(b) The angle (understood in the sense described above) is $\tan^{-1}(1.7/0.65) = 69°$. That is, the first putt must aim in the direction 69° north of east.

59. The vectors can be written as $\vec{a} = a\hat{i}$ and $\vec{b} = b\hat{j}$ where $a, b > 0$.

(a) We are asked to consider

$$\frac{\vec{b}}{d} = \left(\frac{b}{d}\right)\hat{j}$$

in the case $d > 0$. Since the coefficient of $\hat{j}$ is positive, then the vector points in the +y direction.

(b) If, however, $d < 0$, then the coefficient is negative and the vector points in the –y direction.

(c) Since $\cos 90° = 0$, then $\vec{a} \cdot \vec{b} = 0$, using Eq. 3-20.

(d) Since $\vec{b}/d$ is along the y axis, then (by the same reasoning as in the previous part) $\vec{a} \cdot (\vec{b}/d) = 0$.

(e) By the right-hand rule, $\vec{a} \times \vec{b}$ points in the +z-direction.

(f) By the same rule, $\vec{b} \times \vec{a}$ points in the –z-direction. We note that $\vec{b} \times \vec{a} = -\vec{a} \times \vec{b}$ is true in this case and quite generally.

(g) Since $\sin 90° = 1$, Eq. 3-27 gives $|\vec{a} \times \vec{b}| = ab$ where a is the magnitude of $\vec{a}$.

(h) Also, $|\vec{a} \times \vec{b}| = |\vec{b} \times \vec{a}| = ab$.

(i) With $d > 0$, we find that $\vec{a} \times (\vec{b}/d)$ has magnitude ab/d.

(j) The vector $\vec{a} \times (\vec{b}/d)$ points in the $+z$ direction.

60. The vector can be written as $\vec{d} = (2.5 \text{ m})\hat{j}$, where we have taken $\hat{j}$ to be the unit vector pointing north.

(a) The magnitude of the vector $\vec{a} = 4.0\vec{d}$ is $(4.0)(2.5 \text{ m}) = 10$ m.

(b) The direction of the vector $\vec{a} = 4.0\vec{d}$ is the same as the direction of $\vec{d}$ (north).

(c) The magnitude of the vector $\vec{c} = -3.0\vec{d}$ is $(3.0)(2.5 \text{ m}) = 7.5$ m.

(d) The direction of the vector $\vec{c} = -3.0\vec{d}$ is the opposite of the direction of $\vec{d}$. Thus, the direction of $\vec{c}$ is south.

61. We note that the set of choices for unit vector directions has correct orientation (for a right-handed coordinate system). Students sometimes confuse "north" with "up", so it might be necessary to emphasize that these are being treated as the mutually perpendicular directions of our real world, not just some "on the paper" or "on the blackboard" representation of it. Once the terminology is clear, these questions are basic to the definitions of the scalar (dot) and vector (cross) products.

(a) $\hat{i} \cdot \hat{k} = 0$ since $\hat{i} \perp \hat{k}$

(b) $(-\hat{k}) \cdot (-\hat{j}) = 0$ since $\hat{k} \perp \hat{j}$.

(c) $\hat{j} \cdot (-\hat{j}) = -1$.

(d) $\hat{k} \times \hat{j} = -\hat{i}$ (west).

(e) $(-\hat{i}) \times (-\hat{j}) = +\hat{k}$ (upward).

(f) $(-\hat{k}) \times (-\hat{j}) = -\hat{i}$ (west).

62. (a) The vectors should be parallel to achieve a resultant 7 m long (the unprimed case shown below),

(b) anti-parallel (in opposite directions) to achieve a resultant 1 m long (primed case shown),

(c) and perpendicular to achieve a resultant $\sqrt{3^2 + 4^2} = 5$ m long (the double-primed case shown).

In each sketch, the vectors are shown in a "head-to-tail" sketch but the resultant is not shown. The resultant would be a straight line drawn from beginning to end; the beginning is indicated by A (with or without primes, as the case may be) and the end is indicated by B.

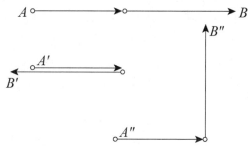

63. A sketch of the displacements is shown. The resultant (not shown) would be a straight line from start (Bank) to finish (Walpole). With a careful drawing, one should find that the resultant vector has length 29.5 km at 35° west of south.

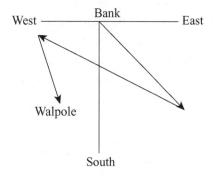

64. The point P is displaced vertically by $2R$, where R is the radius of the wheel. It is displaced horizontally by half the circumference of the wheel, or πR. Since $R = 0.450$ m, the horizontal component of the displacement is 1.414 m and the vertical component of the displacement is 0.900 m. If the x axis is horizontal and the y axis is vertical, the vector displacement (in meters) is $\vec{r} = \left(1.414\ \hat{i} + 0.900\ \hat{j}\right)$. The displacement has a magnitude of

$$|\vec{r}| = \sqrt{(\pi R)^2 + (2R)^2} = R\sqrt{\pi^2 + 4} = 1.68\,\text{m}$$

and an angle of

$$\tan^{-1}\left(\frac{2R}{\pi R}\right) = \tan^{-1}\left(\frac{2}{\pi}\right) = 32.5°$$

above the floor. In physics there are no "exact" measurements, yet that angle computation seemed to yield something *exact*. However, there has to be some uncertainty in the observation that the wheel rolled half of a revolution, which introduces some indefiniteness in our result.

65. Reference to Figure 3-18 (and the accompanying material in that section) is helpful. If we convert $\vec{B}$ to the magnitude-angle notation (as $\vec{A}$ already is) we have $\vec{B} = (14.4 \angle 33.7°)$ (appropriate notation especially if we are using a vector capable calculator in polar mode). Where the length unit is not displayed in the solution, the unit meter should be understood. In the magnitude-angle notation, rotating the axis by +20° amounts to subtracting that angle from the angles previously specified. Thus, $\vec{A} = (12.0 \angle 40.0°)'$ and $\vec{B} = (14.4 \angle 13.7°)'$, where the 'prime' notation indicates that the description is in terms of the new coordinates. Converting these results to (x, y) representations, we obtain

(a) $\vec{A} = (9.19 \text{ m})\,\hat{i}' + (7.71 \text{ m})\,\hat{j}'$.

(b) Similarly, $\vec{B} = (14.0 \text{ m})\,\hat{i}' + (3.41 \text{ m})\,\hat{j}'$.

66. The diagram shows the displacement vectors for the two segments of her walk, labeled $\vec{A}$ and $\vec{B}$, and the total ("final") displacement vector, labeled $\vec{r}$. We take east to be the $+x$ direction and north to be the $+y$ direction. We observe that the angle between $\vec{A}$ and the x axis is 60°. Where the units are not explicitly shown, the distances are understood to be in meters. Thus, the components of $\vec{A}$ are $A_x = 250 \cos 60° = 125$ and $A_y = 250 \sin 60° = 216.5$. The components of $\vec{B}$ are $B_x = 175$ and $B_y = 0$. The components of the total displacement are

$$r_x = A_x + B_x = 125 + 175 = 300$$
$$r_y = A_y + B_y = 216.5 + 0 = 216.5.$$

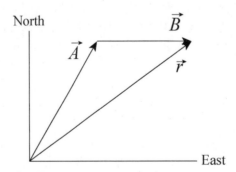

(a) The magnitude of the resultant displacement is

$$|\vec{r}| = \sqrt{r_x^2 + r_y^2} = \sqrt{(300 \text{ m})^2 + (216.5 \text{ m})^2} = 370 \text{ m}.$$

(b) The angle the resultant displacement makes with the $+x$ axis is

$$\tan^{-1}\left(\frac{r_y}{r_x}\right) = \tan^{-1}\left(\frac{216.5 \text{ m}}{300 \text{ m}}\right) = 36°.$$

The direction is 36° north of due east.

(c) The total *distance* walked is $d = 250 \text{ m} + 175 \text{ m} = 425 \text{ m}$.

(d) The total distance walked is greater than the magnitude of the resultant displacement. The diagram shows why: $\vec{A}$ and $\vec{B}$ are not collinear.

67. The three vectors given are

$$\vec{a} = 5.0\,\hat{i} + 4.0\,\hat{j} - 6.0\,\hat{k}$$
$$\vec{b} = -2.0\,\hat{i} + 2.0\,\hat{j} + 3.0\,\hat{k}$$
$$\vec{c} = 4.0\,\hat{i} + 3.0\,\hat{j} + 2.0\,\hat{k}$$

(a) The vector equation $\vec{r} = \vec{a} - \vec{b} + \vec{c}$ is

$$\vec{r} = [5.0 - (-2.0) + 4.0]\hat{i} + (4.0 - 2.0 + 3.0)\hat{j} + (-6.0 - 3.0 + 2.0)\hat{k}$$
$$= 11\hat{i} + 5.0\hat{j} - 7.0\hat{k}.$$

(b) We find the angle from $+z$ by "dotting" (taking the scalar product) $\vec{r}$ with $\hat{k}$. Noting that $r = |\vec{r}| = \sqrt{(11.0)^2 + (5.0)^2 + (-7.0)^2} = 14$, Eq. 3-20 with Eq. 3-23 leads to

$$\vec{r} \cdot \hat{k} = -7.0 = (14)(1)\cos\phi \Rightarrow \phi = 120°.$$

(c) To find the component of a vector in a certain direction, it is efficient to "dot" it (take the scalar product of it) with a unit-vector in that direction. In this case, we make the desired unit-vector by

$$\hat{b} = \frac{\vec{b}}{|\vec{b}|} = \frac{-2.0\hat{i} + 2.0\hat{j} + 3.0\hat{k}}{\sqrt{(-2.0)^2 + (2.0)^2 + (3.0)^2}}.$$

We therefore obtain

$$a_b = \vec{a} \cdot \hat{b} = \frac{(5.0)(-2.0) + (4.0)(2.0) + (-6.0)(3.0)}{\sqrt{(-2.0)^2 + (2.0)^2 + (3.0)^2}} = -4.9.$$

(d) One approach (if all we require is the magnitude) is to use the vector cross product, as the problem suggests; another (which supplies more information) is to subtract the result in part (c) (multiplied by $\hat{b}$) from $\vec{a}$. We briefly illustrate both methods. We note that if

$a\cos\theta$ (where θ is the angle between $\vec{a}$ and $\vec{b}$) gives a_b (the component along $\hat{b}$) then we expect $a\sin\theta$ to yield the orthogonal component:

$$a\sin\theta = \frac{|\vec{a}\times\vec{b}|}{b} = 7.3$$

(alternatively, one might compute θ form part (c) and proceed more directly). The second method proceeds as follows:

$$\vec{a} - a_b\hat{b} = (5.0 - 2.35)\hat{i} + (4.0 - (-2.35))\hat{j} + ((-6.0) - (-3.53))\hat{k}$$
$$= 2.65\hat{i} + 6.35\hat{j} - 2.47\hat{k}$$

This describes the perpendicular part of $\vec{a}$ completely. To find the magnitude of this part, we compute

$$\sqrt{(2.65)^2 + (6.35)^2 + (-2.47)^2} = 7.3$$

which agrees with the first method.

68. The two vectors can be found be solving the simultaneous equations.

(a) If we add the equations, we obtain $2\vec{a} = 6\vec{c}$, which leads to $\vec{a} = 3\vec{c} = 9\hat{i} + 12\hat{j}$.

(b) Plugging this result back in, we find $\vec{b} = \vec{c} = 3\hat{i} + 4\hat{j}$.

69. (a) This is one example of an answer: $(-40\hat{i} - 20\hat{j} + 25\hat{k})$ m, with $\hat{i}$ directed anti-parallel to the first path, $\hat{j}$ directed anti-parallel to the second path and $\hat{k}$ directed upward (in order to have a right-handed coordinate system). Other examples are $(40\hat{i} + 20\hat{j} + 25\hat{k})$ m and $(40\hat{i} - 20\hat{j} - 25\hat{k})$ m (with slightly different interpretations for the unit vectors). Note that the product of the components is positive in each example.

(b) Using Pythagorean theorem, we have $\sqrt{(40\text{ m})^2 + (20\text{ m})^2} = 44.7\text{ m} \approx 45\text{ m}$.

70. The vector $\vec{d}$ (measured in meters) can be represented as $\vec{d} = (3.0\text{ m})(-\hat{j})$, where $-\hat{j}$ is the unit vector pointing south. Therefore,

$$5.0\vec{d} = 5.0(-3.0\text{ m }\hat{j}) = (-15\text{ m})\hat{j}.$$

(a) The positive scalar factor (5.0) affects the magnitude but not the direction. The magnitude of $5.0\vec{d}$ is 15 m.

(b) The new direction of $5\vec{d}$ is the same as the old: south.

The vector $-2.0\vec{d}$ can be written as $-2.0\vec{d} = (6.0 \text{ m})\hat{j}$.

(c) The absolute value of the scalar factor ($|-2.0| = 2.0$) affects the magnitude. The new magnitude is 6.0 m.

(d) The minus sign carried by this scalar factor reverses the direction, so the new direction is $+\hat{j}$, or north.

71. Given: $\vec{A} + \vec{B} = 6.0\,\hat{i} + 1.0\,\hat{j}$ and $\vec{A} - \vec{B} = -4.0\,\hat{i} + 7.0\,\hat{j}$. Solving these simultaneously leads to $\vec{A} = 1.0\,\hat{i} + 4.0\,\hat{j}$. The Pythagorean theorem then leads to $A = \sqrt{(1.0)^2 + (4.0)^2} = 4.1$.

72. The ant's trip consists of three displacements:

$$\vec{d}_1 = (0.40 \text{ m})(\cos 225°\,\hat{i} + \sin 225°\,\hat{j}) = (-0.28 \text{ m})\hat{i} + (-0.28 \text{ m})\hat{j}$$
$$\vec{d}_2 = (0.50 \text{ m})\hat{i}$$
$$\vec{d}_3 = (0.60 \text{ m})(\cos 60°\,\hat{i} + \sin 60°\,\hat{j}) = (0.30 \text{ m})\hat{i} + (0.52 \text{ m})\hat{j},$$

where the angle is measured with respect to the positive x axis. We have taken the positive x and y directions to correspond to east and north, respectively.

(a) The x component of $\vec{d}_1$ is $d_{1x} = (0.40 \text{ m})\cos 225° = -0.28$ m.

(b) The y component of $\vec{d}_1$ is $d_{1y} = (0.40 \text{ m})\sin 225° = -0.28$ m.

(c) The x component of $\vec{d}_2$ is $d_{2x} = 0.50$ m.

(d) The y component of $\vec{d}_2$ is $d_{2y} = 0$ m.

(e) The x component of $\vec{d}_3$ is $d_{3x} = (0.60 \text{ m})\cos 60° = 0.30$ m.

(f) The y component of $\vec{d}_3$ is $d_{3y} = (0.60 \text{ m})\sin 60° = 0.52$ m.

(g) The x component of the net displacement $\vec{d}_{net}$ is

$$d_{net,x} = d_{1x} + d_{2x} + d_{3x} = (-0.28 \text{ m}) + (0.50 \text{ m}) + (0.30 \text{ m}) = 0.52 \text{ m}.$$

(h) The y component of the net displacement $\vec{d}_{net}$ is

$$d_{net,y} = d_{1y} + d_{2y} + d_{3y} = (-0.28 \text{ m}) + (0 \text{ m}) + (0.52 \text{ m}) = 0.24 \text{ m}.$$

(i) The magnitude of the net displacement is

$$d_{net} = \sqrt{d_{net,x}^2 + d_{net,y}^2} = \sqrt{(0.52 \text{ m})^2 + (0.24 \text{ m})^2} = 0.57 \text{ m}.$$

(j) The direction of the net displacement is

$$\theta = \tan^{-1}\left(\frac{d_{net,y}}{d_{net,x}}\right) = \tan^{-1}\left(\frac{0.24 \text{ m}}{0.52 \text{ m}}\right) = 25° \text{ (north of east)}$$

If the ant has to return directly to the starting point, the displacement would be $-\vec{d}_{net}$.

(k) The distance the ant has to travel is $|-\vec{d}_{net}| = 0.57$ m.

(l) The direction the ant has to travel is $25°$ (south of west).

Chapter 4

1. The initial position vector $\vec{r}_0$ satisfies $\vec{r} - \vec{r}_0 = \Delta \vec{r}$, which results in

$$\vec{r}_0 = \vec{r} - \Delta \vec{r} = (3.0\hat{j} - 4.0\hat{k})\text{m} - (2.0\hat{i} - 3.0\hat{j} + 6.0\hat{k})\text{m} = (-2.0 \text{ m})\hat{i} + (6.0 \text{ m})\hat{j} + (-10 \text{ m})\hat{k}.$$

2. (a) The position vector, according to Eq. 4-1, is $\vec{r} = (-5.0 \text{ m})\hat{i} + (8.0 \text{ m})\hat{j}$.

(b) The magnitude is $|\vec{r}| = \sqrt{x^2 + y^2 + z^2} = \sqrt{(-5.0 \text{ m})^2 + (8.0 \text{ m})^2 + (0 \text{ m})^2} = 9.4$ m.

(c) Many calculators have polar $\leftrightarrow$ rectangular conversion capabilities which make this computation more efficient than what is shown below. Noting that the vector lies in the xy plane and using Eq. 3-6, we obtain:

$$\theta = \tan^{-1}\left(\frac{8.0 \text{ m}}{-5.0 \text{ m}}\right) = -58° \text{ or } 122°$$

where the latter possibility (122° measured counterclockwise from the $+x$ direction) is chosen since the signs of the components imply the vector is in the second quadrant.

(d) The sketch is shown on the right. The vector is 122° counterclockwise from the $+x$ direction.

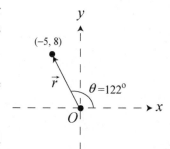

(e) The displacement is $\Delta \vec{r} = \vec{r}' - \vec{r}$ where $\vec{r}$ is given in part (a) and $\vec{r}' = (3.0 \text{ m})\hat{i}$. Therefore, $\Delta \vec{r} = (8.0 \text{ m})\hat{i} - (8.0 \text{ m})\hat{j}$.

(f) The magnitude of the displacement is $|\Delta \vec{r}| = \sqrt{(8.0 \text{ m})^2 + (-8.0 \text{ m})^2} = 11$ m.

(g) The angle for the displacement, using Eq. 3-6, is

$$\tan^{-1}\left(\frac{8.0 \text{ m}}{-8.0 \text{ m}}\right) = -45° \text{ or } 135°$$

where we choose the former possibility (−45°, or 45° measured *clockwise* from $+x$) since the signs of the components imply the vector is in the fourth quadrant. A sketch of $\Delta \vec{r}$ is shown on the right.

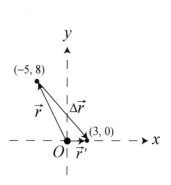

3. (a) The magnitude of $\vec{r}$ is

$$|\vec{r}| = \sqrt{(5.0\text{ m})^2 + (-3.0\text{ m})^2 + (2.0\text{ m})^2} = 6.2\text{ m}.$$

(b) A sketch is shown. The coordinate values are in meters.

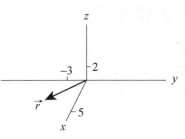

4. We choose a coordinate system with origin at the clock center and $+x$ rightward (towards the "3:00" position) and $+y$ upward (towards "12:00").

(a) In unit-vector notation, we have $\vec{r}_1 = (10\text{ cm})\hat{i}$ and $\vec{r}_2 = (-10\text{ cm})\hat{j}$. Thus, Eq. 4-2 gives

$$\Delta\vec{r} = \vec{r}_2 - \vec{r}_1 = (-10\text{ cm})\hat{i} + (-10\text{ cm})\hat{j}.$$

and the magnitude is given by $|\Delta\vec{r}| = \sqrt{(-10\text{ cm})^2 + (-10\text{ cm})^2} = 14\text{ cm}.$

(b) Using Eq. 3-6, the angle is

$$\theta = \tan^{-1}\left(\frac{-10\text{ cm}}{-10\text{ cm}}\right) = 45° \text{ or } -135°.$$

We choose $-135°$ since the desired angle is in the third quadrant. In terms of the magnitude-angle notation, one may write

$$\Delta\vec{r} = \vec{r}_2 - \vec{r}_1 = (-10\text{ cm})\hat{i} + (-10\text{ cm})\hat{j} \rightarrow (14\text{ cm} \angle -135°).$$

(c) In this case, we have $\vec{r}_1 = (-10\text{ cm})\hat{j}$ and $\vec{r}_2 = (10\text{ cm})\hat{j}$, and $\Delta\vec{r} = (20\text{ cm})\hat{j}$. Thus, $|\Delta\vec{r}| = 20\text{ cm}.$

(d) Using Eq. 3-6, the angle is given by

$$\theta = \tan^{-1}\left(\frac{20\text{ cm}}{0\text{ cm}}\right) = 90°.$$

(e) In a full-hour sweep, the hand returns to its starting position, and the displacement is zero.

(f) The corresponding angle for a full-hour sweep is also zero.

5. Using Eq. 4-3 and Eq. 4-8, we have

$$\vec{v}_{avg} = \frac{(-2.0\hat{i}+8.0\hat{j}-2.0\hat{k})\text{ m} - (5.0\hat{i}-6.0\hat{j}+2.0\hat{k})\text{ m}}{10\text{ s}} = (-0.70\hat{i}+1.40\hat{j}-0.40\hat{k})\text{ m/s}.$$

6. To emphasize the fact that the velocity is a function of time, we adopt the notation $v(t)$ for dx/dt.

(a) Eq. 4-10 leads to

$$v(t) = \frac{d}{dt}(3.00t\hat{i}-4.00t^2\hat{j}+2.00\hat{k}) = (3.00\text{ m/s})\hat{i} - (8.00t\text{ m/s})\hat{j}$$

(b) Evaluating this result at $t = 2.00$ s produces $\vec{v} = (3.00\hat{i} - 16.0\hat{j})$ m/s.

(c) The speed at $t = 2.00$ s is $v = |\vec{v}| = \sqrt{(3.00\text{ m/s})^2 + (-16.0\text{ m/s})^2} = 16.3$ m/s.

(d) The angle of $\vec{v}$ at that moment is

$$\tan^{-1}\left(\frac{-16.0\text{ m/s}}{3.00\text{ m/s}}\right) = -79.4° \text{ or } 101°$$

where we choose the first possibility (79.4° measured *clockwise* from the +x direction, or 281° counterclockwise from +x) since the signs of the components imply the vector is in the fourth quadrant.

7. The average velocity is given by Eq. 4-8. The total displacement $\Delta \vec{r}$ is the sum of three displacements, each result of a (constant) velocity during a given time. We use a coordinate system with +x East and +y North.

(a) In unit-vector notation, the first displacement is given by

$$\Delta \vec{r}_1 = \left(60.0\ \frac{\text{km}}{\text{h}}\right)\left(\frac{40.0\text{ min}}{60\text{ min/h}}\right)\hat{i} = (40.0\text{ km})\hat{i}.$$

The second displacement has a magnitude of $(60.0\ \frac{\text{km}}{\text{h}}) \cdot (\frac{20.0\text{ min}}{60\text{ min/h}}) = 20.0$ km, and its direction is 40° north of east. Therefore,

$$\Delta \vec{r}_2 = (20.0\text{ km})\cos(40.0°)\hat{i} + (20.0\text{ km})\sin(40.0°)\hat{j} = (15.3\text{ km})\hat{i} + (12.9\text{ km})\hat{j}.$$

And the third displacement is

$$\Delta \vec{r}_3 = -\left(60.0\ \frac{\text{km}}{\text{h}}\right)\left(\frac{50.0\text{ min}}{60\text{ min/h}}\right)\hat{i} = (-50.0\text{ km})\hat{i}.$$

The total displacement is

$$\Delta \vec{r} = \Delta \vec{r}_1 + \Delta \vec{r}_2 + \Delta \vec{r}_3 = (40.0 \text{ km})\hat{i} + (15.3 \text{ km})\hat{i} + (12.9 \text{ km})\hat{j} - (50.0 \text{ km})\hat{i}$$
$$= (5.30 \text{ km})\hat{i} + (12.9 \text{ km})\hat{j}.$$

The time for the trip is (40.0 + 20.0 + 50.0) min = 110 min, which is equivalent to 1.83 h. Eq. 4-8 then yields

$$\vec{v}_{avg} = \left(\frac{5.30 \text{ km}}{1.83 \text{ h}}\right)\hat{i} + \left(\frac{12.9 \text{ km}}{1.83 \text{ h}}\right)\hat{j} = (2.90 \text{ km/h})\hat{i} + (7.01 \text{ km/h})\hat{j}.$$

The magnitude is

$$|\vec{v}_{avg}| = \sqrt{(2.90 \text{ km/h})^2 + (7.01 \text{ km/h})^2} = 7.59 \text{ km/h}.$$

(b) The angle is given by

$$\theta = \tan^{-1}\left(\frac{7.01 \text{ km/h}}{2.90 \text{ km/h}}\right) = 67.5° \text{ (north of east)},$$

or 22.5° east of due north.

8. Our coordinate system has $\hat{i}$ pointed east and $\hat{j}$ pointed north. The first displacement is $\vec{r}_{AB} = (483 \text{ km})\hat{i}$ and the second is $\vec{r}_{BC} = (-966 \text{ km})\hat{j}$.

(a) The net displacement is

$$\vec{r}_{AC} = \vec{r}_{AB} + \vec{r}_{BC} = (483 \text{ km})\hat{i} - (966 \text{ km})\hat{j}$$

which yields $|\vec{r}_{AC}| = \sqrt{(483 \text{ km})^2 + (-966 \text{ km})^2} = 1.08 \times 10^3$ km.

(b) The angle is given by

$$\theta = \tan^{-1}\left(\frac{-966 \text{ km}}{483 \text{ km}}\right) = -63.4°.$$

We observe that the angle can be alternatively expressed as 63.4° south of east, or 26.6° east of south.

(c) Dividing the magnitude of $\vec{r}_{AC}$ by the total time (2.25 h) gives

$$\vec{v}_{avg} = \frac{(483 \text{ km})\hat{i} - (966 \text{ km})\hat{j}}{2.25 \text{ h}} = (215 \text{ km/h})\hat{i} - (429 \text{ km/h})\hat{j}.$$

with a magnitude $|\vec{v}_{avg}| = \sqrt{(215 \text{ km/h})^2 + (-429 \text{ km/h})^2} = 480 \text{ km/h}.$

(d) The direction of $\vec{v}_{avg}$ is 26.6° east of south, same as in part (b). In magnitude-angle notation, we would have $\vec{v}_{avg} = (480 \text{ km/h} \angle -63.4°).$

(e) Assuming the AB trip was a straight one, and similarly for the BC trip, then $|\vec{r}_{AB}|$ is the distance traveled during the AB trip, and $|\vec{r}_{BC}|$ is the distance traveled during the BC trip. Since the average speed is the total distance divided by the total time, it equals

$$\frac{483 \text{ km} + 966 \text{ km}}{2.25 \text{ h}} = 644 \text{ km/h}.$$

9. The (x,y) coordinates (in meters) of the points are $A = (15, -15)$, $B = (30, -45)$, $C = (20, -15)$, and $D = (45, 45)$. The respective times are $t_A = 0$, $t_B = 300$ s, $t_C = 600$ s, and $t_D = 900$ s. Average velocity is defined by Eq. 4-8. Each displacement $\Delta \vec{r}$ is understood to originate at point A.

(a) The average velocity having the least magnitude (5.0 m/600 s) is for the displacement ending at point C: $|\vec{v}_{avg}| = 0.0083$ m/s.

(b) The direction of $\vec{v}_{avg}$ is 0° (measured counterclockwise from the $+x$ axis).

(c) The average velocity having the greatest magnitude ($\sqrt{(15 \text{ m})^2 + (30 \text{ m})^2}/300$ s) is for the displacement ending at point B: $|\vec{v}_{avg}| = 0.11$ m/s.

(d) The direction of $\vec{v}_{avg}$ is 297° (counterclockwise from $+x$) or $-63°$ (which is equivalent to measuring 63° *clockwise* from the $+x$ axis).

10. We differentiate $\vec{r} = 5.00t\hat{i} + (et + ft^2)\hat{j}$.

(a) The particle's motion is indicated by the derivative of $\vec{r}$: $\vec{v} = 5.00\hat{i} + (e + 2ft)\hat{j}$. The angle of its direction of motion is consequently

$$\theta = \tan^{-1}(v_y/v_x) = \tan^{-1}[(e + 2ft)/5.00].$$

The graph indicates $\theta_o = 35.0°$ which determines the parameter e:

$$e = (5.00 \text{ m/s}) \tan(35.0°) = 3.50 \text{ m/s}.$$

(b) We note (from the graph) that $\theta = 0$ when $t = 14.0$ s. Thus, $e + 2ft = 0$ at that time. This determines the parameter f:

$$f = \frac{-e}{2t} = \frac{-3.5 \text{ m/s}}{2(14.0 \text{ s})} = -0.125 \text{ m/s}^2.$$

11. We apply Eq. 4-10 and Eq. 4-16.

(a) Taking the derivative of the position vector with respect to time, we have, in SI units (m/s),

$$\vec{v} = \frac{d}{dt}(\hat{i} + 4t^2\hat{j} + t\hat{k}) = 8t\hat{j} + \hat{k}.$$

(b) Taking another derivative with respect to time leads to, in SI units (m/s²),

$$\vec{a} = \frac{d}{dt}(8t\hat{j} + \hat{k}) = 8\hat{j}.$$

12. We use Eq. 4-15 with $\vec{v}_1$ designating the initial velocity and $\vec{v}_2$ designating the later one.

(a) The average acceleration during the $\Delta t = 4$ s interval is

$$\vec{a}_{avg} = \frac{(-2.0\hat{i} - 2.0\hat{j} + 5.0\hat{k}) \text{ m/s} - (4.0\hat{i} - 22\hat{j} + 3.0\hat{k}) \text{ m/s}}{4 \text{ s}} = (-1.5 \text{ m/s}^2)\hat{i} + (0.5 \text{ m/s}^2)\hat{k}.$$

(b) The magnitude of $\vec{a}_{avg}$ is $\sqrt{(-1.5 \text{ m/s}^2)^2 + (0.5 \text{ m/s}^2)^2} = 1.6 \text{ m/s}^2$.

(c) Its angle in the xz plane (measured from the +x axis) is one of these possibilities:

$$\tan^{-1}\left(\frac{0.5 \text{ m/s}^2}{-1.5 \text{ m/s}^2}\right) = -18° \text{ or } 162°$$

where we settle on the second choice since the signs of its components imply that it is in the second quadrant.

13. In parts (b) and (c), we use Eq. 4-10 and Eq. 4-16. For part (d), we find the direction of the velocity computed in part (b), since that represents the asked-for tangent line.

(a) Plugging into the given expression, we obtain

$$\vec{r}\Big|_{t=2.00} = [2.00(8)-5.00(2)]\hat{i} + [6.00-7.00(16)]\hat{j} = (6.00\hat{i} - 106\hat{j})\text{ m}$$

(b) Taking the derivative of the given expression produces

$$\vec{v}(t) = (6.00t^2 - 5.00)\hat{i} - 28.0t^3\hat{j}$$

where we have written $v(t)$ to emphasize its dependence on time. This becomes, at $t = 2.00$ s, $\vec{v} = (19.0\hat{i} - 224\hat{j})$ m/s.

(c) Differentiating the $\vec{v}(t)$ found above, with respect to t produces $12.0t\hat{i} - 84.0t^2\hat{j}$, which yields $\vec{a} = (24.0\hat{i} - 336\hat{j})$ m/s² at $t = 2.00$ s.

(d) The angle of $\vec{v}$, measured from $+x$, is either

$$\tan^{-1}\left(\frac{-224\text{ m/s}}{19.0\text{ m/s}}\right) = -85.2° \text{ or } 94.8°$$

where we settle on the first choice (–85.2°, which is equivalent to 275° measured counterclockwise from the $+x$ axis) since the signs of its components imply that it is in the fourth quadrant.

14. We adopt a coordinate system with $\hat{i}$ pointed east and $\hat{j}$ pointed north; the coordinate origin is the flagpole. We "translate" the given information into unit-vector notation as follows:

$$\vec{r}_o = (40.0\text{ m})\hat{i} \quad \text{and} \quad \vec{v}_o = (-10.0\text{ m/s})\hat{j}$$
$$\vec{r} = (40.0\text{ m})\hat{j} \quad \text{and} \quad \vec{v} = (10.0\text{ m/s})\hat{i}.$$

(a) Using Eq. 4-2, the displacement $\Delta\vec{r}$ is

$$\Delta\vec{r} = \vec{r} - \vec{r}_o = (-40.0\text{ m})\hat{i} + (40.0\text{ m})\hat{j}.$$

with a magnitude $|\Delta\vec{r}| = \sqrt{(-40.0\text{ m})^2 + (40.0\text{ m})^2} = 56.6$ m.

(b) The direction of $\Delta\vec{r}$ is

$$\theta = \tan^{-1}\left(\frac{\Delta y}{\Delta x}\right) = \tan^{-1}\left(\frac{40.0\text{ m}}{-40.0\text{ m}}\right) = -45.0° \text{ or } 135°.$$

Since the desired angle is in the second quadrant, we pick $135°$ ($45°$ north of due west). Note that the displacement can be written as $\Delta \vec{r} = \vec{r} - \vec{r}_o = (56.6 \angle 135°)$ in terms of the magnitude-angle notation.

(c) The magnitude of $\vec{v}_{avg}$ is simply the magnitude of the displacement divided by the time ($\Delta t = 30.0$ s). Thus, the average velocity has magnitude $(56.6 \text{ m})/(30.0 \text{ s}) = 1.89$ m/s.

(d) Eq. 4-8 shows that $\vec{v}_{avg}$ points in the same direction as $\Delta \vec{r}$, i.e, $135°$ ($45°$ north of due west).

(e) Using Eq. 4-15, we have

$$\vec{a}_{avg} = \frac{\vec{v} - \vec{v}_o}{\Delta t} = (0.333 \text{ m/s}^2)\hat{i} + (0.333 \text{ m/s}^2)\hat{j}.$$

The magnitude of the average acceleration vector is therefore equal to $|\vec{a}_{avg}| = \sqrt{(0.333 \text{ m/s}^2)^2 + (0.333 \text{ m/s}^2)^2} = 0.471 \text{ m/s}^2$.

(f) The direction of $\vec{a}_{avg}$ is

$$\theta = \tan^{-1}\left(\frac{0.333 \text{ m/s}^2}{0.333 \text{ m/s}^2}\right) = 45° \text{ or } -135°.$$

Since the desired angle is now in the first quadrant, we choose $45°$, and $\vec{a}_{avg}$ points north of due east.

15. We find t by applying Eq. 2-11 to motion along the y axis (with $v_y = 0$ characterizing $y = y_{max}$):

$$0 = (12 \text{ m/s}) + (-2.0 \text{ m/s}^2)t \implies t = 6.0 \text{ s}.$$

Then, Eq. 2-11 applies to motion along the x axis to determine the answer:

$$v_x = (8.0 \text{ m/s}) + (4.0 \text{ m/s}^2)(6.0 \text{ s}) = 32 \text{ m/s}.$$

Therefore, the velocity of the cart, when it reaches $y = y_{max}$, is $(32 \text{ m/s})\hat{i}$.

16. We find t by solving $\Delta x = x_0 + v_{0x}t + \frac{1}{2}a_x t^2$:

$$12.0 \text{ m} = 0 + (4.00 \text{ m/s})t + \frac{1}{2}(5.00 \text{ m/s}^2)t^2$$

where we have used $\Delta x = 12.0$ m, $v_x = 4.00$ m/s, and $a_x = 5.00$ m/s^2. We use the quadratic formula and find $t = 1.53$ s. Then, Eq. 2-11 (actually, its analog in two dimensions) applies with this value of t. Therefore, its velocity (when $\Delta x = 12.00$ m) is

$$\vec{v} = \vec{v}_0 + \vec{a}t = (4.00 \text{ m/s})\hat{i} + (5.00 \text{ m/s}^2)(1.53 \text{ s})\hat{i} + (7.00 \text{ m/s}^2)(1.53 \text{ s})\hat{j}$$
$$= (11.7 \text{ m/s})\hat{i} + (10.7 \text{ m/s})\hat{j}.$$

Thus, the magnitude of $\vec{v}$ is $|\vec{v}| = \sqrt{(11.7 \text{ m/s})^2 + (10.7 \text{ m/s})^2} = 15.8$ m/s.

(b) The angle of $\vec{v}$, measured from $+x$, is

$$\tan^{-1}\left(\frac{10.7 \text{ m/s}}{11.7 \text{ m/s}}\right) = 42.6°.$$

17. Constant acceleration in both directions (x and y) allows us to use Table 2-1 for the motion along each direction. This can be handled individually (for Δx and Δy) or together with the unit-vector notation (for Δr). Where units are not shown, SI units are to be understood.

(a) The velocity of the particle at any time t is given by $\vec{v} = \vec{v}_0 + \vec{a}t$, where $\vec{v}_0$ is the initial velocity and $\vec{a}$ is the (constant) acceleration. The x component is $v_x = v_{0x} + a_x t = 3.00 - 1.00t$, and the y component is

$$v_y = v_{0y} + a_y t = -0.500t$$

since $v_{0y} = 0$. When the particle reaches its maximum x coordinate at $t = t_m$, we must have $v_x = 0$. Therefore, $3.00 - 1.00t_m = 0$ or $t_m = 3.00$ s. The y component of the velocity at this time is

$$v_y = 0 - 0.500(3.00) = -1.50 \text{ m/s};$$

this is the only nonzero component of $\vec{v}$ at t_m.

(b) Since it started at the origin, the coordinates of the particle at any time t are given by $\vec{r} = \vec{v}_0 t + \frac{1}{2}\vec{a}t^2$. At $t = t_m$ this becomes

$$\vec{r} = (3.00\hat{i})(3.00) + \frac{1}{2}(-1.00\hat{i} - 0.50\hat{j})(3.00)^2 = (4.50\hat{i} - 2.25\hat{j}) \text{ m}.$$

18. We make use of Eq. 4-16.

(a) The acceleration as a function of time is

$$\vec{a} = \frac{d\vec{v}}{dt} = \frac{d}{dt}\left((6.0t - 4.0t^2)\hat{i} + 8.0\hat{j}\right) = (6.0 - 8.0t)\hat{i}$$

in SI units. Specifically, we find the acceleration vector at $t = 3.0$ s to be $(6.0 - 8.0(3.0))\hat{i} = (-18 \text{ m/s}^2)\hat{i}$.

(b) The equation is $\vec{a} = (6.0 - 8.0t)\hat{i} = 0$; we find $t = 0.75$ s.

(c) Since the y component of the velocity, $v_y = 8.0$ m/s, is never zero, the velocity cannot vanish.

(d) Since speed is the magnitude of the velocity, we have

$$v = |\vec{v}| = \sqrt{(6.0t - 4.0t^2)^2 + (8.0)^2} = 10$$

in SI units (m/s). To solve for t, we first square both sides of the above equation, followed by some rearrangement:

$$(6.0t - 4.0t^2)^2 + 64 = 100 \Rightarrow (6.0t - 4.0t^2)^2 = 36$$

Taking the square root of the new expression and making further simplification lead to

$$6.0t - 4.0t^2 = \pm 6.0 \Rightarrow 4.0t^2 - 6.0t \pm 6.0 = 0$$

Finally, using the quadratic formula, we obtain

$$t = \frac{6.0 \pm \sqrt{36 - 4(4.0)(\pm 6.0)}}{2(8.0)}$$

where the requirement of a real positive result leads to the unique answer: $t = 2.2$ s.

19. We make use of Eq. 4-16 and Eq. 4-10.

Using $\vec{a} = 3t\hat{i} + 4t\hat{j}$, we have (in m/s)

$$\vec{v}(t) = \vec{v}_0 + \int_0^t \vec{a}\, dt = (5.00\hat{i} + 2.00\hat{j}) + \int_0^t (3t\hat{i} + 4t\hat{j})\, dt = (5.00 + 3t^2/2)\hat{i} + (2.00 + 2t^2)\hat{j}$$

Integrating using Eq. 4-10 then yields (in metes)

$$\vec{r}(t) = \vec{r}_0 + \int_0^t \vec{v}\,dt = (20.0\hat{i} + 40.0\hat{j}) + \int_0^t [(5.00 + 3t^2/2)\hat{i} + (2.00 + 2t^2)\hat{j}]dt$$
$$= (20.0\hat{i} + 40.0\hat{j}) + (5.00t + t^3/2)\hat{i} + (2.00t + 2t^3/3)\hat{j}$$
$$= (20.0 + 5.00t + t^3/2)\hat{i} + (40.0 + 2.00t + 2t^3/3)\hat{j}$$

(a) At $t = 4.00$ s, we have $\vec{r}(t = 4.00\text{ s}) = (72.0\text{ m})\hat{i} + (90.7\text{ m})\hat{j}$.

(b) $\vec{v}(t = 4.00\text{ s}) = (29.0\text{ m/s})\hat{i} + (34.0\text{ m/s})\hat{j}$. Thus, the angle between the direction of travel and +x, measured counterclockwise, is $\theta = \tan^{-1}[(34.0\text{ m/s})/(29.0\text{ m/s})] = 49.5°$.

20. The acceleration is constant so that use of Table 2-1 (for both the x and y motions) is permitted. Where units are not shown, SI units are to be understood. Collision between particles A and B requires two things. First, the y motion of B must satisfy (using Eq. 2-15 and noting that θ is measured from the y axis)

$$y = \frac{1}{2}a_y t^2 \Rightarrow 30\text{ m} = \frac{1}{2}\left[(0.40\text{ m/s}^2)\cos\theta\right]t^2.$$

Second, the x motions of A and B must coincide:

$$vt = \frac{1}{2}a_x t^2 \Rightarrow (3.0\text{ m/s})t = \frac{1}{2}\left[(0.40\text{ m/s}^2)\sin\theta\right]t^2.$$

We eliminate a factor of t in the last relationship and formally solve for time:

$$t = \frac{2v}{a_x} = \frac{2(3.0\text{ m/s})}{(0.40\text{ m/s}^2)\sin\theta}.$$

This is then plugged into the previous equation to produce

$$30\text{ m} = \frac{1}{2}\left[(0.40\text{ m/s}^2)\cos\theta\right]\left(\frac{2(3.0\text{ m/s})}{(0.40\text{ m/s}^2)\sin\theta}\right)^2$$

which, with the use of $\sin^2\theta = 1 - \cos^2\theta$, simplifies to

$$30 = \frac{9.0}{0.20}\frac{\cos\theta}{1 - \cos^2\theta} \Rightarrow 1 - \cos^2\theta = \frac{9.0}{(0.20)(30)}\cos\theta.$$

We use the quadratic formula (choosing the positive root) to solve for $\cos\theta$:

$$\cos\theta = \frac{-1.5 + \sqrt{1.5^2 - 4(1.0)(-1.0)}}{2} = \frac{1}{2}$$

which yields $\theta = \cos^{-1}\left(\dfrac{1}{2}\right) = 60°$.

21. (a) From Eq. 4-22 (with $\theta_0 = 0$), the time of flight is

$$t = \sqrt{\dfrac{2h}{g}} = \sqrt{\dfrac{2(45.0 \text{ m})}{9.80 \text{ m/s}^2}} = 3.03 \text{ s}.$$

(b) The horizontal distance traveled is given by Eq. 4-21:

$$\Delta x = v_0 t = (250 \text{ m/s})(3.03 \text{ s}) = 758 \text{ m}.$$

(c) And from Eq. 4-23, we find

$$|v_y| = gt = (9.80 \text{ m/s}^2)(3.03 \text{ s}) = 29.7 \text{ m/s}.$$

22. We use Eq. 4-26

$$R_{max} = \left(\dfrac{v_0^2}{g} \sin 2\theta_0\right)_{max} = \dfrac{v_0^2}{g} = \dfrac{(9.50 \text{m/s})^2}{9.80 \text{m/s}^2} = 9.209 \text{ m} \approx 9.21 \text{m}$$

to compare with Powell's long jump; the difference from R_{max} is only $\Delta R = (9.21\text{m} - 8.95\text{m}) = 0.259$ m.

23. Using Eq. (4-26), the take-off speed of the jumper is

$$v_0 = \sqrt{\dfrac{gR}{\sin 2\theta_0}} = \sqrt{\dfrac{(9.80 \text{ m/s}^2)(77.0 \text{ m})}{\sin 2(12.0°)}} = 43.1 \text{ m/s}$$

24. We adopt the positive direction choices used in the textbook so that equations such as Eq. 4-22 are directly applicable.

(a) With the origin at the initial point (edge of table), the y coordinate of the ball is given by $y = -\frac{1}{2}gt^2$. If t is the time of flight and $y = -1.20$ m indicates the level at which the ball hits the floor, then

$$t = \sqrt{\dfrac{2(-1.20 \text{ m})}{-9.80 \text{ m/s}^2}} = 0.495 \text{s}.$$

(b) The initial (horizontal) velocity of the ball is $\vec{v} = v_0 \hat{i}$. Since $x = 1.52$ m is the horizontal position of its impact point with the floor, we have $x = v_0 t$. Thus,

$$v_0 = \frac{x}{t} = \frac{1.52 \text{ m}}{0.495 \text{ s}} = 3.07 \text{ m/s}.$$

25. We adopt the positive direction choices used in the textbook so that equations such as Eq. 4-22 are directly applicable. The initial velocity is horizontal so that $v_{0y} = 0$ and $v_{0x} = v_0 = 10$ m/s.

(a) With the origin at the initial point (where the dart leaves the thrower's hand), the y coordinate of the dart is given by $y = -\frac{1}{2}gt^2$, so that with $y = -PQ$ we have $PQ = \frac{1}{2}(9.8 \text{ m/s}^2)(0.19 \text{ s})^2 = 0.18$ m.

(b) From $x = v_0 t$ we obtain $x = (10 \text{ m/s})(0.19 \text{ s}) = 1.9$ m.

26. (a) Using the same coordinate system assumed in Eq. 4-22, we solve for $y = h$:

$$h = y_0 + v_0 \sin\theta_0 t - \frac{1}{2}gt^2$$

which yields $h = 51.8$ m for $y_0 = 0$, $v_0 = 42.0$ m/s, $\theta_0 = 60.0°$ and $t = 5.50$ s.

(b) The horizontal motion is steady, so $v_x = v_{0x} = v_0 \cos\theta_0$, but the vertical component of velocity varies according to Eq. 4-23. Thus, the speed at impact is

$$v = \sqrt{(v_0 \cos\theta_0)^2 + (v_0 \sin\theta_0 - gt)^2} = 27.4 \text{ m/s}.$$

(c) We use Eq. 4-24 with $v_y = 0$ and $y = H$:

$$H = \frac{(v_0 \sin\theta_0)^2}{2g} = 67.5 \text{ m}.$$

27. We adopt the positive direction choices used in the textbook so that equations such as Eq. 4-22 are directly applicable. The coordinate origin is at ground level directly below the release point. We write $\theta_0 = -30.0°$ since the angle shown in the figure is measured clockwise from horizontal. We note that the initial speed of the decoy is the plane's speed at the moment of release: $v_0 = 290$ km/h, which we convert to SI units: (290)(1000/3600) = 80.6 m/s.

(a) We use Eq. 4-12 to solve for the time:

$$\Delta x = (v_0 \cos\theta_0) t \Rightarrow t = \frac{700 \text{ m}}{(80.6 \text{ m/s})\cos(-30.0°)} = 10.0 \text{ s}.$$

(b) And we use Eq. 4-22 to solve for the initial height y_0:

$$y - y_0 = (v_0 \sin\theta_0)t - \frac{1}{2}gt^2 \Rightarrow 0 - y_0 = (-40.3 \text{ m/s})(10.0 \text{ s}) - \frac{1}{2}(9.80 \text{ m/s}^2)(10.0 \text{ s})^2$$

which yields $y_0 = 897$ m.

28. We adopt the positive direction choices used in the textbook so that equations such as Eq. 4-22 are directly applicable. The coordinate origin is throwing point (the stone's initial position). The x component of its initial velocity is given by $v_{0x} = v_0 \cos\theta_0$ and the y component is given by $v_{0y} = v_0 \sin\theta_0$, where $v_0 = 20$ m/s is the initial speed and $\theta_0 = 40.0°$ is the launch angle.

(a) At $t = 1.10$ s, its x coordinate is

$$x = v_0 t \cos\theta_0 = (20.0 \text{ m/s})(1.10 \text{ s})\cos 40.0° = 16.9 \text{ m}$$

(b) Its y coordinate at that instant is

$$y = v_0 t \sin\theta_0 - \frac{1}{2}gt^2 = (20.0 \text{ m/s})(1.10 \text{ s})\sin 40.0° - \frac{1}{2}(9.80 \text{ m/s}^2)(1.10 \text{ s})^2 = 8.21 \text{ m}.$$

(c) At $t' = 1.80$ s, its x coordinate is $x = (20.0 \text{ m/s})(1.80 \text{ s})\cos 40.0° = 27.6$ m.

(d) Its y coordinate at t' is

$$y = (20.0 \text{ m/s})(1.80 \text{ s})\sin 40.0° - \frac{1}{2}(9.80 \text{ m/s}^2)(1.80 \text{ s}^2) = 7.26 \text{ m}.$$

(e) The stone hits the ground earlier than $t = 5.0$ s. To find the time when it hits the ground solve $y = v_0 t \sin\theta_0 - \frac{1}{2}gt^2 = 0$ for t. We find

$$t = \frac{2v_0}{g}\sin\theta_0 = \frac{2(20.0 \text{ m/s})}{9.8 \text{ m/s}^2}\sin 40° = 2.62 \text{ s}.$$

Its x coordinate on landing is

$$x = v_0 t \cos\theta_0 = (20.0 \text{ m/s})(2.62 \text{ s})\cos 40° = 40.2 \text{ m}.$$

(f) Assuming it stays where it lands, its vertical component at $t = 5.00$ s is $y = 0$.

29. The initial velocity has no vertical component — only an x component equal to +2.00 m/s. Also, $y_0 = +10.0$ m if the water surface is established as $y = 0$.

(a) $x - x_0 = v_x t$ readily yields $x - x_0 = 1.60$ m.

(b) Using $y - y_0 = v_{0y} t - \frac{1}{2} g t^2$, we obtain $y = 6.86$ m when $t = 0.800$ s and $v_{0y} = 0$.

(c) Using the fact that $y = 0$ and $y_0 = 10.0$, the equation $y - y_0 = v_{0y} t - \frac{1}{2} g t^2$ leads to

$$t = \sqrt{2(10.0 \text{ m})/9.80 \text{ m/s}^2} = 1.43 \text{ s}.$$

During this time, the x-displacement of the diver is $x - x_0 = (2.00 \text{ m/s})(1.43 \text{ s}) = 2.86$ m.

30. (a) Since the y-component of the velocity of the stone at the top of its path is zero, its speed is

$$v = \sqrt{v_x^2 + v_y^2} = v_x = v_0 \cos\theta_0 = (28.0 \text{ m/s}) \cos 40.0° = 21.4 \text{ m/s}.$$

(b) Using the fact that $v_y = 0$ at the maximum height y_{max}, the amount of time it takes for the stone to reach y_{max} is given by Eq. 4-23:

$$0 = v_y = v_0 \sin\theta_0 - gt \Rightarrow t = \frac{v_0 \sin\theta_0}{g}.$$

Substituting the above expression into Eq. 4-22, we find the maximum height to be

$$y_{max} = (v_0 \sin\theta_0) t - \frac{1}{2} g t^2 = v_0 \sin\theta_0 \left(\frac{v_0 \sin\theta_0}{g}\right) - \frac{1}{2} g \left(\frac{v_0 \sin\theta_0}{g}\right)^2 = \frac{v_0^2 \sin^2\theta_0}{2g}.$$

To find the time the stone descends to $y = y_{max}/2$, we solve the quadratic equation given in Eq. 4-22:

$$y = \frac{1}{2} y_{max} = \frac{v_0^2 \sin^2\theta_0}{4g} = (v_0 \sin\theta_0) t - \frac{1}{2} g t^2 \Rightarrow t_\pm = \frac{(2 \pm \sqrt{2}) v_0 \sin\theta_0}{2g}.$$

Choosing $t = t_+$ (for descending), we have

$$v_x = v_0 \cos\theta_0 = (28.0 \text{ m/s}) \cos 40.0° = 21.4 \text{ m/s}$$

$$v_y = v_0 \sin\theta_0 - g \frac{(2+\sqrt{2}) v_0 \sin\theta_0}{2g} = -\frac{\sqrt{2}}{2} v_0 \sin\theta_0 = -\frac{\sqrt{2}}{2} (28.0 \text{ m/s}) \sin 40.0° = -12.7 \text{ m/s}$$

Thus, the speed of the stone when $y = y_{max}/2$ is

$$v = \sqrt{v_x^2 + v_y^2} = \sqrt{(21.4 \text{ m/s})^2 + (-12.7 \text{ m/s})^2} = 24.9 \text{ m/s}.$$

(c) The percentage difference is

$$\frac{24.9 \text{ m/s} - 21.4 \text{ m/s}}{21.4 \text{ m/s}} = 0.163 = 16.3\%.$$

31. We adopt the positive direction choices used in the textbook so that equations such as Eq. 4-22 are directly applicable. The coordinate origin is at ground level directly below the release point. We write $\theta_0 = -37.0°$ for the angle measured from +x, since the angle given in the problem is measured from the $-y$ direction. We note that the initial speed of the projectile is the plane's speed at the moment of release.

(a) We use Eq. 4-22 to find v_0:

$$y - y_0 = (v_0 \sin \theta_0) t - \frac{1}{2} gt^2 \Rightarrow 0 - 730 \text{ m} = v_0 \sin(-37.0°)(5.00 \text{ s}) - \frac{1}{2}(9.80 \text{ m/s}^2)(5.00 \text{ s})^2$$

which yields $v_0 = 202$ m/s.

(b) The horizontal distance traveled is $x = v_0 t \cos \theta_0 = (202 \text{ m/s})(5.00 \text{ s})\cos(-37.0°) = 806$ m.

(c) The x component of the velocity (just before impact) is

$$v_x = v_0 \cos \theta_0 = (202 \text{ m/s})\cos(-37.0°) = 161 \text{ m/s}.$$

(d) The y component of the velocity (just before impact) is

$$v_y = v_0 \sin \theta_0 - gt = (202 \text{ m/s}) \sin(-37.0°) - (9.80 \text{ m/s}^2)(5.00 \text{ s}) = -171 \text{ m/s}.$$

32. We adopt the positive direction choices used in the textbook so that equations such as Eq. 4-22 are directly applicable. The coordinate origin is at ground level directly below the point where the ball was hit by the racquet.

(a) We want to know how high the ball is above the court when it is at $x = 12.0$ m. First, Eq. 4-21 tells us the time it is over the fence:

$$t = \frac{x}{v_0 \cos \theta_0} = \frac{12.0 \text{ m}}{(23.6 \text{ m/s}) \cos 0°} = 0.508 \text{ s}.$$

At this moment, the ball is at a height (above the court) of

$$y = y_0 + (v_0 \sin \theta_0) t - \frac{1}{2} gt^2 = 1.10 \text{ m}$$

which implies it does indeed clear the 0.90 m high fence.

(b) At $t = 0.508$ s, the center of the ball is $(1.10 \text{ m} - 0.90 \text{ m}) = 0.20$ m above the net.

(c) Repeating the computation in part (a) with $\theta_0 = -5.0°$ results in $t = 0.510$ s and $y = 0.040$ m, which clearly indicates that it cannot clear the net.

(d) In the situation discussed in part (c), the distance between the top of the net and the center of the ball at $t = 0.510$ s is $0.90 \text{ m} - 0.040 \text{ m} = 0.86$ m.

33. We first find the time it takes for the volleyball to hit the ground. Using Eq. 4-22, we have

$$y - y_0 = (v_0 \sin \theta_0) t - \frac{1}{2} gt^2 \quad \Rightarrow \quad 0 - 2.30 \text{ m} = (-20.0 \text{ m/s}) \sin(18.0°) t - \frac{1}{2}(9.80 \text{ m/s}^2) t^2$$

which gives $t = 0.30$ s. Thus, the range of the volleyball is

$$R = (v_0 \cos \theta_0) t = (20.0 \text{ m/s}) \cos 18.0°(0.30 \text{ s}) = 5.71 \text{ m}$$

On the other hand, when the angle is changed to $\theta_0' = 8.00°$, using the same procedure as shown above, we find

$$y - y_0 = (v_0 \sin \theta_0') t' - \frac{1}{2} gt'^2 \quad \Rightarrow \quad 0 - 2.30 \text{ m} = (-20.0 \text{ m/s}) \sin(8.00°) t' - \frac{1}{2}(9.80 \text{ m/s}^2) t'^2$$

which yields $t' = 0.46$ s, and the range is

$$R' = (v_0 \cos \theta_0) t' = (20.0 \text{ m/s}) \cos 18.0°(0.46 \text{ s}) = 9.06 \text{ m}$$

Thus, the ball travels an extra distance of

$$\Delta R = R' - R = 9.06 \text{ m} - 5.71 \text{ m} = 3.35 \text{ m}$$

34. Although we could use Eq. 4-26 to find where it lands, we choose instead to work with Eq. 4-21 and Eq. 4-22 (for the soccer ball) since these will give information about where *and* when and these are also considered more fundamental than Eq. 4-26. With $\Delta y = 0$, we have

$$\Delta y = (v_0 \sin \theta_0) t - \frac{1}{2} gt^2 \quad \Rightarrow \quad t = \frac{(19.5 \text{ m/s}) \sin 45.0°}{(9.80 \text{ m/s}^2)/2} = 2.81 \text{ s}.$$

Then Eq. 4-21 yields $\Delta x = (v_0 \cos \theta_0)t = 38.7$ m. Thus, using Eq. 4-8, the player must have an average velocity of

$$\vec{v}_{avg} = \frac{\Delta \vec{r}}{\Delta t} = \frac{(38.7 \text{ m})\hat{i} - (55 \text{ m})\hat{i}}{2.81 \text{ s}} = (-5.8 \text{ m/s})\hat{i}$$

which means his average speed (assuming he ran in only one direction) is 5.8 m/s.

35. We adopt the positive direction choices used in the textbook so that equations such as Eq. 4-22 are directly applicable. The coordinate origin is at its initial position (where it is launched). At maximum height, we observe $v_y = 0$ and denote $v_x = v$ (which is also equal to v_{0x}). In this notation, we have $v_0 = 5v$. Next, we observe $v_0 \cos \theta_0 = v_{0x} = v$, so that we arrive at an equation (where $v \neq 0$ cancels) which can be solved for θ_0:

$$(5v)\cos \theta_0 = v \;\Rightarrow\; \theta_0 = \cos^{-1}\left(\frac{1}{5}\right) = 78.5°.$$

36. (a) Solving the quadratic equation Eq. 4-22:

$$y - y_0 = (v_0 \sin \theta_0) t - \frac{1}{2} g t^2 \;\Rightarrow\; 0 - 2.160 \text{ m} = (15.00 \text{ m/s})\sin(45.00°)t - \frac{1}{2}(9.800 \text{ m/s}^2)t^2$$

the total travel time of the shot in the air is found to be $t = 2.352$ s. Therefore, the horizontal distance traveled is

$$R = (v_0 \cos \theta_0)t = (15.00 \text{ m/s})\cos 45.00°(2.352 \text{ s}) = 24.95 \text{ m}.$$

(b) Using the procedure outlined in (a) but for $\theta_0 = 42.00°$, we have

$$y - y_0 = (v_0 \sin \theta_0) t - \frac{1}{2} g t^2 \;\Rightarrow\; 0 - 2.160 \text{ m} = (15.00 \text{ m/s})\sin(42.00°)t - \frac{1}{2}(9.800 \text{ m/s}^2)t^2$$

and the total travel time is $t = 2.245$ s. This gives

$$R = (v_0 \cos \theta_0)t = (15.00 \text{ m/s})\cos 42.00°(2.245 \text{ s}) = 25.02 \text{ m}.$$

37. We designate the given velocity $\vec{v} = (7.6 \text{ m/s})\hat{i} + (6.1 \text{ m/s})\hat{j}$ as $\vec{v}_1$ – as opposed to the velocity when it reaches the max height $\vec{v}_2$ or the velocity when it returns to the ground $\vec{v}_3$ – and take $\vec{v}_0$ as the launch velocity, as usual. The origin is at its launch point on the ground.

(a) Different approaches are available, but since it will be useful (for the rest of the problem) to first find the initial y velocity, that is how we will proceed. Using Eq. 2-16, we have

$$v_{1y}^2 = v_{0y}^2 - 2g\Delta y \quad \Rightarrow \quad (6.1 \text{ m/s})^2 = v_{0y}^2 - 2(9.8 \text{ m/s}^2)(9.1 \text{ m})$$

which yields v_{0y} = 14.7 m/s. Knowing that v_{2y} must equal 0, we use Eq. 2-16 again but now with $\Delta y = h$ for the maximum height:

$$v_{2y}^2 = v_{0y}^2 - 2gh \quad \Rightarrow \quad 0 = (14.7 \text{ m/s})^2 - 2(9.8 \text{ m/s}^2)h$$

which yields h = 11 m.

(b) Recalling the derivation of Eq. 4-26, but using v_{0y} for $v_0 \sin\theta_0$ and v_{0x} for $v_0 \cos\theta_0$, we have

$$0 = v_{0y}t - \frac{1}{2}gt^2, \quad R = v_{0x}t$$

which leads to $R = 2v_{0x}v_{0y}/g$. Noting that $v_{0x} = v_{1x} = 7.6$ m/s, we plug in values and obtain

$$R = 2(7.6 \text{ m/s})(14.7 \text{ m/s})/(9.8 \text{ m/s}^2) = 23 \text{ m}.$$

(c) Since $v_{3x} = v_{1x} = 7.6$ m/s and $v_{3y} = -v_{0y} = -14.7$ m/s, we have

$$v_3 = \sqrt{v_{3x}^2 + v_{3y}^2} = \sqrt{(7.6 \text{ m/s})^2 + (-14.7 \text{ m/s})^2} = 17 \text{ m/s}.$$

(d) The angle (measured from horizontal) for $\vec{v}_3$ is one of these possibilities:

$$\tan^{-1}\left(\frac{-14.7 \text{ m}}{7.6 \text{ m}}\right) = -63° \text{ or } 117°$$

where we settle on the first choice (–63°, which is equivalent to 297°) since the signs of its components imply that it is in the fourth quadrant.

38. We adopt the positive direction choices used in the textbook so that equations such as Eq. 4-22 are directly applicable. The coordinate origin is at the release point (the initial position for the ball as it begins projectile motion in the sense of §4-5), and we let θ_0 be the angle of throw (shown in the figure). Since the horizontal component of the velocity of the ball is $v_x = v_0 \cos 40.0°$, the time it takes for the ball to hit the wall is

$$t = \frac{\Delta x}{v_x} = \frac{22.0 \text{ m}}{(25.0 \text{ m/s})\cos 40.0°} = 1.15 \text{ s}.$$

(a) The vertical distance is

$$\Delta y = (v_0 \sin\theta_0)t - \frac{1}{2}gt^2 = (25.0 \text{ m/s})\sin 40.0°(1.15 \text{ s}) - \frac{1}{2}(9.80 \text{ m/s}^2)(1.15 \text{ s})^2 = 12.0 \text{ m}.$$

(b) The horizontal component of the velocity when it strikes the wall does not change from its initial value: $v_x = v_0 \cos 40.0° = 19.2$ m/s.

(c) The vertical component becomes (using Eq. 4-23)

$$v_y = v_0 \sin\theta_0 - gt = (25.0 \text{ m/s})\sin 40.0° - (9.80 \text{ m/s}^2)(1.15 \text{ s}) = 4.80 \text{ m/s}.$$

(d) Since $v_y > 0$ when the ball hits the wall, it has not reached the highest point yet.

39. We adopt the positive direction choices used in the textbook so that equations such as Eq. 4-22 are directly applicable. The coordinate origin is at the end of the rifle (the initial point for the bullet as it begins projectile motion in the sense of § 4-5), and we let θ_0 be the firing angle. If the target is a distance d away, then its coordinates are $x = d$, $y = 0$. The projectile motion equations lead to $d = v_0 t \cos\theta_0$ and $0 = v_0 t \sin\theta_0 - \frac{1}{2}gt^2$. Eliminating t leads to $2v_0^2 \sin\theta_0 \cos\theta_0 - gd = 0$. Using $\sin\theta_0 \cos\theta_0 = \frac{1}{2}\sin(2\theta_0)$, we obtain

$$v_0^2 \sin(2\theta_0) = gd \Rightarrow \sin(2\theta_0) = \frac{gd}{v_0^2} = \frac{(9.80 \text{ m/s}^2)(45.7 \text{ m})}{(460 \text{ m/s})^2}$$

which yields $\sin(2\theta_0) = 2.11 \times 10^{-3}$ and consequently $\theta_0 = 0.0606°$. If the gun is aimed at a point a distance ℓ above the target, then $\tan\theta_0 = \ell/d$ so that

$$\ell = d \tan\theta_0 = (45.7 \text{ m})\tan(0.0606°) = 0.0484 \text{ m} = 4.84 \text{ cm}.$$

40. We adopt the positive direction choices used in the textbook so that equations such as Eq. 4-22 are directly applicable. The initial velocity is horizontal so that $v_{0y} = 0$ and $v_{0x} = v_0 = 161$ km/h. Converting to SI units, this is $v_0 = 44.7$ m/s.

(a) With the origin at the initial point (where the ball leaves the pitcher's hand), the y coordinate of the ball is given by $y = -\frac{1}{2}gt^2$, and the x coordinate is given by $x = v_0 t$. From the latter equation, we have a simple proportionality between horizontal distance and time, which means the time to travel half the total distance is half the total time. Specifically, if $x = 18.3/2$ m, then $t = (18.3/2 \text{ m})/(44.7 \text{ m/s}) = 0.205$ s.

(b) And the time to travel the next 18.3/2 m must also be 0.205 s. It can be useful to write the horizontal equation as $\Delta x = v_0 \Delta t$ in order that this result can be seen more clearly.

(c) From $y = -\frac{1}{2}gt^2$, we see that the ball has reached the height of $\left| -\frac{1}{2}(9.80 \text{ m/s}^2)(0.205 \text{ s})^2 \right| = 0.205$ m at the moment the ball is halfway to the batter.

(d) The ball's height when it reaches the batter is $-\frac{1}{2}(9.80 \text{ m/s}^2)(0.409 \text{ s})^2 = -0.820$ m, which, when subtracted from the previous result, implies it has fallen another 0.615 m. Since the value of y is not simply proportional to t, we do not expect equal time-intervals to correspond to equal height-changes; in a physical sense, this is due to the fact that the initial y-velocity for the first half of the motion is not the same as the "initial" y-velocity for the second half of the motion.

41. Following the hint, we have the time-reversed problem with the ball thrown from the ground, towards the right, at 60° measured counterclockwise from a rightward axis. We see in this time-reversed situation that it is convenient to use the familiar coordinate system with $+x$ as *rightward* and with positive angles measured counterclockwise.

(a) The x-equation (with $x_0 = 0$ and $x = 25.0$ m) leads to

$$25.0 \text{ m} = (v_0 \cos 60.0°)(1.50 \text{ s}),$$

so that $v_0 = 33.3$ m/s. And with $y_0 = 0$, and $y = h > 0$ at $t = 1.50$ s, we have $y - y_0 = v_{0y}t - \frac{1}{2}gt^2$ where $v_{0y} = v_0 \sin 60.0°$. This leads to $h = 32.3$ m.

(b) We have
$$v_x = v_{0x} = (33.3 \text{ m/s})\cos 60.0° = 16.7 \text{ m/s}$$
$$v_y = v_{0y} - gt = (33.3 \text{ m/s})\sin 60.0° - (9.80 \text{ m/s}^2)(1.50 \text{ s}) = 14.2 \text{ m/s}.$$

The magnitude of $\vec{v}$ is given by

$$|\vec{v}| = \sqrt{v_x^2 + v_y^2} = \sqrt{(16.7 \text{ m/s})^2 + (14.2 \text{ m/s})^2} = 21.9 \text{ m/s}.$$

(c) The angle is

$$\theta = \tan^{-1}\left(\frac{v_y}{v_x}\right) = \tan^{-1}\left(\frac{14.2 \text{ m/s}}{16.7 \text{ m/s}}\right) = 40.4°.$$

(d) We interpret this result ("undoing" the time reversal) as an initial velocity (from the edge of the building) of magnitude 21.9 m/s with angle (down from leftward) of 40.4°.

42. In this projectile motion problem, we have $v_0 = v_x$ = constant, and what is plotted is $v = \sqrt{v_x^2 + v_y^2}$. We infer from the plot that at $t = 2.5$ s, the ball reaches its maximum height, where $v_y = 0$. Therefore, we infer from the graph that $v_x = 19$ m/s.

(a) During $t = 5$ s, the horizontal motion is $x - x_0 = v_x t = 95$ m.

(b) Since $\sqrt{(19 \text{ m/s})^2 + v_{0y}^2} = 31$ m/s (the first point on the graph), we find $v_{0y} = 24.5$ m/s. Thus, with $t = 2.5$ s, we can use $y_{max} - y_0 = v_{0y}t - \frac{1}{2}gt^2$ or $v_y^2 = 0 = v_{0y}^2 - 2g(y_{max} - y_0)$, or $y_{max} - y_0 = \frac{1}{2}(v_y + v_{0y})t$ to solve. Here we will use the latter:

$$y_{max} - y_0 = \frac{1}{2}(v_y + v_{0y})t \Rightarrow y_{max} = \frac{1}{2}(0 + 24.5 \text{m/s})(2.5 \text{ s}) = 31 \text{ m}$$

where we have taken $y_0 = 0$ as the ground level.

43. (a) Let $m = \frac{d_2}{d_1} = 0.600$ be the slope of the ramp, so $y = mx$ there. We choose our coordinate origin at the point of launch and use Eq. 4-25. Thus,

$$y = \tan(50.0°)x - \frac{(9.80 \text{ m/s}^2)x^2}{2(10.0 \text{ m/s})^2(\cos 50.0°)^2} = 0.600x$$

which yields $x = 4.99$ m. This is less than d_1 so the ball *does* land on the ramp.

(b) Using the value of x found in part (a), we obtain $y = mx = 2.99$ m. Thus, the Pythagorean theorem yields a displacement magnitude of $\sqrt{x^2 + y^2} = 5.82$ m.

(c) The angle is, of course, the angle of the ramp: $\tan^{-1}(m) = 31.0°$.

44. (a) Using the fact that the person (as the projectile) reaches the maximum height over the middle wheel located at $x = 23 \text{ m} + (23/2) \text{ m} = 34.5 \text{ m}$, we can deduce the initial launch speed from Eq. 4-26:

$$x = \frac{R}{2} = \frac{v_0^2 \sin 2\theta_0}{2g} \Rightarrow v_0 = \sqrt{\frac{2gx}{\sin 2\theta_0}} = \sqrt{\frac{2(9.8 \text{ m/s}^2)(34.5 \text{ m})}{\sin(2 \cdot 53°)}} = 26.5 \text{ m/s}.$$

Upon substituting the value to Eq. 4-25, we obtain

$$y = y_0 + x \tan \theta_0 - \frac{gx^2}{2v_0^2 \cos^2 \theta_0} = 3.0 \text{ m} + (23 \text{ m})\tan 53° - \frac{(9.8 \text{ m/s}^2)(23 \text{ m})^2}{2(26.5 \text{ m/s})^2(\cos 53°)^2} = 23.3 \text{ m}.$$

Since the height of the wheel is $h_w = 18$ m, the clearance over the first wheel is $\Delta y = y - h_w = 23.3 \text{ m} - 18 \text{ m} = 5.3 \text{ m}$.

(b) The height of the person when he is directly above the second wheel can be found by solving Eq. 4-24. With the second wheel located at $x = 23 \text{ m} + (23/2) \text{ m} = 34.5 \text{ m}$, we have

$$y = y_0 + x \tan\theta_0 - \frac{gx^2}{2v_0^2 \cos^2\theta_0} = 3.0 \text{ m} + (34.5 \text{ m})\tan 53° - \frac{(9.8 \text{ m/s}^2)(34.5 \text{ m})^2}{2(26.52 \text{ m/s})^2(\cos 53°)^2}$$

$$= 25.9 \text{ m}.$$

Therefore, the clearance over the second wheel is $\Delta y = y - h_w = 25.9 \text{ m} - 18 \text{ m} = 7.9 \text{ m}$.

(c) The location of the center of the net is given by

$$0 = y - y_0 = x\tan\theta_0 - \frac{gx^2}{2v_0^2\cos^2\theta_0} \Rightarrow x = \frac{v_0^2 \sin 2\theta_0}{g} = \frac{(26.52 \text{ m/s})^2 \sin(2 \cdot 53°)}{9.8 \text{ m/s}^2} = 69 \text{ m}.$$

45. Using the information given, the position of the insect is given by (with the Archer fish at the origin)

$$x = d\cos\phi = (0.900 \text{ m})\cos 36.0° = 0.728 \text{ m}$$
$$y = d\sin\phi = (0.900 \text{ m})\sin 36.0° = 0.529 \text{ m}$$

Since y corresponds to the maximum height of the parabolic trajectory (see Problem 4-30): $y = y_{\max} = v_0^2 \sin^2\theta_0 / 2g$, the launch angle is found to be

$$\theta_0 = \sin^{-1}\left(\sqrt{\frac{2gy}{v_0^2}}\right) = \sin^{-1}\left(\sqrt{\frac{2(9.8 \text{ m/s}^2)(0.529 \text{ m})}{(3.56 \text{ m/s})^2}}\right) = \sin^{-1}(0.9044) = 64.8°$$

46. Following the hint, we have the time-reversed problem with the ball thrown from the roof, towards the left, at 60° measured clockwise from a leftward axis. We see in this time-reversed situation that it is convenient to take +x as *leftward* with positive angles measured clockwise. Lengths are in meters and time is in seconds.

(a) With $y_0 = 20.0$ m, and $y = 0$ at $t = 4.00$ s, we have $y - y_0 = v_{0y}t - \frac{1}{2}gt^2$ where $v_{0y} = v_0 \sin 60°$. This leads to $v_0 = 16.9$ m/s. This plugs into the x-equation $x - x_0 = v_{0x}t$ (with $x_0 = 0$ and $x = d$) to produce $d = (16.9 \text{ m/s})\cos 60°(4.00 \text{ s}) = 33.7$ m.

(b) We have

$$v_x = v_{0x} = (16.9 \text{ m/s})\cos 60.0° = 8.43 \text{ m/s}$$
$$v_y = v_{0y} - gt = (16.9 \text{ m/s})\sin 60.0° - (9.80 \text{ m/s}^2)(4.00 \text{ s}) = -24.6 \text{ m/s}.$$

The magnitude of $\vec{v}$ is $|\vec{v}| = \sqrt{v_x^2 + v_y^2} = \sqrt{(8.43 \text{ m/s})^2 + (-24.6 \text{ m/s})^2} = 26.0 \text{ m/s}$.

(c) The angle relative to horizontal is

$$\theta = \tan^{-1}\left(\frac{v_y}{v_x}\right) = \tan^{-1}\left(\frac{-24.6 \text{ m/s}}{8.43 \text{ m/s}}\right) = -71.1°.$$

We may convert the result from rectangular components to magnitude-angle representation:

$$\vec{v} = (8.43, -24.6) \rightarrow (26.0 \angle -71.1°)$$

and we now interpret our result ("undoing" the time reversal) as an initial velocity of magnitude 26.0 m/s with angle (up from rightward) of 71.1°.

47. We adopt the positive direction choices used in the textbook so that equations such as Eq. 4-22 are directly applicable. The coordinate origin is at ground level directly below impact point between bat and ball. The *Hint* given in the problem is important, since it provides us with enough information to find v_0 directly from Eq. 4-26.

(a) We want to know how high the ball is from the ground when it is at $x = 97.5$ m, which requires knowing the initial velocity. Using the range information and $\theta_0 = 45°$, we use Eq. 4-26 to solve for v_0:

$$v_0 = \sqrt{\frac{gR}{\sin 2\theta_0}} = \sqrt{\frac{(9.8 \text{ m/s}^2)(107 \text{ m})}{1}} = 32.4 \text{ m/s}.$$

Thus, Eq. 4-21 tells us the time it is over the fence:

$$t = \frac{x}{v_0 \cos \theta_0} = \frac{97.5 \text{ m}}{(32.4 \text{ m/s}) \cos 45°} = 4.26 \text{ s}.$$

At this moment, the ball is at a height (above the ground) of

$$y = y_0 + (v_0 \sin \theta_0)t - \frac{1}{2}gt^2 = 9.88 \text{ m}$$

which implies it does indeed clear the 7.32 m high fence.

(b) At $t = 4.26$ s, the center of the ball is 9.88 m – 7.32 m = 2.56 m above the fence.

48. Using the fact that $v_y = 0$ when the player is at the maximum height y_{max}, the amount of time it takes to reach y_{max} can be solved by using Eq. 4-23:

$$0 = v_y = v_0 \sin \theta_0 - gt \quad \Rightarrow \quad t_{max} = \frac{v_0 \sin \theta_0}{g}.$$

Substituting the above expression into Eq. 4-22, we find the maximum height to be

$$y_{max} = (v_0 \sin\theta_0) t_{max} - \frac{1}{2} g t_{max}^2 = v_0 \sin\theta_0 \left(\frac{v_0 \sin\theta_0}{g}\right) - \frac{1}{2} g \left(\frac{v_0 \sin\theta_0}{g}\right)^2 = \frac{v_0^2 \sin^2\theta_0}{2g}.$$

To find the time when the player is at $y = y_{max}/2$, we solve the quadratic equation given in Eq. 4-22:

$$y = \frac{1}{2} y_{max} = \frac{v_0^2 \sin^2\theta_0}{4g} = (v_0 \sin\theta_0) t - \frac{1}{2} g t^2 \Rightarrow t_\pm = \frac{(2 \pm \sqrt{2}) v_0 \sin\theta_0}{2g}.$$

With $t = t_-$ (for ascending), the amount of time the player spends at a height $y \geq y_{max}/2$ is

$$\Delta t = t_{max} - t_- = \frac{v_0 \sin\theta_0}{g} - \frac{(2-\sqrt{2}) v_0 \sin\theta_0}{2g} = \frac{v_0 \sin\theta_0}{\sqrt{2} g} = \frac{t_{max}}{\sqrt{2}} \Rightarrow \frac{\Delta t}{t_{max}} = \frac{1}{\sqrt{2}} = 0.707.$$

Therefore, the player spends about 70.7% of the time in the upper half of the jump. Note that the ratio $\Delta t / t_{max}$ is independent of v_0 and θ_0, even though Δt and t_{max} depend on these quantities.

49. (a) The skier jumps up at an angle of $\theta_0 = 9.0°$ up from the horizontal and thus returns to the launch level with his velocity vector 9.0° below the horizontal. With the snow surface making an angle of $\alpha = 11.3°$ (downward) with the horizontal, the angle between the slope and the velocity vector is $\phi = \alpha - \theta_0 = 11.3° - 9.0° = 2.3°$.

(b) Suppose the skier lands at a distance d down the slope. Using Eq. 4-25 with $x = d\cos\alpha$ and $y = -d\sin\alpha$ (the edge of the track being the origin), we have

$$-d\sin\alpha = d\cos\alpha \tan\theta_0 - \frac{g(d\cos\alpha)^2}{2v_0^2 \cos^2\theta_0}.$$

Solving for d, we obtain

$$d = \frac{2v_0^2 \cos^2\theta_0}{g \cos^2\alpha}(\cos\alpha \tan\theta_0 + \sin\alpha) = \frac{2v_0^2 \cos\theta_0}{g \cos^2\alpha}(\cos\alpha \sin\theta_0 + \cos\theta_0 \sin\alpha)$$

$$= \frac{2v_0^2 \cos\theta_0}{g \cos^2\alpha} \sin(\theta_0 + \alpha).$$

Substituting the values given, we find

$$d = \frac{2(10 \text{ m/s})^2 \cos(9.0°)}{(9.8 \text{ m/s}^2) \cos^2(11.3°)} \sin(9.0° + 11.3°) = 7.27 \text{ m}.$$

which gives
$$y = -d\sin\alpha = -(7.27\text{ m})\sin(11.3°) = -1.42\text{ m}.$$

Therefore, at landing the skier is approximately 1.4 m below the launch level.

(c) The time it takes for the skier to land is

$$t = \frac{x}{v_x} = \frac{d\cos\alpha}{v_0\cos\theta_0} = \frac{(7.27\text{ m})\cos(11.3°)}{(10\text{ m/s})\cos(9.0°)} = 0.72\text{ s}.$$

Using Eq. 4-23, the x-and y-components of the velocity at landing are

$$v_x = v_0\cos\theta_0 = (10\text{ m/s})\cos(9.0°) = 9.9\text{ m/s}$$
$$v_y = v_0\sin\theta_0 - gt = (10\text{ m/s})\sin(9.0°) - (9.8\text{ m/s}^2)(0.72\text{ s}) = -5.5\text{ m/s}$$

Thus, the direction of travel at landing is

$$\theta = \tan^{-1}\left(\frac{v_y}{v_x}\right) = \tan^{-1}\left(\frac{-5.5\text{ m/s}}{9.9\text{ m/s}}\right) = -29.1°.$$

or 29.1° below the horizontal. The result implies that the angle between the skier's path and the slope is $\phi = 29.1° - 11.3° = 17.8°$, or approximately 18° to two significant figures.

50. From Eq. 4-21, we find $t = x/v_{0x}$. Then Eq. 4-23 leads to

$$v_y = v_{0y} - gt = v_{0y} - \frac{gx}{v_{0x}}.$$

Since the slope of the graph is –0.500, we conclude $\frac{g}{v_{ox}} = \frac{1}{2}$ $\Rightarrow$ v_{ox} = 19.6 m/s. And from the "y intercept" of the graph, we find v_{oy} = 5.00 m/s. Consequently, $\theta_0 = \tan^{-1}(v_{oy}/v_{ox})$ = 14.3°.

51. We adopt the positive direction choices used in the textbook so that equations such as Eq. 4-22 are directly applicable. The coordinate origin is at the point where the ball is kicked. We use x and y to denote the coordinates of ball at the goalpost, and try to find the kicking angle(s) θ_0 so that y = 3.44 m when x = 50 m. Writing the kinematic equations for projectile motion:

$$x = v_0\cos\theta_0,\quad y = v_0 t\sin\theta_0 - \tfrac{1}{2}gt^2,$$

we see the first equation gives $t = x/v_0 \cos\theta_0$, and when this is substituted into the second the result is

$$y = x \tan\theta_0 - \frac{gx^2}{2v_0^2 \cos^2\theta_0}.$$

One may solve this by trial and error: systematically trying values of θ_0 until you find the two that satisfy the equation. A little manipulation, however, will give an algebraic solution: Using the trigonometric identity $1/\cos^2\theta_0 = 1 + \tan^2\theta_0$, we obtain

$$\frac{1}{2}\frac{gx^2}{v_0^2}\tan^2\theta_0 - x\tan\theta_0 + y + \frac{1}{2}\frac{gx^2}{v_0^2} = 0$$

which is a second-order equation for $\tan\theta_0$. To simplify writing the solution, we denote

$$c = \tfrac{1}{2}gx^2/v_0^2 = \tfrac{1}{2}(9.80\text{ m/s}^2)(50\text{ m})^2/(25\text{ m/s})^2 = 19.6\text{m}.$$

Then the second-order equation becomes $c\tan^2\theta_0 - x\tan\theta_0 + y + c = 0$. Using the quadratic formula, we obtain its solution(s).

$$\tan\theta_0 = \frac{x \pm \sqrt{x^2 - 4(y+c)c}}{2c} = \frac{50\text{ m} \pm \sqrt{(50\text{ m})^2 - 4(3.44\text{ m} + 19.6\text{ m})(19.6\text{ m})}}{2(19.6\text{ m})}.$$

The two solutions are given by $\tan\theta_0 = 1.95$ and $\tan\theta_0 = 0.605$. The corresponding (first-quadrant) angles are $\theta_0 = 63°$ and $\theta_0 = 31°$. Thus,

(a) The smallest elevation angle is $\theta_0 = 31°$, and

(b) The greatest elevation angle is $\theta_0 = 63°$.

If kicked at any angle between these two, the ball will travel above the cross bar on the goalposts.

52. For $\Delta y = 0$, Eq. 4-22 leads to $t = 2v_0\sin\theta_0/g$, which immediately implies $t_{max} = 2v_0/g$ (which occurs for the "straight up" case: $\theta_0 = 90°$). Thus,

$$\tfrac{1}{2}t_{max} = v_0/g \Rightarrow \tfrac{1}{2} = \sin\theta_0.$$

Therefore, the half-maximum-time flight is at angle $\theta_0 = 30.0°$. Since the least speed occurs at the top of the trajectory, which is where the velocity is simply the x-component of the initial velocity ($v_0\cos\theta_0 = v_0\cos 30°$ for the half-maximum-time flight), then we need to refer to the graph in order to find v_0 – in order that we may complete the solution.

In the graph, we note that the range is 240 m when $\theta_o = 45.0°$. Eq. 4-26 then leads to $v_o = 48.5$ m/s. The answer is thus $(48.5 \text{ m/s})\cos 30.0° = 42.0$ m/s.

53. We denote h as the height of a step and w as the width. To hit step n, the ball must fall a distance nh and travel horizontally a distance between $(n-1)w$ and nw. We take the origin of a coordinate system to be at the point where the ball leaves the top of the stairway, and we choose the y axis to be positive in the upward direction. The coordinates of the ball at time t are given by $x = v_{0x}t$ and $y = -\frac{1}{2}gt^2$ (since $v_{0y} = 0$). We equate y to $-nh$ and solve for the time to reach the level of step n:

$$t = \sqrt{\frac{2nh}{g}}.$$

The x coordinate then is

$$x = v_{0x}\sqrt{\frac{2nh}{g}} = (1.52 \text{ m/s})\sqrt{\frac{2n(0.203 \text{ m})}{9.8 \text{ m/s}^2}} = (0.309 \text{ m})\sqrt{n}.$$

The method is to try values of n until we find one for which x/w is less than n but greater than $n - 1$. For $n = 1$, $x = 0.309$ m and $x/w = 1.52$, which is greater than n. For $n = 2$, $x = 0.437$ m and $x/w = 2.15$, which is also greater than n. For $n = 3$, $x = 0.535$ m and $x/w = 2.64$. Now, this is less than n and greater than $n - 1$, so the ball hits the third step.

54. We apply Eq. 4-21, Eq. 4-22 and Eq. 4-23.

(a) From $\Delta x = v_{0x}t$, we find $v_{0x} = 40 \text{ m}/2\text{ s} = 20$ m/s.

(b) From $\Delta y = v_{0y}t - \frac{1}{2}gt^2$, we find $v_{0y} = \left(53 \text{ m} + \frac{1}{2}(9.8 \text{ m/s}^2)(2 \text{ s})^2\right)/2 = 36$ m/s.

(c) From $v_y = v_{0y} - gt'$ with $v_y = 0$ as the condition for maximum height, we obtain $t' = (36 \text{ m/s})/(9.8 \text{ m/s}^2) = 3.7$ s. During that time the x-motion is constant, so $x' - x_0 = (20 \text{ m/s})(3.7 \text{ s}) = 74$ m.

55. Let $y_0 = h_0 = 1.00$ m at $x_0 = 0$ when the ball is hit. Let $y_1 = h$ (the height of the wall) and x_1 describe the point where it first rises above the wall one second after being hit; similarly, $y_2 = h$ and x_2 describe the point where it passes back down behind the wall four seconds later. And $y_f = 1.00$ m at $x_f = R$ is where it is caught. Lengths are in meters and time is in seconds.

(a) Keeping in mind that v_x is constant, we have $x_2 - x_1 = 50.0 \text{ m} = v_{1x}(4.00 \text{ s})$, which leads to $v_{1x} = 12.5$ m/s. Thus, applied to the full six seconds of motion:

$$x_f - x_0 = R = v_x(6.00 \text{ s}) = 75.0 \text{ m}.$$

(b) We apply $y - y_0 = v_{0y}t - \frac{1}{2}gt^2$ to the motion above the wall,

$$y_2 - y_1 = 0 = v_{1y}(4.00\text{ s}) - \frac{1}{2}g(4.00\text{ s})^2$$

and obtain v_{1y} = 19.6 m/s. One second earlier, using $v_{1y} = v_{0y} - g(1.00\text{ s})$, we find v_{0y} = 29.4 m/s. Therefore, the velocity of the ball just after being hit is

$$\vec{v} = v_{0x}\hat{i} + v_{0y}\hat{j} = (12.5\text{ m/s})\ \hat{i} + (29.4\text{ m/s})\ \hat{j}$$

Its magnitude is $|\vec{v}| = \sqrt{(12.5\text{ m/s})^2 + (29.4\text{ m/s})^2} = 31.9$ m/s.

(c) The angle is

$$\theta = \tan^{-1}\left(\frac{v_y}{v_x}\right) = \tan^{-1}\left(\frac{29.4\text{ m/s}}{12.5\text{ m/s}}\right) = 67.0°.$$

We interpret this result as a velocity of magnitude 31.9 m/s, with angle (up from rightward) of 67.0°.

(d) During the first 1.00 s of motion, $y = y_0 + v_{0y}t - \frac{1}{2}gt^2$ yields

$$h = 1.0\text{ m} + (29.4\text{ m/s})(1.00\text{ s}) - \frac{1}{2}(9.8\text{ m/s}^2)(1.00\text{ s})^2 = 25.5\text{ m}.$$

56. (a) During constant-speed circular motion, the velocity vector is perpendicular to the acceleration vector at every instant. Thus, $\vec{v} \cdot \vec{a} = 0$.

(b) The acceleration in this vector, at every instant, points towards the center of the circle, whereas the position vector points from the center of the circle to the object in motion. Thus, the angle between $\vec{r}$ and $\vec{a}$ is 180° at every instant, so $\vec{r} \times \vec{a} = 0$.

57. (a) Since the wheel completes 5 turns each minute, its period is one-fifth of a minute, or 12 s.

(b) The magnitude of the centripetal acceleration is given by $a = v^2/R$, where R is the radius of the wheel, and v is the speed of the passenger. Since the passenger goes a distance $2\pi R$ for each revolution, his speed is

$$v = \frac{2\pi(15\text{ m})}{12\text{ s}} = 7.85\text{ m/s}$$

and his centripetal acceleration is $a = \dfrac{(7.85 \text{ m/s})^2}{15 \text{ m}} = 4.1 \text{ m/s}^2$.

(c) When the passenger is at the highest point, his centripetal acceleration is downward, toward the center of the orbit.

(d) At the lowest point, the centripetal acceleration is $a = 4.1 \text{ m/s}^2$, same as part (b).

(e) The direction is up, toward the center of the orbit.

58. The magnitude of the acceleration is

$$a = \dfrac{v^2}{r} = \dfrac{(10 \text{ m/s})^2}{25 \text{ m}} = 4.0 \text{ m/s}^2.$$

59. We apply Eq. 4-35 to solve for speed v and Eq. 4-34 to find centripetal acceleration a.

(a) $v = 2\pi r/T = 2\pi(20 \text{ km})/1.0 \text{ s} = 126 \text{ km/s} = 1.3 \times 10^5 \text{ m/s}$.

(b) The magnitude of the acceleration is

$$a = \dfrac{v^2}{r} = \dfrac{(126 \text{ km/s})^2}{20 \text{ km}} = 7.9 \times 10^5 \text{ m/s}^2.$$

(c) Clearly, both v and a will increase if T is reduced.

60. We apply Eq. 4-35 to solve for speed v and Eq. 4-34 to find acceleration a.

(a) Since the radius of Earth is 6.37×10^6 m, the radius of the satellite orbit is

$$r = (6.37 \times 10^6 + 640 \times 10^3) \text{ m} = 7.01 \times 10^6 \text{ m}.$$

Therefore, the speed of the satellite is

$$v = \dfrac{2\pi r}{T} = \dfrac{2\pi(7.01 \times 10^6 \text{ m})}{(98.0 \text{ min})(60 \text{ s/min})} = 7.49 \times 10^3 \text{ m/s}.$$

(b) The magnitude of the acceleration is

$$a = \dfrac{v^2}{r} = \dfrac{(7.49 \times 10^3 \text{ m/s})^2}{7.01 \times 10^6 \text{ m}} = 8.00 \text{ m/s}^2.$$

61. The magnitude of centripetal acceleration ($a = v^2/r$) and its direction (towards the center of the circle) form the basis of this problem.

(a) If a passenger at this location experiences $\vec{a} = 1.83 \text{ m/s}^2$ east, then the center of the circle is *east* of this location. The distance is $r = v^2/a = (3.66 \text{ m/s})^2/(1.83 \text{ m/s}^2) = 7.32$ m.

(b) Thus, relative to the center, the passenger at that moment is located 7.32 m toward the west.

(c) If the direction of $\vec{a}$ experienced by the passenger is now *south*—indicating that the center of the merry-go-round is south of him, then relative to the center, the passenger at that moment is located 7.32 m toward the north.

62. (a) The circumference is $c = 2\pi r = 2\pi(0.15 \text{ m}) = 0.94$ m.

(b) With $T = (60 \text{ s})/1200 = 0.050$ s, the speed is $v = c/T = (0.94 \text{ m})/(0.050 \text{ s}) = 19$ m/s. This is equivalent to using Eq. 4-35.

(c) The magnitude of the acceleration is $a = v^2/r = (19 \text{ m/s})^2/(0.15 \text{ m}) = 2.4 \times 10^3 \text{ m/s}^2$.

(d) The period of revolution is $(1200 \text{ rev/min})^{-1} = 8.3 \times 10^{-4}$ min which becomes, in SI units, $T = 0.050$ s $= 50$ ms.

63. Since the period of a uniform circular motion is $T = 2\pi r / v$, where r is the radius and v is the speed, the centripetal acceleration can be written as

$$a = \frac{v^2}{r} = \frac{1}{r}\left(\frac{2\pi r}{T}\right)^2 = \frac{4\pi^2 r}{T^2}.$$

Based on this expression, we compare the (magnitudes) of the wallet and purse accelerations, and find their ratio is the ratio of r values. Therefore, $a_{\text{wallet}} = 1.50 \, a_{\text{purse}}$. Thus, the wallet acceleration vector is

$$\vec{a} = 1.50[(2.00 \text{ m/s}^2)\hat{i} + (4.00 \text{ m/s}^2)\hat{j}] = (3.00 \text{ m/s}^2)\hat{i} + (6.00 \text{ m/s}^2)\hat{j}.$$

64. The fact that the velocity is in the +y direction, and the acceleration is in the +x direction at $t_1 = 4.00$ s implies that the motion is clockwise. The position corresponds to the "9:00 position." On the other hand, the position at $t_2 = 10.0$ s is in the "6:00 position" since the velocity points in the -x direction and the acceleration is in the +y direction. The time interval $\Delta t = 10.0 \text{ s} - 4.00 \text{ s} = 6.00$ s is equal to 3/4 of a period:

$$6.00 \text{ s} = \frac{3}{4}T \implies T = 8.00 \text{ s}.$$

Eq. 4-35 then yields

$$r = \frac{vT}{2\pi} = \frac{(3.00 \text{ m/s})(8.00 \text{ s})}{2\pi} = 3.82 \text{ m}.$$

(a) The x coordinate of the center of the circular path is $x = 5.00 \text{ m} + 3.82 \text{ m} = 8.82 \text{ m}$.

(b) The y coordinate of the center of the circular path is $y = 6.00 \text{ m}$.

In other words, the center of the circle is at $(x,y) = (8.82 \text{ m}, 6.00 \text{ m})$.

65. We first note that $\vec{a}_1$ (the acceleration at $t_1 = 2.00$ s) is perpendicular to $\vec{a}_2$ (the acceleration at $t_2 = 5.00$ s), by taking their scalar (dot) product.:

$$\vec{a}_1 \cdot \vec{a}_2 = [(6.00 \text{ m/s}^2)\hat{i} + (4.00 \text{ m/s}^2)\hat{j}] \cdot [(4.00 \text{ m/s}^2)\hat{i} + (-6.00 \text{ m/s}^2)\hat{j}] = 0.$$

Since the acceleration vectors are in the (negative) radial directions, then the two positions (at t_1 and t_2) are a quarter-circle apart (or three-quarters of a circle, depending on whether one measures clockwise or counterclockwise). A quick sketch leads to the conclusion that if the particle is moving counterclockwise (as the problem states) then it travels three-quarters of a circumference in moving from the position at time t_1 to the position at time t_2. Letting T stand for the period, then $t_2 - t_1 = 3.00$ s $= 3T/4$. This gives $T = 4.00$ s. The magnitude of the acceleration is

$$a = \sqrt{a_x^2 + a_y^2} = \sqrt{(6.00 \text{ m/s}^2)^2 + (4.00 \text{ m/s}^2)^2} = 7.21 \text{ m/s}^2.$$

Using Eq. 4-34 and 4-35, we have $a = 4\pi^2 r/T^2$, which yields

$$r = \frac{aT^2}{4\pi^2} = \frac{(7.21 \text{ m/s}^2)(4.00 \text{ s})^2}{4\pi^2} = 2.92 \text{ m}.$$

66. When traveling in circular motion with constant speed, the instantaneous acceleration vector necessarily points towards the center. Thus, the center is "straight up" from the cited point.

(a) Since the center is "straight up" from (4.00 m, 4.00 m), the x coordinate of the center is 4.00 m.

(b) To find out "how far up" we need to know the radius. Using Eq. 4-34 we find

$$r = \frac{v^2}{a} = \frac{(5.00 \text{ m/s})^2}{12.5 \text{ m/s}^2} = 2.00 \text{ m}.$$

Thus, the y coordinate of the center is 2.00 m + 4.00 m = 6.00 m. Thus, the center may be written as $(x, y) = (4.00 \text{ m}, 6.00 \text{ m})$.

67. To calculate the centripetal acceleration of the stone, we need to know its speed during its circular motion (this is also its initial speed when it flies off). We use the kinematic equations of projectile motion (discussed in §4-6) to find that speed. Taking the $+y$ direction to be upward and placing the origin at the point where the stone leaves its circular orbit, then the coordinates of the stone during its motion as a projectile are given by $x = v_0 t$ and $y = -\frac{1}{2}gt^2$ (since $v_{0y} = 0$). It hits the ground at $x = 10$ m and $y = -2.0$ m. Formally solving the second equation for the time, we obtain $t = \sqrt{-2y/g}$, which we substitute into the first equation:

$$v_0 = x\sqrt{-\frac{g}{2y}} = (10 \text{ m})\sqrt{-\frac{9.8 \text{ m/s}^2}{2(-2.0 \text{ m})}} = 15.7 \text{ m/s}.$$

Therefore, the magnitude of the centripetal acceleration is

$$a = \frac{v^2}{r} = \frac{(15.7 \text{ m/s})^2}{1.5 \text{ m}} = 160 \text{ m/s}^2.$$

68. We note that after three seconds have elapsed ($t_2 - t_1 = 3.00$ s) the velocity (for this object in circular motion of period T) is reversed; we infer that it takes three seconds to reach the opposite side of the circle. Thus, $T = 2(3.00 \text{ s}) = 6.00$ s.

(a) Using Eq. 4-35, $r = vT/2\pi$, where $v = \sqrt{(3.00 \text{ m/s})^2 + (4.00 \text{ m/s})^2} = 5.00$ m/s, we obtain $r = 4.77$ m. The magnitude of the object's centripetal acceleration is therefore $a = v^2/r = 5.24 \text{ m/s}^2$.

(b) The average acceleration is given by Eq. 4-15:

$$\vec{a}_{avg} = \frac{\vec{v}_2 - \vec{v}_1}{t_2 - t_1} = \frac{(-3.00\hat{i} - 4.00\hat{j}) \text{ m/s} - (3.00\hat{i} + 4.00\hat{j}) \text{ m/s}}{5.00 \text{ s} - 2.00 \text{ s}} = (-2.00 \text{ m/s}^2)\hat{i} + (-2.67 \text{ m/s}^2)\hat{j}$$

which implies $|\vec{a}_{avg}| = \sqrt{(-2.00 \text{ m/s}^2)^2 + (-2.67 \text{ m/s}^2)^2} = 3.33 \text{ m/s}^2$.

69. We use Eq. 4-15 first using velocities relative to the truck (subscript t) and then using velocities relative to the ground (subscript g). We work with SI units, so 20 km/h → 5.6 m/s, 30 km/h → 8.3 m/s, and 45 km/h → 12.5 m/s. We choose east as the $+\hat{i}$ direction.

(a) The velocity of the cheetah (subscript c) at the end of the 2.0 s interval is (from Eq. 4-44)

$$\vec{v}_{ct} = \vec{v}_{cg} - \vec{v}_{tg} = (12.5 \text{ m/s})\,\hat{i} - (-5.6 \text{ m/s})\,\hat{i} = (18.1 \text{ m/s})\,\hat{i}$$

relative to the truck. Since the velocity of the cheetah relative to the truck at the beginning of the 2.0 s interval is $(-8.3 \text{ m/s})\hat{i}$, the (average) acceleration vector relative to the cameraman (in the truck) is

$$\vec{a}_{avg} = \frac{(18.1 \text{ m/s})\hat{i} - (-8.3 \text{ m/s})\hat{i}}{2.0 \text{ s}} = (13 \text{ m/s}^2)\hat{i},$$

or $|\vec{a}_{avg}| = 13 \text{ m/s}^2$.

(b) The direction of $\vec{a}_{avg}$ is $+\hat{i}$, or eastward.

(c) The velocity of the cheetah at the start of the 2.0 s interval is (from Eq. 4-44)

$$\vec{v}_{0cg} = \vec{v}_{0ct} + \vec{v}_{0tg} = (-8.3 \text{ m/s})\hat{i} + (-5.6 \text{ m/s})\hat{i} = (-13.9 \text{ m/s})\hat{i}$$

relative to the ground. The (average) acceleration vector relative to the crew member (on the ground) is

$$\vec{a}_{avg} = \frac{(12.5 \text{ m/s})\hat{i} - (-13.9 \text{ m/s})\hat{i}}{2.0 \text{ s}} = (13 \text{ m/s}^2)\hat{i}, \quad |\vec{a}_{avg}| = 13 \text{ m/s}^2$$

identical to the result of part (a).

(d) The direction of $\vec{a}_{avg}$ is $+\hat{i}$, or eastward.

70. We use Eq. 4-44, noting that the upstream corresponds to the $+\hat{i}$ direction.

(a) The subscript b is for the boat, w is for the water, and g is for the ground.

$$\vec{v}_{bg} = \vec{v}_{bw} + \vec{v}_{wg} = (14 \text{ km/h})\,\hat{i} + (-9 \text{ km/h})\,\hat{i} = (5 \text{ km/h})\,\hat{i}.$$

Thus, the magnitude is $|\vec{v}_{bg}| = 5 \text{ km/h}$.

(b) The direction of $\vec{v}_{bg}$ is $+x$, or upstream.

(c) We use the subscript c for the child, and obtain

$$\vec{v}_{cg} = \vec{v}_{cb} + \vec{v}_{bg} = (-6 \text{ km/h})\,\hat{i} + (5 \text{ km/h})\,\hat{i} = (-1 \text{ km/h})\,\hat{i}.$$

The magnitude is $|\vec{v}_{cg}| = 1$ km/h.

(d) The direction of $\vec{v}_{cg}$ is $-x$, or downstream.

71. While moving in the same direction as the sidewalk's motion (covering a distance d relative to the ground in time $t_1 = 2.50$ s), Eq. 4-44 leads to

$$v_{\text{sidewalk}} + v_{\text{man running}} = \frac{d}{t_1}.$$

While he runs back (taking time $t_2 = 10.0$ s) we have

$$v_{\text{sidewalk}} - v_{\text{man running}} = -\frac{d}{t_2}.$$

Dividing these equations and solving for the desired ratio, we get $\frac{12.5}{7.5} = \frac{5}{3} = 1.67$.

72. We denote the velocity of the player with $\vec{v}_{PF}$ and the relative velocity between the player and the ball be $\vec{v}_{BP}$. Then the velocity $\vec{v}_{BF}$ of the ball relative to the field is given by $\vec{v}_{BF} = \vec{v}_{PF} + \vec{v}_{BP}$. The smallest angle $\theta_{\min}$ corresponds to the case when $\vec{v}_{BF} \perp \vec{v}_{PF}$. Hence,

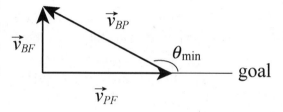

$$\theta_{\min} = 180° - \cos^{-1}\left(\frac{|\vec{v}_{PF}|}{|\vec{v}_{BP}|}\right) = 180° - \cos^{-1}\left(\frac{4.0 \text{ m/s}}{6.0 \text{ m/s}}\right) = 130°.$$

73. The velocity vectors (relative to the shore) for ships A and B are given by

$$\vec{v}_A = -(v_A \cos 45°)\,\hat{i} + (v_A \sin 45°)\,\hat{j}$$
$$\vec{v}_B = -(v_B \sin 40°)\,\hat{i} - (v_B \cos 40°)\,\hat{j},$$

with $v_A = 24$ knots and $v_B = 28$ knots. We take east as $+\hat{i}$ and north as $\hat{j}$.

(a) Their relative velocity is

$$\vec{v}_{AB} = \vec{v}_A - \vec{v}_B = (v_B \sin 40° - v_A \cos 45°)\,\hat{i} + (v_B \cos 40° + v_A \sin 45°)\,\hat{j}$$

the magnitude of which is $|\vec{v}_{AB}| = \sqrt{(1.03 \text{ knots})^2 + (38.4 \text{ knots})^2} \approx 38$ knots.

(b) The angle θ which $\vec{v}_{AB}$ makes with north is given by

$$\theta = \tan^{-1}\left(\frac{v_{AB,x}}{v_{AB,y}}\right) = \tan^{-1}\left(\frac{1.03 \text{ knots}}{38.4 \text{ knots}}\right) = 1.5°$$

which is to say that $\vec{v}_{AB}$ points 1.5° east of north.

(c) Since they started at the same time, their relative velocity describes at what rate the distance between them is increasing. Because the rate is steady, we have

$$t = \frac{|\Delta r_{AB}|}{|\vec{v}_{AB}|} = \frac{160}{38.4} = 4.2 \text{ h}.$$

(d) The velocity $\vec{v}_{AB}$ does not change with time in this problem, and $\vec{r}_{AB}$ is in the same direction as $\vec{v}_{AB}$ since they started at the same time. Reversing the points of view, we have $\vec{v}_{AB} = -\vec{v}_{BA}$ so that $\vec{r}_{AB} = -\vec{r}_{BA}$ (i.e., they are 180° opposite to each other). Hence, we conclude that B stays at a bearing of 1.5° west of south relative to A during the journey (neglecting the curvature of Earth).

74. The destination is $\vec{D} = 800 \text{ km } \hat{j}$ where we orient axes so that $+y$ points north and $+x$ points east. This takes two hours, so the (constant) velocity of the plane (relative to the ground) is $\vec{v}_{pg} = (400 \text{ km/h}) \hat{j}$. This must be the vector sum of the plane's velocity with respect to the air which has (x,y) components $(500\cos 70°, 500\sin 70°)$ and the velocity of the air (*wind*) relative to the ground $\vec{v}_{ag}$. Thus,

$$(400 \text{ km/h}) \hat{j} = (500 \text{ km/h}) \cos 70° \hat{i} + (500 \text{ km/h}) \sin 70° \hat{j} + \vec{v}_{ag}$$

which yields

$$\vec{v}_{ag} = (-171 \text{ km/h})\hat{i} - (70.0 \text{ km/h})\hat{j}.$$

(a) The magnitude of $\vec{v}_{ag}$ is $|\vec{v}_{ag}| = \sqrt{(-171 \text{ km/h})^2 + (-70.0 \text{ km/h})^2} = 185 \text{ km/h}$.

(b) The direction of $\vec{v}_{ag}$ is

$$\theta = \tan^{-1}\left(\frac{-70.0 \text{ km/h}}{-171 \text{ km/h}}\right) = 22.3° \text{ (south of west)}.$$

75. Relative to the car the velocity of the snowflakes has a vertical component of 8.0 m/s and a horizontal component of 50 km/h = 13.9 m/s. The angle θ from the vertical is found from

$$\tan \theta = \frac{v_h}{v_v} = \frac{13.9 \text{ m/s}}{8.0 \text{ m/s}} = 1.74$$

which yields $\theta = 60°$.

76. Velocities are taken to be constant; thus, the velocity of the plane relative to the ground is $\vec{v}_{PG} = (55 \text{ km})/(1/4 \text{ hour}) \hat{j} = (220 \text{ km/h})\hat{j}$. In addition,

$$\vec{v}_{AG} = (42 \text{ km/h})(\cos 20° \hat{i} - \sin 20° \hat{j}) = (39 \text{ km/h})\hat{i} - (14 \text{ km/h})\hat{j}.$$

Using $\vec{v}_{PG} = \vec{v}_{PA} + \vec{v}_{AG}$, we have

$$\vec{v}_{PA} = \vec{v}_{PG} - \vec{v}_{AG} = -(39 \text{ km/h})\hat{i} + (234 \text{ km/h})\hat{j}.$$

which implies $|\vec{v}_{PA}| = 237$ km/h, or 240 km/h (to two significant figures.)

77. Since the raindrops fall vertically relative to the train, the horizontal component of the velocity of a raindrop is $v_h = 30$ m/s, the same as the speed of the train. If v_v is the vertical component of the velocity and θ is the angle between the direction of motion and the vertical, then $\tan \theta = v_h/v_v$. Thus $v_v = v_h/\tan \theta = (30 \text{ m/s})/\tan 70° = 10.9$ m/s. The speed of a raindrop is

$$v = \sqrt{v_h^2 + v_v^2} = \sqrt{(30 \text{ m/s})^2 + (10.9 \text{ m/s})^2} = 32 \text{ m/s}.$$

78. This is a classic problem involving two-dimensional relative motion. We align our coordinates so that *east* corresponds to $+x$ and *north* corresponds to $+y$. We write the vector addition equation as $\vec{v}_{BG} = \vec{v}_{BW} + \vec{v}_{WG}$. We have $\vec{v}_{WG} = (2.0 \angle 0°)$ in the magnitude-angle notation (with the unit m/s understood), or $\vec{v}_{WG} = 2.0\hat{i}$ in unit-vector notation. We also have $\vec{v}_{BW} = (8.0 \angle 120°)$ where we have been careful to phrase the angle in the 'standard' way (measured counterclockwise from the $+x$ axis), or $\vec{v}_{BW} = (-4.0\hat{i} + 6.9\hat{j})$ m/s.

(a) We can solve the vector addition equation for $\vec{v}_{BG}$:

$$\vec{v}_{BG} = \vec{v}_{BW} + \vec{v}_{WG} = (2.0 \text{ m/s})\hat{i} + (-4.0\hat{i} + 6.9\hat{j}) \text{ m/s} = (-2.0 \text{ m/s})\hat{i} + (6.9 \text{ m/s})\hat{j}.$$

Thus, we find $|\vec{v}_{BG}| = 7.2$ m/s.

(b) The direction of $\vec{v}_{BG}$ is $\theta = \tan^{-1}[(6.9 \text{ m/s})/(-2.0 \text{ m/s})] = 106°$ (measured counterclockwise from the $+x$ axis), or 16° west of north.

(c) The velocity is constant, and we apply $y - y_0 = v_y t$ in a reference frame. Thus, in the *ground* reference frame, we have $(200 \text{ m}) = (7.2 \text{ m/s})\sin(106°)t \rightarrow t = 29$ s. Note: if a student obtains "28 s", then the student has probably neglected to take the y component properly (a common mistake).

79. We denote the police and the motorist with subscripts p and m, respectively. The coordinate system is indicated in Fig. 4-49.

(a) The velocity of the motorist with respect to the police car is

$$\vec{v}_{mp} = \vec{v}_m - \vec{v}_p = (-60 \text{ km/h})\hat{j} - (-80 \text{ km/h})\hat{i} = (80 \text{ km/h})\hat{i} - (60 \text{ km/h})\hat{j}.$$

(b) $\vec{v}_{mp}$ does happen to be along the line of sight. Referring to Fig. 4-49, we find the vector pointing from one car to another is $\vec{r} = (800 \text{ m})\hat{i} - (600 \text{ m})\hat{j}$ (from M to P). Since the ratio of components in $\vec{r}$ is the same as in $\vec{v}_{mp}$, they must point the same direction.

(c) No, they remain unchanged.

80. We make use of Eq. 4-44 and Eq. 4-45.

The velocity of Jeep P relative to A at the instant is

$$\vec{v}_{PA} = (40.0 \text{ m/s})(\cos 60°\hat{i} + \sin 60°\hat{j}) = (20.0 \text{ m/s})\hat{i} + (34.6 \text{ m/s})\hat{j}.$$

Similarly, the velocity of Jeep B relative to A at the instant is

$$\vec{v}_{BA} = (20.0 \text{ m/s})(\cos 30°\hat{i} + \sin 30°\hat{j}) = (17.3 \text{ m/s})\hat{i} + (10.0 \text{ m/s})\hat{j}.$$

Thus, the velocity of P relative to B is

$$\vec{v}_{PB} = \vec{v}_{PA} - \vec{v}_{BA} = (20.0\hat{i} + 34.6\hat{j}) \text{ m/s} - (17.3\hat{i} + 10.0\hat{j}) \text{ m/s} = (2.68 \text{ m/s})\hat{i} + (24.6 \text{ m/s})\hat{j}.$$

(a) The magnitude of $\vec{v}_{PB}$ is $|\vec{v}_{PB}| = \sqrt{(2.68 \text{ m/s})^2 + (24.6 \text{ m/s})^2} = 24.8 \text{ m/s}.$

(b) The direction of $\vec{v}_{PB}$ is $\theta = \tan^{-1}[(24.6 \text{ m/s})/(2.68 \text{ m/s})] = 83.8°$ north of east (or 6.2° east of north).

(c) The acceleration of P is

$$\vec{a}_{PA} = (0.400 \text{ m/s}^2)(\cos 60.0°\hat{i} + \sin 60.0°\hat{j}) = (0.200 \text{ m/s}^2)\hat{i} + (0.346 \text{ m/s}^2)\hat{j},$$

and $\vec{a}_{PA} = \vec{a}_{PB}$. Thus, we have $|\vec{a}_{PB}| = 0.400 \text{ m/s}^2.$

(d) The direction is 60.0° north of east (or 30.0° east of north).

81. Here, the subscript W refers to the water. Our coordinates are chosen with $+x$ being *east* and $+y$ being *north*. In these terms, the angle specifying *east* would be $0°$ and the angle specifying *south* would be $-90°$ or $270°$. Where the length unit is not displayed, km is to be understood.

(a) We have $\vec{v}_{AW} = \vec{v}_{AB} + \vec{v}_{BW}$, so that

$$\vec{v}_{AB} = (22 \angle -90°) - (40 \angle 37°) = (56 \angle -125°)$$

in the magnitude-angle notation (conveniently done with a vector-capable calculator in polar mode). Converting to rectangular components, we obtain

$$\vec{v}_{AB} = (-32 \text{km/h}) \hat{i} - (46 \text{ km/h}) \hat{j}.$$

Of course, this could have been done in unit-vector notation from the outset.

(b) Since the velocity-components are constant, integrating them to obtain the position is straightforward ($\vec{r} - \vec{r}_0 = \int \vec{v} \, dt$)

$$\vec{r} = (2.5 - 32t) \hat{i} + (4.0 - 46t) \hat{j}$$

with lengths in kilometers and time in hours.

(c) The magnitude of this $\vec{r}$ is $r = \sqrt{(2.5 - 32t)^2 + (4.0 - 46t)^2}$. We minimize this by taking a derivative and requiring it to equal zero — which leaves us with an equation for t

$$\frac{dr}{dt} = \frac{1}{2} \frac{6286t - 528}{\sqrt{(2.5 - 32t)^2 + (4.0 - 46t)^2}} = 0$$

which yields $t = 0.084$ h.

(d) Plugging this value of t back into the expression for the distance between the ships (r), we obtain $r = 0.2$ km. Of course, the calculator offers more digits ($r = 0.225...$), but they are not significant; in fact, the uncertainties implicit in the given data, here, should make the ship captains worry.

82. We construct a right triangle starting from the clearing on the south bank, drawing a line (200 m long) due north (*upward* in our sketch) across the river, and then a line due west (upstream, leftward in our sketch) along the north bank for a distance $(82 \text{ m}) + (1.1 \text{ m/s})t$, where the t-dependent contribution is the distance that the river will carry the boat downstream during time t.

The hypotenuse of this right triangle (the arrow in our sketch) also depends on t and on the boat's speed (relative to the water), and we set it equal to the Pythagorean "sum" of the triangle's sides:

$$(4.0)t = \sqrt{200^2 + (82 + 1.1t)^2}$$

which leads to a quadratic equation for t

$$46724 + 180.4t - 14.8t^2 = 0.$$

We solve this and find a positive value: $t = 62.6$ s.

The angle between the northward (200 m) leg of the triangle and the hypotenuse (which is measured "west of north") is then given by

$$\theta = \tan^{-1}\left(\frac{82 + 1.1t}{200}\right) = \tan^{-1}\left(\frac{151}{200}\right) = 37°.$$

83. Using displacement = velocity × time (for each constant-velocity part of the trip), along with the fact that 1 hour = 60 minutes, we have the following vector addition exercise (using notation appropriate to many vector capable calculators):

(1667 m $\angle$ 0°) + (1333 m $\angle$ −90°) + (333 m $\angle$ 180°) + (833 m $\angle$ −90°) + (667 m $\angle$ 180°) + (417 m $\angle$ −90°) = (2668 m $\angle$ −76°).

(a) Thus, the magnitude of the net displacement is 2.7 km.

(b) Its direction is 76° clockwise (relative to the initial direction of motion).

84. We compute the coordinate pairs (x, y) from $x = (v_0 \cos\theta)t$ and $y = v_0 \sin\theta t - \frac{1}{2}gt^2$ for $t = 20$ s and the speeds and angles given in the problem.

(a) We obtain

$(x_A, y_A) = (10.1 \text{ km}, 0.56 \text{ km})$ $(x_B, y_B) = (12.1 \text{ km}, 1.51 \text{ km})$
$(x_C, y_C) = (14.3 \text{ km}, 2.68 \text{ km})$ $(x_D, y_D) = (16.4 \text{ km}, 3.99 \text{ km})$

and $(x_E, y_E) = (18.5 \text{ km}, 5.53 \text{ km})$ which we plot in the next part.

(b) The vertical (y) and horizontal (x) axes are in kilometers. The graph does not start at the origin. The curve to "fit" the data is not shown, but is easily imagined (forming the "curtain of death").

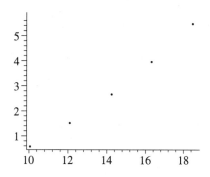

85. Let $v_o = 2\pi(0.200 \text{ m})/(0.00500 \text{ s}) \approx 251$ m/s (using Eq. 4-35) be the speed it had in circular motion and $\theta_o = (1 \text{ hr})(360°/12 \text{ hr [for full rotation]}) = 30.0°$. Then Eq. 4-25 leads to

$$y = (2.50 \text{ m})\tan 30.0° - \frac{(9.8 \text{ m/s}^2)(2.50 \text{ m})^2}{2(251 \text{ m/s})^2(\cos 30.0°)^2} \approx 1.44 \text{ m}$$

which means its height above the floor is 1.44 m + 1.20 m = 2.64 m.

86. For circular motion, we must have $\vec{v}$ with direction perpendicular to $\vec{r}$ and (since the speed is constant) magnitude $v = 2\pi r/T$ where $r = \sqrt{(2.00 \text{ m})^2 + (-3.00 \text{ m})^2}$ and $T = 7.00$ s. The $\vec{r}$ (given in the problem statement) specifies a point in the fourth quadrant, and since the motion is clockwise then the velocity must have both components negative. Our result, satisfying these three conditions, (using unit-vector notation which makes it easy to double-check that $\vec{r} \cdot \vec{v} = 0$) for $\vec{v} = (-2.69 \text{ m/s})\hat{i} + (-1.80 \text{ m/s})\hat{j}$.

87. Using Eq. 2-16, we obtain $v^2 = v_0^2 - 2gh$, or $h = (v_0^2 - v^2)/2g$.

(a) Since $v = 0$ at the maximum height of an upward motion, with $v_0 = 7.00$ m/s, we have $h = (7.00 \text{ m/s})^2/2(9.80 \text{ m/s}^2) = 2.50$ m.

(b) The relative speed is $v_r = v_0 - v_c = 7.00 \text{ m/s} - 3.00 \text{ m/s} = 4.00$ m/s with respect to the floor. Using the above equation we obtain $h = (4.00 \text{ m/s})^2/2(9.80 \text{ m/s}^2) = 0.82$ m.

(c) The acceleration, or the rate of change of speed of the ball with respect to the ground is 9.80 m/s^2 (downward).

(d) Since the elevator cab moves at constant velocity, the rate of change of speed of the ball with respect to the cab floor is also 9.80 m/s^2 (downward).

88. Relative to the sled, the launch velocity is $\vec{v}_{o\,rel} = v_{ox}\hat{i} + v_{oy}\hat{j}$. Since the sled's motion is in the negative direction with speed v_s (note that we are treating v_s as a positive number, so the sled's velocity is actually $-v_s\hat{i}$), then the launch velocity relative to the

ground is $\vec{v}_o = (v_{ox} - v_s)\hat{i} + v_{oy}\hat{j}$. The horizontal and vertical displacement (relative to the ground) are therefore

$$x_{land} - x_{launch} = \Delta x_{bg} = (v_{ox} - v_s) t_{flight}$$

$$y_{land} - y_{launch} = 0 = v_{oy} t_{flight} + \tfrac{1}{2}(-g)(t_{flight})^2 .$$

Combining these equations leads to

$$\Delta x_{bg} = \frac{2 v_{ox} v_{oy}}{g} - \left(\frac{2 v_{oy}}{g}\right) v_s.$$

The first term corresponds to the "y intercept" on the graph, and the second term (in parentheses) corresponds to the magnitude of the "slope." From Figure 4-54, we have

$$\Delta x_{bg} = 40 - 4 v_s.$$

This implies $v_{oy} = (4.0 \text{ s})(9.8 \text{ m/s}^2)/2 = 19.6$ m/s, and that furnishes enough information to determine v_{ox}.

(a) $v_{ox} = 40 g / 2 v_{oy} = (40 \text{ m})(9.8 \text{ m/s}^2)/(39.2 \text{ m/s}) = 10$ m/s.

(b) As noted above, $v_{oy} = 19.6$ m/s.

(c) Relative to the sled, the displacement Δx_{bs} does not depend on the sled's speed, so $\Delta x_{bs} = v_{ox} t_{flight} = 40$ m.

(d) As in (c), relative to the sled, the displacement Δx_{bs} does not depend on the sled's speed, and $\Delta x_{bs} = v_{ox} t_{flight} = 40$ m.

89. We establish coordinates with $\hat{i}$ pointing to the far side of the river (perpendicular to the current) and $\hat{j}$ pointing in the direction of the current. We are told that the magnitude (presumed constant) of the velocity of the boat relative to the water is $|\vec{v}_{bw}| = 6.4$ km/h. Its angle, relative to the x axis is θ. With km and h as the understood units, the velocity of the water (relative to the ground) is $\vec{v}_{wg} = (3.2 \text{ km/h})\hat{j}$.

(a) To reach a point "directly opposite" means that the velocity of her boat relative to ground must be $\vec{v}_{bg} = v_{bg}\hat{i}$ where $v_{bg} > 0$ is unknown. Thus, all $\hat{j}$ components must cancel in the vector sum $\vec{v}_{bw} + \vec{v}_{wg} = \vec{v}_{bg}$, which means the $\vec{v}_{bw} \sin\theta = (-3.2 \text{ km/h})\hat{j}$, so

$$\theta = \sin^{-1}[(-3.2 \text{ km/h})/(6.4 \text{ km/h})] = -30°.$$

(b) Using the result from part (a), we find $v_{bg} = v_{bw}\cos\theta = 5.5$ km/h. Thus, traveling a distance of $\ell = 6.4$ km requires a time of $(6.4 \text{ km})/(5.5 \text{ km/h}) = 1.15$ h or 69 min.

(c) If her motion is completely along the y axis (as the problem implies) then with $v_{wg} = 3.2$ km/h (the water speed) we have

$$t_{total} = \frac{D}{v_{bw} + v_{wg}} + \frac{D}{v_{bw} - v_{wg}} = 1.33 \text{ h}$$

where $D = 3.2$ km. This is equivalent to 80 min.

(d) Since

$$\frac{D}{v_{bw}+v_{wg}} + \frac{D}{v_{bw}-v_{wg}} = \frac{D}{v_{bw}-v_{wg}} + \frac{D}{v_{bw}+v_{wg}}$$

the answer is the same as in the previous part, i.e., $t_{total} = 80$ min.

(e) The shortest-time path should have $\theta = 0°$. This can also be shown by noting that the case of general θ leads to

$$\vec{v}_{bg} = \vec{v}_{bw} + \vec{v}_{wg} = v_{bw}\cos\theta\,\hat{i} + (v_{bw}\sin\theta + v_{wg})\,\hat{j}$$

where the x component of $\vec{v}_{bg}$ must equal l/t. Thus,

$$t = \frac{l}{v_{bw}\cos\theta}$$

which can be minimized using $dt/d\theta = 0$.

(f) The above expression leads to $t = (6.4 \text{ km})/(6.4 \text{ km/h}) = 1.0$ h, or 60 min.

90. We use a coordinate system with $+x$ eastward and $+y$ upward.

(a) We note that 123° is the angle between the initial position and later position vectors, so that the angle from $+x$ to the later position vector is $40° + 123° = 163°$. In unit-vector notation, the position vectors are

$$\vec{r}_1 = (360 \text{ m})\cos(40°)\,\hat{i} + (360 \text{ m})\sin(40°)\,\hat{j} = (276 \text{ m})\hat{i} + (231 \text{ m})\hat{j}$$
$$\vec{r}_2 = (790 \text{ m})\cos(163°)\,\hat{i} + (790 \text{ m})\sin(163°)\,\hat{j} = (-755 \text{ m})\hat{i} + (231 \text{ m})\hat{j}$$

respectively. Consequently, we plug into Eq. 4-3

$$\Delta \vec{r} = [(-755 \text{ m}) - (276 \text{ m})]\hat{i} + (231 \text{ m} - 231 \text{ m})\hat{j} = -(1031 \text{ m})\hat{i}.$$

The magnitude of the displacement $\Delta \vec{r}$ is $|\Delta \vec{r}| = 1031$ m.

(b) The direction of $\Delta \vec{r}$ is $-\hat{i}$, or westward.

91. We adopt the positive direction choices used in the textbook so that equations such as Eq. 4-22 are directly applicable.

(a) With the origin at the firing point, the y coordinate of the bullet is given by $y = -\frac{1}{2}gt^2$. If t is the time of flight and $y = -0.019$ m indicates where the bullet hits the target, then

$$t = \sqrt{\frac{2(0.019 \text{ m})}{9.8 \text{ m/s}^2}} = 6.2 \times 10^{-2} \text{ s}.$$

(b) The muzzle velocity is the initial (horizontal) velocity of the bullet. Since $x = 30$ m is the horizontal position of the target, we have $x = v_0 t$. Thus,

$$v_0 = \frac{x}{t} = \frac{30 \text{ m}}{6.3 \times 10^{-2} \text{ s}} = 4.8 \times 10^2 \text{ m/s}.$$

92. Eq. 4-34 describes an inverse proportionality between r and a, so that a large acceleration results from a small radius. Thus, an upper limit for a corresponds to a lower limit for r.

(a) The minimum turning radius of the train is given by

$$r_{min} = \frac{v^2}{a_{max}} = \frac{(216 \text{ km/h})^2}{(0.050)(9.8 \text{ m/s}^2)} = 7.3 \times 10^3 \text{ m}.$$

(b) The speed of the train must be reduced to no more than

$$v = \sqrt{a_{max} r} = \sqrt{0.050(9.8 \text{ m/s}^2)(1.00 \times 10^3 \text{ m})} = 22 \text{ m/s}$$

which is roughly 80 km/h.

93. (a) With $r = 0.15$ m and $a = 3.0 \times 10^{14}$ m/s^2, Eq. 4-34 gives

$$v = \sqrt{ra} = 6.7 \times 10^6 \text{ m/s}.$$

(b) The period is given by Eq. 4-35:

$$T = \frac{2\pi r}{v} = 1.4 \times 10^{-7}\,\text{s}.$$

94. We use Eq. 4-2 and Eq. 4-3.

(a) With the initial position vector as $\vec{r}_1$ and the later vector as $\vec{r}_2$, Eq. 4-3 yields

$$\Delta r = [(-2.0\,\text{m}) - 5.0\,\text{m}]\hat{i} + [(6.0\,\text{m}) - (-6.0\,\text{m})]\hat{j} + (2.0\,\text{m} - 2.0\,\text{m})\hat{k} = (-7.0\,\text{m})\hat{i} + (12\,\text{m})\hat{j}$$

for the displacement vector in unit-vector notation.

(b) Since there is no z component (that is, the coefficient of $\hat{k}$ is zero), the displacement vector is in the xy plane.

95. We write our magnitude-angle results in the form $(R \angle \theta)$ with SI units for the magnitude understood (m for distances, m/s for speeds, m/s^2 for accelerations). All angles θ are measured counterclockwise from $+x$, but we will occasionally refer to angles ϕ which are measured counterclockwise from the vertical line between the circle-center and the coordinate origin and the line drawn from the circle-center to the particle location (see r in the figure). We note that the speed of the particle is $v = 2\pi r/T$ where $r = 3.00$ m and $T = 20.0$ s; thus, $v = 0.942$ m/s. The particle is moving counterclockwise in Fig. 4-56.

(a) At $t = 5.0$ s, the particle has traveled a fraction of

$$\frac{t}{T} = \frac{5.00\,\text{s}}{20.0\,\text{s}} = \frac{1}{4}$$

of a full revolution around the circle (starting at the origin). Thus, relative to the circle-center, the particle is at

$$\phi = \frac{1}{4}(360°) = 90°$$

measured from vertical (as explained above). Referring to Fig. 4-56, we see that this position (which is the "3 o'clock" position on the circle) corresponds to $x = 3.0$ m and $y = 3.0$ m relative to the coordinate origin. In our magnitude-angle notation, this is expressed as $(R \angle \theta) = (4.2 \angle 45°)$. Although this position is easy to analyze without resorting to trigonometric relations, it is useful (for the computations below) to note that these values of x and y relative to coordinate origin can be gotten from the angle ϕ from the relations

$$x = r\sin\phi, \quad y = r - r\cos\phi.$$

Of course, $R = \sqrt{x^2 + y^2}$ and θ comes from choosing the appropriate possibility from $\tan^{-1}(y/x)$ (or by using particular functions of vector-capable calculators).

(b) At $t = 7.5$ s, the particle has traveled a fraction of $7.5/20 = 3/8$ of a revolution around the circle (starting at the origin). Relative to the circle-center, the particle is therefore at $\phi = 3/8 \, (360°) = 135°$ measured from vertical in the manner discussed above. Referring to Fig. 4-56, we compute that this position corresponds to

$$x = (3.00 \text{ m})\sin 135° = 2.1 \text{ m}$$
$$y = (3.0 \text{ m}) - (3.0 \text{ m})\cos 135° = 5.1 \text{ m}$$

relative to the coordinate origin. In our magnitude-angle notation, this is expressed as $(R \angle \theta) = (5.5 \angle 68°)$.

(c) At $t = 10.0$ s, the particle has traveled a fraction of $10/20 = 1/2$ of a revolution around the circle. Relative to the circle-center, the particle is at $\phi = 180°$ measured from vertical (see explanation, above). Referring to Fig. 4-56, we see that this position corresponds to $x = 0$ and $y = 6.0$ m relative to the coordinate origin. In our magnitude-angle notation, this is expressed as $(R \angle \theta) = (6.0 \angle 90°)$.

(d) We subtract the position vector in part (a) from the position vector in part (c):

$$(6.0 \angle 90°) - (4.2 \angle 45°) = (4.2 \angle 135°)$$

using magnitude-angle notation (convenient when using vector-capable calculators). If we wish instead to use unit-vector notation, we write

$$\Delta \vec{R} = (0 - 3.0 \text{ m}) \hat{i} + (6.0 \text{ m} - 3.0 \text{ m}) \hat{j} = (-3.0 \text{ m})\hat{i} + (3.0 \text{ m})\hat{j}$$

which leads to $|\Delta \vec{R}| = 4.2$ m and $\theta = 135°$.

(e) From Eq. 4-8, we have $\vec{v}_{avg} = \Delta \vec{R} / \Delta t$. With $\Delta t = 5.0$ s, we have

$$\vec{v}_{avg} = (-0.60 \text{ m/s}) \hat{i} + (0.60 \text{ m/s}) \hat{j}$$

in unit-vector notation or $(0.85 \angle 135°)$ in magnitude-angle notation.

(f) The speed has already been noted ($v = 0.94$ m/s), but its direction is best seen by referring again to Fig. 4-56. The velocity vector is tangent to the circle at its "3 o'clock position" (see part (a)), which means $\vec{v}$ is vertical. Thus, our result is $(0.94 \angle 90°)$.

(g) Again, the speed has been noted above ($v = 0.94$ m/s), but its direction is best seen by referring to Fig. 4-56. The velocity vector is tangent to the circle at its "12 o'clock position" (see part (c)), which means $\vec{v}$ is horizontal. Thus, our result is $(0.94 \angle 180°)$.

(h) The acceleration has magnitude $a = v^2/r = 0.30$ m/s^2, and at this instant (see part (a)) it is horizontal (towards the center of the circle). Thus, our result is $(0.30 \angle 180°)$.

(i) Again, $a = v^2/r = 0.30$ m/s^2, but at this instant (see part (c)) it is vertical (towards the center of the circle). Thus, our result is $(0.30 \angle 270°)$.

96. Noting that $\vec{v}_2 = 0$, then, using Eq. 4-15, the average acceleration is

$$\vec{a}_{avg} = \frac{\Delta \vec{v}}{\Delta t} = \frac{0 - (6.30\hat{i} - 8.42\hat{j}) \text{ m/s}}{3 \text{ s}} = (-2.1\hat{i} + 2.8\hat{j}) \text{ m/s}^2$$

97. (a) The magnitude of the displacement vector $\Delta \vec{r}$ is given by

$$|\Delta \vec{r}| = \sqrt{(21.5 \text{ km})^2 + (9.7 \text{ km})^2 + (2.88 \text{ km})^2} = 23.8 \text{ km}.$$

Thus,

$$|\vec{v}_{avg}| = \frac{|\Delta \vec{r}|}{\Delta t} = \frac{23.8 \text{ km}}{3.50 \text{ h}} = 6.79 \text{ km/h}.$$

(b) The angle θ in question is given by

$$\theta = \tan^{-1}\left(\frac{2.88 \text{ km}}{\sqrt{(21.5 \text{ km})^2 + (9.7 \text{ km})^2}}\right) = 6.96°.$$

98. The initial velocity has magnitude v_0 and because it is horizontal, it is equal to v_x the horizontal component of velocity at impact. Thus, the speed at impact is

$$\sqrt{v_0^2 + v_y^2} = 3v_0$$

where $v_y = \sqrt{2gh}$ and we have used Eq. 2-16 with Δx replaced with $h = 20$ m. Squaring both sides of the first equality and substituting from the second, we find

$$v_0^2 + 2gh = (3v_0)^2$$

which leads to $gh = 4v_0^2$ and therefore to $v_0 = \sqrt{(9.8 \text{ m/s}^2)(20 \text{ m})/2} = 7.0$ m/s.

99. We choose horizontal x and vertical y axes such that both components of $\vec{v}_0$ are positive. Positive angles are counterclockwise from $+x$ and negative angles are clockwise from it. In unit-vector notation, the velocity at each instant during the projectile motion is

$$\vec{v} = v_0 \cos\theta_0\, \hat{i} + (v_0 \sin\theta_0 - gt)\,\hat{j}.$$

(a) With $v_0 = 30$ m/s and $\theta_0 = 60°$, we obtain $\vec{v} = (15\hat{i} + 6.4\hat{j})$ m/s, for $t = 2.0$ s. The magnitude of $\vec{v}$ is $|\vec{v}| = \sqrt{(15\text{ m/s})^2 + (6.4\text{ m/s})^2} = 16$ m/s.

(b) The direction of $\vec{v}$ is
$$\theta = \tan^{-1}[(6.4\text{ m/s})/(15\text{ m/s})] = 23°,$$

measured *counterclockwise* from $+x$.

(c) Since the angle is positive, it is above the horizontal.

(d) With $t = 5.0$ s, we find $\vec{v} = (15\hat{i} - 23\hat{j})$ m/s, which yields

$$|\vec{v}| = \sqrt{(15\text{ m/s})^2 + (-23\text{ m/s})^2} = 27 \text{ m/s}.$$

(e) The direction of $\vec{v}$ is $\theta = \tan^{-1}[(-23\text{ m/s})/(15\text{ m/s})] = -57°$, or $57°$ measured *clockwise* from $+x$.

(f) Since the angle is negative, it is below the horizontal.

100. The velocity of Larry is v_1 and that of Curly is v_2. Also, we denote the length of the corridor by L. Now, Larry's time of passage is $t_1 = 150$ s (which must equal L/v_1), and Curly's time of passage is $t_2 = 70$ s (which must equal L/v_2). The time Moe takes is therefore

$$t = \frac{L}{v_1 + v_2} = \frac{1}{v_1/L + v_2/L} = \frac{1}{\frac{1}{150\text{ s}} + \frac{1}{70\text{ s}}} = 48\text{s}.$$

101. We adopt the positive direction choices used in the textbook so that equations such as Eq. 4-22 are directly applicable. The coordinate origin is at the initial position for the football as it begins projectile motion in the sense of §4-5), and we let θ_0 be the angle of its initial velocity measured from the $+x$ axis.

(a) $x = 46$ m and $y = -1.5$ m are the coordinates for the landing point; it lands at time $t = 4.5$ s. Since $x = v_{0x}t$,

$$v_{0x} = \frac{x}{t} = \frac{46\text{ m}}{4.5\text{ s}} = 10.2 \text{ m/s}.$$

Since $y = v_{0y}t - \frac{1}{2}gt^2$,

$$v_{0y} = \frac{y + \frac{1}{2}gt^2}{t} = \frac{(-1.5\text{ m}) + \frac{1}{2}(9.8\text{ m/s}^2)(4.5\text{ s})^2}{4.5\text{ s}} = 21.7\text{ m/s}.$$

The magnitude of the initial velocity is

$$v_0 = \sqrt{v_{0x}^2 + v_{0y}^2} = \sqrt{(10.2\text{ m/s})^2 + (21.7\text{ m/s})^2} = 24\text{ m/s}.$$

(b) The initial angle satisfies $\tan\theta_0 = v_{0y}/v_{0x}$. Thus, $\theta_0 = \tan^{-1}[(21.7\text{ m/s})/(10.2\text{ m/s})] = 65°$.

102. We assume the ball's initial velocity is perpendicular to the plane of the net. We choose coordinates so that $(x_0, y_0) = (0, 3.0)$ m, and $v_x > 0$ (note that $v_{0y} = 0$).

(a) To (barely) clear the net, we have

$$y - y_0 = v_{0y}t - \frac{1}{2}gt^2 \implies 2.24\text{ m} - 3.0\text{ m} = 0 - \frac{1}{2}(9.8\text{ m/s}^2)t^2$$

which gives $t = 0.39$ s for the time it is passing over the net. This is plugged into the x-equation to yield the (minimum) initial velocity $v_x = (8.0\text{ m})/(0.39\text{ s}) = 20.3$ m/s.

(b) We require $y = 0$ and find t from $y - y_0 = v_{0y}t - \frac{1}{2}gt^2$. This value $\left(t = \sqrt{2(3.0\text{ m})/(9.8\text{ m/s}^2)} = 0.78\text{ s}\right)$ is plugged into the x-equation to yield the (maximum) initial velocity $v_x = (17.0\text{ m})/(0.78\text{ s}) = 21.7$ m/s.

103. (a) With $\Delta x = 8.0$ m, $t = \Delta t_1$, $a = a_x$, and $v_{ox} = 0$, Eq. 2-15 gives

$$8.0\text{ m} = \tfrac{1}{2}a_x(\Delta t_1)^2,$$

and the corresponding expression for motion along the y axis leads to

$$\Delta y = 12\text{ m} = \tfrac{1}{2}a_y(\Delta t_1)^2.$$

Dividing the second expression by the first leads to $a_y/a_x = 3/2 = 1.5$.

(b) Letting $t = 2\Delta t_1$, then Eq. 2-15 leads to $\Delta x = (8.0\text{ m})(2)^2 = 32$ m, which implies that its x coordinate is now $(4.0 + 32)$ m $= 36$ m. Similarly, $\Delta y = (12\text{ m})(2)^2 = 48$ m, which means its y coordinate has become $(6.0 + 48)$ m $= 54$ m.

104. We apply Eq. 4-34 to solve for speed v and Eq. 4-35 to find the period T.

(a) We obtain
$$v = \sqrt{ra} = \sqrt{(5.0 \text{ m})(7.0)(9.8 \text{ m/s}^2)} = 19 \text{ m/s}.$$

(b) The time to go around once (the period) is $T = 2\pi r/v = 1.7$ s. Therefore, in one minute ($t = 60$ s), the astronaut executes

$$\frac{t}{T} = \frac{60 \text{ s}}{1.7 \text{ s}} = 35$$

revolutions. Thus, 35 rev/min is needed to produce a centripetal acceleration of $7g$ when the radius is 5.0 m.

(c) As noted above, $T = 1.7$ s.

105. The radius of Earth may be found in Appendix C.

(a) The speed of an object at Earth's equator is $v = 2\pi R/T$, where R is the radius of Earth (6.37×10^6 m) and T is the length of a day (8.64×10^4 s):

$$v = 2\pi(6.37 \times 10^6 \text{ m})/(8.64 \times 10^4 \text{ s}) = 463 \text{ m/s}.$$

The magnitude of the acceleration is given by

$$a = \frac{v^2}{R} = \frac{(463 \text{ m/s})^2}{6.37 \times 10^6 \text{ m}} = 0.034 \text{ m/s}^2.$$

(b) If T is the period, then $v = 2\pi R/T$ is the speed and the magnitude of the acceleration is

$$a = \frac{v^2}{R} = \frac{(2\pi R/T)^2}{R} = \frac{4\pi^2 R}{T^2}.$$

Thus,

$$T = 2\pi\sqrt{\frac{R}{a}} = 2\pi\sqrt{\frac{6.37 \times 10^6 \text{ m}}{9.8 \text{ m/s}^2}} = 5.1 \times 10^3 \text{ s} = 84 \text{ min}.$$

106. When the escalator is stalled the speed of the person is $v_p = \ell/t$, where ℓ is the length of the escalator and t is the time the person takes to walk up it. This is $v_p = (15 \text{ m})/(90 \text{ s}) = 0.167$ m/s. The escalator moves at $v_e = (15 \text{ m})/(60 \text{ s}) = 0.250$ m/s. The speed of the person walking up the moving escalator is

$$v = v_p + v_e = 0.167 \text{ m/s} + 0.250 \text{ m/s} = 0.417 \text{ m/s}$$

and the time taken to move the length of the escalator is

$$t = \ell/v = (15 \text{ m})/(0.417 \text{ m/s}) = 36 \text{ s}.$$

If the various times given are independent of the escalator length, then the answer does not depend on that length either. In terms of ℓ (in meters) the speed (in meters per second) of the person walking on the stalled escalator is $\ell/90$, the speed of the moving escalator is $\ell/60$, and the speed of the person walking on the moving escalator is $v = (\ell/90) + (\ell/60) = 0.0278\ell$. The time taken is $t = \ell/v = \ell/0.0278\ell = 36$ s and is independent of ℓ.

107. (a) Eq. 2-15 can be applied to the vertical (y axis) motion related to reaching the maximum height (when $t = 3.0$ s and $v_y = 0$):

$$y_{\max} - y_0 = v_y t - \frac{1}{2}gt^2.$$

With ground level chosen so $y_0 = 0$, this equation gives the result $y_{\max} = \frac{1}{2}g(3.0 \text{ s})^2 = 44$ m.

(b) After the moment it reached maximum height, it is falling; at $t = 2.5$ s, it will have fallen an amount given by Eq. 2-18:

$$y_{\text{fence}} - y_{\max} = (0)(2.5 \text{ s}) - \frac{1}{2}g(2.5 \text{ s})^2$$

which leads to $y_{\text{fence}} = 13$ m.

(c) Either the *range* formula, Eq. 4-26, can be used or one can note that after passing the fence, it will strike the ground in 0.5 s (so that the total "fall-time" equals the "rise-time"). Since the horizontal component of velocity in a projectile-motion problem is constant (neglecting air friction), we find the original x-component from 97.5 m = $v_{0x}(5.5$ s) and then apply it to that final 0.5 s. Thus, we find $v_{0x} = 17.7$ m/s and that after the fence

$$\Delta x = (17.7 \text{ m/s})(0.5 \text{ s}) = 8.9 \text{ m}.$$

108. With $g_B = 9.8128$ m/s^2 and $g_M = 9.7999$ m/s^2, we apply Eq. 4-26:

$$R_M - R_B = \frac{v_0^2 \sin 2\theta_0}{g_M} - \frac{v_0^2 \sin 2\theta_0}{g_B} = \frac{v_0^2 \sin 2\theta_0}{g_B}\left(\frac{g_B}{g_M} - 1\right)$$

which becomes

$$R_M - R_B = R_B\left(\frac{9.8128 \text{ m/s}^2}{9.7999 \text{ m/s}^2} - 1\right)$$

and yields (upon substituting $R_B = 8.09$ m) $R_M - R_B = 0.01$ m = 1 cm.

109. We make use of Eq. 4-25.

(a) By rearranging Eq. 4-25, we obtain the initial speed:

$$v_0 = \frac{x}{\cos\theta_0}\sqrt{\frac{g}{2(x\tan\theta_0 - y)}}$$

which yields $v_0 = 255.5 \approx 2.6 \times 10^2$ m/s for $x = 9400$ m, $y = -3300$ m, and $\theta_0 = 35°$.

(b) From Eq. 4-21, we obtain the time of flight:

$$t = \frac{x}{v_0 \cos\theta_0} = \frac{9400 \text{ m}}{(255.5 \text{ m/s})\cos 35°} = 45 \text{ s}.$$

(c) We expect the air to provide resistance but no appreciable lift to the rock, so we would need a greater launching speed to reach the same target.

110. When moving in the same direction as the jet stream (of speed v_s), the time is

$$t_1 = \frac{d}{v_{ja} + v_s},$$

where $d = 4000$ km is the distance and v_{ja} is the speed of the jet relative to the air (1000 km/h). When moving against the jet stream, the time is

$$t_2 = \frac{d}{v_{ja} - v_s},$$

where $t_2 - t_1 = \frac{70}{60}$ h. Combining these equations and using the quadratic formula to solve gives $v_s = 143$ km/h.

111. Since the x and y components of the acceleration are constants, we can use Table 2-1 for the motion along both axes. This can be handled individually (for Δx and Δy) or together with the unit-vector notation (for Δr). Where units are not shown, SI units are to be understood.

(a) Since $\vec{r}_0 = 0$, the position vector of the particle is (adapting Eq. 2-15)

$$\vec{r} = \vec{v}_0 t + \frac{1}{2}\vec{a}t^2 = (8.0\hat{j})t + \frac{1}{2}(4.0\hat{i} + 2.0\hat{j})t^2 = (2.0t^2)\hat{i} + (8.0t + 1.0t^2)\hat{j}.$$

Therefore, we find when $x = 29$ m, by solving $2.0t^2 = 29$, which leads to $t = 3.8$ s. The y coordinate at that time is $y = (8.0 \text{ m/s})(3.8 \text{ s}) + (1.0 \text{ m/s}^2)(3.8 \text{ s})^2 = 45$ m.

(b) Adapting Eq. 2-11, the velocity of the particle is given by

$$\vec{v} = \vec{v}_0 + \vec{a}t.$$

Thus, at $t = 3.8$ s, the velocity is

$$\vec{v} = (8.0 \text{ m/s})\hat{j} + \left((4.0 \text{ m/s}^2)\hat{i} + (2.0 \text{ m/s}^2)\hat{j}\right)(3.8 \text{ s}) = (15.2 \text{ m/s})\hat{i} + (15.6 \text{ m/s})\hat{j}$$

which has a magnitude of

$$v = \sqrt{v_x^2 + v_y^2} = \sqrt{(15.2 \text{ m/s})^2 + (15.6 \text{ m/s})^2} = 22 \text{ m/s}.$$

112. We make use of Eq. 4-34 and Eq. 4-35.

(a) The track radius is given by

$$r = \frac{v^2}{a} = \frac{(9.2 \text{ m/s})^2}{3.8 \text{ m/s}^2} = 22 \text{ m}.$$

(b) The period of the circular motion is $T = 2\pi(22 \text{ m})/(9.2 \text{ m/s}) = 15$ s.

113. Since this problem involves constant downward acceleration of magnitude a, similar to the projectile motion situation, we use the equations of §4-6 as long as we substitute a for g. We adopt the positive direction choices used in the textbook so that equations such as Eq. 4-22 are directly applicable. The initial velocity is horizontal so that $v_{0y} = 0$ and

$$v_{0x} = v_0 = 1.00 \times 10^9 \text{ cm/s}.$$

(a) If ℓ is the length of a plate and t is the time an electron is between the plates, then $\ell = v_0 t$, where v_0 is the initial speed. Thus

$$t = \frac{\ell}{v_0} = \frac{2.00 \text{ cm}}{1.00 \times 10^9 \text{ cm/s}} = 2.00 \times 10^{-9} \text{ s}.$$

(b) The vertical displacement of the electron is

$$y = -\frac{1}{2}at^2 = -\frac{1}{2}\left(1.00 \times 10^{17} \text{ cm/s}^2\right)\left(2.00 \times 10^{-9} \text{ s}\right)^2 = -0.20 \text{ cm} = -2.00 \text{ mm},$$

or $|y| = 2.00$ mm.

(c) The x component of velocity does not change: $v_x = v_0 = 1.00 \times 10^9$ cm/s $= 1.00 \times 10^7$ m/s.

(d) The y component of the velocity is

$$v_y = a_y t = (1.00 \times 10^{17} \text{ cm/s}^2)(2.00 \times 10^{-9} \text{ s}) = 2.00 \times 10^8 \text{ cm/s} = 2.00 \times 10^6 \text{ m/s}.$$

114. We neglect air resistance, which justifies setting $a = -g = -9.8$ m/s^2 (taking *down* as the $-y$ direction) for the duration of the motion of the shot ball. We are allowed to use Table 2-1 (with Δy replacing Δx) because the ball has constant acceleration motion. We use primed variables (except t) with the constant-velocity elevator (so $v' = 10$ m/s), and unprimed variables with the ball (with initial velocity $v_0 = v' + 20 = 30$ m/s, relative to the ground). SI units are used throughout.

(a) Taking the time to be zero at the instant the ball is shot, we compute its maximum height y (relative to the ground) with $v^2 = v_0^2 - 2g(y - y_0)$, where the highest point is characterized by $v = 0$. Thus,

$$y = y_o + \frac{v_0^2}{2g} = 76 \text{ m}$$

where $y_o = y'_o + 2 = 30$ m (where $y'_o = 28$ m is given in the problem) and $v_0 = 30$ m/s relative to the ground as noted above.

(b) There are a variety of approaches to this question. One is to continue working in the frame of reference adopted in part (a) (which treats the ground as motionless and "fixes" the coordinate origin to it); in this case, one describes the elevator motion with $y' = y'_o + v't$ and the ball motion with Eq. 2-15, and solves them for the case where they reach the same point at the same time. Another is to work in the frame of reference of the elevator (the boy in the elevator might be oblivious to the fact the elevator is moving since it isn't accelerating), which is what we show here in detail:

$$\Delta y_e = v_{0_e} t - \frac{1}{2} g t^2 \quad \Rightarrow \quad t = \frac{v_{0_e} + \sqrt{v_{0_e}^2 - 2g \Delta y_e}}{g}$$

where $v_{0e} = 20$ m/s is the initial velocity of the ball relative to the elevator and $\Delta y_e = -2.0$ m is the ball's displacement relative to the floor of the elevator. The positive root is chosen to yield a positive value for t; the result is $t = 4.2$ s.

115. (a) With $v = c/10 = 3 \times 10^7$ m/s and $a = 20g = 196$ m/s^2, Eq. 4-34 gives

$$r = v^2/a = 4.6 \times 10^{12} \text{ m}.$$

(b) The period is given by Eq. 4-35: $T = 2\pi r / v = 9.6 \times 10^5$ s. Thus, the time to make a quarter-turn is $T/4 = 2.4 \times 10^5$ s or about 2.8 days.

116. Using the same coordinate system assumed in Eq. 4-25, we rearrange that equation to solve for the initial speed:

$$v_0 = \frac{x}{\cos\theta_0}\sqrt{\frac{g}{2(x\tan\theta_0 - y)}}$$

which yields $v_0 = 23$ ft/s for $g = 32$ ft/s^2, $x = 13$ ft, $y = 3$ ft and $\theta_0 = 55°$.

117. The (box)car has velocity $\vec{v}_{cg} = v_1\,\hat{i}$ relative to the ground, and the bullet has velocity

$$\vec{v}_{0bg} = v_2\cos\theta\,\hat{i} + v_2\sin\theta\,\hat{j}$$

relative to the ground before entering the car (we are neglecting the effects of gravity on the bullet). While in the car, its velocity relative to the outside ground is $\vec{v}_{bg} = 0.8 v_2 \cos\theta\,\hat{i} + 0.8 v_2 \sin\theta\,\hat{j}$ (due to the 20% reduction mentioned in the problem). The problem indicates that the velocity of the bullet in the car *relative to the car* is (with v_3 unspecified) $\vec{v}_{bc} = v_3\,\hat{j}$. Now, Eq. 4-44 provides the condition

$$\vec{v}_{bg} = \vec{v}_{bc} + \vec{v}_{cg}$$
$$0.8v_2\cos\theta\,\hat{i} + 0.8v_2\sin\theta\,\hat{j} = v_3\,\hat{j} + v_1\,\hat{i}$$

so that equating x components allows us to find θ. If one wished to find v_3 one could also equate the y components, and from this, if the car width were given, one could find the time spent by the bullet in the car, but this information is not asked for (which is why the width is irrelevant). Therefore, examining the x components in SI units leads to

$$\theta = \cos^{-1}\left(\frac{v_1}{0.8v_2}\right) = \cos^{-1}\left(\frac{85\text{ km/h}\left(\frac{1000\text{ m/km}}{3600\text{ s/h}}\right)}{0.8\,(650\text{ m/s})}\right)$$

which yields 87° for the direction of $\vec{v}_{bg}$ (measured from $\hat{i}$, which is the direction of motion of the car). The problem asks, "from what direction was it fired?" — which means the answer is not 87° but rather its supplement 93° (measured from the direction of motion). Stating this more carefully, in the coordinate system we have adopted in our solution, the bullet velocity vector is in the first quadrant, at 87° measured counterclockwise from the $+x$ direction (the direction of train motion), which means that the direction from which the bullet came (where the sniper is) is in the third quadrant, at $-93°$ (that is, 93° measured clockwise from $+x$).

118. Since $v_y^2 = v_{0y}^2 - 2g\Delta y$, and $v_y=0$ at the target, we obtain

$$v_{0y} = \sqrt{2(9.80 \text{ m/s}^2)(5.00 \text{ m})} = 9.90 \text{ m/s}$$

(a) Since $v_0 \sin \theta_0 = v_{0y}$, with $v_0 = 12.0$ m/s, we find $\theta_0 = 55.6°$.

(b) Now, $v_y = v_{0y} - gt$ gives $t = (9.90 \text{ m/s})/(9.80 \text{ m/s}^2) = 1.01$ s. Thus, $\Delta x = (v_0 \cos \theta_0)t = 6.85$ m.

(c) The velocity at the target has only the v_x component, which is equal to $v_{0x} = v_0 \cos \theta_0 = 6.78$ m/s.

119. From the figure, the three displacements can be written as

$$\vec{d}_1 = d_1(\cos \theta_1 \hat{i} + \sin \theta_1 \hat{j}) = (5.00 \text{ m})(\cos 30° \hat{i} + \sin 30° \hat{j}) = (4.33 \text{ m})\hat{i} + (2.50 \text{ m})\hat{j}$$

$$\vec{d}_2 = d_2[\cos(180° + \theta_1 - \theta_2)\hat{i} + \sin(180° + \theta_1 - \theta_2)\hat{j}] = (8.00 \text{ m})(\cos 160° \hat{i} + \sin 160° \hat{j})$$
$$= (-7.52 \text{ m})\hat{i} + (2.74 \text{ m})\hat{j}$$

$$\vec{d}_3 = d_3[\cos(360° - \theta_3 - \theta_2 + \theta_1)\hat{i} + \sin(360° - \theta_3 - \theta_2 + \theta_1)\hat{j}] = (12.0 \text{ m})(\cos 260° \hat{i} + \sin 260° \hat{j})$$
$$= (-2.08 \text{ m})\hat{i} - (11.8 \text{ m})\hat{j}$$

where the angles are measured from the +x axis. The net displacement is

$$\vec{d} = \vec{d}_1 + \vec{d}_2 + \vec{d}_3 = (-5.27 \text{ m})\hat{i} - (6.58 \text{ m})\hat{j}.$$

(a) The magnitude of the net displacement is

$$|\vec{d}| = \sqrt{(-5.27 \text{ m})^2 + (-6.58 \text{ m})^2} = 8.43 \text{ m}.$$

(b) The direction of $\vec{d}$ is

$$\theta = \tan^{-1}\left(\frac{d_y}{d_x}\right) = \tan^{-1}\left(\frac{-6.58 \text{ m}}{-5.27 \text{ m}}\right) = 51.3° \text{ or } 231°.$$

We choose 231° (measured counterclockwise from +x) since the desired angle is in the third quadrant. An equivalent answer is −129° (measured clockwise from +x).

120. With $v_0 = 30.0$ m/s and $R = 20.0$ m, Eq. 4-26 gives

$$\sin 2\theta_0 = \frac{gR}{v_0^2} = 0.218.$$

Because sin ϕ = sin ($180° - \phi$), there are two roots of the above equation:

$$2\theta_0 = \sin^{-1}(0.218) = 12.58° \text{ and } 167.4°.$$

which correspond to the two possible launch angles that will hit the target (in the absence of air friction and related effects).

(a) The smallest angle is $\theta_0 = 6.29°$.

(b) The greatest angle is and $\theta_0 = 83.7°$.

An alternative approach to this problem in terms of Eq. 4-25 (with $y = 0$ and $1/\cos^2 = 1 + \tan^2$) is possible — and leads to a quadratic equation for $\tan\theta_0$ with the roots providing these two possible θ_0 values.

121. On the one hand, we could perform the vector addition of the displacements with a vector-capable calculator in polar mode $((75 \angle 37°) + (65 \angle -90°) = (63 \angle -18°))$, but in keeping with Eq. 3-5 and Eq. 3-6 we will show the details in unit-vector notation. We use a 'standard' coordinate system with $+x$ East and $+y$ North. Lengths are in kilometers and times are in hours.

(a) We perform the vector addition of individual displacements to find the net displacement of the camel.

$$\Delta\vec{r}_1 = (75 \text{ km})\cos(37°)\hat{i} + (75 \text{ km})\sin(37°)\hat{j}$$
$$\Delta\vec{r}_2 = (-65 \text{ km})\hat{j}$$
$$\Delta\vec{r} = \Delta\vec{r}_1 + \Delta\vec{r}_2 = (60 \text{ km})\hat{i} - (20 \text{ km})\hat{j}.$$

If it is desired to express this in magnitude-angle notation, then this is equivalent to a vector of length $|\Delta\vec{r}| = \sqrt{(60 \text{ km})^2 + (-20 \text{ km})^2} = 63 \text{ km}$.

(b) The direction of $\Delta\vec{r}$ is $\theta = \tan^{-1}[(-20 \text{ km})/(60 \text{ km})] = -18°$, or $18°$ south of east.

(c) We use the result from part (a) in Eq. 4-8 along with the fact that $\Delta t = 90$ h. In unit vector notation, we obtain

$$\vec{v}_{avg} = \frac{(60\hat{i} - 20\hat{j}) \text{ km}}{90 \text{ h}} = (0.67\hat{i} - 0.22\hat{j}) \text{ km/h}.$$

This leads to $|\vec{v}_{avg}| = 0.70$ km/h.

(d) The direction of $\vec{v}_{avg}$ is $\theta = \tan^{-1}[(-0.22 \text{ km/h})/(0.67 \text{ km/h})] = -18°$, or 18° south of east.

(e) The average speed is distinguished from the magnitude of average velocity in that it depends on the total distance as opposed to the net displacement. Since the camel travels 140 km, we obtain (140 km)/(90 h) = 1.56 km/h ≈ 1.6 km/h.

(f) The net displacement is required to be the 90 km East from A to B. The displacement from the resting place to B is denoted $\Delta \vec{r}_3$. Thus, we must have

$$\Delta \vec{r}_1 + \Delta \vec{r}_2 + \Delta \vec{r}_3 = (90 \text{ km})\hat{i}$$

which produces $\Delta \vec{r}_3 = (30 \text{ km})\hat{i} + (20 \text{ km})\hat{j}$ in unit-vector notation, or $(36 \angle 33°)$ in magnitude-angle notation. Therefore, using Eq. 4-8 we obtain

$$|\vec{v}_{avg}| = \frac{36 \text{ km}}{(120-90) \text{ h}} = 1.2 \text{ km/h}.$$

(g) The direction of $\vec{v}_{avg}$ is the same as $\vec{r}_3$ (that is, 33° north of east).

122. We make use of Eq. 4-21 and Eq. 4-22.

(a) With $v_0 = 16$ m/s, we square Eq. 4-21 and Eq. 4-22 and add them, then (using Pythagoras' theorem) take the square root to obtain r:

$$r = \sqrt{(x-x_0)^2 + (y-y_0)^2} = \sqrt{(v_0 \cos\theta_0 t)^2 + (v_0 \sin\theta_0 t - gt^2/2)^2}$$

$$= t\sqrt{v_0^2 - v_0 g \sin\theta_0 t + g^2 t^2/4}$$

Below we plot r as a function of time for $\theta_0 = 40.0°$:

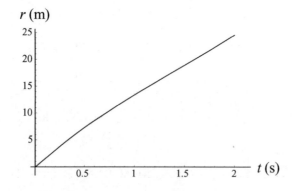

(b) For this next graph for r versus t we set $\theta_0 = 80.0°$.

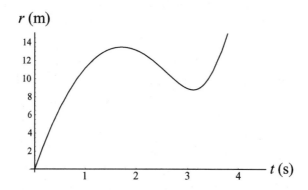

(c) Differentiating r with respect to t, we obtain

$$\frac{dr}{dt} = \frac{v_0^2 - 3v_0 gt \sin\theta_0/2 + g^2 t^2/2}{\sqrt{v_0^2 - v_0 g \sin\theta_0 t + g^2 t^2/4}}$$

Setting $dr/dt = 0$, with $v_0 = 16.0$ m/s and $\theta_0 = 40.0°$, we have $256 - 151t + 48t^2 = 0$. The equation has no real solution. This means that the maximum is reached at the end of the flight, with

$$t_{total} = 2v_0 \sin\theta_0/g = 2(16.0 \text{ m/s})\sin(40.0°)/(9.80 \text{ m/s}^2) = 2.10 \text{ s}.$$

(d) The value of r is given by

$$r = (2.10)\sqrt{(16.0)^2 - (16.0)(9.80)\sin 40.0°(2.10) + (9.80)^2(2.10)^2/4} = 25.7 \text{ m}.$$

(e) The horizontal distance is $r_x = v_0 \cos\theta_0 t = (16.0 \text{ m/s})\cos 40.0°(2.10 \text{ s}) = 25.7$ m.

(f) The vertical distance is $r_y = 0$.

(g) For the $\theta_0 = 80°$ launch, the condition for maximum r is $256 - 232t + 48t^2 = 0$, or $t = 1.71$ s (the other solution, $t = 3.13$ s, corresponds to a minimum.)

(h) The distance traveled is

$$r = (1.71)\sqrt{(16.0)^2 - (16.0)(9.80)\sin 80.0°(1.71) + (9.80)^2(1.71)^2/4} = 13.5 \text{ m}.$$

(i) The horizontal distance is

$$r_x = v_0 \cos\theta_0 t = (16.0 \text{ m/s})\cos 80.0°(1.71 \text{ s}) = 4.75 \text{ m}.$$

(j) The vertical distance is

$$r_y = v_0 \sin\theta_0 t - \frac{gt^2}{2} = (16.0 \text{ m/s})\sin 80°(1.71\text{ s}) - \frac{(9.80 \text{ m/s}^2)(1.71\text{ s})^2}{2} = 12.6 \text{ m}.$$

123. Using the same coordinate system assumed in Eq. 4-25, we find x for the elevated cannon from

$$y = x\tan\theta_0 - \frac{gx^2}{2(v_0\cos\theta_0)^2} \quad \text{where } y = -30 \text{ m}.$$

Using the quadratic formula (choosing the positive root), we find

$$x = v_0\cos\theta_0 \left(\frac{v_0\sin\theta_0 + \sqrt{(v_0\sin\theta_0)^2 - 2gy}}{g}\right)$$

which yields $x = 715$ m for $v_0 = 82$ m/s and $\theta_0 = 45°$. This is 29 m longer than the 686 m found in that Sample Problem. Since the "9" in 29 m is not reliable, due to the low level of precision in the given data, we write the answer as 3×10^1 m.

124. (a) Using the same coordinate system assumed in Eq. 4-25, we find

$$y = x\tan\theta_0 - \frac{gx^2}{2(v_0\cos\theta_0)^2} = -\frac{gx^2}{2v_0^2} \quad \text{if } \theta_0 = 0.$$

Thus, with $v_0 = 3.0 \times 10^6$ m/s and $x = 1.0$ m, we obtain $y = -5.4 \times 10^{-13}$ m which is not practical to measure (and suggests why gravitational processes play such a small role in the fields of atomic and subatomic physics).

(b) It is clear from the above expression that $|y|$ decreases as v_0 is increased.

125. At maximum height, the y-component of a projectile's velocity vanishes, so the given 10 m/s is the (constant) x-component of velocity.

(a) Using v_{0y} to denote the y-velocity 1.0 s before reaching the maximum height, then (with $v_y = 0$) the equation $v_y = v_{0y} - gt$ leads to $v_{0y} = 9.8$ m/s. The magnitude of the velocity vector (or *speed*) at that moment is therefore

$$\sqrt{v_x^2 + v_{0y}^2} = \sqrt{(10 \text{ m/s})^2 + (9.8 \text{ m/s})^2} = 14 \text{ m/s}.$$

(b) It is clear from the symmetry of the problem that the speed is the same 1.0 s after reaching the top, as it was 1.0 s before (14 m/s again). This may be verified by using $v_y =$

$v_{0y} - gt$ again but now "starting the clock" at the highest point so that $v_{0y} = 0$ (and $t = 1.0$ s). This leads to $v_y = -9.8$ m/s and $\sqrt{(10 \text{ m/s})^2 + (-9.8 \text{ m/s})^2} = 14$ m/s.

(c) The x_0 value may be obtained from $x = 0 = x_0 + (10 \text{ m/s})(1.0\text{s})$, which yields $x_0 = -10$ m.

(d) With $v_{0y} = 9.8$ m/s denoting the y-component of velocity one second before the top of the trajectory, then we have $y = 0 = y_0 + v_{0y}t - \frac{1}{2}gt^2$ where $t = 1.0$ s. This yields $y_0 = -4.9$ m.

(e) By using $x - x_0 = (10 \text{ m/s})(1.0 \text{ s})$ where $x_0 = 0$, we obtain $x = 10$ m.

(f) Let $t = 0$ at the top with $y_0 = v_{0y} = 0$. From $y - y_0 = v_{0y}t - \frac{1}{2}gt^2$, we have, for $t = 1.0$ s,

$$y = -(9.8 \text{ m/s}^2)(1.0 \text{ s})^2/2 = -4.9 \text{ m}.$$

126. With no acceleration in the x direction yet a constant acceleration of 1.4 m/s^2 in the y direction, the position (in meters) as a function of time (in seconds) must be

$$\vec{r} = (6.0t)\hat{i} + \left(\frac{1}{2}(1.4)t^2\right)\hat{j}$$

and $\vec{v}$ is its derivative with respect to t.

(a) At $t = 3.0$ s, therefore, $\vec{v} = (6.0\hat{i} + 4.2\hat{j})$ m/s.

(b) At $t = 3.0$ s, the position is $\vec{r} = (18\hat{i} + 6.3\hat{j})$ m.

127. We note that
$$\vec{v}_{PG} = \vec{v}_{PA} + \vec{v}_{AG}$$

describes a right triangle, with one leg being $\vec{v}_{PG}$ (east), another leg being $\vec{v}_{AG}$ (magnitude = 20, direction = south), and the hypotenuse being $\vec{v}_{PA}$ (magnitude = 70). Lengths are in kilometers and time is in hours. Using the Pythagorean theorem, we have

$$|\vec{v}_{PA}| = \sqrt{|\vec{v}_{PG}|^2 + |\vec{v}_{AG}|^2} \Rightarrow 70 \text{ km/h} = \sqrt{|\vec{v}_{PG}|^2 + (20 \text{ km/h})^2}$$

which is easily solved for the ground speed: $|\vec{v}_{PG}| = 67$ km/h.

128. The figure offers many interesting points to analyze, and others are easily inferred (such as the point of maximum height). The focus here, to begin with, will be the final

point shown (1.25 s after the ball is released) which is when the ball returns to its original height. In English units, $g = 32$ ft/s².

(a) Using $x - x_0 = v_x t$ we obtain $v_x = (40 \text{ ft})/(1.25 \text{ s}) = 32$ ft/s. And $y - y_0 = 0 = v_{0y} t - \frac{1}{2} g t^2$ yields $v_{0y} = \frac{1}{2}(32 \text{ ft/s}^2)(1.25 \text{ s}) = 20$ ft/s. Thus, the initial speed is

$$v_0 = |\vec{v}_0| = \sqrt{(32 \text{ ft/s})^2 + (20 \text{ ft/s})^2} = 38 \text{ ft/s}.$$

(b) Since $v_y = 0$ at the maximum height and the horizontal velocity stays constant, then the speed at the top is the same as $v_x = 32$ ft/s.

(c) We can infer from the figure (or compute from $v_y = 0 = v_{0y} - gt$) that the time to reach the top is 0.625 s. With this, we can use $y - y_0 = v_{0y} t - \frac{1}{2} g t^2$ to obtain 9.3 ft (where $y_0 = 3$ ft has been used). An alternative approach is to use $v_y^2 = v_{0y}^2 - 2g(y - y_0)$.

129. We denote $\vec{v}_{PG}$ as the velocity of the plane relative to the ground, $\vec{v}_{AG}$ as the velocity of the air relative to the ground, and $\vec{v}_{PA}$ as the velocity of the plane relative to the air.

(a) The vector diagram is shown on the right: $\vec{v}_{PG} = \vec{v}_{PA} + \vec{v}_{AG}$. Since the magnitudes v_{PG} and v_{PA} are equal the triangle is isosceles, with two sides of equal length.

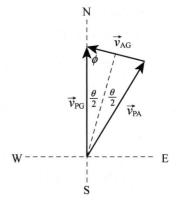

Consider either of the right triangles formed when the bisector of θ is drawn (the dashed line). It bisects $\vec{v}_{AG}$, so

$$\sin(\theta/2) = \frac{v_{AG}}{2 v_{PG}} = \frac{70.0 \text{ mi/h}}{2(135 \text{ mi/h})}$$

which leads to $\theta = 30.1°$. Now $\vec{v}_{AG}$ makes the same angle with the E-W line as the dashed line does with the N-S line. The wind is blowing in the direction 15.0° north of west. Thus, it is blowing *from* 75.0° east of south.

(b) The plane is headed along $\vec{v}_{PA}$, in the direction 30.0° east of north. There is another solution, with the plane headed 30.0° west of north and the wind blowing 15° north of east (that is, from 75° west of south).

130. Taking derivatives of $\vec{r} = 2t\hat{i} + 2\sin(\pi t/4)\hat{j}$ (with lengths in meters, time in seconds and angles in radians) provides expressions for velocity and acceleration:

$$\vec{v} = \frac{d\vec{r}}{dt} = 2\hat{i} + \frac{\pi}{2}\cos\left(\frac{\pi t}{4}\right)\hat{j}$$

$$\vec{a} = \frac{d\vec{v}}{dt} = -\frac{\pi^2}{8}\sin\left(\frac{\pi t}{4}\right)\hat{j}.$$

Thus, we obtain:

			time t	0.0	1.0	2.0	3.0	4.0
(a)	$\vec{r}$ position	x		0.0	2.0	4.0	6.0	8.0
		y		0.0	1.4	2.0	1.4	0.0
(b)	$\vec{v}$ velocity	v_x			2.0	2.0	2.0	
		v_y			1.1	0.0	−1.1	
(c)	$\vec{a}$ acceleration	a_x			0.0	0.0	0.0	
		a_y			−0.87	−1.2	−0.87	

And the path of the particle in the xy plane is shown in the following graph. The arrows indicating the velocities are not shown here, but they would appear as tangent-lines, as expected.

131. We make use of Eq. 4-24 and Eq. 4-25.

(a) With $x = 180$ m, $\theta_0 = 30°$, and $v_o = 43$ m/s, we obtain

$$y = \tan(30°)(180 \text{ m}) - \frac{(9.8 \text{ m/s}^2)(180 \text{ m})^2}{2(43 \text{ m/s})^2(\cos 30°)^2} = -11 \text{ m}$$

or $|y| = 11$ m. This implies the rise is roughly eleven meters above the fairway.

(b) The horizontal component (in the absence of air friction) is unchanged, but the vertical component increases (see Eq. 4-24). The Pythagorean theorem then gives the magnitude of final velocity (right before striking the ground): 45 m/s.

132. We let g_p denote the magnitude of the gravitational acceleration on the planet. A number of the points on the graph (including some "inferred" points — such as the max height point at $x = 12.5$ m and $t = 1.25$ s) can be analyzed profitably; for future reference, we label (with subscripts) the first $((x_0, y_0) = (0, 2)$ at $t_0 = 0)$ and last ("final") points $((x_f, y_f) = (25, 2)$ at $t_f = 2.5)$, with lengths in meters and time in seconds.

(a) The x-component of the initial velocity is found from $x_f - x_0 = v_{0x} t_f$. Therefore, $v_{0x} = 25/2.5 = 10$ m/s. And we try to obtain the y-component from

$y_f - y_0 = 0 = v_{0y}t_f - \frac{1}{2}g_p t_f^2$. This gives us $v_{0y} = 1.25 g_p$, and we see we need another equation (by analyzing another point, say, the next-to-last one) $y - y_0 = v_{0y}t - \frac{1}{2}g_p t^2$ with $y = 6$ and $t = 2$; this produces our second equation $v_{0y} = 2 + g_p$. Simultaneous solution of these two equations produces results for v_{0y} and g_p (relevant to part (b)). Thus, our complete answer for the initial velocity is $\vec{v} = (10 \text{ m/s})\hat{i} + (10 \text{ m/s})\hat{j}$.

(b) As a by-product of the part (a) computations, we have $g_p = 8.0 \text{ m/s}^2$.

(c) Solving for t_g (the time to reach the ground) in $y_g = 0 = y_0 + v_{0y}t_g - \frac{1}{2}g_p t_g^2$ leads to a positive answer: $t_g = 2.7$ s.

(d) With $g = 9.8 \text{ m/s}^2$, the method employed in part (c) would produce the quadratic equation $-4.9 t_g^2 + 10 t_g + 2 = 0$ and then the positive result $t_g = 2.2$ s.

Chapter 5

1. We apply Newton's second law (specifically, Eq. 5-2).

(a) We find the x component of the force is

$$F_x = ma_x = ma\cos 20.0° = (1.00\,\text{kg})(2.00\,\text{m/s}^2)\cos 20.0° = 1.88\,\text{N}.$$

(b) The y component of the force is

$$F_y = ma_y = ma\sin 20.0° = (1.0\,\text{kg})(2.00\,\text{m/s}^2)\sin 20.0° = 0.684\,\text{N}.$$

(c) In unit-vector notation, the force vector is

$$\vec{F} = F_x\hat{i} + F_y\hat{j} = (1.88\,\text{N})\hat{i} + (0.684\,\text{N})\hat{j}.$$

2. We apply Newton's second law (Eq. 5-1 or, equivalently, Eq. 5-2). The net force applied on the chopping block is $\vec{F}_{net} = \vec{F}_1 + \vec{F}_2$, where the vector addition is done using unit-vector notation. The acceleration of the block is given by $\vec{a} = (\vec{F}_1 + \vec{F}_2)/m$.

(a) In the first case

$$\vec{F}_1 + \vec{F}_2 = \left[(3.0\,\text{N})\hat{i} + (4.0\,\text{N})\hat{j}\right] + \left[(-3.0\,\text{N})\hat{i} + (-4.0\,\text{N})\hat{j}\right] = 0$$

so $\vec{a} = 0$.

(b) In the second case, the acceleration $\vec{a}$ equals

$$\frac{\vec{F}_1 + \vec{F}_2}{m} = \frac{\left((3.0\,\text{N})\hat{i} + (4.0\,\text{N})\hat{j}\right) + \left((-3.0\,\text{N})\hat{i} + (4.0\,\text{N})\hat{j}\right)}{2.0\,\text{kg}} = (4.0\,\text{m/s}^2)\hat{j}.$$

(c) In this final situation, $\vec{a}$ is

$$\frac{\vec{F}_1 + \vec{F}_2}{m} = \frac{\left((3.0\,\text{N})\hat{i} + (4.0\,\text{N})\hat{j}\right) + \left((3.0\,\text{N})\hat{i} + (-4.0\,\text{N})\hat{j}\right)}{2.0\,\text{kg}} = (3.0\,\text{m/s}^2)\hat{i}.$$

3. We are only concerned with horizontal forces in this problem (gravity plays no direct role). We take East as the $+x$ direction and North as $+y$. This calculation is efficiently implemented on a vector-capable calculator, using magnitude-angle notation (with SI units understood).

$$\vec{a} = \frac{\vec{F}}{m} = \frac{(9.0 \angle 0°) + (8.0 \angle 118°)}{3.0} = (2.9 \angle 53°)$$

Therefore, the acceleration has a magnitude of 2.9 m/s^2.

4. We note that $m\vec{a} = (-16 \text{ N})\hat{i} + (12 \text{ N})\hat{j}$. With the other forces as specified in the problem, then Newton's second law gives the third force as

$$\vec{F}_3 = m\vec{a} - \vec{F}_1 - \vec{F}_2 = (-34 \text{ N})\hat{i} - (12 \text{ N})\hat{j}.$$

5. We denote the two forces $\vec{F}_1$ and $\vec{F}_2$. According to Newton's second law, $\vec{F}_1 + \vec{F}_2 = m\vec{a}$, so $\vec{F}_2 = m\vec{a} - \vec{F}_1$.

(a) In unit vector notation $\vec{F}_1 = (20.0 \text{ N})\hat{i}$ and

$$\vec{a} = -(12.0 \sin 30.0° \text{ m/s}^2)\hat{i} - (12.0 \cos 30.0° \text{ m/s}^2)\hat{j} = -(6.00 \text{ m/s}^2)\hat{i} - (10.4 \text{ m/s}^2)\hat{j}.$$

Therefore,

$$\vec{F}_2 = (2.00 \text{ kg})(-6.00 \text{ m/s}^2)\hat{i} + (2.00 \text{ kg})(-10.4 \text{ m/s}^2)\hat{j} - (20.0 \text{ N})\hat{i}$$
$$= (-32.0 \text{ N})\hat{i} - (20.8 \text{ N})\hat{j}.$$

(b) The magnitude of $\vec{F}_2$ is

$$|\vec{F}_2| = \sqrt{F_{2x}^2 + F_{2y}^2} = \sqrt{(-32.0 \text{ N})^2 + (-20.8 \text{ N})^2} = 38.2 \text{ N}.$$

(c) The angle that $\vec{F}_2$ makes with the positive x axis is found from

$$\tan \theta = (F_{2y}/F_{2x}) = [(-20.8 \text{ N})/(-32.0 \text{ N})] = 0.656.$$

Consequently, the angle is either $33.0°$ or $33.0° + 180° = 213°$. Since both the x and y components are negative, the correct result is $213°$. An alternative answer is $213° - 360° = -147°$.

6. Since $\vec{v}$ = constant, we have $\vec{a} = 0$, which implies

$$\vec{F}_{net} = \vec{F}_1 + \vec{F}_2 = m\vec{a} = 0.$$

Thus, the other force must be

$$\vec{F}_2 = -\vec{F}_1 = (-2\text{ N})\hat{i} + (6\text{ N})\hat{j}.$$

7. The net force applied on the chopping block is $\vec{F}_{net} = \vec{F}_1 + \vec{F}_2 + \vec{F}_3$, where the vector addition is done using unit-vector notation. The acceleration of the block is given by $\vec{a} = (\vec{F}_1 + \vec{F}_2 + \vec{F}_3)/m$.

(a) The forces exerted by the three astronauts can be expressed in unit-vector notation as follows:

$$\vec{F}_1 = (32\text{ N})(\cos 30°\hat{i} + \sin 30°\hat{j}) = (27.7\text{ N})\hat{i} + (16\text{ N})\hat{j}$$
$$\vec{F}_2 = (55\text{ N})(\cos 0°\hat{i} + \sin 0°\hat{j}) = (55\text{ N})\hat{i}$$
$$\vec{F}_3 = (41\text{ N})(\cos(-60°)\hat{i} + \sin(-60°)\hat{j}) = (20.5\text{ N})\hat{i} - (35.5\text{ N})\hat{j}.$$

The resultant acceleration of the asteroid of mass $m = 120$ kg is therefore

$$\vec{a} = \frac{(27.7\hat{i} + 16\hat{j})\text{ N} + (55\hat{i})\text{ N} + (20.5\hat{i} - 35.5\hat{j})\text{ N}}{120\text{ kg}} = (0.86\text{m/s}^2)\hat{i} - (0.16\text{m/s}^2)\hat{j}.$$

(b) The magnitude of the acceleration vector is

$$|\vec{a}| = \sqrt{a_x^2 + a_y^2} = \sqrt{(0.86\text{ m/s}^2)^2 + (-0.16\text{ m/s}^2)^2} = 0.88\text{ m/s}^2.$$

(c) The vector $\vec{a}$ makes an angle θ with the $+x$ axis, where

$$\theta = \tan^{-1}\left(\frac{a_y}{a_x}\right) = \tan^{-1}\left(\frac{-0.16\text{ m/s}^2}{0.86\text{ m/s}^2}\right) = -11°.$$

8. Since the tire remains stationary, by Newton's second law, the net force must be zero:

$$\vec{F}_{net} = \vec{F}_A + \vec{F}_B + \vec{F}_C = m\vec{a} = 0.$$

From the free-body diagram shown on the right, we have

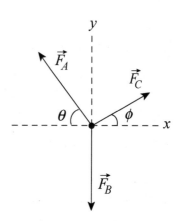

$$0 = \sum F_{\text{net},x} = F_C \cos\phi - F_A \cos\theta$$
$$0 = \sum F_{\text{net},y} = F_A \sin\theta + F_C \sin\phi - F_B$$

To solve for F_B, we first compute ϕ. With $F_A = 220$ N, $F_C = 170$ N and $\theta = 47°$, we get

$$\cos\phi = \frac{F_A \cos\theta}{F_C} = \frac{(220\text{ N})\cos 47.0°}{170\text{ N}} = 0.883 \Rightarrow \phi = 28.0°$$

Substituting the value into the second force equation, we find

$$F_B = F_A \sin\theta + F_C \sin\phi = (220\text{ N})\sin 47.0° + (170\text{ N})\sin 28.0 = 241\text{ N}.$$

9. The velocity is the derivative (with respect to time) of given function x, and the acceleration is the derivative of the velocity. Thus, $a = 2c - 3(2.0)(2.0)t$, which we use in Newton's second law: $F = (2.0\text{ kg})a = 4.0c - 24t$ (with SI units understood). At $t = 3.0$ s, we are told that $F = -36$ N. Thus, $-36 = 4.0c - 24(3.0)$ can be used to solve for c. The result is $c = +9.0$ m/s^2.

10. To solve the problem, we note that acceleration is the second time derivative of the position function, and the net force is related to the acceleration via Newton's second law. Thus, differentiating

$$x(t) = -13.00 + 2.00t + 4.00t^2 - 3.00t^3$$

twice with respect to t, we get

$$\frac{dx}{dt} = 2.00 + 8.00t - 9.00t^2, \quad \frac{d^2x}{dt^2} = 8.00 - 18.0t$$

The net force acting on the particle at $t = 3.40$ s is

$$\vec{F} = m\frac{d^2x}{dt^2}\hat{i} = (0.150)[8.00 - 18.0(3.40)]\hat{i} = (-7.98\text{ N})\hat{i}$$

11. To solve the problem, we note that acceleration is the second time derivative of the position function; it is a vector and can be determined from its components. The net force is related to the acceleration via Newton's second law. Thus, differentiating $x(t) = -15.0 + 2.00t + 4.00t^3$ twice with respect to t, we get

$$\frac{dx}{dt} = 2.00 - 12.0t^2, \quad \frac{d^2x}{dt^2} = -24.0t$$

Similarly, differentiating $y(t) = 25.0 + 7.00t - 9.00t^2$ twice with respect to t yields

$$\frac{dy}{dt} = 7.00 - 18.0t, \quad \frac{d^2y}{dt^2} = -18.0$$

(a) The acceleration is

$$\vec{a} = a_x\hat{i} + a_y\hat{j} = \frac{d^2x}{dt^2}\hat{i} + \frac{d^2y}{dt^2}\hat{j} = (-24.0t)\hat{i} + (-18.0)\hat{j}.$$

At $t = 0.700$ s, we have $\vec{a} = (-16.8)\hat{i} + (-18.0)\hat{j}$ with a magnitude of

$$a = |\vec{a}| = \sqrt{(-16.8)^2 + (-18.0)^2} = 24.6 \text{ m/s}^2.$$

Thus, the magnitude of the force is $F = ma = (0.34 \text{ kg})(24.6 \text{ m/s}^2) = 8.37$ N.

(b) The angle $\vec{F}$ or $\vec{a} = \vec{F}/m$ makes with $+x$ is

$$\theta = \tan^{-1}\left(\frac{a_y}{a_x}\right) = \tan^{-1}\left(\frac{-18.0 \text{ m/s}^2}{-16.8 \text{ m/s}^2}\right) = 47.0° \text{ or } -133°.$$

We choose the latter ($-133°$) since $\vec{F}$ is in the third quadrant.

(c) The direction of travel is the direction of a tangent to the path, which is the direction of the velocity vector:

$$\vec{v}(t) = v_x\hat{i} + v_y\hat{j} = \frac{dx}{dt}\hat{i} + \frac{dy}{dt}\hat{j} = (2.00 - 12.0t^2)\hat{i} + (7.00 - 18.0t)\hat{j}.$$

At $t = 0.700$ s, we have $\vec{v}(t = 0.700 \text{ s}) = (-3.88 \text{ m/s})\hat{i} + (-5.60 \text{ m/s})\hat{j}$. Therefore, the angle $\vec{v}$ makes with $+x$ is

$$\theta_v = \tan^{-1}\left(\frac{v_y}{v_x}\right) = \tan^{-1}\left(\frac{-5.60 \text{ m/s}}{-3.88 \text{ m/s}}\right) = 55.3° \text{ or } -125°.$$

We choose the latter ($-125°$) since $\vec{v}$ is in the third quadrant.

12. From the slope of the graph we find $a_x = 3.0$ m/s^2. Applying Newton's second law to the x axis (and taking θ to be the angle between F_1 and F_2), we have

$$F_1 + F_2\cos\theta = ma_x \quad \Rightarrow \quad \theta = 56°.$$

13. (a) – (c) In all three cases the scale is not accelerating, which means that the two cords exert forces of equal magnitude on it. The scale reads the magnitude of either of these forces. In each case the tension force of the cord attached to the salami must be the

same in magnitude as the weight of the salami because the salami is not accelerating. Thus the scale reading is mg, where m is the mass of the salami. Its value is $(11.0 \text{ kg})(9.8 \text{ m/s}^2) = 108 \text{ N}$.

14. Three vertical forces are acting on the block: the earth pulls down on the block with gravitational force 3.0 N; a spring pulls up on the block with elastic force 1.0 N; and, the surface pushes up on the block with normal force F_N. There is no acceleration, so

$$\sum F_y = 0 = F_N + (1.0 \text{ N}) + (-3.0 \text{ N})$$

yields $F_N = 2.0$ N.

(a) By Newton's third law, the force exerted by the block on the surface has that same magnitude but opposite direction: 2.0 N.

(b) The direction is down.

15. (a) From the fact that $T_3 = 9.8$ N, we conclude the mass of disk D is 1.0 kg. Both this and that of disk C cause the tension $T_2 = 49$ N, which allows us to conclude that disk C has a mass of 4.0 kg. The weights of these two disks plus that of disk B determine the tension $T_1 = 58.8$ N, which leads to the conclusion that $m_B = 1.0$ kg. The weights of all the disks must add to the 98 N force described in the problem; therefore, disk A has mass 4.0 kg.

(b) $m_B = 1.0$ kg, as found in part (a).

(c) $m_C = 4.0$ kg, as found in part (a).

(d) $m_D = 1.0$ kg, as found in part (a).

16. (a) There are six legs, and the vertical component of the tension force in each leg is $T \sin \theta$ where $\theta = 40°$. For vertical equilibrium (zero acceleration in the y direction) then Newton's second law leads to

$$6T \sin \theta = mg \Rightarrow T = \frac{mg}{6 \sin \theta}$$

which (expressed as a multiple of the bug's weight mg) gives roughly $T/mg \approx 0.26 \, 0$.

(b) The angle θ is measured from horizontal, so as the insect "straightens out the legs" θ will increase (getting closer to 90°), which causes $\sin \theta$ to increase (getting closer to 1) and consequently (since $\sin \theta$ is in the denominator) causes T to decrease.

17. (a) The coin undergoes free fall. Therefore, with respect to ground, its acceleration is

$$\vec{a}_{\text{coin}} = \vec{g} = (-9.8 \text{ m/s}^2)\hat{j}.$$

(b) Since the customer is being pulled down with an acceleration of $\vec{a}'_{customer} = 1.24\vec{g} = (-12.15 \text{ m/s}^2)\hat{j}$, the acceleration of the coin with respect to the customer is

$$\vec{a}_{rel} = \vec{a}_{coin} - \vec{a}'_{customer} = (-9.8 \text{ m/s}^2)\hat{j} - (-12.15 \text{ m/s}^2)\hat{j} = (+2.35 \text{ m/s}^2)\hat{j}.$$

(c) The time it takes for the coin to reach the ceiling is

$$t = \sqrt{\frac{2h}{a_{rel}}} = \sqrt{\frac{2(2.20 \text{ m})}{2.35 \text{ m/s}^2}} = 1.37 \text{ s}.$$

(d) Since gravity is the only force acting on the coin, the actual force on the coin is

$$\vec{F}_{coin} = m\vec{a}_{coin} = m\vec{g} = (0.567 \times 10^{-3} \text{ kg})(-9.8 \text{ m/s}^2)\hat{j} = (-5.56 \times 10^{-3} \text{ N})\hat{j}.$$

(e) In the customer's frame, the coin travels upward at a constant acceleration. Therefore, the apparent force on the coin is

$$\vec{F}_{app} = m\vec{a}_{rel} = (0.567 \times 10^{-3} \text{ kg})(+2.35 \text{ m/s}^2)\hat{j} = (+1.33 \times 10^{-3} \text{ N})\hat{j}.$$

18. We note that the rope is 22.0° from vertical – and therefore 68.0° from horizontal.

(a) With $T = 760$ N, then its components are

$$\vec{T} = T\cos 68.0° \hat{i} + T\sin 68.0° \hat{j} = (285 \text{N})\hat{i} + (705 \text{N})\hat{j}.$$

(b) No longer in contact with the cliff, the only other force on Tarzan is due to earth's gravity (his weight). Thus,

$$\vec{F}_{net} = \vec{T} + \vec{W} = (285 \text{ N})\hat{i} + (705 \text{ N})\hat{j} - (820 \text{ N})\hat{j} = (285 \text{N})\hat{i} - (115 \text{ N})\hat{j}.$$

(c) In a manner that is efficiently implemented on a vector-capable calculator, we convert from rectangular (x, y) components to magnitude-angle notation:

$$\vec{F}_{net} = (285, -115) \rightarrow (307 \angle -22.0°)$$

so that the net force has a magnitude of 307 N.

(d) The angle (see part (c)) has been found to be −22.0°, or 22.0° below horizontal (away from cliff).

(e) Since $\vec{a} = \vec{F}_{net}/m$ where $m = W/g = 83.7$ kg, we obtain $\vec{a} = 3.67$ m/s^2.

(f) Eq. 5-1 requires that $\vec{a} \parallel \vec{F}_{net}$ so that the angle is also $-22.0°$, or $22.0°$ below horizontal (away from cliff).

19. (a) Since the acceleration of the block is zero, the components of the Newton's second law equation yield
$$T - mg \sin\theta = 0$$
$$F_N - mg \cos\theta = 0.$$

Solving the first equation for the tension in the string, we find
$$T = mg \sin\theta = (8.5 \text{ kg})(9.8 \text{ m/s}^2) \sin 30° = 42 \text{ N}.$$

(b) We solve the second equation in part (a) for the normal force F_N:
$$F_N = mg \cos\theta = (8.5 \text{ kg})(9.8 \text{ m/s}^2) \cos 30° = 72 \text{ N}.$$

(c) When the string is cut, it no longer exerts a force on the block and the block accelerates. The x component of the second law becomes $-mg\sin\theta = ma$, so the acceleration becomes
$$a = -g \sin\theta = -(9.8 \text{ m/s}^2)\sin 30° = -4.9 \text{ m/s}^2.$$

The negative sign indicates the acceleration is down the plane. The magnitude of the acceleration is 4.9 m/s².

20. We take rightwards as the $+x$ direction. Thus, $\vec{F}_1 = (20 \text{ N})\hat{i}$. In each case, we use Newton's second law $\vec{F}_1 + \vec{F}_2 = m\vec{a}$ where $m = 2.0$ kg.

(a) If $\vec{a} = (+10 \text{ m/s}^2)\hat{i}$, then the equation above gives $\vec{F}_2 = 0$.

(b) If, $\vec{a} = (+20 \text{ m/s}^2)\hat{i}$, then that equation gives $\vec{F}_2 = (20 \text{ N})\hat{i}$.

(c) If $\vec{a} = 0$, then the equation gives $\vec{F}_2 = (-20 \text{ N})\hat{i}$.

(d) If $\vec{a} = (-10 \text{ m/s}^2)\hat{i}$, the equation gives $\vec{F}_2 = (-40 \text{ N})\hat{i}$.

(e) If $\vec{a} = (-20 \text{ m/s}^2)\hat{i}$, the equation gives $\vec{F}_2 = (-60 \text{ N})\hat{i}$.

21. (a) The slope of each graph gives the corresponding component of acceleration. Thus, we find $a_x = 3.00$ m/s² and $a_y = -5.00$ m/s². The magnitude of the acceleration

vector is therefore $a = \sqrt{(3.00 \text{ m/s}^2)^2 + (-5.00 \text{ m/s}^2)^2} = 5.83 \text{ m/s}^2$, and the force is obtained from this by multiplying with the mass ($m = 2.00$ kg). The result is $F = ma$ =11.7 N.

(b) The direction of the force is the same as that of the acceleration:

$$\theta = \tan^{-1}[(-5.00 \text{ m/s}^2)/(3.00 \text{ m/s}^2)] = -59.0°.$$

22. The free-body diagram of the cars is shown on the right. The force exerted by John Massis is

$$F = 2.5mg = 2.5(80 \text{ kg})(9.8 \text{ m/s}^2) = 1960 \text{ N}.$$

Since the motion is along the horizontal x-axis, using Newton's second law, we have $Fx = F\cos\theta = Ma_x$, where M is the total mass of the railroad cars. Thus, the acceleration of the cars is

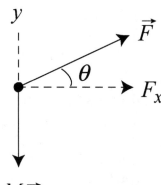

$$a_x = \frac{F\cos\theta}{M} = \frac{(1960 \text{ N})\cos 30°}{(7.0\times 10^5 \text{ N}/9.8 \text{ m/s}^2)} = 0.024 \text{ m/s}^2.$$

Using Eq. 2-16, the speed of the car at the end of the pull is

$$v_x = \sqrt{2a_x \Delta x} = \sqrt{2(0.024 \text{ m/s}^2)(1.0 \text{ m})} = 0.22 \text{ m/s}.$$

23. (a) The acceleration is

$$a = \frac{F}{m} = \frac{20 \text{ N}}{900 \text{ kg}} = 0.022 \text{ m/s}^2.$$

(b) The distance traveled in 1 day (= 86400 s) is

$$s = \frac{1}{2}at^2 = \frac{1}{2}(0.0222 \text{ m/s}^2)(86400 \text{ s})^2 = 8.3 \times 10^7 \text{ m}.$$

(c) The speed it will be traveling is given by

$$v = at = (0.0222 \text{ m/s}^2)(86400 \text{ s}) = 1.9 \times 10^3 \text{ m/s}.$$

24. Some assumptions (not so much for realism but rather in the interest of using the given information efficiently) are needed in this calculation: we assume the fishing line and the path of the salmon are horizontal. Thus, the weight of the fish contributes only (via Eq. 5-12) to information about its mass ($m = W/g = 8.7$ kg). Our +x axis is in the direction of the salmon's velocity (away from the fisherman), so that its acceleration

("deceleration") is negative-valued and the force of tension is in the $-x$ direction: $\vec{T} = -T$. We use Eq. 2-16 and SI units (noting that $v = 0$).

$$v^2 = v_0^2 + 2a\Delta x \Rightarrow a = -\frac{v_0^2}{2\Delta x} = -\frac{(2.8 \text{ m/s})^2}{2(0.11 \text{ m})} = -36 \text{ m/s}^2.$$

Assuming there are no significant horizontal forces other than the tension, Eq. 5-1 leads to

$$\vec{T} = m\vec{a} \Rightarrow -T = (8.7 \text{ kg})(-36 \text{ m/s}^2)$$

which results in $T = 3.1 \times 10^2$ N.

25. In terms of magnitudes, Newton's second law is $F = ma$, where $F = |\vec{F}_{net}|$, $a = |\vec{a}|$, and m is the (always positive) mass. The magnitude of the acceleration can be found using constant acceleration kinematics (Table 2-1). Solving $v = v_0 + at$ for the case where it starts from rest, we have $a = v/t$ (which we interpret in terms of magnitudes, making specification of coordinate directions unnecessary). The velocity is

$$v = (1600 \text{ km/h})(1000 \text{ m/km})/(3600 \text{ s/h}) = 444 \text{ m/s},$$

so

$$F = ma = m\frac{v}{t} = (500 \text{ kg})\frac{444 \text{ m/s}}{1.8 \text{ s}} = 1.2 \times 10^5 \text{ N}.$$

26. The stopping force $\vec{F}$ and the path of the passenger are horizontal. Our $+x$ axis is in the direction of the passenger's motion, so that the passenger's acceleration ("deceleration") is negative-valued and the stopping force is in the $-x$ direction: $\vec{F} = -F\hat{i}$. Using Eq. 2-16 with

$$v_0 = (53 \text{ km/h})(1000 \text{ m/km})/(3600 \text{ s/h}) = 14.7 \text{ m/s}$$

and $v = 0$, the acceleration is found to be

$$v^2 = v_0^2 + 2a\Delta x \Rightarrow a = -\frac{v_0^2}{2\Delta x} = -\frac{(14.7 \text{ m/s})^2}{2(0.65 \text{ m})} = -167 \text{ m/s}^2.$$

Assuming there are no significant horizontal forces other than the stopping force, Eq. 5-1 leads to

$$\vec{F} = m\vec{a} \Rightarrow -F = (41 \text{ kg})(-167 \text{ m/s}^2)$$

which results in $F = 6.8 \times 10^3$ N.

27. We choose up as the $+y$ direction, so $\vec{a} = (-3.00 \text{ m/s}^2)\hat{j}$ (which, without the unit-vector, we denote as a since this is a 1-dimensional problem in which Table 2-1 applies). From Eq. 5-12, we obtain the firefighter's mass: $m = W/g = 72.7$ kg.

(a) We denote the force exerted by the pole on the firefighter $\vec{F}_{fp} = F_{fp}\hat{j}$ and apply Eq. 5-1. Since $\vec{F}_{net} = m\vec{a}$, we have

$$F_{fp} - F_g = ma \Rightarrow F_{fp} - 712 \text{ N} = (72.7 \text{ kg})(-3.00 \text{ m/s}^2)$$

which yields $F_{fp} = 494$ N.

(b) The fact that the result is positive means $\vec{F}_{fp}$ points up.

(c) Newton's third law indicates $\vec{F}_{fp} = -\vec{F}_{pf}$, which leads to the conclusion that $|\vec{F}_{pf}| = 494$ N.

(d) The direction of $\vec{F}_{pf}$ is down.

28. The stopping force $\vec{F}$ and the path of the toothpick are horizontal. Our $+x$ axis is in the direction of the toothpick's motion, so that the toothpick's acceleration ("deceleration") is negative-valued and the stopping force is in the $-x$ direction: $\vec{F} = -F\hat{i}$. Using Eq. 2-16 with $v_0 = 220$ m/s and $v = 0$, the acceleration is found to be

$$v^2 = v_0^2 + 2a\Delta x \Rightarrow a = -\frac{v_0^2}{2\Delta x} = -\frac{(220 \text{ m/s})^2}{2(0.015 \text{ m})} = -1.61 \times 10^6 \text{ m/s}^2.$$

Thus, the magnitude of the force exerted by the branch on the toothpick is

$$F = m|a| = (1.3 \times 10^{-4} \text{ kg})(1.61 \times 10^6 \text{ m/s}^2) = 2.1 \times 10^2 \text{ N}.$$

29. The acceleration of the electron is vertical and for all practical purposes the only force acting on it is the electric force. The force of gravity is negligible. We take the $+x$ axis to be in the direction of the initial velocity and the $+y$ axis to be in the direction of the electrical force, and place the origin at the initial position of the electron. Since the force and acceleration are constant, we use the equations from Table 2-1: $x = v_0 t$ and

$$y = \frac{1}{2}at^2 = \frac{1}{2}\left(\frac{F}{m}\right)t^2.$$

The time taken by the electron to travel a distance x (= 30 mm) horizontally is $t = x/v_0$ and its deflection in the direction of the force is

$$y = \frac{1}{2}\frac{F}{m}\left(\frac{x}{v_0}\right)^2 = \frac{1}{2}\left(\frac{4.5\times 10^{-16}\text{ N}}{9.11\times 10^{-31}\text{ kg}}\right)\left(\frac{30\times 10^{-3}\text{ m}}{1.2\times 10^{7}\text{ m/s}}\right)^2 = 1.5\times 10^{-3}\text{ m}.$$

30. The stopping force $\vec{F}$ and the path of the car are horizontal. Thus, the weight of the car contributes only (via Eq. 5-12) to information about its mass ($m = W/g = 1327$ kg). Our $+x$ axis is in the direction of the car's velocity, so that its acceleration ("deceleration") is negative-valued and the stopping force is in the $-x$ direction: $\vec{F} = -F\hat{i}$.

(a) We use Eq. 2-16 and SI units (noting that $v = 0$ and $v_0 = 40(1000/3600) = 11.1$ m/s).

$$v^2 = v_0^2 + 2a\Delta x \quad\Rightarrow\quad a = -\frac{v_0^2}{2\Delta x} = -\frac{(11.1\text{ m/s})^2}{2(15\text{ m})}$$

which yields $a = -4.12$ m/s^2. Assuming there are no significant horizontal forces other than the stopping force, Eq. 5-1 leads to

$$\vec{F} = m\vec{a} \quad\Rightarrow\quad -F = (1327\text{ kg})(-4.12\text{ m/s}^2)$$

which results in $F = 5.5\times 10^3$ N.

(b) Eq. 2-11 readily yields $t = -v_0/a = 2.7$ s.

(c) Keeping F the same means keeping a the same, in which case (since $v = 0$) Eq. 2-16 expresses a direct proportionality between Δx and v_0^2. Therefore, doubling v_0 means quadrupling Δx. That is, the new over the old stopping distances is a factor of 4.0.

(d) Eq. 2-11 illustrates a direct proportionality between t and v_0 so that doubling one means doubling the other. That is, the new time of stopping is a factor of 2.0 greater than the one found in part (b).

31. The acceleration vector as a function of time is

$$\vec{a} = \frac{d\vec{v}}{dt} = \frac{d}{dt}\left(8.00t\,\hat{i} + 3.00t^2\,\hat{j}\right)\text{ m/s} = (8.00\,\hat{i} + 6.00t\,\hat{j})\text{ m/s}^2.$$

(a) The magnitude of the force acting on the particle is

$$F = ma = m|\vec{a}| = (3.00)\sqrt{(8.00)^2 + (6.00t)^2} = (3.00)\sqrt{64.0 + 36.0\,t^2}\text{ N}.$$

Thus, $F = 35.0$ N corresponds to $t = 1.415$ s, and the acceleration vector at this instant is

$$\vec{a} = [8.00\,\hat{i} + 6.00(1.415)\,\hat{j}]\,\text{m/s}^2 = (8.00\,\text{m/s}^2)\,\hat{i} + (8.49\,\text{m/s}^2)\,\hat{j}.$$

The angle $\vec{a}$ makes with +x is

$$\theta_a = \tan^{-1}\left(\frac{a_y}{a_x}\right) = \tan^{-1}\left(\frac{8.49\,\text{m/s}^2}{8.00\,\text{m/s}^2}\right) = 46.7°.$$

(b) The velocity vector at $t = 1.415$ s is

$$\vec{v} = \left[8.00(1.415)\,\hat{i} + 3.00(1.415)^2\,\hat{j}\right]\,\text{m/s} = (11.3\,\text{m/s})\,\hat{i} + (6.01\,\text{m/s})\,\hat{j}.$$

Therefore, the angle $\vec{v}$ makes with +x is

$$\theta_v = \tan^{-1}\left(\frac{v_y}{v_x}\right) = \tan^{-1}\left(\frac{6.01\,\text{m/s}}{11.3\,\text{m/s}}\right) = 28.0°.$$

32. We resolve this horizontal force into appropriate components.

(a) Newton's second law applied to the x-axis produces

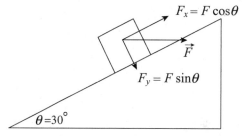

$$F\cos\theta - mg\sin\theta = ma.$$

For $a = 0$, this yields $F = 566$ N.

(b) Applying Newton's second law to the y axis (where there is no acceleration), we have

$$F_N - F\sin\theta - mg\cos\theta = 0$$

which yields the normal force $F_N = 1.13 \times 10^3$ N.

33. (a) Since friction is negligible the force of the girl is the only horizontal force on the sled. The vertical forces (the force of gravity and the normal force of the ice) sum to zero. The acceleration of the sled is

$$a_s = \frac{F}{m_s} = \frac{5.2\,\text{N}}{8.4\,\text{kg}} = 0.62\,\text{m/s}^2.$$

(b) According to Newton's third law, the force of the sled on the girl is also 5.2 N. Her acceleration is

$$a_g = \frac{F}{m_g} = \frac{5.2\,\text{N}}{40\,\text{kg}} = 0.13\,\text{m/s}^2.$$

(c) The accelerations of the sled and girl are in opposite directions. Assuming the girl starts at the origin and moves in the +x direction, her coordinate is given by $x_g = \frac{1}{2}a_g t^2$. The sled starts at $x_0 = 15$ m and moves in the $-x$ direction. Its coordinate is given by $x_s = x_0 - \frac{1}{2}a_s t^2$. They meet when $x_g = x_s$, or

$$\frac{1}{2}a_g t^2 = x_0 - \frac{1}{2}a_s t^2.$$

This occurs at time

$$t = \sqrt{\frac{2x_0}{a_g + a_s}}.$$

By then, the girl has gone the distance

$$x_g = \frac{1}{2}a_g t^2 = \frac{x_0 a_g}{a_g + a_s} = \frac{(15\,\text{m})(0.13\,\text{m/s}^2)}{0.13\,\text{m/s}^2 + 0.62\,\text{m/s}^2} = 2.6\,\text{m}.$$

34. (a) Using notation suitable to a vector capable calculator, the $\vec{F}_{\text{net}} = 0$ condition becomes

$$\vec{F}_1 + \vec{F}_2 + \vec{F}_3 = (6.00 \angle 150°) + (7.00 \angle -60.0°) + \vec{F}_3 = 0.$$

Thus, $\vec{F}_3 = (1.70\,\text{N})\,\hat{i} + (3.06\,\text{N})\hat{j}$.

(b) A constant velocity condition requires zero acceleration, so the answer is the same.

(c) Now, the acceleration is $\vec{a} = (13.0\,\text{m/s}^2)\hat{i} - (14.0\,\text{m/s}^2)\hat{j}$. Using $\vec{F}_{\text{net}} = m\vec{a}$ (with $m = 0.025$ kg) we now obtain

$$\vec{F}_3 = (2.02\,\text{N})\,\hat{i} + (2.71\,\text{N})\,\hat{j}.$$

35. The free-body diagram is shown next. $\vec{F}_N$ is the normal force of the plane on the block and $m\vec{g}$ is the force of gravity on the block. We take the +x direction to be down the incline, in the direction of the acceleration, and the +y direction to be in the direction of the normal force exerted by the incline on the block. The x component of Newton's second law is then $mg \sin\theta = ma$; thus, the acceleration is $a = g \sin\theta$.

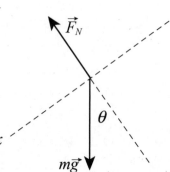

(a) Placing the origin at the bottom of the plane, the kinematic equations (Table 2-1) for motion along the x axis which we will use are $v^2 = v_0^2 + 2ax$ and $v = v_0 + at$. The block momentarily stops at its highest point, where $v = 0$; according to the second equation, this occurs at time $t = -v_0/a$. The position where it stops is

$$x = -\frac{1}{2}\frac{v_0^2}{a} = -\frac{1}{2}\left(\frac{(-3.50 \text{ m/s})^2}{(9.8 \text{ m/s}^2)\sin 32.0°}\right) = -1.18 \text{ m},$$

or $|x| = 1.18$ m.

(b) The time is

$$t = \frac{v_0}{a} = -\frac{v_0}{g \sin \theta} = -\frac{-3.50 \text{ m/s}}{(9.8 \text{ m/s}^2)\sin 32.0°} = 0.674 \text{ s}.$$

(c) That the return-speed is identical to the initial speed is to be expected since there are no dissipative forces in this problem. In order to prove this, one approach is to set $x = 0$ and solve $x = v_0 t + \tfrac{1}{2}at^2$ for the total time (up and back down) t. The result is

$$t = -\frac{2v_0}{a} = -\frac{2v_0}{g \sin \theta} = -\frac{2(-3.50 \text{ m/s})}{(9.8 \text{ m/s}^2)\sin 32.0°} = 1.35 \text{ s}.$$

The velocity when it returns is therefore

$$v = v_0 + at = v_0 + gt \sin \theta = -3.50 \text{ m/s} + (9.8 \text{ m/s}^2)(1.35 \text{ s})\sin 32° = 3.50 \text{ m/s}.$$

36. We label the 40 kg skier "m" which is represented as a block in the figure shown. The force of the wind is denoted $\vec{F}_w$ and might be either "uphill" or "downhill" (it is shown uphill in our sketch). The incline angle θ is 10°. The $-x$ direction is downhill.

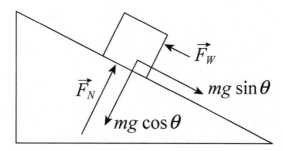

(a) Constant velocity implies zero acceleration; thus, application of Newton's second law along the x axis leads to

$$mg \sin \theta - F_w = 0.$$

This yields $F_w = 68$ N (uphill).

(b) Given our coordinate choice, we have $a = |a| = 1.0$ m/s^2. Newton's second law

$$mg \sin \theta - F_w = ma$$

now leads to $F_w = 28$ N (uphill).

(c) Continuing with the forces as shown in our figure, the equation

$$mg \sin \theta - F_w = ma$$

will lead to $F_w = -12$ N when $|a| = 2.0$ m/s^2. This simply tells us that the wind is opposite to the direction shown in our sketch; in other words, $\vec{F}_w = 12$ N *downhill*.

37. The solutions to parts (a) and (b) have been combined here. The free-body diagram is shown below, with the tension of the string $\vec{T}$, the force of gravity $m\vec{g}$, and the force of the air $\vec{F}$. Our coordinate system is shown. Since the sphere is motionless the net force on it is zero, and the x and the y components of the equations are:

$$T \sin \theta - F = 0$$
$$T \cos \theta - mg = 0,$$

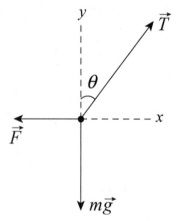

where $\theta = 37°$. We answer the questions in the reverse order. Solving $T \cos \theta - mg = 0$ for the tension, we obtain

$$T = mg/\cos \theta = (3.0 \times 10^{-4} \text{ kg})(9.8 \text{ m/s}^2)/\cos 37° = 3.7 \times 10^{-3} \text{ N}.$$

Solving $T \sin \theta - F = 0$ for the force of the air:

$$F = T \sin \theta = (3.7 \times 10^{-3} \text{ N}) \sin 37° = 2.2 \times 10^{-3} \text{ N}.$$

38. The acceleration of an object (neither pushed nor pulled by any force other than gravity) on a smooth inclined plane of angle θ is $a = -g\sin\theta$. The slope of the graph shown with the problem statement indicates $a = -2.50$ m/s^2. Therefore, we find $\theta = 14.8°$. Examining the forces perpendicular to the incline (which must sum to zero since there is no component of acceleration in this direction) we find $F_N = mg\cos\theta$, where $m = 5.00$ kg. Thus, the normal (perpendicular) force exerted at the box/ramp interface is 47.4 N.

39. The free-body diagram is shown below. Let $\vec{T}$ be the tension of the cable and $m\vec{g}$ be the force of gravity. If the upward direction is positive, then Newton's second law is $T - mg = ma$, where a is the acceleration.

Thus, the tension is $T = m(g + a)$. We use constant acceleration kinematics (Table 2-1) to find the acceleration (where $v = 0$ is the final velocity, $v_0 = -12$ m/s is the initial velocity, and $y = -42$ m is the coordinate at the stopping point). Consequently, $v^2 = v_0^2 + 2ay$ leads to

$$a = -\frac{v_0^2}{2y} = -\frac{(-12 \text{ m/s})^2}{2(-42 \text{ m})} = 1.71 \text{ m/s}^2.$$

We now return to calculate the tension:

$$T = m(g + a)$$
$$= (1600 \text{ kg})(9.8 \text{ m/s}^2 + 1.71 \text{ m/s}^2)$$
$$= 1.8 \times 10^4 \text{ N}.$$

40. (a) Constant velocity implies zero acceleration, so the "uphill" force must equal (in magnitude) the "downhill" force: $T = mg \sin \theta$. Thus, with $m = 50$ kg and $\theta = 8.0°$, the tension in the rope equals 68 N.

(b) With an uphill acceleration of 0.10 m/s², Newton's second law (applied to the x axis) yields

$$T - mg \sin \theta = ma \Rightarrow T - (50 \text{ kg})(9.8 \text{ m/s}^2) \sin 8.0° = (50 \text{ kg})(0.10 \text{ m/s}^2)$$

which leads to $T = 73$ N.

41. (a) The mass of the elevator is $m = (27800/9.80) = 2837$ kg and (with $+y$ upward) the acceleration is $a = +1.22$ m/s². Newton's second law leads to

$$T - mg = ma \Rightarrow T = m(g + a)$$

which yields $T = 3.13 \times 10^4$ N for the tension.

(b) The term "deceleration" means the acceleration vector is in the direction opposite to the velocity vector (which the problem tells us is upward). Thus (with $+y$ upward) the acceleration is now $a = -1.22$ m/s², so that the tension is

$$T = m(g + a) = 2.43 \times 10^4 \text{ N}.$$

42. (a) The term "deceleration" means the acceleration vector is in the direction opposite to the velocity vector (which the problem tells us is downward). Thus (with $+y$ upward) the acceleration is $a = +2.4$ m/s^2. Newton's second law leads to

$$T - mg = ma \Rightarrow m = \frac{T}{g + a}$$

which yields $m = 7.3$ kg for the mass.

(b) Repeating the above computation (now to solve for the tension) with $a = +2.4$ m/s^2 will, of course, lead us right back to $T = 89$ N. Since the direction of the velocity did not enter our computation, this is to be expected.

43. The mass of the bundle is $m = (449 \text{ N})/(9.80 \text{ m/s}^2) = 45.8$ kg and we choose $+y$ upward.

(a) Newton's second law, applied to the bundle, leads to

$$T - mg = ma \Rightarrow a = \frac{387 \text{ N} - 449 \text{ N}}{45.8 \text{ kg}}$$

which yields $a = -1.4$ m/s^2 (or $|a| = 1.4$ m/s^2) for the acceleration. The minus sign in the result indicates the acceleration vector points down. Any downward acceleration of magnitude greater than this is also acceptable (since that would lead to even smaller values of tension).

(b) We use Eq. 2-16 (with Δx replaced by $\Delta y = -6.1$ m). We assume $v_0 = 0$.

$$|v| = \sqrt{2a\Delta y} = \sqrt{2(-1.35 \text{ m/s}^2)(-6.1 \text{ m})} = 4.1 \text{ m/s}.$$

For downward accelerations greater than 1.4 m/s^2, the speeds at impact will be larger than 4.1 m/s.

44. With a_{ce} meaning "the acceleration of the coin relative to the elevator" and a_{eg} meaning "the acceleration of the elevator relative to the ground", we have

$$a_{ce} + a_{eg} = a_{cg} \Rightarrow -8.00 \text{ m/s}^2 + a_{eg} = -9.80 \text{ m/s}^2$$

which leads to $a_{eg} = -1.80$ m/s^2. We have chosen upward as the positive y direction. Then Newton's second law (in the "ground" reference frame) yields $T - mg = m a_{eg}$, or

$$T = mg + m a_{eg} = m(g + a_{eg}) = (2000 \text{ kg})(8.00 \text{ m/s}^2) = 16.0 \text{ kN}.$$

45. (a) The links are numbered from bottom to top. The forces on the bottom link are the force of gravity $m\vec{g}$, downward, and the force $\vec{F}_{2\text{on}1}$ of link 2, upward. Take the positive direction to be upward. Then Newton's second law for this link is $F_{2\text{on}1} - mg = ma$. Thus,

$$F_{2\text{on}1} = m(a + g) = (0.100 \text{ kg})(2.50 \text{ m/s}^2 + 9.80 \text{ m/s}^2) = 1.23 \text{ N}.$$

(b) The forces on the second link are the force of gravity $m\vec{g}$, downward, the force $\vec{F}_{1\text{on}2}$ of link 1, downward, and the force $\vec{F}_{3\text{on}2}$ of link 3, upward. According to Newton's third law $\vec{F}_{1\text{on}2}$ has the same magnitude as $\vec{F}_{2\text{on}1}$. Newton's second law for the second link is $F_{3\text{on}2} - F_{1\text{on}2} - mg = ma$, so

$$F_{3\text{on}2} = m(a + g) + F_{1\text{on}2} = (0.100 \text{ kg})(2.50 \text{ m/s}^2 + 9.80 \text{ m/s}^2) + 1.23 \text{ N} = 2.46 \text{ N}.$$

(c) Newton's second for link 3 is $F_{4\text{on}3} - F_{2\text{on}3} - mg = ma$, so

$$F_{4\text{on}3} = m(a + g) + F_{2\text{on}3} = (0.100 \text{ N})(2.50 \text{ m/s}^2 + 9.80 \text{ m/s}^2) + 2.46 \text{ N} = 3.69 \text{ N},$$

where Newton's third law implies $F_{2\text{on}3} = F_{3\text{on}2}$ (since these are magnitudes of the force vectors).

(d) Newton's second law for link 4 is $F_{5\text{on}4} - F_{3\text{on}4} - mg = ma$, so

$$F_{5\text{on}4} = m(a + g) + F_{3\text{on}4} = (0.100 \text{ kg})(2.50 \text{ m/s}^2 + 9.80 \text{ m/s}^2) + 3.69 \text{ N} = 4.92 \text{ N},$$

where Newton's third law implies $F_{3\text{on}4} = F_{4\text{on}3}$.

(e) Newton's second law for the top link is $F - F_{4\text{on}5} - mg = ma$, so

$$F = m(a + g) + F_{4\text{on}5} = (0.100 \text{ kg})(2.50 \text{ m/s}^2 + 9.80 \text{ m/s}^2) + 4.92 \text{ N} = 6.15 \text{ N},$$

where $F_{4\text{on}5} = F_{5\text{on}4}$ by Newton's third law.

(f) Each link has the same mass and the same acceleration, so the same net force acts on each of them:

$$F_{\text{net}} = ma = (0.100 \text{ kg})(2.50 \text{ m/s}^2) = 0.250 \text{ N}.$$

46. Applying Newton's second law to cab B (of mass m) we have $a = \dfrac{T}{m} - g = 4.89 \text{ m/s}^2$. Next, we apply it to the box (of mass m_b) to find the normal force:

$$F_N = m_b(g + a) = 176 \text{ N}.$$

47. The free-body diagram (not to scale) for the block is shown below. $\vec{F}_N$ is the normal force exerted by the floor and $m\vec{g}$ is the force of gravity.

(a) The x component of Newton's second law is $F\cos\theta = ma$, where m is the mass of the block and a is the x component of its acceleration. We obtain

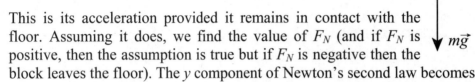

$$a = \frac{F\cos\theta}{m} = \frac{(12.0\,\text{N})\cos 25.0°}{5.00\,\text{kg}} = 2.18\,\text{m/s}^2.$$

This is its acceleration provided it remains in contact with the floor. Assuming it does, we find the value of F_N (and if F_N is positive, then the assumption is true but if F_N is negative then the block leaves the floor). The y component of Newton's second law becomes

$$F_N + F\sin\theta - mg = 0,$$

so

$$F_N = mg - F\sin\theta = (5.00\,\text{kg})(9.80\,\text{m/s}^2) - (12.0\,\text{N})\sin 25.0° = 43.9\,\text{N}.$$

Hence the block remains on the floor and its acceleration is $a = 2.18\,\text{m/s}^2$.

(b) If F is the minimum force for which the block leaves the floor, then $F_N = 0$ and the y component of the acceleration vanishes. The y component of the second law becomes

$$F\sin\theta - mg = 0 \quad\Rightarrow\quad F = \frac{mg}{\sin\theta} = \frac{(5.00\,\text{kg})(9.80\,\text{m/s}^2)}{\sin 25.0°} = 116\,\text{N}.$$

(c) The acceleration is still in the x direction and is still given by the equation developed in part (a):

$$a = \frac{F\cos\theta}{m} = \frac{(116\,\text{N})\cos 25.0°}{5.00\,\text{kg}} = 21.0\,\text{m/s}^2.$$

48. The direction of motion (the direction of the barge's acceleration) is $+\hat{i}$, and $+\hat{j}$ is chosen so that the pull $\vec{F}_h$ from the horse is in the first quadrant. The components of the unknown force of the water are denoted simply F_x and F_y.

(a) Newton's second law applied to the barge, in the x and y directions, leads to

$$(7900\,\text{N})\cos 18° + F_x = ma$$
$$(7900\,\text{N})\sin 18° + F_y = 0$$

respectively. Plugging in $a = 0.12$ m/s^2 and $m = 9500$ kg, we obtain $F_x = -6.4 \times 10^3$ N and $F_y = -2.4 \times 10^3$ N. The magnitude of the force of the water is therefore

$$F_{\text{water}} = \sqrt{F_x^2 + F_y^2} = 6.8 \times 10^3 \text{ N}.$$

(b) Its angle measured from $+\hat{i}$ is either

$$\tan^{-1}\left(\frac{F_y}{F_x}\right) = +21° \text{ or } 201°.$$

The signs of the components indicate the latter is correct, so $\vec{F}_{\text{water}}$ is at 201° measured counterclockwise from the line of motion (+x axis).

49. Using Eq. 4-26, the launch speed of the projectile is

$$v_0 = \sqrt{\frac{gR}{\sin 2\theta}} = \sqrt{\frac{(9.8 \text{ m/s}^2)(69 \text{ m})}{\sin 2(53°)}} = 26.52 \text{ m/s}.$$

The horizontal and vertical components of the speed are

$$v_x = v_0 \cos\theta = (26.52 \text{ m/s})\cos 53° = 15.96 \text{ m/s}$$
$$v_y = v_0 \sin\theta = (26.52 \text{ m/s})\sin 53° = 21.18 \text{ m/s}.$$

Since the acceleration is constant, we can use Eq. 2-16 to analyze the motion. The component of the acceleration in the horizontal direction is

$$a_x = \frac{v_x^2}{2x} = \frac{(15.96 \text{ m/s})^2}{2(5.2 \text{ m})\cos 53°} = 40.7 \text{ m/s}^2,$$

and the force component is $F_x = ma_x = (85 \text{ kg})(40.7 \text{ m/s}^2) = 3460$ N. Similarly, in the vertical direction, we have

$$a_y = \frac{v_y^2}{2y} = \frac{(21.18 \text{ m/s})^2}{2(5.2 \text{ m})\sin 53°} = 54.0 \text{ m/s}^2.$$

and the force component is

$$F_y = ma_y + mg = (85 \text{ kg})(54.0 \text{ m/s}^2 + 9.80 \text{ m/s}^2) = 5424 \text{ N}.$$

Thus, the magnitude of the force is

$$F = \sqrt{F_x^2 + F_y^2} = \sqrt{(3460 \text{ N})^2 + (5424 \text{ N})^2} = 6434 \text{ N} \approx 6.4 \times 10^3 \text{ N},$$

to two significant figures.

50. First, we consider all the penguins (1 through 4, counting left to right) as one system, to which we apply Newton's second law:

$$T_4 = (m_1 + m_2 + m_3 + m_4)a \implies 222\text{N} = (12\text{kg} + m_2 + 15\text{kg} + 20\text{kg})a.$$

Second, we consider penguins 3 and 4 as one system, for which we have

$$T_4 - T_2 = (m_3 + m_4)a$$
$$111\text{N} = (15 \text{ kg} + 20\text{kg})a \implies a = 3.2 \text{ m/s}^2.$$

Substituting the value, we obtain $m_2 = 23$ kg.

51. We apply Newton's second law first to the three blocks as a single system and then to the individual blocks. The +x direction is to the right in Fig. 5-49.

(a) With $m_{\text{sys}} = m_1 + m_2 + m_3 = 67.0$ kg, we apply Eq. 5-2 to the x motion of the system – in which case, there is only one force $\vec{T}_3 = +T_3 \, \hat{\text{i}}$. Therefore,

$$T_3 = m_{\text{sys}} a \implies 65.0 \text{N} = (67.0\text{kg})a$$

which yields $a = 0.970$ m/s² for the system (and for each of the blocks individually).

(b) Applying Eq. 5-2 to block 1, we find

$$T_1 = m_1 a = (12.0\text{kg})(0.970\text{m/s}^2) = 11.6\text{N}.$$

(c) In order to find T_2, we can either analyze the forces on block 3 or we can treat blocks 1 and 2 as a system and examine its forces. We choose the latter.

$$T_2 = (m_1 + m_2)a = (12.0 \text{ kg} + 24.0 \text{ kg})(0.970 \text{ m/s}^2) = 34.9 \text{ N}.$$

52. Both situations involve the same applied force and the same total mass, so the accelerations must be the same in both figures.

(a) The (direct) force causing B to have this acceleration in the first figure is twice as big as the (direct) force causing A to have that acceleration. Therefore, B has the twice the mass of A. Since their total is given as 12.0 kg then B has a mass of $m_B = 8.00$ kg and A has mass $m_A = 4.00$ kg. Considering the first figure, (20.0 N)/(8.00 kg) = 2.50 m/s². Of course, the same result comes from considering the second figure ((10.0 N)/(4.00 kg) = 2.50 m/s²).

(b) $F_a = (12.0 \text{ kg})(2.50 \text{ m/s}^2) = 30.0 \text{ N}$

53. The free-body diagrams for part (a) are shown below. $\vec{F}$ is the applied force and $\vec{f}$ is the force exerted by block 1 on block 2. We note that $\vec{F}$ is applied directly to block 1 and that block 2 exerts the force $-\vec{f}$ on block 1 (taking Newton's third law into account).

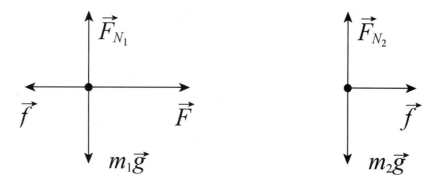

(a) Newton's second law for block 1 is $F - f = m_1 a$, where a is the acceleration. The second law for block 2 is $f = m_2 a$. Since the blocks move together they have the same acceleration and the same symbol is used in both equations. From the second equation we obtain the expression $a = f/m_2$, which we substitute into the first equation to get $F - f = m_1 f/m_2$. Therefore,

$$f = \frac{Fm_2}{m_1 + m_2} = \frac{(3.2 \text{ N})(1.2 \text{ kg})}{2.3 \text{ kg} + 1.2 \text{ kg}} = 1.1 \text{ N}.$$

(b) If $\vec{F}$ is applied to block 2 instead of block 1 (and in the opposite direction), the force of contact between the blocks is

$$f = \frac{Fm_1}{m_1 + m_2} = \frac{(3.2 \text{ N})(2.3 \text{ kg})}{2.3 \text{ kg} + 1.2 \text{ kg}} = 2.1 \text{ N}.$$

(c) We note that the acceleration of the blocks is the same in the two cases. In part (a), the force f is the only horizontal force on the block of mass m_2 and in part (b) f is the only horizontal force on the block with $m_1 > m_2$. Since $f = m_2 a$ in part (a) and $f = m_1 a$ in part (b), then for the accelerations to be the same, f must be larger in part (b).

54. (a) The net force on the *system* (of total mass $M = 80.0$ kg) is the force of gravity acting on the total overhanging mass ($m_{BC} = 50.0$ kg). The magnitude of the acceleration is therefore $a = (m_{BC}\, g)/M = 6.125$ m/s². Next we apply Newton's second law to block C itself (choosing *down* as the +y direction) and obtain

$$m_C g - T_{BC} = m_C a.$$

This leads to $T_{BC} = 36.8$ N.

(b) We use Eq. 2-15 (choosing *rightward* as the +x direction): $\Delta x = 0 + \frac{1}{2}at^2 = 0.191$ m.

55. The free-body diagrams for m_1 and m_2 are shown in the figures below. The only forces on the blocks are the upward tension $\vec{T}$ and the downward gravitational forces $\vec{F}_1 = m_1 g$ and $\vec{F}_2 = m_2 g$. Applying Newton's second law, we obtain:

$$T - m_1 g = m_1 a$$

$$m_2 g - T = m_2 a$$

which can be solved to yield

$$a = \left(\frac{m_2 - m_1}{m_2 + m_1}\right)g$$

Substituting the result back, we have

$$T = \left(\frac{2m_1 m_2}{m_1 + m_2}\right)g$$

(a) With $m_1 = 1.3$ kg and $m_2 = 2.8$ kg, the acceleration becomes

$$a = \left(\frac{2.80 \text{ kg} - 1.30 \text{ kg}}{2.80 \text{ kg} + 1.30 \text{ kg}}\right)(9.80 \text{ m/s}^2) = 3.59 \text{ m/s}^2.$$

(b) Similarly, the tension in the cord is

$$T = \frac{2(1.30 \text{ kg})(2.80 \text{ kg})}{1.30 \text{ kg} + 2.80 \text{ kg}}(9.80 \text{ m/s}^2) = 17.4 \text{ N}.$$

56. To solve the problem, we note that the acceleration along the slanted path depends on only the force components along the path, not the components perpendicular to the path. (a) From the free-body diagram shown, we see that the net force on the putting shot along the +x-axis is

$$F_{\text{net},x} = F - mg \sin\theta = 380.0 \text{ N} - (7.260 \text{ kg})(9.80 \text{ m/s}^2)\sin 30° = 344.4 \text{ N},$$

which in turn gives

$$a_x = F_{net,x}/m = (344.4 \text{ N})/(7.260 \text{ kg}) = 47.44 \text{ m/s}^2.$$

Using Eq. 2-16 for constant-acceleration motion, the speed of the shot at the end of the acceleration phase is

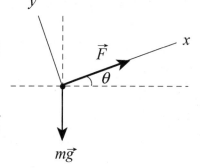

$$v = \sqrt{v_0^2 + 2a_x \Delta x} = \sqrt{(2.500 \text{ m/s})^2 + 2(47.44 \text{ m/s}^2)(1.650 \text{ m})}$$
$$= 12.76 \text{ m/s}.$$

(b) If $\theta = 42°$, then

$$a_x = \frac{F_{net,x}}{m} = \frac{F - mg \sin\theta}{m} = \frac{380.0 \text{ N} - (7.260 \text{ kg})(9.80 \text{ m/s}^2)\sin 42.00°}{7.260 \text{ kg}} = 45.78 \text{ m/s}^2,$$

and the final (launch) speed is

$$v = \sqrt{v_0^2 + 2a_x \Delta x} = \sqrt{(2.500 \text{ m/s})^2 + 2(45.78 \text{ m/s}^2)(1.650 \text{ m})} = 12.54 \text{ m/s}.$$

(c) The decrease in launch speed when changing the angle from 30.00° to 42.00° is

$$\frac{12.76 \text{ m/s} - 12.54 \text{ m/s}}{12.76 \text{ m/s}} = 0.0169 = 16.9\%.$$

57. We take $+y$ to be up for both the monkey and the package.

(a) The force the monkey pulls downward on the rope has magnitude F. According to Newton's third law, the rope pulls upward on the monkey with a force of the same magnitude, so Newton's second law for forces acting on the monkey leads to

$$F - m_m g = m_m a_m,$$

where m_m is the mass of the monkey and a_m is its acceleration. Since the rope is massless $F = T$ is the tension in the rope. The rope pulls upward on the package with a force of magnitude F, so Newton's second law for the package is

$$F + F_N - m_p g = m_p a_p,$$

where m_p is the mass of the package, a_p is its acceleration, and F_N is the normal force exerted by the ground on it. Now, if F is the minimum force required to lift the package, then $F_N = 0$ and $a_p = 0$. According to the second law equation for the package, this means $F = m_p g$. Substituting $m_p g$ for F in the equation for the monkey, we solve for a_m:

$$a_m = \frac{F - m_m g}{m_m} = \frac{(m_p - m_m)g}{m_m} = \frac{(15 \text{ kg} - 10 \text{ kg})(9.8 \text{ m/s}^2)}{10 \text{ kg}} = 4.9 \text{ m/s}^2.$$

(b) As discussed, Newton's second law leads to $F - m_p g = m_p a_p$ for the package and $F - m_m g = m_m a_m$ for the monkey. If the acceleration of the package is downward, then the acceleration of the monkey is upward, so $a_m = -a_p$. Solving the first equation for F

$$F = m_p(g + a_p) = m_p(g - a_m)$$

and substituting this result into the second equation, we solve for a_m:

$$a_m = \frac{(m_p - m_m)g}{m_p + m_m} = \frac{(15 \text{ kg} - 10 \text{ kg})(9.8 \text{ m/s}^2)}{15 \text{ kg} + 10 \text{ kg}} = 2.0 \text{ m/s}^2.$$

(c) The result is positive, indicating that the acceleration of the monkey is upward.

(d) Solving the second law equation for the package, we obtain

$$F = m_p(g - a_m) = (15 \text{ kg})(9.8 \text{ m/s}^2 - 2.0 \text{ m/s}^2) = 120 \text{ N}.$$

58. Referring to Fig. 5-10(c) is helpful. In this case, viewing the man-rope-sandbag as a system means that we should be careful to choose a consistent positive direction of motion (though there are other ways to proceed – say, starting with individual application of Newton's law to each mass). We take *down* as positive for the man's motion and *up* as positive for the sandbag's motion and, without ambiguity, denote their acceleration as *a*. The net force on the system is the different between the weight of the man and that of the sandbag. The system mass is $m_{sys} = 85 \text{ kg} + 65 \text{ kg} = 150 \text{ kg}$. Thus, Eq. 5-1 leads to

$$(85 \text{ kg})(9.8 \text{ m/s}^2) - (65 \text{ kg})(9.8 \text{ m/s}^2) = m_{sys} a$$

which yields $a = 1.3 \text{ m/s}^2$. Since the system starts from rest, Eq. 2-16 determines the speed (after traveling $\Delta y = 10 \text{ m}$) as follows:

$$v = \sqrt{2a\Delta y} = \sqrt{2(1.3 \text{ m/s}^2)(10 \text{ m})} = 5.1 \text{ m/s}.$$

59. The free-body diagram for each block is shown below. T is the tension in the cord and $\theta = 30°$ is the angle of the incline. For block 1, we take the $+x$ direction to be up the incline and the $+y$ direction to be in the direction of the normal force $\vec{F}_N$ that the plane

exerts on the block. For block 2, we take the +y direction to be down. In this way, the accelerations of the two blocks can be represented by the same symbol a, without ambiguity. Applying Newton's second law to the x and y axes for block 1 and to the y axis of block 2, we obtain

$$T - m_1 g \sin \theta = m_1 a$$
$$F_N - m_1 g \cos \theta = 0$$
$$m_2 g - T = m_2 a$$

respectively. The first and third of these equations provide a simultaneous set for obtaining values of a and T. The second equation is not needed in this problem, since the normal force is neither asked for nor is it needed as part of some further computation (such as can occur in formulas for friction).

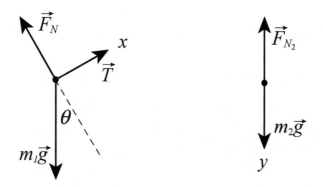

(a) We add the first and third equations above:

$$m_2 g - m_1 g \sin \theta = m_1 a + m_2 a.$$

Consequently, we find

$$a = \frac{(m_2 - m_1 \sin \theta) g}{m_1 + m_2} = \frac{[2.30 \text{ kg} - (3.70 \text{ kg}) \sin 30.0°](9.80 \text{ m/s}^2)}{3.70 \text{ kg} + 2.30 \text{ kg}} = 0.735 \text{ m/s}^2.$$

(b) The result for a is positive, indicating that the acceleration of block 1 is indeed up the incline and that the acceleration of block 2 is vertically down.

(c) The tension in the cord is

$$T = m_1 a + m_1 g \sin \theta = (3.70 \text{ kg})(0.735 \text{ m/s}^2) + (3.70 \text{ kg})(9.80 \text{ m/s}^2) \sin 30.0° = 20.8 \text{ N}.$$

60. The motion of the man-and-chair is positive if upward.

(a) When the man is grasping the rope, pulling with a force equal to the tension T in the rope, the total upward force on the man-and-chair due its two contact points with the rope is $2T$. Thus, Newton's second law leads to

$$2T - mg = ma$$

so that when $a = 0$, the tension is $T = 466$ N.

(b) When $a = +1.30$ m/s^2 the equation in part (a) predicts that the tension will be $T = 527$ N.

(c) When the man is not holding the rope (instead, the co-worker attached to the ground is pulling on the rope with a force equal to the tension T in it), there is only one contact point between the rope and the man-and-chair, and Newton's second law now leads to

$$T - mg = ma$$

so that when $a = 0$, the tension is $T = 931$ N.

(d) When $a = +1.30$ m/s^2, the equation in (c) yields $T = 1.05 \times 10^3$ N.

(e) The rope comes into contact (pulling down in each case) at the left edge and the right edge of the pulley, producing a total downward force of magnitude $2T$ on the ceiling. Thus, in part (a) this gives $2T = 931$ N.

(f) In part (b) the downward force on the ceiling has magnitude $2T = 1.05 \times 10^3$ N.

(g) In part (c) the downward force on the ceiling has magnitude $2T = 1.86 \times 10^3$ N.

(h) In part (d) the downward force on the ceiling has magnitude $2T = 2.11 \times 10^3$ N.

61. The forces on the balloon are the force of gravity $m\vec{g}$ (down) and the force of the air $\vec{F}_a$ (up). We take the $+y$ to be up, and use a to mean the *magnitude* of the acceleration (which is not its usual use in this chapter). When the mass is M (before the ballast is thrown out) the acceleration is downward and Newton's second law is

$$F_a - Mg = -Ma.$$

After the ballast is thrown out, the mass is $M - m$ (where m is the mass of the ballast) and the acceleration is upward. Newton's second law leads to

$$F_a - (M - m)g = (M - m)a.$$

The previous equation gives $F_a = M(g - a)$, and this plugs into the new equation to give

$$M(g - a) - (M - m)g = (M - m)a \quad \Rightarrow \quad m = \frac{2Ma}{g + a}.$$

62. The horizontal component of the acceleration is determined by the net horizontal force.

(a) If the rate of change of the angle is

$$\frac{d\theta}{dt} = (2.00\times 10^{-2})°/s = (2.00\times 10^{-2})°/s \cdot \left(\frac{\pi \text{ rad}}{180°}\right) = 3.49\times 10^{-4} \text{rad/s},$$

then, using $F_x = F\cos\theta$, we find the rate of change of acceleration to be

$$\frac{da_x}{dt} = \frac{d}{dt}\left(\frac{F\cos\theta}{m}\right) = -\frac{F\sin\theta}{m}\frac{d\theta}{dt} = -\frac{(20.0 \text{ N})\sin 25.0°}{5.00 \text{ kg}}(3.49\times 10^{-4}\text{ rad/s})$$
$$= -5.90\times 10^{-4} \text{ m/s}^3.$$

(b) If the rate of change of the angle is

$$\frac{d\theta}{dt} = -(2.00\times 10^{-2})°/s = -(2.00\times 10^{-2})°/s \cdot \left(\frac{\pi \text{ rad}}{180°}\right) = -3.49\times 10^{-4} \text{rad/s},$$

then the rate of change of acceleration would be

$$\frac{da_x}{dt} = \frac{d}{dt}\left(\frac{F\cos\theta}{m}\right) = -\frac{F\sin\theta}{m}\frac{d\theta}{dt} = -\frac{(20.0 \text{ N})\sin 25.0°}{5.00 \text{ kg}}(-3.49\times 10^{-4}\text{ rad/s})$$
$$= +5.90\times 10^{-4} \text{ m/s}^3.$$

63. The free-body diagrams for m_1 and m_2 are shown in the figures below. The only forces on the blocks are the upward tension $\vec{T}$ and the downward gravitational forces $\vec{F}_1 = m_1 g$ and $\vec{F}_2 = m_2 g$. Applying Newton's second law, we obtain:

$$T - m_1 g = m_1 a$$
$$m_2 g - T = m_2 a$$

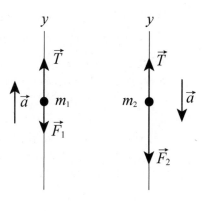

which can be solved to give

$$a = \left(\frac{m_2 - m_1}{m_2 + m_1}\right)g$$

(a) At $t = 0$, $m_{10} = 1.30$ kg. With $dm_1/dt = -0.200$ kg/s, we find the rate of change of acceleration to be

204   CHAPTER 5

$$\frac{da}{dt} = \frac{da}{dm_1}\frac{dm_1}{dt} = -\frac{2m_2 g}{(m_2+m_{10})^2}\frac{dm_1}{dt} = -\frac{2(2.80\text{ kg})(9.80\text{ m/s}^2)}{(2.80\text{ kg}+1.30\text{ kg})^2}(-0.200\text{ kg/s}) = 0.653\text{ m/s}^3.$$

(b) At $t = 3.00$ s, $m_1 = m_{10} + (dm_1/dt)t = 1.30\text{ kg} + (-0.200\text{ kg/s})(3.00\text{ s}) = 0.700\text{ kg}$, and the rate of change of acceleration is

$$\frac{da}{dt} = \frac{da}{dm_1}\frac{dm_1}{dt} = -\frac{2m_2 g}{(m_2+m_1)^2}\frac{dm_1}{dt} = -\frac{2(2.80\text{ kg})(9.80\text{ m/s}^2)}{(2.80\text{ kg}+0.700\text{ kg})^2}(-0.200\text{ kg/s}) = 0.896\text{ m/s}^3.$$

(c) The acceleration reaches its maximum value when

$$0 = m_1 = m_{10} + (dm_1/dt)t = 1.30\text{ kg} + (-0.200\text{ kg/s})t,$$

or $t = 6.50$ s.

64. We first use Eq. 4-26 to solve for the launch speed of the shot:

$$y - y_0 = (\tan\theta)x - \frac{gx^2}{2(v'\cos\theta)^2}.$$

With $\theta = 34.10°$, $y_0 = 2.11$ m and $(x, y) = (15.90\text{ m}, 0)$, we find the launch speed to be $v' = 11.85$ m/s. During this phase, the acceleration is

$$a = \frac{v'^2 - v_0^2}{2L} = \frac{(11.85\text{ m/s})^2 - (2.50\text{ m/s})^2}{2(1.65\text{ m})} = 40.63\text{ m/s}^2.$$

Since the acceleration along the slanted path depends on only the force components along the path, not the components perpendicular to the path, the average force on the shot during the acceleration phase is

$$F = m(a + g\sin\theta) = (7.260\text{ kg})\left[40.63\text{ m/s}^2 + (9.80\text{ m/s}^2)\sin 34.10°\right] = 334.8\text{ N}.$$

65. First we analyze the entire *system* with "clockwise" motion considered positive (that is, downward is positive for block C, rightward is positive for block B, and upward is positive for block A): $m_C g - m_A g = Ma$ (where $M =$ mass of the *system* $= 24.0$ kg). This yields an acceleration of

$$a = g(m_C - m_A)/M = 1.63\text{ m/s}^2.$$

Next we analyze the forces just on block C: $m_C g - T = m_C a$. Thus the tension is

$$T = m_C g(2m_A + m_B)/M = 81.7\text{ N}.$$

66. The $+x$ direction for $m_2=1.0$ kg is "downhill" and the $+x$ direction for $m_1=3.0$ kg is rightward; thus, they accelerate with the same sign.

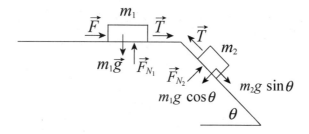

(a) We apply Newton's second law to the x axis of each box:

$$m_2 g \sin\theta - T = m_2 a$$
$$F + T = m_1 a$$

Adding the two equations allows us to solve for the acceleration:

$$a = \frac{m_2 g \sin\theta + F}{m_1 + m_2}$$

With $F = 2.3$ N and $\theta = 30°$, we have $a = 1.8$ m/s². We plug back and find $T = 3.1$ N.

(b) We consider the "critical" case where the F has reached the *max* value, causing the tension to vanish. The first of the equations in part (a) shows that $a = g \sin 30°$ in this case; thus, $a = 4.9$ m/s². This implies (along with $T = 0$ in the second equation in part (a)) that

$$F = (3.0 \text{ kg})(4.9 \text{ m/s}^2) = 14.7 \text{ N} \approx 15 \text{ N}$$

in the critical case.

67. (a) The acceleration (which equals F/m in this problem) is the derivative of the velocity. Thus, the velocity is the integral of F/m, so we find the "area" in the graph (15 units) and divide by the mass (3) to obtain $v - v_0 = 15/3 = 5$. Since $v_0 = 3.0$ m/s, then $v = 8.0$ m/s.

(b) Our positive answer in part (a) implies $\vec{v}$ points in the $+x$ direction.

68. The free-body diagram is shown on the right. Newton's second law for the mass m for the x direction leads to

$$T_1 - T_2 - mg \sin\theta = ma$$

which gives the difference in the tension in the pull cable:

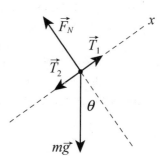

$$T_1 - T_2 = m(g\sin\theta + a) = (2800 \text{ kg})\left[(9.8 \text{ m/s}^2)\sin 35° + 0.81 \text{ m/s}^2\right]$$
$$= 1.8 \times 10^4 \text{ N}.$$

69. (a) We quote our answers to many figures – probably more than are truly "significant." Here (7682 L)("1.77 kg/L") = 13597 kg. The quotation marks around the 1.77 are due to the fact that this was believed (by the flight crew) to be a legitimate conversion factor (it is not).

(b) The amount they felt should be added was 22300 kg – 13597 kg = 87083 kg, which they believed to be equivalent to (87083 kg)/("1.77 kg/L") = 4917 L.

(c) Rounding to 4 figures as instructed, the conversion factor is 1.77 lb/L → 0.8034 kg/L, so the amount on board was (7682 L)(0.8034 kg/L) = 6172 kg.

(d) The implication is that what as needed was 22300 kg – 6172 kg = 16128 kg, so the request should have been for (16128 kg)/(0.8034 kg/L) = 20075 L.

(e) The percentage of the required fuel was

$$\frac{7682 \text{ L (on board)} + 4917 \text{ L (added)}}{(22300 \text{ kg required}) /(0.8034 \text{ kg/L})} = 45\%.$$

70. We are only concerned with horizontal forces in this problem (gravity plays no direct role). Without loss of generality, we take one of the forces along the +x direction and the other at 80° (measured counterclockwise from the x axis). This calculation is efficiently implemented on a vector capable calculator in polar mode, as follows (using magnitude-angle notation, with angles understood to be in degrees):

$$\vec{F}_{net} = (20 \angle 0) + (35 \angle 80) = (43 \angle 53) \Rightarrow |\vec{F}_{net}| = 43 \text{ N}.$$

Therefore, the mass is $m = (43 \text{ N})/(20 \text{ m/s}^2) = 2.2$ kg.

71. The goal is to arrive at the least magnitude of $\vec{F}_{net}$, and as long as the magnitudes of $\vec{F}_2$ and $\vec{F}_3$ are (in total) less than or equal to $|\vec{F}_1|$ then we should orient them opposite to the direction of $\vec{F}_1$ (which is the +x direction).

(a) We orient both $\vec{F}_2$ and $\vec{F}_3$ in the –x direction. Then, the magnitude of the net force is 50 – 30 – 20 = 0, resulting in zero acceleration for the tire.

(b) We again orient $\vec{F}_2$ and $\vec{F}_3$ in the negative x direction. We obtain an acceleration along the +x axis with magnitude

$$a = \frac{F_1 - F_2 - F_3}{m} = \frac{50\,\text{N} - 30\,\text{N} - 10\,\text{N}}{12\,\text{kg}} = 0.83\,\text{m/s}^2.$$

(c) In this case, the forces $\vec{F}_2$ and $\vec{F}_3$ are collectively strong enough to have y components (one positive and one negative) which cancel each other and still have enough x contributions (in the $-x$ direction) to cancel $\vec{F}_1$. Since $|\vec{F}_2| = |\vec{F}_3|$, we see that the angle above the $-x$ axis to one of them should equal the angle below the $-x$ axis to the other one (we denote this angle θ). We require

$$-50\,\text{N} = F_{2x} + F_{3x} = -(30\,\text{N})\cos\theta - (30\,\text{N})\cos\theta$$

which leads to

$$\theta = \cos^{-1}\left(\frac{50\,\text{N}}{60\,\text{N}}\right) = 34°.$$

72. (a) A small segment of the rope has mass and is pulled down by the gravitational force of the Earth. Equilibrium is reached because neighboring portions of the rope pull up sufficiently on it. Since tension is a force *along* the rope, at least one of the neighboring portions must slope up away from the segment we are considering. Then, the tension has an upward component which means the rope sags.

(b) The only force acting with a horizontal component is the applied force $\vec{F}$. Treating the block and rope as a single object, we write Newton's second law for it: $F = (M + m)a$, where a is the acceleration and the positive direction is taken to be to the right. The acceleration is given by $a = F/(M + m)$.

(c) The force of the rope F_r is the only force with a horizontal component acting on the block. Then Newton's second law for the block gives

$$F_r = Ma = \frac{MF}{M + m}$$

where the expression found above for a has been used.

(d) Treating the block and half the rope as a single object, with mass $M + \tfrac{1}{2}m$, where the horizontal force on it is the tension T_m at the midpoint of the rope, we use Newton's second law:

$$T_m = \left(M + \frac{1}{2}m\right)a = \frac{(M + m/2)F}{(M + m)} = \frac{(2M + m)F}{2(M + m)}.$$

73. Although the full specification of $\vec{F}_{\text{net}} = m\vec{a}$ in this situation involves both x and y axes, only the x-application is needed to find what this particular problem asks for. We

note that $a_y = 0$ so that there is no ambiguity denoting a_x simply as a. We choose $+x$ to the right and $+y$ up. We also note that the x component of the rope's tension (acting on the crate) is

$$F_x = F\cos\theta = (450 \text{ N}) \cos 38° = 355 \text{ N},$$

and the resistive force (pointing in the $-x$ direction) has magnitude $f = 125$ N.

(a) Newton's second law leads to

$$F_x - f = ma \Rightarrow a = \frac{355 \text{ N} - 125 \text{ N}}{310 \text{ kg}} = 0.74 \text{ m/s}^2.$$

(b) In this case, we use Eq. 5-12 to find the mass: $m = W/g = 31.6$ kg. Now, Newton's second law leads to

$$T_x - f = ma \Rightarrow a = \frac{355 \text{ N} - 125 \text{ N}}{31.6 \text{ kg}} = 7.3 \text{ m/s}^2.$$

74. Since the velocity of the particle does not change, it undergoes no acceleration and must therefore be subject to zero net force. Therefore,

$$\vec{F}_{net} = \vec{F}_1 + \vec{F}_2 + \vec{F}_3 = 0.$$

Thus, the third force $\vec{F}_3$ is given by

$$\vec{F}_3 = -\vec{F}_1 - \vec{F}_2 = -\left(2\hat{i} + 3\hat{j} - 2\hat{k}\right)\text{N} - \left(-5\hat{i} + 8\hat{j} - 2\hat{k}\right)\text{N} = \left(3\hat{i} - 11\hat{j} + 4\hat{k}\right)\text{N}.$$

The specific value of the velocity is not used in the computation.

75. (a) Since the performer's weight is (52 kg)(9.8 m/s^2) = 510 N, the rope breaks.

(b) Setting $T = 425$ N in Newton's second law (with $+y$ upward) leads to

$$T - mg = ma \Rightarrow a = \frac{T}{m} - g$$

which yields $|a| = 1.6$ m/s^2.

76. (a) For the 0.50 meter drop in "free-fall", Eq. 2-16 yields a speed of 3.13 m/s. Using this as the "initial speed" for the final motion (over 0.02 meter) during which his motion slows at rate "a", we find the magnitude of his average acceleration from when his feet first touch the patio until the moment his body stops moving is $a = 245$ m/s^2.

(b) We apply Newton's second law: $F_{stop} - mg = ma \Rightarrow F_{stop} = 20.4$ kN.

77. We begin by examining a slightly different problem: similar to this figure but without the string. The motivation is that if (without the string) block A is found to accelerate faster (or exactly as fast) as block B then (returning to the original problem) the tension in the string is trivially zero. In the absence of the string,

$$a_A = F_A/m_A = 3.0 \text{ m/s}^2$$
$$a_B = F_B/m_B = 4.0 \text{ m/s}^2$$

so the trivial case does not occur. We now (with the string) consider the net force on the system: $Ma = F_A + F_B = 36$ N. Since $M = 10$ kg (the total mass of the system) we obtain $a = 3.6$ m/s^2. The two forces on block A are F_A and T (in the same direction), so we have

$$m_A a = F_A + T \quad \Rightarrow \quad T = 2.4 \text{ N}.$$

78. With SI units understood, the net force on the box is

$$\vec{F}_{net} = (3.0 + 14\cos 30° - 11)\hat{i} + (14\sin 30° + 5.0 - 17)\hat{j}$$

which yields $\vec{F}_{net} = (4.1 \text{ N})\hat{i} - (5.0 \text{ N})\hat{j}$.

(a) Newton's second law applied to the $m = 4.0$ kg box leads to

$$\vec{a} = \frac{\vec{F}_{net}}{m} = (1.0 \text{ m/s}^2)\hat{i} - (1.3 \text{ m/s}^2)\hat{j}.$$

(b) The magnitude of $\vec{a}$ is $a = \sqrt{(1.0 \text{ m/s}^2)^2 + (-1.3 \text{ m/s}^2)^2} = 1.6 \text{ m/s}^2$.

(c) Its angle is $\tan^{-1}[(-1.3 \text{ m/s}^2)/(1.0 \text{ m/s}^2)] = -50°$ (that is, 50° measured clockwise from the rightward axis).

79. The "certain force" denoted F is assumed to be the net force on the object when it gives m_1 an acceleration $a_1 = 12$ m/s^2 and when it gives m_2 an acceleration $a_2 = 3.3$ m/s^2. Thus, we substitute $m_1 = F/a_1$ and $m_2 = F/a_2$ in appropriate places during the following manipulations.

(a) Now we seek the acceleration a of an object of mass $m_2 - m_1$ when F is the net force on it. Thus,

$$a = \frac{F}{m_2 - m_1} = \frac{F}{(F/a_2) - (F/a_1)} = \frac{a_1 a_2}{a_1 - a_2}$$

which yields $a = 4.6$ m/s^2.

(b) Similarly for an object of mass $m_2 + m_1$:

$$a = \frac{F}{m_2 + m_1} = \frac{F}{(F/a_2)+(F/a_1)} = \frac{a_1 a_2}{a_1 + a_2}$$

which yields $a = 2.6$ m/s^2.

80. We use the notation g as the acceleration due to gravity near the surface of Callisto, m as the mass of the landing craft, a as the acceleration of the landing craft, and F as the rocket thrust. We take down to be the positive direction. Thus, Newton's second law takes the form $mg - F = ma$. If the thrust is F_1 (= 3260 N), then the acceleration is zero, so $mg - F_1 = 0$. If the thrust is F_2 (= 2200 N), then the acceleration is a_2 (= 0.39 m/s^2), so $mg - F_2 = ma_2$.

(a) The first equation gives the weight of the landing craft: $mg = F_1 = 3260$ N.

(b) The second equation gives the mass:

$$m = \frac{mg - F_2}{a_2} = \frac{3260 \text{ N} - 2200 \text{ N}}{0.39 \text{ m/s}^2} = 2.7 \times 10^3 \text{ kg}.$$

(c) The weight divided by the mass gives the acceleration due to gravity:

$$g = (3260 \text{ N})/(2.7 \times 10^3 \text{ kg}) = 1.2 \text{ m/s}^2.$$

81. From the reading when the elevator was at rest, we know the mass of the object is $m = (65 \text{ N})/(9.8 \text{ m/s}^2) = 6.6$ kg. We choose $+y$ upward and note there are two forces on the object: mg downward and T upward (in the cord that connects it to the balance; T is the reading on the scale by Newton's third law).

(a) "Upward at constant speed" means constant velocity, which means no acceleration. Thus, the situation is just as it was at rest: $T = 65$ N.

(b) The term "deceleration" is used when the acceleration vector points in the direction opposite to the velocity vector. We're told the velocity is upward, so the acceleration vector points downward ($a = -2.4$ m/s^2). Newton's second law gives

$$T - mg = ma \implies T = (6.6 \text{ kg})(9.8 \text{ m/s}^2 - 2.4 \text{ m/s}^2) = 49 \text{ N}.$$

82. We take $+x$ uphill for the $m_2 = 1.0$ kg box and $+x$ rightward for the $m_1 = 3.0$ kg box (so the accelerations of the two boxes have the same magnitude and the same sign). The uphill force on m_2 is F and the downhill forces on it are T and $m_2 g \sin \theta$, where $\theta = 37°$. The only horizontal force on m_1 is the rightward-pointed tension. Applying Newton's second law to each box, we find

$$F - T - m_2 g \sin\theta = m_2 a$$
$$T = m_1 a$$

which can be added to obtain $F - m_2 g \sin\theta = (m_1 + m_2)a$. This yields the acceleration

$$a = \frac{12 \text{ N} - (1.0 \text{ kg})(9.8 \text{ m/s}^2)\sin 37°}{1.0 \text{ kg} + 3.0 \text{ kg}} = 1.53 \text{ m/s}^2.$$

Thus, the tension is $T = m_1 a = (3.0 \text{ kg})(1.53 \text{ m/s}^2) = 4.6 \text{ N}$.

83. We apply Eq. 5-12.

(a) The mass is $m = W/g = (22 \text{ N})/(9.8 \text{ m/s}^2) = 2.2$ kg. At a place where $g = 4.9$ m/s², the mass is still 2.2 kg but the gravitational force is $F_g = mg = (2.2 \text{ kg})(4.0 \text{ m/s}^2) = 11$ N.

(b) As noted, $m = 2.2$ kg.

(c) At a place where $g = 0$ the gravitational force is zero.

(d) The mass is still 2.2 kg.

84. We use $W_p = mg_p$, where W_p is the weight of an object of mass m on the surface of a certain planet p, and g_p is the acceleration of gravity on that planet.

(a) The weight of the space ranger on Earth is

$$W_e = mg_e = (75 \text{ kg})(9.8 \text{ m/s}^2) = 7.4 \times 10^2 \text{ N}.$$

(b) The weight of the space ranger on Mars is

$$W_m = mg_m = (75 \text{ kg})(3.7 \text{ m/s}^2) = 2.8 \times 10^2 \text{ N}.$$

(c) The weight of the space ranger in interplanetary space is zero, where the effects of gravity are negligible.

(d) The mass of the space ranger remains the same, $m=75$ kg, at all the locations.

85. (a) When $\vec{F}_{net} = 3F - mg = 0$, we have

$$F = \frac{1}{3}mg = \frac{1}{3}(1400 \text{ kg})(9.8 \text{ m/s}^2) = 4.6 \times 10^3 \text{ N}$$

for the force exerted by each bolt on the engine.

(b) The force on each bolt now satisfies $3F - mg = ma$, which yields

$$F = \frac{1}{3}m(g+a) = \frac{1}{3}(1400 \text{ kg})(9.8 \text{ m/s}^2 + 2.6 \text{ m/s}^2) = 5.8 \times 10^3 \text{ N}.$$

86. We take the down to be the $+y$ direction.

(a) The first diagram (shown below left) is the free-body diagram for the person and parachute, considered as a single object with a mass of 80 kg + 5.0 kg = 85 kg.

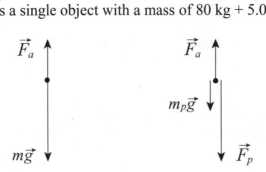

$\vec{F}_a$ is the force of the air on the parachute and $m\vec{g}$ is the force of gravity. Application of Newton's second law produces $mg - F_a = ma$, where a is the acceleration. Solving for F_a we find

$$F_a = m(g-a) = (85 \text{ kg})(9.8 \text{ m/s}^2 - 2.5 \text{ m/s}^2) = 620 \text{ N}.$$

(b) The second diagram (above right) is the free-body diagram for the parachute alone. $\vec{F}_a$ is the force of the air, $m_p\vec{g}$ is the force of gravity, and $\vec{F}_p$ is the force of the person. Now, Newton's second law leads to

$$m_p g + F_p - F_a = m_p a.$$

Solving for F_p, we obtain

$$F_p = m_p(a-g) + F_a = (5.0 \text{ kg})(2.5 \text{ m/s}^2 - 9.8 \text{ m/s}^2) + 620 \text{ N} = 580 \text{ N}.$$

87. (a) Intuition readily leads to the conclusion that the heavier block should be the hanging one, for largest acceleration. The force that "drives" the system into motion is the weight of the hanging block (gravity acting on the block on the table has no effect on the dynamics, so long as we ignore friction). Thus, $m = 4.0$ kg.

The acceleration of the system and the tension in the cord can be readily obtained by solving

$$mg - T = ma$$
$$T = Ma.$$

(b) The acceleration is given by

$$a = \left(\frac{m}{m+M}\right)g = 6.5 \text{ m/s}^2.$$

(c) The tension is

$$T = Ma = \left(\frac{Mm}{m+M}\right)g = 13 \text{ N}.$$

88. We assume the direction of motion is $+x$ and assume the refrigerator starts from rest (so that the speed being discussed is the velocity $\vec{v}$ which results from the process). The only force along the x axis is the x component of the applied force $\vec{F}$.

(a) Since $v_0 = 0$, the combination of Eq. 2-11 and Eq. 5-2 leads simply to

$$F_x = m\left(\frac{v}{t}\right) \Rightarrow v_i = \left(\frac{F\cos\theta_i}{m}\right)t$$

for $i = 1$ or 2 (where we denote $\theta_1 = 0$ and $\theta_2 = \theta$ for the two cases). Hence, we see that the ratio v_2 over v_1 is equal to $\cos\theta$.

(b) Since $v_0 = 0$, the combination of Eq. 2-16 and Eq. 5-2 leads to

$$F_x = m\left(\frac{v^2}{2\Delta x}\right) \Rightarrow v_i = \sqrt{2\left(\frac{F\cos\theta_i}{m}\right)\Delta x}$$

for $i = 1$ or 2 (again, $\theta_1 = 0$ and $\theta_2 = \theta$ is used for the two cases). In this scenario, we see that the ratio v_2 over v_1 is equal to $\sqrt{\cos\theta}$.

89. The mass of the pilot is $m = 735/9.8 = 75$ kg. Denoting the upward force exerted by the spaceship (his seat, presumably) on the pilot as $\vec{F}$ and choosing upward the $+y$ direction, then Newton's second law leads to

$$F - mg_{moon} = ma \Rightarrow F = (75 \text{ kg})(1.6 \text{ m/s}^2 + 1.0 \text{ m/s}^2) = 195 \text{ N}.$$

90. We denote the thrust as T and choose $+y$ upward. Newton's second law leads to

$$T - Mg = Ma \Rightarrow a = \frac{2.6\times10^5 \text{ N}}{1.3\times10^4 \text{ kg}} - 9.8 \text{ m/s}^2 = 10 \text{ m/s}^2.$$

91. (a) The bottom cord is only supporting $m_2 = 4.5$ kg against gravity, so its tension is

$$T_2 = m_2 g = (4.5 \text{ kg})(9.8 \text{ m/s}^2) = 44 \text{ N}.$$

(b) The top cord is supporting a total mass of $m_1 + m_2 = (3.5 \text{ kg} + 4.5 \text{ kg}) = 8.0$ kg against gravity, so the tension there is

$$T_1 = (m_1 + m_2)g = (8.0 \text{ kg})(9.8 \text{ m/s}^2) = 78 \text{ N}.$$

(c) In the second picture, the lowest cord supports a mass of $m_5 = 5.5$ kg against gravity and consequently has a tension of $T_5 = (5.5 \text{ kg})(9.8 \text{ m/s}^2) = 54$ N.

(d) The top cord, we are told, has tension $T_3 = 199$ N which supports a total of $(199 \text{ N})/(9.80 \text{ m/s}^2) = 20.3$ kg, 10.3 kg of which is already accounted for in the figure. Thus, the unknown mass in the middle must be $m_4 = 20.3 \text{ kg} - 10.3 \text{ kg} = 10.0$ kg, and the tension in the cord above it must be enough to support

$$m_4 + m_5 = (10.0 \text{ kg} + 5.50 \text{ kg}) = 15.5 \text{ kg},$$

so $T_4 = (15.5 \text{ kg})(9.80 \text{ m/s}^2) = 152$ N. Another way to analyze this is to examine the forces on m_3; one of the downward forces on it is T_4.

92. (a) With SI units understood, the net force is

$$\vec{F}_{net} = \vec{F}_1 + \vec{F}_2 = (3.0 \text{ N} + (-2.0 \text{ N}))\hat{i} + (4.0 \text{ N} + (-6.0 \text{ N}))\hat{j}$$

which yields $\vec{F}_{net} = (1.0 \text{ N})\hat{i} - (2.0 \text{ N})\hat{j}$.

(b) The magnitude of $\vec{F}_{net}$ is $F_{net} = \sqrt{(1.0 \text{ N})^2 + (-2.0 \text{ N})^2} = 2.2$ N.

(c) The angle of $\vec{F}_{net}$ is

$$\theta = \tan^{-1}\left(\frac{-2.0 \text{ N}}{1.0 \text{ N}}\right) = -63°.$$

(d) The magnitude of $\vec{a}$ is

$$a = F_{net}/m = (2.2 \text{ N})/(1.0 \text{ kg}) = 2.2 \text{ m/s}^2.$$

(e) Since $\vec{F}_{net}$ is equal to $\vec{a}$ multiplied by mass m, which is a positive scalar that cannot affect the direction of the vector it multiplies, $\vec{a}$ has the same angle as the net force, i.e, $\theta = -63°$. In magnitude-angle notation, we may write $\vec{a} = (2.2 \text{ m/s}^2 \angle -63°)$.

93. According to Newton's second law, the magnitude of the force is given by $F = ma$, where a is the magnitude of the acceleration of the neutron. We use kinematics (Table 2-1) to find the acceleration that brings the neutron to rest in a distance d. Assuming the acceleration is constant, then $v^2 = v_0^2 + 2ad$ produces the value of a:

$$a = \frac{(v^2 - v_0^2)}{2d} = \frac{-(1.4 \times 10^7 \text{ m/s})^2}{2(1.0 \times 10^{-14} \text{ m})} = -9.8 \times 10^{27} \text{ m/s}^2.$$

The magnitude of the force is consequently

$$F = ma = (1.67 \times 10^{-27} \text{ kg})(9.8 \times 10^{27} \text{ m/s}^2) = 16 \text{ N}.$$

94. Making separate free-body diagrams for the helicopter and the truck, one finds there are two forces on the truck ($\vec{T}$ upward, caused by the tension, which we'll think of as that of a single cable, and $m\vec{g}$ downward, where m = 4500 kg) and three forces on the helicopter ($\vec{T}$ downward, $\vec{F}_{\text{lift}}$ upward, and $M\vec{g}$ downward, where M = 15000 kg). With $+y$ upward, then $a = +1.4$ m/s² for both the helicopter and the truck.

(a) Newton's law applied to the helicopter and truck separately gives

$$F_{\text{lift}} - T - Mg = Ma$$
$$T - mg = ma$$

which we add together to obtain

$$F_{\text{lift}} - (M+m)g = (M+m)a.$$

From this equation, we find $F_{\text{lift}} = 2.2 \times 10^5$ N.

(b) From the truck equation $T - mg = ma$ we obtain $T = 5.0 \times 10^4$ N.

95. The free-body diagrams is shown on the right. Note that F_{m, r_y} and F_{m, r_x}, respectively, and thought of as the y and x components of the force $\vec{F}_{m, r}$ exerted by the motorcycle on the rider.

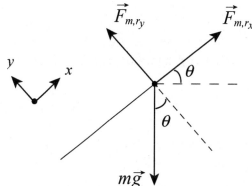

(a) Since the net force equals ma, then the magnitude of the net force on the rider is
(60.0 kg) (3.0 m/s²) = 1.8×10^2 N.

(b) We apply Newton's second law to the x axis:

$$F_{m, r_x} - mg \sin\theta = ma$$

where m = 60.0 kg, a = 3.0 m/s², and θ = 10°. Thus, F_{m, r_x} = 282 N Applying it to the y axis (where there is no acceleration), we have

$$F_{m,r_y} - mg\cos\theta = 0$$

which produces $F_{m,r_y} = 579$ N. Using the Pythagorean theorem, we find

$$\sqrt{F_{m,r_x}^2 + F_{m,r_y}^2} = 644 \text{ N}.$$

Now, the magnitude of the force exerted on the rider by the motorcycle is the same magnitude of force exerted by the rider on the motorcycle, so the answer is 6.4×10^2 N, to two significant figures.

96. We write the length unit light-month, the distance traveled by light in one month, as c·month in this solution.

(a) The magnitude of the required acceleration is given by

$$a = \frac{\Delta v}{\Delta t} = \frac{(0.10)(3.0 \times 10^8 \text{ m/s})}{(3.0 \text{ days})(86400 \text{ s/day})} = 1.2 \times 10^2 \text{ m/s}^2.$$

(b) The acceleration in terms of g is

$$a = \left(\frac{a}{g}\right)g = \left(\frac{1.2 \times 10^2 \text{ m/s}^2}{9.8 \text{ m/s}^2}\right)g = 12g.$$

(c) The force needed is

$$F = ma = (1.20 \times 10^6 \text{ kg})(1.2 \times 10^2 \text{ m/s}^2) = 1.4 \times 10^8 \text{ N}.$$

(d) The spaceship will travel a distance $d = 0.1$ c·month during one month. The time it takes for the spaceship to travel at constant speed for 5.0 light-months is

$$t = \frac{d}{v} = \frac{5.0 \text{ c·months}}{0.1c} = 50 \text{ months} \approx 4.2 \text{ years}.$$

97. The coordinate choices are made in the problem statement.

(a) We write the velocity of the armadillo as $\vec{v} = v_x \hat{i} + v_y \hat{j}$. Since there is no net force exerted on it in the x direction, the x component of the velocity of the armadillo is a constant: $v_x = 5.0$ m/s. In the y direction at $t = 3.0$ s, we have (using Eq. 2-11 with $v_{0y} = 0$)

$$v_y = v_{0y} + a_y t = v_{0y} + \left(\frac{F_y}{m}\right)t = \left(\frac{17\text{ N}}{12\text{ kg}}\right)(3.0\text{ s}) = 4.3\text{ m/s}.$$

Thus, $\vec{v} = (5.0\text{ m/s})\hat{i} + (4.3\text{ m/s})\hat{j}$.

(b) We write the position vector of the armadillo as $\vec{r} = r_x\,\hat{i} + r_y\,\hat{j}$. At $t = 3.0$ s we have $r_x = (5.0\text{ m/s})(3.0\text{ s}) = 15$ m and (using Eq. 2-15 with $v_{0y} = 0$)

$$r_y = v_{0y}t + \frac{1}{2}a_y t^2 = \frac{1}{2}\left(\frac{F_y}{m}\right)t^2 = \frac{1}{2}\left(\frac{17\text{ N}}{12\text{ kg}}\right)(3.0\text{ s})^2 = 6.4\text{ m}.$$

The position vector at $t = 3.0$ s is therefore

$$\vec{r} = (15\text{ m})\hat{i} + (6.4\text{ m})\hat{j}.$$

98. (a) From Newton's second law, the magnitude of the maximum force on the passenger from the floor is given by

$$F_{max} - mg = ma \quad \text{where} \quad a = a_{max} = 2.0\text{ m/s}^2$$

we obtain $F_N = 590$ N for $m = 50$ kg.

(b) The direction is upward.

(c) Again, we use Newton's second law, the magnitude of the minimum force on the passenger from the floor is given by

$$F_{min} - mg = ma \quad \text{where} \quad a = a_{min} = -3.0\text{ m/s}^2.$$

Now, we obtain $F_N = 340$ N.

(d) The direction is upward.

(e) Returning to part (a), we use Newton's third law, and conclude that the force exerted by the passenger on the floor is $|\vec{F}_{PF}| = 590$ N.

(f) The direction is downward.

99. The $+x$ axis is "uphill" for $m_1 = 3.0$ kg and "downhill" for $m_2 = 2.0$ kg (so they both accelerate with the same sign). The x components of the two masses along the x axis are given by $w_{1x} = m_1 g \sin\theta_1$ and $w_{2x} = m_2 g \sin\theta_2$, respectively.

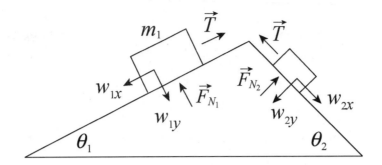

Applying Newton's second law, we obtain

$$T - m_1 g \sin \theta_1 = m_1 a$$
$$m_2 g \sin \theta_2 - T = m_2 a$$

Adding the two equations allows us to solve for the acceleration:

$$a = \left(\frac{m_2 \sin \theta_2 - m_1 \sin \theta_1}{m_2 + m_1} \right) g$$

With $\theta_1 = 30°$ and $\theta_2 = 60°$, we have $a = 0.45$ m/s^2. This value is plugged back into either of the two equations to yield the tension $T = 16$ N.

100. (a) In unit vector notation,

$$m\vec{a} = (-3.76 \text{ N})\hat{i} + (1.37 \text{ N})\hat{j}.$$

Thus, Newton's second law leads to

$$\vec{F}_2 = m\vec{a} - \vec{F}_1 = (-6.26 \text{ N})\hat{i} - (3.23 \text{ N})\hat{j}.$$

(b) The magnitude of $\vec{F}_2$ is $F_2 = \sqrt{(-6.26 \text{ N})^2 + (-3.23 \text{ N})^2} = 7.04$ N.

(c) Since $\vec{F}_2$ is in the third quadrant, the angle is

$$\theta = \tan^{-1}\left(\frac{-3.23 \text{ N}}{-6.26 \text{ N}} \right) = 207°.$$

counterclockwise from positive direction of x axis (or 153° *clockwise* from $+x$).

101. We first analyze the forces on $m_1 = 1.0$ kg.

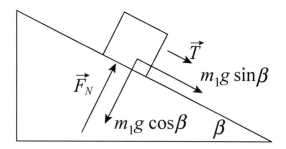

The $+x$ direction is "downhill" (parallel to $\vec{T}$).

With the acceleration (5.5 m/s^2) in the positive x direction for m_1, then Newton's second law, applied to the x axis, becomes

$$T + m_1 g \sin\beta = m_1 \left(5.5 \text{m/s}^2\right)$$

But for $m_2 = 2.0$ kg, using the more familiar vertical y axis (with *up* as the positive direction), we have the acceleration in the negative direction:

$$F + T - m_2 g = m_2 \left(-5.5 \text{m/s}^2\right)$$

where the tension comes in as an upward force (the cord can pull, not push).

(a) From the equation for m_2, with $F = 6.0$ N, we find the tension $T = 2.6$ N.

(b) From the equation for m, using the result from part (a), we obtain the angle $\beta = 17°$.

102. (a) The word "hovering" is taken to imply that the upward (thrust) force is equal in magnitude to the downward (gravitational) force: $mg = 4.9 \times 10^5$ N.

(b) Now the thrust must exceed the answer of part (a) by $ma = 10 \times 10^5$ N, so the thrust must be 1.5×10^6 N.

103. (a) Choosing the direction of motion as $+x$, Eq. 2-11 gives

$$a = \frac{88.5 \text{ km/h} - 0}{6.0 \text{ s}} = 15 \text{ km/h/s}.$$

Converting to SI, this is $a = 4.1$ m/s^2.

(b) With mass $m = 2000/9.8 = 204$ kg, Newton's second law gives $\vec{F} = m\vec{a} = 836$ N in the $+x$ direction.

104. (a) With $v_0 = 0$, Eq. 2-16 leads to

$$a = \frac{v^2}{2\Delta x} = \frac{(6.0 \times 10^6 \text{ m/s})^2}{2(0.015 \text{ m})} = 1.2 \times 10^{15} \text{ m/s}^2.$$

The force responsible for producing this acceleration is

$$F = ma = (9.11 \times 10^{-31} \text{ kg})(1.2 \times 10^{15} \text{ m/s}^2) = 1.1 \times 10^{-15} \text{ N}.$$

(b) The weight is $mg = 8.9 \times 10^{-30}$ N, many orders of magnitude smaller than the result of part (a). As a result, gravity plays a negligible role in most atomic and subatomic processes.

Chapter 6

1. We do not consider the possibility that the bureau might tip, and treat this as a purely horizontal motion problem (with the person's push $\vec{F}$ in the $+x$ direction). Applying Newton's second law to the x and y axes, we obtain

$$F - f_{s,\,max} = ma$$
$$F_N - mg = 0$$

respectively. The second equation yields the normal force $F_N = mg$, whereupon the maximum static friction is found to be (from Eq. 6-1) $f_{s,max} = \mu_s mg$. Thus, the first equation becomes

$$F - \mu_s mg = ma = 0$$

where we have set $a = 0$ to be consistent with the fact that the static friction is still (just barely) able to prevent the bureau from moving.

(a) With $\mu_s = 0.45$ and $m = 45$ kg, the equation above leads to $F = 198$ N. To bring the bureau into a state of motion, the person should push with any force greater than this value. Rounding to two significant figures, we can therefore say the minimum required push is $F = 2.0 \times 10^2$ N.

(b) Replacing $m = 45$ kg with $m = 28$ kg, the reasoning above leads to roughly $F = 1.2 \times 10^2$ N.

2. To maintain the stone's motion, a horizontal force (in the $+x$ direction) is needed that cancels the retarding effect due to kinetic friction. Applying Newton's second to the x and y axes, we obtain

$$F - f_k = ma$$
$$F_N - mg = 0$$

respectively. The second equation yields the normal force $F_N = mg$, so that (using Eq. 6-2) the kinetic friction becomes $f_k = \mu_k mg$. Thus, the first equation becomes

$$F - \mu_k mg = ma = 0$$

where we have set $a = 0$ to be consistent with the idea that the horizontal velocity of the stone should remain constant. With $m = 20$ kg and $\mu_k = 0.80$, we find $F = 1.6 \times 10^2$ N.

3. We denote $\vec{F}$ as the horizontal force of the person exerted on the crate (in the $+x$ direction), $\vec{f}_k$ is the force of kinetic friction (in the $-x$ direction), F_N is the vertical normal force exerted by the floor (in the $+y$ direction), and $m\vec{g}$ is the force of gravity. The magnitude of the force of friction is given by $f_k = \mu_k F_N$ (Eq. 6-2). Applying Newton's second law to the x and y axes, we obtain

$$F - f_k = ma$$
$$F_N - mg = 0$$

respectively.

(a) The second equation yields the normal force $F_N = mg$, so that the friction is

$$f_k = \mu_k mg = (0.35)(55 \text{ kg})(9.8 \text{ m/s}^2) = 1.9 \times 10^2 \text{ N}.$$

(b) The first equation becomes

$$F - \mu_k mg = ma$$

which (with $F = 220$ N) we solve to find

$$a = \frac{F}{m} - \mu_k g = 0.56 \text{ m/s}^2.$$

4. The free-body diagram for the player is shown next. $\vec{F}_N$ is the normal force of the ground on the player, $m\vec{g}$ is the force of gravity, and $\vec{f}$ is the force of friction. The force of friction is related to the normal force by $f = \mu_k F_N$. We use Newton's second law applied to the vertical axis to find the normal force. The vertical component of the acceleration is zero, so we obtain $F_N - mg = 0$; thus, $F_N = mg$. Consequently,

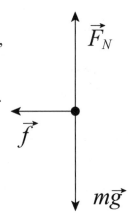

$$\mu_k = \frac{f}{F_N} = \frac{470 \text{ N}}{(79 \text{ kg})(9.8 \text{ m/s}^2)} = 0.61.$$

5. The greatest deceleration (of magnitude a) is provided by the maximum friction force (Eq. 6-1, with $F_N = mg$ in this case). Using Newton's second law, we find

$$a = f_{s,\max}/m = \mu_s g.$$

Eq. 2-16 then gives the shortest distance to stop: $|\Delta x| = v^2/2a = 36$ m. In this calculation, it is important to first convert v to 13 m/s.

6. We first analyze the forces on the pig of mass m. The incline angle is θ.

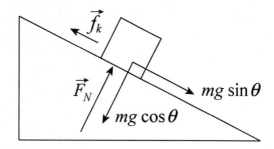

The +x direction is "downhill."

Application of Newton's second law to the x- and y-axes leads to

$$mg \sin \theta - f_k = ma$$
$$F_N - mg \cos \theta = 0.$$

Solving these along with Eq. 6-2 ($f_k = \mu_k F_N$) produces the following result for the pig's downhill acceleration:

$$a = g(\sin \theta - \mu_k \cos \theta).$$

To compute the time to slide from rest through a downhill distance ℓ, we use Eq. 2-15:

$$\ell = v_0 t + \frac{1}{2} at^2 \Rightarrow t = \sqrt{\frac{2\ell}{a}}.$$

We denote the frictionless ($\mu_k = 0$) case with a prime and set up a ratio:

$$\frac{t}{t'} = \frac{\sqrt{2\ell/a}}{\sqrt{2\ell/a'}} = \sqrt{\frac{a'}{a}}$$

which leads us to conclude that if $t/t' = 2$ then $a' = 4a$. Putting in what we found out above about the accelerations, we have

$$g \sin \theta = 4g(\sin \theta - \mu_k \cos \theta).$$

Using $\theta = 35°$, we obtain $\mu_k = 0.53$.

7. We choose $+x$ horizontally rightwards and $+y$ upwards and observe that the 15 N force has components $F_x = F \cos \theta$ and $F_y = -F \sin \theta$.

(a) We apply Newton's second law to the y axis:

$$F_N - F \sin \theta - mg = 0 \Rightarrow F_N = (15 \text{ N}) \sin 40° + (3.5 \text{ kg})(9.8 \text{ m/s}^2) = 44 \text{ N}.$$

With $\mu_k = 0.25$, Eq. 6-2 leads to $f_k = 11$ N.

(b) We apply Newton's second law to the x axis:

$$F\cos\theta - f_k = ma \Rightarrow a = \frac{(15\text{ N})\cos 40° - 11\text{ N}}{3.5\text{ kg}} = 0.14\text{ m/s}^2.$$

Since the result is positive-valued, then the block is accelerating in the $+x$ (rightward) direction.

8. In addition to the forces already shown in Fig. 6-21, a free-body diagram would include an upward normal force $\vec{F}_N$ exerted by the floor on the block, a downward $m\vec{g}$ representing the gravitational pull exerted by Earth, and an assumed-leftward $\vec{f}$ for the kinetic or static friction. We choose $+x$ rightwards and $+y$ upwards. We apply Newton's second law to these axes:

$$F - f = ma$$
$$P + F_N - mg = 0$$

where $F = 6.0$ N and $m = 2.5$ kg is the mass of the block.

(a) In this case, $P = 8.0$ N leads to

$$F_N = (2.5\text{ kg})(9.8\text{ m/s}^2) - 8.0\text{ N} = 16.5\text{ N}.$$

Using Eq. 6-1, this implies $f_{s,\max} = \mu_s F_N = 6.6$ N, which is larger than the 6.0 N rightward force – so the block (which was initially at rest) does not move. Putting $a = 0$ into the first of our equations above yields a static friction force of $f = P = 6.0$ N.

(b) In this case, $P = 10$ N, the normal force is $F_N = (2.5\text{ kg})(9.8\text{ m/s}^2) - 10\text{ N} = 14.5$ N. Using Eq. 6-1, this implies $f_{s,\max} = \mu_s F_N = 5.8$ N, which is less than the 6.0 N rightward force – so the block does move. Hence, we are dealing not with static but with kinetic friction, which Eq. 6-2 reveals to be $f_k = \mu_k F_N = 3.6$ N.

(c) In this last case, $P = 12$ N leads to $F_N = 12.5$ N and thus to $f_{s,\max} = \mu_s F_N = 5.0$ N, which (as expected) is less than the 6.0 N rightward force – so the block moves. The kinetic friction force, then, is $f_k = \mu_k F_N = 3.1$ N.

9. Applying Newton's second law to the horizontal motion, we have $F - \mu_k mg = ma$, where we have used Eq. 6-2, assuming that $F_N = mg$ (which is equivalent to assuming that the vertical force from the broom is negligible). Eq. 2-16 relates the distance traveled and the final speed to the acceleration: $v^2 = 2a\Delta x$. This gives $a = 1.4$ m/s^2. Returning to the force equation, we find (with $F = 25$ N and $m = 3.5$ kg) that $\mu_k = 0.58$.

10. There is no acceleration, so the (upward) static friction forces (there are four of them, one for each thumb and one for each set of opposing fingers) equals the magnitude of the (downward) pull of gravity. Using Eq. 6-1, we have

$$4\mu_s F_N = mg = (79 \text{ kg})(9.8 \text{ m/s}^2)$$

which, with $\mu_s = 0.70$, yields $F_N = 2.8 \times 10^2$ N.

11. We denote the magnitude of 110 N force exerted by the worker on the crate as F. The magnitude of the static frictional force can vary between zero and $f_{s,\max} = \mu_s F_N$.

(a) In this case, application of Newton's second law in the vertical direction yields $F_N = mg$. Thus,

$$f_{s,\max} = \mu_s F_N = \mu_s mg = (0.37)(35\text{kg})(9.8\text{m/s}^2) = 1.3\times 10^2 \text{ N}$$

which is greater than F.

(b) The block, which is initially at rest, stays at rest since $F < f_{s,\max}$. Thus, it does not move.

(c) By applying Newton's second law to the horizontal direction, that the magnitude of the frictional force exerted on the crate is $f_s = 1.1 \times 10^2$ N.

(d) Denoting the upward force exerted by the second worker as F_2, then application of Newton's second law in the vertical direction yields $F_N = mg - F_2$, which leads to

$$f_{s,\max} = \mu_s F_N = \mu_s (mg - F_2).$$

In order to move the crate, F must satisfy the condition $F > f_{s,\max} = \mu_s (mg - F_2)$

or

$$110\text{N} > (0.37)\left[(35\text{kg})(9.8\text{m/s}^2) - F_2\right].$$

The minimum value of F_2 that satisfies this inequality is a value slightly bigger than 45.7 N, so we express our answer as $F_{2,\min} = 46$ N.

(e) In this final case, moving the crate requires a greater horizontal push from the worker than static friction (as computed in part (a)) can resist. Thus, Newton's law in the horizontal direction leads to

$$F + F_2 > f_{s,\max} \quad \Rightarrow \quad 110 \text{ N} + F_2 > 126.9 \text{ N}$$

which leads (after appropriate rounding) to $F_{2, min} = 17$ N.

12. (a) Using the result obtained in Sample Problem 6-2, the maximum angle for which static friction applies is

$$\theta_{max} = \tan^{-1} \mu_s = \tan^{-1} 0.63 \approx 32°.$$

This is greater than the dip angle in the problem, so the block does not slide.

(b) We analyze forces in a manner similar to that shown in Sample Problem 6-3, but with the addition of a downhill force F.

$$F + mg \sin \theta - f_{s, max} = ma = 0$$
$$F_N - mg \cos \theta = 0.$$

Along with Eq. 6-1 ($f_{s, max} = \mu_s F_N$) we have enough information to solve for F. With $\theta = 24°$ and $m = 1.8 \times 10^7$ kg, we find

$$F = mg(\mu_s \cos \theta - \sin \theta) = 3.0 \times 10^7 \text{ N}.$$

13. (a) The free-body diagram for the crate is shown on the right. $\vec{T}$ is the tension force of the rope on the crate, $\vec{F}_N$ is the normal force of the floor on the crate, $m\vec{g}$ is the force of gravity, and $\vec{f}$ is the force of friction. We take the $+x$ direction to be horizontal to the right and the $+y$ direction to be up. We assume the crate is motionless. The equations for the x and the y components of the force according to Newton's second law are:

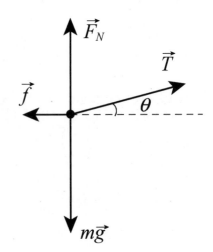

$$T \cos \theta - f = 0$$
$$T \sin \theta + F_N - mg = 0$$

where $\theta = 15°$ is the angle between the rope and the horizontal. The first equation gives $f = T \cos \theta$ and the second gives $F_N = mg - T \sin \theta$. If the crate is to remain at rest, f must be less than $\mu_s F_N$, or $T \cos \theta < \mu_s (mg - T \sin \theta)$. When the tension force is sufficient to just start the crate moving, we must have

$$T \cos \theta = \mu_s (mg - T \sin \theta).$$

We solve for the tension:

$$T = \frac{\mu_s mg}{\cos\theta + \mu_s \sin\theta} = \frac{(0.50)(68\text{ kg})(9.8\text{ m/s}^2)}{\cos 15° + 0.50 \sin 15°} = 304\text{ N} \approx 3.0\times 10^2\text{ N}.$$

(b) The second law equations for the moving crate are

$$T\cos\theta - f = ma$$
$$F_N + T\sin\theta - mg = 0.$$

Now $f = \mu_k F_N$, and the second equation gives $F_N = mg - T\sin\theta$, which yields $f = \mu_k(mg - T\sin\theta)$. This expression is substituted for f in the first equation to obtain

$$T\cos\theta - \mu_k(mg - T\sin\theta) = ma,$$

so the acceleration is

$$a = \frac{T(\cos\theta + \mu_k \sin\theta)}{m} - \mu_k g.$$

Numerically, it is given by

$$a = \frac{(304\text{ N})(\cos 15° + 0.35\sin 15°)}{68\text{ kg}} - (0.35)(9.8\text{ m/s}^2) = 1.3\text{ m/s}^2.$$

14. (a) The free-body diagram for the block is shown on the right, with $\vec{F}$ being the force applied to the block, $\vec{F}_N$ the normal force of the floor on the block, $m\vec{g}$ the force of gravity, and $\vec{f}$ the force of friction. We take the $+x$ direction to be horizontal to the right and the $+y$ direction to be up. The equations for the x and the y components of the force according to Newton's second law are:

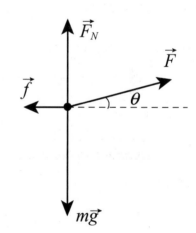

$$F_x = F\cos\theta - f = ma$$
$$F_y = F\sin\theta + F_N - mg = 0$$

Now $f = \mu_k F_N$, and the second equation gives $F_N = mg - F\sin\theta$, which yields $f = \mu_k(mg - F\sin\theta)$. This expression is substituted for f in the first equation to obtain

$$F\cos\theta - \mu_k(mg - F\sin\theta) = ma,$$

so the acceleration is

$$a = \frac{F}{m}(\cos\theta + \mu_k \sin\theta) - \mu_k g.$$

(a) If $\mu_s = 0.600$ and $\mu_k = 0.500$, then the magnitude of $\vec{f}$ has a maximum value of

$$f_{s,\max} = \mu_s F_N = (0.600)(mg - 0.500mg \sin 20°) = 0.497mg.$$

On the other hand, $F \cos\theta = 0.500mg \cos 20° = 0.470mg$. Therefore, $F \cos\theta < f_{s,\max}$ and the block remains stationary with $a = 0$.

(b) If $\mu_s = 0.400$ and $\mu_k = 0.300$, then the magnitude of $\vec{f}$ has a maximum value of

$$f_{s,\max} = \mu_s F_N = (0.400)(mg - 0.500mg \sin 20°) = 0.332mg.$$

In this case, $F \cos\theta = 0.500mg \cos 20° = 0.470mg > f_{s,\max}$. Therefore, the acceleration of the block is

$$\begin{aligned}a &= \frac{F}{m}(\cos\theta + \mu_k \sin\theta) - \mu_k g \\ &= (0.500)(9.80 \text{ m/s}^2)[\cos 20° + (0.300)\sin 20°] - (0.300)(9.80 \text{ m/s}^2) \\ &= 2.17 \text{ m/s}^2.\end{aligned}$$

15. An excellent discussion and equation development related to this problem is given in Sample Problem 6-2. We merely quote (and apply) their main result:

$$\theta = \tan^{-1}\mu_s = \tan^{-1} 0.04 \approx 2°.$$

16. (a) We apply Newton's second law to the "downhill" direction:

$$mg \sin\theta - f = ma,$$

where, using Eq. 6-11,

$$f = f_k = \mu_k F_N = \mu_k mg \cos\theta.$$

Thus, with $\mu_k = 0.600$, we have

$$a = g\sin\theta - \mu_k \cos\theta = -3.72 \text{ m/s}^2$$

which means, since we have chosen the positive direction in the direction of motion (down the slope) then the acceleration vector points "uphill"; it is decelerating. With $v_0 = 18.0$ m/s and $\Delta x = d = 24.0$ m, Eq. 2-16 leads to

$$v = \sqrt{v_0^2 + 2ad} = 12.1 \text{ m/s}.$$

(b) In this case, we find $a = +1.1$ m/s^2, and the speed (when impact occurs) is 19.4 m/s.

17. (a) The free-body diagram for the block is shown below. $\vec{F}$ is the applied force, $\vec{F}_N$ is the normal force of the wall on the block, $\vec{f}$ is the force of friction, and $m\vec{g}$ is the force of gravity. To determine if the block falls, we find the magnitude f of the force of friction required to hold it without accelerating and also find the normal force of the wall on the block. We compare f and $\mu_s F_N$. If $f < \mu_s F_N$, the block does not slide on the wall but if $f > \mu_s F_N$, the block does slide. The horizontal component of Newton's second law is $F - F_N = 0$, so $F_N = F = 12$ N and

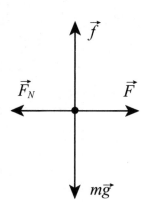

$$\mu_s F_N = (0.60)(12 \text{ N}) = 7.2 \text{ N}.$$

The vertical component is $f - mg = 0$, so $f = mg = 5.0$ N. Since $f < \mu_s F_N$ the block does not slide.

(b) Since the block does not move $f = 5.0$ N and $F_N = 12$ N. The force of the wall on the block is

$$\vec{F}_w = -F_N \hat{i} + f \hat{j} = -(12\text{N})\hat{i} + (5.0\text{N})\hat{j}$$

where the axes are as shown on Fig. 6-26 of the text.

18. We find the acceleration from the slope of the graph (recall Eq. 2-11): $a = 4.5$ m/s^2. Thus, Newton's second law leads to

$$F - \mu_k mg = ma,$$

where $F = 40.0$ N is the constant horizontal force applied. With $m = 4.1$ kg, we arrive at $\mu_k = 0.54$.

19. Fig. 6-4 in the textbook shows a similar situation (using ϕ for the unknown angle) along with a free-body diagram. We use the same coordinate system as in that figure.

(a) Thus, Newton's second law leads to

$$x: \quad T\cos\phi - f = ma$$
$$y: \quad T\sin\phi + F_N - mg = 0$$

Setting $a = 0$ and $f = f_{s,\max} = \mu_s F_N$, we solve for the mass of the box-and-sand (as a function of angle):

$$m = \frac{T}{g}\left(\sin\phi + \frac{\cos\phi}{\mu_s}\right)$$

which we will solve with calculus techniques (to find the angle ϕ_m corresponding to the maximum mass that can be pulled).

$$\frac{dm}{dt} = \frac{T}{g}\left(\cos\phi_m - \frac{\sin\phi_m}{\mu_s}\right) = 0$$

This leads to $\tan\phi_m = \mu_s$ which (for $\mu_s = 0.35$) yields $\phi_m = 19°$.

(b) Plugging our value for ϕ_m into the equation we found for the mass of the box-and-sand yields $m = 340$ kg. This corresponds to a weight of $mg = 3.3 \times 10^3$ N.

20. (a) In this situation, we take $\vec{f}_s$ to point uphill and to be equal to its maximum value, in which case $f_{s,\,max} = \mu_s F_N$ applies, where $\mu_s = 0.25$. Applying Newton's second law to the block of mass $m = W/g = 8.2$ kg, in the x and y directions, produces

$$F_{min\,1} - mg\sin\theta + f_{s,\,max} = ma = 0$$
$$F_N - mg\cos\theta = 0$$

which (with $\theta = 20°$) leads to

$$F_{min\,1} - mg\left(\sin\theta + \mu_s\cos\theta\right) = 8.6 \text{ N}.$$

(b) Now we take $\vec{f}_s$ to point downhill and to be equal to its maximum value, in which case $f_{s,\,max} = \mu_s F_N$ applies, where $\mu_s = 0.25$. Applying Newton's second law to the block of mass $m = W/g = 8.2$ kg, in the x and y directions, produces

$$F_{min\,2} = mg\sin\theta - f_{s,\,max} = ma = 0$$
$$F_N - mg\cos\theta = 0$$

which (with $\theta = 20°$) leads to

$$F_{min\,2} = mg\left(\sin\theta + \mu_s\cos\theta\right) = 46 \text{ N}.$$

A value slightly larger than the "exact" result of this calculation is required to make it accelerate uphill, but since we quote our results here to two significant figures, 46 N is a "good enough" answer.

(c) Finally, we are dealing with kinetic friction (pointing downhill), so that

$$0 = F - mg\sin\theta - f_k = ma$$
$$0 = F_N - mg\cos\theta$$

along with $f_k = \mu_k F_N$ (where $\mu_k = 0.15$) brings us to

$$F = mg\left(\sin\theta + \mu_k \cos\theta\right) = 39 \text{ N}.$$

21. If the block is sliding then we compute the kinetic friction from Eq. 6-2; if it is not sliding, then we determine the extent of static friction from applying Newton's law, with zero acceleration, to the x axis (which is parallel to the incline surface). The question of whether or not it is sliding is therefore crucial, and depends on the maximum static friction force, as calculated from Eq. 6-1. The forces are resolved in the incline plane coordinate system in Figure 6-5 in the textbook. The acceleration, if there is any, is along the x axis, and we are taking uphill as $+x$. The net force along the y axis, then, is certainly zero, which provides the following relationship:

$$\sum F_y = 0 \Rightarrow F_N = W\cos\theta$$

where $W = mg = 45$ N is the weight of the block, and $\theta = 15°$ is the incline angle. Thus, $F_N = 43.5$ N, which implies that the maximum static friction force should be

$$f_{s,\max} = (0.50)(43.5 \text{ N}) = 21.7 \text{ N}.$$

(a) For $\vec{P} = (-5.0 \text{ N})\hat{i}$, Newton's second law, applied to the x axis becomes

$$f - |P| - mg\sin\theta = ma.$$

Here we are assuming $\vec{f}$ is pointing uphill, as shown in Figure 6-5, and if it turns out that it points downhill (which *is* a possibility), then the result for f_s will be negative. If $f = f_s$ then $a = 0$, we obtain

$$f_s = |P| + mg\sin\theta = 5.0 \text{ N} + (43.5 \text{ N})\sin15° = 17 \text{ N},$$

or $\vec{f}_s = (17 \text{ N})\hat{i}$. This is clearly allowed since f_s is less than $f_{s,\max}$.

(b) For $\vec{P} = (-8.0 \text{ N})\hat{i}$, we obtain (from the same equation) $\vec{f}_s = (20 \text{ N})\hat{i}$, which is still allowed since it is less than $f_{s,\max}$.

(c) But for $\vec{P} = (-15 \text{ N})\hat{i}$, we obtain (from the same equation) $f_s = 27$ N, which is not allowed since it is larger than $f_{s,\max}$. Thus, we conclude that it is the kinetic friction instead of the static friction that is relevant in this case. The result is

$$\vec{f}_k = \mu_k F_N \hat{i} = (0.34)(43.5 \text{ N})\hat{i} = (15 \text{ N})\hat{i}.$$

22. Treating the two boxes as a single system of total mass $m_C + m_W = 1.0 + 3.0 = 4.0$ kg, subject to a total (leftward) friction of magnitude $2.0 \text{ N} + 4.0 \text{ N} = 6.0$ N, we apply Newton's second law (with $+x$ rightward):

$$F - f_{total} = m_{total}\, a \quad \Rightarrow \quad 12.0\text{ N} - 6.0\text{ N} = (4.0\text{ kg})a$$

which yields the acceleration $a = 1.5$ m/s². We have treated F as if it were known to the nearest tenth of a Newton so that our acceleration is "good" to two significant figures. Turning our attention to the larger box (the Wheaties box of mass $m_W = 3.0$ kg) we apply Newton's second law to find the contact force F' exerted by the Cheerios box on it.

$$F' - f_W = m_W a \quad \Rightarrow \quad F' - 4.0\text{ N} = (3.0\text{ kg})(1.5\text{ m/s}^2).$$

From the above equation, we find the contact force to be $F' = 8.5$ N.

23. The free-body diagrams for block B and for the knot just above block A are shown next. $\vec{T}_1$ is the tension force of the rope pulling on block B or pulling on the knot (as the case may be), $\vec{T}_2$ is the tension force exerted by the second rope (at angle $\theta = 30°$) on the knot, $\vec{f}$ is the force of static friction exerted by the horizontal surface on block B, $\vec{F}_N$ is normal force exerted by the surface on block B, W_A is the weight of block A (W_A is the magnitude of $m_A \vec{g}$), and W_B is the weight of block B ($W_B = 711$ N is the magnitude of $m_B \vec{g}$).

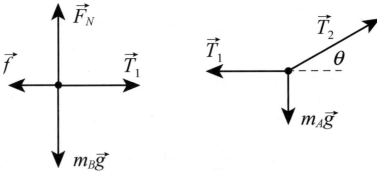

For each object we take $+x$ horizontally rightward and $+y$ upward. Applying Newton's second law in the x and y directions for block B and then doing the same for the knot results in four equations:

$$T_1 - f_{s,\max} = 0$$
$$F_N - W_B = 0$$
$$T_2 \cos\theta - T_1 = 0$$
$$T_2 \sin\theta - W_A = 0$$

where we assume the static friction to be at its maximum value (permitting us to use Eq. 6-1). Solving these equations with $\mu_s = 0.25$, we obtain $W_A = 103\text{ N} \approx 1.0 \times 10^2$ N.

24. The free-body diagram for the block is shown below, with $\vec{F}$ being the force applied to the block, $\vec{F}_N$ the normal force of the floor on the block, $m\vec{g}$ the force of gravity, and

$\vec{f}$ the force of friction. We take the +x direction to be horizontal to the right and the +y direction to be up. The equations for the x and the y components of the force according to Newton's second law are:

$$F_x = F\cos\theta - f = ma$$
$$F_y = F_N - F\sin\theta - mg = 0$$

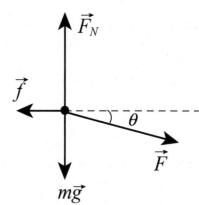

Now $f = \mu_k F_N$, and the second equation gives $F_N = mg + F\sin\theta$, which yields

$$f = \mu_k(mg + F\sin\theta).$$

This expression is substituted for f in the first equation to obtain

$$F\cos\theta - \mu_k(mg + F\sin\theta) = ma,$$

so the acceleration is

$$a = \frac{F}{m}(\cos\theta - \mu_k \sin\theta) - \mu_k g.$$

From Fig. 6-32, we see that $a = 3.0 \text{ m/s}^2$ when $\mu_k = 0$. This implies

$$3.0 \text{ m/s}^2 = \frac{F}{m}\cos\theta.$$

We also find $a = 0$ when $\mu_k = 0.20$:

$$0 = \frac{F}{m}(\cos\theta - (0.20)\sin\theta) - (0.20)(9.8 \text{ m/s}^2) = 3.00 \text{ m/s}^2 - 0.20\frac{F}{m}\sin\theta - 1.96 \text{ m/s}^2$$
$$= 1.04 \text{ m/s}^2 - 0.20\frac{F}{m}\sin\theta$$

which yields $5.2 \text{ m/s}^2 = \frac{F}{m}\sin\theta$. Combining the two results, we get

$$\tan\theta = \left(\frac{5.2 \text{ m/s}^2}{3.0 \text{ m/s}^2}\right) = 1.73 \Rightarrow \theta = 60°.$$

25. Let the tensions on the strings connecting m_2 and m_3 be T_{23}, and that connecting m_2 and m_1 be T_{12}, respectively. Applying Newton's second law (and Eq. 6-2, with $F_N = m_2 g$ in this case) to the *system* we have

$$m_3 g - T_{23} = m_3 a$$
$$T_{23} - \mu_k m_2 g - T_{12} = m_2 a$$
$$T_{12} - m_1 g = m_1 a$$

Adding up the three equations and using $m_1 = M, m_2 = m_3 = 2M$, we obtain

$$2Mg - 2\mu_k Mg - Mg = 5Ma.$$

With $a = 0.500$ m/s^2 this yields $\mu_k = 0.372$. Thus, the coefficient of kinetic friction is roughly $\mu_k = 0.37$.

26. The free-body diagram for the sled is shown on the right, with $\vec{F}$ being the force applied to the sled, $\vec{F}_N$ the normal force of the inclined plane on the sled, $m\vec{g}$ the force of gravity, and $\vec{f}$ the force of friction. We take the $+x$ direction to be along the inclined plane and the $+y$ direction to be in its normal direction. The equations for the x and the y components of the force according to Newton's second law are:

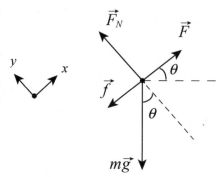

$$F_x = F - f - mg\sin\theta = ma = 0$$
$$F_y = F_N - mg\cos\theta = 0$$

Now $f = \mu F_N$, and the second equation gives $F_N = mg\cos\theta$, which yields $f = \mu mg\cos\theta$. This expression is substituted for f in the first equation to obtain

$$F = mg(\sin\theta + \mu\cos\theta)$$

From Fig. 6-34, we see that $F = 2.0$ N when $\mu = 0$. This implies $mg\sin\theta = 2.0$ N. Similarly, we also find $F = 5.0$ N when $\mu = 0.5$:

$$5.0\text{ N} = mg(\sin\theta + 0.50\cos\theta) = 2.0\text{ N} + 0.50 mg\cos\theta$$

which yields $mg\cos\theta = 6.0$ N. Combining the two results, we get

$$\tan\theta = \frac{2}{6} = \frac{1}{3} \quad\Rightarrow\quad \theta = 18°.$$

27. The free-body diagrams for the two blocks are shown next. T is the magnitude of the tension force of the string, $\vec{F}_{NA}$ is the normal force on block A (the leading block), $\vec{F}_{NB}$ is the normal force on block B, $\vec{f}_A$ is kinetic friction force on block A, $\vec{f}_B$ is kinetic friction force on block B. Also, m_A is the mass of block A (where $m_A = W_A/g$ and $W_A = 3.6$ N), and m_B is the mass of block B (where $m_B = W_B/g$ and $W_B = 7.2$ N). The angle of the incline is $\theta = 30°$.

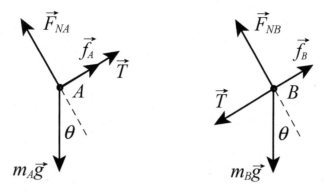

For each block we take $+x$ downhill (which is toward the lower-left in these diagrams) and $+y$ in the direction of the normal force. Applying Newton's second law to the x and y directions of both blocks A and B, we arrive at four equations:

$$W_A \sin\theta - f_A - T = m_A a$$
$$F_{NA} - W_A \cos\theta = 0$$
$$W_B \sin\theta - f_B + T = m_B a$$
$$F_{NB} - W_B \cos\theta = 0$$

which, when combined with Eq. 6-2 ($f_A = \mu_{kA} F_{NA}$ where $\mu_{k\,A} = 0.10$ and $f_B = \mu_{kB} F_{NB} f_B$ where $\mu_{k\,B} = 0.20$), fully describe the dynamics of the system so long as the blocks have the same acceleration and $T > 0$.

(a) From these equations, we find the acceleration to be

$$a = g\left(\sin\theta - \left(\frac{\mu_{kA} W_A + \mu_{kB} W_B}{W_A + W_B}\right)\cos\theta\right) = 3.5 \text{ m/s}^2.$$

(b) We solve the above equations for the tension and obtain

$$T = \left(\frac{W_A W_B}{W_A + W_B}\right)(\mu_{kB} - \mu_{kA})\cos\theta = 0.21 \text{ N}.$$

Simply returning the value for a found in part (a) into one of the above equations is certainly fine, and probably easier than solving for T algebraically as we have done, but the algebraic form does illustrate the $\mu_{k\,B} - \mu_{k\,A}$ factor which aids in the understanding of the next part.

28. (a) Applying Newton's second law to the *system* (of total mass $M = 60.0$ kg) and using Eq. 6-2 (with $F_N = Mg$ in this case) we obtain

$$F - \mu_k Mg = Ma \Rightarrow a = 0.473 \text{ m/s}^2.$$

Next, we examine the forces just on m_3 and find $F_{32} = m_3(a + \mu_k g) = 147$ N. If the algebra steps are done more systematically, one ends up with the interesting relationship: $F_{32} = (m_3/M)F$ (which is independent of the friction!).

(b) As remarked at the end of our solution to part (a), the result does not depend on the frictional parameters. The answer here is the same as in part (a).

29. First, we check to see if the bodies start to move. We assume they remain at rest and compute the force of (static) friction which holds them there, and compare its magnitude with the maximum value $\mu_s F_N$. The free-body diagrams are shown below. T is the magnitude of the tension force of the string, f is the magnitude of the force of friction on body A, F_N is the magnitude of the normal force of the plane on body A, $m_A \vec{g}$ is the force of gravity on body A (with magnitude $W_A = 102$ N), and $m_B \vec{g}$ is the force of gravity on body B (with magnitude $W_B = 32$ N). $\theta = 40°$ is the angle of incline. We are told the direction of $\vec{f}$ but we assume it is downhill. If we obtain a negative result for f, then we know the force is actually up the plane.

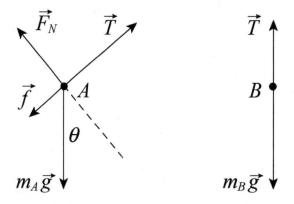

(a) For A we take the $+x$ to be uphill and $+y$ to be in the direction of the normal force. The x and y components of Newton's second law become

$$T - f - W_A \sin \theta = 0$$
$$F_N - W_A \cos \theta = 0.$$

Taking the positive direction to be *downward* for body B, Newton's second law leads to $W_B - T = 0$. Solving these three equations leads to

$$f = W_B - W_A \sin \theta = 32 \text{ N} - (102 \text{ N}) \sin 40° = -34 \text{ N}$$

(indicating that the force of friction is *uphill*) and to

$$F_N = W_A \cos \theta = (102 \text{ N}) \cos 40° = 78 \text{N}$$

which means that

$$f_{s,\text{max}} = \mu_s F_N = (0.56)(78\text{ N}) = 44\text{ N}.$$

Since the magnitude f of the force of friction that holds the bodies motionless is less than $f_{s,\text{max}}$ the bodies remain at rest. The acceleration is zero.

(b) Since A is moving up the incline, the force of friction is downhill with magnitude $f_k = \mu_k F_N$. Newton's second law, using the same coordinates as in part (a), leads to

$$T - f_k - W_A \sin\theta = m_A a$$
$$F_N - W_A \cos\theta = 0$$
$$W_B - T = m_B a$$

for the two bodies. We solve for the acceleration:

$$a = \frac{W_B - W_A \sin\theta - \mu_k W_A \cos\theta}{m_B + m_A} = \frac{32\text{N} - (102\text{N})\sin 40° - (0.25)(102\text{N})\cos 40°}{(32\text{N} + 102\text{N})/(9.8\text{ m/s}^2)}$$
$$= -3.9\text{ m/s}^2.$$

The acceleration is down the plane, i.e., $\vec{a} = (-3.9\text{ m/s}^2)\hat{i}$, which is to say (since the initial velocity was uphill) that the objects are slowing down. We note that $m = W/g$ has been used to calculate the masses in the calculation above.

(c) Now body A is initially moving down the plane, so the force of friction is uphill with magnitude $f_k = \mu_k F_N$. The force equations become

$$T + f_k - W_A \sin\theta = m_A a$$
$$F_N - W_A \cos\theta = 0$$
$$W_B - T = m_B a$$

which we solve to obtain

$$a = \frac{W_B - W_A \sin\theta + \mu_k W_A \cos\theta}{m_B + m_A} = \frac{32\text{N} - (102\text{N})\sin 40° + (0.25)(102\text{N})\cos 40°}{(32\text{N} + 102\text{N})/(9.8\text{ m/s}^2)}$$
$$= -1.0\text{ m/s}^2.$$

The acceleration is again downhill the plane, i.e., $\vec{a} = (-1.0\text{ m/s}^2)\hat{i}$. In this case, the objects are speeding up.

30. The free-body diagrams are shown below. T is the magnitude of the tension force of the string, f is the magnitude of the force of friction on block A, F_N is the magnitude of the normal force of the plane on block A, $m_A \vec{g}$ is the force of gravity on body A (where

$m_A = 10$ kg), and $m_B \vec{g}$ is the force of gravity on block B. $\theta = 30°$ is the angle of incline. For A we take the $+x$ to be uphill and $+y$ to be in the direction of the normal force; the positive direction is chosen *downward* for block B.

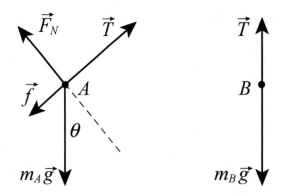

Since A is moving down the incline, the force of friction is uphill with magnitude $f_k = \mu_k F_N$ (where $\mu_k = 0.20$). Newton's second law leads to

$$T - f_k + m_A g \sin\theta = m_A a = 0$$
$$F_N - m_A g \cos\theta = 0$$
$$m_B g - T = m_B a = 0$$

for the two bodies (where $a = 0$ is a consequence of the velocity being constant). We solve these for the mass of block B.

$$m_B = m_A \left(\sin\theta - \mu_k \cos\theta \right) = 3.3 \text{ kg}.$$

31. (a) Free-body diagrams for the blocks A and C, considered as a single object, and for the block B are shown below. T is the magnitude of the tension force of the rope, F_N is the magnitude of the normal force of the table on block A, f is the magnitude of the force of friction, W_{AC} is the combined weight of blocks A and C (the magnitude of force $\vec{F}_{g\,AC}$ shown in the figure), and W_B is the weight of block B (the magnitude of force $\vec{F}_{g\,B}$ shown). Assume the blocks are not moving. For the blocks on the table we take the x axis to be to the right and the y axis to be upward. From Newton's second law, we have

x component: $\quad T - f = 0$

y component: $\quad F_N - W_{AC} = 0.$

For block B take the downward direction to be positive. Then Newton's second law for that block is $W_B - T = 0$. The third equation gives $T = W_B$ and the first gives $f = T = W_B$. The second equation gives $F_N = W_{AC}$. If sliding is not to occur, f must be less than $\mu_s F_N$, or $W_B < \mu_s W_{AC}$. The smallest that W_{AC} can be with the blocks still at rest is

$$W_{AC} = W_B/\mu_s = (22 \text{ N})/(0.20) = 110 \text{ N}.$$

Since the weight of block A is 44 N, the least weight for C is (110 − 44) N = 66 N.

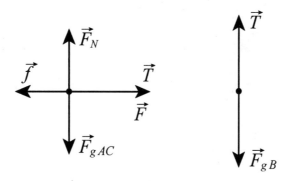

(b) The second law equations become

$$T - f = (W_A/g)a$$
$$F_N - W_A = 0$$
$$W_B - T = (W_B/g)a.$$

In addition, $f = \mu_k F_N$. The second equation gives $F_N = W_A$, so $f = \mu_k W_A$. The third gives $T = W_B - (W_B/g)a$. Substituting these two expressions into the first equation, we obtain

$$W_B - (W_B/g)a - \mu_k W_A = (W_A/g)a.$$

Therefore,

$$a = \frac{g(W_B - \mu_k W_A)}{W_A + W_B} = \frac{(9.8 \text{ m/s}^2)(22 \text{ N} - (0.15)(44 \text{ N}))}{44 \text{ N} + 22 \text{ N}} = 2.3 \text{ m/s}^2.$$

32. We use the familiar horizontal and vertical axes for x and y directions, with rightward and upward positive, respectively. The rope is assumed massless so that the force exerted by the child $\vec{F}$ is identical to the tension uniformly through the rope. The x and y components of $\vec{F}$ are $F\cos\theta$ and $F\sin\theta$, respectively. The static friction force points leftward.

(a) Newton's Law applied to the y-axis, where there is presumed to be no acceleration, leads to

$$F_N + F\sin\theta - mg = 0$$

which implies that the maximum static friction is $\mu_s(mg - F\sin\theta)$. If $f_s = f_{s,\text{max}}$ is assumed, then Newton's second law applied to the x axis (which also has $a = 0$ even though it is "verging" on moving) yields

$$F\cos\theta - f_s = ma \Rightarrow F\cos\theta - \mu_s(mg - F\sin\theta) = 0$$

which we solve, for $\theta = 42°$ and $\mu_s = 0.42$, to obtain $F = 74$ N.

(b) Solving the above equation algebraically for F, with W denoting the weight, we obtain

$$F = \frac{\mu_s W}{\cos\theta + \mu_s \sin\theta} = \frac{(0.42)(180\text{ N})}{\cos\theta + (0.42)\sin\theta} = \frac{76\text{ N}}{\cos\theta + (0.42)\sin\theta}.$$

(c) We minimize the above expression for F by working through the condition:

$$\frac{dF}{d\theta} = \frac{\mu_s W(\sin\theta - \mu_s \cos\theta)}{(\cos\theta + \mu_s \sin\theta)^2} = 0$$

which leads to the result $\theta = \tan^{-1}\mu_s = 23°$.

(d) Plugging $\theta = 23°$ into the above result for F, with $\mu_s = 0.42$ and $W = 180$ N, yields $F = 70$ N.

33. The free-body diagrams for the two blocks, treated individually, are shown below (first m and then M). F' is the contact force between the two blocks, and the static friction force $\vec{f}_s$ is at its maximum value (so Eq. 6-1 leads to $f_s = f_{s,\max} = \mu_s F'$ where $\mu_s = 0.38$).

Treating the two blocks together as a single system (sliding across a frictionless floor), we apply Newton's second law (with $+x$ rightward) to find an expression for the acceleration:

$$F = m_{\text{total}}\, a \quad \Rightarrow \quad a = \frac{F}{m + M}$$

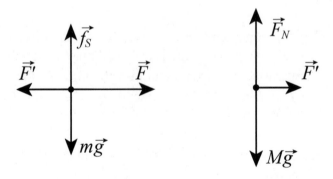

This is equivalent to having analyzed the two blocks individually and then combined their equations. Now, when we analyze the small block individually, we apply Newton's second law to the x and y axes, substitute in the above expression for a, and use Eq. 6-1.

$$F - F' = ma \quad \Rightarrow \quad F' = F - m\left(\frac{F}{m+M}\right)$$

$$f_s - mg = 0 \quad \Rightarrow \quad \mu_s F' - mg = 0$$

These expressions are combined (to eliminate F') and we arrive at

$$F = \frac{mg}{\mu_s\left(1 - \dfrac{m}{m+M}\right)}$$

which we find to be $F = 4.9 \times 10^2$ N.

34. The free-body diagrams for the slab and block are shown below.

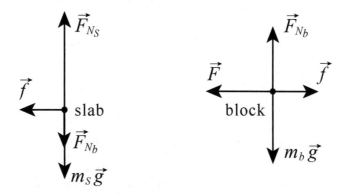

$\vec{F}$ is the 100 N force applied to the block, $\vec{F}_{Ns}$ is the normal force of the floor on the slab, F_{Nb} is the magnitude of the normal force between the slab and the block, $\vec{f}$ is the force of friction between the slab and the block, m_s is the mass of the slab, and m_b is the mass of the block. For both objects, we take the $+x$ direction to be to the right and the $+y$ direction to be up.

Applying Newton's second law for the x and y axes for (first) the slab and (second) the block results in four equations:

$$-f = m_s a_s$$
$$F_{Ns} - F_{Ns}' - m_s g = 0$$
$$f - F = m_b a_b$$
$$F_{Nb} - m_b g = 0$$

from which we note that the maximum possible static friction magnitude would be

$$\mu_s F_{Nb} = \mu_s m_b g = (0.60)(10 \text{ kg})(9.8 \text{ m/s}^2) = 59 \text{ N}.$$

We check to see if the block slides on the slab. Assuming it does not, then $a_s = a_b$ (which we denote simply as a) and we solve for f:

$$f = \frac{m_s F}{m_s + m_b} = \frac{(40 \text{ kg})(100 \text{ N})}{40 \text{ kg} + 10 \text{ kg}} = 80 \text{ N}$$

which is greater than $f_{s,\text{max}}$ so that we conclude the block is sliding across the slab (their accelerations are different).

(a) Using $f = \mu_k F_{Nb}$ the above equations yield

$$a_b = \frac{\mu_k m_b g - F}{m_b} = \frac{(0.40)(10 \text{ kg})(9.8 \text{ m/s}^2) - 100 \text{ N}}{10 \text{ kg}} = -6.1 \text{ m/s}^2.$$

The negative sign means that the acceleration is leftward. That is, $\vec{a}_b = (-6.1 \text{ m/s}^2)\hat{i}$

(b) We also obtain

$$a_s = -\frac{\mu_k m_b g}{m_s} = -\frac{(0.40)(10 \text{ kg})(9.8 \text{ m/s}^2)}{40 \text{ kg}} = -0.98 \text{ m/s}^2.$$

As mentioned above, this means it accelerates to the left. That is, $\vec{a}_s = (-0.98 \text{ m/s}^2)\hat{i}$

35. We denote the magnitude of the frictional force αv, where $\alpha = 70 \text{ N} \cdot \text{s/m}$. We take the direction of the boat's motion to be positive. Newton's second law gives

$$-\alpha v = m \frac{dv}{dt}.$$

Thus,

$$\int_{v_0}^{v} \frac{dv}{v} = -\frac{\alpha}{m} \int_0^t dt$$

where v_0 is the velocity at time zero and v is the velocity at time t. The integrals are evaluated with the result

$$\ln\left(\frac{v}{v_0}\right) = -\frac{\alpha t}{m}$$

We take $v = v_0/2$ and solve for time:

$$t = \frac{m}{\alpha} \ln 2 = \frac{1000 \text{ kg}}{70 \text{ N} \cdot \text{s/m}} \ln 2 = 9.9 \text{ s}.$$

36. Using Eq. 6-16, we solve for the area

$$A \frac{2m\,g}{C\rho v_t^2}$$

which illustrates the inverse proportionality between the area and the speed-squared. Thus, when we set up a ratio of areas – of the slower case to the faster case – we obtain

$$\frac{A_{slow}}{A_{fast}} = \left(\frac{310 \text{ km/h}}{160 \text{ km/h}}\right)^2 = 3.75.$$

37. For the passenger jet $D_j = \frac{1}{2} C \rho_1 A v_j^2$, and for the prop-driven transport $D_t = \frac{1}{2} C \rho_2 A v_t^2$, where ρ_1 and ρ_2 represent the air density at 10 km and 5.0 km, respectively. Thus the ratio in question is

$$\frac{D_j}{D_t} = \frac{\rho_1 v_j^2}{\rho_2 v_t^2} = \frac{(0.38 \text{ kg/m}^3)(1000 \text{ km/h})^2}{(0.67 \text{ kg/m}^3)(500 \text{ km/h})^2} = 2.3.$$

38. This problem involves Newton's second law for motion along the slope.

(a) The force along the slope is given by

$$F_g = mg \sin\theta - \mu F_N = mg \sin\theta - \mu mg \cos\theta = mg(\sin\theta - \mu \cos\theta)$$
$$= (85.0 \text{ kg})(9.80 \text{ m/s}^2)[\sin 40.0° - (0.04000)\cos 40.0°]$$
$$= 510 \text{ N}.$$

Thus, the terminal speed of the skier is

$$v_t = \sqrt{\frac{2F_g}{C\rho A}} = \sqrt{\frac{2(510 \text{ N})}{(0.150)(1.20 \text{ kg/m}^3)(1.30 \text{ m}^2)}} = 66.0 \text{ m/s}.$$

(b) Differentiating v_t with respect to C, we obtain

$$dv_t = -\frac{1}{2}\sqrt{\frac{2F_g}{\rho A}} C^{-3/2} dC = -\frac{1}{2}\sqrt{\frac{2(510 \text{ N})}{(1.20 \text{ kg/m}^3)(1.30 \text{ m}^2)}} (0.150)^{-3/2} dC$$
$$= -(2.20 \times 10^2 \text{ m/s}) dC.$$

39. In the solution to exercise 4, we found that the force provided by the wind needed to equal $F = 157$ N (where that last figure is not "significant").

(a) Setting $F = D$ (for Drag force) we use Eq. 6-14 to find the wind speed V along the ground (which actually is relative to the moving stone, but we assume the stone is moving slowly enough that this does not invalidate the result):

$$V = \sqrt{\frac{2F}{C\rho A}} = \sqrt{\frac{2(157 \text{ N})}{(0.80)(1.21 \text{ kg/m}^3)(0.040 \text{ m}^2)}} = 90 \text{ m/s} = 3.2 \times 10^2 \text{ km/h}.$$

(b) Doubling our previous result, we find the reported speed to be 6.5×10^2 km/h.

(c) The result is not reasonable for a terrestrial storm. A category 5 hurricane has speeds on the order of 2.6×10^2 m/s.

40. (a) From Table 6-1 and Eq. 6-16, we have

$$v_t = \sqrt{\frac{2F_g}{C\rho A}} \Rightarrow C\rho A = 2\frac{mg}{v_t^2}$$

where $v_t = 60$ m/s. We estimate the pilot's mass at about $m = 70$ kg. Now, we convert $v = 1300(1000/3600) \approx 360$ m/s and plug into Eq. 6-14:

$$D = \frac{1}{2}C\rho A v^2 = \frac{1}{2}\left(2\frac{mg}{v_t^2}\right)v^2 = mg\left(\frac{v}{v_t}\right)^2$$

which yields $D = (70 \text{ kg})(9.8 \text{ m/s}^2)(360/60)^2 \approx 2 \times 10^4$ N.

(b) We assume the mass of the ejection seat is roughly equal to the mass of the pilot. Thus, Newton's second law (in the horizontal direction) applied to this system of mass $2m$ gives the magnitude of acceleration:

$$|a| = \frac{D}{2m} = \frac{g}{2}\left(\frac{v}{v_t}\right)^2 = 18g.$$

41. The magnitude of the acceleration of the cyclist as it rounds the curve is given by v^2/R, where v is the speed of the cyclist and R is the radius of the curve. Since the road is horizontal, only the frictional force of the road on the tires makes this acceleration possible. The horizontal component of Newton's second law is $f = mv^2/R$. If F_N is the normal force of the road on the bicycle and m is the mass of the bicycle and rider, the vertical component of Newton's second law leads to $F_N = mg$. Thus, using Eq. 6-1, the maximum value of static friction is $f_{s,\max} = \mu_s F_N = \mu_s mg$. If the bicycle does not slip, $f \leq \mu_s mg$. This means

$$\frac{v^2}{R} \leq \mu_s g \Rightarrow R \geq \frac{v^2}{\mu_s g}.$$

Consequently, the minimum radius with which a cyclist moving at 29 km/h = 8.1 m/s can round the curve without slipping is

$$R_{min} = \frac{v^2}{\mu_s g} = \frac{(8.1 \text{ m/s})^2}{(0.32)(9.8 \text{ m/s}^2)} = 21 \text{ m}.$$

42. With $v = 96.6$ km/h $= 26.8$ m/s, Eq. 6-17 readily yields

$$a = \frac{v^2}{R} = \frac{(26.8 \text{ m/s})^2}{7.6 \text{ m}} = 94.7 \text{ m/s}^2$$

which we express as a multiple of g:

$$a = \left(\frac{a}{g}\right) g = \left(\frac{94.7 \text{ m/s}^2}{9.80 \text{ m/s}^2}\right) g = 9.7 g.$$

43. Perhaps surprisingly, the equations pertaining to this situation are exactly those in Sample Problem 6-9, although the logic is a little different. In the Sample Problem, the car moves along a (stationary) road, whereas in this problem the cat is stationary relative to the merry-go-around platform. But the static friction plays the same role in both cases since the bottom-most point of the car tire is instantaneously at rest with respect to the race track, just as static friction applies to the contact surface between cat and platform. Using Eq. 6-23 with Eq. 4-35, we find

$$\mu_s = (2\pi R/T)^2/gR = 4\pi^2 R/gT^2.$$

With $T = 6.0$ s and $R = 5.4$ m, we obtain $\mu_s = 0.60$.

44. The magnitude of the acceleration of the car as it rounds the curve is given by v^2/R, where v is the speed of the car and R is the radius of the curve. Since the road is horizontal, only the frictional force of the road on the tires makes this acceleration possible. The horizontal component of Newton's second law is $f = mv^2/R$. If F_N is the normal force of the road on the car and m is the mass of the car, the vertical component of Newton's second law leads to $F_N = mg$. Thus, using Eq. 6-1, the maximum value of static friction is

$$f_{s,\max} = \mu_s F_N = \mu_s mg.$$

If the car does not slip, $f \leq \mu_s mg$. This means

$$\frac{v^2}{R} \leq \mu_s g \quad \Rightarrow \quad v \leq \sqrt{\mu_s R g}.$$

Consequently, the maximum speed with which the car can round the curve without slipping is

$$v_{max} = \sqrt{\mu_s R g} = \sqrt{(0.60)(30.5 \text{ m})(9.8 \text{ m/s}^2)} = 13 \text{ m/s} \approx 48 \text{ km/h}.$$

45. (a) Eq. 4-35 gives $T = 2\pi R/v = 2\pi(10 \text{ m})/(6.1 \text{ m/s}) = 10$ s.

(b) The situation is similar to that of Sample Problem 6-7 but with the normal force direction reversed. Adapting Eq. 6-19, we find

$$F_N = m(g - v^2/R) = 486 \text{ N} \approx 4.9 \times 10^2 \text{ N}.$$

(c) Now we reverse both the normal force direction and the acceleration direction (from what is shown in Sample Problem 6-7) and adapt Eq. 6-19 accordingly. Thus we obtain

$$F_N = m(g + v^2/R) = 1081 \text{ N} \approx 1.1 \text{ kN}.$$

46. We will start by assuming that the normal force (on the car from the rail) points up. Note that gravity points down, and the y axis is chosen positive upwards. Also, the direction to the center of the circle (the direction of centripetal acceleration) is down. Thus, Newton's second law leads to

$$F_N - mg = m\left(-\frac{v^2}{r}\right).$$

(a) When $v = 11$ m/s, we obtain $F_N = 3.7 \times 10^3$ N.

(b) $\vec{F}_N$ points upward.

(c) When $v = 14$ m/s, we obtain $F_N = -1.3 \times 10^3$ N, or $|F_N| = 1.3 \times 10^3$ N.

(d) The fact that this answer is negative means that $\vec{F}_N$ points opposite to what we had assumed. Thus, the magnitude of $\vec{F}_N$ is $|\vec{F}_N| = 1.3$ kN and its direction is *down*.

47. At the top of the hill, the situation is similar to that of Sample Problem 6-7 but with the normal force direction reversed. Adapting Eq. 6-19, we find

$$F_N = m(g - v^2/R).$$

Since $F_N = 0$ there (as stated in the problem) then $v^2 = gR$. Later, at the bottom of the valley, we reverse both the normal force direction and the acceleration direction (from what is shown in Sample Problem 6-7) and adapt Eq. 6-19 accordingly. Thus we obtain

$$F_N = m(g + v^2/R) = 2mg = 1372 \text{ N} \approx 1.37 \times 10^3 \text{ N}.$$

48. (a) We note that the speed 80.0 km/h in SI units is roughly 22.2 m/s. The horizontal force that keeps her from sliding must equal the centripetal force (Eq. 6-18), and the upward force on her must equal mg. Thus,

$$F_{net} = \sqrt{(mg)^2 + (mv^2/R)^2} = 547 \text{ N}.$$

(b) The angle is $\tan^{-1}[(mv^2/R)/(mg)] = \tan^{-1}(v^2/gR) = 9.53°$ (as measured from a vertical axis).

49. (a) At the top (the highest point in the circular motion) the seat pushes up on the student with a force of magnitude $F_N = 556$ N. Earth pulls down with a force of magnitude $W = 667$ N. The seat is pushing up with a force that is smaller than the student's weight, and we say the student experiences a decrease in his "apparent weight" at the highest point. Thus, he feels "light."

(b) Now F_N is the magnitude of the upward force exerted by the seat when the student is at the lowest point. The net force toward the center of the circle is $F_b - W = mv^2/R$ (note that we are now choosing upward as the positive direction). The Ferris wheel is "steadily rotating" so the value mv^2/R is the same as in part (a). Thus,

$$F_N = \frac{mv^2}{R} + W = 111 \text{ N} + 667 \text{ N} = 778 \text{ N}.$$

(c) If the speed is doubled, mv^2/R increases by a factor of 4, to 444 N. Therefore, at the highest point we have $W - F_N = mv^2/R$, which leads to

$$F_N = 667 \text{ N} - 444 \text{ N} = 223 \text{ N}.$$

(d) Similarly, the normal force at the lowest point is now found to be

$$F_N = 667 \text{ N} + 444 \text{ N} \approx 1.11 \text{ kN}.$$

50. The situation is somewhat similar to that shown in the "loop-the-loop" example done in the textbook (see Figure 6-10) except that, instead of a downward normal force, we are dealing with the force of the boom $\vec{F}_B$ on the car – which is capable of pointing any direction. We will assume it to be upward as we apply Newton's second law to the car (of total weight 5000 N): $F_B - W = ma$ where $m = W/g$ and $a = -v^2/r$. Note that the centripetal acceleration is downward (our choice for negative direction) for a body at the top of its circular trajectory.

(a) If $r = 10$ m and $v = 5.0$ m/s, we obtain $F_B = 3.7 \times 10^3$ N = 3.7 kN.

(b) The direction of $\vec{F}_B$ is up.

(c) If $r = 10$ m and $v = 12$ m/s, we obtain $F_B = -2.3 \times 10^3$ N $= -2.3$ kN, or $|F_B| = 2.3$ kN.

(d) The minus sign indicates that $\vec{F}_B$ points downward.

51. The free-body diagram (for the hand straps of mass m) is the view that a passenger might see if she was looking forward and the streetcar was curving towards the right (so $\vec{a}$ points rightwards in the figure). We note that $|\vec{a}| = v^2/R$ where $v = 16$ km/h $= 4.4$ m/s.

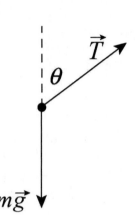

Applying Newton's law to the axes of the problem ($+x$ is rightward and $+y$ is upward) we obtain

$$T \sin\theta = m\frac{v^2}{R}$$
$$T \cos\theta = mg.$$

We solve these equations for the angle:

$$\theta = \tan^{-1}\left(\frac{v^2}{Rg}\right)$$

which yields $\theta = 12°$.

52. The centripetal force on the passenger is $F = mv^2/r$.

(a) The variation of F with respect to r while holding v constant is

$$dF = -\frac{mv^2}{r^2}dr.$$

(b) The variation of F with respect to v while holding r constant is

$$dF = \frac{2mv}{r}dv.$$

(c) The period of the circular ride is $T = 2\pi r/v$. Thus,

$$F = \frac{mv^2}{r} = \frac{m}{r}\left(\frac{2\pi r}{T}\right)^2 = \frac{4\pi^2 mr}{T^2},$$

and the variation of F with respect to T while holding r constant is

$$dF = -\frac{8\pi^2 mr}{T^3}dT = -8\pi^2 mr\left(\frac{v}{2\pi r}\right)^3 dT = -\left(\frac{mv^3}{\pi r^2}\right)dT.$$

53. The free-body diagram (for the airplane of mass m) is shown below. We note that $\vec{F}_\ell$ is the force of aerodynamic lift and $\vec{a}$ points rightwards in the figure. We also note that $|\vec{a}| = v^2/R$ where $v = 480$ km/h $= 133$ m/s.

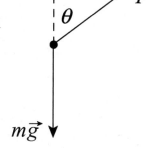

Applying Newton's law to the axes of the problem ($+x$ rightward and $+y$ upward) we obtain

$$F_\ell \sin\theta = m\frac{v^2}{R}$$
$$F_\ell \cos\theta = mg$$

where $\theta = 40°$. Eliminating mass from these equations leads to

$$\tan\theta = \frac{v^2}{gR}$$

which yields $R = v^2/g \tan\theta = 2.2 \times 10^3$ m.

54. The centripetal force on the passenger is $F = mv^2/r$.

(a) The slope of the plot at $v = 8.30$ m/s is

$$\left.\frac{dF}{dv}\right|_{v=8.30\,\text{m/s}} = \left.\frac{2mv}{r}\right|_{v=8.30\,\text{m/s}} = \frac{2(85.0\text{ kg})(8.30\text{ m/s})}{3.50\text{ m}} = 403\text{ N}\cdot\text{s/m}.$$

(b) The period of the circular ride is $T = 2\pi r/v$. Thus,

$$F = \frac{mv^2}{r} = \frac{m}{r}\left(\frac{2\pi r}{T}\right)^2 = \frac{4\pi^2 mr}{T^2},$$

and the variation of F with respect to T while holding r constant is

$$dF = -\frac{8\pi^2 mr}{T^3}dT.$$

The slope of the plot at $T = 2.50$ s is

$$\left.\frac{dF}{dT}\right|_{T=2.50\,\text{s}} = -\left.\frac{8\pi^2 mr}{T^3}\right|_{T=2.50\,\text{s}} = -\frac{8\pi^2(85.0\text{ kg})(3.50\text{ m})}{(2.50\text{ s})^3} = -1.50\times 10^3\text{ N/s}.$$

55. For the puck to remain at rest the magnitude of the tension force T of the cord must equal the gravitational force Mg on the cylinder. The tension force supplies the

centripetal force that keeps the puck in its circular orbit, so $T = mv^2/r$. Thus $Mg = mv^2/r$. We solve for the speed:

$$v = \sqrt{\frac{Mgr}{m}} = \sqrt{\frac{(2.50\text{ kg})(9.80\text{ m/s}^2)(0.200\text{ m})}{1.50\text{ kg}}} = 1.81\text{ m/s}.$$

56. (a) Using the kinematic equation given in Table 2-1, the deceleration of the car is

$$v^2 = v_0^2 + 2ad \Rightarrow 0 = (35\text{ m/s})^2 + 2a(107\text{ m})$$

which gives $a = -5.72\text{ m/s}^2$. Thus, the force of friction required to stop by car is

$$f = m|a| = (1400\text{ kg})(5.72\text{ m/s}^2) \approx 8.0\times10^3\text{ N}.$$

(b) The maximum possible static friction is

$$f_{s,\max} = \mu_s mg = (0.50)(1400\text{ kg})(9.80\text{ m/s}^2) \approx 6.9\times10^3\text{ N}.$$

(c) If $\mu_k = 0.40$, then $f_k = \mu_k mg$ and the deceleration is $a = -\mu_k g$. Therefore, the speed of the car when it hits the wall is

$$v = \sqrt{v_0^2 + 2ad} = \sqrt{(35\text{ m/s})^2 - 2(0.40)(9.8\text{ m/s}^2)(107\text{ m})} \approx 20\text{ m/s}.$$

(d) The force required to keep the motion circular is

$$F_r = \frac{mv_0^2}{r} = \frac{(1400\text{ kg})(35.0\text{ m/s})^2}{107\text{ m}} = 1.6\times10^4\text{ N}.$$

(e) Since $F_r > f_{s,\max}$, no circular path is possible.

57. We note that the period T is eight times the time between flashes ($\frac{1}{2000}$ s), so $T = 0.0040$ s. Combining Eq. 6-18 with Eq. 4-35 leads to

$$F = \frac{4m\pi^2 R}{T^2} = \frac{4(0.030\text{ kg})\pi^2(0.035\text{ m})}{(0.0040\text{ s})^2} = 2.6\times10^3\text{ N}.$$

58. We refer the reader to Sample Problem 6-10, and use the result Eq. 6-26:

$$\theta = \tan^{-1}\left(\frac{v^2}{gR}\right)$$

with $v = 60(1000/3600) = 17$ m/s and $R = 200$ m. The banking angle is therefore $\theta = 8.1°$. Now we consider a vehicle taking this banked curve at $v' = 40(1000/3600) = 11$ m/s. Its (horizontal) acceleration is $a' = v'^2/R$, which has components parallel the incline and perpendicular to it:

$$a_\| = a'\cos\theta = \frac{v'^2 \cos\theta}{R}$$

$$a_\perp = a'\sin\theta = \frac{v'^2 \sin\theta}{R}.$$

These enter Newton's second law as follows (choosing downhill as the $+x$ direction and away-from-incline as $+y$):

$$mg\sin\theta - f_s = ma_\|$$
$$F_N - mg\cos\theta = ma_\perp$$

and we are led to

$$\frac{f_s}{F_N} = \frac{mg\sin\theta - mv'^2 \cos\theta/R}{mg\cos\theta + mv'^2 \sin\theta/R}.$$

We cancel the mass and plug in, obtaining $f_s/F_N = 0.078$. The problem implies we should set $f_s = f_{s,\max}$ so that, by Eq. 6-1, we have $\mu_s = 0.078$.

59. The free-body diagram for the ball is shown below. $\vec{T}_u$ is the tension exerted by the upper string on the ball, $\vec{T}_\ell$ is the tension force of the lower string, and m is the mass of the ball. Note that the tension in the upper string is greater than the tension in the lower string. It must balance the downward pull of gravity and the force of the lower string.

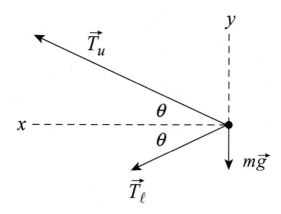

(a) We take the $+x$ direction to be leftward (toward the center of the circular orbit) and $+y$ upward. Since the magnitude of the acceleration is $a = v^2/R$, the x component of Newton's second law is

$$T_u \cos\theta + T_\ell \cos\theta = \frac{mv^2}{R},$$

where v is the speed of the ball and R is the radius of its orbit. The y component is

$$T_u \sin\theta - T_\ell \sin\theta - mg = 0.$$

The second equation gives the tension in the lower string: $T_\ell = T_u - mg/\sin\theta$. Since the triangle is equilateral $\theta = 30.0°$. Thus

$$T_\ell = 35.0\,\text{N} - \frac{(1.34\,\text{kg})(9.80\,\text{m/s}^2)}{\sin 30.0°} = 8.74\,\text{N}.$$

(b) The net force has magnitude

$$F_{\text{net,str}} = (T_u + T_\ell)\cos\theta = (35.0\,\text{N} + 8.74\,\text{N})\cos 30.0° = 37.9\,\text{N}.$$

(c) The radius of the path is

$$R = ((1.70\,\text{m})/2)\tan 30.0° = 1.47\,\text{m}.$$

Using $F_{\text{net,str}} = mv^2/R$, we find that the speed of the ball is

$$v = \sqrt{\frac{RF_{\text{net,str}}}{m}} = \sqrt{\frac{(1.47\,\text{m})(37.9\,\text{N})}{1.34\,\text{kg}}} = 6.45\,\text{m/s}.$$

(d) The direction of $\vec{F}_{\text{net,str}}$ is leftward ("radially inward").

60. (a) We note that R (the horizontal distance from the bob to the axis of rotation) is the circumference of the circular path divided by 2π; therefore, $R = 0.94/2\pi = 0.15$ m. The angle that the cord makes with the horizontal is now easily found:

$$\theta = \cos^{-1}(R/L) = \cos^{-1}(0.15\,\text{m}/0.90\,\text{m}) = 80°.$$

The vertical component of the force of tension in the string is $T\sin\theta$ and must equal the downward pull of gravity (mg). Thus,

$$T = \frac{mg}{\sin\theta} = 0.40\,\text{N}.$$

Note that we are using T for tension (not for the period).

(b) The horizontal component of that tension must supply the centripetal force (Eq. 6-18), so we have $T\cos\theta = mv^2/R$. This gives speed $v = 0.49$ m/s. This divided into the circumference gives the time for one revolution: $0.94/0.49 = 1.9$ s.

61. The layer of ice has a mass of

$$m_{ice} = (917 \text{ kg/m}^3)(400 \text{ m} \times 500 \text{ m} \times 0.0040 \text{ m}) = 7.34 \times 10^5 \text{ kg}.$$

This added to the mass of the hundred stones (at 20 kg each) comes to $m = 7.36 \times 10^5$ kg.

(a) Setting $F = D$ (for Drag force) we use Eq. 6-14 to find the wind speed v along the ground (which actually is relative to the moving stone, but we assume the stone is moving slowly enough that this does not invalidate the result):

$$v = \sqrt{\frac{\mu_k mg}{4 C_{ice} \rho A_{ice}}} = \sqrt{\frac{(0.10)(7.36 \times 10^5 \text{ kg})(9.8 \text{ m/s}^2)}{4(0.002)(1.21 \text{ kg/m}^3)(400 \times 500 \text{ m}^2)}} = 19 \text{ m/s} \approx 69 \text{ km/h}.$$

(b) Doubling our previous result, we find the reported speed to be 139 km/h.

(c) The result is reasonable for storm winds. (A category-5 hurricane has speeds on the order of 2.6×10^2 m/s.)

62. (a) To be on the verge of sliding out means that the force of static friction is acting "down the bank" (in the sense explained in the problem statement) with maximum possible magnitude. We first consider the vector sum $\vec{F}$ of the (maximum) static friction force and the normal force. Due to the facts that they are perpendicular and their magnitudes are simply proportional (Eq. 6-1), we find $\vec{F}$ is at angle (measured from the vertical axis) $\phi = \theta + \theta_s$, where $\tan\theta_s = \mu_s$ (compare with Eq. 6-13), and θ is the bank angle (as stated in the problem). Now, the vector sum of $\vec{F}$ and the vertically downward pull (mg) of gravity must be equal to the (horizontal) centripetal force (mv^2/R), which leads to a surprisingly simple relationship:

$$\tan\phi = \frac{mv^2/R}{mg} = \frac{v^2}{Rg}.$$

Writing this as an expression for the maximum speed, we have

$$v_{max} = \sqrt{Rg \tan(\theta + \tan^{-1}\mu_s)} = \sqrt{\frac{Rg(\tan\theta + \mu_s)}{1 - \mu_s \tan\theta}}$$

(b) The graph is shown below (with θ in radians):

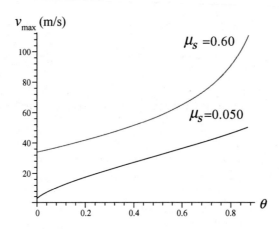

(c) Either estimating from the graph (μ_s = 0.60, upper curve) or calculated it more carefully leads to v = 41.3 m/s = 149 km/h when θ = 10° = 0.175 radian.

(d) Similarly (for μ_s = 0.050, the lower curve) we find v = 21.2 m/s = 76.2 km/h when θ = 10° = 0.175 radian.

63. (a) With θ = 60°, we apply Newton's second law to the "downhill" direction:

$$mg\sin\theta - f = ma$$
$$f = f_k = \mu_k F_N = \mu_k mg\cos\theta.$$

Thus,

$$a = g(\sin\theta - \mu_k \cos\theta) = 7.5 \text{ m/s}^2.$$

(b) The direction of the acceleration $\vec{a}$ is down the slope.

(c) Now the friction force is in the "downhill" direction (which is our positive direction) so that we obtain

$$a = g(\sin\theta + \mu_k \cos\theta) = 9.5 \text{ m/s}^2.$$

(d) The direction is down the slope.

64. Note that since no static friction coefficient is mentioned, we assume f_s is not relevant to this computation. We apply Newton's second law to each block's x axis, which for m_1 is positive rightward and for m_2 is positive downhill:

$$T - f_k = m_1 a$$
$$m_2 g \sin\theta - T = m_2 a$$

Adding the equations, we obtain the acceleration:

$$a = \frac{m_2 g \sin\theta - f_k}{m_1 + m_2}$$

For $f_k = \mu_k F_N = \mu_k m_1 g$, we obtain

$$a = \frac{(3.0 \text{ kg})(9.8 \text{ m/s}^2)\sin 30° - (0.25)(2.0 \text{ kg})(9.8 \text{ m/s}^2)}{3.0 \text{ kg} + 2.0 \text{ kg}} = 1.96 \text{ m/s}^2.$$

Returning this value to either of the above two equations, we find $T = 8.8$ N.

65. (a) Using $F = \mu_s mg$, the coefficient of static friction for the surface between the two blocks is $\mu_s = (12 \text{ N})/(39.2 \text{ N}) = 0.31$, where $m_t g = (4.0 \text{ kg})(9.8 \text{ m/s}^2) = 39.2$ N is the weight of the top block. Let $M = m_t + m_b = 9.0$ kg be the total *system* mass, then the maximum horizontal force has a magnitude $Ma = M\mu_s g = 27$ N.

(b) The acceleration (in the maximal case) is $a = \mu_s g = 3.0 \text{ m/s}^2$.

66. With $\theta = 40°$, we apply Newton's second law to the "downhill" direction:

$$mg\sin\theta - f = ma,$$
$$f = f_k = \mu_k F_N = \mu_k mg\cos\theta$$

using Eq. 6-12. Thus,

$$a = 0.75 \text{ m/s}^2 = g(\sin\theta - \mu_k \cos\theta)$$

determines the coefficient of kinetic friction: $\mu_k = 0.74$.

67. (a) To be "on the verge of sliding" means the applied force is equal to the maximum possible force of static friction (Eq. 6-1, with $F_N = mg$ in this case):

$$f_{s,\max} = \mu_s mg = 35.3 \text{ N}.$$

(b) In this case, the applied force $\vec{F}$ indirectly decreases the maximum possible value of friction (since its y component causes a reduction in the normal force) as well as directly opposing the friction force itself (because of its x component). The normal force turns out to be

$$F_N = mg - F\sin\theta$$

where $\theta = 60°$, so that the horizontal equation (the x application of Newton's second law) becomes

$$F\cos\theta - f_{s,\max} = F\cos\theta - \mu_s(mg - F\sin\theta) = 0 \quad \Rightarrow F = 39.7 \text{ N}.$$

(c) Now, the applied force $\vec{F}$ indirectly increases the maximum possible value of friction (since its y component causes a reduction in the normal force) as well as directly opposing the friction force itself (because of its x component). The normal force in this case turns out to be

$$F_N = mg + F\sin\theta,$$

where $\theta = 60°$, so that the horizontal equation becomes

$$F\cos\theta - f_{s,\max} = F\cos\theta - \mu_s(mg + F\sin\theta) = 0 \Rightarrow F = 320 \text{ N}.$$

68. The free-body diagrams for the two boxes are shown below. T is the magnitude of the force in the rod (when $T > 0$ the rod is said to be in tension and when $T < 0$ the rod is under compression), $\vec{F}_{N2}$ is the normal force on box 2 (the uncle box), $\vec{F}_{N1}$ is the the normal force on the aunt box (box 1), $\vec{f}_1$ is kinetic friction force on the aunt box, and $\vec{f}_2$ is kinetic friction force on the uncle box. Also, $m_1 = 1.65$ kg is the mass of the aunt box and $m_2 = 3.30$ kg is the mass of the uncle box (which is a lot of ants!).

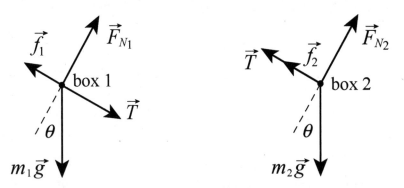

For each block we take $+x$ downhill (which is toward the lower-right in these diagrams) and $+y$ in the direction of the normal force. Applying Newton's second law to the x and y directions of first box 2 and next box 1, we arrive at four equations:

$$m_2 g \sin\theta - f_2 - T = m_2 a$$
$$F_{N2} - m_2 g \cos\theta = 0$$
$$m_1 g \sin\theta - f_1 + T = m_1 a$$
$$F_{N1} - m_1 g \cos\theta = 0$$

which, when combined with Eq. 6-2 ($f_1 = \mu_1 F_{N1}$ where $\mu_1 = 0.226$ and $f_2 = \mu_2 F_{N2}$ where $\mu_2 = 0.113$), fully describe the dynamics of the system.

(a) We solve the above equations for the tension and obtain

$$T = \left(\frac{m_2 m_1 g}{m_2 + m_1}\right)(\mu_1 - \mu_2)\cos\theta = 1.05 \text{ N}.$$

(b) These equations lead to an acceleration equal to

$$a = g\left[\sin\theta - \left(\frac{\mu_2 m_2 + \mu_1 m_1}{m_2 + m_1}\right)\cos\theta\right] = 3.62 \text{ m/s}^2.$$

(c) Reversing the blocks is equivalent to switching the labels. We see from our algebraic result in part (a) that this gives a negative value for T (equal in magnitude to the result we got before). Thus, the situation is as it was before except that the rod is now in a state of compression.

69. Each side of the trough exerts a normal force on the crate. The first diagram shows the view looking in toward a cross section. The net force is along the dashed line. Since each of the normal forces makes an angle of 45° with the dashed line, the magnitude of the resultant normal force is given by

$$F_{Nr} = 2F_N \cos 45° = \sqrt{2} F_N.$$

The second diagram is the free-body diagram for the crate (from a "side" view, similar to that shown in the first picture in Fig. 6-53). The force of gravity has magnitude mg, where m is the mass of the crate, and the magnitude of the force of friction is denoted by f. We take the $+x$ direction to be down the incline and $+y$ to be in the direction of $\vec{F}_{Nr}$. Then the x and the y components of Newton's second law are

x: $mg \sin\theta - f = ma$
y: $F_{Nr} - mg \cos\theta = 0.$

Since the crate is moving, each side of the trough exerts a force of kinetic friction, so the total frictional force has magnitude

$$f = 2\mu_k F_N = 2\mu_k F_{Nr}/\sqrt{2} = \sqrt{2}\mu_k F_{Nr}.$$

Combining this expression with $F_{Nr} = mg \cos\theta$ and substituting into the x component equation, we obtain

$$mg \sin\theta - \sqrt{2} mg \cos\theta = ma.$$

Therefore $a = g(\sin\theta - \sqrt{2}\mu_k \cos\theta)$.

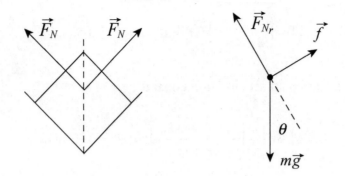

70. (a) The coefficient of static friction is $\mu_s = \tan(\theta_{\text{slip}}) = 0.577 \approx 0.58$.

(b) Using
$$mg\sin\theta - f = ma$$
$$f = f_k = \mu_k F_N = \mu_k mg\cos\theta$$

and $a = 2d/t^2$ (with $d = 2.5$ m and $t = 4.0$ s), we obtain $\mu_k = 0.54$.

71. We may treat all 25 cars as a single object of mass $m = 25 \times 5.0 \times 10^4$ kg and (when the speed is 30 km/h = 8.3 m/s) subject to a friction force equal to

$$f = 25 \times 250 \times 8.3 = 5.2 \times 10^4 \text{ N}.$$

(a) Along the level track, this object experiences a "forward" force T exerted by the locomotive, so that Newton's second law leads to

$$T - f = ma \;\Rightarrow\; T = 5.2\times 10^4 + (1.25\times 10^6)(0.20) = 3.0\times 10^5 \text{ N}.$$

(b) The free-body diagram is shown next, with θ as the angle of the incline. The $+x$ direction (which is the only direction to which we will be applying Newton's second law) is uphill (to the upper right in our sketch).

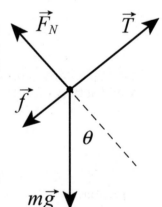

Thus, we obtain
$$T - f - mg\sin\theta = ma$$

where we set $a = 0$ (implied by the problem statement) and solve for the angle. We obtain $\theta = 1.2°$.

72. An excellent discussion and equation development related to this problem is given in Sample Problem 6-2. Using the result, we obtain

$$\theta = \tan^{-1}\mu_s = \tan^{-1} 0.50 = 27°$$

which implies that the angle through which the slope should be *reduced* is

$$\phi = 45° - 27° \approx 20°.$$

73. We make use of Eq. 6-16 which yields

$$\sqrt{\frac{2mg}{C\rho\pi R^2}} = \sqrt{\frac{2(6)(9.8)}{(1.6)(1.2)\pi(0.03)^2}} = 147 \text{ m/s}.$$

74. (a) The upward force exerted by the car on the passenger is equal to the downward force of gravity ($W = 500$ N) on the passenger. So the *net* force does not have a vertical contribution; it only has the contribution from the horizontal force (which is necessary for maintaining the circular motion). Thus $|\vec{F}_{net}| = F = 210$ N.

(b) Using Eq. 6-18, we have

$$v = \sqrt{\frac{FR}{m}} = \sqrt{\frac{(210 \text{ N})(470 \text{ m})}{51.0 \text{ kg}}} = 44.0 \text{ m/s}.$$

75. (a) We note that $F_N = mg$ in this situation, so

$$f_{s,max} = \mu_s mg = (0.52)(11 \text{ kg})(9.8 \text{ m/s}^2) = 56 \text{ N}.$$

Consequently, the horizontal force $\vec{F}$ needed to initiate motion must be (at minimum) slightly more than 56 N.

(b) Analyzing vertical forces when $\vec{F}$ is at nonzero θ yields

$$F\sin\theta + F_N = mg \Rightarrow f_{s,max} = \mu_s(mg - F\sin\theta).$$

Now, the horizontal component of $\vec{F}$ needed to initiate motion must be (at minimum) slightly more than this, so

$$F\cos\theta = \mu_s(mg - F\sin\theta) \Rightarrow F = \frac{\mu_s mg}{\cos\theta + \mu_s \sin\theta}$$

which yields $F = 59$ N when $\theta = 60°$.

(c) We now set $\theta = -60°$ and obtain

$$F = \frac{(0.52)(11 \text{ kg})(9.8 \text{ m/s}^2)}{\cos(-60°) + (0.52)\sin(-60°)} = 1.1 \times 10^3 \text{ N}.$$

76. We use Eq. 6-14, $D = \frac{1}{2} C\rho A v^2$, where ρ is the air density, A is the cross-sectional area of the missile, v is the speed of the missile, and C is the drag coefficient. The area is given by $A = \pi R^2$, where $R = 0.265$ m is the radius of the missile. Thus

$$D = \frac{1}{2}(0.75)(1.2 \text{ kg/m}^3)\pi(0.265 \text{ m})^2 (250 \text{ m/s})^2 = 6.2 \times 10^3 \text{ N}.$$

77. The magnitude of the acceleration of the cyclist as it moves along the horizontal circular path is given by v^2/R, where v is the speed of the cyclist and R is the radius of the curve.

(a) The horizontal component of Newton's second law is $f = mv^2/R$, where f is the static friction exerted horizontally by the ground on the tires. Thus,

$$f = \frac{(85.0 \text{ kg})(9.00 \text{ m/s})^2}{25.0 \text{ m}} = 275 \text{ N}.$$

(b) If F_N is the vertical force of the ground on the bicycle and m is the mass of the bicycle and rider, the vertical component of Newton's second law leads to $F_N = mg = 833$ N. The magnitude of the force exerted by the ground on the bicycle is therefore

$$\sqrt{f^2 + F_N^2} = \sqrt{(275 \text{ N})^2 + (833 \text{ N})^2} = 877 \text{ N}.$$

78. The free-body diagram for the puck is shown below. $\vec{F}_N$ is the normal force of the ice on the puck, $\vec{f}$ is the force of friction (in the $-x$ direction), and $m\vec{g}$ is the force of gravity.

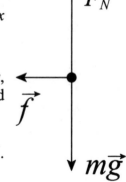

(a) The horizontal component of Newton's second law gives $-f = ma$, and constant acceleration kinematics (Table 2-1) can be used to find the acceleration.

Since the final velocity is zero, $v^2 = v_0^2 + 2ax$ leads to $a = -v_0^2/2x$. This is substituted into the Newton's law equation to obtain

$$f = \frac{mv_0^2}{2x} = \frac{(0.110 \text{ kg})(6.0 \text{ m/s})^2}{2(15 \text{ m})} = 0.13 \text{ N}.$$

(b) The vertical component of Newton's second law gives $F_N - mg = 0$, so $F_N = mg$ which implies (using Eq. 6-2) $f = \mu_k mg$. We solve for the coefficient:

$$\mu_k = \frac{f}{mg} = \frac{0.13 \text{ N}}{(0.110 \text{ kg})(9.8 \text{ m/s}^2)} = 0.12.$$

79. (a) The free-body diagram for the person (shown as an L-shaped block) is shown below. The force that she exerts on the rock slabs is not directly shown (since the diagram should only show forces exerted on her), but it is related by Newton's third law) to the normal forces $\vec{F}_{N1}$ and $\vec{F}_{N2}$ exerted horizontally by the slabs onto her shoes and back, respectively. We will show in part (b) that $F_{N1} = F_{N2}$ so that we there is no ambiguity in saying that the magnitude of her push is F_{N2}. The total upward force due to (maximum) static friction is $\vec{f} = \vec{f}_1 + \vec{f}_2$ where $f_1 = \mu_{s1} F_{N1}$ and $f_2 = \mu_{s2} F_{N2}$. The problem gives the values $\mu_{s1} = 1.2$ and $\mu_{s2} = 0.8$.

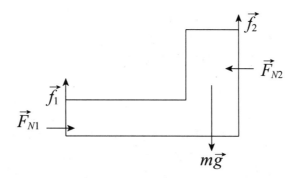

(b) We apply Newton's second law to the x and y axes (with $+x$ rightward and $+y$ upward and there is no acceleration in either direction).

$$F_{N1} - F_{N2} = 0$$
$$f_1 + f_2 - mg = 0$$

The first equation tells us that the normal forces are equal $F_{N1} = F_{N2} = F_N$. Consequently, from Eq. 6-1,

$$f_1 = \mu_{s1} F_N$$
$$f_2 = \mu_{s2} F_N$$

we conclude that

$$f_1 = \left(\frac{\mu_{s1}}{\mu_{s2}}\right) f_2 .$$

Therefore, $f_1 + f_2 - mg = 0$ leads to

$$\left(\frac{\mu_{s1}}{\mu_{s2}} + 1\right) f_2 = mg$$

which (with $m = 49$ kg) yields $f_2 = 192$ N. From this we find $F_N = f_2 / \mu_{s2} = 240$ N. This is equal to the magnitude of the push exerted by the rock climber.

(c) From the above calculation, we find $f_1 = \mu_{s1} F_N = 288$ N which amounts to a fraction

$$\frac{f_1}{W} = \frac{288}{(49)(9.8)} = 0.60$$

or 60% of her weight.

80. The free-body diagram for the stone is shown on the right, with $\vec{F}$ being the force applied to the stone, $\vec{F}_N$ the *downward* normal force of the ceiling on the stone, $m\vec{g}$ the force of gravity, and $\vec{f}$ the force of friction. We take the +x direction to be horizontal to the right and the +y direction to be up. The equations for the x and the y components of the force according to Newton's second law are:

$$F_x = F\cos\theta - f = ma$$
$$F_y = F\sin\theta - F_N - mg = 0$$

Now $f = \mu_k F_N$, and the second equation gives $F_N = F\sin\theta - mg$, which yields $f = \mu_k(F\sin\theta - mg)$. This expression is substituted for f in the first equation to obtain

$$F\cos\theta - \mu_k(F\sin\theta - mg) = ma.$$

For $a = 0$, the force is

$$F = \frac{-\mu_k mg}{\cos\theta - \mu_k \sin\theta}.$$

With $\mu_k = 0.65$, $m = 5.0$ kg, and $\theta = 70°$, we obtain $F = 118$ N.

81. (a) If we choose "downhill" positive, then Newton's law gives

$$m_A g \sin\theta - f_A - T = m_A a$$

for block A (where $\theta = 30°$). For block B we choose leftward as the positive direction and write $T - f_B = m_B a$. Now

$$f_A = \mu_{k,\text{incline}} F_{NA} = \mu' m_A g \cos\theta$$

using Eq. 6-12 applies to block A, and

$$f_B = \mu_k F_{NB} = \mu_k m_B g.$$

In this particular problem, we are asked to set $\mu' = 0$, and the resulting equations can be straightforwardly solved for the tension: $T = 13$ N.

(b) Similarly, finding the value of a is straightforward:

$$a = g(m_A \sin\theta - \mu_k m_B)/(m_A + m_B) = 1.6 \text{ m/s}^2.$$

82. (a) If the skier covers a distance L during time t with zero initial speed and a constant acceleration a, then $L = at^2/2$, which gives the acceleration a_1 for the first (old) pair of skis:

$$a_1 = \frac{2L}{t_1^2} = \frac{2(200\text{ m})}{(61\text{ s})^2} = 0.11 \text{ m/s}^2.$$

(b) The acceleration a_2 for the second (new) pair is

$$a_2 = \frac{2L}{t_2^2} = \frac{2(200\text{ m})}{(42\text{ s})^2} = 0.23 \text{ m/s}^2.$$

(c) The net force along the slope acting on the skier of mass m is

$$F_{net} = mg\sin\theta - f_k = mg(\sin\theta - \mu_k \cos\theta) = ma$$

which we solve for μ_{k1} for the first pair of skis:

$$\mu_{k1} = \tan\theta - \frac{a_1}{g\cos\theta} = \tan 3.0° - \frac{0.11 \text{ m/s}^2}{(9.8 \text{ m/s}^2)\cos 3.0°} = 0.041$$

(d) For the second pair, we have

$$\mu_{k2} = \tan\theta - \frac{a_2}{g\cos\theta} = \tan 3.0° - \frac{0.23 \text{ m/s}^2}{(9.8 \text{ m/s}^2)\cos 3.0°} = 0.029.$$

83. If we choose "downhill" positive, then Newton's law gives

$$mg\sin\theta - f_k = ma$$

for the sliding child. Now using Eq. 6-12

$$f_k = \mu_k F_N = \mu_k mg,$$

so we obtain $a = g(\sin\theta - \mu_k \cos\theta) = -0.5 \text{ m/s}^2$ (note that the problem gives the direction of the acceleration vector as uphill, even though the child is sliding downhill, so it is a deceleration). With $\theta = 35°$, we solve for the coefficient and find $\mu_k = 0.76$.

84. At the top of the hill the vertical forces on the car are the upward normal force exerted by the ground and the downward pull of gravity. Designating $+y$ downward, we have

$$mg - F_N = \frac{mv^2}{R}$$

from Newton's second law. To find the greatest speed without leaving the hill, we set $F_N = 0$ and solve for v:

$$v = \sqrt{gR} = \sqrt{(9.8 \text{ m/s}^2)(250 \text{ m})} = 49.5 \text{ m/s} = 49.5(3600/1000) \text{ km/h} = 178 \text{ km/h}.$$

85. The mass of the car is $m = (10700/9.80)$ kg $= 1.09 \times 10^3$ kg. We choose "inward" (horizontally towards the center of the circular path) as the positive direction.

(a) With $v = 13.4$ m/s and $R = 61$ m, Newton's second law (using Eq. 6-18) leads to

$$f_s = \frac{mv^2}{R} = 3.21 \times 10^3 \text{ N}.$$

(b) Noting that $F_N = mg$ in this situation, the maximum possible static friction is found to be

$$f_{s,\max} = \mu_s mg = (0.35)(10700 \text{ N}) = 3.75 \times 10^3 \text{ N}$$

using Eq. 6-1. We see that the static friction found in part (a) is less than this, so the car rolls (no skidding) and successfully negotiates the curve.

86. (a) Our $+x$ direction is horizontal and is chosen (as we also do with $+y$) so that the components of the 100 N force $\vec{F}$ are non-negative. Thus, $F_x = F \cos\theta = 100$ N, which the textbook denotes F_h in this problem.

(b) Since there is no vertical acceleration, application of Newton's second law in the y direction gives

$$F_N + F_y = mg \Rightarrow F_N = mg - F\sin\theta$$

where $m = 25.0$ kg. This yields $F_N = 245$ N in this case ($\theta = 0°$).

(c) Now, $F_x = F_h = F \cos\theta = 86.6$ N for $\theta = 30.0°$.

(d) And $F_N = mg - F \sin\theta = 195$ N.

(e) We find $F_x = F_h = F \cos\theta = 50.0$ N for $\theta = 60.0°$.

(f) And $F_N = mg - F \sin\theta = 158$ N.

(g) The condition for the chair to slide is

$$F_x > f_{s,\max} = \mu_s F_N \quad \text{where} \quad \mu_s = 0.42.$$

For $\theta = 0°$, we have

$$F_x = 100 \text{ N} < f_{s,\max} = (0.42)(245 \text{ N}) = 103 \text{ N}$$

so the crate remains at rest.

(h) For $\theta = 30.0°$, we find

$$F_x = 86.6 \text{ N} > f_{s,\max} = (0.42)(195 \text{ N}) = 81.9 \text{ N}$$

so the crate slides.

(i) For $\theta = 60°$, we get

$$F_x = 50.0 \text{ N} < f_{s,\max} = (0.42)(158 \text{ N}) = 66.4 \text{ N}$$

which means the crate must remain at rest.

87. For simplicity, we denote the 70° angle as θ and the magnitude of the push (80 N) as P. The vertical forces on the block are the downward normal force exerted on it by the ceiling, the downward pull of gravity (of magnitude mg) and the vertical component of $\vec{P}$ (which is upward with magnitude $P \sin \theta$). Since there is no acceleration in the vertical direction, we must have

$$F_N = P \sin \theta - mg$$

in which case the leftward-pointed kinetic friction has magnitude

$$f_k = \mu_k (P \sin \theta - mg).$$

Choosing $+x$ rightward, Newton's second law leads to

$$P \cos \theta - f_k = ma \implies a = \frac{P \cos \theta - u_k (P \sin \theta - mg)}{m}$$

which yields $a = 3.4$ m/s^2 when $\mu_k = 0.40$ and $m = 5.0$ kg.

88. (a) The intuitive conclusion, that the tension is greatest at the bottom of the swing, is certainly supported by application of Newton's second law there:

$$T - mg = \frac{mv^2}{R} \Rightarrow T = m\left(g + \frac{v^2}{R}\right)$$

where Eq. 6-18 has been used. Increasing the speed eventually leads to the tension at the bottom of the circle reaching that breaking value of 40 N.

(b) Solving the above equation for the speed, we find

$$v = \sqrt{R\left(\frac{T}{m} - g\right)} = \sqrt{(0.91 \text{ m})\left(\frac{40 \text{ N}}{0.37 \text{ kg}} - 9.8 \text{ m/s}^2\right)}$$

which yields $v = 9.5$ m/s.

89. (a) The push (to get it moving) must be at least as big as $f_{s,\max} = \mu_s F_N$ (Eq. 6-1, with $F_N = mg$ in this case), which equals $(0.51)(165 \text{ N}) = 84.2$ N.

(b) While in motion, constant velocity (zero acceleration) is maintained if the push is equal to the kinetic friction force $f_k = \mu_k F_N = \mu_k mg = 52.8$ N.

(c) We note that the mass of the crate is $165/9.8 = 16.8$ kg. The acceleration, using the push from part (a), is

$$a = (84.2 \text{ N} - 52.8 \text{ N})/(16.8 \text{ kg}) \approx 1.87 \text{ m/s}^2.$$

90. In the figure below, $m = 140/9.8 = 14.3$ kg is the mass of the child. We use $\vec{w}_x$ and $\vec{w}_y$ as the components of the gravitational pull of Earth on the block; their magnitudes are $w_x = mg \sin \theta$ and $w_y = mg \cos \theta$.

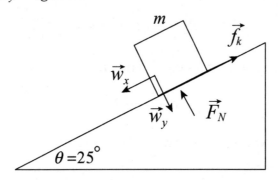

(a) With the x axis directed up along the incline (so that $a = -0.86$ m/s^2), Newton's second law leads to

$$f_k - 140 \sin 25° = m(-0.86)$$

which yields $f_k = 47$ N. We also apply Newton's second law to the y axis (perpendicular to the incline surface), where the acceleration-component is zero:

$$F_N - 140\cos 25° = 0 \quad \Rightarrow \quad F_N = 127 \text{ N}.$$

Therefore, $\mu_k = f_k/F_N = 0.37$.

(b) Returning to our first equation in part (a), we see that if the downhill component of the weight force were insufficient to overcome static friction, the child would not slide at all. Therefore, we require $140 \sin 25° > f_{s,max} = \mu_s F_N$, which leads to $\tan 25° = 0.47 > \mu_s$. The minimum value of μ_s equals μ_k and is more subtle; reference to §6-1 is recommended. If μ_k exceeded μ_s then when static friction were overcome (as the incline is raised) then it should start to move – which is impossible if f_k is large enough to cause deceleration! The bounds on μ_s are therefore given by $0.47 > \mu_s > 0.37$.

91. We apply Newton's second law (as $F_{push} - f = ma$). If we find $F_{push} < f_{max}$, we conclude "no, the cabinet does not move" (which means a is actually 0 and $f = F_{push}$), and if we obtain $a > 0$ then it is moves (so $f = f_k$). For f_{max} and f_k we use Eq. 6-1 and Eq. 6-2 (respectively), and in those formulas we set the magnitude of the normal force equal to 556 N. Thus, $f_{max} = 378$ N and $f_k = 311$ N.

(a) Here we find $F_{push} < f_{max}$ which leads to $f = F_{push} = 222$ N.

(b) Again we find $F_{push} < f_{max}$ which leads to $f = F_{push} = 334$ N.

(c) Now we have $F_{push} > f_{max}$ which means it moves and $f = f_k = 311$ N.

(d) Again we have $F_{push} > f_{max}$ which means it moves and $f = f_k = 311$ N.

(e) The cabinet moves in (c) and (d).

92. (a) The tension will be the greatest at the lowest point of the swing. Note that there is no substantive difference between the tension T in this problem and the normal force F_N in Sample Problem 6-7. Eq. 6-19 of that Sample Problem examines the situation at the top of the circular path (where F_N is the least), and rewriting that for the bottom of the path leads to

$$T = mg + mv^2/r$$

where F_N is at its greatest value.

(b) At the breaking point $T = 33$ N $= m(g + v^2/r)$ where $m = 0.26$ kg and $r = 0.65$ m. Solving for the speed, we find that the cord should break when the speed (at the lowest point) reaches 8.73 m/s.

93. (a) The component of the weight along the incline (with downhill understood as the positive direction) is $mg\sin\theta$ where $m = 630$ kg and $\theta = 10.2°$. With $f = 62.0$ N, Newton's second law leads to

$$mg\sin\theta - f = ma$$

which yields $a = 1.64$ m/s². Using Eq. 2-15, we have

$$80.0 \text{ m} = \left(6.20\frac{\text{m}}{\text{s}}\right)t + \frac{1}{2}\left(1.64\frac{\text{m}}{\text{s}^2}\right)t^2.$$

This is solved using the quadratic formula. The positive root is $t = 6.80$ s.

(b) Running through the calculation of part (a) with $f = 42.0$ N instead of $f = 62$ N results in $t = 6.76$ s.

94. (a) The x component of $\vec{F}$ tries to move the crate while its y component indirectly contributes to the inhibiting effects of friction (by increasing the normal force). Newton's second law implies

$$x \text{ direction: } F\cos\theta - f_s = 0$$

$$y \text{ direction: } F_N - F\sin\theta - mg = 0.$$

To be "on the verge of sliding" means $f_s = f_{s,\max} = \mu_s F_N$ (Eq. 6-1). Solving these equations for F (actually, for the ratio of F to mg) yields

$$\frac{F}{mg} = \frac{\mu_s}{\cos\theta - \mu_s \sin\theta}.$$

This is plotted on the right (θ in degrees).

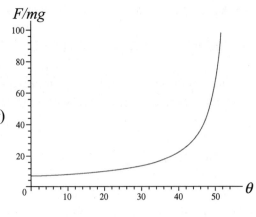

(b) The denominator of our expression (for F/mg) vanishes when

$$\cos\theta - \mu_s \sin\theta = 0 \quad \Rightarrow \quad \theta_{\inf} = \tan^{-1}\left(\frac{1}{\mu_s}\right)$$

For $\mu_s = 0.70$, we obtain $\theta_{\inf} = \tan^{-1}\left(\frac{1}{\mu_s}\right) = 55°$.

(c) Reducing the coefficient means increasing the angle by the condition in part (b).

(d) For $\mu_s = 0.60$ we have $\theta_{inf} = \tan^{-1}\left(\dfrac{1}{\mu_s}\right) = 59°$.

95. The car is in "danger of sliding" down when

$$\mu_s = \tan\theta = \tan 35.0° = 0.700.$$

This value represents a 3.4% decrease from the given 0.725 value.

96. For the $m_2 = 1.0$ kg block, application of Newton's laws result in

$$\begin{aligned} F\cos\theta - T - f_k &= m_2 a & x\text{ axis} \\ F_N - F\sin\theta - m_2 g &= 0 & y\text{ axis} \end{aligned}$$

Since $f_k = \mu_k F_N$, these equations can be combined into an equation to solve for a:

$$F(\cos\theta - \mu_k \sin\theta) - T - \mu_k m_2 g = m_2 a$$

Similarly (but without the applied push) we analyze the $m_1 = 2.0$ kg block:

$$\begin{aligned} T - f'_k &= m_1 a & x\text{ axis} \\ F'_N - m_1 g &= 0 & y\text{ axis} \end{aligned}$$

Using $f_k = \mu_k F'_N$, the equations can be combined:

$$T - \mu_k m_1 g = m_1 a$$

Subtracting the two equations for a and solving for the tension, we obtain

$$T = \dfrac{m_1(\cos\theta - \mu_k \sin\theta)}{m_1 + m_2} F = \dfrac{(2.0\text{ kg})[\cos 35° - (0.20)\sin 35°]}{2.0\text{ kg} + 1.0\text{ kg}}(20\text{ N}) = 9.4\text{ N}.$$

97. (a) The x component of $\vec{F}$ contributes to the motion of the crate while its y component indirectly contributes to the inhibiting effects of friction (by increasing the normal force). Along the y direction, we have $F_N - F\cos\theta - mg = 0$ and along the x direction we have $F\sin\theta - f_k = 0$ (since it is not accelerating, according to the problem). Also, Eq. 6-2 gives $f_k = \mu_k F_N$. Solving these equations for F yields

$$F = \dfrac{\mu_k mg}{\sin\theta - \mu_k \cos\theta}.$$

(b) When $\theta < \theta_0 = \tan^{-1}\mu_s$, F will not be able to move the mop head.

98. Consider that the car is "on the verge of sliding out" – meaning that the force of static friction is acting "down the bank" (or "downhill" from the point of view of an ant on the banked curve) with maximum possible magnitude. We first consider the vector sum $\vec{F}$ of the (maximum) static friction force and the normal force. Due to the facts that they are perpendicular and their magnitudes are simply proportional (Eq. 6-1), we find $\vec{F}$ is at angle (measured from the vertical axis) $\phi = \theta + \theta_s$ where $\tan \theta_s = \mu_s$ (compare with Eq. 6-13), and θ is the bank angle. Now, the vector sum of $\vec{F}$ and the vertically downward pull (mg) of gravity must be equal to the (horizontal) centripetal force (mv^2/R), which leads to a surprisingly simple relationship:

$$\tan\phi = \frac{mv^2/R}{mg} = \frac{v^2}{Rg}.$$

Writing this as an expression for the maximum speed, we have

$$v_{max} = \sqrt{Rg\tan(\theta + \tan^{-1}\mu_s)} = \sqrt{\frac{Rg(\tan\theta + \mu_s)}{1 - \mu_s \tan\theta}}.$$

(a) We note that the given speed is (in SI units) roughly 17 m/s. If we do not want the cars to "depend" on the static friction to keep from sliding out (that is, if we want the component "down the back" of gravity to be sufficient), then we can set $\mu_s = 0$ in the above expression and obtain $v = \sqrt{Rg\tan\theta}$. With $R = 150$ m, this leads to $\theta = 11°$.

(b) If, however, the curve is not banked (so $\theta = 0$) then the above expression becomes

$$v = \sqrt{Rg\tan(\tan^{-1}\mu_s)} = \sqrt{Rg\mu_s}$$

Solving this for the coefficient of static friction $\mu_s = 0.19$.

99. Replace f_s with f_k in Fig. 6-5(b) to produce the appropriate force diagram for the first part of this problem (when it is sliding downhill with zero acceleration). This amounts to replacing the static coefficient with the kinetic coefficient in Eq. 6-13: $\mu_k = \tan\theta$. Now (for the second part of the problem, with the block projected uphill) the friction direction is reversed from what is shown in Fig. 6-5(b). Newton's second law for the uphill motion (and Eq. 6-12) leads to

$$-mg\sin\theta - \mu_k mg\cos\theta = ma.$$

Canceling the mass and substituting what we found earlier for the coefficient, we have

$$-g\sin\theta - \tan\theta g\cos\theta = a.$$

This simplifies to $-2g\sin\theta = a$. Eq. 2-16 then gives the distance to stop: $\Delta x = -v_0^2/2a$.

(a) Thus, the distance up the incline traveled by the block is $\Delta x = v_0^2/(4g\sin\theta)$.

(b) We usually expect $\mu_s > \mu_k$ (see the discussion in section 6-1). Sample Problem 6-2 treats the "angle of repose" (the minimum angle necessary for a stationary block to start sliding downhill): $\mu_s = \tan(\theta_{repose})$. Therefore, we expect $\theta_{repose} > \theta$ found in part (a). Consequently, when the block comes to rest, the incline is not steep enough to cause it to start slipping down the incline again.

100. Analysis of forces in the horizontal direction (where there can be no acceleration) leads to the conclusion that $F = F_N$; the magnitude of the normal force is 60 N. The maximum possible static friction force is therefore $\mu_s F_N = 33$ N, and the kinetic friction force (when applicable) is $\mu_k F_N = 23$ N.

(a) In this case, $\vec{P} = 34$ N upward. Assuming $\vec{f}$ points down, then Newton's second law for the y leads to

$$P - mg - f = ma \ .$$

if we assume $f = f_s$ and $a = 0$, we obtain $f = (34 - 22)$ N $= 12$ N. This is less than $f_{s,\,max}$, which shows the consistency of our assumption. The answer is: $\vec{f_s} = 12$ N down.

(b) In this case, $\vec{P} = 12$ N upward. The above equation, with the same assumptions as in part (a), leads to $f = (12 - 22)$ N $= -10$ N. Thus, $|f_s| < f_{s,\,max}$, justifying our assumption that the block is stationary, but its negative value tells us that our initial assumption about the direction of $\vec{f}$ is incorrect in this case. Thus, the answer is: $\vec{f_s} = 10$ N up.

(c) In this case, $\vec{P} = 48$ N upward. The above equation, with the same assumptions as in part (a), leads to $f = (48 - 22)$ N $= 26$ N. Thus, we again have $f_s < f_{s,\,max}$, and our answer is: $\vec{f_s} = 26$ N down.

(d) In this case, $\vec{P} = 62$ N upward. The above equation, with the same assumptions as in part (a), leads to $f = (62 - 22)$ N $= 40$ N, which is larger than $f_{s,\,max}$, -- invalidating our assumptions. Therefore, we take $f = f_k$ and $a \neq 0$ in the above equation; if we wished to find the value of a we would find it to be positive, as we should expect. The answer is: $\vec{f_k} = 23$ N down.

(e) In this case, $\vec{P} = 10$ N downward. The above equation (but with P replaced with $-P$) with the same assumptions as in part (a), leads to $f = (-10 - 22)$ N $= -32$ N. Thus, we have $|f_s| < f_{s,\,max}$, justifying our assumption that the block is stationary, but its negative value tells us that our initial assumption about the direction of $\vec{f}$ is incorrect in this case. Thus, the answer is: $\vec{f_s}$ = 32 N up.

(f) In this case, $\vec{P}$ = 18 N downward. The above equation (but with P replaced with –P) with the same assumptions as in part (a), leads to $f = (-18 - 22)$ N = –40 N, which is larger (in absolute value) than $f_{s,\,max}$, -- invalidating our assumptions. Therefore, we take $f = f_k$ and $a \neq 0$ in the above equation; if we wished to find the value of a we would find it to be negative, as we should expect. The answer is: $\vec{f_k}$ = 23 N up.

(g) The block moves up the wall in case (d) where $a > 0$.

(h) The block moves down the wall in case (f) where $a < 0$.

(i) The frictional force $\vec{f_s}$ is directed down in cases (a), (c) and (d).

101. (a) The distance traveled by the coin in 3.14 s is $3(2\pi r) = 6\pi(0.050) = 0.94$ m. Thus, its speed is $v = 0.94/3.14 = 0.30$ m/s.

(b) The centripetal acceleration is given by Eq. 6-17:

$$a = \frac{v^2}{r} = \frac{(0.30 \text{ m/s})^2}{0.050 \text{ m}} = 1.8 \text{ m/s}^2 \ .$$

(c) The acceleration vector (at any instant) is horizontal and points from the coin towards the center of the turntable.

(d) The only horizontal force acting on the coin is static friction f_s and must be large enough to supply the acceleration of part (b) for the $m = 0.0020$ kg coin. Using Newton's second law,

$$f_s = ma = (0.0020 \text{ kg})(1.8 \text{ m/s}^2) = 3.6 \times 10^{-3} \text{ N} \ .$$

(e) The static friction f_s must point in the same direction as the acceleration (towards the center of the turntable).

(f) We note that the normal force exerted upward on the coin by the turntable must equal the coin's weight (since there is no vertical acceleration in the problem). We also note that if we repeat the computations in parts (a) and (b) for $r' = 0.10$ m, then we obtain $v' = 0.60$ m/s and $a' = 3.6$ m/s^2. Now, if friction is at its maximum at $r = r'$, then, by Eq. 6-1, we obtain

$$\mu_s = \frac{f_{s,\text{max}}}{mg} = \frac{ma'}{mg} = 0.37 \ .$$

102. (a) The distance traveled in one revolution is $2\pi R = 2\pi(4.6 \text{ m}) = 29$ m. The (constant) speed is consequently $v = (29 \text{ m})/(30 \text{ s}) = 0.96$ m/s.

(b) Newton's second law (using Eq. 6-17 for the magnitude of the acceleration) leads to

$$f_s = m\left(\frac{v^2}{R}\right) = m(0.20)$$

in SI units. Noting that $F_N = mg$ in this situation, the maximum possible static friction is $f_{s,\text{max}} = \mu_s mg$ using Eq. 6-1. Equating this with $f_s = m(0.20)$ we find the mass m cancels and we obtain $\mu_s = 0.20/9.8 = 0.021$.

103. (a) The box doesn't move until $t = 2.8$ s, which is when the applied force $\vec{F}$ reaches a magnitude of $F = (1.8)(2.8) = 5.0$ N, implying therefore that $f_{s,\text{max}} = 5.0$ N. Analysis of the vertical forces on the block leads to the observation that the normal force magnitude equals the weight $F_N = mg = 15$ N. Thus, $\mu_s = f_{s,\text{max}}/F_N = 0.34$.

(b) We apply Newton's second law to the horizontal x axis (positive in the direction of motion):

$$F - f_k = ma \Rightarrow 1.8t - f_k = (1.5)(1.2t - 2.4)$$

Thus, we find $f_k = 3.6$ N. Therefore, $\mu_k = f_k/F_N = 0.24$.

104. We note that $F_N = mg$ in this situation, so $f_k = \mu_k mg = (0.32)(220 \text{ N}) = 70.4$ N and $f_{s,\text{max}} = \mu_s mg = (0.41)(220 \text{ N}) = 90.2$ N.

(a) The person needs to push at least as hard as the static friction maximum if he hopes to start it moving. Denoting his force as P, this means a value of P slightly larger than 90.2 N is sufficient. Rounding to two figures, we obtain $P = 90$ N.

(b) Constant velocity (zero acceleration) implies the push equals the kinetic friction, so $P = 70$ N.

(c) Applying Newton's second law, we have

$$P - f_k = ma \Rightarrow a = \frac{\mu_s mg - \mu_k mg}{m}$$

which simplifies to $a = g(\mu_s - \mu_k) = 0.88$ m/s^2.

105. Probably the most appropriate picture in the textbook to represent the situation in this problem is in the previous chapter: Fig. 5-9. We adopt the familiar axes with $+x$ rightward and $+y$ upward, and refer to the 85 N horizontal push of the worker as P (and

assume it to be rightward). Applying Newton's second law to the x axis and y axis, respectively, produces

$$P - f_k = ma$$
$$F_N - mg = 0.$$

Using $v^2 = v_0^2 + 2a\Delta x$ we find $a = 0.36$ m/s^2. Consequently, we obtain $f_k = 71$ N and $F_N = 392$ N. Therefore, $\mu_k = f_k/F_N = 0.18$.

106. (a) The centripetal force is given by Eq. 6-18:

$$F = \frac{mv^2}{R} = \frac{(1.00 \text{ kg})(465 \text{ m/s})^2}{6.40 \times 10^6 \text{ m}} = 0.0338 \text{ N}.$$

(b) Calling downward (towards the center of Earth) the positive direction, Newton's second law leads to

$$mg - T = ma$$

where $mg = 9.80$ N and $ma = 0.034$ N, calculated in part (a). Thus, the tension in the cord by which the body hangs from the balance is $T = 9.80$ N $-$ 0.03 N $= 9.77$ N. Thus, this is the reading for a standard kilogram mass, of the scale at the equator of the spinning Earth.

107. Except for replacing f_s with f_k, Fig 6-5 in the textbook is appropriate. With that figure in mind, we choose uphill as the $+x$ direction. Applying Newton's second law to the x axis, we have

$$f_k - W \sin\theta = ma \quad \text{where } m = \frac{W}{g},$$

and where $W = 40$ N, $a = +0.80$ m/s^2 and $\theta = 25°$. Thus, we find $f_k = 20$ N. Along the y axis, we have

$$\sum \vec{F}_y = 0 \Rightarrow F_N = W \cos\theta$$

so that $\mu_k = f_k/F_N = 0.56$.

108. The assumption that there is no slippage indicates that we are dealing with static friction f_s, and it is this force that is responsible for "pushing" the luggage along as the belt moves. Thus, Fig. 6-5 in the textbook is appropriate for this problem -- *if* one reverses the arrow indicating the direction of motion (and removes the word "impending"). The mass of the box is $m = 69/9.8 = 7.0$ kg. Applying Newton's law to the x axis leads to

$$f_s - mg \sin\theta = ma$$

where $\theta = 2.5°$ and uphill is the positive direction.

(a) Interpreting "temporarily at rest" (which is not meant to be the same thing as "momentarily at rest") to mean that the box is at equilibrium, we have $a = 0$ and, consequently, $f_s = mg \sin \theta = 3.0$ N. It is positive and therefore pointed uphill.

(b) Constant speed in a one-dimensional setting implies that the velocity is constant -- thus, $a = 0$ again. We recover the answer $f_s = 3.0$ N uphill, which we obtained in part (a).

(c) Early in the problem, the direction of motion of the luggage was given: downhill. Thus, an increase in that speed indicates a downhill acceleration $a = -0.20$ m/s^2. We now solve for the friction and obtain

$$f_s = ma + mg \sin \theta = 1.6 \text{ N},$$

which is positive -- therefore, uphill.

(d) A decrease in the (downhill) speed indicates the acceleration vector points uphill; thus, $a = +0.20$ m/s^2. We solve for the friction and obtain

$$f_s = ma + mg \sin \theta = 4.4 \text{ N},$$

which is positive -- therefore, uphill.

(e) The situation is similar to the one described in part (c), but with $a = -0.57$ m/s^2. Now,

$$f_s = ma + mg \sin \theta = -1.0 \text{ N},$$

or $|f_s| = 1.0$ N. Since f_s is negative, the direction is downhill.

(f) From the above, the only case where f_s is directed downhill is (e).

109. We resolve this horizontal force into appropriate components.

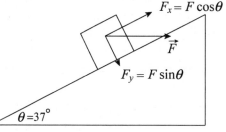

(a) Applying Newton's second law to the x (directed uphill) and y (directed away from the incline surface) axes, we obtain

$$F \cos \theta - f_k - mg \sin \theta = ma$$
$$F_N - F \sin \theta - mg \cos \theta = 0.$$

Using $f_k = \mu_k F_N$, these equations lead to

$$a = \frac{F}{m}(\cos \theta - \mu_k \sin \theta) - g(\sin \theta + \mu_k \cos \theta)$$

which yields $a = -2.1$ m/s^2, or $|a| = 2.1$ m/s^2, for $\mu_k = 0.30$, $F = 50$ N and $m = 5.0$ kg.

(b) The direction of $\vec{a}$ is down the plane.

(c) With $v_0 = +4.0$ m/s and $v = 0$, Eq. 2-16 gives

$$\Delta x = -\frac{(4.0 \text{ m/s})^2}{2(-2.1 \text{ m/s}^2)} = 3.9 \text{ m}.$$

(d) We expect $\mu_s \geq \mu_k$; otherwise, an object started into motion would immediately start decelerating (before it gained any speed)! In the minimal expectation case, where $\mu_s = 0.30$, the maximum possible (downhill) static friction is, using Eq. 6-1,

$$f_{s,\max} = \mu_s F_N = \mu_s (F \sin\theta + mg \cos\theta)$$

which turns out to be 21 N. But in order to have no acceleration along the x axis, we must have

$$f_s = F \cos\theta - mg \sin\theta = 10 \text{ N}$$

(the fact that this is positive reinforces our suspicion that $\vec{f}_s$ points downhill). Since the f_s needed to remain at rest is less than $f_{s,\max}$ then it stays at that location.

Chapter 7

1. (a) The change in kinetic energy for the meteorite would be

$$\Delta K = K_f - K_i = -K_i = -\frac{1}{2}m_i v_i^2 = -\frac{1}{2}(4\times10^6 \text{ kg})(15\times10^3 \text{ m/s})^2 = -5\times10^{14} \text{ J},$$

or $|\Delta K| = 5\times10^{14}$ J. The negative sign indicates that kinetic energy is lost.

(b) The energy loss in units of megatons of TNT would be

$$-\Delta K = (5\times10^{14} \text{ J})\left(\frac{1 \text{ megaton TNT}}{4.2\times10^{15} \text{ J}}\right) = 0.1 \text{ megaton TNT}.$$

(c) The number of bombs N that the meteorite impact would correspond to is found by noting that megaton = 1000 kilotons and setting up the ratio:

$$N = \frac{0.1\times1000 \text{ kiloton TNT}}{13 \text{ kiloton TNT}} = 8.$$

2. With speed $v = 11200$ m/s, we find

$$K = \frac{1}{2}mv^2 = \frac{1}{2}(2.9\times10^5 \text{ kg})(11200 \text{ m/s})^2 = 1.8\times10^{13} \text{ J}.$$

3. (a) From Table 2-1, we have $v^2 = v_0^2 + 2a\Delta x$. Thus,

$$v = \sqrt{v_0^2 + 2a\Delta x} = \sqrt{(2.4\times10^7 \text{ m/s})^2 + 2(3.6\times10^{15} \text{ m/s}^2)(0.035 \text{ m})} = 2.9\times10^7 \text{ m/s}.$$

(b) The initial kinetic energy is

$$K_i = \frac{1}{2}mv_0^2 = \frac{1}{2}(1.67\times10^{-27} \text{ kg})(2.4\times10^7 \text{ m/s})^2 = 4.8\times10^{-13} \text{ J}.$$

The final kinetic energy is

$$K_f = \frac{1}{2}mv^2 = \frac{1}{2}(1.67\times10^{-27} \text{ kg})(2.9\times10^7 \text{ m/s})^2 = 6.9\times10^{-13} \text{ J}.$$

278 CHAPTER 7

The change in kinetic energy is $\Delta K = 6.9 \times 10^{-13}$ J $- 4.8 \times 10^{-13}$ J $= 2.1 \times 10^{-13}$ J.

4. The work done by the applied force $\vec{F}_a$ is given by $W = \vec{F}_a \cdot \vec{d} = F_a d \cos\phi$. From Fig. 7-24, we see that $W = 25$ J when $\phi = 0$ and $d = 5.0$ cm. This yields the magnitude of $\vec{F}_a$:

$$F_a = \frac{W}{d} = \frac{25 \text{ J}}{0.050 \text{ m}} = 5.0 \times 10^2 \text{ N}.$$

(a) For $\phi = 64°$, we have $W = F_a d \cos\phi = (5.0 \times 10^2 \text{ N})(0.050 \text{ m})\cos 64° = 11$ J.

(b) For $\phi = 147°$, we have $W = F_a d \cos\phi = (5.0 \times 10^2 \text{ N})(0.050 \text{ m})\cos 147° = -21$ J.

5. We denote the mass of the father as m and his initial speed v_i. The initial kinetic energy of the father is

$$K_i = \frac{1}{2} K_{\text{son}}$$

and his final kinetic energy (when his speed is $v_f = v_i + 1.0$ m/s) is $K_f = K_{\text{son}}$. We use these relations along with Eq. 7-1 in our solution.

(a) We see from the above that $K_i = \frac{1}{2} K_f$ which (with SI units understood) leads to

$$\frac{1}{2} m v_i^2 = \frac{1}{2} \left[\frac{1}{2} m \left(v_i + 1.0 \text{ m/s} \right)^2 \right].$$

The mass cancels and we find a second-degree equation for v_i:

$$\frac{1}{2} v_i^2 - v_i - \frac{1}{2} = 0.$$

The positive root (from the quadratic formula) yields $v_i = 2.4$ m/s.

(b) From the first relation above $\left(K_i = \frac{1}{2} K_{\text{son}}\right)$, we have

$$\frac{1}{2} m v_i^2 = \frac{1}{2} \left(\frac{1}{2} (m/2) v_{\text{son}}^2 \right)$$

and (after canceling m and one factor of 1/2) are led to $v_{\text{son}} = 2 v_i = 4.8$ m/s.

6. We apply the equation $x(t) = x_0 + v_0 t + \frac{1}{2}at^2$, found in Table 2-1. Since at $t = 0$ s, $x_0 = 0$ and $v_0 = 12$ m/s, the equation becomes (in unit of meters)

$$x(t) = 12t + \tfrac{1}{2}at^2.$$

With $x = 10$ m when $t = 1.0$ s, the acceleration is found to be $a = -4.0$ m/s^2. The fact that $a < 0$ implies that the bead is decelerating. Thus, the position is described by $x(t) = 12t - 2.0t^2$. Differentiating x with respect to t then yields

$$v(t) = \frac{dx}{dt} = 12 - 4.0t.$$

Indeed at $t = 3.0$ s, $v(t = 3.0) = 0$ and the bead stops momentarily. The speed at $t = 10$ s is $v(t = 10) = -28$ m/s, and the corresponding kinetic energy is

$$K = \frac{1}{2}mv^2 = \frac{1}{2}(1.8 \times 10^{-2}\,\text{kg})(-28\,\text{m/s})^2 = 7.1\,\text{J}.$$

7. By the work-kinetic energy theorem,

$$W = \Delta K = \frac{1}{2}mv_f^2 - \frac{1}{2}mv_i^2 = \frac{1}{2}(2.0\,\text{kg})\left((6.0\,\text{m/s})^2 - (4.0\,\text{m/s})^2\right) = 20\,\text{J}.$$

We note that the *directions* of $\vec{v}_f$ and $\vec{v}_i$ play no role in the calculation.

8. Eq. 7-8 readily yields

$$W = F_x \Delta x + F_y \Delta y = (2.0\,\text{N})\cos(100°)(3.0\,\text{m}) + (2.0\,\text{N})\sin(100°)(4.0\,\text{m}) = 6.8\,\text{J}.$$

9. Since this involves constant-acceleration motion, we can apply the equations of Table 2-1, such as $x = v_0 t + \frac{1}{2}at^2$ (where $x_0 = 0$). We choose to analyze the third and fifth points, obtaining

$$0.2\,\text{m} = v_0(1.0\,\text{s}) + \frac{1}{2}a\,(1.0\,\text{s})^2$$

$$0.8\,\text{m} = v_0(2.0\,\text{s}) + \frac{1}{2}a\,(2.0\,\text{s})^2$$

Simultaneous solution of the equations leads to $v_0 = 0$ and $a = 0.40\,\text{m/s}^2$. We now have two ways to finish the problem. One is to compute force from $F = ma$ and then obtain the work from Eq. 7-7. The other is to find ΔK as a way of computing W (in accordance with Eq. 7-10). In this latter approach, we find the velocity at $t = 2.0$ s from $v = v_0 + at$ (so $v = 0.80$ m/s). Thus,

$$W = \Delta K = \frac{1}{2}(3.0\,\text{kg})(0.80\,\text{m/s})^2 = 0.96\,\text{J}.$$

10. Using Eq. 7-8 (and Eq. 3-23), we find the work done by the water on the ice block:

$$W = \vec{F}\cdot\vec{d} = \left[(210\,\text{N})\hat{i} - (150\,\text{N})\hat{j}\right]\cdot\left[(15\,\text{m})\hat{i} - (12\,\text{m})\hat{j}\right] = (210\,\text{N})(15\,\text{m}) + (-150\,\text{N})(-12\,\text{m})$$
$$= 5.0\times 10^3\,\text{J}.$$

11. We choose $+x$ as the direction of motion (so $\vec{a}$ and $\vec{F}$ are negative-valued).

(a) Newton's second law readily yields $\vec{F} = (85\,\text{kg})(-2.0\,\text{m/s}^2)$ so that

$$F = |\vec{F}| = 1.7\times 10^2\,\text{N}.$$

(b) From Eq. 2-16 (with $v = 0$) we have

$$0 = v_0^2 + 2a\Delta x \quad\Rightarrow\quad \Delta x = -\frac{(37\,\text{m/s})^2}{2(-2.0\,\text{m/s}^2)} = 3.4\times 10^2\,\text{m}.$$

Alternatively, this can be worked using the work-energy theorem.

(c) Since $\vec{F}$ is opposite to the direction of motion (so the angle ϕ between $\vec{F}$ and $\vec{d} = \Delta x$ is 180°) then Eq. 7-7 gives the work done as $W = -F\Delta x = -5.8\times 10^4\,\text{J}$.

(d) In this case, Newton's second law yields $\vec{F} = (85\,\text{kg})(-4.0\,\text{m/s}^2)$ so that $F = |\vec{F}| = 3.4\times 10^2\,\text{N}$.

(e) From Eq. 2-16, we now have

$$\Delta x = -\frac{(37\,\text{m/s})^2}{2(-4.0\,\text{m/s}^2)} = 1.7\times 10^2\,\text{m}.$$

(f) The force $\vec{F}$ is again opposite to the direction of motion (so the angle ϕ is again 180°) so that Eq. 7-7 leads to $W = -F\Delta x = -5.8\times 10^4\,\text{J}$. The fact that this agrees with the result of part (c) provides insight into the concept of work.

12. The change in kinetic energy can be written as

$$\Delta K = \frac{1}{2}m(v_f^2 - v_i^2) = \frac{1}{2}m(2a\Delta x) = ma\Delta x$$

where we have used $v_f^2 = v_i^2 + 2a\Delta x$ from Table 2-1. From Fig. 7-27, we see that $\Delta K = (0-30)$ J $= -30$ J when $\Delta x = +5$ m. The acceleration can then be obtained as

$$a = \frac{\Delta K}{m\Delta x} = \frac{(-30 \text{ J})}{(8.0 \text{ kg})(5.0 \text{ m})} = -0.75 \text{ m/s}^2.$$

The negative sign indicates that the mass is decelerating. From the figure, we also see that when $x = 5$ m the kinetic energy becomes zero, implying that the mass comes to rest momentarily. Thus,

$$v_0^2 = v^2 - 2a\Delta x = 0 - 2(-0.75 \text{ m/s}^2)(5.0 \text{ m}) = 7.5 \text{ m}^2/\text{s}^2,$$

or $v_0 = 2.7$ m/s. The speed of the object when $x = -3.0$ m is

$$v = \sqrt{v_0^2 + 2a\Delta x} = \sqrt{7.5 \text{ m}^2/\text{s}^2 + 2(-0.75 \text{ m/s}^2)(-3.0 \text{ m})} = \sqrt{12} \text{ m/s} = 3.5 \text{ m/s}.$$

13. (a) The forces are constant, so the work done by any one of them is given by $W = \vec{F} \cdot \vec{d}$, where $\vec{d}$ is the displacement. Force $\vec{F}_1$ is in the direction of the displacement, so

$$W_1 = F_1 d \cos\phi_1 = (5.00 \text{ N})(3.00 \text{ m})\cos 0° = 15.0 \text{ J}.$$

Force $\vec{F}_2$ makes an angle of 120° with the displacement, so

$$W_2 = F_2 d \cos\phi_2 = (9.00 \text{ N})(3.00 \text{ m})\cos 120° = -13.5 \text{ J}.$$

Force $\vec{F}_3$ is perpendicular to the displacement, so

$$W_3 = F_3 d \cos\phi_3 = 0 \text{ since } \cos 90° = 0.$$

The net work done by the three forces is

$$W = W_1 + W_2 + W_3 = 15.0 \text{ J} - 13.5 \text{ J} + 0 = +1.50 \text{ J}.$$

(b) If no other forces do work on the box, its kinetic energy increases by 1.50 J during the displacement.

14. (a) From Eq. 7-6, $F = W/x = 3.00$ N (this is the slope of the graph).

(b) Eq. 7-10 yields $K = K_i + W = 3.00$ J $+ 6.00$ J $= 9.00$ J.

15. Using the work-kinetic energy theorem, we have

$$\Delta K = W = \vec{F} \cdot \vec{d} = Fd\cos\phi$$

In addition, $F = 12$ N and $d = \sqrt{(2.00\text{ m})^2 + (-4.00\text{ m})^2 + (3.00\text{ m})^2} = 5.39$ m.

(a) If $\Delta K = +30.0$ J, then

$$\phi = \cos^{-1}\left(\frac{\Delta K}{Fd}\right) = \cos^{-1}\left(\frac{30.0\text{ J}}{(12.0\text{ N})(5.39\text{ m})}\right) = 62.3°.$$

(b) $\Delta K = -30.0$ J, then

$$\phi = \cos^{-1}\left(\frac{\Delta K}{Fd}\right) = \cos^{-1}\left(\frac{-30.0\text{ J}}{(12.0\text{ N})(5.39\text{ m})}\right) = 118°$$

16. The forces are all constant, so the total work done by them is given by $W = F_{net}\Delta x$, where F_{net} is the magnitude of the net force and Δx is the magnitude of the displacement. We add the three vectors, finding the x and y components of the net force:

$$F_{net\,x} = -F_1 - F_2\sin 50.0° + F_3\cos 35.0° = -3.00\text{ N} - (4.00\text{ N})\sin 35.0° + (10.0\text{ N})\cos 35.0°$$
$$= 2.13\text{ N}$$

$$F_{net\,y} = -F_2\cos 50.0° + F_3\sin 35.0° = -(4.00\text{ N})\cos 50.0° + (10.0\text{ N})\sin 35.0°$$
$$= 3.17\text{ N}.$$

The magnitude of the net force is

$$F_{net} = \sqrt{F_{net\,x}^2 + F_{net\,y}^2} = \sqrt{(2.13\text{ N})^2 + (3.17\text{ N})^2} = 3.82\text{ N}.$$

The work done by the net force is

$$W = F_{net}d = (3.82\text{ N})(4.00\text{ m}) = 15.3\text{ J}$$

where we have used the fact that $\vec{d} \parallel \vec{F}_{net}$ (which follows from the fact that the canister started from rest and moved horizontally under the action of horizontal forces — the resultant effect of which is expressed by $\vec{F}_{net}$).

17. (a) We use $\vec{F}$ to denote the upward force exerted by the cable on the astronaut. The force of the cable is upward and the force of gravity is mg downward. Furthermore, the acceleration of the astronaut is $g/10$ upward. According to Newton's second law, $F - mg$

$= mg/10$, so $F = 11\,mg/10$. Since the force $\vec{F}$ and the displacement $\vec{d}$ are in the same direction, the work done by $\vec{F}$ is

$$W_F = Fd = \frac{11mgd}{10} = \frac{11\,(72\text{ kg})(9.8\text{ m/s}^2)(15\text{ m})}{10} = 1.164 \times 10^4\text{ J}$$

which (with respect to significant figures) should be quoted as 1.2×10^4 J.

(b) The force of gravity has magnitude mg and is opposite in direction to the displacement. Thus, using Eq. 7-7, the work done by gravity is

$$W_g = -mgd = -(72\text{ kg})(9.8\text{ m/s}^2)(15\text{ m}) = -1.058 \times 10^4\text{ J}$$

which should be quoted as -1.1×10^4 J.

(c) The total work done is $W = 1.164 \times 10^4\text{ J} - 1.058 \times 10^4\text{ J} = 1.06 \times 10^3\text{ J}$. Since the astronaut started from rest, the work-kinetic energy theorem tells us that this (which we round to 1.1×10^3 J) is her final kinetic energy.

(d) Since $K = \frac{1}{2}mv^2$, her final speed is

$$v = \sqrt{\frac{2K}{m}} = \sqrt{\frac{2(1.06 \times 10^3\text{ J})}{72\text{ kg}}} = 5.4\text{ m/s}.$$

18. In both cases, there is no acceleration, so the lifting force is equal to the weight of the object.

(a) Eq. 7-8 leads to $W = \vec{F} \cdot \vec{d} = (360\text{ kN})(0.10\text{ m}) = 36$ kJ.

(b) In this case, we find $W = (4000\text{ N})(0.050\text{ m}) = 2.0 \times 10^2$ J.

19. (a) We use F to denote the magnitude of the force of the cord on the block. This force is upward, opposite to the force of gravity (which has magnitude Mg). The acceleration is $\vec{a} = g/4$ downward. Taking the downward direction to be positive, then Newton's second law yields

$$\vec{F}_{net} = m\vec{a} \Rightarrow Mg - F = M\left(\frac{g}{4}\right)$$

so $F = 3Mg/4$. The displacement is downward, so the work done by the cord's force is, using Eq. 7-7,

$$W_F = -Fd = -3Mgd/4.$$

(b) The force of gravity is in the same direction as the displacement, so it does work $W_g = Mgd$.

(c) The total work done on the block is $-3Mgd/4 + Mgd = Mgd/4$. Since the block starts from rest, we use Eq. 7-15 to conclude that this $(Mgd/4)$ is the block's kinetic energy K at the moment it has descended the distance d.

(d) Since $K = \frac{1}{2}Mv^2$, the speed is

$$v = \sqrt{\frac{2K}{M}} = \sqrt{\frac{2(Mgd/4)}{M}} = \sqrt{\frac{gd}{2}}$$

at the moment the block has descended the distance d.

20. (a) Using notation common to many vector capable calculators, we have (from Eq. 7-8) $W = \text{dot}([20.0,0] + [0, -(3.00)(9.8)], [0.500 \angle 30.0°]) = +1.31$ J.

(b) Eq. 7-10 (along with Eq. 7-1) then leads to

$$v = \sqrt{2(1.31 \text{ J})/(3.00 \text{ kg})} = 0.935 \text{ m/s}.$$

21. The fact that the applied force $\vec{F}_a$ causes the box to move up a frictionless ramp at a constant speed implies that there is no net change in the kinetic energy: $\Delta K = 0$. Thus, the work done by $\vec{F}_a$ must be equal to the negative work done by gravity: $W_a = -W_g$. Since the box is displaced vertically upward by $h = 0.150$ m, we have

$$W_a = +mgh = (3.00 \text{ kg})(9.80 \text{ m/s}^2)(0.150 \text{ m}) = 4.41 \text{ J}$$

22. From the figure, one may write the kinetic energy (in units of J) as a function of x as

$$K = K_s - 20x = 40 - 20x$$

Since $W = \Delta K = \vec{F}_x \cdot \Delta x$, the component of the force along the force along $+x$ is $F_x = dK/dx = -20$ N. The normal force on the block is $F_N = F_y$, which is related to the gravitational force by

$$mg = \sqrt{F_x^2 + (-F_y)^2}.$$

(Note that F_N points in the opposite direction of the component of the gravitational force.) With an initial kinetic energy $K_s = 40.0$ J and $v_0 = 4.00$ m/s, the mass of the block is

$$m = \frac{2K_s}{v_0^2} = \frac{2(40.0 \text{ J})}{(4.00 \text{ m/s})^2} = 5.00 \text{ kg}.$$

Thus, the normal force is

$$F_y = \sqrt{(mg)^2 - F_x^2} = \sqrt{(5.0 \text{ kg})^2 (9.8 \text{ m/s}^2)^2 - (20 \text{ N})^2} = 44.7 \text{ N} \approx 45 \text{ N}.$$

23. Eq. 7-15 applies, but the wording of the problem suggests that it is only necessary to examine the contribution from the rope (which would be the "W_a" term in Eq. 7-15):

$$W_a = -(50 \text{ N})(0.50 \text{ m}) = -25 \text{ J}$$

(the minus sign arises from the fact that the pull from the rope is anti-parallel to the direction of motion of the block). Thus, the kinetic energy would have been 25 J greater if the rope had not been attached (given the same displacement).

24. We use d to denote the magnitude of the spelunker's displacement during each stage. The mass of the spelunker is $m = 80.0$ kg. The work done by the lifting force is denoted W_i where $i = 1, 2, 3$ for the three stages. We apply the work-energy theorem, Eq. 17-15.

(a) For stage 1, $W_1 - mgd = \Delta K_1 = \frac{1}{2} mv_1^2$, where $v_1 = 5.00$ m/s. This gives

$$W_1 = mgd + \frac{1}{2} mv_1^2 = (80.0 \text{ kg})(9.80 \text{ m/s}^2)(10.0 \text{ m}) + \frac{1}{2}(80.0 \text{ kg})(5.00 \text{ m/s})^2 = 8.84 \times 10^3 \text{ J}.$$

(b) For stage 2, $W_2 - mgd = \Delta K_2 = 0$, which leads to

$$W_2 = mgd = (80.0 \text{ kg})(9.80 \text{ m/s}^2)(10.0 \text{ m}) = 7.84 \times 10^3 \text{ J}.$$

(c) For stage 3, $W_3 - mgd = \Delta K_3 = -\frac{1}{2} mv_1^2$. We obtain

$$W_3 = mgd - \frac{1}{2} mv_1^2 = (80.0 \text{ kg})(9.80 \text{ m/s}^2)(10.0 \text{ m}) - \frac{1}{2}(80.0 \text{ kg})(5.00 \text{ m/s})^2 = 6.84 \times 10^3 \text{ J}.$$

25. (a) The net upward force is given by

$$F + F_N - (m + M)g = (m + M)a$$

where $m = 0.250$ kg is the mass of the cheese, $M = 900$ kg is the mass of the elevator cab, F is the force from the cable, and $F_N = 3.00$ N is the normal force on the cheese. On the cheese alone, we have

$$F_N - mg = ma \Rightarrow a = \frac{3.00 \text{ N} - (0.250 \text{ kg})(9.80 \text{ m/s}^2)}{0.250 \text{ kg}} = 2.20 \text{ m/s}^2.$$

Thus the force from the cable is $F = (m+M)(a+g) - F_N = 1.08 \times 10^4 \text{ N}$, and the work done by the cable on the cab is

$$W = Fd_1 = (1.80 \times 10^4 \text{ N})(2.40 \text{ m}) = 2.59 \times 10^4 \text{ J}.$$

(b) If $W = 92.61$ kJ and $d_2 = 10.5$ m, the magnitude of the normal force is

$$F_N = (m+M)g - \frac{W}{d_2} = (0.250 \text{ kg} + 900 \text{ kg})(9.80 \text{ m/s}^2) - \frac{9.261 \times 10^4 \text{ J}}{10.5 \text{ m}} = 2.45 \text{ N}.$$

26. The spring constant is k = 100 N/m and the maximum elongation is x_i = 5.00 m. Using Eq. 7-25 with x_f = 0, the work is found to be

$$W = \frac{1}{2}kx_i^2 = \frac{1}{2}(100 \text{ N/m})(5.00 \text{ m})^2 = 1.25 \times 10^3 \text{ J}.$$

27. From Eq. 7-25, we see that the work done by the spring force is given by

$$W_s = \frac{1}{2}k(x_i^2 - x_f^2).$$

The fact that 360 N of force must be applied to pull the block to x = + 4.0 cm implies that the spring constant is

$$k = \frac{360 \text{ N}}{4.0 \text{ cm}} = 90 \text{ N/cm} = 9.0 \times 10^3 \text{ N/m}.$$

(a) When the block moves from $x_i = +5.0$ cm to $x = +3.0$ cm, we have

$$W_s = \frac{1}{2}(9.0 \times 10^3 \text{ N/m})[(0.050 \text{ m})^2 - (0.030 \text{ m})^2] = 7.2 \text{ J}.$$

(b) Moving from $x_i = +5.0$ cm to $x = -3.0$ cm, we have

$$W_s = \frac{1}{2}(9.0 \times 10^3 \text{ N/m})[(0.050 \text{ m})^2 - (-0.030 \text{ m})^2] = 7.2 \text{ J}.$$

(c) Moving from $x_i = +5.0$ cm to $x = -5.0$ cm, we have

$$W_s = \frac{1}{2}(9.0 \times 10^3 \text{ N/m})[(0.050 \text{ m})^2 - (-0.050 \text{ m})^2] = 0 \text{ J}.$$

(d) Moving from $x_i = +5.0$ cm to $x = -9.0$ cm, we have

$$W_s = \frac{1}{2}(9.0 \times 10^3 \text{ N/m})[(0.050 \text{ m})^2 - (-0.090 \text{ m})^2] = -25 \text{ J}.$$

28. We make use of Eq. 7-25 and Eq. 7-28 since the block is stationary before and after the displacement. The work done by the applied force can be written as

$$W_a = -W_s = \frac{1}{2}k(x_f^2 - x_i^2).$$

The spring constant is $k = (80 \text{ N})/(2.0 \text{ cm}) = 4.0 \times 10^3 \text{ N/m}$. With $W_a = 4.0$ J, and $x_i = -2.0$ cm, we have

$$x_f = \pm\sqrt{\frac{2W_a}{k} + x_i^2} = \pm\sqrt{\frac{2(4.0 \text{ J})}{(4.0 \times 10^3 \text{ N/m})} + (-0.020 \text{ m})^2} = \pm 0.049 \text{ m} = \pm 4.9 \text{ cm}.$$

29. (a) As the body moves along the x axis from $x_i = 3.0$ m to $x_f = 4.0$ m the work done by the force is

$$W = \int_{x_i}^{x_f} F_x \, dx = \int_{x_i}^{x_f} -6x \, dx = -3(x_f^2 - x_i^2) = -3(4.0^2 - 3.0^2) = -21 \text{ J}.$$

According to the work-kinetic energy theorem, this gives the change in the kinetic energy:

$$W = \Delta K = \frac{1}{2}m(v_f^2 - v_i^2)$$

where v_i is the initial velocity (at x_i) and v_f is the final velocity (at x_f). The theorem yields

$$v_f = \sqrt{\frac{2W}{m} + v_i^2} = \sqrt{\frac{2(-21 \text{ J})}{2.0 \text{ kg}} + (8.0 \text{ m/s})^2} = 6.6 \text{ m/s}.$$

(b) The velocity of the particle is $v_f = 5.0$ m/s when it is at $x = x_f$. The work-kinetic energy theorem is used to solve for x_f. The net work done on the particle is $W = -3(x_f^2 - x_i^2)$, so the theorem leads to

$$-3(x_f^2 - x_i^2) = \frac{1}{2}m(v_f^2 - v_i^2).$$

Thus,

$$x_f = \sqrt{-\frac{m}{6}(v_f^2 - v_i^2) + x_i^2} = \sqrt{-\frac{2.0 \text{ kg}}{6 \text{ N/m}}\left((5.0 \text{ m/s})^2 - (8.0 \text{ m/s})^2\right) + (3.0 \text{ m})^2} = 4.7 \text{ m}.$$

30. The work done by the spring force is given by Eq. 7-25:

$$W_s = \frac{1}{2}k(x_i^2 - x_f^2).$$

Since $F_x = -kx$, the slope in Fig. 7-36 corresponds to the spring constant k. Its value is given by $k = 80$ N/cm$=8.0 \times 10^3$ N/m.

(a) When the block moves from $x_i = +8.0$ cm to $x = +5.0$ cm, we have

$$W_s = \frac{1}{2}(8.0 \times 10^3 \text{ N/m})[(0.080 \text{ m})^2 - (0.050 \text{ m})^2] = 15.6 \text{ J} \approx 16 \text{ J}.$$

(b) Moving from $x_i = +8.0$ cm to $x = -5.0$ cm, we have

$$W_s = \frac{1}{2}(8.0 \times 10^3 \text{ N/m})[(0.080 \text{ m})^2 - (-0.050 \text{ m})^2] = 15.6 \text{ J} \approx 16 \text{ J}.$$

(c) Moving from $x_i = +8.0$ cm to $x = -8.0$ cm, we have

$$W_s = \frac{1}{2}(8.0 \times 10^3 \text{ N/m})[(0.080 \text{ m})^2 - (-0.080 \text{ m})^2] = 0 \text{ J}.$$

(d) Moving from $x_i = +8.0$ cm to $x = -10.0$ cm, we have

$$W_s = \frac{1}{2}(8.0 \times 10^3 \text{ N/m})[(0.080 \text{ m})^2 - (-0.10 \text{ m})^2] = -14.4 \text{ J} \approx -14 \text{ J}.$$

31. The work done by the spring force is given by Eq. 7-25: $W_s = \frac{1}{2}k(x_i^2 - x_f^2)$.

The spring constant k can be deduced from Fig. 7-37 which shows the amount of work done to pull the block from 0 to $x = 3.0$ cm. The parabola $W_a = kx^2/2$ contains (0,0), (2.0 cm, 0.40 J) and (3.0 cm, 0.90 J). Thus, we may infer from the data that $k = 2.0 \times 10^3$ N/m.

(a) When the block moves from $x_i = +5.0$ cm to $x = +4.0$ cm, we have

$$W_s = \frac{1}{2}(2.0\times10^3 \text{ N/m})[(0.050 \text{ m})^2 - (0.040 \text{ m})^2] = 0.90 \text{ J}.$$

(b) Moving from $x_i = +5.0$ cm to $x = -2.0$ cm, we have

$$W_s = \frac{1}{2}(2.0\times10^3 \text{ N/m})[(0.050 \text{ m})^2 - (-0.020 \text{ m})^2] = 2.1 \text{ J}.$$

(c) Moving from $x_i = +5.0$ cm to $x = -5.0$ cm, we have

$$W_s = \frac{1}{2}(2.0\times10^3 \text{ N/m})[(0.050 \text{ m})^2 - (-0.050 \text{ m})^2] = 0 \text{ J}.$$

32. Hooke's law and the work done by a spring is discussed in the chapter. We apply work-kinetic energy theorem, in the form of $\Delta K = W_a + W_s$, to the points in Figure 7-38 at $x = 1.0$ m and $x = 2.0$ m, respectively. The "applied" work W_a is that due to the constant force $\vec{P}$.

$$4 \text{ J} = P(1.0 \text{ m}) - \frac{1}{2}k(1.0 \text{ m})^2$$

$$0 = P(2.0 \text{ m}) - \frac{1}{2}k(2.0 \text{ m})^2$$

(a) Simultaneous solution leads to $P = 8.0$ N.

(b) Similarly, we find $k = 8.0$ N/m.

33. (a) This is a situation where Eq. 7-28 applies, so we have

$$Fx = \tfrac{1}{2}kx^2 \Rightarrow (3.0 \text{ N}) x = \tfrac{1}{2}(50 \text{ N/m})x^2$$

which (other than the trivial root) gives $x = (3.0/25)$ m $= 0.12$ m.

(b) The work done by the applied force is $W_a = Fx = (3.0 \text{ N})(0.12 \text{ m}) = 0.36$ J.

(c) Eq. 7-28 immediately gives $W_s = -W_a = -0.36$ J.

(d) With $K_f = K$ considered variable and $K_i = 0$, Eq. 7-27 gives $K = Fx - \tfrac{1}{2}kx^2$. We take the derivative of K with respect to x and set the resulting expression equal to zero, in order to find the position x_c which corresponds to a maximum value of K:

$$x_c = \frac{F}{k} = (3.0/50) \text{ m} = 0.060 \text{ m}.$$

We note that x_c is also the point where the applied and spring forces "balance."

(e) At x_c we find $K = K_{max} = 0.090$ J.

34. From Eq. 7-32, we see that the "area" in the graph is equivalent to the work done. Finding that area (in terms of rectangular [length × width] and triangular [$\frac{1}{2}$ base × height] areas) we obtain

$$W = W_{0<x<2} + W_{2<x<4} + W_{4<x<6} + W_{6<x<8} = (20+10+0-5) \text{ J} = 25 \text{ J}.$$

35. (a) The graph shows F as a function of x assuming x_0 is positive. The work is negative as the object moves from $x = 0$ to $x = x_0$ and positive as it moves from $x = x_0$ to $x = 2x_0$.

Since the area of a triangle is (base)(altitude)/2, the work done from $x = 0$ to $x = x_0$ is $-(x_0)(F_0)/2$ and the work done from $x = x_0$ to $x = 2x_0$ is

$$(2x_0 - x_0)(F_0)/2 = (x_0)(F_0)/2$$

The total work is the sum, which is zero.

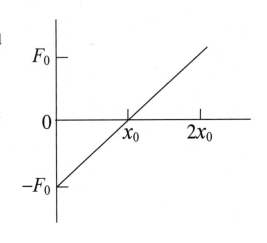

(b) The integral for the work is

$$W = \int_0^{2x_0} F_0 \left(\frac{x}{x_0} - 1 \right) dx = F_0 \left(\frac{x^2}{2x_0} - x \right) \Big|_0^{2x_0} = 0.$$

36. According to the graph the acceleration a varies linearly with the coordinate x. We may write $a = \alpha x$, where α is the slope of the graph. Numerically,

$$\alpha = \frac{20 \text{ m/s}^2}{8.0 \text{ m}} = 2.5 \text{ s}^{-2}.$$

The force on the brick is in the positive x direction and, according to Newton's second law, its magnitude is given by $F = ma = m\alpha x$. If x_f is the final coordinate, the work done by the force is

$$W = \int_0^{x_f} F\, dx = m\alpha \int_0^{x_f} x\, dx = \frac{m\alpha}{2} x_f^2 = \frac{(10 \text{ kg})(2.5 \text{ s}^{-2})}{2}(8.0 \text{ m})^2 = 8.0 \times 10^2 \text{ J}.$$

37. We choose to work this using Eq. 7-10 (the work-kinetic energy theorem). To find the initial and final kinetic energies, we need the speeds, so

$$v = \frac{dx}{dt} = 3.0 - 8.0t + 3.0t^2$$

in SI units. Thus, the initial speed is $v_i = 3.0$ m/s and the speed at $t = 4$ s is $v_f = 19$ m/s. The change in kinetic energy for the object of mass $m = 3.0$ kg is therefore

$$\Delta K = \frac{1}{2}m\left(v_f^2 - v_i^2\right) = 528 \text{ J}$$

which we round off to two figures and (using the work-kinetic energy theorem) conclude that the work done is $W = 5.3 \times 10^2$ J.

38. Using Eq. 7-32, we find

$$W = \int_{0.25}^{1.25} e^{-4x^2}\, dx = 0.21 \text{ J}$$

where the result has been obtained numerically. Many modern calculators have that capability, as well as most math software packages that a great many students have access to.

39. (a) We first multiply the vertical axis by the mass, so that it becomes a graph of the applied force. Now, adding the triangular and rectangular "areas" in the graph (for $0 \le x \le 4$) gives 42 J for the work done.

(b) Counting the "areas" under the axis as negative contributions, we find (for $0 \le x \le 7$) the work to be 30 J at $x = 7.0$ m.

(c) And at $x = 9.0$ m, the work is 12 J.

(d) Eq. 7-10 (along with Eq. 7-1) leads to speed $v = 6.5$ m/s at $x = 4.0$ m. Returning to the original graph (where a was plotted) we note that (since it started from rest) it has received acceleration(s) (up to this point) only in the $+x$ direction and consequently must have a velocity vector pointing in the $+x$ direction at $x = 4.0$ m.

(e) Now, using the result of part (b) and Eq. 7-10 (along with Eq. 7-1) we find the speed is 5.5 m/s at $x = 7.0$ m. Although it has experienced some deceleration during the $0 \le x \le 7$ interval, its velocity vector still points in the $+x$ direction.

(f) Finally, using the result of part (c) and Eq. 7-10 (along with Eq. 7-1) we find its speed $v = 3.5$ m/s at $x = 9.0$ m. It certainly has experienced a significant amount of deceleration during the $0 \le x \le 9$ interval; nonetheless, its velocity vector *still* points in the $+x$ direction.

40. (a) Using the work-kinetic energy theorem

$$K_f = K_i + \int_0^{2.0} (2.5 - x^2)\, dx = 0 + (2.5)(2.0) - \frac{1}{3}(2.0)^3 = 2.3 \text{ J}.$$

(b) For a variable end-point, we have K_f as a function of x, which could be differentiated to find the extremum value, but we recognize that this is equivalent to solving $F = 0$ for x:

$$F = 0 \Rightarrow 2.5 - x^2 = 0.$$

Thus, K is extremized at $x = \sqrt{2.5} \approx 1.6$ m and we obtain

$$K_f = K_i + \int_0^{\sqrt{2.5}} (2.5 - x^2)\, dx = 0 + (2.5)(\sqrt{2.5}) - \frac{1}{3}(\sqrt{2.5})^3 = 2.6 \text{ J}.$$

Recalling our answer for part (a), it is clear that this extreme value is a maximum.

41. As the body moves along the x axis from $x_i = 0$ m to $x_f = 3.00$ m the work done by the force is

$$W = \int_{x_i}^{x_f} F_x\, dx = \int_{x_i}^{x_f} (cx - 3.00 x^2)\, dx = \left(\frac{c}{2} x^2 - x^3\right)\bigg|_0^3 = \frac{c}{2}(3.00)^2 - (3.00)^3$$
$$= 4.50c - 27.0.$$

However, $W = \Delta K = (11.0 - 20.0) = -9.00$ J from the work-kinetic energy theorem. Thus,
$$4.50c - 27.0 = -9.00$$

or $c = 4.00$ N/m.

42. We solve the problem using the work-kinetic energy theorem which states that the change in kinetic energy is equal to the work done by the applied force, $\Delta K = W$. In our problem, the work done is $W = Fd$, where F is the tension in the cord and d is the length of the cord pulled as the cart slides from x_1 to x_2. From Fig. 7-42, we have

$$d = \sqrt{x_1^2 + h^2} - \sqrt{x_2^2 + h^2} = \sqrt{(3.00 \text{ m})^2 + (1.20 \text{ m})^2} - \sqrt{(1.00 \text{ m})^2 + (1.20 \text{ m})^2}$$
$$= 3.23 \text{ m} - 1.56 \text{ m} = 1.67 \text{ m}$$

which yields $\Delta K = Fd = (25.0 \text{ N})(1.67 \text{ m}) = 41.7$ J.

43. The power associated with force $\vec{F}$ is given by $P = \vec{F} \cdot \vec{v}$, where $\vec{v}$ is the velocity of the object on which the force acts. Thus,

$$P = \vec{F} \cdot \vec{v} = Fv\cos\phi = (122 \text{ N})(5.0 \text{ m/s})\cos 37° = 4.9 \times 10^2 \text{ W}.$$

44. Recognizing that the force in the cable must equal the total weight (since there is no acceleration), we employ Eq. 7-47:

$$P = Fv \cos\theta = mg\left(\frac{\Delta x}{\Delta t}\right)$$

where we have used the fact that $\theta = 0°$ (both the force of the cable and the elevator's motion are upward). Thus,

$$P = (3.0\times10^3 \text{ kg})(9.8 \text{ m/s}^2)\left(\frac{210 \text{ m}}{23 \text{ s}}\right) = 2.7\times10^5 \text{ W}.$$

45. (a) The power is given by $P = Fv$ and the work done by $\vec{F}$ from time t_1 to time t_2 is given by

$$W = \int_{t_1}^{t_2} P\, dt = \int_{t_1}^{t_2} Fv\, dt.$$

Since $\vec{F}$ is the net force, the magnitude of the acceleration is $a = F/m$, and, since the initial velocity is $v_0 = 0$, the velocity as a function of time is given by $v = v_0 + at = (F/m)t$. Thus

$$W = \int_{t_1}^{t_2} (F^2/m)t\, dt = \frac{1}{2}(F^2/m)(t_2^2 - t_1^2).$$

For $t_1 = 0$ and $t_2 = 1.0\text{ s}$,

$$W = \frac{1}{2}\left(\frac{(5.0 \text{ N})^2}{15 \text{ kg}}\right)(1.0 \text{ s})^2 = 0.83 \text{ J}.$$

(b) For $t_1 = 1.0\text{ s}$, and $t_2 = 2.0\text{ s}$,

$$W = \frac{1}{2}\left(\frac{(5.0 \text{ N})^2}{15 \text{ kg}}\right)[(2.0 \text{ s})^2 - (1.0 \text{ s})^2] = 2.5 \text{ J}.$$

(c) For $t_1 = 2.0\text{ s}$ and $t_2 = 3.0\text{ s}$,

$$W = \frac{1}{2}\left(\frac{(5.0 \text{ N})^2}{15 \text{ kg}}\right)[(3.0 \text{ s})^2 - (2.0 \text{ s})^2] = 4.2 \text{ J}.$$

(d) Substituting $v = (F/m)t$ into $P = Fv$ we obtain $P = F^2 t/m$ for the power at any time t. At the end of the third second

$$P = \left(\frac{(5.0 \text{ N})^2 (3.0 \text{ s})}{15 \text{ kg}}\right) = 5.0 \text{ W}.$$

46. (a) Since constant speed implies $\Delta K = 0$, we require $W_a = -W_g$, by Eq. 7-15. Since W_g is the same in both cases (same weight and same path), then $W_a = 9.0 \times 10^2$ J just as it was in the first case.

(b) Since the speed of 1.0 m/s is constant, then 8.0 meters is traveled in 8.0 seconds. Using Eq. 7-42, and noting that average power is *the* power when the work is being done at a steady rate, we have

$$P = \frac{W}{\Delta t} = \frac{900 \text{ J}}{8.0 \text{ s}} = 1.1 \times 10^2 \text{ W}.$$

(c) Since the speed of 2.0 m/s is constant, 8.0 meters is traveled in 4.0 seconds. Using Eq. 7-42, with *average power* replaced by *power*, we have

$$P = \frac{W}{\Delta t} = \frac{900 \text{ J}}{4.0 \text{ s}} = 225 \text{ W} \approx 2.3 \times 10^2 \text{ W}.$$

47. The total work is the sum of the work done by gravity on the elevator, the work done by gravity on the counterweight, and the work done by the motor on the system:

$$W_T = W_e + W_c + W_s.$$

Since the elevator moves at constant velocity, its kinetic energy does not change and according to the work-kinetic energy theorem the total work done is zero. This means $W_e + W_c + W_s = 0$. The elevator moves upward through 54 m, so the work done by gravity on it is

$$W_e = -m_e g d = -(1200 \text{ kg})(9.80 \text{ m/s}^2)(54 \text{ m}) = -6.35 \times 10^5 \text{ J}.$$

The counterweight moves downward the same distance, so the work done by gravity on it is

$$W_c = m_c g d = (950 \text{ kg})(9.80 \text{ m/s}^2)(54 \text{ m}) = 5.03 \times 10^5 \text{ J}.$$

Since $W_T = 0$, the work done by the motor on the system is

$$W_s = -W_e - W_c = 6.35 \times 10^5 \text{ J} - 5.03 \times 10^5 \text{ J} = 1.32 \times 10^5 \text{ J}.$$

This work is done in a time interval of $\Delta t = 3.0$ min $= 180$ s, so the power supplied by the motor to lift the elevator is

$$P = \frac{W_s}{\Delta t} = \frac{1.32 \times 10^5 \text{ J}}{180 \text{ s}} = 7.4 \times 10^2 \text{ W}.$$

48. (a) Using Eq. 7-48 and Eq. 3-23, we obtain

$$P = \vec{F} \cdot \vec{v} = (4.0 \text{ N})(-2.0 \text{ m/s}) + (9.0 \text{ N})(4.0 \text{ m/s}) = 28 \text{ W}.$$

(b) We again use Eq. 7-48 and Eq. 3-23, but with a one-component velocity: $\vec{v} = v\hat{j}$.

$$P = \vec{F} \cdot \vec{v} \Rightarrow -12 \text{ W} = (-2.0 \text{ N})v.$$

which yields $v = 6$ m/s.

49. (a) Eq. 7-8 yields

$$W = F_x \Delta x + F_y \Delta y + F_z \Delta z$$
$$= (2.00 \text{ N})(7.5 \text{ m} - 0.50 \text{ m}) + (4.00 \text{ N})(12.0 \text{ m} - 0.75 \text{ m}) + (6.00 \text{ N})(7.2 \text{m} - 0.20 \text{ m})$$
$$= 101 \text{ J} \approx 1.0 \times 10^2 \text{ J}.$$

(b) Dividing this result by 12 s (see Eq. 7-42) yields $P = 8.4$ W.

50. (a) Since the force exerted by the spring on the mass is zero when the mass passes through the equilibrium position of the spring, the rate at which the spring is doing work on the mass at this instant is also zero.

(b) The rate is given by $P = \vec{F} \cdot \vec{v} = -Fv$, where the minus sign corresponds to the fact that $\vec{F}$ and $\vec{v}$ are anti-parallel to each other. The magnitude of the force is given by

$$F = kx = (500 \text{ N/m})(0.10 \text{ m}) = 50 \text{ N},$$

while v is obtained from conservation of energy for the spring-mass system:

$$E = K + U = 10 \text{ J} = \frac{1}{2}mv^2 + \frac{1}{2}kx^2 = \frac{1}{2}(0.30 \text{ kg})v^2 + \frac{1}{2}(500 \text{ N/m})(0.10 \text{ m})^2$$

which gives $v = 7.1$ m/s. Thus,

$$P = -Fv = -(50 \text{ N})(7.1 \text{ m/s}) = -3.5 \times 10^2 \text{ W}.$$

51. (a) The object's displacement is

$$\vec{d} = \vec{d}_f - \vec{d}_i = (-8.00 \text{ m})\hat{i} + (6.00 \text{ m})\hat{j} + (2.00 \text{ m})\hat{k}.$$

Thus, Eq. 7-8 gives

$$W = \vec{F} \cdot \vec{d} = (3.00 \text{ N})(-8.00 \text{ m}) + (7.00 \text{ N})(6.00 \text{ m}) + (7.00 \text{ N})(2.00 \text{ m}) = 32.0 \text{ J}.$$

(b) The average power is given by Eq. 7-42:

$$P_{avg} = \frac{W}{t} = \frac{32.0}{4.00} = 8.00 \text{ W}.$$

(c) The distance from the coordinate origin to the initial position is

$$d_i = \sqrt{(3.00 \text{ m})^2 + (-2.00 \text{ m})^2 + (5.00 \text{ m})^2} = 6.16 \text{ m},$$

and the magnitude of the distance from the coordinate origin to the final position is

$$d_f = \sqrt{(-5.00 \text{ m})^2 + (4.00 \text{ m})^2 + (7.00 \text{ m})^2} = 9.49 \text{ m}.$$

Their scalar (dot) product is

$$\vec{d}_i \cdot \vec{d}_f = (3.00 \text{ m})(-5.00 \text{ m}) + (-2.00 \text{ m})(4.00 \text{ m}) + (5.00 \text{ m})(7.00 \text{ m}) = 12.0 \text{ m}^2.$$

Thus, the angle between the two vectors is

$$\phi = \cos^{-1}\left(\frac{\vec{d}_i \cdot \vec{d}_f}{d_i d_f}\right) = \cos^{-1}\left(\frac{12.0}{(6.16)(9.49)}\right) = 78.2°.$$

52. According to the problem statement, the power of the car is

$$P = \frac{dW}{dt} = \frac{d}{dt}\left(\frac{1}{2}mv^2\right) = mv\frac{dv}{dt} = \text{constant}.$$

The condition implies $dt = mv\, dv / P$, which can be integrated to give

$$\int_0^T dt = \int_0^{v_T} \frac{mv\, dv}{P} \quad \Rightarrow \quad T = \frac{mv_T^2}{2P}$$

where v_T is the speed of the car at $t = T$. On the other hand, the total distance traveled can be written as

$$L = \int_0^T v\, dt = \int_0^{v_T} v\frac{mv\, dv}{P} = \frac{m}{P}\int_0^{v_T} v^2\, dv = \frac{mv_T^3}{3P}.$$

By squaring the expression for L and substituting the expression for T, we obtain

$$L^2 = \left(\frac{mv_T^3}{3P}\right)^2 = \frac{8P}{9m}\left(\frac{mv_T^2}{2P}\right)^3 = \frac{8PT^3}{9m}$$

which implies that

$$PT^3 = \frac{9}{8}mL^2 = \text{constant.}$$

Differentiating the above equation gives $dPT^3 + 3PT^2 dT = 0$, or $dT = -\frac{T}{3P}dP$.

53. (a) We set up the ratio

$$\frac{50 \text{ km}}{1 \text{ km}} = \left(\frac{E}{1 \text{ megaton}}\right)^{1/3}$$

and find $E = 50^3 \approx 1 \times 10^5$ megatons of TNT.

(b) We note that 15 kilotons is equivalent to 0.015 megatons. Dividing the result from part (a) by 0.013 yields about ten million bombs.

54. (a) The compression of the spring is $d = 0.12$ m. The work done by the force of gravity (acting on the block) is, by Eq. 7-12,

$$W_1 = mgd = (0.25 \text{ kg})(9.8 \text{ m/s}^2)(0.12 \text{ m}) = 0.29 \text{ J.}$$

(b) The work done by the spring is, by Eq. 7-26,

$$W_2 = -\frac{1}{2}kd^2 = -\frac{1}{2}(250 \text{ N/m})(0.12 \text{ m})^2 = -1.8 \text{ J.}$$

(c) The speed v_i of the block just before it hits the spring is found from the work-kinetic energy theorem (Eq. 7-15):

$$\Delta K = 0 - \frac{1}{2}mv_i^2 = W_1 + W_2$$

which yields

$$v_i = \sqrt{\frac{(-2)(W_1 + W_2)}{m}} = \sqrt{\frac{(-2)(0.29 \text{ J} - 1.8 \text{ J})}{0.25 \text{ kg}}} = 3.5 \text{ m/s.}$$

(d) If we instead had $v_i' = 7$ m/s, we reverse the above steps and solve for d'. Recalling the theorem used in part (c), we have

$$0 - \frac{1}{2}mv_i'^2 = W_1' + W_2' = mgd' - \frac{1}{2}kd'^2$$

which (choosing the positive root) leads to

$$d' = \frac{mg + \sqrt{m^2g^2 + mkv_i'^2}}{k}$$

which yields $d' = 0.23$ m. In order to obtain this result, we have used more digits in our intermediate results than are shown above (so $v_i = \sqrt{12.048}$ m/s $= 3.471$ m/s and $v_i' = 6.942$ m/s).

55. One approach is to assume a "path" from $\vec{r}_i$ to $\vec{r}_f$ and do the line-integral accordingly. Another approach is to simply use Eq. 7-36, which we demonstrate:

$$W = \int_{x_i}^{x_f} F_x dx + \int_{y_i}^{y_f} F_y dy = \int_{2}^{-4} (2x) dx + \int_{3}^{-3} (3) \, dy$$

with SI units understood. Thus, we obtain $W = 12$ J $- 18$ J $= -6$ J.

56. (a) The force of the worker on the crate is constant, so the work it does is given by $W_F = \vec{F} \cdot \vec{d} = Fd \cos\phi$, where $\vec{F}$ is the force, $\vec{d}$ is the displacement of the crate, and ϕ is the angle between the force and the displacement. Here $F = 210$ N, $d = 3.0$ m, and $\phi = 20°$. Thus,

$$W_F = (210 \text{ N})(3.0 \text{ m}) \cos 20° = 590 \text{ J}.$$

(b) The force of gravity is downward, perpendicular to the displacement of the crate. The angle between this force and the displacement is 90° and cos 90° = 0, so the work done by the force of gravity is zero.

(c) The normal force of the floor on the crate is also perpendicular to the displacement, so the work done by this force is also zero.

(d) These are the only forces acting on the crate, so the total work done on it is 590 J.

57. There is no acceleration, so the lifting force is equal to the weight of the object. We note that the person's pull $\vec{F}$ is equal (in magnitude) to the tension in the cord.

(a) As indicated in the *hint*, tension contributes twice to the lifting of the canister: $2T = mg$. Since $|\vec{F}| = T$, we find $|\vec{F}| = 98$ N.

(b) To rise 0.020 m, two segments of the cord (see Fig. 7-44) must shorten by that amount. Thus, the amount of string pulled down at the left end (this is the magnitude of $\vec{d}$, the downward displacement of the hand) is $d = 0.040$ m.

(c) Since (at the left end) both $\vec{F}$ and $\vec{d}$ are downward, then Eq. 7-7 leads to

$$W = \vec{F} \cdot \vec{d} = (98 \text{ N})(0.040 \text{ m}) = 3.9 \text{ J}.$$

(d) Since the force of gravity $\vec{F}_g$ (with magnitude mg) is opposite to the displacement $\vec{d}_c = 0.020$ m (up) of the canister, Eq. 7-7 leads to

$$W = \vec{F}_g \cdot \vec{d}_c = -(196 \text{ N})(0.020 \text{ m}) = -3.9 \text{ J}.$$

This is consistent with Eq. 7-15 since there is no change in kinetic energy.

58. With SI units understood, Eq. 7-8 leads to $W = (4.0)(3.0) - c(2.0) = 12 - 2c$.

(a) If $W = 0$, then $c = 6.0$ N.

(b) If $W = 17$ J, then $c = -2.5$ N.

(c) If $W = -18$ J, then $c = 15$ N.

59. Using Eq. 7-8, we find

$$W = \vec{F} \cdot \vec{d} = (F\cos\theta\, \hat{i} + F\sin\theta\, \hat{j}) \cdot (x\hat{i} + y\hat{j}) = Fx\cos\theta + Fy\sin\theta$$

where $x = 2.0$ m, $y = -4.0$ m, $F = 10$ N, and $\theta = 150°$. Thus, we obtain $W = -37$ J. Note that the given mass value (2.0 kg) is not used in the computation.

60. The acceleration is constant, so we may use the equations in Table 2-1. We choose the direction of motion as $+x$ and note that the displacement is the same as the distance traveled, in this problem. We designate the force (assumed singular) along the x direction acting on the $m = 2.0$ kg object as F.

(a) With $v_0 = 0$, Eq. 2-11 leads to $a = v/t$. And Eq. 2-17 gives $\Delta x = \frac{1}{2}vt$. Newton's second law yields the force $F = ma$. Eq. 7-8, then, gives the work:

$$W = F\Delta x = m\left(\frac{v}{t}\right)\left(\frac{1}{2}vt\right) = \frac{1}{2}mv^2$$

as we expect from the work-kinetic energy theorem. With $v = 10$ m/s, this yields $W = 1.0 \times 10^2$ J.

(b) Instantaneous power is defined in Eq. 7-48. With $t = 3.0$ s, we find

$$P = Fv = m\left(\frac{v}{t}\right)v = 67 \text{ W}.$$

(c) The velocity at $t' = 1.5$ s is $v' = at' = 5.0$ m/s. Thus, $P' = Fv' = 33$ W.

61. The total weight is $(100)(660 \text{ N}) = 6.60 \times 10^4$ N, and the words "raises ... at constant speed" imply zero acceleration, so the lift-force is equal to the total weight. Thus

$$P = Fv = (6.60 \times 10^4)(150 \text{ m}/60.0 \text{ s}) = 1.65 \times 10^5 \text{ W}.$$

62. (a) The force $\vec{F}$ of the incline is a combination of normal and friction force which is serving to "cancel" the tendency of the box to fall downward (due to its 19.6 N weight). Thus, $\vec{F} = mg$ upward. In this part of the problem, the angle ϕ between the belt and $\vec{F}$ is 80°. From Eq. 7-47, we have

$$P = Fv \cos\phi = (19.6 \text{ N})(0.50 \text{ m/s}) \cos 80° = 1.7 \text{ W}.$$

(b) Now the angle between the belt and $\vec{F}$ is 90°, so that $P = 0$.

(c) In this part, the angle between the belt and $\vec{F}$ is 100°, so that

$$P = (19.6 \text{ N})(0.50 \text{ m/s}) \cos 100° = -1.7 \text{ W}.$$

63. (a) In 10 min the cart moves

$$d = \left(6.0 \ \frac{\text{mi}}{\text{h}}\right)\left(\frac{5280 \text{ ft/mi}}{60 \text{ min/h}}\right)(10 \text{ min}) = 5280 \text{ ft}$$

so that Eq. 7-7 yields

$$W = Fd\cos\phi = (40 \text{ lb})(5280 \text{ ft}) \cos 30° = 1.8 \times 10^5 \text{ ft} \cdot \text{lb}.$$

(b) The average power is given by Eq. 7-42, and the conversion to horsepower (hp) can be found on the inside back cover. We note that 10 min is equivalent to 600 s.

$$P_{\text{avg}} = \frac{1.8 \times 10^5 \text{ ft} \cdot \text{lb}}{600 \text{ s}} = 305 \text{ ft} \cdot \text{lb/s}$$

which (upon dividing by 550) converts to $P_{\text{avg}} = 0.55$ hp.

64. Using Eq. 7-7, we have $W = Fd \cos\phi = 1504$ J. Then, by the work-kinetic energy theorem, we find the kinetic energy $K_f = K_i + W = 0 + 1504$ J. The answer is therefore 1.5 kJ.

65. (a) To hold the crate at equilibrium in the final situation, $\vec{F}$ must have the same magnitude as the horizontal component of the rope's tension $T \sin \theta$, where θ is the angle between the rope (in the final position) and vertical:

$$\theta = \sin^{-1}\left(\frac{4.00}{12.0}\right) = 19.5°.$$

But the vertical component of the tension supports against the weight: $T \cos \theta = mg$. Thus, the tension is

$$T = (230 \text{ kg})(9.80 \text{ m/s}^2)/\cos 19.5° = 2391 \text{ N}$$

and $F = (2391 \text{ N}) \sin 19.5° = 797 \text{ N}$.

An alternative approach based on drawing a vector triangle (of forces) in the final situation provides a quick solution.

(b) Since there is no change in kinetic energy, the net work on it is zero.

(c) The work done by gravity is $W_g = \vec{F}_g \cdot \vec{d} = -mgh$, where $h = L(1 - \cos \theta)$ is the vertical component of the displacement. With $L = 12.0$ m, we obtain $W_g = -1547$ J which should be rounded to three figures: -1.55 kJ.

(d) The tension vector is everywhere perpendicular to the direction of motion, so its work is zero (since $\cos 90° = 0$).

(e) The implication of the previous three parts is that the work due to $\vec{F}$ is $-W_g$ (so the net work turns out to be zero). Thus, $W_F = -W_g = 1.55$ kJ.

(f) Since $\vec{F}$ does not have constant magnitude, we cannot expect Eq. 7-8 to apply.

66. From Eq. 7-32, we see that the "area" in the graph is equivalent to the work done. We find the area in terms of rectangular [length × width] and triangular [$\frac{1}{2}$ base × height] areas and use the work-kinetic energy theorem appropriately. The initial point is taken to be $x = 0$, where $v_0 = 4.0$ m/s.

(a) With $K_i = \frac{1}{2} m v_0^2 = 16$ J, we have

$$K_3 - K_0 = W_{0<x<1} + W_{1<x<2} + W_{2<x<3} = -4.0 \text{ J}$$

so that K_3 (the kinetic energy when $x = 3.0$ m) is found to equal 12 J.

(b) With SI units understood, we write $W_{3<x<x_f}$ as $F_x \Delta x = (-4.0 \text{ N})(x_f - 3.0 \text{ m})$ and apply the work-kinetic energy theorem:

$$K_{x_f} - K_3 = W_{3<x<x_f}$$
$$K_{xf} - 12 = (-4)(x_f - 3.0)$$

so that the requirement $K_{xf} = 8.0$ J leads to $x_f = 4.0$ m.

(c) As long as the work is positive, the kinetic energy grows. The graph shows this situation to hold until $x = 1.0$ m. At that location, the kinetic energy is

$$K_1 = K_0 + W_{0<x<1} = 16 \text{ J} + 2.0 \text{ J} = 18 \text{ J}.$$

67. (a) Noting that the x component of the third force is $F_{3x} = (4.00 \text{ N})\cos(60°)$, we apply Eq. 7-8 to the problem:

$$W = [5.00 \text{ N} - 1.00 \text{ N} + (4.00 \text{ N})\cos 60°](0.20 \text{ m}) = 1.20 \text{ J}.$$

(b) Eq. 7-10 (along with Eq. 7-1) then yields $v = \sqrt{2W/m} = 1.10$ m/s.

68. (a) In the work-kinetic energy theorem, we include both the work due to an applied force W_a and work done by gravity W_g in order to find the latter quantity.

$$\Delta K = W_a + W_g \quad \Rightarrow \quad 30 \text{ J} = (100 \text{ N})(1.8 \text{ m})\cos 180° + W_g$$

leading to $W_g = 2.1 \times 10^2$ J.

(b) The value of W_g obtained in part (a) still applies since the weight and the path of the child remain the same, so $\Delta K = W_g = 2.1 \times 10^2$ J.

69. (a) Eq. 7-6 gives $W_a = Fd = (209 \text{ N})(1.50 \text{ m}) \approx 314$ J.

(b) Eq. 7-12 leads to $W_g = (25.0 \text{ kg})(9.80 \text{ m/s}^2)(1.50 \text{ m})\cos(115°) \approx -155$ J.

(c) The angle between the normal force and the direction of motion remains 90° at all times, so the work it does is zero.

(d) The total work done on the crate is $W_T = 314 \text{ J} - 155 \text{ J} = 158$ J.

70. After converting the speed to meters-per-second, we find

$$K = \tfrac{1}{2}mv^2 = 667 \text{ kJ}.$$

71. (a) Hooke's law and the work done by a spring is discussed in the chapter. Taking absolute values, and writing that law in terms of differences ΔF and Δx, we analyze the first two pictures as follows:

$$|\Delta F| = k|\Delta x|$$
$$240 \text{ N} - 110 \text{ N} = k(60 \text{ mm} - 40 \text{ mm})$$

which yields $k = 6.5$ N/mm. Designating the relaxed position (as read by that scale) as x_o we look again at the first picture:

$$110 \text{ N} = k(40 \text{ mm} - x_o)$$

which (upon using the above result for k) yields $x_o = 23$ mm.

(b) Using the results from part (a) to analyze that last picture, we find

$$W = k(30 \text{ mm} - x_o) = 45 \text{ N}.$$

72. (a) Using Eq. 7-8 and SI units, we find

$$W = \vec{F} \cdot \vec{d} = (2\hat{i} - 4\hat{j}) \cdot (8\hat{i} + c\hat{j}) = 16 - 4c$$

which, if equal zero, implies $c = 16/4 = 4$ m.

(b) If $W > 0$ then $16 > 4c$, which implies $c < 4$ m.

(c) If $W < 0$ then $16 < 4c$, which implies $c > 4$ m.

73. A convenient approach is provided by Eq. 7-48.

$$P = Fv = (1800 \text{ kg} + 4500 \text{ kg})(9.8 \text{ m/s}^2)(3.80 \text{ m/s}) = 235 \text{ kW}.$$

Note that we have set the applied force equal to the weight in order to maintain constant velocity (zero acceleration).

74. (a) The component of the force of gravity exerted on the ice block (of mass m) along the incline is $mg \sin \theta$, where $\theta = \sin^{-1}(0.91/1.5)$ gives the angle of inclination for the inclined plane. Since the ice block slides down with uniform velocity, the worker must exert a force $\vec{F}$ "uphill" with a magnitude equal to $mg \sin \theta$. Consequently,

$$F = mg \sin \theta = (45 \text{ kg})(9.8 \text{ m/s}^2)\left(\frac{0.91 \text{ m}}{1.5 \text{ m}}\right) = 2.7 \times 10^2 \text{ N}.$$

(b) Since the "downhill" displacement is opposite to $\vec{F}$, the work done by the worker is

$$W_1 = -(2.7 \times 10^2 \, \text{N})(1.5 \, \text{m}) = -4.0 \times 10^2 \, \text{J}.$$

(c) Since the displacement has a vertically downward component of magnitude 0.91 m (in the same direction as the force of gravity), we find the work done by gravity to be

$$W_2 = (45 \, \text{kg})(9.8 \, \text{m/s}^2)(0.91 \, \text{m}) = 4.0 \times 10^2 \, \text{J}.$$

(d) Since $\vec{F}_N$ is perpendicular to the direction of motion of the block, and $\cos 90° = 0$, work done by the normal force is $W_3 = 0$ by Eq. 7-7.

(e) The resultant force $\vec{F}_{net}$ is zero since there is no acceleration. Thus, its work is zero, as can be checked by adding the above results $W_1 + W_2 + W_3 = 0$.

75. (a) The plot of the function (with SI units understood) is shown below.

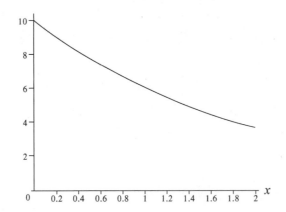

Estimating the area under the curve allows for a range of answers. Estimates from 11 J to 14 J are typical.

(b) Evaluating the work analytically (using Eq. 7-32), we have

$$W = \int_0^2 10 e^{-x/2} dx = -20 e^{-x/2} \Big|_0^2 = 12.6 \, \text{J} \approx 13 \, \text{J}.$$

76. (a) Eq. 7-10 (along with Eq. 7-1 and Eq. 7-7) leads to

$$v_f = (2 \frac{d}{m} F \cos\theta)^{1/2} = (\cos\theta)^{1/2},$$

where we have substituted $F = 2.0$ N, $m = 4.0$ kg and $d = 1.0$ m.

(b) With $v_i = 1$, those same steps lead to $v_f = (1 + \cos\theta)^{1/2}$.

(c) Replacing θ with $180° - \theta$, and still using $v_i = 1$, we find

$$v_f = [1 + \cos(180° - \theta)]^{1/2} = (1 - \cos\theta)^{1/2}.$$

(d) The graphs are shown on the right. Note that as θ is increased in parts (a) and (b) the force provides less and less of a positive acceleration, whereas in part (c) the force provides less and less of a deceleration (as its θ value increases). The highest curve (which slowly decreases from 1.4 to 1) is the curve for part (b); the other decreasing curve (starting at 1 and ending at 0) is for part (a). The rising curve is for part (c); it is equal to 1 where $\theta = 90°$.

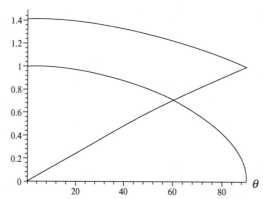

77. (a) We can easily fit the curve to a concave-downward parabola: $x = \frac{1}{10}t(10 - t)$, from which (by taking two derivatives) we find the acceleration to be $a = -0.20$ m/s^2. The (constant) force is therefore $F = ma = -0.40$ N, with a corresponding work given by $W = Fx = \frac{2}{50}t(t - 10)$. It also follows from the x expression that $v_0 = 1.0$ m/s. This means that $K_i = \frac{1}{2}mv^2 = 1.0$ J. Therefore, when $t = 1.0$ s, Eq. 7-10 gives $K = K_i + W = 0.64$ J ≈ 0.6 J, where the second significant figure is not to be taken too seriously.

(b) At $t = 5.0$ s, the above method gives $K = 0$.

(c) Evaluating the $W = \frac{2}{50}t(t - 10)$ expression at $t = 5.0$ s and $t = 1.0$ s, and subtracting, yields -0.6 J. This can also be inferred from the answers for parts (a) and (b).

78. The problem indicates that SI units are understood, so the result (of Eq. 7-23) is in Joules. Done numerically, using features available on many modern calculators, the result is roughly 0.47 J. For the interested student it might be worthwhile to quote the "exact" answer (in terms of the "error function"):

$$\int_{.15}^{1.2} e^{-2x^2} dx = \tfrac{1}{4}\sqrt{2\pi}\,[\mathrm{erf}(6\sqrt{2}/5) - \mathrm{erf}(3\sqrt{2}/20)]\,.$$

79. (a) To estimate the area under the curve between $x = 1$ m and $x = 3$ m (which should yield the value for the work done), one can try "counting squares" (or half-squares or thirds of squares) between the curve and the axis. Estimates between 5 J and 8 J are typical for this (crude) procedure.

(b) Eq. 7-32 gives

$$\int_1^3 \frac{a}{x^2}\,dx = \frac{a}{3} - \frac{a}{1} = 6\text{ J}$$

where $a = -9$ N·m^2 is given in the problem statement.

80. (a) Using Eq. 7-32, the work becomes $W = \frac{9}{2}x^2 - x^3$ (SI units understood). The plot is shown below:

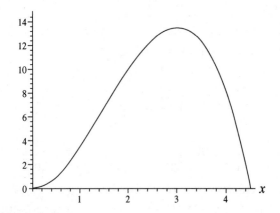

(b) We see from the graph that its peak value occurs at $x = 3.00$ m. This can be verified by taking the derivative of W and setting equal to zero, or simply by noting that this is where the force vanishes.

(c) The maximum value is $W = \frac{9}{2}(3.00)^2 - (3.00)^3 = 13.50$ J.

(d) We see from the graph (or from our analytic expression) that $W = 0$ at $x = 4.50$ m.

(e) The case is at rest when $v = 0$. Since $W = \Delta K = mv^2/2$, the condition implies $W = 0$. This happens at $x = 4.50$ m.

Chapter 8

1. (a) Noting that the vertical displacement is 10.0 m – 1.50 m = 8.50 m downward (same direction as $\vec{F}_g$), Eq. 7-12 yields

$$W_g = mgd \cos\phi = (2.00 \text{ kg})(9.80 \text{ m/s}^2)(8.50 \text{ m})\cos 0° = 167 \text{ J}.$$

(b) One approach (which is fairly trivial) is to use Eq. 8-1, but we feel it is instructive to instead calculate this as ΔU where $U = mgy$ (with upwards understood to be the $+y$ direction). The result is

$$\Delta U = mg(y_f - y_i) = (2.00 \text{ kg})(9.80 \text{ m/s}^2)(1.50 \text{ m} - 10.0 \text{ m}) = -167 \text{ J}.$$

(c) In part (b) we used the fact that $U_i = mgy_i = 196$ J.

(d) In part (b), we also used the fact $U_f = mgy_f = 29$ J.

(e) The computation of W_g does not use the new information (that $U = 100$ J at the ground), so we again obtain $W_g = 167$ J.

(f) As a result of Eq. 8-1, we must again find $\Delta U = -W_g = -167$ J.

(g) With this new information (that $U_0 = 100$ J where $y = 0$) we have

$$U_i = mgy_i + U_0 = 296 \text{ J}.$$

(h) With this new information (that $U_0 = 100$ J where $y = 0$) we have

$$U_f = mgy_f + U_0 = 129 \text{ J}.$$

We can check part (f) by subtracting the new U_i from this result.

2. (a) The only force that does work on the ball is the force of gravity; the force of the rod is perpendicular to the path of the ball and so does no work. In going from its initial position to the lowest point on its path, the ball moves vertically through a distance equal to the length L of the rod, so the work done by the force of gravity is

$$W = mgL = (0.341 \text{ kg})(9.80 \text{ m/s}^2)(0.452 \text{ m}) = 1.51 \text{ J}.$$

(b) In going from its initial position to the highest point on its path, the ball moves vertically through a distance equal to L, but this time the displacement is upward, opposite the direction of the force of gravity. The work done by the force of gravity is

$$W = -mgL = -(0.341 \text{ kg})(9.80 \text{ m/s}^2)(0.452 \text{ m}) = -1.51 \text{ J}.$$

(c) The final position of the ball is at the same height as its initial position. The displacement is horizontal, perpendicular to the force of gravity. The force of gravity does no work during this displacement.

(d) The force of gravity is conservative. The change in the gravitational potential energy of the ball-Earth system is the negative of the work done by gravity:

$$\Delta U = -mgL = -(0.341 \text{ kg})(9.80 \text{ m/s}^2)(0.452 \text{ m}) = -1.51 \text{ J}$$

as the ball goes to the lowest point.

(e) Continuing this line of reasoning, we find

$$\Delta U = +mgL = (0.341 \text{ kg})(9.80 \text{ m/s}^2)(0.452 \text{ m}) = 1.51 \text{ J}$$

as it goes to the highest point.

(f) Continuing this line of reasoning, we have $\Delta U = 0$ as it goes to the point at the same height.

(g) The change in the gravitational potential energy depends only on the initial and final positions of the ball, not on its speed anywhere. The change in the potential energy is the *same* since the initial and final positions are the same.

3. (a) The force of gravity is constant, so the work it does is given by $W = \vec{F} \cdot \vec{d}$, where $\vec{F}$ is the force and $\vec{d}$ is the displacement. The force is vertically downward and has magnitude mg, where m is the mass of the flake, so this reduces to $W = mgh$, where h is the height from which the flake falls. This is equal to the radius r of the bowl. Thus

$$W = mgr = (2.00 \times 10^{-3} \text{ kg})(9.8 \text{ m/s}^2)(22.0 \times 10^{-2} \text{ m}) = 4.31 \times 10^{-3} \text{ J}.$$

(b) The force of gravity is conservative, so the change in gravitational potential energy of the flake-Earth system is the negative of the work done: $\Delta U = -W = -4.31 \times 10^{-3}$ J.

(c) The potential energy when the flake is at the top is greater than when it is at the bottom by $|\Delta U|$. If $U = 0$ at the bottom, then $U = +4.31 \times 10^{-3}$ J at the top.

(d) If $U = 0$ at the top, then $U = -4.31 \times 10^{-3}$ J at the bottom.

(e) All the answers are proportional to the mass of the flake. If the mass is doubled, all answers are doubled.

4. We use Eq. 7-12 for W_g and Eq. 8-9 for U.

(a) The displacement between the initial point and A is horizontal, so $\phi = 90.0°$ and $W_g = 0$ (since $\cos 90.0° = 0$).

(b) The displacement between the initial point and B has a vertical component of $h/2$ downward (same direction as $\vec{F}_g$), so we obtain

$$W_g = \vec{F}_g \cdot \vec{d} = \frac{1}{2}mgh = \frac{1}{2}(825 \text{ kg})(9.80 \text{ m/s}^2)(42.0 \text{ m}) = 1.70 \times 10^5 \text{ J}.$$

(c) The displacement between the initial point and C has a vertical component of h downward (same direction as $\vec{F}_g$), so we obtain

$$W_g = \vec{F}_g \cdot \vec{d} = mgh = (825 \text{ kg})(9.80 \text{ m/s}^2)(42.0 \text{ m}) = 3.40 \times 10^5 \text{ J}.$$

(d) With the reference position at C, we obtain

$$U_B = \frac{1}{2}mgh = \frac{1}{2}(825 \text{ kg})(9.80 \text{ m/s}^2)(42.0 \text{ m}) = 1.70 \times 10^5 \text{ J}$$

(e) Similarly, we find

$$U_A = mgh = (825 \text{ kg})(9.80 \text{ m/s}^2)(42.0 \text{ m}) = 3.40 \times 10^5 \text{ J}$$

(f) All the answers are proportional to the mass of the object. If the mass is doubled, all answers are doubled.

5. The potential energy stored by the spring is given by $U = \frac{1}{2}kx^2$, where k is the spring constant and x is the displacement of the end of the spring from its position when the spring is in equilibrium. Thus

$$k = \frac{2U}{x^2} = \frac{2(25 \text{ J})}{(0.075 \text{ m})^2} = 8.9 \times 10^3 \text{ N/m}.$$

6. (a) The force of gravity is constant, so the work it does is given by $W = \vec{F} \cdot \vec{d}$, where $\vec{F}$ is the force and $\vec{d}$ is the displacement. The force is vertically downward and has magnitude mg, where m is the mass of the snowball. The expression for the work reduces to $W = mgh$, where h is the height through which the snowball drops. Thus

$$W = mgh = (1.50 \text{ kg})(9.80 \text{ m/s}^2)(12.5 \text{ m}) = 184 \text{ J}.$$

(b) The force of gravity is conservative, so the change in the potential energy of the snowball-Earth system is the negative of the work it does: $\Delta U = -W = -184$ J.

(c) The potential energy when it reaches the ground is less than the potential energy when it is fired by $|\Delta U|$, so $U = -184$ J when the snowball hits the ground.

7. The main challenge for students in this type of problem seems to be working out the trigonometry in order to obtain the height of the ball (relative to the low point of the swing) $h = L - L \cos \theta$ (for angle θ measured from vertical as shown in Fig. 8-34). Once this relation (which we will not derive here since we have found this to be most easily illustrated at the blackboard) is established, then the principal results of this problem follow from Eq. 7-12 (for W_g) and Eq. 8-9 (for U).

(a) The vertical component of the displacement vector is downward with magnitude h, so we obtain

$$W_g = \vec{F}_g \cdot \vec{d} = mgh = mgL(1 - \cos \theta)$$
$$= (5.00 \text{ kg})(9.80 \text{ m/s}^2)(2.00 \text{ m})(1 - \cos 30°) = 13.1 \text{ J}$$

(b) From Eq. 8-1, we have $\Delta U = -W_g = -mgL(1 - \cos \theta) = -13.1$ J.

(c) With $y = h$, Eq. 8-9 yields $U = mgL(1 - \cos \theta) = 13.1$ J.

(d) As the angle increases, we intuitively see that the height h increases (and, less obviously, from the mathematics, we see that $\cos \theta$ decreases so that $1 - \cos \theta$ increases), so the answers to parts (a) and (c) increase, and the absolute value of the answer to part (b) also increases.

8. We use Eq. 7-12 for W_g and Eq. 8-9 for U.

(a) The displacement between the initial point and Q has a vertical component of $h - R$ downward (same direction as $\vec{F}_g$), so (with $h = 5R$) we obtain

$$W_g = \vec{F}_g \cdot \vec{d} = 4mgR = 4(3.20 \times 10^{-2} \text{ kg})(9.80 \text{ m/s}^2)(0.12 \text{ m}) = 0.15 \text{ J}.$$

(b) The displacement between the initial point and the top of the loop has a vertical component of $h - 2R$ downward (same direction as $\vec{F}_g$), so (with $h = 5R$) we obtain

$$W_g = \vec{F}_g \cdot \vec{d} = 3mgR = 3(3.20 \times 10^{-2} \text{ kg})(9.80 \text{ m/s}^2)(0.12 \text{ m}) = 0.11 \text{ J}.$$

(c) With $y = h = 5R$, at P we find

$$U = 5mgR = 5(3.20 \times 10^{-2} \text{ kg})(9.80 \text{ m/s}^2)(0.12 \text{ m}) = 0.19 \text{ J}.$$

(d) With $y = R$, at Q we have

$$U = mgR = (3.20 \times 10^{-2} \text{ kg})(9.80 \text{ m/s}^2)(0.12 \text{ m}) = 0.038 \text{ J}$$

(e) With $y = 2R$, at the top of the loop, we find

$$U = 2mgR = 2(3.20 \times 10^{-2} \text{ kg})(9.80 \text{ m/s}^2)(0.12 \text{ m}) = 0.075 \text{ J}$$

(f) The new information $(v_i \neq 0)$ is not involved in any of the preceding computations; the above results are unchanged.

9. We neglect any work done by friction. We work with SI units, so the speed is converted: $v = 130(1000/3600) = 36.1$ m/s.

(a) We use Eq. 8-17: $K_f + U_f = K_i + U_i$ with $U_i = 0$, $U_f = mgh$ and $K_f = 0$. Since $K_i = \frac{1}{2}mv^2$, where v is the initial speed of the truck, we obtain

$$\frac{1}{2}mv^2 = mgh \quad \Rightarrow \quad h = \frac{v^2}{2g} = \frac{(36.1 \text{ m/s})^2}{2(9.8 \text{ m/s}^2)} = 66.5 \text{ m}.$$

If L is the length of the ramp, then $L \sin 15° = 66.5$ m so that $L = (66.5 \text{ m})/\sin 15° = 257$ m. Therefore, the ramp must be about 2.6×10^2 m long if friction is negligible.

(b) The answers do not depend on the mass of the truck. They remain the same if the mass is reduced.

(c) If the speed is decreased, h and L both decrease (note that h is proportional to the square of the speed and that L is proportional to h).

10. We use Eq. 8-17, representing the conservation of mechanical energy (which neglects friction and other dissipative effects).

(a) In the solution to exercise 2 (to which this problem refers), we found $U_i = mgy_i = 196$J and $U_f = mgy_f = 29.0$ J (assuming the reference position is at the ground). Since $K_i = 0$ in this case, we have

$$0 + 196 \text{ J} = K_f + 29.0 \text{ J}$$

which gives $K_f = 167$ J and thus leads to

$$v = \sqrt{\frac{2K_f}{m}} = \sqrt{\frac{2(167\text{ J})}{2.00\text{ kg}}} = 12.9\text{ m/s}.$$

(b) If we proceed algebraically through the calculation in part (a), we find $K_f = -\Delta U = mgh$ where $h = y_i - y_f$ and is positive-valued. Thus,

$$v = \sqrt{\frac{2K_f}{m}} = \sqrt{2gh}$$

as we might also have derived from the equations of Table 2-1 (particularly Eq. 2-16). The fact that the answer is independent of mass means that the answer to part (b) is identical to that of part (a), i.e., $v = 12.9$ m/s.

(c) If $K_i \neq 0$, then we find $K_f = mgh + K_i$ (where K_i is necessarily positive-valued). This represents a larger value for K_f than in the previous parts, and thus leads to a larger value for v.

11. (a) If K_i is the kinetic energy of the flake at the edge of the bowl, K_f is its kinetic energy at the bottom, U_i is the gravitational potential energy of the flake-Earth system with the flake at the top, and U_f is the gravitational potential energy with it at the bottom, then $K_f + U_f = K_i + U_i$.

Taking the potential energy to be zero at the bottom of the bowl, then the potential energy at the top is $U_i = mgr$ where $r = 0.220$ m is the radius of the bowl and m is the mass of the flake. $K_i = 0$ since the flake starts from rest. Since the problem asks for the speed at the bottom, we write $\frac{1}{2}mv^2$ for K_f. Energy conservation leads to

$$W_g = \vec{F}_g \cdot \vec{d} = mgh = mgL(1-\cos\theta).$$

The speed is $v = \sqrt{2gr} = 2.08$ m/s.

(b) Since the expression for speed does not contain the mass of the flake, the speed would be the same, 2.08 m/s, regardless of the mass of the flake.

(c) The final kinetic energy is given by $K_f = K_i + U_i - U_f$. Since K_i is greater than before, K_f is greater. This means the final speed of the flake is greater.

12. We use Eq. 8-18, representing the conservation of mechanical energy (which neglects friction and other dissipative effects).

(a) In the solution to Problem 4 we found $\Delta U = mgL$ as it goes to the highest point. Thus, we have

$$\Delta K + \Delta U = 0$$
$$K_{\text{top}} - K_0 + mgL = 0$$

which, upon requiring $K_{\text{top}} = 0$, gives $K_0 = mgL$ and thus leads to

$$v_0 = \sqrt{\frac{2K_0}{m}} = \sqrt{2gL} = \sqrt{2(9.80 \text{ m/s}^2)(0.452 \text{ m})} = 2.98 \text{ m/s}.$$

(b) We also found in the Problem 4 that the potential energy change is $\Delta U = -mgL$ in going from the initial point to the lowest point (the bottom). Thus,

$$\Delta K + \Delta U = 0$$
$$K_{\text{bottom}} - K_0 - mgL = 0$$

which, with $K_0 = mgL$, leads to $K_{\text{bottom}} = 2mgL$. Therefore,

$$v_{\text{bottom}} = \sqrt{\frac{2K_{\text{bottom}}}{m}} = \sqrt{4gL} = \sqrt{4(9.80 \text{ m/s}^2)(0.452 \text{ m})} = 4.21 \text{ m/s}.$$

(c) Since there is no change in height (going from initial point to the rightmost point), then $\Delta U = 0$, which implies $\Delta K = 0$. Consequently, the speed is the same as what it was initially,

$$v_{\text{right}} = v_0 = 2.98 \text{ m/s}.$$

(d) It is evident from the above manipulations that the results do not depend on mass. Thus, a different mass for the ball must lead to the same results.

13. We use Eq. 8-17, representing the conservation of mechanical energy (which neglects friction and other dissipative effects).

(a) In Problem 4, we found $U_A = mgh$ (with the reference position at C). Referring again to Fig. 8-33, we see that this is the same as U_0 which implies that $K_A = K_0$ and thus that

$$v_A = v_0 = 17.0 \text{ m/s}.$$

(b) In the solution to Problem 4, we also found $U_B = mgh/2$. In this case, we have

$$K_0 + U_0 = K_B + U_B$$
$$\frac{1}{2}mv_0^2 + mgh = \frac{1}{2}mv_B^2 + mg\left(\frac{h}{2}\right)$$

which leads to

$$v_B = \sqrt{v_0^2 + gh} = \sqrt{(17.0 \text{ m/s})^2 + (9.80 \text{ m/s}^2)(42.0 \text{ m})} = 26.5 \text{ m/s}.$$

(c) Similarly,

$$v_C = \sqrt{v_0^2 + 2gh} = \sqrt{(17.0 \text{ m/s})^2 + 2(9.80 \text{ m/s}^2)(42.0 \text{ m})} = 33.4 \text{ m/s}.$$

(d) To find the "final" height, we set $K_f = 0$. In this case, we have

$$K_0 + U_0 = K_f + U_f$$

$$\frac{1}{2}mv_0^2 + mgh = 0 + mgh_f$$

which yields $h_f = h + \dfrac{v_0^2}{2g} = 42.0 \text{ m} + \dfrac{(17.0 \text{ m/s})^2}{2(9.80 \text{ m/s}^2)} = 56.7 \text{ m}.$

(e) It is evident that the above results do not depend on mass. Thus, a different mass for the coaster must lead to the same results.

14. We use Eq. 8-18, representing the conservation of mechanical energy. We choose the reference position for computing U to be at the ground below the cliff; it is also regarded as the "final" position in our calculations.

(a) Using Eq. 8-9, the initial potential energy is given by $U_i = mgh$ where $h = 12.5$ m and $m = 1.50$ kg. Thus, we have

$$K_i + U_i = K_f + U_f$$

$$\frac{1}{2}mv_i^2 + mgh = \frac{1}{2}mv^2 + 0$$

which leads to the speed of the snowball at the instant before striking the ground:

$$v = \sqrt{\frac{2}{m}\left(\frac{1}{2}mv_i^2 + mgh\right)} = \sqrt{v_i^2 + 2gh}$$

where $v_i = 14.0$ m/s is the magnitude of its initial velocity (not just one component of it). Thus we find $v = 21.0$ m/s.

(b) As noted above, v_i is the magnitude of its initial velocity and not just one component of it; therefore, there is no dependence on launch angle. The answer is again 21.0 m/s.

(c) It is evident that the result for v in part (a) does not depend on mass. Thus, changing the mass of the snowball does not change the result for v.

15. We take the reference point for gravitational potential energy at the position of the marble when the spring is compressed.

(a) The gravitational potential energy when the marble is at the top of its motion is $U_g = mgh$, where $h = 20$ m is the height of the highest point. Thus,

$$U_g = (5.0 \times 10^{-3} \text{ kg})(9.8 \text{ m/s}^2)(20 \text{ m}) = 0.98 \text{ J}.$$

(b) Since the kinetic energy is zero at the release point and at the highest point, then conservation of mechanical energy implies $\Delta U_g + \Delta U_s = 0$, where ΔU_s is the change in the spring's elastic potential energy. Therefore, $\Delta U_s = -\Delta U_g = -0.98$ J.

(c) We take the spring potential energy to be zero when the spring is relaxed. Then, our result in the previous part implies that its initial potential energy is $U_s = 0.98$ J. This must be $\frac{1}{2}kx^2$, where k is the spring constant and x is the initial compression. Consequently,

$$k = \frac{2U_s}{x^2} = \frac{2(0.98 \text{ J})}{(0.080 \text{ m})^2} = 3.1 \times 10^2 \text{ N/m} = 3.1 \text{ N/cm}.$$

16. We use Eq. 8-18, representing the conservation of mechanical energy. The reference position for computing U is the lowest point of the swing; it is also regarded as the "final" position in our calculations.

(a) In the solution to problem 7, we found $U = mgL(1 - \cos\theta)$ at the position shown in Fig. 8-34 (which we consider to be the initial position). Thus, we have

$$K_i + U_i = K_f + U_f$$

$$0 + mgL(1 - \cos\theta) = \frac{1}{2}mv^2 + 0$$

which leads to

$$v = \sqrt{\frac{2mgL(1 - \cos\theta)}{m}} = \sqrt{2gL(1 - \cos\theta)}.$$

Plugging in $L = 2.00$ m and $\theta = 30.0°$ we find $v = 2.29$ m/s.

(b) It is evident that the result for v does not depend on mass. Thus, a different mass for the ball must not change the result.

17. We use Eq. 8-18, representing the conservation of mechanical energy (which neglects friction and other dissipative effects). The reference position for computing U (and height

h) is the lowest point of the swing; it is also regarded as the "final" position in our calculations.

(a) Careful examination of the figure leads to the trigonometric relation $h = L - L\cos\theta$ when the angle is measured from vertical as shown. Thus, the gravitational potential energy is $U = mgL(1 - \cos\theta_0)$ at the position shown in Fig. 8-34 (the initial position). Thus, we have

$$K_0 + U_0 = K_f + U_f$$

$$\frac{1}{2}mv_0^2 + mgL(1-\cos\theta_0) = \frac{1}{2}mv^2 + 0$$

which leads to

$$v = \sqrt{\frac{2}{m}\left[\frac{1}{2}mv_0^2 + mgL(1-\cos\theta_0)\right]} = \sqrt{v_0^2 + 2gL(1-\cos\theta_0)}$$

$$= \sqrt{(8.00 \text{ m/s})^2 + 2(9.80 \text{ m/s}^2)(1.25 \text{ m})(1-\cos 40°)} = 8.35 \text{ m/s}.$$

(b) We look for the initial speed required to barely reach the horizontal position — described by $v_h = 0$ and $\theta = 90°$ (or $\theta = -90°$, if one prefers, but since $\cos(-\phi) = \cos\phi$, the sign of the angle is not a concern).

$$K_0 + U_0 = K_h + U_h$$

$$\frac{1}{2}mv_0^2 + mgL(1-\cos\theta_0) = 0 + mgL$$

which yields

$$v_0 = \sqrt{2gL\cos\theta_0} = \sqrt{2(9.80 \text{ m/s}^2)(1.25 \text{ m})\cos 40°} = 4.33 \text{ m/s}.$$

(c) For the cord to remain straight, then the centripetal force (at the top) must be (at least) equal to gravitational force:

$$\frac{mv_t^2}{r} = mg \Rightarrow mv_t^2 = mgL$$

where we recognize that $r = L$. We plug this into the expression for the kinetic energy (at the top, where $\theta = 180°$).

$$K_0 + U_0 = K_t + U_t$$

$$\frac{1}{2}mv_0^2 + mgL(1-\cos\theta_0) = \frac{1}{2}mv_t^2 + mg(1-\cos 180°)$$

$$\frac{1}{2}mv_0^2 + mgL(1-\cos\theta_0) = \frac{1}{2}(mgL) + mg(2L)$$

which leads to

$$v_0 = \sqrt{gL(3+2\cos\theta_0)} = \sqrt{(9.80 \text{ m/s}^2)(1.25 \text{ m})(3+2\cos 40°)} = 7.45 \text{ m/s}.$$

(d) The more initial potential energy there is, the less initial kinetic energy there needs to be, in order to reach the positions described in parts (b) and (c). Increasing θ_0 amounts to increasing U_0, so we see that a greater value of θ_0 leads to smaller results for v_0 in parts (b) and (c).

18. We place the reference position for evaluating gravitational potential energy at the relaxed position of the spring. We use x for the spring's compression, measured positively downwards (so $x > 0$ means it is compressed).

(a) With $x = 0.190$ m, Eq. 7-26 gives

$$W_s = -\frac{1}{2}kx^2 = -7.22 \text{ J} \approx -7.2 \text{ J}$$

for the work done by the spring force. Using Newton's third law, we see that the work done on the spring is 7.2 J.

(b) As noted above, $W_s = -7.2$ J.

(c) Energy conservation leads to

$$K_i + U_i = K_f + U_f$$

$$mgh_0 = -mgx + \frac{1}{2}kx^2$$

which (with $m = 0.70$ kg) yields $h_0 = 0.86$ m.

(d) With a new value for the height $h_0' = 2h_0 = 1.72$ m, we solve for a new value of x using the quadratic formula (taking its positive root so that $x > 0$).

$$mgh_0' = -mgx + \frac{1}{2}kx^2 \Rightarrow x = \frac{mg + \sqrt{(mg)^2 + 2mgkh_0'}}{k}$$

which yields $x = 0.26$ m.

19. (a) At Q the block (which is in circular motion at that point) experiences a centripetal acceleration v^2/R leftward. We find v^2 from energy conservation:

$$K_P + U_P = K_Q + U_Q$$

$$0 + mgh = \frac{1}{2}mv^2 + mgR$$

318   CHAPTER 8

Using the fact that $h = 5R$, we find $mv^2 = 8mgR$. Thus, the horizontal component of the net force on the block at Q is

$$F = mv^2/R = 8mg = 8(0.032 \text{ kg})(9.8 \text{ m/s}^2) = 2.5 \text{ N}.$$

and points left (in the same direction as $\vec{a}$).

(b) The downward component of the net force on the block at Q is the downward force of gravity

$$F = mg = (0.032 \text{ kg})(9.8 \text{ m/s}^2) = 0.31 \text{ N}.$$

(c) To barely make the top of the loop, the centripetal force there must equal the force of gravity:

$$\frac{mv_t^2}{R} = mg \implies mv_t^2 = mgR$$

This requires a different value of h than was used above.

$$K_P + U_P = K_t + U_t$$

$$0 + mgh = \frac{1}{2}mv_t^2 + mgh_t$$

$$mgh = \frac{1}{2}(mgR) + mg(2R)$$

Consequently, $h = 2.5R = (2.5)(0.12 \text{ m}) = 0.30$ m.

(d) The normal force F_N, for speeds v_t greater than $\sqrt{gR}$ (which are the only possibilities for non-zero F_N — see the solution in the previous part), obeys

$$F_N = \frac{mv_t^2}{R} - mg$$

from Newton's second law. Since v_t^2 is related to h by energy conservation

$$K_P + U_P = K_t + U_t \implies gh = \frac{1}{2}v_t^2 + 2gR$$

then the normal force, as a function for h (so long as $h \geq 2.5R$ — see solution in previous part), becomes

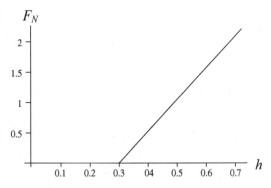

$$F_N = \frac{2mgh}{R} - 5mg$$

Thus, the graph for $h \geq 2.5R$ consists of a straight line of positive slope $2mg/R$ (which can be set to some convenient values for graphing purposes).

Note that for $h \leq 2.5R$, the normal force is zero.

20. (a) With energy in Joules and length in meters, we have

$$\Delta U = U(x) - U(0) = -\int_0^x (6x' - 12)dx'.$$

Therefore, with $U(0) = 27$ J, we obtain $U(x)$ (written simply as U) by integrating and rearranging:

$$U = 27 + 12x - 3x^2.$$

(b) We can maximize the above function by working through the $dU/dx = 0$ condition, or we can treat this as a force equilibrium situation — which is the approach we show.

$$F = 0 \Rightarrow 6x_{eq} - 12 = 0$$

Thus, $x_{eq} = 2.0$ m, and the above expression for the potential energy becomes $U = 39$ J.

(c) Using the quadratic formula or using the polynomial solver on an appropriate calculator, we find the negative value of x for which $U = 0$ to be $x = -1.6$ m.

(d) Similarly, we find the positive value of x for which $U = 0$ to be $x = 5.6$ m

21. (a) As the string reaches its lowest point, its original potential energy $U = mgL$ (measured relative to the lowest point) is converted into kinetic energy. Thus,

$$mgL = \frac{1}{2}mv^2 \Rightarrow v = \sqrt{2gL}.$$

With $L = 1.20$ m we obtain $v = 4.85$ m/s.

(b) In this case, the total mechanical energy is shared between kinetic $\frac{1}{2}mv_b^2$ and potential mgy_b. We note that $y_b = 2r$ where $r = L - d = 0.450$ m. Energy conservation leads to

$$mgL = \frac{1}{2}mv_b^2 + mgy_b$$

which yields $v_b = \sqrt{2gL - 2g(2r)} = 2.42$ m/s.

22. We denote m as the mass of the block, $h = 0.40$ m as the height from which it dropped (measured from the relaxed position of the spring), and x the compression of the spring (measured downward so that it yields a positive value). Our reference point for the gravitational potential energy is the initial position of the block. The block drops a total distance $h + x$, and the final gravitational potential energy is $-mg(h + x)$. The spring potential energy is $\frac{1}{2}kx^2$ in the final situation, and the kinetic energy is zero both at the beginning and end. Since energy is conserved

$$K_i + U_i = K_f + U_f$$
$$0 = -mg(h+x) + \frac{1}{2}kx^2$$

which is a second degree equation in x. Using the quadratic formula, its solution is

$$x = \frac{mg \pm \sqrt{(mg)^2 + 2mghk}}{k}.$$

Now $mg = 19.6$ N, $h = 0.40$ m, and $k = 1960$ N/m, and we choose the positive root so that $x > 0$.

$$x = \frac{19.6 + \sqrt{19.6^2 + 2(19.6)(0.40)(1960)}}{1960} = 0.10 \text{ m}.$$

23. Since time does not directly enter into the energy formulations, we return to Chapter 4 (or Table 2-1 in Chapter 2) to find the change of height during this $t = 6.0$ s flight.

$$\Delta y = v_{0y}t - \frac{1}{2}gt^2$$

This leads to $\Delta y = -32$ m. Therefore $\Delta U = mg\Delta y = -318$ J $\approx -3.2 \times 10^{-2}$ J.

24. From Chapter 4, we know the height h of the skier's jump can be found from $v_y^2 = 0 = v_{0y}^2 - 2gh$ where $v_{0y} = v_0 \sin 28°$ is the upward component of the skier's "launch velocity." To find v_0 we use energy conservation.

(a) The skier starts at rest $y = 20$ m above the point of "launch" so energy conservation leads to

$$mgy = \frac{1}{2}mv^2 \Rightarrow v = \sqrt{2gy} = 20 \text{ m/s}$$

which becomes the initial speed v_0 for the launch. Hence, the above equation relating h to v_0 yields

$$h = \frac{(v_0 \sin 28°)^2}{2g} = 4.4 \text{ m}.$$

(b) We see that all reference to mass cancels from the above computations, so a new value for the mass will yield the same result as before.

25. (a) To find out whether or not the vine breaks, it is sufficient to examine it at the moment Tarzan swings through the lowest point, which is when the vine — if it didn't break — would have the greatest tension. Choosing upward positive, Newton's second law leads to

$$T - mg = m\frac{v^2}{r}$$

where $r = 18.0$ m and $m = W/g = 688/9.8 = 70.2$ kg. We find the v^2 from energy conservation (where the reference position for the potential energy is at the lowest point).

$$mgh = \frac{1}{2}mv^2 \Rightarrow v^2 = 2gh$$

where $h = 3.20$ m. Combining these results, we have

$$T = mg + m\frac{2gh}{r} = mg\left(1 + \frac{2h}{r}\right)$$

which yields 933 N. Thus, the vine does not break.

(b) Rounding to an appropriate number of significant figures, we see the maximum tension is roughly 9.3×10^2 N.

26. (a) We take the reference point for gravitational energy to be at the lowest point of the swing. Let θ be the angle measured from vertical. Then the height y of the pendulum "bob" (the object at the end of the pendulum, which i this problem is the stone) is given by $L(1 - \cos\theta) = y$. Hence, the gravitational potential energy is

$$mgy = mgL(1 - \cos\theta).$$

When $\theta = 0°$ (the string at its lowest point) we are told that its speed is 8.0 m/s; its kinetic energy there is therefore 64 J (using Eq. 7-1). At $\theta = 60°$ its mechanical energy is

$$E_{mech} = \frac{1}{2}mv^2 + mgL(1 - \cos\theta).$$

Energy conservation (since there is no friction) requires that this be equal to 64 J. Solving for the speed, we find $v = 5.0$ m/s.

(b) We now set the above expression again equal to 64 J (with θ being the unknown) but with zero speed (which gives the condition for the maximum point, or "turning point" that it reaches). This leads to $\theta_{max} = 79°$.

(c) As observed in our solution to part (a), the total mechanical energy is 64 J.

27. We convert to SI units and choose upward as the $+y$ direction. Also, the relaxed position of the top end of the spring is the origin, so the initial compression of the spring (defining an equilibrium situation between the spring force and the force of gravity) is $y_0 = -0.100$ m and the additional compression brings it to the position $y_1 = -0.400$ m.

(a) When the stone is in the equilibrium ($a = 0$) position, Newton's second law becomes

$$\vec{F}_{net} = ma$$
$$F_{spring} - mg = 0$$
$$-k(-0.100) - (8.00)(9.8) = 0$$

where Hooke's law (Eq. 7-21) has been used. This leads to a spring constant equal to $k = 784$ N/m.

(b) With the additional compression (and release) the acceleration is no longer zero, and the stone will start moving upwards, turning some of its elastic potential energy (stored in the spring) into kinetic energy. The amount of elastic potential energy at the moment of release is, using Eq. 8-11,

$$U = \frac{1}{2} k y_1^2 = \frac{1}{2}(784)(-0.400)^2 = 62.7 \text{ J}.$$

(c) Its maximum height y_2 is beyond the point that the stone separates from the spring (entering free-fall motion). As usual, it is characterized by having (momentarily) zero speed. If we choose the y_1 position as the reference position in computing the gravitational potential energy, then

$$K_1 + U_1 = K_2 + U_2$$
$$0 + \frac{1}{2} k y_1^2 = 0 + mgh$$

where $h = y_2 - y_1$ is the height above the release point. Thus, mgh (the gravitational potential energy) is seen to be equal to the previous answer, 62.7 J, and we proceed with the solution in the next part.

(d) We find $h = k y_1^2 / 2mg = 0.800$ m, or 80.0 cm.

28. We take the original height of the box to be the $y = 0$ reference level and observe that, in general, the height of the box (when the box has moved a distance d downhill) is $y = -d \sin 40°$.

(a) Using the conservation of energy, we have

$$K_i + U_i = K + U \Rightarrow 0 + 0 = \frac{1}{2}mv^2 + mgy + \frac{1}{2}kd^2.$$

Therefore, with $d = 0.10$ m, we obtain $v = 0.81$ m/s.

(b) We look for a value of $d \neq 0$ such that $K = 0$.

$$K_i + U_i = K + U \Rightarrow 0 + 0 = 0 + mgy + \frac{1}{2}kd^2.$$

Thus, we obtain $mgd \sin 40° = \frac{1}{2}kd^2$ and find $d = 0.21$ m.

(c) The uphill force is caused by the spring (Hooke's law) and has magnitude $kd = 25.2$ N. The downhill force is the component of gravity $mg \sin 40° = 12.6$ N. Thus, the net force on the box is $(25.2 - 12.6)$ N $= 12.6$ N uphill, with $a = F/m = (12.6$ N$)/(2.0$ kg$) = 6.3$ m/s^2.

(d) The acceleration is up the incline.

29. The reference point for the gravitational potential energy U_g (and height h) is at the block when the spring is maximally compressed. When the block is moving to its highest point, it is first accelerated by the spring; later, it separates from the spring and finally reaches a point where its speed v_f is (momentarily) zero. The x axis is along the incline, pointing uphill (so x_0 for the initial compression is negative-valued); its origin is at the relaxed position of the spring. We use SI units, so $k = 1960$ N/m and $x_0 = -0.200$ m.

(a) The elastic potential energy is $\frac{1}{2}kx_0^2 = 39.2$ J.

(b) Since initially $U_g = 0$, the change in U_g is the same as its final value mgh where $m = 2.00$ kg. That this must equal the result in part (a) is made clear in the steps shown in the next part. Thus, $\Delta U_g = U_g = 39.2$ J.

(c) The principle of mechanical energy conservation leads to

$$K_0 + U_0 = K_f + U_f$$
$$0 + \frac{1}{2}kx_0^2 = 0 + mgh$$

which yields $h = 2.00$ m. The problem asks for the distance *along the incline*, so we have $d = h/\sin 30° = 4.00$ m.

30. From the slope of the graph, we find the spring constant

$$k = \frac{\Delta F}{\Delta x} = 0.10 \, \text{N/cm} = 10 \, \text{N/m}.$$

(a) Equating the potential energy of the compressed spring to the kinetic energy of the cork at the moment of release, we have

$$\frac{1}{2}kx^2 = \frac{1}{2}mv^2 \Rightarrow v = x\sqrt{\frac{k}{m}}$$

which yields $v = 2.8$ m/s for $m = 0.0038$ kg and $x = 0.055$ m.

(b) The new scenario involves some potential energy at the moment of release. With $d = 0.015$ m, energy conservation becomes

$$\frac{1}{2}kx^2 = \frac{1}{2}mv^2 + \frac{1}{2}kd^2 \Rightarrow v = \sqrt{\frac{k}{m}(x^2 - d^2)}$$

which yields $v = 2.7$ m/s.

31. We refer to its starting point as A, the point where it first comes into contact with the spring as B, and the point where the spring is compressed $|x| = 0.055$ m as C. Point C is our reference point for computing gravitational potential energy. Elastic potential energy (of the spring) is zero when the spring is relaxed. Information given in the second sentence allows us to compute the spring constant. From Hooke's law, we find

$$k = \frac{F}{x} = \frac{270 \, \text{N}}{0.02 \, \text{m}} = 1.35 \times 10^4 \, \text{N/m}.$$

(a) The distance between points A and B is $\vec{F}_g$ and we note that the total sliding distance $\ell + |x|$ is related to the initial height h of the block (measured relative to C) by

$$\frac{h}{\ell + |x|} = \sin\theta$$

where the incline angle θ is 30°. Mechanical energy conservation leads to

$$K_A + U_A = K_C + U_C$$
$$0 + mgh = 0 + \frac{1}{2}kx^2$$

which yields

$$h = \frac{kx^2}{2mg} = \frac{(1.35 \times 10^4 \text{ N/m})(0.055 \text{ m})^2}{2(12 \text{ kg})(9.8 \text{ m/s}^2)} = 0.174 \text{ m}.$$

Therefore,

$$\ell + |x| = \frac{h}{\sin 30°} = \frac{0.174 \text{ m}}{\sin 30°} = 0.35 \text{ m}.$$

(b) From this result, we find $\ell = 0.35 - 0.055 = 0.29$ m, which means that $\Delta y = -\ell \sin\theta = -0.15$ m in sliding from point A to point B. Thus, Eq. 8-18 gives

$$\Delta K + \Delta U = 0$$
$$\frac{1}{2}mv_B^2 + mg\Delta h = 0$$

which yields $v_B = \sqrt{-2g\Delta h} = \sqrt{-(9.8)(-0.15)} = 1.7$ m/s.

32. The work required is the change in the gravitational potential energy as a result of the chain being pulled onto the table. Dividing the hanging chain into a large number of infinitesimal segments, each of length dy, we note that the mass of a segment is $(m/L)\, dy$ and the change in potential energy of a segment when it is a distance $|y|$ below the table top is

$$dU = (m/L)g|y|\, dy = -(m/L)gy\, dy$$

since y is negative-valued (we have $+y$ upward and the origin is at the tabletop). The total potential energy change is

$$\Delta U = -\frac{mg}{L}\int_{-L/4}^{0} y\, dy = \frac{1}{2}\frac{mg}{L}(L/4)^2 = mgL/32.$$

The work required to pull the chain onto the table is therefore

$$W = \Delta U = mgL/32 = (0.012 \text{ kg})(9.8 \text{ m/s}^2)(0.28 \text{ m})/32 = 0.0010 \text{ J}.$$

33. All heights h are measured from the lower end of the incline (which is our reference position for computing gravitational potential energy mgh). Our x axis is along the incline, with $+x$ being uphill (so spring compression corresponds to $x > 0$) and its origin being at the relaxed end of the spring. The height that corresponds to the canister's initial position

(with spring compressed amount $x = 0.200$ m) is given by $h_1 = (D + x)\sin\theta$, where $\theta = 37°$.

(a) Energy conservation leads to

$$K_1 + U_1 = K_2 + U_2 \quad \Rightarrow \quad 0 + mg(D+x)\sin\theta + \frac{1}{2}kx^2 = \frac{1}{2}mv_2^2 + mgD\sin\theta$$

which yields, using the data $m = 2.00$ kg and $k = 170$ N/m,

$$v_2 = \sqrt{2gx\sin\theta + kx^2/m} = 2.40 \text{ m/s}.$$

(b) In this case, energy conservation leads to

$$K_1 + U_1 = K_3 + U_3$$
$$0 + mg(D+x)\sin\theta + \frac{1}{2}kx^2 = \frac{1}{2}mv_3^2 + 0$$

which yields $v_3 = \sqrt{2g(D+x)\sin\theta + kx^2/m} = 4.19$ m/s.

34. The distance the marble travels is determined by its initial speed (and the methods of Chapter 4), and the initial speed is determined (using energy conservation) by the original compression of the spring. We denote h as the height of the table, and x as the horizontal distance to the point where the marble lands. Then $x = v_0 t$ and $h = \frac{1}{2}gt^2$ (since the vertical component of the marble's "launch velocity" is zero). From these we find $x = v_0\sqrt{2h/g}$. We note from this that the distance to the landing point is directly proportional to the initial speed. We denote v_{01} be the initial speed of the first shot and $D_1 = (2.20 - 0.27)$ m $= 1.93$ m be the horizontal distance to its landing point; similarly, v_{02} is the initial speed of the second shot and $D = 2.20$ m is the horizontal distance to its landing spot. Then

$$\frac{v_{02}}{v_{01}} = \frac{D}{D_1} \quad \Rightarrow \quad v_{02} = \frac{D}{D_1}v_{01}$$

When the spring is compressed an amount ℓ, the elastic potential energy is $\frac{1}{2}k\ell^2$. When the marble leaves the spring its kinetic energy is $\frac{1}{2}mv_0^2$. Mechanical energy is conserved: $\frac{1}{2}mv_0^2 = \frac{1}{2}k\ell^2$, and we see that the initial speed of the marble is directly proportional to the original compression of the spring. If ℓ_1 is the compression for the first shot and ℓ_2 is the compression for the second, then $v_{02} = (\ell_2/\ell_1)v_{01}$. Relating this to the previous result, we obtain

$$\ell_2 = \frac{D}{D_1}\ell_1 = \left(\frac{2.20 \text{ m}}{1.93 \text{ m}}\right)(1.10 \text{ cm}) = 1.25 \text{ cm}.$$

35. Consider a differential element of length dx at a distance x from one end (the end which remains stuck) of the cord. As the cord turns vertical, its change in potential energy is given by

$$dU = -(\lambda dx)gx$$

where $\lambda = m/h$ is the mass/unit length and the negative sign indicates that the potential energy decreases. Integrating over the entire length, we obtain the total change in the potential energy:

$$\Delta U = \int dU = -\int_0^h \lambda gx\, dx = -\frac{1}{2}\lambda gh^2 = -\frac{1}{2}mgh.$$

With $m = 15$ g and $h = 25$ cm, we have $\Delta U = -0.018$ J.

36. Let $\vec{F}_N$ be the normal force of the ice on him and m is his mass. The net inward force is $mg \cos\theta - F_N$ and, according to Newton's second law, this must be equal to mv^2/R, where v is the speed of the boy. At the point where the boy leaves the ice $F_N = 0$, so $g \cos\theta = v^2/R$. We wish to find his speed. If the gravitational potential energy is taken to be zero when he is at the top of the ice mound, then his potential energy at the time shown is

$$U = -mgR(1 - \cos\theta).$$

He starts from rest and his kinetic energy at the time shown is $\frac{1}{2}mv^2$. Thus conservation of energy gives

$$0 = \frac{1}{2}mv^2 - mgR(1 - \cos\theta),$$

or $v^2 = 2gR(1 - \cos\theta)$. We substitute this expression into the equation developed from the second law to obtain $g \cos\theta = 2g(1 - \cos\theta)$. This gives $\cos\theta = 2/3$. The height of the boy above the bottom of the mound is

$$h = R \cos\theta = 2R/3 = 2(13.8 \text{ m})/3 = 9.20 \text{ m}.$$

37. (a) The (final) elastic potential energy is

$$U = \frac{1}{2}kx^2 = \frac{1}{2}(431 \text{ N/m})(0.210 \text{ m})^2 = 9.50 \text{ J}.$$

Ultimately this must come from the original (gravitational) energy in the system mgy (where we are measuring y from the lowest "elevation" reached by the block, so $y = (d + x)\sin(30°)$. Thus,

$$mg(d+x)\sin(30°) = 9.50 \text{ J} \quad \Rightarrow \quad d = 0.396 \text{ m}.$$

(b) The block is still accelerating (due to the component of gravity along the incline, $mg\sin(30°)$) for a few moments after coming into contact with the spring (which exerts the Hooke's law force kx), until the Hooke's law force is strong enough to cause the block to being decelerating. This point is reached when

$$kx = mg\sin 30°$$

which leads to $x = 0.0364$ m $= 3.64$ cm; this is long before the block finally stops (36.0 cm before it stops).

38. (a) The force at the equilibrium position $r = r_{eq}$ is

$$F = -\frac{dU}{dr}\bigg|_{r=r_{eq}} = 0 \quad \Rightarrow \quad -\frac{12A}{r_{eq}^{13}} + \frac{6B}{r_{eq}^{7}} = 0$$

which leads to the result

$$r_{eq} = \left(\frac{2A}{B}\right)^{\frac{1}{6}} = 1.12\left(\frac{A}{B}\right)^{\frac{1}{6}}.$$

(b) This defines a minimum in the potential energy curve (as can be verified either by a graph or by taking another derivative and verifying that it is concave upward at this point), which means that for values of r slightly smaller than r_{eq} the slope of the curve is negative (so the force is positive, repulsive).

(c) And for values of r slightly larger than r_{eq} the slope of the curve must be positive (so the force is negative, attractive).

39. From Fig. 8-50, we see that at $x = 4.5$ m, the potential energy is $U_1 = 15$ J. If the speed is $v = 7.0$ m/s, then the kinetic energy is

$$K_1 = mv^2/2 = (0.90 \text{ kg})(7.0 \text{ m/s})^2/2 = 22 \text{ J}.$$

The total energy is $E_1 = U_1 + K_1 = (15 + 22)$ J $= 37$ J.

(a) At $x = 1.0$ m, the potential energy is $U_2 = 35$ J. From energy conservation, we have $K_2 = 2.0$ J > 0. This means that the particle can reach there with a corresponding speed

$$v_2 = \sqrt{\frac{2K_2}{m}} = \sqrt{\frac{2(2.0 \text{ J})}{0.90 \text{ kg}}} = 2.1 \text{ m/s}.$$

(b) The force acting on the particle is related to the potential energy by the negative of the slope:

$$F_x = -\frac{\Delta U}{\Delta x}$$

From the figure we have $F_x = -\dfrac{35\text{ J}-15\text{ J}}{2\text{ m}-4\text{ m}} = +10\text{ N}$.

(c) Since the magnitude $F_x > 0$, the force points in the +x direction.

(d) At $x = 7.0$ m, the potential energy is $U_3 = 45$ J which exceeds the initial total energy E_1. Thus, the particle can never reach there. At the turning point, the kinetic energy is zero. Between $x = 5$ and 6 m, the potential energy is given by

$$U(x) = 15 + 30(x-5), \quad 5 \le x \le 6.$$

Thus, the turning point is found by solving $37 = 15 + 30(x-5)$, which yields $x = 5.7$ m.

(e) At x =5.0 m, the force acting on the particle is

$$F_x = -\frac{\Delta U}{\Delta x} = -\frac{(45-15)\text{ J}}{(6-5)\text{ m}} = -30\text{ N}$$

The magnitude is $|F_x| = 30$ N.

(f) The fact that $F_x < 0$ indicated that the force points in the –x direction.

40. In this problem, the mechanical energy (the sum of K and U) remains constant as the particle moves.

(a) Since mechanical energy is conserved, $U_B + K_B = U_A + K_A$, the kinetic energy of the particle in region A (3.00 m $\le x \le$ 4.00 m) is

$$K_A = U_B - U_A + K_B = 12.0\text{ J} - 9.00\text{ J} + 4.00\text{ J} = 7.00\text{ J}.$$

With $K_A = mv_A^2/2$, the speed of the particle at $x = 3.5$ m (within region A) is

$$v_A = \sqrt{\frac{2K_A}{m}} = \sqrt{\frac{2(7.00\text{ J})}{0.200\text{ kg}}} = 8.37\text{ m/s}.$$

(b) At $x = 6.5$ m, $U = 0$ and $K = U_B + K_B = 12.0\text{ J} + 4.00\text{ J} = 16.0$ J by mechanical energy conservation. Therefore, the speed at this point is

$$v = \sqrt{\frac{2K}{m}} = \sqrt{\frac{2(16.0 \text{ J})}{0.200 \text{ kg}}} = 12.6 \text{ m/s}.$$

(c) At the turning point, the speed of the particle is zero. Let the position of the right turning point be x_R. From the figure shown on the right, we find x_R to be

$$\frac{16.00 \text{ J} - 0}{x_R - 7.00 \text{ m}} = \frac{24.00 \text{ J} - 16.00 \text{ J}}{8.00 \text{ m} - x_R} \Rightarrow x_R = 7.67 \text{ m}.$$

(d) Let the position of the left turning point be x_L. From the figure shown, we find x_L to be

$$\frac{16.00 \text{ J} - 20.00 \text{ J}}{x_L - 1.00 \text{ m}} = \frac{9.00 \text{ J} - 16.00 \text{ J}}{3.00 \text{ m} - x_L} \Rightarrow x_L = 1.73 \text{ m}.$$

41. (a) The energy at $x = 5.0$ m is $E = K + U = 2.0 \text{ J} - 5.7 \text{ J} = -3.7 \text{ J}$.

(b) A plot of the potential energy curve (SI units understood) and the energy E (the horizontal line) is shown for $0 \le x \le 10$ m.

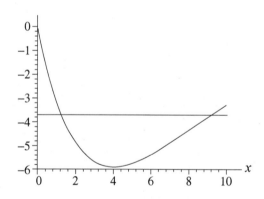

(c) The problem asks for a graphical determination of the turning points, which are the points on the curve corresponding to the total energy computed in part (a). The result for the smallest turning point (determined, to be honest, by more careful means) is $x = 1.3$ m.

(d) And the result for the largest turning point is $x = 9.1$ m.

(e) Since $K = E - U$, then maximizing K involves finding the minimum of U. A graphical determination suggests that this occurs at $x = 4.0$ m, which plugs into the expression $E - U = -3.7 - (-4xe^{-x/4})$ to give $K = 2.16 \text{ J} \approx 2.2 \text{ J}$. Alternatively, one can measure from the graph from the minimum of the U curve up to the level representing the total energy E and thereby obtain an estimate of K at that point.

(f) As mentioned in the previous part, the minimum of the U curve occurs at $x = 4.0$ m.

(g) The force (understood to be in newtons) follows from the potential energy, using Eq. 8-20 (and Appendix E if students are unfamiliar with such derivatives).

$$F = \frac{dU}{dx} = (4-x)e^{-x/4}$$

(h) This revisits the considerations of parts (d) and (e) (since we are returning to the minimum of $U(x)$) — but now with the advantage of having the analytic result of part (g). We see that the location which produces $F = 0$ is exactly $x = 4.0$ m.

42. Since the velocity is constant, $\vec{a} = 0$ and the horizontal component of the worker's push $F \cos\theta$ (where $\theta = 32°$) must equal the friction force magnitude $f_k = \mu_k F_N$. Also, the vertical forces must cancel, implying

$$W_{\text{applied}} = (8.0\text{N})(0.70\text{m}) = 5.6 \text{ J}$$

which is solved to find $F = 71$ N.

(a) The work done on the block by the worker is, using Eq. 7-7,

$$W = Fd\cos\theta = (71\text{ N})(9.2\text{ m})\cos 32° = 5.6 \times 10^2 \text{ J}.$$

(b) Since $f_k = \mu_k (mg + F \sin\theta)$, we find $\Delta E_{\text{th}} = f_k d = (60\text{ N})(9.2\text{ m}) = 5.6 \times 10^2$ J.

43. (a) Using Eq. 7-8, we have

$$W_{\text{applied}} = (8.0\text{ N})(0.70\text{ m}) = 5.6 \text{ J}.$$

(b) Using Eq. 8-31, the thermal energy generated is

$$\Delta E_{\text{th}} = f_k d = (5.0\text{ N})(0.70\text{ m}) = 3.5 \text{ J}.$$

44. (a) The work is $W = Fd = (35.0\text{ N})(3.00\text{ m}) = 105$ J.

(b) The total amount of energy that has gone to thermal forms is (see Eq. 8-31 and Eq. 6-2)

$$\Delta E_{\text{th}} = \mu_k mgd = (0.600)(4.00\text{ kg})(9.80\text{ m/s}^2)(3.00\text{ m}) = 70.6 \text{ J}.$$

If 40.0 J has gone to the block then $(70.6 - 40.0)$ J $= 30.6$ J has gone to the floor.

(c) Much of the work (105 J) has been "wasted" due to the 70.6 J of thermal energy generated, but there still remains (105 − 70.6) J = 34.4 J which has gone into increasing the kinetic energy of the block. (It has not gone into increasing the potential energy of the block because the floor is presumed to be horizontal.)

45. (a) The work done on the block by the force in the rope is, using Eq. 7-7,

$$W = Fd\cos\theta = (7.68\,\text{N})(4.06\,\text{m})\cos 15.0° = 30.1\,\text{J}.$$

(b) Using f for the magnitude of the kinetic friction force, Eq. 8-29 reveals that the increase in thermal energy is

$$\Delta E_{th} = fd = (7.42\,\text{N})(4.06\,\text{m}) = 30.1\,\text{J}.$$

(c) We can use Newton's second law of motion to obtain the frictional and normal forces, then use $\mu_k = f/F_N$ to obtain the coefficient of friction. Place the x axis along the path of the block and the y axis normal to the floor. The x and the y component of Newton's second law are

$$x:\quad F\cos\theta - f = 0$$
$$y:\quad F_N + F\sin\theta - mg = 0,$$

where m is the mass of the block, F is the force exerted by the rope, and θ is the angle between that force and the horizontal. The first equation gives

$$f = F\cos\theta = (7.68\,\text{N})\cos 15.0° = 7.42\,\text{N}$$

and the second gives

$$F_N = mg - F\sin\theta = (3.57\,\text{kg})(9.8\,\text{m/s}^2) - (7.68\,\text{N})\sin 15.0° = 33.0\,\text{N}.$$

Thus,

$$\mu_k = \frac{f}{F_N} = \frac{7.42\,\text{N}}{33.0\,\text{N}} = 0.225.$$

46. Equation 8-33 provides $\Delta E_{th} = -\Delta E_{mec}$ for the energy "lost" in the sense of this problem. Thus,

$$\Delta E_{th} = \frac{1}{2}m(v_i^2 - v_f^2) + mg(y_i - y_f) = \frac{1}{2}(60\,\text{kg})[(24\,\text{m/s})^2 - (22\,\text{m/s})^2] + (60\,\text{kg})(9.8\,\text{m/s}^2)(14\,\text{m})$$
$$= 1.1\times 10^4\,\text{J}.$$

That the angle of 25° is nowhere used in this calculation is indicative of the fact that energy is a scalar quantity.

47. (a) We take the initial gravitational potential energy to be $U_i = 0$. Then the final gravitational potential energy is $U_f = -mgL$, where L is the length of the tree. The change is

$$U_f - U_i = -mgL = -(25 \text{ kg})(9.8 \text{ m/s}^2)(12 \text{ m}) = -2.9 \times 10^3 \text{ J}.$$

(b) The kinetic energy is $K = \frac{1}{2}mv^2 = \frac{1}{2}(25 \text{ kg})(5.6 \text{ m/s})^2 = 3.9 \times 10^2 \text{ J}$.

(c) The changes in the mechanical and thermal energies must sum to zero. The change in thermal energy is $\Delta E_{th} = fL$, where f is the magnitude of the average frictional force; therefore,

$$f = -\frac{\Delta K + \Delta U}{L} = -\frac{3.9 \times 10^2 \text{ J} - 2.9 \times 10^3 \text{ J}}{12 \text{ m}} = 2.1 \times 10^2 \text{ N}.$$

48. We work this using the English units (with $g = 32$ ft/s), but for consistency we convert the weight to pounds

$$mg = (9.0) \text{ oz} \left(\frac{1 \text{ lb}}{16 \text{ oz}} \right) = 0.56 \text{ lb}$$

which implies $m = 0.018 \text{ lb} \cdot \text{s}^2/\text{ft}$ (which can be phrased as 0.018 slug as explained in Appendix D). And we convert the initial speed to feet-per-second

$$v_i = (81.8 \text{ mi/h}) \left(\frac{5280 \text{ ft/mi}}{3600 \text{ s/h}} \right) = 120 \text{ ft/s}$$

or a more "direct" conversion from Appendix D can be used. Equation 8-30 provides $\Delta E_{th} = -\Delta E_{mec}$ for the energy "lost" in the sense of this problem. Thus,

$$\Delta E_{th} = \frac{1}{2}m(v_i^2 - v_f^2) + mg(y_i - y_f) = \frac{1}{2}(0.018)(120^2 - 110^2) + 0 = 20 \text{ ft} \cdot \text{lb}.$$

49. We use SI units so $m = 0.075$ kg. Equation 8-33 provides $\Delta E_{th} = -\Delta E_{mec}$ for the energy "lost" in the sense of this problem. Thus,

$$\begin{aligned}\Delta E_{th} &= \frac{1}{2}m(v_i^2 - v_f^2) + mg(y_i - y_f) \\ &= \frac{1}{2}(0.075 \text{ kg})[(12 \text{ m/s})^2 - (10.5 \text{ m/s})^2] + (0.075 \text{ kg})(9.8 \text{ m/s}^2)(1.1 \text{ m} - 2.1 \text{ m}) \\ &= 0.53 \text{ J}.\end{aligned}$$

50. We use Eq. 8-31 to obtain

$$\Delta E_{th} = f_k d = (10\,\text{N})(5.0\,\text{m}) = 50\,\text{J}$$

and Eq. 7-8 to get

$$W = Fd = (2.0\,\text{N})(5.0\,\text{m}) = 10\,\text{J}.$$

Similarly, Eq. 8-31 gives

$$W = \Delta K + \Delta U + \Delta E_{th}$$
$$10 = 35 + \Delta U + 50$$

which yields $\Delta U = -75$ J. By Eq. 8-1, then, the work done by gravity is $W = -\Delta U = 75$ J.

51. (a) The initial potential energy is

$$U_i = mgy_i = (520\,\text{kg})(9.8\,\text{m/s}^2)(300\,\text{m}) = 1.53\times 10^6\,\text{J}$$

where $+y$ is upward and $y = 0$ at the bottom (so that $U_f = 0$).

(b) Since $f_k = \mu_k F_N = \mu_k mg\cos\theta$ we have $\Delta E_{th} = f_k d = \mu_k mgd\cos\theta$ from Eq. 8-31. Now, the hillside surface (of length $d = 500$ m) is treated as an hypotenuse of a 3-4-5 triangle, so $\cos\theta = x/d$ where $x = 400$ m. Therefore,

$$\Delta E_{th} = \mu_k mgd\frac{x}{d} = \mu_k mgx = (0.25)(520)(9.8)(400) = 5.1\times 10^5\,\text{J}.$$

(c) Using Eq. 8-31 (with $W = 0$) we find

$$K_f = K_i + U_i - U_f - \Delta E_{th}$$
$$= 0 + 1.53\times 10^6 - 0 - 5.1\times 10^5$$
$$= 0 + 1.02\times 10^6\,\text{J}.$$

(d) From $K_f = mv^2/2$, we obtain $v = 63$ m/s.

52. Energy conservation, as expressed by Eq. 8-33 (with $W = 0$) leads to

$$\Delta E_{th} = K_i - K_f + U_i - U_f \;\Rightarrow\; f_k d = 0 - 0 + \frac{1}{2}kx^2 - 0$$

$$\Rightarrow \mu_k mgd = \frac{1}{2}(200\,\text{N/m})(0.15\,\text{m})^2 \;\Rightarrow\; \mu_k(2.0\,\text{kg})(9.8\,\text{m/s}^2)(0.75\,\text{m}) = 2.25\,\text{J}$$

which yields $\mu_k = 0.15$ as the coefficient of kinetic friction.

53. Since the valley is frictionless, the only reason for the speed being less when it reaches the higher level is the gain in potential energy $\Delta U = mgh$ where $h = 1.1$ m. Sliding along the rough surface of the higher level, the block finally stops since its remaining kinetic energy has turned to thermal energy $\Delta E_{th} = f_k d = \mu mgd$, where $\mu = 0.60$. Thus, Eq. 8-33 (with $W = 0$) provides us with an equation to solve for the distance d:

$$K_i = \Delta U + \Delta E_{th} = mg(h + \mu d)$$

where $K_i = mv_i^2 / 2$ and $v_i = 6.0$ m/s. Dividing by mass and rearranging, we obtain

$$d = \frac{v_i^2}{2\mu g} - \frac{h}{\mu} = 1.2 \text{ m}.$$

54. (a) An appropriate picture (once friction is included) for this problem is Figure 8-3 in the textbook. We apply equation 8-31, $\Delta E_{th} = f_k d$, and relate initial kinetic energy K_i to the "resting" potential energy U_r:

$$K_i + U_i = f_k d + K_r + U_r \quad \Rightarrow \quad 20.0 \text{ J} + 0 = f_k d + 0 + \frac{1}{2}kd^2$$

where $f_k = 10.0$ N and $k = 400$ N/m. We solve the equation for d using the quadratic formula or by using the polynomial solver on an appropriate calculator, with $d = 0.292$ m being the only positive root.

(b) We apply equation 8-31 again and relate U_r to the "second" kinetic energy K_s it has at the unstretched position.

$$K_r + U_r = f_k d + K_s + U_s \quad \Rightarrow \quad \frac{1}{2}kd^2 = f_k d + K_s + 0$$

Using the result from part (a), this yields $K_s = 14.2$ J.

55. (a) The vertical forces acting on the block are the normal force, upward, and the force of gravity, downward. Since the vertical component of the block's acceleration is zero, Newton's second law requires $F_N = mg$, where m is the mass of the block. Thus $f = \mu_k F_N = \mu_k mg$. The increase in thermal energy is given by $\Delta E_{th} = fd = \mu_k mgD$, where D is the distance the block moves before coming to rest. Using Eq. 8-29, we have

$$\Delta E_{th} = (0.25)(3.5 \text{ kg})(9.8 \text{ m/s}^2)(7.8 \text{ m}) = 67 \text{ J}.$$

(b) The block has its maximum kinetic energy K_{max} just as it leaves the spring and enters the region where friction acts. Therefore, the maximum kinetic energy equals the thermal energy generated in bringing the block back to rest, 67 J.

(c) The energy that appears as kinetic energy is originally in the form of potential energy in the compressed spring. Thus, $K_{max} = U_i = \frac{1}{2}kx^2$, where k is the spring constant and x is the compression. Thus,

$$x = \sqrt{\frac{2K_{max}}{k}} = \sqrt{\frac{2(67\,\text{J})}{640\,\text{N/m}}} = 0.46\,\text{m}.$$

56. We look for the distance along the incline d which is related to the height ascended by $\Delta h = d \sin \theta$. By a force analysis of the style done in Ch. 6, we find the normal force has magnitude $F_N = mg \cos \theta$ which means $f_k = \mu_k mg \cos\theta$. Thus, Eq. 8-33 (with $W = 0$) leads to

$$0 = K_f - K_i + \Delta U + \Delta E_{th}$$
$$= 0 - K_i + mgd \sin \theta + \mu_k mgd \cos \theta$$

which leads to

$$d = \frac{K_i}{mg(\sin \theta + \mu_k \cos \theta)} = \frac{128}{(4.0)(9.8)(\sin 30° + 0.30 \cos 30°)} = 4.3\,\text{m}.$$

57. Before the launch, the mechanical energy is $\Delta E_{mech,0} = 0$. At the maximum height h where the speed of the beetle vanishes, the mechanical energy is $\Delta E_{mech,1} = mgh$. The change of the mechanical energy is related to the external force by

$$\Delta E_{mech} = \Delta E_{mech,1} - \Delta E_{mech,0} = mgh = F_{avg} d \cos \phi,$$

where F_{avg} is the average magnitude of the external force on the beetle.

(a) From the above equation, we have

$$F_{avg} = \frac{mgh}{d \cos \phi} = \frac{(4.0 \times 10^{-6}\,\text{kg})(9.80\,\text{m/s}^2)(0.30\,\text{m})}{(7.7 \times 10^{-4}\,\text{m})(\cos 0°)} = 1.5 \times 10^{-2}\,\text{N}.$$

(b) Dividing the above result by the mass of the beetle, we obtain

$$a = \frac{F_{avg}}{m} = \frac{h}{d \cos \phi} g = \frac{(0.30\,\text{m})}{(7.7 \times 10^{-4}\,\text{m})(\cos 0°)} g = 3.8 \times 10^2 g.$$

58. (a) Using the force analysis shown in Chapter 6, we find the normal force $F_N = mg \cos \theta$ (where $mg = 267$ N) which means $f_k = \mu_k F_N = \mu_k mg \cos \theta$. Thus, Eq. 8-31 yields

$$\Delta E_{th} = f_k d = \mu_k mgd \cos\theta = (0.10)(267)(6.1)\cos 20° = 1.5\times 10^2 \text{ J}.$$

(b) The potential energy change is

$$\Delta U = mg(-d \sin\theta) = (267 \text{ N})(-6.1 \text{ m}) \sin 20° = -5.6 \times 10^2 \text{ J}.$$

The initial kinetic energy is

$$K_i = \frac{1}{2}mv_i^2 = \frac{1}{2}\left(\frac{267 \text{ N}}{9.8 \text{ m/s}^2}\right)(0.457 \text{ m/s}^2) = 2.8 \text{ J}.$$

Therefore, using Eq. 8-33 (with $W = 0$), the final kinetic energy is

$$K_f = K_i - \Delta U - \Delta E_{th} = 2.8 - (-5.6\times 10^2) - 1.5\times 10^2 = 4.1\times 10^2 \text{ J}.$$

Consequently, the final speed is $v_f = \sqrt{2K_f/m} = 5.5$ m/s.

59. (a) With $x = 0.075$ m and $k = 320$ N/m, Eq. 7-26 yields $W_s = -\frac{1}{2}kx^2 = -0.90$ J. For later reference, this is equal to the negative of ΔU.

(b) Analyzing forces, we find $F_N = mg$ which means $f_k = \mu_k F_N = \mu_k mg$. With $d = x$, Eq. 8-31 yields

$$\Delta E_{th} = f_k d = \mu_k mgx = (0.25)(2.5)(9.8)(0.075) = 0.46 \text{ J}.$$

(c) Eq. 8-33 (with $W = 0$) indicates that the initial kinetic energy is

$$K_i = \Delta U + \Delta E_{th} = 0.90 + 0.46 = 1.36 \text{ J}$$

which leads to $v_i = \sqrt{2K_i/m} = 1.0$ m/s.

60. This can be worked entirely by the methods of Chapters 2–6, but we will use energy methods in as many steps as possible.

(a) By a force analysis of the style done in Ch. 6, we find the normal force has magnitude $F_N = mg \cos\theta$ (where $\theta = 40°$) which means $f_k = \mu_k F_N = \mu_k mg \cos\theta$ where $\mu_k = 0.15$. Thus, Eq. 8-31 yields

$$\Delta E_{th} = f_k d = \mu_k mgd \cos\theta.$$

Also, elementary trigonometry leads us to conclude that $\Delta U = mgd \sin\theta$. Eq. 8-33 (with $W = 0$ and $K_f = 0$) provides an equation for determining d:

$$K_i = \Delta U + \Delta E_{th}$$

$$\frac{1}{2}mv_i^2 = mgd(\sin\theta + \mu_k \cos\theta)$$

where $v_i = 1.4 \text{ m/s}$. Dividing by mass and rearranging, we obtain

$$d = \frac{v_i^2}{2g(\sin\theta + \mu_k \cos\theta)} = 0.13 \text{ m}.$$

(b) Now that we know where on the incline it stops ($d' = 0.13 + 0.55 = 0.68$ m from the bottom), we can use Eq. 8-33 again (with $W = 0$ and now with $K_i = 0$) to describe the final kinetic energy (at the bottom):

$$K_f = -\Delta U - \Delta E_{th}$$

$$\frac{1}{2}mv^2 = mgd'(\sin\theta - \mu_k \cos\theta)$$

which — after dividing by the mass and rearranging — yields

$$v = \sqrt{2gd'(\sin\theta - \mu_k \cos\theta)} = 2.7 \text{ m/s}.$$

(c) In part (a) it is clear that d increases if μ_k decreases — both mathematically (since it is a positive term in the denominator) and intuitively (less friction — less energy "lost"). In part (b), there are two terms in the expression for v which imply that it should increase if μ_k were smaller: the increased value of $d' = d_0 + d$ and that last factor $\sin\theta - \mu_k \cos\theta$ which indicates that less is being subtracted from $\sin\theta$ when μ_k is less (so the factor itself increases in value).

61. (a) The maximum height reached is h. The thermal energy generated by air resistance as the stone rises to this height is $\Delta E_{th} = fh$ by Eq. 8-31. We use energy conservation in the form of Eq. 8-33 (with $W = 0$):

$$K_f + U_f + \Delta E_{th} = K_i + U_i$$

and we take the potential energy to be zero at the throwing point (ground level). The initial kinetic energy is $K_i = \frac{1}{2}mv_0^2$, the initial potential energy is $U_i = 0$, the final kinetic energy is $K_f = 0$, and the final potential energy is $U_f = wh$, where $w = mg$ is the weight of the stone. Thus, $wh + fh = \frac{1}{2}mv_0^2$, and we solve for the height:

$$h = \frac{mv_0^2}{2(w+f)} = \frac{v_0^2}{2g(1+f/w)}.$$

Numerically, we have, with $m = (5.29 \text{ N})/(9.80 \text{ m/s}^2) = 0.54$ kg,

$$h = \frac{(20.0 \text{ m/s})^2}{2(9.80 \text{ m/s}^2)(1+0.265/5.29)} = 19.4 \text{ m/s}.$$

(b) We notice that the force of the air is downward on the trip up and upward on the trip down, since it is opposite to the direction of motion. Over the entire trip the increase in thermal energy is $\Delta E_{th} = 2fh$. The final kinetic energy is $K_f = \frac{1}{2}mv^2$, where v is the speed of the stone just before it hits the ground. The final potential energy is $U_f = 0$. Thus, using Eq. 8-31 (with $W = 0$), we find

$$\frac{1}{2}mv^2 + 2fh = \frac{1}{2}mv_0^2.$$

We substitute the expression found for h to obtain

$$\frac{2fv_0^2}{2g(1+f/w)} = \frac{1}{2}mv^2 - \frac{1}{2}mv_0^2$$

which leads to

$$v^2 = v_0^2 - \frac{2fv_0^2}{mg(1+f/w)} = v_0^2 - \frac{2fv_0^2}{w(1+f/w)} = v_0^2\left(1 - \frac{2f}{w+f}\right) = v_0^2 \frac{w-f}{w+f}$$

where w was substituted for mg and some algebraic manipulations were carried out. Therefore,

$$v = v_0\sqrt{\frac{w-f}{w+f}} = (20.0 \text{ m/s})\sqrt{\frac{5.29 \text{ N} - 0.265 \text{ N}}{5.29 \text{ N} + 0.265 \text{ N}}} = 19.0 \text{ m/s}.$$

62. In the absence of friction, we have a simple conversion (as it moves along the inclined ramps) of energy between the kinetic form (Eq. 7-1) and the potential form (Eq. 8-9). Along the horizontal plateaus, however, there is friction which causes some of the kinetic energy to dissipate in accordance with Eq. 8-31 (along with Eq. 6-2 where $\mu_k = 0.50$ and $F_N = mg$ in this situation). Thus, after it slides down a (vertical) distance d it has gained $K = \frac{1}{2}mv^2 = mgd$, some of which ($\Delta E_{th} = \mu_k mgd$) is dissipated, so that the value of kinetic energy at the end of the first plateau (just before it starts descending towards the lowest plateau) is $K = mgd - \mu_k mgd = 0.5mgd$. In its descent to the lowest plateau, it gains $mgd/2$ more kinetic energy, but as it slides across it "loses" $\mu_k mgd/2$ of it. Therefore, as it starts its climb up the right ramp, it has kinetic energy equal to

$$K = 0.5mgd + mgd/2 - \mu_k mgd/2 = 3\, mgd/4.$$

Setting this equal to Eq. 8-9 (to find the height to which it climbs) we get $H = \tfrac{3}{4}d$. Thus, the block (momentarily) stops on the inclined ramp at the right, at a height of

$$H = 0.75d = 0.75\,(40\text{ cm}) = 30\text{ cm}$$

measured from the lowest plateau.

63. The initial and final kinetic energies are zero, and we set up energy conservation in the form of Eq. 8-33 (with $W = 0$) according to our assumptions. Certainly, it can only come to a permanent stop somewhere in the flat part, but the question is whether this occurs during its first pass through (going rightward) or its second pass through (going leftward) or its third pass through (going rightward again), and so on. If it occurs during its first pass through, then the thermal energy generated is $\Delta E_{\text{th}} = f_k d$ where $d \leq L$ and $f_k = \mu_k mg$. If it occurs during its second pass through, then the total thermal energy is $\Delta E_{\text{th}} = \mu_k mg(L + d)$ where we again use the symbol d for how far through the level area it goes during that last pass (so $0 \leq d \leq L$). Generalizing to the nth pass through, we see that

$$\Delta E_{\text{th}} = \mu_k mg[(n-1)L + d].$$

In this way, we have

$$mgh = \mu_k mg\big((n-1)L + d\big)$$

which simplifies (when $h = L/2$ is inserted) to

$$\frac{d}{L} = 1 + \frac{1}{2\mu_k} - n.$$

The first two terms give $1 + 1/2\mu_k = 3.5$, so that the requirement $0 \leq d/L \leq 1$ demands that $n = 3$. We arrive at the conclusion that $d/L = \tfrac{1}{2}$, or

$$d = \frac{1}{2}L = \frac{1}{2}(40\text{ cm}) = 20\text{ cm}$$

and that this occurs on its third pass through the flat region.

64. We will refer to the point where it first encounters the "rough region" as point C (this is the point at a height h above the reference level). From Eq. 8-17, we find the speed it has at point C to be

$$v_C = \sqrt{v_A^2 - 2gh} = \sqrt{(8.0)^2 - 2(9.8)(2.0)} = 4.980 \approx 5.0\text{ m/s}.$$

Thus, we see that its kinetic energy right at the beginning of its "rough slide" (heading uphill towards B) is

$$K_C = \frac{1}{2} m(4.980 \text{ m/s})^2 = 12.4m$$

(with SI units understood). Note that we "carry along" the mass (as if it were a known quantity); as we will see, it will cancel out, shortly. Using Eq. 8-37 (and Eq. 6-2 with $F_N = mg\cos\theta$) and $y = d\sin\theta$, we note that if $d < L$ (the block does not reach point B), this kinetic energy will turn entirely into thermal (and potential) energy

$$K_C = mgy + f_k d \quad \Rightarrow \quad 12.4m = mgd\sin\theta + \mu_k mgd\cos\theta.$$

With $\mu_k = 0.40$ and $\theta = 30°$, we find $d = 1.49$ m, which is greater than L (given in the problem as 0.75 m), so our assumption that $d < L$ is incorrect. What is its kinetic energy as it reaches point B? The calculation is similar to the above, but with d replaced by L and the final v^2 term being the unknown (instead of assumed zero):

$$\frac{1}{2} mv^2 = K_C - (mgL\sin\theta + \mu_k mgL\cos\theta).$$

This determines the speed with which it arrives at point B:

$$v_B = \sqrt{v_C^2 - 2gL(\sin\theta + \mu_k \cos\theta)}$$
$$= \sqrt{(4.98 \text{ m/s})^2 - 2(9.80 \text{ m/s}^2)(0.75 \text{ m})(\sin 30° + 0.4\cos 30°)} = 3.5 \text{ m/s}.$$

65. We observe that the last line of the problem indicates that static friction is not to be considered a factor in this problem. The friction force of magnitude $f = 4400$ N mentioned in the problem is kinetic friction and (as mentioned) is constant (and directed upward), and the thermal energy change associated with it is $\Delta E_{th} = fd$ (Eq. 8-31) where $d = 3.7$ m in part (a) (but will be replaced by x, the spring compression, in part (b)).

(a) With $W = 0$ and the reference level for computing $U = mgy$ set at the top of the (relaxed) spring, Eq. 8-33 leads to

$$U_i = K + \Delta E_{th} \Rightarrow v = \sqrt{2d\left(g - \frac{f}{m}\right)}$$

which yields $v = 7.4$ m/s for $m = 1800$ kg.

(b) We again utilize Eq. 8-33 (with $W = 0$), now relating its kinetic energy at the moment it makes contact with the spring to the system energy at the bottom-most point. Using the same reference level for computing $U = mgy$ as we did in part (a), we end up with gravitational potential energy equal to $mg(-x)$ at that bottom-most point, where the spring (with spring constant $k = 1.5 \times 10^5$ N/m) is fully compressed.

$$K = mg(-x) + \frac{1}{2}kx^2 + fx$$

where $K = \frac{1}{2}mv^2 = 4.9 \times 10^4$ J using the speed found in part (a). Using the abbreviation $\xi = mg - f = 1.3 \times 10^4$ N, the quadratic formula yields

$$x = \frac{\xi \pm \sqrt{\xi^2 + 2kK}}{k} = 0.90 \text{ m}$$

where we have taken the positive root.

(c) We relate the energy at the bottom-most point to that of the highest point of rebound (a distance d' above the relaxed position of the spring). We assume $d' > x$. We now use the bottom-most point as the reference level for computing gravitational potential energy.

$$\frac{1}{2}kx^2 = mgd' + fd' \Rightarrow d' = \frac{kx^2}{2(mg+d)} = 2.8 \text{ m}.$$

(d) The non-conservative force (§8-1) is friction, and the energy term associated with it is the one that keeps track of the total distance traveled (whereas the potential energy terms, coming as they do from conservative forces, depend on positions — but not on the paths that led to them). We assume the elevator comes to final rest at the equilibrium position of the spring, with the spring compressed an amount d_{eq} given by

$$mg = kd_{eq} \Rightarrow d_{eq} = \frac{mg}{k} = 0.12 \text{ m}.$$

In this part, we use that final-rest point as the reference level for computing gravitational potential energy, so the original $U = mgy$ becomes $mg(d_{eq} + d)$. In that final position, then, the gravitational energy is zero and the spring energy is $kd_{eq}^2/2$. Thus, Eq. 8-33 becomes

$$mg(d_{eq} + d) = \frac{1}{2}kd_{eq}^2 + fd_{total}$$

$$(1800)(9.8)(0.12 + 3.7) = \frac{1}{2}(1.5 \times 10^5)(0.12)^2 + (4400)d_{total}$$

which yields d_{total} = 15 m.

66. (a) Since the speed of the crate of mass m increases from 0 to 1.20 m/s relative to the factory ground, the kinetic energy supplied to it is

$$K = \frac{1}{2}mv^2 = \frac{1}{2}(300\,\text{kg})(120\,\text{m/s})^2 = 216\,\text{J}.$$

(b) The magnitude of the kinetic frictional force is

$$f = \mu F_N = \mu mg = (0.400)(300\,\text{kg})(9.8\,\text{m/s}^2) = 1.18 \times 10^3\,\text{N}.$$

(c) Let the distance the crate moved relative to the conveyor belt before it stops slipping be d, then from Eq. 2-16 ($v^2 = 2ad = 2(f/m)d$) we find

$$\Delta E_{\text{th}} = fd = \frac{1}{2}mv^2 = K.$$

Thus, the total energy that must be supplied by the motor is

$$W = K + \Delta E_{\text{th}} = 2K = (2)(216\,\text{J}) = 432\,\text{J}.$$

(d) The energy supplied by the motor is the work W it does on the system, and must be greater than the kinetic energy gained by the crate computed in part (b). This is due to the fact that part of the energy supplied by the motor is being used to compensate for the energy dissipated ΔE_{th} while it was slipping.

67. (a) The assumption is that the slope of the bottom of the slide is horizontal, like the ground. A useful analogy is that of the pendulum of length $R = 12$ m that is pulled leftward to an angle θ (corresponding to being at the top of the slide at height $h = 4.0$ m) and released so that the pendulum swings to the lowest point (zero height) gaining speed $v = 6.2$ m/s. Exactly as we would analyze the trigonometric relations in the pendulum problem, we find

$$h = R(1-\cos\theta) \Rightarrow \theta = \cos^{-1}\left(1 - \frac{h}{R}\right) = 48°$$

or 0.84 radians. The slide, representing a circular arc of length $s = R\theta$, is therefore (12 m)(0.84) = 10 m long.

(b) To find the magnitude f of the frictional force, we use Eq. 8-31 (with $W = 0$):

$$0 = \Delta K + \Delta U + \Delta E_{\text{th}}$$
$$= \frac{1}{2}mv^2 - mgh + fs$$

so that (with $m = 25$ kg) we obtain $f = 49$ N.

(c) The assumption is no longer that the slope of the bottom of the slide is horizontal, but rather that the slope of the top of the slide is vertical (and 12 m to the left of the center of curvature). Returning to the pendulum analogy, this corresponds to releasing the pendulum from horizontal (at $\theta_1 = 90°$ measured from vertical) and taking a snapshot of its motion a few moments later when it is at angle θ_2 with speed $v = 6.2$ m/s. The difference in height between these two positions is (just as we would figure for the pendulum of length R)

$$\Delta h = R(1-\cos\theta_2) - R(1-\cos\theta_1) = -R\cos\theta_2$$

where we have used the fact that $\cos\theta_1 = 0$. Thus, with $\Delta h = -4.0$ m, we obtain $\theta_2 = 70.5°$ which means the arc subtends an angle of $|\Delta\theta| = 19.5°$ or 0.34 radians. Multiplying this by the radius gives a slide length of $s' = 4.1$ m.

(d) We again find the magnitude f' of the frictional force by using Eq. 8-31 (with $W = 0$):

$$0 = \Delta K + \Delta U + \Delta E_{th}$$
$$= \frac{1}{2}mv^2 - mgh + f's'$$

so that we obtain $f' = 1.2 \times 10^2$ N.

68. We use conservation of mechanical energy: the mechanical energy must be the same at the top of the swing as it is initially. Newton's second law is used to find the speed, and hence the kinetic energy, at the top. There the tension force T of the string and the force of gravity are both downward, toward the center of the circle. We notice that the radius of the circle is $r = L - d$, so the law can be written

$$T + mg = mv^2/(L-d),$$

where v is the speed and m is the mass of the ball. When the ball passes the highest point with the least possible speed, the tension is zero. Then

$$mg = m\frac{v^2}{L-d} \Rightarrow v = \sqrt{g(L-d)}.$$

We take the gravitational potential energy of the ball-Earth system to be zero when the ball is at the bottom of its swing. Then the initial potential energy is mgL. The initial kinetic energy is zero since the ball starts from rest. The final potential energy, at the top of the swing, is $2mg(L-d)$ and the final kinetic energy is $\frac{1}{2}mv^2 = \frac{1}{2}mg(L-d)$ using the above result for v. Conservation of energy yields

$$mgL = 2mg(L-d) + \frac{1}{2}mg(L-d) \Rightarrow d = 3L/5.$$

With $L = 1.20$ m, we have $d = 0.60(1.20\text{ m}) = 0.72$ m.

Notice that if d is greater than this value, so the highest point is lower, then the speed of the ball is greater as it reaches that point and the ball passes the point. If d is less, the ball cannot go around. Thus the value we found for d is a lower limit.

69. There is the same potential energy change in both circumstances, so we can equate the kinetic energy changes as well:

$$\Delta K_2 = \Delta K_1 \quad \Rightarrow \quad \frac{1}{2}mv_B^2 - \frac{1}{2}m(4.00\text{ m/s})^2 = \frac{1}{2}m(2.60\text{ m/s})^2 - \frac{1}{2}m(2.00\text{ m/s})^2$$

which leads to $v_B = 4.33$ m/s.

70. (a) To stretch the spring an external force, equal in magnitude to the force of the spring but opposite to its direction, is applied. Since a spring stretched in the positive x direction exerts a force in the negative x direction, the applied force must be $F = 52.8x + 38.4x^2$, in the $+x$ direction. The work it does is

$$W = \int_{0.50}^{1.00} (52.8x + 38.4x^2)dx = \left(\frac{52.8}{2}x^2 + \frac{38.4}{3}x^3\right)\Bigg|_{0.50}^{1.00} = 31.0\text{ J}.$$

(b) The spring does 31.0 J of work and this must be the increase in the kinetic energy of the particle. Its speed is then

$$v = \sqrt{\frac{2K}{m}} = \sqrt{\frac{2(31.0\text{ J})}{2.17\text{ kg}}} = 5.35\text{ m/s}.$$

(c) The force is conservative since the work it does as the particle goes from any point x_1 to any other point x_2 depends only on x_1 and x_2, not on details of the motion between x_1 and x_2.

71. This can be worked entirely by the methods of Chapters 2–6, but we will use energy methods in as many steps as possible.

(a) By a force analysis in the style of Chapter 6, we find the normal force has magnitude $F_N = mg\cos\theta$ (where $\theta = 39°$) which means $f_k = \mu_k mg\cos\theta$ where $\mu_k = 0.28$. Thus, Eq. 8-31 yields

$$\Delta E_{\text{th}} = f_k d = \mu_k mgd\cos\theta.$$

Also, elementary trigonometry leads us to conclude that $\Delta U = -mgd\sin\theta$ where $d = 3.7$ m. Since $K_i = 0$, Eq. 8-33 (with $W = 0$) indicates that the final kinetic energy is

$$K_f = -\Delta U - \Delta E_{th} = mgd(\sin\theta - \mu_k \cos\theta)$$

which leads to the speed at the bottom of the ramp

$$v = \sqrt{\frac{2K_f}{m}} = \sqrt{2gd(\sin\theta - \mu_k \cos\theta)} = 5.5 \text{ m/s}.$$

(b) This speed begins its horizontal motion, where $f_k = \mu_k mg$ and $\Delta U = 0$. It slides a distance d' before it stops. According to Eq. 8-31 (with $W = 0$),

$$0 = \Delta K + \Delta U + \Delta E_{th}$$
$$= 0 - \frac{1}{2}mv^2 + 0 + \mu_k mgd'$$
$$= -\frac{1}{2}(2gd(\sin\theta - \mu_k \cos\theta)) + \mu_k gd'$$

where we have divided by mass and substituted from part (a) in the last step. Therefore,

$$d' = \frac{d(\sin\theta - \mu_k \cos\theta)}{\mu_k} = 5.4 \text{ m}.$$

(c) We see from the algebraic form of the results, above, that the answers do not depend on mass. A 90 kg crate should have the same speed at the bottom and sliding distance across the floor, to the extent that the friction relations in Ch. 6 are accurate. Interestingly, since g does not appear in the relation for d', the sliding distance would seem to be the same if the experiment were performed on Mars!

72. (a) At B the speed is (from Eq. 8-17)

$$v = \sqrt{v_0^2 + 2gh_1} = \sqrt{(7.0 \text{ m/s})^2 + 2(9.8 \text{ m/s}^2)(6.0 \text{ m})} = 13 \text{ m/s}.$$

(a) Here what matters is the difference in heights (between A and C):

$$v = \sqrt{v_0^2 + 2g(h_1 - h_2)} = \sqrt{(7.0 \text{ m/s})^2 + 2(9.8 \text{ m/s}^2)(4.0 \text{ m})} = 11.29 \text{ m/s} \approx 11 \text{ m/s}.$$

(c) Using the result from part (b), we see that its kinetic energy right at the beginning of its "rough slide" (heading horizontally towards D) is $\frac{1}{2}m(11.29 \text{ m/s})^2 = 63.7m$ (with SI units understood). Note that we "carry along" the mass (as if it were a known quantity); as we will see, it will cancel out, shortly. Using Eq. 8-31 (and Eq. 6-2 with $F_N = mg$) we note that this kinetic energy will turn entirely into thermal energy

$$63.7m = \mu_k mgd$$

if $d < L$. With $\mu_k = 0.70$, we find $d = 9.3$ m, which is indeed less than L (given in the problem as 12 m). We conclude that the block stops before passing out of the "rough" region (and thus does not arrive at point D).

73. (a) By mechanical energy conversation, the kinetic energy as it reaches the floor (which we choose to be the $U = 0$ level) is the sum of the initial kinetic and potential energies:

$$K = K_i + U_i = \frac{1}{2}(2.50 \text{ kg})(3.00 \text{ m/s})^2 + (2.50 \text{ kg})(9.80 \text{ m/s}^2)(4.00 \text{ m}) = 109 \text{ J}.$$

For later use, we note that the speed with which it reaches the ground is

$$v = \sqrt{2K/m} = 9.35 \text{ m/s}.$$

(b) When the drop in height is 2.00 m instead of 4.00 m, the kinetic energy is

$$K = \frac{1}{2}(2.50 \text{ kg})(3.00 \text{ m/s})^2 + (2.50 \text{ kg})(9.80 \text{ m/s}^2)(2.00 \text{ m}) = 60.3 \text{ J}.$$

(c) A simple way to approach this is to imagine the can is *launched* from the ground at $t = 0$ with speed 9.35 m/s (see above) and ask of its height and speed at $t = 0.200$ s, using Eq. 2-15 and Eq. 2-11:

$$y = (9.35 \text{ m/s})(0.200 \text{ s}) - \frac{1}{2}(9.80 \text{ m/s}^2)(0.200 \text{ s})^2 = 1.67 \text{ m},$$

$$v = 9.35 \text{ m/s} - (9.80 \text{ m/s}^2)(0.200 \text{ s}) = 7.39 \text{ m/s}.$$

The kinetic energy is

$$K = \frac{1}{2}(2.50 \text{ kg})(7.39 \text{ m/s})^2 = 68.2 \text{ J}.$$

(d) The gravitational potential energy

$$U = mgy = (2.5 \text{ kg})(9.8 \text{ m/s}^2)(1.67 \text{ m}) = 41.0 \text{ J}$$

74. (a) The initial kinetic energy is $K_i = \frac{1}{2}(1.5)(3)^2 = 6.75 \text{ J}$.

(b) The work of gravity is the negative of its change in potential energy. At the highest point, all of K_i has converted into U (if we neglect air friction) so we conclude the work of gravity is -6.75 J.

(c) And we conclude that $\Delta U = 6.75$ J.

348

(d) The potential energy there is $U_f = U_i + \Delta U = 6.75$ J.

(e) If $U_f = 0$, then $U_i = U_f - \Delta U = -6.75$ J.

(f) Since $mg\Delta y = \Delta U$, we obtain $\Delta y = 0.459$ m.

75. We note that if the larger mass (block B, $m_B = 2$ kg) falls $d = 0.25$ m, then the smaller mass (blocks A, $m_A = 1$ kg) must increase its height by $h = d \sin 30°$. Thus, by mechanical energy conservation, the kinetic energy of the system is

$$K_{total} = m_B g d - m_A g h = 3.7 \text{ J}.$$

76. (a) At the point of maximum height, where $y = 140$ m, the vertical component of velocity vanishes but the horizontal component remains what it was when it was launched (if we neglect air friction). Its kinetic energy at that moment is

$$K = \frac{1}{2}(0.55 \text{ kg})v_x^2.$$

Also, its potential energy (with the reference level chosen at the level of the cliff edge) at that moment is $U = mgy = 755$ J. Thus, by mechanical energy conservation,

$$K = K_i - U = 1550 - 755 \Rightarrow v_x = \sqrt{\frac{2(1550 - 755)}{0.55}} = 54 \text{ m/s}.$$

(b) As mentioned $v_x = v_{ix}$ so that the initial kinetic energy

$$K_i = \frac{1}{2}m\left(v_{ix}^2 + v_{iy}^2\right)$$

can be used to find v_{iy}. We obtain $v_{iy} = 52$ m/s.

(c) Applying Eq. 2-16 to the vertical direction (with $+y$ upward), we have

$$v_y^2 = v_{iy}^2 - 2g\Delta y \Rightarrow (65 \text{ m/s})^2 = (52 \text{ m/s})^2 - 2(9.8 \text{ m/s}^2)\Delta y$$

which yields $\Delta y = -76$ m. The minus sign tells us it is below its launch point.

77. The work done by $\vec{F}$ is the negative of its potential energy change (see Eq. 8-6), so $U_B = U_A - 25 = 15$ J.

78. The free-body diagram for the trunk is shown.

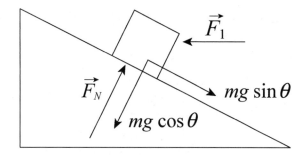

The x and y applications of Newton's second law provide two equations:

$$F_1 \cos\theta - f_k - mg\sin\theta = ma$$
$$F_N - F_1 \sin\theta - mg\cos\theta = 0.$$

(a) The trunk is moving up the incline at constant velocity, so $a = 0$. Using $f_k = \mu_k F_N$, we solve for the push-force F_1 and obtain

$$F_1 = \frac{mg(\sin\theta + \mu_k \cos\theta)}{\cos\theta - \mu_k \sin\theta}.$$

The work done by the push-force $\vec{F}_1$ as the trunk is pushed through a distance ℓ up the inclined plane is therefore

$$W_1 = F_1 \ell \cos\theta = \frac{(mg\ell\cos\theta)(\sin\theta + \mu_k \cos\theta)}{\cos\theta - \mu_k \sin\theta}$$

$$= \frac{(50\text{ kg})(9.8\text{ m/s}^2)(6.0\text{ m})(\cos 30°)(\sin 30° + (0.20)\cos 30°)}{\cos 30° - (0.20)\sin 30°}$$

$$= 2.2 \times 10^3 \text{ J}.$$

(b) The increase in the gravitational potential energy of the trunk is

$$\Delta U = mg\ell \sin\theta = (50\text{ kg})(9.8\text{ m/s}^2)(6.0\text{ m})\sin 30° = 1.5 \times 10^3 \text{ J}.$$

Since the speed (and, therefore, the kinetic energy) of the trunk is unchanged, Eq. 8-33 leads to

$$W_1 = \Delta U + \Delta E_{th}.$$

Thus, using more precise numbers than are shown above, the increase in thermal energy (generated by the kinetic friction) is 2.24×10^3 J $- 1.47 \times 10^3$ J $= 7.7 \times 10^2$ J. An alternate way to this result is to use $\Delta E_{th} = f_k \ell$ (Eq. 8-31).

79. The initial height of the $2M$ block, shown in Fig. 8-64, is the $y = 0$ level in our computations of its value of U_g. As that block drops, the spring stretches accordingly. Also, the kinetic energy K_{sys} is evaluated for the *system* -- that is, for a total moving mass of $3M$.

(a) The conservation of energy, Eq. 8-17, leads to

$$K_i + U_i = K_{sys} + U_{sys} \implies 0 + 0 = K_{sys} + (2M)g(-0.090) + \frac{1}{2} k(0.090)^2 .$$

Thus, with $M = 2.0$ kg, we obtain $K_{sys} = 2.7$ J.

(b) The kinetic energy of the $2M$ block represents a fraction of the total kinetic energy:

$$\frac{K_{2M}}{K_{sys}} = \frac{(2M)v^2/2}{(3M)v^2/2} = \frac{2}{3}.$$

Therefore, $K_{2M} = \frac{2}{3}(2.7 \text{ J}) = 1.8$ J.

(c) Here we let $y = -d$ and solve for d.

$$K_i + U_i = K_{sys} + U_{sys} \implies 0 + 0 = 0 + (2M)g(-d) + \frac{1}{2} kd^2 .$$

Thus, with $M = 2.0$ kg, we obtain $d = 0.39$ m.

80. Sample Problem 8-3 illustrates simple energy conservation in a similar situation, and derives the frequently encountered relationship: $v = \sqrt{2gh}$. In our present problem, the height is related to the distance (on the $\theta = 10°$ slope) $d = 920$ m by the trigonometric relation $h = d \sin\theta$. Thus,

$$v = \sqrt{2(9.8 \text{ m/s}^2)(920 \text{ m})\sin 10°} = 56 \text{ m/s}.$$

81. Eq. 8-33 gives $mgy_f = K_i + mgy_i - \Delta E_{th}$, or

$$(0.50 \text{ kg})(9.8 \text{ m/s}^2)(0.80 \text{ m}) = \frac{1}{2}(0.50 \text{ kg})(4.00 \text{ /s})^2 + (0.50 \text{ kg})(9.8 \text{ m/s}^2)(0) - \Delta E_{th}$$

which yields $\Delta E_{th} = 4.00$ J $- 3.92$ J $= 0.080$ J.

81. (a) The loss of the initial $K = \frac{1}{2} mv^2 = \frac{1}{2}(70 \text{ kg})(10 \text{ m/s})^2$ is 3500 J, or 3.5 kJ.

(b) This is dissipated as thermal energy; $\Delta E_{th} = 3500$ J $= 3.5$ kJ.

83. The initial height shown in the figure is the $y = 0$ level in our computations of U_g, and in parts (a) and (b) the heights are $y_a = (0.80 \text{ m})\sin 40° = 0.51$ m and $y_b = (1.00 \text{ m}) \sin 40° = 0.64$ m, respectively.

(a) The conservation of energy, Eq. 8-17, leads to

$$K_i + U_i = K_a + U_a \quad \Rightarrow \quad 16 \text{ J} + 0 = K_a + mgy_a + \frac{1}{2}k(0.20 \text{ m})^2$$

from which we obtain $K_a = (16 - 5.0 - 4.0)$ J $= 7.0$ J.

(b) Again we use the conservation of energy

$$K_i + U_i = K_b + U_b \quad \Rightarrow \quad K_i + 0 = 0 + mgy_b + \frac{1}{2} k(0.40 \text{ m})^2$$

from which we obtain $K_i = 6.0$ J $+ 16$ J $= 22$ J.

84. (a) Eq. 8-9 gives $U = (3.2 \text{ kg})(9.8 \text{ m/s}^2)(3.0 \text{ m}) = 94$ J.

(b) The mechanical energy is conserved, so $K = 94$ J.

(c) The speed (from solving Eq. 7-1) is $v = \sqrt{2(94 \text{ J})/(32 \text{ kg})} = 7.7$ m/s.

85. (a) Resolving the gravitational force into components and applying Newton's second law (as well as Eq. 6-2), we find

$$F_{\text{machine}} - mg\sin\theta - \mu_k mg\cos\theta = ma.$$

In the situation described in the problem, we have $a = 0$, so

$$F_{\text{machine}} = mg\sin\theta + \mu_k mg\cos\theta = 372 \text{ N}.$$

Thus, the work done by the machine is $F_{\text{machine}}d = 744$ J $= 7.4 \times 10^2$ J.

(b) The thermal energy generated is $\mu_k mg\cos\theta \, d = 240$ J $= 2.4 \times 10^2$ J.

86. We use $P = Fv$ to compute the force:

$$F = \frac{P}{v} = \frac{92 \times 10^6 \text{ W}}{(32.5 \text{ knot})\left(1.852 \frac{\text{km/h}}{\text{knot}}\right)\left(\frac{1000 \text{ m/km}}{3600 \text{ s/h}}\right)} = 5.5 \times 10^6 \text{ N}.$$

87. Since the speed is constant $\Delta K = 0$ and Eq. 8-33 (an application of the energy conservation concept) implies

$$W_{\text{applied}} = \Delta E_{\text{th}} = \Delta E_{\text{th(cube)}} + \Delta E_{\text{th(floor)}}.$$

Thus, if $W_{\text{applied}} = (15 \text{ N})(3.0 \text{ m}) = 45 \text{ J}$, and we are told that $\Delta E_{\text{th (cube)}} = 20 \text{ J}$, then we conclude that $\Delta E_{\text{th (floor)}} = 25 \text{ J}$.

88. (a) We take the gravitational potential energy of the skier-Earth system to be zero when the skier is at the bottom of the peaks. The initial potential energy is $U_i = mgH$, where m is the mass of the skier, and H is the height of the higher peak. The final potential energy is $U_f = mgh$, where h is the height of the lower peak. The skier initially has a kinetic energy of $K_i = 0$, and the final kinetic energy is $K_f = \frac{1}{2}mv^2$, where v is the speed of the skier at the top of the lower peak. The normal force of the slope on the skier does no work and friction is negligible, so mechanical energy is conserved:

$$U_i + K_i = U_f + K_f \quad \Rightarrow \quad mgH = mgh + \frac{1}{2}mv^2$$

Thus,

$$v = \sqrt{2g(H-h)} = \sqrt{2(9.8 \text{ m/s}^2)(850 \text{ m} - 750 \text{ m})} = 44 \text{ m/s}.$$

(b) We recall from analyzing objects sliding down inclined planes that the normal force of the slope on the skier is given by $F_N = mg \cos \theta$, where θ is the angle of the slope from the horizontal, 30° for each of the slopes shown. The magnitude of the force of friction is given by $f = \mu_k F_N = \mu_k mg \cos \theta$. The thermal energy generated by the force of friction is $fd = \mu_k mgd \cos \theta$, where d is the total distance along the path. Since the skier gets to the top of the lower peak with no kinetic energy, the increase in thermal energy is equal to the decrease in potential energy. That is, $\mu_k mgd \cos \theta = mg(H-h)$. Consequently,

$$\mu_k = \frac{H-h}{d \cos \theta} = \frac{(850 \text{ m} - 750 \text{ m})}{(3.2 \times 10^3 \text{ m}) \cos 30°} = 0.036.$$

89. To swim at constant velocity the swimmer must push back against the water with a force of 110 N. Relative to him the water is going at 0.22 m/s toward his rear, in the same direction as his force. Using Eq. 7-48, his power output is obtained:

$$P = \vec{F} \cdot \vec{v} = Fv = (110 \text{ N})(0.22 \text{ m/s}) = 24 \text{ W}.$$

90. The initial kinetic energy of the automobile of mass m moving at speed v_i is $K_i = \frac{1}{2}mv_i^2$, where $m = 16400/9.8 = 1673$ kg. Using Eq. 8-31 and Eq. 8-33, this relates to

the effect of friction force f in stopping the auto over a distance d by $K_i = fd$, where the road is assumed level (so $\Delta U = 0$). With

$$v_i = (113 \text{ km/h}) = (113 \text{ km/h})(1000 \text{ m/km})(1 \text{ h}/3600 \text{ s}) = 31.4 \text{ m/s},$$

we obtain

$$d = \frac{K_i}{f} = \frac{mv_i^2}{2f} = \frac{(1673 \text{ kg})(31.4 \text{ m/s})^2}{2(8230 \text{ N})} = 100 \text{ m}.$$

91. With the potential energy reference level set at the point of throwing, we have (with SI units understood)

$$\Delta E = mgh - \frac{1}{2}mv_0^2 = m\left((9.8)(8.1) - \frac{1}{2}(14)^2\right)$$

which yields $\Delta E = -12$ J for $m = 0.63$ kg. This "loss" of mechanical energy is presumably due to air friction.

92. (a) The (internal) energy the climber must convert to gravitational potential energy is

$$\Delta U = mgh = (90 \text{ kg})(9.80 \text{ m/s}^2)(8850 \text{ m}) = 7.8 \times 10^6 \text{ J}.$$

(b) The number of candy bars this corresponds to is

$$N = \frac{7.8 \times 10^6 \text{ J}}{1.25 \times 10^6 \text{ J/bar}} \approx 6.2 \text{ bars}.$$

93. (a) The acceleration of the sprinter is (using Eq. 2-15)

$$a = \frac{2\Delta x}{t^2} = \frac{(2)(7.0 \text{ m})}{(1.6 \text{ s})^2} = 5.47 \text{ m/s}^2.$$

Consequently, the speed at $t = 1.6$ s is $v = at = (5.47 \text{ m/s}^2)(1.6 \text{ s}) = 8.8 \text{ m/s}$. Alternatively, Eq. 2-17 could be used.

(b) The kinetic energy of the sprinter (of weight w and mass $m = w/g$) is

$$K = \frac{1}{2}mv^2 = \frac{1}{2}\left(\frac{w}{g}\right)v^2 = \frac{1}{2}(670 \text{ N}/(9.8 \text{ m/s}^2))(8.8 \text{ m/s})^2 = 2.6 \times 10^3 \text{ J}.$$

(c) The average power is

354   CHAPTER 8

$$P_{avg} = \frac{\Delta K}{\Delta t} = \frac{2.6 \times 10^3 \text{ J}}{1.6 \text{ s}} = 1.6 \times 10^3 \text{ W}.$$

94. We note that in one second, the block slides $d = 1.34$ m up the incline, which means its height increase is $h = d \sin\theta$ where

$$\theta = \tan^{-1}\left(\frac{30}{40}\right) = 37°.$$

We also note that the force of kinetic friction in this inclined plane problem is $f_k = \mu_k mg \cos\theta$, where $\mu_k = 0.40$ and $m = 1400$ kg. Thus, using Eq. 8-31 and Eq. 8-33, we find

$$W = mgh + f_k d = mgd(\sin\theta + \mu_k \cos\theta)$$

or $W = 1.69 \times 10^4$ J for this one-second interval. Thus, the power associated with this is

$$P = \frac{1.69 \times 10^4 \text{ J}}{1 \text{ s}} = 1.69 \times 10^4 \text{ W} \approx 1.7 \times 10^4 \text{ W}.$$

95. (a) The initial kinetic energy is $K_i = (1.5 \text{ kg})(20 \text{ m/s})^2 / 2 = 300$ J.

(b) At the point of maximum height, the vertical component of velocity vanishes but the horizontal component remains what it was when it was "shot" (if we neglect air friction). Its kinetic energy at that moment is

$$K = \frac{1}{2}(1.5 \text{ kg})[(20 \text{ m/s})\cos 34°]^2 = 206 \text{ J}.$$

Thus, $\Delta U = K_i - K = 300$ J $- 206$ J $= 93.8$ J.

(c) Since $\Delta U = mg \Delta y$, we obtain

$$\Delta y = \frac{94 \text{ J}}{(1.5 \text{ kg})(9.8 \text{ m/s}^2)} = 6.38 \text{ m}.$$

96. From Eq. 8-6, we find (with SI units understood)

$$U(\xi) = -\int_0^\xi (-3x - 5x^2)\, dx = \frac{3}{2}\xi^2 + \frac{5}{3}\xi^3.$$

(a) Using the above formula, we obtain $U(2) \approx 19$ J.

(b) When its speed is $v = 4$ m/s, its mechanical energy is $\frac{1}{2}mv^2 + U(5)$. This must equal the energy at the origin:

$$\frac{1}{2}mv^2 + U(5) = \frac{1}{2}mv_o^2 + U(0)$$

so that the speed at the origin is

$$v_o = \sqrt{v^2 + \frac{2}{m}(U(5) - U(0))}.$$

Thus, with $U(5) = 246$ J, $U(0) = 0$ and $m = 20$ kg, we obtain $v_o = 6.4$ m/s.

(c) Our original formula for U is changed to

$$U(x) = -8 + \tfrac{3}{2}x^2 + \tfrac{5}{3}x^3$$

in this case. Therefore, $U(2) = 11$ J. But we still have $v_o = 6.4$ m/s since that calculation only depended on the difference of potential energy values (specifically, $U(5) - U(0)$).

97. Eq. 8-8 leads directly to $\Delta y = \dfrac{68000 \text{ J}}{(9.4 \text{ kg})(9.8 \text{ m/s}^2)} = 738$ m.

98. Since the period T is $(2.5 \text{ rev/s})^{-1} = 0.40$ s, then Eq. 4-33 leads to $v = 3.14$ m/s. The frictional force has magnitude (using Eq. 6-2)

$$f = \mu_k F_N = (0.320)(180 \text{ N}) = 57.6 \text{ N}.$$

The power dissipated by the friction must equal that supplied by the motor, so Eq. 7-48 gives $P = (57.6 \text{ N})(3.14 \text{ m/s}) = 181$ W.

99. (a) In the initial situation, the elongation was (using Eq. 8-11)

$$x_i = \sqrt{2(1.44)/3200} = 0.030 \text{ m (or 3.0 cm)}.$$

In the next situation, the elongation is only 2.0 cm (or 0.020 m), so we now have less stored energy (relative to what we had initially). Specifically,

$$\Delta U = \frac{1}{2}(3200 \text{ N/m})(0.020 \text{ m})^2 - 1.44 \text{ J} = -0.80 \text{ J}.$$

(b) The elastic stored energy for $|x| = 0.020$ m, does not depend on whether this represents a stretch or a compression. The answer is the same as in part (a), $\Delta U = -0.80$ J.

(c) Now we have $|x| = 0.040$ m which is greater than x_i, so this represents an increase in the potential energy (relative to what we had initially). Specifically,

$$\Delta U = \frac{1}{2}(3200 \text{ N/m})(0.040 \text{ m})^2 - 1.44 \text{ J} = +1.12 \text{ J} \approx 1.1 \text{ J}.$$

100. (a) At the highest point, the velocity $v = v_x$ is purely horizontal and is equal to the horizontal component of the launch velocity (see section 4-6): $v_{ox} = v_o \cos\theta$, where $\theta = 30°$ in this problem. Eq. 8-17 relates the kinetic energy at the highest point to the launch kinetic energy:

$$K_o = mgy + \frac{1}{2}mv^2 = \frac{1}{2}mv_{ox}^2 + \frac{1}{2}mv_{oy}^2.$$

with $y = 1.83$ m. Since the $mv_{ox}^2/2$ term on the left-hand side cancels the $mv^2/2$ term on the right-hand side, this yields $v_{oy} = \sqrt{2gy} \approx 6$ m/s. With $v_{oy} = v_o \sin\theta$, we obtain

$$v_o = 11.98 \text{ m/s} \approx 12 \text{ m/s}.$$

(b) Energy conservation (including now the energy stored elastically in the spring, Eq. 8-11) also applies to the motion along the muzzle (through a distance d which corresponds to a vertical height increase of $d\sin\theta$):

$$\frac{1}{2}kd^2 = K_o + mg\,d\sin\theta \quad \Rightarrow \quad d = 0.11 \text{ m}.$$

101. (a) We implement Eq. 8-37 as

$$K_f = K_i + mgy_i - f_k d = 0 + (60 \text{ kg})(9.8 \text{ m/s}^2)(4.0 \text{ m}) - 0 = 2.35 \times 10^3 \text{ J}.$$

(b) Now it applies with a nonzero thermal term:

$$K_f = K_i + mgy_i - f_k d = 0 + (60 \text{ kg})(9.8 \text{ m/s}^2)(4.0 \text{ m}) - (500 \text{ N})(4.0 \text{ m}) = 352 \text{ J}.$$

102. (a) We assume his mass is between $m_1 = 50$ kg and $m_2 = 70$ kg (corresponding to a weight between 110 lb and 154 lb). His increase in gravitational potential energy is therefore in the range

$$m_1 gh \leq \Delta U \leq m_2 gh \quad \Rightarrow \quad 2\times 10^5 \leq \Delta U \leq 3\times 10^5$$

in SI units (J), where $h = 443$ m.

(b) The problem only asks for the amount of internal energy which converts into gravitational potential energy, so this result is the same as in part (a). But if we were to consider his *total* internal energy "output" (much of which converts to heat) we can expect that external climb is quite different from taking the stairs.

103. We use SI units so $m = 0.030$ kg and $d = 0.12$ m.

(a) Since there is no change in height (and we assume no changes in elastic potential energy), then $\Delta U = 0$ and we have

$$\Delta E_{mech} = \Delta K = -\frac{1}{2} m v_0^2 = -3.8 \times 10^3 \text{ J}.$$

where $v_0 = 500$ m/s and the final speed is zero.

(b) By Eq. 8-33 (with $W = 0$) we have $\Delta E_{th} = 3.8 \times 10^3$ J, which implies

$$f = \frac{\Delta E_{th}}{d} = 3.1 \times 10^4 \text{ N}$$

using Eq. 8-31 with f_k replaced by f (effectively generalizing that equation to include a greater variety of dissipative forces than just those obeying Eq. 6-2).

104. We work this in SI units and convert to horsepower in the last step. Thus,

$$v = (80 \text{ km/h}) \left(\frac{1000 \text{ m/km}}{3600 \text{ s/h}} \right) = 22.2 \text{ m/s}.$$

The force F_P needed to propel the car (of weight w and mass $m = w/g$) is found from Newton's second law:

$$F_{net} = F_P - F = ma = \frac{wa}{g}$$

where $F = 300 + 1.8 v^2$ in SI units. Therefore, the power required is

$$P = \vec{F}_P \cdot \vec{v} = \left(F + \frac{wa}{g} \right) v = \left(300 + 1.8(22.2)^2 + \frac{(12000)(0.92)}{9.8} \right)(22.2) = 5.14 \times 10^4 \text{ W}$$

$$= (5.14 \times 10^4 \text{ W}) \left(\frac{1 \text{ hp}}{746 \text{ W}} \right) = 69 \text{ hp}.$$

105. (a) With $P = 1.5$ MW $= 1.5 \times 10^6$ W (assumed constant) and $t = 6.0$ min $= 360$ s, the work-kinetic energy theorem becomes

$$W = Pt = \Delta K = \frac{1}{2} m \left(v_f^2 - v_i^2 \right).$$

The mass of the locomotive is then

$$m = \frac{2Pt}{v_f^2 - v_i^2} = \frac{(2)(1.5 \times 10^6 \text{ W})(360 \text{ s})}{(25 \text{ m/s})^2 - (10 \text{ m/s})^2} = 2.1 \times 10^6 \text{ kg}.$$

(b) With t arbitrary, we use $Pt = \frac{1}{2}m(v^2 - v_i^2)$ to solve for the speed $v = v(t)$ as a function of time and obtain

$$v(t) = \sqrt{v_i^2 + \frac{2Pt}{m}} = \sqrt{(10)^2 + \frac{(2)(1.5 \times 10^6)t}{2.1 \times 10^6}} = \sqrt{100 + 1.5t}$$

in SI units (v in m/s and t in s).

(c) The force $F(t)$ as a function of time is

$$F(t) = \frac{P}{v(t)} = \frac{1.5 \times 10^6}{\sqrt{100 + 1.5t}}$$

in SI units (F in N and t in s).

(d) The distance d the train moved is given by

$$d = \int_0^t v(t')dt' = \int_0^{360}\left(100 + \frac{3}{2}t\right)^{1/2} dt = \frac{4}{9}\left(100 + \frac{3}{2}t\right)^{3/2}\Bigg|_0^{360} = 6.7 \times 10^3 \text{ m}.$$

106. We take the bottom of the incline to be the $y = 0$ reference level. The incline angle is $\theta = 30°$. The distance along the incline d (measured from the bottom) is related to height y by the relation $y = d \sin \theta$.

(a) Using the conservation of energy, we have

$$K_0 + U_0 = K_{top} + U_{top} \Rightarrow \frac{1}{2}mv_0^2 + 0 = 0 + mgy$$

with $v_0 = 5.0 \text{ m/s}$. This yields $y = 1.3$ m, from which we obtain $d = 2.6$ m.

(b) An analysis of forces in the manner of Chapter 6 reveals that the magnitude of the friction force is $f_k = \mu_k mg \cos \theta$. Now, we write Eq. 8-33 as

$$K_0 + U_0 = K_{top} + U_{top} + f_k d$$

$$\frac{1}{2}mv_0^2 + 0 = 0 + mgy + f_k d$$

$$\frac{1}{2}mv_0^2 = mgd\sin\theta + \mu_k mgd\cos\theta$$

which — upon canceling the mass and rearranging — provides the result for d:

$$d = \frac{v_0^2}{2g(\mu_k \cos\theta + \sin\theta)} = 1.5 \text{ m}.$$

(c) The thermal energy generated by friction is $f_k d = \mu_k mgd \cos\theta = 26$ J.

(d) The slide back down, from the height $y = 1.5 \sin 30°$ is also described by Eq. 8-33. With ΔE_{th} again equal to 26 J, we have

$$K_{top} + U_{top} = K_{bot} + U_{bot} + f_k d \Rightarrow 0 + mgy = \frac{1}{2}mv_{bot}^2 + 0 + 26$$

from which we find $v_{bot} = 2.1$ m/s.

107. (a) The effect of a (sliding) friction is described in terms of energy dissipated as shown in Eq. 8-31. We have

$$\Delta E = K + \frac{1}{2}k(0.08)^2 - \frac{1}{2}k(0.10)^2 = -f_k(0.02)$$

where distances are in meters and energies are in Joules. With $k = 4000$ N/m and $f_k = 80$ N, we obtain $K = 5.6$ J.

(b) In this case, we have $d = 0.10$ m. Thus,

$$\Delta E = K + 0 - \frac{1}{2}k(0.10)^2 = -f_k(0.10)$$

which leads to $K = 12$ J.

(c) We can approach this two ways. One way is to examine the dependence of energy on the variable d:

$$\Delta E = K + \frac{1}{2}k(d_0 - d)^2 - \frac{1}{2}kd_0^2 = -f_k d$$

where $d_0 = 0.10$ m, and solving for K as a function of d:

$$K = -\frac{1}{2}kd^2 + (kd_0)d - f_k d.$$

In this first approach, we could work through the $\frac{dK}{dd} = 0$ condition (or with the special capabilities of a graphing calculator) to obtain the answer $K_{max} = \frac{1}{2k}(kd_0 - f_k)^2$. In the second (and perhaps easier) approach, we note that K is maximum where v is maximum — which is where $a = 0 \Rightarrow$ equilibrium of forces. Thus, the second approach simply solves for the equilibrium position

$$|F_{spring}| = f_k \Rightarrow kx = 80.$$

Thus, with $k = 4000$ N/m we obtain $x = 0.02$ m. But $x = d_0 - d$ so this corresponds to $d = 0.08$ m. Then the methods of part (a) lead to the answer $K_{max} = 12.8$ J ≈ 13 J.

108. We assume his initial kinetic energy (when he jumps) is negligible. Then, his initial gravitational potential energy measured relative to where he momentarily stops is what becomes the elastic potential energy of the stretched net (neglecting air friction). Thus,

$$U_{net} = U_{grav} = mgh$$

where $h = 11.0$ m $+ 1.5$ m $= 12.5$ m. With $m = 70$ kg, we obtain $U_{net} = 8580$ J $\approx 8.6 \times 10^3$ J.

109. The connection between angle θ (measured from vertical) and height h (measured from the lowest point, which is our choice of reference position in computing the gravitational potential energy mgh) is given by $h = L(1 - \cos \theta)$ where L is the length of the pendulum.

(a) Using this formula (or simply using intuition) we see the initial height is $h_1 = 2L$, and of course $h_2 = 0$. We use energy conservation in the form of Eq. 8-17.

$$K_1 + U_1 = K_2 + U_2$$
$$0 + mg(2L) = \frac{1}{2}mv^2 + 0$$

This leads to $v = 2\sqrt{gL}$. With $L = 0.62$ m, we have

$$v = 2\sqrt{(9.8 \text{ m/s}^2)(0.62 \text{ m})} = 4.9 \text{ m/s}.$$

(b) The ball is in circular motion with the center of the circle above it, so $\vec{a} = v^2/r$ upward, where $r = L$. Newton's second law leads to

$$T - mg = m\frac{v^2}{r} \Rightarrow T = m\left(g + \frac{4gL}{L}\right) = 5\,mg.$$

With $m = 0.092$ kg, the tension is given by $T = 4.5$ N.

(c) The pendulum is now started (with zero speed) at $\theta_i = 90°$ (that is, $h_i = L$), and we look for an angle θ such that $T = mg$. When the ball is moving through a point at angle θ, then Newton's second law applied to the axis along the rod yields

$$T - mg\cos\theta = m\frac{v^2}{r}$$

which (since $r = L$) implies $v^2 = gL(1 - \cos\theta)$ at the position we are looking for. Energy conservation leads to

$$K_i + U_i = K + U$$
$$0 + mgL = \frac{1}{2}mv^2 + mgL(1 - \cos\theta)$$
$$gL = \frac{1}{2}(gL(1 - \cos\theta)) + gL(1 - \cos\theta)$$

where we have divided by mass in the last step. Simplifying, we obtain

$$\theta = \cos^{-1}\left(\frac{1}{3}\right) = 71°.$$

(d) Since the angle found in (c) is independent of the mass, the result remains the same if the mass of the ball is changed.

110. We take her original elevation to be the $y = 0$ reference level and observe that the top of the hill must consequently have $y_A = R(1 - \cos 20°) = 1.2$ m, where R is the radius of the hill. The mass of the skier is $600/9.8 = 61$ kg.

(a) Applying energy conservation, Eq. 8-17, we have

$$K_B + U_B = K_A + U_A \Rightarrow K_B + 0 = K_A + mgy_A.$$

Using $K_B = \frac{1}{2}(61\,\text{kg})(8.0\,\text{m/s})^2$, we obtain $K_A = 1.2 \times 10^3$ J. Thus, we find the speed at the hilltop is

$$v = \sqrt{2K/m} = 6.4\,\text{m/s}.$$

Note: one might wish to check that the skier stays in contact with the hill — which is indeed the case, here. For instance, at A we find $v^2/r \approx 2$ m/s^2 which is considerably less than g.

(b) With $K_A = 0$, we have

$$K_B + U_B = K_A + U_A \Rightarrow K_B + 0 = 0 + mgy_A$$

which yields $K_B = 724$ J, and the corresponding speed is $v = \sqrt{2K/m} = 4.9$ m/s.

(c) Expressed in terms of mass, we have

$$K_B + U_B = K_A + U_A \Rightarrow$$
$$\frac{1}{2}mv_B^2 + mgy_B = \frac{1}{2}mv_A^2 + mgy_A.$$

Thus, the mass m cancels, and we observe that solving for speed does not depend on the value of mass (or weight).

111. (a) At the top of its flight, the vertical component of the velocity vanishes, and the horizontal component (neglecting air friction) is the same as it was when it was thrown. Thus,

$$K_{\text{top}} = \frac{1}{2}mv_x^2 = \frac{1}{2}(0.050\,\text{kg})\bigl((8.0\,\text{m/s})\cos 30°\bigr)^2 = 1.2\text{ J}.$$

(b) We choose the point 3.0 m below the window as the reference level for computing the potential energy. Thus, equating the mechanical energy when it was thrown to when it is at this reference level, we have (with SI units understood)

$$mgy_0 + K_0 = K$$
$$m(9.8)(3.0) + \frac{1}{2}m(8.0)^2 = \frac{1}{2}mv^2$$

which yields (after canceling m and simplifying) $v = 11$ m/s.

(c) As mentioned, m cancels — and is therefore not relevant to that computation.

(d) The v in the kinetic energy formula is the magnitude of the velocity vector; it does not depend on the direction.

112. (a) The rate of change of the gravitational potential energy is

$$\frac{dU}{dt} = mg\frac{dy}{dt} = -mg|v| = -(68)(9.8)(59) = -3.9 \times 10^4 \text{ J/s}.$$

Thus, the gravitational energy is being reduced at the rate of 3.9×10^4 W.

(b) Since the velocity is constant, the rate of change of the kinetic energy is zero. Thus the rate at which the mechanical energy is being dissipated is the same as that of the gravitational potential energy (3.9×10^4 W).

113. The water has gained

$$\Delta K = \frac{1}{2}(10 \text{ kg})(13 \text{ m/s})^2 - \frac{1}{2}(10 \text{ kg})(3.2 \text{ m/s})^2 = 794 \text{ J}$$

of kinetic energy, and it has lost $\Delta U = (10 \text{ kg})(9.8 \text{ m/s}^2)(15 \text{ m}) = 1470$ J.

of potential energy (the lack of agreement between these two values is presumably due to transfer of energy into thermal forms). The ratio of these values is 0.54 = 54%. The mass of the water cancels when we take the ratio, so that the assumption (stated at the end of the problem: $m = 10$ kg) is not needed for the final result.

114. (a) The integral (see Eq. 8-6, where the value of U at $x = \infty$ is required to vanish) is straightforward. The result is $U(x) = -Gm_1m_2/x$.

(b) One approach is to use Eq. 8-5, which means that we are effectively doing the integral of part (a) all over again. Another approach is to use our result from part (a) (and thus use Eq. 8-1). Either way, we arrive at

$$W = \frac{Gm_1m_2}{x_1} - \frac{Gm_1m_2}{x_1+d} = \frac{Gm_1m_2 d}{x_1(x_1+d)}.$$

115. (a) During one second, the decrease in potential energy is

$$-\Delta U = mg(-\Delta y) = (5.5 \times 10^6 \text{ kg})(9.8 \text{ m/s}^2)(50 \text{ m}) = 2.7 \times 10^9 \text{ J}$$

where $+y$ is upward and $\Delta y = y_f - y_i$.

(b) The information relating mass to volume is not needed in the computation. By Eq. 8-40 (and the SI relation W = J/s), the result follows:

$$P = (2.7 \times 10^9 \text{ J})/(1 \text{ s}) = 2.7 \times 10^9 \text{ W}.$$

(c) One year is equivalent to $24 \times 365.25 = 8766$ h which we write as 8.77 kh. Thus, the energy supply rate multiplied by the cost and by the time is

$$(2.7 \times 10^9 \text{ W})(8.77 \text{ kh})\left(\frac{1 \text{ cent}}{1 \text{ kWh}}\right) = 2.4 \times 10^{10} \text{ cents} = \$2.4 \times 10^8.$$

116. (a) The kinetic energy K of the automobile of mass m at $t = 30$ s is

$$K = \frac{1}{2}mv^2 = \frac{1}{2}(1500\,\text{kg})\left[(72\,\text{km/h})\left(\frac{1000\,\text{m/km}}{3600\,\text{s/h}}\right)\right]^2 = 3.0\times 10^5\,\text{J}.$$

(b) The average power required is

$$P_{\text{avg}} = \frac{\Delta K}{\Delta t} = \frac{3.0\times 10^5\,\text{J}}{30\,\text{s}} = 1.0\times 10^4\,\text{W}.$$

(c) Since the acceleration a is constant, the power is $P = Fv = mav = ma(at) = ma^2 t$ using Eq. 2-11. By contrast, from part (b), the average power is $P_{\text{avg}} = \frac{mv^2}{2t}$ which becomes $\frac{1}{2}ma^2 t$ when $v = at$ is again utilized. Thus, the instantaneous power at the end of the interval is twice the average power during it: $P = 2P_{\text{avg}} = (2)(1.0\times 10^4\,\text{W}) = 2.0\times 10^4\,\text{W}$.

117. (a) The remark in the problem statement that the forces can be associated with potential energies is illustrated as follows: the work from $x = 3.00$ m to $x = 2.00$ m is

$$W = F_2 \Delta x = (5.00\,\text{N})(-1.00\,\text{m}) = -5.00\,\text{J},$$

so the potential energy at $x = 2.00$ m is $U_2 = +5.00$ J.

(b) Now, it is evident from the problem statement that $E_{\text{max}} = 14.0$ J, so the kinetic energy at $x = 2.00$ m is

$$K_2 = E_{\text{max}} - U_2 = 14.0 - 5.00 = 9.00\,\text{J}.$$

(c) The work from $x = 2.00$ m to $x = 0$ is $W = F_1 \Delta x = (3.00\,\text{N})(-2.00\,\text{m}) = -6.00$ J, so the potential energy at $x = 0$ is

$$U_0 = 6.00\,\text{J} + U_2 = (6.00 + 5.00)\,\text{J} = 11.0\,\text{J}.$$

(d) Similar reasoning to that presented in part (a) then gives

$$K_0 = E_{\text{max}} - U_0 = (14.0 - 11.0)\,\text{J} = 3.00\,\text{J}.$$

(e) The work from $x = 8.00$ m to $x = 11.0$ m is $W = F_3 \Delta x = (-4.00\,\text{N})(3.00\,\text{m}) = -12.0$ J, so the potential energy at $x = 11.0$ m is $U_{11} = 12.0$ J.

(f) The kinetic energy at $x = 11.0$ m is therefore

$$K_{11} = E_{max} - U_{11} = (14.0 - 12.0) \text{ J} = 2.00 \text{ J}.$$

(g) Now we have $W = F_4 \Delta x = (-1.00 \text{ N})(1.00 \text{ m}) = -1.00 \text{ J}$, so the potential energy at $x = 12.0$ m is

$$U_{12} = 1.00 \text{ J} + U_{11} = (1.00 + 12.0) \text{ J} = 13.0 \text{ J}.$$

(h) Thus, the kinetic energy at $x = 12.0$ m is

$$K_{12} = E_{max} - U_{12} = (14.0 - 13.0) = 1.00 \text{ J}.$$

(i) There is no work done in this interval (from $x = 12.0$ m to $x = 13.0$ m) so the answers are the same as in part (g): $U_{12} = 13.0$ J.

(j) There is no work done in this interval (from $x = 12.0$ m to $x = 13.0$ m) so the answers are the same as in part (h): $K_{12} = 1.00$ J.

(k) Although the plot is not shown here, it would look like a "potential well" with piecewise-sloping sides: from $x = 0$ to $x = 2$ (SI units understood) the graph if U is a decreasing line segment from 11 to 5, and from $x = 2$ to $x = 3$, it then heads down to zero, where it stays until $x = 8$, where it starts increasing to a value of 12 (at $x = 11$), and then in another positive-slope line segment it increases to a value of 13 (at $x = 12$). For $x > 12$ its value does not change (this is the "top of the well").

(l) The particle can be thought of as "falling" down the $0 < x < 3$ slopes of the well, gaining kinetic energy as it does so, and certainly is able to reach $x = 5$. Since $U = 0$ at $x = 5$, then it's initial potential energy (11 J) has completely converted to kinetic: now $K = 11.0$ J.

(m) This is not sufficient to climb up and out of the well on the large x side ($x > 8$), but does allow it to reach a "height" of 11 at $x = 10.8$ m. As discussed in section 8-5, this is a "turning point" of the motion.

(n) Next it "falls" back down and rises back up the small x slopes until it comes back to its original position. Stating this more carefully, when it is (momentarily) stopped at $x = 10.8$ m it is accelerated to the left by the force $\vec{F}_3$; it gains enough speed as a result that it eventually is able to return to $x = 0$, where it stops again.

118. (a) At $x = 5.00$ m the potential energy is zero, and the kinetic energy is

$$K = \frac{1}{2} mv^2 = \frac{1}{2} (2.00 \text{ kg})(3.45 \text{ m/s})^2 = 11.9 \text{ J}.$$

The total energy, therefore, is great enough to reach the point $x = 0$ where $U = 11.0$ J, with a little "left over" (11.9 J – 11.0 J = 0.9025 J). This is the kinetic energy at $x = 0$, which means the speed there is

$$v = \sqrt{2(0.9025 \text{ J})/(2 \text{ kg})} = 0.950 \text{ m/s}.$$

It has now come to a stop, therefore, so it has not encountered a turning point.

(b) The total energy (11.9 J) is equal to the potential energy (in the scenario where it is initially moving rightward) at $x = 10.9756 \approx 11.0$ m. This point may be found by interpolation or simply by using the work-kinetic-energy theorem:

$$K_f = K_i + W = 0 \quad \Rightarrow \quad 11.9025 + (-4)d = 0 \quad \Rightarrow \quad d = 2.9756 \approx 2.98$$

(which when added to $x = 8.00$ [the point where F_3 begins to act] gives the correct result). This provides a turning point for the particle's motion.

119. (a) During the final $d = 12$ m of motion, we use

$$K_1 + U_1 = K_2 + U_2 + f_k d$$
$$\frac{1}{2}mv^2 + 0 = 0 + 0 + f_k d$$

where $v = 4.2$ m/s. This gives $f_k = 0.31$ N. Therefore, the thermal energy change is $f_k d = 3.7$ J.

(b) Using $f_k = 0.31$ N we obtain $f_k d_{\text{total}} = 4.3$ J for the thermal energy generated by friction; here, $d_{\text{total}} = 14$ m.

(c) During the initial $d' = 2$ m of motion, we have

$$K_0 + U_0 + W_{\text{app}} = K_1 + U_1 + f_k d' \Rightarrow 0 + 0 + W_{\text{app}} = \frac{1}{2}mv^2 + 0 + f_k d'$$

which essentially combines Eq. 8-31 and Eq. 8-33. This leads to the result $W_{\text{app}} = 4.3$ J, and — reasonably enough — is the same as our answer in part (b).

120. (a) The table shows that the force is $+(3.0 \text{ N})\hat{i}$ while the displacement is in the $+x$ direction ($\vec{d} = +(3.0 \text{ m})\hat{i}$), and it is $-(3.0 \text{ N})\hat{i}$ while the displacement is in the $-x$ direction. Using Eq. 7-8 for each part of the trip, and adding the results, we find the work done is 18 J. This is not a conservative force field; if it had been, then the net work done would have been zero (since it returned to where it started).

(b) This, however, is a conservative force field, as can be easily verified by calculating that the net work done here is zero.

(c) The two integrations that need to be performed are each of the form $\int 2x\,dx$ so that we are adding two equivalent terms, where each equals x^2 (evaluated at $x = 4$, minus its value at $x = 1$). Thus, the work done is $2(4^2 - 1^2) = 30$ J.

(d) This is another conservative force field, as can be easily verified by calculating that the net work done here is zero.

(e) The forces in (b) and (d) are conservative.

121. We use Eq. 8-20.

(a) The force at $x = 2.0$ m is

$$F = -\frac{dU}{dx} \approx -\frac{-(17.5\text{ J}) - (-2.8\text{ J})}{4.0\text{ m} - 1.0\text{ m}} = 4.9\text{ N}.$$

(b) The force points in the $+x$ direction (but there is some uncertainty in reading the graph which makes the last digit not very significant).

(c) The total mechanical energy at $x = 2.0$ m is

$$E = \frac{1}{2}mv^2 + U \approx \frac{1}{2}(2.0)(-1.5)^2 - 7.7 = -5.5$$

in SI units (Joules). Again, there is some uncertainty in reading the graph which makes the last digit not very significant. At that level (–5.5 J) on the graph, we find two points where the potential energy curve has that value — at $x \approx 1.5$ m and $x \approx 13.5$ m. Therefore, the particle remains in the region $1.5 < x < 13.5$ m. The left boundary is at $x = 1.5$ m.

(d) From the above results, the right boundary is at $x = 13.5$ m.

(e) At $x = 7.0$ m, we read $U \approx -17.5$ J. Thus, if its total energy (calculated in the previous part) is $E \approx -5.5$ J, then we find

$$\frac{1}{2}mv^2 = E - U \approx 12\text{ J} \Rightarrow v = \sqrt{\frac{2}{m}(E-U)} \approx 3.5\text{ m/s}$$

where there is certainly room for disagreement on that last digit for the reasons cited above.

122. The connection between angle θ (measured from vertical) and height h (measured from the lowest point, which is our choice of reference position in computing the gravitational potential energy) is given by $h = L(1 - \cos\theta)$ where L is the length of the pendulum.

(a) We use energy conservation in the form of Eq. 8-17.

$$K_1 + U_1 = K_2 + U_2$$

$$0 + mgL(1 - \cos\theta_1) = \frac{1}{2}mv_2^2 + mgL(1 - \cos\theta_2)$$

With $L = 1.4$ m, $\theta_1 = 30°$, and $\theta_2 = 20°$, we have

$$v_2 = \sqrt{2gL(\cos\theta_2 - \cos\theta_1)} = 1.4 \text{ m/s}.$$

(b) The maximum speed v_3 is at the lowest point. Our formula for h gives $h_3 = 0$ when $\theta_3 = 0°$, as expected. From

$$K_1 + U_1 = K_3 + U_3$$

$$0 + mgL(1 - \cos\theta_1) = \frac{1}{2}mv_3^2 + 0$$

we obtain $v_3 = 1.9$ m/s.

(c) We look for an angle θ_4 such that the speed there is $v_4 = v_3/3$. To be as accurate as possible, we proceed algebraically (substituting $v_3^2 = 2gL(1 - \cos\theta_1)$ at the appropriate place) and plug numbers in at the end. Energy conservation leads to

$$K_1 + U_1 = K_4 + U_4$$

$$0 + mgL(1 - \cos\theta_1) = \frac{1}{2}mv_4^2 + mgL(1 - \cos\theta_4)$$

$$mgL(1 - \cos\theta_1) = \frac{1}{2}m\frac{v_3^2}{9} + mgL(1 - \cos\theta_4)$$

$$-gL\cos\theta_1 = \frac{1}{2}\frac{2gL(1 - \cos\theta_1)}{9} - gL\cos\theta_4$$

where in the last step we have subtracted out mgL and then divided by m. Thus, we obtain

$$\theta_4 = \cos^{-1}\left(\frac{1}{9} + \frac{8}{9}\cos\theta_1\right) = 28.2° \approx 28°.$$

123. Converting to SI units, $v_0 = 8.3$ m/s and $v = 11.1$ m/s. The incline angle is $\theta = 5.0°$. The height difference between the car's highest and lowest points is (50 m) sin $\theta = 4.4$ m. We take the lowest point (the car's final reported location) to correspond to the $y = 0$ reference level.

(a) Using Eq. 8-31 and Eq. 8-33, we find

$$f_k d = -\Delta K - \Delta U \Rightarrow f_k d = \frac{1}{2} m \left(v_0^2 - v^2 \right) + mgy_0 .$$

Therefore, the mechanical energy reduction (due to friction) is $f_k d = 2.4 \times 10^4$ J.

(b) With $d = 50$ m, we solve for f_k and obtain 4.7×10^2 N.

124. Equating the mechanical energy at his initial position (as he emerges from the canon, where we set the reference level for computing potential energy) to his energy as he lands, we obtain

$$K_i = K_f + U_f$$
$$\frac{1}{2}(60\,\text{kg})(16\,\text{m/s})^2 = K_f + (60\,\text{kg})(9.8\,\text{m/s}^2)(3.9\,\text{m})$$

which leads to $K_f = 5.4 \times 10^3$ J.

125. (a) The compression is "spring-like" so the maximum force relates to the distance x by Hooke's law:

$$F_x = kx \Rightarrow x = \frac{750}{2.5 \times 10^5} = 0.0030\,\text{m}.$$

(b) The work is what produces the "spring-like" potential energy associated with the compression. Thus, using Eq. 8-11,

$$W = \frac{1}{2} kx^2 = \frac{1}{2}(2.5 \times 10^5)(0.0030)^2 = 1.1\,\text{J}.$$

(c) By Newton's third law, the force F exerted by the tooth is equal and opposite to the "spring-like" force exerted by the licorice, so the graph of F is a straight line of slope k. We plot F (in newtons) versus x (in millimeters); both are taken as positive.

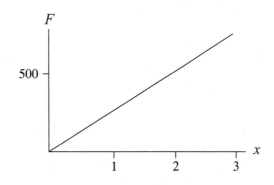

(d) As mentioned in part (b), the spring potential energy expression is relevant. Now, whether or not we can ignore dissipative processes is a deeper question. In other words, it seems unlikely that — if the tooth at any moment were to reverse its motion — that the licorice could "spring back" to its original shape. Still, to the extent that $U = \frac{1}{2}kx^2$ applies, the graph is a parabola (not shown here) which has its vertex at the origin and is either concave upward or concave downward depending on how one wishes to define the sign of F (the connection being $F = -dU/dx$).

(e) As a crude estimate, the area under the curve is roughly half the area of the entire plotting-area (8000 N by 12 mm). This leads to an approximate work of

$\frac{1}{2}$ (8000 N) (0.012 m) $\approx$ 50 J. Estimates in the range $40 \leq W \leq 50$ J are acceptable.

(f) Certainly dissipative effects dominate this process, and we cannot assign it a meaningful potential energy.

126. (a) This part is essentially a free-fall problem, which can be easily done with Chapter 2 methods. Instead, choosing energy methods, we take $y = 0$ to be the ground level.

$$K_i + U_i = K + U \Rightarrow 0 + mgy_i = \frac{1}{2}mv^2 + 0$$

Therefore $v = \sqrt{2gy_i} = 9.2$ m/s, where $y_i = 4.3$ m.

(b) Eq. 8-29 provides $\Delta E_{th} = f_k d$ for thermal energy generated by the kinetic friction force. We apply Eq. 8-31:

$$K_i + U_i = K + U \Rightarrow 0 + mgy_i = \frac{1}{2}mv^2 + 0 + f_k d.$$

With $d = y_i$, $m = 70$ kg and $f_k = 500$ N, this yields $v = 4.8$ m/s.

127. (a) When there is no change in potential energy, Eq. 8-24 leads to

$$W_{app} = \Delta K = \frac{1}{2}m\left(v^2 - v_0^2\right).$$

Therefore, $\Delta E = 6.0 \times 10^3$ J.

(b) From the above manipulation, we see $W_{app} = 6.0 \times 10^3$ J. Also, from Chapter 2, we know that $\Delta t = \Delta v/a = 10$ s. Thus, using Eq. 7-42,

$$P_{avg} = \frac{W}{\Delta t} = \frac{6.0 \times 10^3}{10} = 600 \text{ W}.$$

(c) and (d) The constant applied force is $ma = 30$ N and clearly in the direction of motion, so Eq. 7-48 provides the results for instantaneous power

$$P = \vec{F} \cdot \vec{v} = \begin{cases} 300 \text{ W} & \text{for } v = 10 \text{ m/s} \\ 900 \text{ W} & \text{for } v = 30 \text{ m/s} \end{cases}$$

We note that the average of these two values agrees with the result in part (b).

128. The distance traveled up the incline can be figured with Chapter 2 techniques: $v^2 = v_0^2 + 2a\Delta x \rightarrow \Delta x = 200$ m. This corresponds to an increase in height equal to $y = (200 \text{ m}) \sin\theta = 17$ m, where $\theta = 5.0°$. We take its initial height to be $y = 0$.

(a) Eq. 8-24 leads to

$$W_{app} = \Delta E = \frac{1}{2}m(v^2 - v_0^2) + mgy.$$

Therefore, $\Delta E = 8.6 \times 10^3$ J.

(b) From the above manipulation, we see $W_{app} = 8.6 \times 10^3$ J. Also, from Chapter 2, we know that $\Delta t = \Delta v/a = 10$ s. Thus, using Eq. 7-42,

$$P_{avg} = \frac{W}{\Delta t} = \frac{8.6 \times 10^3}{10} = 860 \text{ W}$$

where the answer has been rounded off (from the 856 value that is provided by the calculator).

(c) and (d) Taking into account the component of gravity along the incline surface, the applied force is $ma + mg \sin\theta = 43$ N and clearly in the direction of motion, so Eq. 7-48 provides the results for instantaneous power

$$P = \vec{F} \cdot \vec{v} = \begin{cases} 430 \text{ W} & \text{for } v = 10 \text{ m/s} \\ 1300 \text{ W} & \text{for } v = 30 \text{ m/s} \end{cases}$$

where these answers have been rounded off (from 428 and 1284, respectively). We note that the average of these two values agrees with the result in part (b).

129. We want to convert (at least in theory) the water that falls through $h = 500$ m into electrical energy. The problem indicates that in one year, a volume of water equal to $A\Delta z$ lands in the form of rain on the country, where $A = 8 \times 10^{12}$ m^2 and $\Delta z = 0.75$ m. Multiplying this volume by the density $\rho = 1000$ kg/m^3 leads to

$$m_{total} = \rho A \Delta z = (1000)(8 \times 10^{12})(0.75) = 6 \times 10^{15} \text{ kg}$$

for the mass of rainwater. One-third of this "falls" to the ocean, so it is $m = 2 \times 10^{15}$ kg that we want to use in computing the gravitational potential energy mgh (which will turn into electrical energy during the year). Since a year is equivalent to 3.2×10^7 s, we obtain

$$P_{\text{avg}} = \frac{(2 \times 10^{15})(9.8)(500)}{3.2 \times 10^7} = 3.1 \times 10^{11} \text{ W}.$$

130. The spring is relaxed at $y = 0$, so the elastic potential energy (Eq. 8-11) is $U_{\text{el}} = \frac{1}{2}ky^2$. The total energy is conserved, and is zero (determined by evaluating it at its initial position). We note that U is the same as ΔU in these manipulations. Thus, we have

$$0 = K + U_g + U_e \quad \Rightarrow \quad K = -U_g - U_e$$

where $U_g = mgy = (20 \text{ N})y$ with y in meters (so that the energies are in Joules). We arrange the results in a table:

position y	−0.05	−0.10	−0.15	−0.20
K	(a) 0.75	(d) 1.0	(g) 0.75	(j) 0
U_g	(b) −1.0	(e) −2.0	(h) −3.0	(k) −4.0
U_e	(c) 0.25	(f) 1.0	(i) 2.25	(l) 4.0

131. The power generation (assumed constant, so average power is the same as instantaneous power) is

$$P = \frac{mgh}{t} = \frac{(3/4)(1200 \text{ m}^3)(10^3 \text{ kg/m}^3)(9.8 \text{ m/s}^2)(100 \text{ m})}{1.0 \text{ s}} = 8.80 \times 10^8 \text{ W}.$$

132. The style of reasoning used here is presented in §8-5.

(a) The horizontal line representing E_1 intersects the potential energy curve at a value of $r \approx 0.07$ nm and seems not to intersect the curve at larger r (though this is somewhat unclear since $U(r)$ is graphed only up to $r = 0.4$ nm). Thus, if m were propelled towards M from large r with energy E_1 it would "turn around" at 0.07 nm and head back in the direction from which it came.

(b) The line representing E_2 has two intersection points $r_1 \approx 0.16$ nm and $r_2 \approx 0.28$ nm with the $U(r)$ plot. Thus, if m starts in the region $r_1 < r < r_2$ with energy E_2 it will bounce back and forth between these two points, presumably forever.

(c) At $r = 0.3$ nm, the potential energy is roughly $U = -1.1 \times 10^{-19}$ J.

(d) With $M >> m$, the kinetic energy is essentially just that of m. Since $E = 1 \times 10^{-19}$ J, its kinetic energy is $K = E - U \approx 2.1 \times 10^{-19}$ J.

(e) Since force is related to the slope of the curve, we must (crudely) estimate $|F| \approx 1 \times 10^{-9}$ N at this point. The sign of the slope is positive, so by Eq. 8-20, the force is negative-valued. This is interpreted to mean that the atoms are attracted to each other.

(f) Recalling our remarks in the previous part, we see that the sign of F is positive (meaning it's repulsive) for $r < 0.2$ nm.

(g) And the sign of F is negative (attractive) for $r > 0.2$ nm.

(h) At $r = 0.2$ nm, the slope (hence, F) vanishes.

133. (a) Sample Problem 8-3 illustrates simple energy conservation in a similar situation, and derives the frequently encountered relationship: $v = \sqrt{2gh}$. In our present problem, the height change is equal to the rod length L. Thus, using the suggested notation for the speed, we have $v_o = \sqrt{2gL}$.

(b) At B the speed is (from Eq. 8-17)

$$v = \sqrt{v_0^2 + 2gL} = \sqrt{4gL}.$$

The direction of the centripetal acceleration ($v^2/r = 4gL/L = 4g$) is upward (at that moment), as is the tension force. Thus, Newton's second law gives

$$T - mg = m(4g) \;\Rightarrow\; T = 5mg.$$

(c) The difference in height between C and D is L, so the "loss" of mechanical energy (which goes into thermal energy) is $-mgL$.

(d) The difference in height between B and D is $2L$, so the total "loss" of mechanical energy (which all goes into thermal energy) is $-2mgL$.

134. (a) The force (SI units understood) from Eq. 8-20 is plotted in the graph below.

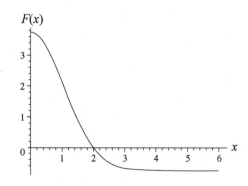

(b) The potential energy $U(x)$ and the kinetic energy $K(x)$ are shown in the next. The potential energy curve begins at 4 and drops (until about $x = 2$); the kinetic energy curve is the one that starts at zero and rises (until about $x = 2$).

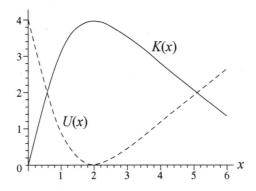

135. Let the amount of stretch of the spring be x. For the object to be in equilibrium

$$kx - mg = 0 \Rightarrow x = mg/k.$$

Thus the gain in elastic potential energy for the spring is

$$\Delta U_e = \frac{1}{2}kx^2 = \frac{1}{2}k\left(\frac{mg}{k}\right)^2 = \frac{m^2g^2}{2k}$$

while the loss in the gravitational potential energy of the system is

$$-\Delta U_g = mgx = mg\left(\frac{mg}{k}\right) = \frac{m^2g^2}{k}$$

which we see (by comparing with the previous expression) is equal to $2\Delta U_e$. The reason why $|\Delta U_g| \neq \Delta U_e$ is that, since the object is slowly lowered, an upward external force (e.g., due to the hand) must have been exerted on the object during the lowering process, preventing it from accelerating downward. This force does *negative* work on the object, reducing the total mechanical energy of the system.

Chapter 9

1. We use Eq. 9-5 to solve for (x_3, y_3).

(a) The x coordinates of the system's center of mass is:

$$x_{com} = \frac{m_1 x_1 + m_2 x_2 + m_3 x_3}{m_1 + m_2 + m_3} = \frac{(2.00 \text{ kg})(-1.20 \text{ m}) + (4.00 \text{ kg})(0.600 \text{ m}) + (3.00 \text{ kg}) x_3}{2.00 \text{ kg} + 4.00 \text{ kg} + 3.00 \text{ kg}}$$
$$= -0.500 \text{ m}.$$

Solving the equation yields $x_3 = -1.50$ m.

(b) The y coordinates of the system's center of mass is:

$$y_{com} = \frac{m_1 y_1 + m_2 y_2 + m_3 y_3}{m_1 + m_2 + m_3} = \frac{(2.00 \text{ kg})(0.500 \text{ m}) + (4.00 \text{ kg})(-0.750 \text{ m}) + (3.00 \text{ kg}) y_3}{2.00 \text{ kg} + 4.00 \text{ kg} + 3.00 \text{ kg}}$$
$$= -0.700 \text{ m}.$$

Solving the equation yields $y_3 = -1.43$ m.

2. Our notation is as follows: $x_1 = 0$ and $y_1 = 0$ are the coordinates of the $m_1 = 3.0$ kg particle; $x_2 = 2.0$ m and $y_2 = 1.0$ m are the coordinates of the $m_2 = 4.0$ kg particle; and, $x_3 = 1.0$ m and $y_3 = 2.0$ m are the coordinates of the $m_3 = 8.0$ kg particle.

(a) The x coordinate of the center of mass is

$$x_{com} = \frac{m_1 x_1 + m_2 x_2 + m_3 x_3}{m_1 + m_2 + m_3} = \frac{0 + (4.0 \text{ kg})(2.0 \text{ m}) + (8.0 \text{ kg})(1.0 \text{ m})}{3.0 \text{ kg} + 4.0 \text{ kg} + 8.0 \text{ kg}} = 1.1 \text{ m}.$$

(b) The y coordinate of the center of mass is

$$y_{com} = \frac{m_1 y_1 + m_2 y_2 + m_3 y_3}{m_1 + m_2 + m_3} = \frac{0 + (4.0 \text{ kg})(1.0 \text{ m}) + (8.0 \text{ kg})(2.0 \text{ m})}{3.0 \text{ kg} + 4.0 \text{ kg} + 8.0 \text{ kg}} = 1.3 \text{ m}.$$

(c) As the mass of m_3, the topmost particle, is increased, the center of mass shifts toward that particle. As we approach the limit where m_3 is infinitely more massive than the others, the center of mass becomes infinitesimally close to the position of m_3.

3. Since the plate is uniform, we can split it up into three rectangular pieces, with the mass of each piece being proportional to its area and its center of mass being at its geometric center. We'll refer to the large 35 cm × 10 cm piece (shown to the left of the y axis in Fig. 9-38) as section 1; it has 63.6% of the total area and its center of mass is at $(x_1,y_1) = (-5.0$ cm, -2.5 cm). The top 20 cm × 5 cm piece (section 2, in the first quadrant) has 18.2% of the total area; its center of mass is at $(x_2,y_2) = (10$ cm, 12.5 cm). The bottom 10 cm × 10 cm piece (section 3) also has 18.2% of the total area; its center of mass is at $(x_3,y_3) = (5$ cm, -15 cm).

(a) The x coordinate of the center of mass for the plate is

$$x_{com} = (0.636)x_1 + (0.182)x_2 + (0.182)x_3 = -0.45 \text{ cm}.$$

(b) The y coordinate of the center of mass for the plate is

$$y_{com} = (0.636)y_1 + (0.182)y_2 + (0.182)y_3 = -2.0 \text{ cm}.$$

4. We will refer to the arrangement as a "table." We locate the coordinate origin at the left end of the tabletop (as shown in Fig. 9-37). With $+x$ rightward and $+y$ upward, then the center of mass of the right leg is at $(x,y) = (+L, -L/2)$, the center of mass of the left leg is at $(x,y) = (0, -L/2)$, and the center of mass of the tabletop is at $(x,y) = (L/2, 0)$.

(a) The x coordinate of the (whole table) center of mass is

$$x_{com} = \frac{M(+L) + M(0) + 3M(+L/2)}{M + M + 3M} = 0.5L.$$

With $L = 22$ cm, we have $x_{com} = 11$ cm.

(b) The y coordinate of the (whole table) center of mass is

$$y_{com} = \frac{M(-L/2) + M(-L/2) + 3M(0)}{M + M + 3M} = -\frac{L}{5},$$

or $y_{com} = -4.4$ cm.

From the coordinates, we see that the whole table center of mass is a small distance 4.4 cm directly below the middle of the tabletop.

5. (a) By symmetry the center of mass is located on the axis of symmetry of the molecule – the y axis. Therefore $x_{com} = 0$.

(b) To find y_{com}, we note that $3m_H y_{com} = m_N(y_N - y_{com})$, where y_N is the distance from the nitrogen atom to the plane containing the three hydrogen atoms:

$$y_N = \sqrt{(10.14 \times 10^{-11} \text{ m})^2 - (9.4 \times 10^{-11} \text{ m})^2} = 3.803 \times 10^{-11} \text{ m}.$$

Thus,

$$y_{com} = \frac{m_N y_N}{m_N + 3m_H} = \frac{(14.0067)(3.803 \times 10^{-11} \text{ m})}{14.0067 + 3(1.00797)} = 3.13 \times 10^{-11} \text{ m}$$

where Appendix F has been used to find the masses.

6. The centers of mass (with centimeters understood) for each of the five sides are as follows:

$(x_1, y_1, z_1) = (0, 20, 20)$ for the side in the yz plane
$(x_2, y_2, z_2) = (20, 0, 20)$ for the side in the xz plane
$(x_3, y_3, z_3) = (20, 20, 0)$ for the side in the xy plane
$(x_4, y_4, z_4) = (40, 20, 20)$ for the remaining side parallel to side 1
$(x_5, y_5, z_5) = (20, 40, 20)$ for the remaining side parallel to side 2

Recognizing that all sides have the same mass m, we plug these into Eq. 9-5 to obtain the results (the first two being expected based on the symmetry of the problem).

(a) The x coordinate of the center of mass is

$$x_{com} = \frac{mx_1 + mx_2 + mx_3 + mx_4 + mx_5}{5m} = \frac{0 + 20 + 20 + 40 + 20}{5} = 20 \text{ cm}$$

(b) The y coordinate of the center of mass is

$$y_{com} = \frac{my_1 + my_2 + my_3 + my_4 + my_5}{5m} = \frac{20 + 0 + 20 + 20 + 40}{5} = 20 \text{ cm}$$

(c) The z coordinate of the center of mass is

$$z_{com} = \frac{mz_1 + mz_2 + mz_3 + mz_4 + mz_5}{5m} = \frac{20 + 20 + 0 + 20 + 20}{5} = 16 \text{ cm}$$

7. We use Eq. 9-5 to locate the coordinates.

(a) By symmetry $x_{com} = -d_1/2 = -(13 \text{ cm})/2 = -6.5$ cm. The negative value is due to our choice of the origin.

(b) We find y_{com} as

$$y_{com} = \frac{m_i y_{com,i} + m_a y_{com,a}}{m_i + m_a} = \frac{\rho_i V_i y_{com,i} + \rho_a V_a y_{cm,a}}{\rho_i V_i + \rho_a V_a}$$

$$= \frac{(11 \text{ cm}/2)(7.85 \text{ g/cm}^3) + 3(11 \text{ cm}/2)(2.7 \text{ g/cm}^3)}{7.85 \text{ g/cm}^3 + 2.7 \text{ g/cm}^3} = 8.3 \text{ cm}.$$

(c) Again by symmetry, we have $z_{com} = (2.8 \text{ cm})/2 = 1.4$ cm.

8. (a) Since the can is uniform, its center of mass is at its geometrical center, a distance $H/2$ above its base. The center of mass of the soda alone is at its geometrical center, a distance $x/2$ above the base of the can. When the can is full this is $H/2$. Thus the center of mass of the can and the soda it contains is a distance

$$h = \frac{M(H/2) + m(H/2)}{M + m} = \frac{H}{2}$$

above the base, on the cylinder axis. With $H = 12$ cm, we obtain $h = 6.0$ cm.

(b) We now consider the can alone. The center of mass is $H/2 = 6.0$ cm above the base, on the cylinder axis.

(c) As x decreases the center of mass of the soda in the can at first drops, then rises to $H/2 = 6.0$ cm again.

(d) When the top surface of the soda is a distance x above the base of the can, the mass of the soda in the can is $m_p = m(x/H)$, where m is the mass when the can is full ($x = H$). The center of mass of the soda alone is a distance $x/2$ above the base of the can. Hence

$$h = \frac{M(H/2) + m_p(x/2)}{M + m_p} = \frac{M(H/2) + m(x/H)(x/2)}{M + (mx/H)} = \frac{MH^2 + mx^2}{2(MH + mx)}.$$

We find the lowest position of the center of mass of the can and soda by setting the derivative of h with respect to x equal to 0 and solving for x. The derivative is

$$\frac{dh}{dx} = \frac{2mx}{2(MH + mx)} - \frac{(MH^2 + mx^2)m}{2(MH + mx)^2} = \frac{m^2 x^2 + 2MmHx - MmH^2}{2(MH + mx)^2}.$$

The solution to $m^2 x^2 + 2MmHx - MmH^2 = 0$ is

$$x = \frac{MH}{m}\left(-1 + \sqrt{1 + \frac{m}{M}}\right).$$

The positive root is used since x must be positive. Next, we substitute the expression found for x into $h = (MH^2 + mx^2)/2(MH + mx)$. After some algebraic manipulation we obtain

$$h = \frac{HM}{m}\left(\sqrt{1+\frac{m}{M}}-1\right) = \frac{(12 \text{ cm})(0.14 \text{ kg})}{1.31 \text{ kg}}\left(\sqrt{1+\frac{1.31 \text{ kg}}{0.14 \text{ kg}}}-1\right) = 2.8 \text{ cm}.$$

9. The implication in the problem regarding $\vec{v}_0$ is that the olive and the nut start at rest. Although we could proceed by analyzing the forces on each object, we prefer to approach this using Eq. 9-14. The total force on the nut-olive system is $\vec{F}_o + \vec{F}_n = (-\hat{i}+\hat{j})$ N. Thus, Eq. 9-14 becomes

$$(-\hat{i}+\hat{j}) \text{ N} = M\vec{a}_{com}$$

where $M = 2.0$ kg. Thus, $\vec{a}_{com} = (-\tfrac{1}{2}\hat{i}+\tfrac{1}{2}\hat{j})$ m/s^2. Each component is constant, so we apply the equations discussed in Chapters 2 and 4 and obtain

$$\Delta\vec{r}_{com} = \frac{1}{2}\vec{a}_{com}t^2 = (-4.0 \text{ m})\hat{i}+(4.0 \text{ m})\hat{j}$$

when $t = 4.0$ s. It is perhaps instructive to work through this problem the *long way* (separate analysis for the olive and the nut and then application of Eq. 9-5) since it helps to point out the computational advantage of Eq. 9-14.

10. Since the center of mass of the two-skater system does not move, both skaters will end up at the center of mass of the system. Let the center of mass be a distance x from the 40-kg skater, then

$$(65 \text{ kg})(10 \text{ m}-x) = (40 \text{ kg})x \Rightarrow x = 6.2 \text{ m}.$$

Thus the 40-kg skater will move by 6.2 m.

11. We use the constant-acceleration equations of Table 2-1 (with $+y$ downward and the origin at the release point), Eq. 9-5 for y_{com} and Eq. 9-17 for $\vec{v}_{com}$.

(a) The location of the first stone (of mass m_1) at $t = 300 \times 10^{-3}$ s is

$$y_1 = (1/2)gt^2 = (1/2)(9.8 \text{ m/s}^2)(300 \times 10^{-3} \text{ s})^2 = 0.44 \text{ m},$$

and the location of the second stone (of mass $m_2 = 2m_1$) at $t = 300 \times 10^{-3}$ s is

$$y_2 = (1/2)gt^2 = (1/2)(9.8 \text{ m/s}^2)(300 \times 10^{-3} \text{ s} - 100 \times 10^{-3} \text{ s})^2 = 0.20 \text{ m}.$$

Thus, the center of mass is at

$$y_{com} = \frac{m_1 y_1 + m_2 y_2}{m_1 + m_2} = \frac{m_1(0.44 \text{ m}) + 2m_1(0.20 \text{ m})}{m_1 + 2m_1} = 0.28 \text{ m}.$$

(b) The speed of the first stone at time t is $v_1 = gt$, while that of the second stone is

$$v_2 = g(t - 100 \times 10^{-3} \text{ s}).$$

Thus, the center-of-mass speed at $t = 300 \times 10^{-3}$ s is

$$v_{com} = \frac{m_1 v_1 + m_2 v_2}{m_1 + m_2} = \frac{m_1(9.8 \text{ m/s}^2)(300 \times 10^{-3} \text{ s}) + 2m_1(9.8 \text{ m/s}^2)(300 \times 10^{-3} \text{ s} - 100 \times 10^{-3} \text{ s})}{m_1 + 2m_1}$$
$$= 2.3 \text{ m/s}.$$

12. We use the constant-acceleration equations of Table 2-1 (with the origin at the traffic light), Eq. 9-5 for x_{com} and Eq. 9-17 for $\vec{v}_{com}$. At $t = 3.0$ s, the location of the automobile (of mass m_1) is

$$x_1 = \tfrac{1}{2} at^2 = \tfrac{1}{2}(4.0 \text{ m/s}^2)(3.0 \text{ s})^2 = 18 \text{ m},$$

while that of the truck (of mass m_2) is $x_2 = vt = (8.0 \text{ m/s})(3.0 \text{ s}) = 24$ m. The speed of the automobile then is $v_1 = at = (4.0 \text{ m/s}^2)(3.0 \text{ s}) = 12$ m/s, while the speed of the truck remains $v_2 = 8.0$ m/s.

(a) The location of their center of mass is

$$x_{com} = \frac{m_1 x_1 + m_2 x_2}{m_1 + m_2} = \frac{(1000 \text{ kg})(18 \text{ m}) + (2000 \text{ kg})(24 \text{ m})}{1000 \text{ kg} + 2000 \text{ kg}} = 22 \text{ m}.$$

(b) The speed of the center of mass is

$$v_{com} = \frac{m_1 v_1 + m_2 v_2}{m_1 + m_2} = \frac{(1000 \text{ kg})(12 \text{ m/s}) + (2000 \text{ kg})(8.0 \text{ m/s})}{1000 \text{ kg} + 2000 \text{ kg}} = 9.3 \text{ m/s}.$$

13. (a) The net force on the *system* (of total mass $m_1 + m_2$) is $m_2 g$. Thus, Newton's second law leads to $a = g(m_2/(m_1 + m_2)) = 0.4g$. For block1, this acceleration is to the right (the $\hat{i}$ direction), and for block 2 this is an acceleration downward (the $-\hat{j}$ direction). Therefore, Eq. 9-18 gives

$$\vec{a}_{com} = \frac{m_1 \vec{a}_1 + m_2 \vec{a}_2}{m_1 + m_2} = \frac{(0.6)(0.4g\hat{i}) + (0.4)(-0.4g\hat{j})}{0.6 + 0.4} = (2.35\,\hat{i} - 1.57\,\hat{j}) \text{ m/s}^2.$$

(b) Integrating Eq. 4-16, we obtain

$$\vec{v}_{com} = (2.35\,\hat{i} - 1.57\hat{j})\,t$$

(with SI units understood), since it started at rest. We note that the *ratio* of the y-component to the x-component (for the velocity vector) does not change with time, and it is that ratio which determines the angle of the velocity vector (by Eq. 3-6), and thus the direction of motion for the center of mass of the system.

(c) The last sentence of our answer for part (b) implies that the path of the center-of-mass is a straight line.

(d) Eq. 3-6 leads to $\theta = -34°$. The path of the center of mass is therefore straight, at downward angle 34°.

14. (a) The phrase (in the problem statement) "such that it [particle 2] always stays directly above particle 1 during the flight" means that the shadow (as if a light were directly above the particles shining down on them) of particle 2 coincides with the position of particle 1, at each moment. We say, in this case, that they are vertically aligned. Because of that alignment, $v_{2x} = v_1 = 10.0$ m/s. Because the initial value of v_2 is given as 20.0 m/s, then (using the Pythagorean theorem) we must have

$$v_{2y} = \sqrt{v_2^2 - v_{2x}^2} = \sqrt{300}\ \text{m/s}$$

for the initial value of the y component of particle 2's velocity. Eq. 2-16 (or conservation of energy) readily yields $y_{max} = 300/19.6 = 15.3$ m. Thus, we obtain

$$H_{max} = m_2 y_{max}/m_{total} = (3.00\ \text{g})(15.3\ \text{m})/(8.00\ \text{g}) = 5.74\ \text{m}.$$

(b) Since both particles have the same horizontal velocity, and particle 2's vertical component of velocity vanishes at that highest point, then the center of mass velocity then is simply $(10.0\ \text{m/s})\hat{i}$ (as one can verify using Eq. 9-17).

(c) Only particle 2 experiences any acceleration (the free fall acceleration downward), so Eq. 9-18 (or Eq. 9-19) leads to

$$a_{com} = m_2 g/m_{total} = (3.00\ \text{g})(9.8\ \text{m/s}^2)/(8.00\ \text{g}) = 3.68\ \text{m/s}^2$$

for the magnitude of the downward acceleration of the center of mass of this system. Thus, $\vec{a}_{com} = (-3.68\ \text{m/s}^2)\hat{j}$.

15. We need to find the coordinates of the point where the shell explodes and the velocity of the fragment that does not fall straight down. The coordinate origin is at the firing point, the +x axis is rightward, and the +y direction is upward. The y component of the velocity is given by $v = v_{0y} - gt$ and this is zero at time $t = v_{0y}/g = (v_0/g)\sin\theta_0$, where v_0

is the initial speed and θ_0 is the firing angle. The coordinates of the highest point on the trajectory are

$$x = v_{0x}t = v_0 t\cos\theta_0 = \frac{v_0^2}{g}\sin\theta_0\cos\theta_0 = \frac{(20\text{ m/s})^2}{9.8\text{ m/s}^2}\sin 60°\cos 60° = 17.7\text{ m}$$

and

$$y = v_{0y}t - \frac{1}{2}gt^2 = \frac{1}{2}\frac{v_0^2}{g}\sin^2\theta_0 = \frac{1}{2}\frac{(20\text{ m/s})^2}{9.8\text{ m/s}^2}\sin^2 60° = 15.3\text{ m}.$$

Since no horizontal forces act, the horizontal component of the momentum is conserved. Since one fragment has a velocity of zero after the explosion, the momentum of the other equals the momentum of the shell before the explosion. At the highest point the velocity of the shell is $v_0\cos\theta_0$, in the positive x direction. Let M be the mass of the shell and let V_0 be the velocity of the fragment. Then $Mv_0\cos\theta_0 = MV_0/2$, since the mass of the fragment is $M/2$. This means

$$V_0 = 2v_0\cos\theta_0 = 2(20\text{ m/s})\cos 60° = 20\text{ m/s}.$$

This information is used in the form of initial conditions for a projectile motion problem to determine where the fragment lands. Resetting our clock, we now analyze a projectile launched horizontally at time $t = 0$ with a speed of 20 m/s from a location having coordinates $x_0 = 17.7$ m, $y_0 = 15.3$ m. Its y coordinate is given by $y = y_0 - \frac{1}{2}gt^2$, and when it lands this is zero. The time of landing is $t = \sqrt{2y_0/g}$ and the x coordinate of the landing point is

$$x = x_0 + V_0 t = x_0 + V_0\sqrt{\frac{2y_0}{g}} = 17.7\text{ m} + (20\text{ m/s})\sqrt{\frac{2(15.3\text{ m})}{9.8\text{ m/s}^2}} = 53\text{ m}.$$

16. We denote the mass of Ricardo as M_R and that of Carmelita as M_C. Let the center of mass of the two-person system (assumed to be closer to Ricardo) be a distance x from the middle of the canoe of length L and mass m. Then

$$M_R(L/2 - x) = mx + M_C(L/2 + x).$$

Now, after they switch positions, the center of the canoe has moved a distance $2x$ from its initial position. Therefore, $x = 40$ cm$/2 = 0.20$ m, which we substitute into the above equation to solve for M_C:

$$M_C = \frac{M_R(L/2 - x) - mx}{L/2 + x} = \frac{(80)(\frac{3.0}{2} - 0.20) - (30)(0.20)}{(3.0/2) + 0.20} = 58\text{ kg}.$$

17. There is no net horizontal force on the dog-boat system, so their center of mass does not move. Therefore by Eq. 9-16, $M\Delta x_{com} = 0 = m_b\Delta x_b + m_d\Delta x_d$, which implies

$$|\Delta x_b| = \frac{m_d}{m_b}|\Delta x_d|.$$

Now we express the geometrical condition that *relative to the boat* the dog has moved a distance $d = 2.4$ m:

$$|\Delta x_b| + |\Delta x_d| = d$$

which accounts for the fact that the dog moves one way and the boat moves the other. We substitute for $|\Delta x_b|$ from above:

$$\frac{m_d}{m_b}|(\Delta x_d)| + |\Delta x_d| = d$$

which leads to $|\Delta x_d| = \dfrac{d}{1 + m_d/m_b} = \dfrac{2.4 \text{ m}}{1 + (4.5/18)} = 1.92$ m.

The dog is therefore 1.9 m closer to the shore than initially (where it was $D = 6.1$ m from it). Thus, it is now $D - |\Delta x_d| = 4.2$ m from the shore.

18. The magnitude of the ball's momentum change is

$$\Delta p = |mv_i - mv_f| = (0.70 \text{ kg})|5.0 \text{ m/s} - (-2.0 \text{ m/s})| = 4.9 \text{ kg} \cdot \text{m/s}.$$

19. (a) The change in kinetic energy is

$$\Delta K = \frac{1}{2}mv_f^2 - \frac{1}{2}mv_i^2 = \frac{1}{2}(2100 \text{ kg})\left((51 \text{ km/h})^2 - (41 \text{ km/h})^2\right)$$
$$= 9.66 \times 10^4 \text{ kg} \cdot (\text{km/h})^2 \left((10^3 \text{ m/km})(1 \text{ h}/3600 \text{ s})\right)^2$$
$$= 7.5 \times 10^4 \text{ J}.$$

(b) The magnitude of the change in velocity is

$$|\Delta \vec{v}| = \sqrt{(-v_i)^2 + (v_f)^2} = \sqrt{(-41 \text{ km/h})^2 + (51 \text{ km/h})^2} = 65.4 \text{ km/h}$$

so the magnitude of the change in momentum is

$$|\Delta \vec{p}| = m|\Delta \vec{v}| = (2100 \text{ kg})(65.4 \text{ km/h})\left(\frac{1000 \text{ m/km}}{3600 \text{ s/h}}\right) = 3.8 \times 10^4 \text{ kg} \cdot \text{m/s}.$$

(c) The vector $\Delta \vec{p}$ points at an angle θ south of east, where

$$\theta = \tan^{-1}\left(\frac{v_i}{v_f}\right) = \tan^{-1}\left(\frac{41 \text{ km/h}}{51 \text{ km/h}}\right) = 39°.$$

20. (a) Since the force of impact on the ball is in the y direction, p_x is conserved:

$$p_{xi} = mv_i \sin\theta_1 = p_{xf} = mv_i \sin\theta_2.$$

With $\theta_1 = 30.0°$, we find $\theta_2 = 30.0°$.

(b) The momentum change is

$$\Delta\vec{p} = mv_i \cos\theta_2 \left(-\hat{j}\right) - mv_i \cos\theta_2 \left(+\hat{j}\right) = -2(0.165 \text{ kg})(2.00 \text{ m/s})(\cos 30°)\hat{j}$$
$$= (-0.572 \text{ kg} \cdot \text{m/s})\hat{j}.$$

21. We use coordinates with $+x$ horizontally toward the pitcher and $+y$ upward. Angles are measured counterclockwise from the $+x$ axis. Mass, velocity and momentum units are SI. Thus, the initial momentum can be written $\vec{p}_0 = (4.5 \angle 215°)$ in magnitude-angle notation.

(a) In magnitude-angle notation, the momentum change is

$$(6.0 \angle -90°) - (4.5 \angle 215°) = (5.0 \angle -43°)$$

(efficiently done with a vector-capable calculator in polar mode). The magnitude of the momentum change is therefore 5.0 kg·m/s.

(b) The momentum change is $(6.0 \angle 0°) - (4.5 \angle 215°) = (10 \angle 15°)$. Thus, the magnitude of the momentum change is 10 kg·m/s.

22. We infer from the graph that the horizontal component of momentum p_x is 4.0 kg·m/s. Also, its initial magnitude of momentum p_0 is 6.0 kg·m/s. Thus,

$$\cos\theta_0 = \frac{p_x}{p_0} \Rightarrow \theta_0 = 48°.$$

23. The initial direction of motion is in the $+x$ direction. The magnitude of the average force F_{avg} is given by

$$F_{\text{avg}} = \frac{J}{\Delta t} = \frac{32.4 \text{ N} \cdot \text{s}}{2.70 \times 10^{-2} \text{ s}} = 1.20 \times 10^3 \text{ N}$$

The force is in the negative direction. Using the linear momentum-impulse theorem stated in Eq. 9-31, we have

$$-F_{\text{avg}}\Delta t = mv_f - mv_i.$$

where m is the mass, v_i the initial velocity, and v_f the final velocity of the ball. Thus,

$$v_f = \frac{mv_i - F_{avg}\Delta t}{m} = \frac{(0.40\,\text{kg})(14\,\text{m/s}) - (1200\,\text{N})(27\times 10^{-3}\,\text{s})}{0.40\,\text{kg}} = -67\,\text{m/s}.$$

(a) The final speed of the ball is $|v_f| = 67$ m/s.

(b) The negative sign indicates that the velocity is in the $-x$ direction, which is opposite to the initial direction of travel.

(c) From the above, the average magnitude of the force is $F_{avg} = 1.20\times 10^3$ N.

(d) The direction of the impulse on the ball is $-x$, same as the applied force.

24. (a) By energy conservation, the speed of the victim when he falls to the floor is

$$\frac{1}{2}mv^2 = mgh \;\Rightarrow\; v = \sqrt{2gh} = \sqrt{2(9.8\,\text{m/s}^2)(0.50\,\text{m})} = 3.1\,\text{m/s}.$$

Thus, the magnitude of the impulse is

$$J = |\Delta p| = m|\Delta v| = mv = (70\,\text{kg})(3.1\,\text{m/s}) \approx 2.2\times 10^2\,\text{N}\cdot\text{s}.$$

(b) With duration of $\Delta t = 0.082$ s for the collision, the average force is

$$F_{avg} = \frac{J}{\Delta t} = \frac{2.2\times 10^2\,\text{N}\cdot\text{s}}{0.082\,\text{s}} \approx 2.7\times 10^3\,\text{N}.$$

25. We estimate his mass in the neighborhood of 70 kg and compute the upward force F of the water from Newton's second law: $F - mg = ma$, where we have chosen $+y$ upward, so that $a > 0$ (the acceleration is upward since it represents a deceleration of his downward motion through the water). His speed when he arrives at the surface of the water is found either from Eq. 2-16 or from energy conservation: $v = \sqrt{2gh}$, where $h = 12$ m, and since the deceleration a reduces the speed to zero over a distance $d = 0.30$ m we also obtain $v = \sqrt{2ad}$. We use these observations in the following.

Equating our two expressions for v leads to $a = gh/d$. Our force equation, then, leads to

$$F = mg + m\left(g\frac{h}{d}\right) = mg\left(1 + \frac{h}{d}\right)$$

386   CHAPTER 9

which yields $F \approx 2.8 \times 10^4$ kg. Since we are not at all certain of his mass, we express this as a guessed-at range (in kN) $25 < F < 30$.

Since $F \gg mg$, the impulse $\vec{J}$ due to the net force (while he is in contact with the water) is overwhelmingly caused by the upward force of the water: $\int F\,dt = \vec{J}$ to a good approximation. Thus, by Eq. 9-29,

$$\int F\,dt = \vec{p}_f - \vec{p}_i = 0 - m\left(-\sqrt{2gh}\right)$$

(the minus sign with the initial velocity is due to the fact that downward is the negative direction) which yields $(70\text{ kg})\sqrt{2(9.8\text{ m/s}^2)(12\text{ m})} = 1.1 \times 10^3$ kg·m/s. Expressing this as a range we estimate

$$1.0 \times 10^3 \text{ kg·m/s} < \int F\,dt < 1.2 \times 10^3 \text{ kg·m/s}.$$

26. We choose $+y$ upward, which implies $a > 0$ (the acceleration is upward since it represents a deceleration of his downward motion through the snow).

(a) The maximum deceleration a_{max} of the paratrooper (of mass m and initial speed $v = 56$ m/s) is found from Newton's second law

$$F_{snow} - mg = ma_{max}$$

where we require $F_{snow} = 1.2 \times 10^5$ N. Using Eq. 2-15 $v^2 = 2a_{max}d$, we find the minimum depth of snow for the man to survive:

$$d = \frac{v^2}{2a_{max}} = \frac{mv^2}{2(F_{snow} - mg)} \approx \frac{(85\text{kg})(56\text{m/s})^2}{2(1.2 \times 10^5 \text{N})} = 1.1 \text{ m}.$$

(b) His short trip through the snow involves a change in momentum

$$\Delta\vec{p} = \vec{p}_f - \vec{p}_i = 0 - (85\text{kg})(-56\text{m/s}) = -4.8 \times 10^3 \text{ kg·m/s},$$

or $|\Delta\vec{p}| = 4.8 \times 10^3$ kg·m/s. The negative value of the initial velocity is due to the fact that downward is the negative direction. By the impulse-momentum theorem, this equals the impulse due to the net force $F_{snow} - mg$, but since $F_{snow} \gg mg$ we can approximate this as the impulse on him just from the snow.

27. We choose $+y$ upward, which means $\vec{v}_i = -25$ m/s and $\vec{v}_f = +10$ m/s. During the collision, we make the reasonable approximation that the net force on the ball is equal to F_{avg} – the average force exerted by the floor up on the ball.

(a) Using the impulse momentum theorem (Eq. 9-31) we find

$$\vec{J} = m\vec{v}_f - m\vec{v}_i = (1.2)(10) - (1.2)(-25) = 42 \text{ kg} \cdot \text{m/s}.$$

(b) From Eq. 9-35, we obtain

$$\vec{F}_{avg} = \frac{J}{\Delta t} = \frac{42}{0.020} = 2.1 \times 10^3 \text{ N}.$$

28. (a) The magnitude of the impulse is

$$J = |\Delta p| = m|\Delta v| = mv = (0.70 \text{ kg})(13 \text{ m/s}) \approx 9.1 \text{ kg} \cdot \text{m/s} = 9.1 \text{ N} \cdot \text{s}.$$

(b) With duration of $\Delta t = 5.0 \times 10^{-3}$ s for the collision, the average force is

$$F_{avg} = \frac{J}{\Delta t} = \frac{9.1 \text{ N} \cdot \text{s}}{5.0 \times 10^{-3} \text{ s}} \approx 1.8 \times 10^3 \text{ N}.$$

29. We choose the positive direction in the direction of rebound so that $\vec{v}_f > 0$ and $\vec{v}_i < 0$. Since they have the same speed v, we write this as $\vec{v}_f = v$ and $\vec{v}_i = -v$. Therefore, the change in momentum for each bullet of mass m is $\Delta \vec{p} = m\Delta v = 2mv$. Consequently, the total change in momentum for the 100 bullets (each minute) $\Delta \vec{P} = 100 \Delta \vec{p} = 200 mv$. The average force is then

$$\vec{F}_{avg} = \frac{\Delta \vec{P}}{\Delta t} = \frac{(200)(3 \times 10^{-3} \text{ kg})(500 \text{ m/s})}{(1 \text{ min})(60 \text{ s/min})} \approx 5 \text{ N}.$$

30. (a) By the impulse-momentum theorem (Eq. 9-31) the change in momentum must equal the "area" under the $F(t)$ curve. Using the facts that the area of a triangle is $\frac{1}{2}$ (base)(height), and that of a rectangle is (height)(width), we find the momentum at $t = 4$ s to be $(30 \text{ kg} \cdot \text{m/s})\hat{i}$.

(b) Similarly (but keeping in mind that areas beneath the axis are counted negatively) we find the momentum at $t = 7$ s is $(38 \text{ kg} \cdot \text{m/s})\hat{i}$.

(c) At $t = 9$ s, we obtain $\vec{p} = (6.0 \text{ m/s})\hat{i}$.

31. We use coordinates with $+x$ rightward and $+y$ upward, with the usual conventions for measuring the angles (so that the initial angle becomes $180 + 35 = 215°$). Using SI units and magnitude-angle notation (efficient to work with when using a vector-capable calculator), the change in momentum is

$$\vec{J} = \Delta \vec{p} = \vec{p}_f - \vec{p}_i = (3.00 \angle 90°) - (3.60 \angle 215°) = (5.86 \angle 59.8°).$$

(a) The magnitude of the impulse is $J = \Delta p = 5.86 \text{ kg} \cdot \text{m/s} = 5.86 \text{ N} \cdot \text{s}$.

(b) The direction of $\vec{J}$ is 59.8° measured counterclockwise from the +x axis.

(c) Eq. 9-35 leads to

$$J = F_{avg} \Delta t = 5.86 \text{ N} \cdot \text{s} \implies F_{avg} = \frac{5.86 \text{ N} \cdot \text{s}}{2.00 \times 10^{-3} \text{ s}} \approx 2.93 \times 10^3 \text{ N}.$$

We note that this force is very much larger than the weight of the ball, which justifies our (implicit) assumption that gravity played no significant role in the collision.

(d) The direction of $\vec{F}_{avg}$ is the same as $\vec{J}$, 59.8° measured counterclockwise from the +x axis.

32. (a) Choosing upward as the positive direction, the momentum change of the foot is

$$\Delta \vec{p} = 0 - m_{foot} \vec{v}_i = -(0.003 \text{ kg})(-1.50 \text{ m/s}) = 4.50 \times 10^{-3} \text{ N} \cdot \text{s}.$$

(b) Using Eq. 9-35 and now treating *downward* as the positive direction, we have

$$\vec{J} = \vec{F}_{avg} \Delta t = m_{lizard} g \, \Delta t = (0.090 \text{ kg})(9.80 \text{ m/s}^2)(0.60 \text{ s}) = 0.529 \text{ N} \cdot \text{s}.$$

(c) Push is what provides the primary support.

33. (a) By energy conservation, the speed of the passenger when the elevator hits the floor is

$$\frac{1}{2} m v^2 = mgh \implies v = \sqrt{2gh} = \sqrt{2(9.8 \text{ m/s}^2)(36 \text{ m})} = 26.6 \text{ m/s}.$$

Thus, the magnitude of the impulse is

$$J = |\Delta p| = m|\Delta v| = mv = (90 \text{ kg})(26.6 \text{ m/s}) \approx 2.39 \times 10^3 \text{ N} \cdot \text{s}.$$

(b) With duration of $\Delta t = 5.0 \times 10^{-3}$ s for the collision, the average force is

$$F_{avg} = \frac{J}{\Delta t} = \frac{2.39 \times 10^3 \text{ N} \cdot \text{s}}{5.0 \times 10^{-3} \text{ s}} \approx 4.78 \times 10^5 \text{ N}.$$

(c) If the passenger were to jump upward with a speed of $v' = 7.0$ m/s, then the resulting downward velocity would be

$$v'' = v - v' = 26.6 \text{ m/s} - 7.0 \text{ m/s} = 19.6 \text{ m/s},$$

and the magnitude of the impulse becomes

$$J'' = |\Delta p''| = m|\Delta v''| = mv'' = (90 \text{ kg})(19.6 \text{ m/s}) \approx 1.76 \times 10^3 \text{ N} \cdot \text{s}.$$

(d) The corresponding average force would be

$$F''_{avg} = \frac{J''}{\Delta t} = \frac{1.76 \times 10^3 \text{ N} \cdot \text{s}}{5.0 \times 10^{-3} \text{ s}} \approx 3.52 \times 10^5 \text{ N}.$$

34. (a) By Eq. 9-30, impulse can be determined from the "area" under the $F(t)$ curve. Keeping in mind that the area of a triangle is $\frac{1}{2}$(base)(height), we find the impulse in this case is 1.00 N·s.

(b) By definition (of the average of function, in the calculus sense) the average force must be the result of part (a) divided by the time (0.010 s). Thus, the average force is found to be 100 N.

(c) Consider ten hits. Thinking of ten hits as 10 $F(t)$ triangles, our total time interval is 10(0.050 s) = 0.50 s, and the total area is 10(1.0 N·s). We thus obtain an average force of 10/0.50 = 20.0 N. One could consider 15 hits, 17 hits, and so on, and still arrive at this same answer.

35. (a) We take the force to be in the positive direction, at least for earlier times. Then the impulse is

$$J = \int_0^{3.0 \times 10^{-3}} F\, dt = \int_0^{3.0 \times 10^{-3}} \left[(6.0 \times 10^6)t - (2.0 \times 10^9)t^2\right] dt$$

$$= \left[\frac{1}{2}(6.0 \times 10^6)t^2 - \frac{1}{3}(2.0 \times 10^9)t^3\right]\bigg|_0^{3.0 \times 10^{-3}}$$

$$= 9.0 \text{ N} \cdot \text{s}.$$

(b) Since $J = F_{avg} \Delta t$, we find

$$F_{avg} = \frac{J}{\Delta t} = \frac{9.0 \text{ N} \cdot \text{s}}{3.0 \times 10^{-3} \text{ s}} = 3.0 \times 10^3 \text{ N}.$$

(c) To find the time at which the maximum force occurs, we set the derivative of F with respect to time equal to zero – and solve for t. The result is $t = 1.5 \times 10^{-3}$ s. At that time the force is

$$F_{max} = (6.0\times10^6)(1.5\times10^{-3}) - (2.0\times10^9)(1.5\times10^{-3})^2 = 4.5\times10^3 \text{ N}.$$

(d) Since it starts from rest, the ball acquires momentum equal to the impulse from the kick. Let m be the mass of the ball and v its speed as it leaves the foot. Then,

$$v = \frac{p}{m} = \frac{J}{m} = \frac{9.0 \text{ N}\cdot\text{s}}{0.45 \text{ kg}} = 20 \text{ m/s}.$$

36. From Fig. 9-55, $+y$ corresponds to the direction of the rebound (directly away from the wall) and $+x$ towards the right. Using unit-vector notation, the ball's initial and final velocities are

$$\vec{v}_i = v\cos\theta\,\hat{i} - v\sin\theta\,\hat{j} = 5.2\,\hat{i} - 3.0\,\hat{j}$$
$$\vec{v}_f = v\cos\theta\,\hat{i} + v\sin\theta\,\hat{j} = 5.2\,\hat{i} + 3.0\,\hat{j}$$

respectively (with SI units understood).

(a) With $m = 0.30$ kg, the impulse-momentum theorem (Eq. 9-31) yields

$$\vec{J} = m\vec{v}_f - m\vec{v}_i = 2(0.30 \text{ kg})(3.0 \text{ m/s }\hat{j}) = (1.8 \text{ N}\cdot\text{s})\hat{j}$$

(b) Using Eq. 9-35, the force on the ball by the wall is $\vec{J}/\Delta t = (1.8/0.010)\hat{j} = (180\text{ N})\hat{j}$. By Newton's third law, the force on the wall by the ball is $(-180 \text{ N})\hat{j}$ (that is, its magnitude is 180 N and its direction is directly into the wall, or "down" in the view provided by Fig. 9-55).

37. We choose our positive direction in the direction of the rebound (so the ball's initial velocity is negative-valued). We evaluate the integral $J = \int F\,dt$ by adding the appropriate areas (of a triangle, a rectangle, and another triangle) shown in the graph (but with the t converted to seconds). With $m = 0.058$ kg and $v = 34$ m/s, we apply the impulse-momentum theorem:

$$\int F_{wall}\,dt = m\vec{v}_f - m\vec{v}_i \Rightarrow \int_0^{0.002} F\,dt + \int_{0.002}^{0.004} F\,dt + \int_{0.004}^{0.006} F\,dt = m(+v) - m(-v)$$

$$\Rightarrow \frac{1}{2}F_{max}(0.002\text{ s}) + F_{max}(0.002\text{ s}) + \frac{1}{2}F_{max}(0.002\text{ s}) = 2mv$$

which yields $F_{max}(0.004\text{ s}) = 2(0.058\text{ kg})(34\text{ m/s}) = 9.9\times10^2$ N.

38. (a) Performing the integral (from time a to time b) indicated in Eq. 9-30, we obtain

$$\int_a^b (12 - 3t^2)\,dt = 12(b-a) - (b^3 - a^3)$$

in SI units. If $b = 1.25$ s and $a = 0.50$ s, this gives 7.17 N·s.

(b) This integral (the impulse) relates to the change of momentum in Eq. 9-31. We note that the force is zero at $t = 2.00$ s. Evaluating the above expression for $a = 0$ and $b = 2.00$ gives an answer of 16.0 kg·m/s.

39. No external forces with horizontal components act on the man-stone system and the vertical forces sum to zero, so the total momentum of the system is conserved. Since the man and the stone are initially at rest, the total momentum is zero both before and after the stone is kicked. Let m_s be the mass of the stone and v_s be its velocity after it is kicked; let m_m be the mass of the man and v_m be his velocity after he kicks the stone. Then

$$m_s v_s + m_m v_m = 0 \rightarrow v_m = -m_s v_s / m_m.$$

We take the axis to be positive in the direction of motion of the stone. Then

$$v_m = -\frac{(0.068 \text{ kg})(4.0 \text{ m/s})}{91 \text{ kg}} = -3.0 \times 10^{-3} \text{ m/s},$$

or $|v_m| = 3.0 \times 10^{-3}$ m/s. The negative sign indicates that the man moves in the direction opposite to the direction of motion of the stone.

40. Our notation is as follows: the mass of the motor is M; the mass of the module is m; the initial speed of the system is v_0; the relative speed between the motor and the module is v_r; and, the speed of the module relative to the Earth is v after the separation. Conservation of linear momentum requires

$$(M + m)v_0 = mv + M(v - v_r).$$

Therefore,

$$v = v_0 + \frac{Mv_r}{M + m} = 4300 \text{ km/h} + \frac{(4m)(82 \text{ km/h})}{4m + m} = 4.4 \times 10^3 \text{ km/h}.$$

41. With $\vec{v}_0 = (9.5\,\hat{i} + 4.0\,\hat{j})$ m/s, the initial speed is

$$v_0 = \sqrt{v_{x0}^2 + v_{y0}^2} = \sqrt{(9.5 \text{ m/s})^2 + (4.0 \text{ m/s})^2} = 10.31 \text{ m/s}$$

and the takeoff angle of the athlete is

$$\theta_0 = \tan^{-1}\left(\frac{v_{y0}}{v_{x0}}\right) = \tan^{-1}\left(\frac{4.0}{9.5}\right) = 22.8°.$$

Using Equation 4-26, the range of the athlete without using halteres is

$$R_0 = \frac{v_0^2 \sin 2\theta_0}{g} = \frac{(10.31 \text{ m/s})^2 \sin 2(22.8°)}{9.8 \text{ m/s}^2} = 7.75 \text{ m}.$$

On the other hand, if two halteres of mass $m = 5.50$ kg were thrown at the maximum height, then, by momentum conservation, the subsequent speed of the athlete would be

$$(M + 2m)v_{x0} = Mv'_x \Rightarrow v'_x = \frac{M + 2m}{M} v_{x0}$$

Thus, the change in the x-component of the velocity is

$$\Delta v_x = v'_x - v_{x0} = \frac{M + 2m}{M} v_{x0} - v_{x0} = \frac{2m}{M} v_{x0} = \frac{2(5.5 \text{ kg})}{78 \text{ kg}} (9.5 \text{ m/s}) = 1.34 \text{ m/s}.$$

The maximum height is attained when $v_y = v_{y0} - gt = 0$, or

$$t = \frac{v_{y0}}{g} = \frac{4.0 \text{ m/s}}{9.8 \text{ m/s}^2} = 0.41 \text{ s}.$$

Therefore, the increase in range with use of halteres is

$$\Delta R = (\Delta v'_x)t = (1.34 \text{ m/s})(0.41 \text{ s}) = 0.55 \text{ m}.$$

42. Our $+x$ direction is east and $+y$ direction is north. The linear momenta for the two $m = 2.0$ kg parts are then

$$\vec{p}_1 = m\vec{v}_1 = mv_1 \hat{j}$$

where $v_1 = 3.0$ m/s, and

$$\vec{p}_2 = m\vec{v}_2 = m(v_{2x} \hat{i} + v_{2y} \hat{j}) = mv_2(\cos\theta \hat{i} + \sin\theta \hat{j})$$

where $v_2 = 5.0$ m/s and $\theta = 30°$. The combined linear momentum of both parts is then

$$\vec{P} = \vec{p}_1 + \vec{p}_2 = mv_1 \hat{j} + mv_2 (\cos\theta \hat{i} + \sin\theta \hat{j}) = (mv_2 \cos\theta) \hat{i} + (mv_1 + mv_2 \sin\theta) \hat{j}$$

$$= (2.0 \text{ kg})(5.0 \text{ m/s})(\cos 30°) \hat{i} + (2.0 \text{ kg})(3.0 \text{ m/s} + (5.0 \text{ m/s})(\sin 30°)) \hat{j}$$

$$= (8.66 \hat{i} + 11 \hat{j}) \text{ kg·m/s}.$$

From conservation of linear momentum we know that this is also the linear momentum of the whole kit before it splits. Thus the speed of the 4.0-kg kit is

$$v = \frac{P}{M} = \frac{\sqrt{P_x^2 + P_y^2}}{M} = \frac{\sqrt{(8.66 \text{ kg} \cdot \text{m/s})^2 + (11 \text{ kg} \cdot \text{m/s})^2}}{4.0 \text{ kg}} = 3.5 \text{ m/s}.$$

43. (a) With SI units understood, the velocity of block L (in the frame of reference indicated in the figure that goes with the problem) is $(v_1 - 3)\hat{i}$. Thus, momentum conservation (for the explosion at $t = 0$) gives

$$m_L(v_1 - 3) + (m_C + m_R)v_1 = 0$$

which leads to

$$v_1 = \frac{3 m_L}{m_L + m_C + m_R} = \frac{3(2 \text{ kg})}{10 \text{ kg}} = 0.60 \text{ m/s}.$$

Next, at $t = 0.80$ s, momentum conservation (for the second explosion) gives

$$m_C v_2 + m_R(v_2 + 3) = (m_C + m_R)v_1 = (8 \text{ kg})(0.60 \text{ m/s}) = 4.8 \text{ kg·m/s}.$$

This yields $v_2 = -0.15$. Thus, the velocity of block C after the second explosion is

$$v_2 = -(0.15 \text{ m/s})\hat{i}.$$

(b) Between $t = 0$ and $t = 0.80$ s, the block moves $v_1 \Delta t = (0.60 \text{ m/s})(0.80 \text{ s}) = 0.48$ m. Between $t = 0.80$ s and $t = 2.80$ s, it moves an additional

$$v_2 \Delta t = (-0.15 \text{ m/s})(2.00 \text{ s}) = -0.30 \text{ m}.$$

Its net displacement since $t = 0$ is therefore $0.48 \text{ m} - 0.30 \text{ m} = 0.18 \text{ m}$.

44. Our notation (and, implicitly, our choice of coordinate system) is as follows: the mass of the original body is m; its initial velocity is $\vec{v}_0 = v\hat{i}$; the mass of the less massive piece is m_1; its velocity is $\vec{v}_1 = 0$; and, the mass of the more massive piece is m_2. We note that the conditions $m_2 = 3m_1$ (specified in the problem) and $m_1 + m_2 = m$ generally assumed in classical physics (before Einstein) lead us to conclude

$$m_1 = \frac{1}{4}m \text{ and } m_2 = \frac{3}{4}m.$$

Conservation of linear momentum requires

$$m\vec{v}_0 = m_1 \vec{v}_1 + m_2 \vec{v}_2 \Rightarrow mv\hat{i} = 0 + \frac{3}{4}m\vec{v}_2$$

which leads to $\vec{v}_2 = \frac{4}{3}v\hat{i}$. The increase in the system's kinetic energy is therefore

$$\Delta K = \frac{1}{2}m_1v_1^2 + \frac{1}{2}m_2v_2^2 - \frac{1}{2}mv_0^2 = 0 + \frac{1}{2}\left(\frac{3}{4}m\right)\left(\frac{4}{3}v\right)^2 - \frac{1}{2}mv^2 = \frac{1}{6}mv^2.$$

45. Our notation (and, implicitly, our choice of coordinate system) is as follows: the mass of one piece is $m_1 = m$; its velocity is $\vec{v}_1 = (-30 \text{ m/s})\hat{i}$; the mass of the second piece is $m_2 = m$; its velocity is $\vec{v}_2 = (-30 \text{ m/s})\hat{j}$; and, the mass of the third piece is $m_3 = 3m$.

(a) Conservation of linear momentum requires

$$m\vec{v}_0 = m_1\vec{v}_1 + m_2\vec{v}_2 + m_3\vec{v}_3 \Rightarrow 0 = m(-30\hat{i}) + m(-30\hat{j}) + 3m\vec{v}_3$$

which leads to $\vec{v}_3 = (10\hat{i} + 10\hat{j})$ m/s. Its magnitude is $v_3 = 10\sqrt{2} \approx 14$ m/s.

(b) The direction is 45° *counterclockwise* from +x (in this system where we have m_1 flying off in the –x direction and m_2 flying off in the –y direction).

46. We can think of the sliding-until-stopping as an example of kinetic energy converting into thermal energy (see Eq. 8-29 and Eq. 6-2, with $F_N = mg$). This leads to $v^2 = 2\mu gd$ being true separately for each piece. Thus we can set up a ratio:

$$\left(\frac{v_L}{v_R}\right)^2 = \frac{2\mu_L g d_L}{2\mu_R g d_R} = \frac{12}{25}.$$

But (by the conservation of momentum) the ratio of speeds must be inversely proportional to the ratio of masses (since the initial momentum – before the explosion – was zero). Consequently,

$$\left(\frac{m_R}{m_L}\right)^2 = \frac{12}{25} \Rightarrow m_R = \frac{2}{5}\sqrt{3}\ m_L = 1.39 \text{ kg}.$$

Therefore, the total mass is $m_R + m_L \approx 3.4$ kg.

47. Our notation is as follows: the mass of the original body is $M = 20.0$ kg; its initial velocity is $\vec{v}_0 = (200 \text{ m/s})\hat{i}$; the mass of one fragment is $m_1 = 10.0$ kg; its velocity is $\vec{v}_1 = (-100 \text{ m/s})\hat{j}$; the mass of the second fragment is $m_2 = 4.0$ kg; its velocity is $\vec{v}_2 = (-500 \text{ m/s})\hat{i}$; and, the mass of the third fragment is $m_3 = 6.00$ kg.

(a) Conservation of linear momentum requires $M\vec{v}_0 = m_1\vec{v}_1 + m_2\vec{v}_2 + m_3\vec{v}_3$, which (using the above information) leads to

$$\vec{v}_3 = (1.00\times10^3\,\hat{i} - 0.167\times10^3\,\hat{j})\text{ m/s}.$$

The magnitude of $\vec{v}_3$ is $v_3 = \sqrt{(1000\text{ m/s})^2 + (-167\text{ m/s})^2} = 1.01\times10^3$ m/s. It points at $\theta = \tan^{-1}(-167/1000) = -9.48°$ (that is, at 9.5° measured clockwise from the $+x$ axis).

(b) We are asked to calculate ΔK or

$$\left(\frac{1}{2}m_1v_1^2 + \frac{1}{2}m_2v_2^2 + \frac{1}{2}m_3v_3^2\right) - \frac{1}{2}Mv_0^2 = 3.23\times10^6 \text{ J}.$$

48. This problem involves both mechanical energy conservation $U_i = K_1 + K_2$, where $U_i = 60$ J, and momentum conservation

$$0 = m_1\vec{v}_1 + m_2\vec{v}_2$$

where $m_2 = 2m_1$. From the second equation, we find $|\vec{v}_1| = 2|\vec{v}_2|$ which in turn implies (since $v_1 = |\vec{v}_1|$ and likewise for v_2)

$$K_1 = \frac{1}{2}m_1v_1^2 = \frac{1}{2}\left(\frac{1}{2}m_2\right)(2v_2)^2 = 2\left(\frac{1}{2}m_2v_2^2\right) = 2K_2.$$

(a) We substitute $K_1 = 2K_2$ into the energy conservation relation and find

$$U_i = 2K_2 + K_2 \Rightarrow K_2 = \frac{1}{3}U_i = 20 \text{ J}.$$

(b) And we obtain $K_1 = 2(20) = 40$ J.

49. We refer to the discussion in the textbook (see Sample Problem 9-8, which uses the same notation that we use here) for many of the important details in the reasoning. Here we only present the primary computational step (using SI units):

$$v = \frac{m+M}{m}\sqrt{2gh} = \frac{2.010}{0.010}\sqrt{2(9.8)(0.12)} = 3.1\times10^2 \text{ m/s}.$$

50. (a) We choose $+x$ along the initial direction of motion and apply momentum conservation:

$$m_{\text{bullet}}\vec{v}_i = m_{\text{bullet}}\vec{v}_1 + m_{\text{block}}\vec{v}_2$$

$$(5.2 \text{ g})(672 \text{ m/s}) = (5.2 \text{ g})(428 \text{ m/s}) + (700 \text{ g})\vec{v}_2$$

which yields $v_2 = 1.81$ m/s.

(b) It is a consequence of momentum conservation that the velocity of the center of mass is unchanged by the collision. We choose to evaluate it before the collision:

$$\vec{v}_{com} = \frac{m_{bullet}\vec{v}_i}{m_{bullet}+m_{block}} = \frac{(5.2 \text{ g})(672 \text{ m/s})}{5.2 \text{ g}+700 \text{ g}} = 4.96 \text{ m/s}.$$

51. With an initial speed of v_i, the initial kinetic energy of the car is $K_i = m_c v_i^2/2$. After a totally inelastic collision with a moose of mass m_m, by momentum conservation, the speed of the combined system is

$$m_c v_i = (m_c + m_m)v_f \Rightarrow v_f = \frac{m_c v_i}{m_c + m_m},$$

with final kinetic energy

$$K_f = \frac{1}{2}(m_c + m_m)v_f^2 = \frac{1}{2}(m_c + m_m)\left(\frac{m_c v_i}{m_c + m_m}\right)^2 = \frac{1}{2}\frac{m_c^2}{m_c + m_m}v_i^2.$$

(a) The percentage loss of kinetic energy due to collision is

$$\frac{\Delta K}{K_i} = \frac{K_i - K_f}{K_i} = 1 - \frac{K_f}{K_i} = 1 - \frac{m_c}{m_c + m_m} = \frac{m_m}{m_c + m_m} = \frac{500 \text{ kg}}{1000 \text{ kg}+500 \text{ kg}} = \frac{1}{3} = 33.3\%.$$

(b) If the collision were with a camel of mass $m_{camel} = 300$ kg, then the percentage loss of kinetic energy would be

$$\frac{\Delta K}{K_i} = \frac{m_{camel}}{m_c + m_{camel}} = \frac{300 \text{ kg}}{1000 \text{ kg}+300 \text{ kg}} = \frac{3}{13} = 23\%.$$

(c) As the animal mass decreases, the percentage loss of kinetic energy also decreases.

52. (a) The magnitude of the deceleration of each of the cars is $a = f/m = \mu_k mg/m = \mu_k g$. If a car stops in distance d, then its speed v just after impact is obtained from Eq. 2-16:

$$v^2 = v_0^2 + 2ad \Rightarrow v = \sqrt{2ad} = \sqrt{2\mu_k gd}$$

since $v_0 = 0$ (this could alternatively have been derived using Eq. 8-31). Thus,

$$v_A = \sqrt{2\mu_k gd_A} = \sqrt{2(0.13)(9.8 \text{ m/s}^2)(8.2 \text{ m})} = 4.6 \text{ m/s}.$$

(b) Similarly, $v_B = \sqrt{2\mu_k gd_B} = \sqrt{2(0.13)(9.8 \text{ m/s}^2)(6.1 \text{ m})} = 3.9$ m/s.

(c) Let the speed of car B be v just before the impact. Conservation of linear momentum gives $m_B v = m_A v_A + m_B v_B$, or

$$v = \frac{(m_A v_A + m_B v_B)}{m_B} = \frac{(1100)(4.6)+(1400)(3.9)}{1400} = 7.5 \text{ m/s}.$$

(d) The conservation of linear momentum during the impact depends on the fact that the only significant force (during impact of duration Δt) is the force of contact between the bodies. In this case, that implies that the force of friction exerted by the road on the cars is neglected during the brief Δt. This neglect would introduce some error in the analysis. Related to this is the assumption we are making that the transfer of momentum occurs at one location – that the cars do not slide appreciably during Δt – which is certainly an approximation (though probably a good one). Another source of error is the application of the friction relation Eq. 6-2 for the sliding portion of the problem (after the impact); friction is a complex force that Eq. 6-2 only partially describes.

53. In solving this problem, our $+x$ direction is to the right (so all velocities are positive-valued).

(a) We apply momentum conservation to relate the situation just before the bullet strikes the second block to the situation where the bullet is embedded within the block.

$$(0.0035 \text{ kg})v = (1.8035 \text{ kg})(1.4 \text{ m/s}) \Rightarrow v = 721 \text{ m/s}.$$

(b) We apply momentum conservation to relate the situation just before the bullet strikes the first block to the instant it has passed through it (having speed v found in part (a)).

$$(0.0035 \text{ kg})v_0 = (1.20 \text{ kg})(0.630 \text{ m/s}) + (0.00350 \text{ kg})(721 \text{ m/s})$$

which yields $v_0 = 937$ m/s.

54. We think of this as having two parts: the first is the collision itself – where the bullet passes through the block so quickly that the block has not had time to move through any distance yet – and then the subsequent "leap" of the block into the air (up to height h measured from its initial position). The first part involves momentum conservation (with $+y$ upward):

$$(0.01 \text{ kg})(1000 \text{ m/s}) = (5.0 \text{ kg})\bar{v} + (0.01 \text{ kg})(400 \text{ m/s})$$

which yields $\bar{v} = 1.2$ m/s. The second part involves either the free-fall equations from Ch. 2 (since we are ignoring air friction) or simple energy conservation from Ch. 8. Choosing the latter approach, we have

$$\frac{1}{2}(5.0\,\text{kg})(1.2\,\text{m/s})^2 = (5.0\,\text{kg})(9.8\,\text{m/s}^2)h$$

which gives the result $h = 0.073$ m.

55. (a) Let v be the final velocity of the ball-gun system. Since the total momentum of the system is conserved $mv_i = (m + M)v$. Therefore,

$$v = \frac{mv_i}{m+M} = \frac{(60\,\text{g})(22\,\text{m/s})}{60\,\text{g} + 240\,\text{g}} = 4.4\,\text{m/s}.$$

(b) The initial kinetic energy is $K_i = \frac{1}{2}mv_i^2$ and the final kinetic energy is $K_f = \frac{1}{2}(m+M)v^2 = \frac{1}{2}m^2v_i^2/(m+M)$. The problem indicates $\Delta E_{\text{th}} = 0$, so the difference $K_i - K_f$ must equal the energy U_s stored in the spring:

$$U_s = \frac{1}{2}mv_i^2 - \frac{1}{2}\frac{m^2v_i^2}{(m+M)} = \frac{1}{2}mv_i^2\left(1 - \frac{m}{m+M}\right) = \frac{1}{2}mv_i^2\frac{M}{m+M}.$$

Consequently, the fraction of the initial kinetic energy that becomes stored in the spring is

$$\frac{U_s}{K_i} = \frac{M}{m+M} = \frac{240}{60+240} = 0.80.$$

56. The total momentum immediately before the collision (with $+x$ upward) is

$$p_i = (3.0\,\text{kg})(20\,\text{m/s}) + (2.0\,\text{kg})(-12\,\text{m/s}) = 36\,\text{kg·m/s}.$$

Their momentum immediately after, when they constitute a combined mass of $M = 5.0$ kg, is $p_f = (5.0\,\text{kg})\vec{v}$. By conservation of momentum, then, we obtain $\vec{v} = 7.2$ m/s, which becomes their "initial" velocity for their subsequent free-fall motion. We can use Ch. 2 methods or energy methods to analyze this subsequent motion; we choose the latter. The level of their collision provides the reference ($y = 0$) position for the gravitational potential energy, and we obtain

$$K_0 + U_0 = K + U \Rightarrow \frac{1}{2}Mv_0^2 + 0 = 0 + Mgy_{\text{max}}.$$

Thus, with $v_0 = 7.2$ m/s, we find $y_{\text{max}} = 2.6$ m.

57. We choose $+x$ in the direction of (initial) motion of the blocks, which have masses $m_1 = 5$ kg and $m_2 = 10$ kg. Where units are not shown in the following, SI units are to be understood.

(a) Momentum conservation leads to

$$m_1\vec{v}_{1i} + m_2\vec{v}_{2i} = m_1\vec{v}_{1f} + m_2\vec{v}_{2f}$$
$$(5\text{ kg})(3.0\text{ m/s}) + (10\text{ kg})(2.0\text{ m/s}) = (5\text{ kg})\vec{v}_{1f} + (10\text{ kg})(2.5\text{ m/s})$$

which yields $\vec{v}_{1f} = 2.0$ m/s. Thus, the speed of the 5.0 kg block immediately after the collision is 2.0 m/s.

(b) We find the reduction in total kinetic energy:

$$K_i - K_f = \frac{1}{2}(5\text{ kg})(3\text{ m/s})^2 + \frac{1}{2}(10\text{ kg})(2\text{ m/s})^2 - \frac{1}{2}(5\text{ kg})(2\text{ m/s})^2 - \frac{1}{2}(10\text{ kg})(2.5\text{ m/s})^2$$
$$= -1.25\text{ J} \approx -1.3\text{ J}.$$

(c) In this new scenario where $\vec{v}_{2f} = 4.0$ m/s, momentum conservation leads to $\vec{v}_{1f} = -1.0$ m/s and we obtain $\Delta K = +40$ J.

(d) The creation of additional kinetic energy is possible if, say, some gunpowder were on the surface where the impact occurred (initially stored chemical energy would then be contributing to the result).

58. We think of this as having two parts: the first is the collision itself – where the blocks "join" so quickly that the 1.0-kg block has not had time to move through any distance yet – and then the subsequent motion of the 3.0 kg system as it compresses the spring to the maximum amount x_m. The first part involves momentum conservation (with $+x$ rightward):

$$m_1 v_1 = (m_1 + m_2)v \quad \Rightarrow \quad (2.0\text{ kg})(4.0\text{ m/s}) = (3.0\text{ kg})\vec{v}$$

which yields $\vec{v} = 2.7$ m/s. The second part involves mechanical energy conservation:

$$\frac{1}{2}(3.0\text{ kg})(2.7\text{ m/s})^2 = \frac{1}{2}(200\text{ N/m})x_m^2$$

which gives the result $x_m = 0.33$ m.

59. As hinted in the problem statement, the velocity v of the system as a whole – when the spring reaches the maximum compression x_m – satisfies

$$m_1 v_{1i} + m_2 v_{2i} = (m_1 + m_2)v.$$

The change in kinetic energy of the system is therefore

$$\Delta K = \frac{1}{2}(m_1+m_2)v^2 - \frac{1}{2}m_1 v_{1i}^2 - \frac{1}{2}m_2 v_{2i}^2 = \frac{(m_1 v_{1i}+m_2 v_{2i})^2}{2(m_1+m_2)} - \frac{1}{2}m_1 v_{1i}^2 - \frac{1}{2}m_2 v_{2i}^2$$

which yields $\Delta K = -35$ J. (Although it is not necessary to do so, still it is worth noting that algebraic manipulation of the above expression leads to $|\Delta K| = \frac{1}{2}\left(\frac{m_1 m_2}{m_1+m_2}\right) v_{rel}^2$ where $v_{rel} = v_1 - v_2$). Conservation of energy then requires

$$\frac{1}{2}k x_m^2 = -\Delta K \Rightarrow x_m = \sqrt{\frac{-2\Delta K}{k}} = \sqrt{\frac{-2(-35\text{ J})}{1120\text{ N/m}}} = 0.25 \text{ m}.$$

60. (a) Let m_1 be the mass of one sphere, v_{1i} be its velocity before the collision, and v_{1f} be its velocity after the collision. Let m_2 be the mass of the other sphere, v_{2i} be its velocity before the collision, and v_{2f} be its velocity after the collision. Then, according to Eq. 9-75,

$$v_{1f} = \frac{m_1 - m_2}{m_1 + m_2} v_{1i} + \frac{2 m_2}{m_1 + m_2} v_{2i}.$$

Suppose sphere 1 is originally traveling in the positive direction and is at rest after the collision. Sphere 2 is originally traveling in the negative direction. Replace v_{1i} with v, v_{2i} with $-v$, and v_{1f} with zero to obtain $0 = m_1 - 3m_2$. Thus,

$$m_2 = m_1/3 = (300 \text{ g})/3 = 100 \text{ g}.$$

(b) We use the velocities before the collision to compute the velocity of the center of mass:

$$v_{com} = \frac{m_1 v_{1i} + m_2 v_{2i}}{m_1 + m_2} = \frac{(300 \text{ g})(2.00 \text{ m/s}) + (100 \text{ g})(-2.00 \text{ m/s})}{300 \text{ g} + 100 \text{ g}} = 1.00 \text{ m/s}.$$

61. (a) Let m_1 be the mass of the cart that is originally moving, v_{1i} be its velocity before the collision, and v_{1f} be its velocity after the collision. Let m_2 be the mass of the cart that is originally at rest and v_{2f} be its velocity after the collision. Then, according to Eq. 9-67,

$$v_{1f} = \frac{m_1 - m_2}{m_1 + m_2} v_{1i}.$$

Using SI units (so $m_1 = 0.34$ kg), we obtain

$$m_2 = \frac{v_{1i} - v_{1f}}{v_{1i} + v_{1f}} m_1 = \left(\frac{1.2 \text{ m/s} - 0.66 \text{ m/s}}{1.2 \text{ m/s} + 0.66 \text{ m/s}}\right)(0.34 \text{ kg}) = 0.099 \text{ kg}.$$

(b) The velocity of the second cart is given by Eq. 9-68:

$$v_{2f} = \frac{2m_1}{m_1+m_2}v_{1i} = \left(\frac{2(0.34 \text{ kg})}{0.34 \text{ kg}+0.099 \text{ kg}}\right)(1.2 \text{ m/s}) = 1.9 \text{ m/s}.$$

(c) The speed of the center of mass is

$$v_{com} = \frac{m_1v_{1i}+m_2v_{2i}}{m_1+m_2} = \frac{(0.34)(1.2)+0}{0.34+0.099} = 0.93 \text{ m/s}.$$

Values for the initial velocities were used but the same result is obtained if values for the final velocities are used.

62. (a) Let m_A be the mass of the block on the left, v_{Ai} be its initial velocity, and v_{Af} be its final velocity. Let m_B be the mass of the block on the right, v_{Bi} be its initial velocity, and v_{Bf} be its final velocity. The momentum of the two-block system is conserved, so

$$m_A v_{Ai} + m_B v_{Bi} = m_A v_{Af} + m_B v_{Bf}$$

and

$$v_{Af} = \frac{m_A v_{Ai} + m_B v_{Bi} - m_B v_{Bf}}{m_A} = \frac{(1.6 \text{ kg})(5.5 \text{ m/s})+(2.4 \text{ kg})(2.5 \text{ m/s})-(2.4 \text{ kg})(4.9 \text{ m/s})}{1.6 \text{ kg}}$$
$$= 1.9 \text{ m/s}.$$

(b) The block continues going to the right after the collision.

(c) To see if the collision is elastic, we compare the total kinetic energy before the collision with the total kinetic energy after the collision. The total kinetic energy before is

$$K_i = \frac{1}{2}m_A v_{Ai}^2 + \frac{1}{2}m_B v_{Bi}^2 = \frac{1}{2}(1.6 \text{ kg})(5.5 \text{ m/s})^2 + \frac{1}{2}(2.4 \text{ kg})(2.5 \text{ m/s})^2 = 31.7 \text{ J}.$$

The total kinetic energy after is

$$K_f = \frac{1}{2}m_A v_{Af}^2 + \frac{1}{2}m_B v_{Bf}^2 = \frac{1}{2}(1.6 \text{ kg})(1.9 \text{ m/s})^2 + \frac{1}{2}(2.4 \text{ kg})(4.9 \text{ m/s})^2 = 31.7 \text{ J}.$$

Since $K_i = K_f$ the collision is found to be elastic.

63. (a) Let m_1 be the mass of the body that is originally moving, v_{1i} be its velocity before the collision, and v_{1f} be its velocity after the collision. Let m_2 be the mass of the body that is originally at rest and v_{2f} be its velocity after the collision. Then, according to Eq. 9-67,

$$v_{1f} = \frac{m_1-m_2}{m_1+m_2}v_{1i}.$$

We solve for m_2 to obtain

$$m_2 = \frac{v_{1i} - v_{1f}}{v_{1f} + v_{1i}} m_1 .$$

We combine this with $v_{1f} = v_{1i}/4$ to obtain $m_2 = 3m_1/5 = 3(2.0 \text{ kg})/5 = 1.2 \text{ kg}$.

(b) The speed of the center of mass is

$$v_{com} = \frac{m_1 v_{1i} + m_2 v_{2i}}{m_1 + m_2} = \frac{(2.0 \text{ kg})(4.0 \text{ m/s})}{2.0 \text{ kg} + 1.2 \text{ kg}} = 2.5 \text{ m/s} .$$

64. This is a completely inelastic collision, but Eq. 9-53 ($V = \frac{m_1}{m_1 + m_2} v_{1i}$) is not easily applied since that equation is designed for use when the struck particle is initially stationary. To deal with this case (where particle 2 is already in motion), we return to the principle of momentum conservation:

$$m_1 \vec{v}_1 + m_2 \vec{v}_2 = (m_1 + m_2) \vec{V} \quad \Rightarrow \quad \vec{V} = \frac{2(4\hat{i} - 5\hat{j}) + 4(6\hat{i} - 2\hat{j})}{2 + 4} .$$

(a) In unit-vector notation, then,

$$\vec{V} = (2.67 \text{ m/s})\hat{i} + (-3.00 \text{ m/s})\hat{j} .$$

(b) The magnitude of $\vec{V}$ is $|\vec{V}| = 4.01$ m/s

(c) The direction of $\vec{V}$ is 48.4° (measured *clockwise* from the +x axis).

65. We use Eq 9-67 and 9-68 to find the velocities of the particles after their first collision (at $x = 0$ and $t = 0$):

$$v_{1f} = \frac{m_1 - m_2}{m_1 + m_2} v_{1i} = \frac{-0.1 \text{ kg}}{0.7 \text{ kg}} (2.0 \text{ m/s}) = \frac{-2}{7} \text{ m/s}$$

$$v_{2f} = \frac{2m_1}{m_1 + m_2} v_{1i} = \frac{0.6 \text{ kg}}{0.7 \text{ kg}} (2.0 \text{ m/s}) = \frac{12}{7} \text{ m/s} \approx 1.7 \text{ m/s}.$$

At a rate of motion of 1.7 m/s, $2x_w = 140$ cm (the distance to the wall and back to $x = 0$) will be traversed by particle 2 in 0.82 s. At $t = 0.82$ s, particle 1 is located at

$$x = (-2/7)(0.82) = -23 \text{ cm},$$

and particle 2 is "gaining" at a rate of (10/7) m/s leftward; this is their relative velocity at that time. Thus, this "gap" of 23 cm between them will be closed after an additional time of (0.23 m)/(10/7 m/s) = 0.16 s has passed. At this time ($t = 0.82 + 0.16 = 0.98$ s) the two particles are at $x = (-2/7)(0.98) = -28$ cm.

66. First, we find the speed v of the ball of mass m_1 right before the collision (just as it reaches its lowest point of swing). Mechanical energy conservation (with $h = 0.700$ m) leads to

$$m_1 gh = \frac{1}{2} m_1 v^2 \Rightarrow v = \sqrt{2gh} = 3.7 \text{ m/s}.$$

(a) We now treat the elastic collision using Eq. 9-67:

$$v_{1f} = \frac{m_1 - m_2}{m_1 + m_2} v = \frac{0.5 \text{ kg} - 2.5 \text{ kg}}{0.5 \text{ kg} + 2.5 \text{ kg}} (3.7 \text{ m/s}) = -2.47 \text{ m/s}$$

which means the final speed of the ball is 2.47 m/s.

(b) Finally, we use Eq. 9-68 to find the final speed of the block:

$$v_{2f} = \frac{2m_1}{m_1 + m_2} v = \frac{2(0.5 \text{ kg})}{0.5 \text{ kg} + 2.5 \text{ kg}} (3.7 \text{ m/s}) = 1.23 \text{ m/s}.$$

67. (a) The center of mass velocity does not change in the absence of external forces. In this collision, only forces of one block on the other (both being part of the same system) are exerted, so the center of mass velocity is 3.00 m/s before and after the collision.

(b) We can find the velocity v_{1i} of block 1 before the collision (when the velocity of block 2 is known to be zero) using Eq. 9-17:

$$(m_1 + m_2) v_{com} = m_1 v_{1i} + 0 \Rightarrow v_{1i} = 12.0 \text{ m/s}.$$

Now we use Eq. 9-68 to find v_{2f}:

$$v_{2f} = \frac{2m_1}{m_1 + m_2} v_{1i} = 6.00 \text{ m/s}.$$

68. (a) If the collision is perfectly elastic, then Eq. 9-68 applies

$$v_2 = \frac{2m_1}{m_1 + m_2} v_{1i} = \frac{2m_1}{m_1 + (2.00)m_1} \sqrt{2gh} = \frac{2}{3}\sqrt{2gh}$$

where we have used the fact (found most easily from energy conservation) that the speed of block 1 at the bottom of the frictionless ramp is $\sqrt{2gh}$ (where $h = 2.50$ m). Next, for block 2's "rough slide" we use Eq. 8-37:

$$\tfrac{1}{2}m_2 v_2^2 = \Delta E_{\text{th}} = f_k d = \mu_k m_2 g d.$$

where $\mu_k = 0.500$. Solving for the sliding distance d, we find that m_2 cancels out and we obtain $d = 2.22$ m.

(b) In a completely inelastic collision, we apply Eq. 9-53: $v_2 = \dfrac{m_1}{m_1 + m_2} v_{1i}$ (where, as above, $v_{1i} = \sqrt{2gh}$). Thus, in this case we have $v_2 = \sqrt{2gh}/3$. Now, Eq. 8-37 (using the total mass since the blocks are now joined together) leads to a sliding distance of $d = 0.556$ m (one-fourth of the part (a) answer).

69. (a) We use conservation of mechanical energy to find the speed of either ball after it has fallen a distance h. The initial kinetic energy is zero, the initial gravitational potential energy is Mgh, the final kinetic energy is $\tfrac{1}{2}Mv^2$, and the final potential energy is zero. Thus $Mgh = \tfrac{1}{2}Mv^2$ and $v = \sqrt{2gh}$. The collision of the ball of M with the floor is an elastic collision of a light object with a stationary massive object. The velocity of the light object reverses direction without change in magnitude. After the collision, the ball is traveling upward with a speed of $\sqrt{2gh}$. The ball of mass m is traveling downward with the same speed. We use Eq. 9-75 to find an expression for the velocity of the ball of mass M after the collision:

$$v_{Mf} = \frac{M-m}{M+m} v_{Mi} + \frac{2m}{M+m} v_{mi} = \frac{M-m}{M+m}\sqrt{2gh} - \frac{2m}{M+m}\sqrt{2gh} = \frac{M-3m}{M+m}\sqrt{2gh}.$$

For this to be zero, $m = M/3$. With $M = 0.63$ kg, we have $m = 0.21$ kg.

(b) We use the same equation to find the velocity of the ball of mass m after the collision:

$$v_{mf} = -\frac{m-M}{M+m}\sqrt{2gh} + \frac{2M}{M+m}\sqrt{2gh} = \frac{3M-m}{M+m}\sqrt{2gh}$$

which becomes (upon substituting $M = 3m$) $v_{mf} = 2\sqrt{2gh}$. We next use conservation of mechanical energy to find the height h' to which the ball rises. The initial kinetic energy is $\tfrac{1}{2}mv_{mf}^2$, the initial potential energy is zero, the final kinetic energy is zero, and the final potential energy is mgh'. Thus,

$$\frac{1}{2}mv_{mf}^2 = mgh' \Rightarrow h' = \frac{v_{mf}^2}{2g} = 4h.$$

With $h = 1.8$ m, we have $h' = 7.2$ m.

70. We use Eqs. 9-67, 9-68 and 4-21 for the elastic collision and the subsequent projectile motion. We note that both pucks have the same time-of-fall t (during their projectile motions). Thus, we have

$$\Delta x_2 = v_2 t \quad \text{where } \Delta x_2 = d \text{ and } v_2 = \frac{2m_1}{m_1 + m_2} v_{1i}$$

$$\Delta x_1 = v_1 t \quad \text{where } \Delta x_1 = -2d \text{ and } v_1 = \frac{m_1 - m_2}{m_1 + m_2} v_{1i}.$$

Dividing the first equation by the second, we arrive at

$$\frac{d}{-2d} = \frac{\frac{2m_1}{m_1 + m_2} v_{1i} t}{\frac{m_1 - m_2}{m_1 + m_2} v_{1i} t}.$$

After canceling v_{1i}, t and d, and solving, we obtain $m_2 = 1.0$ kg.

71. We orient our $+x$ axis along the initial direction of motion, and specify angles in the "standard" way — so $\theta = +60°$ for the proton (1) which is assumed to scatter into the first quadrant and $\phi = -30°$ for the target proton (2) which scatters into the fourth quadrant (recall that the problem has told us that this is perpendicular to θ). We apply the conservation of linear momentum to the x and y axes respectively.

$$m_1 v_1 = m_1 v'_1 \cos\theta + m_2 v'_2 \cos\phi$$
$$0 = m_1 v'_1 \sin\theta + m_2 v'_2 \sin\phi$$

We are given $v_1 = 500$ m/s, which provides us with two unknowns and two equations, which is sufficient for solving. Since $m_1 = m_2$ we can cancel the mass out of the equations entirely.

(a) Combining the above equations and solving for v'_2 we obtain

$$v'_2 = \frac{v_1 \sin\theta}{\sin(\theta - \phi)} = \frac{(500 \text{ m/s}) \sin(60°)}{\sin(90°)} = 433 \text{ m/s}.$$

We used the identity $\sin\theta \cos\phi - \cos\theta \sin\phi = \sin(\theta - \phi)$ in simplifying our final expression.

(b) In a similar manner, we find

$$v'_1 = \frac{v_1 \sin\theta}{\sin(\phi - \theta)} = \frac{(500 \text{ m/s}) \sin(-30°)}{\sin(-90°)} = 250 \text{ m/s}.$$

72. (a) Conservation of linear momentum implies

$$m_A \vec{v}_A + m_B \vec{v}_B = m_A \vec{v}'_A + m_B \vec{v}'_B.$$

Since $m_A = m_B = m = 2.0$ kg, the masses divide out and we obtain

$$\vec{v}'_B = \vec{v}_A + \vec{v}_B - \vec{v}'_A = (15\hat{i} + 30\hat{j}) \text{ m/s} + (-10\hat{i} + 5\hat{j}) \text{ m/s} - (-5\hat{i} + 20\hat{j}) \text{ m/s}$$
$$= (10\hat{i} + 15\hat{j}) \text{ m/s}.$$

(b) The final and initial kinetic energies are

$$K_f = \frac{1}{2}mv'^2_A + \frac{1}{2}mv'^2_B = \frac{1}{2}(2.0)\left((-5)^2 + 20^2 + 10^2 + 15^2\right) = 8.0 \times 10^2 \text{ J}$$

$$K_i = \frac{1}{2}mv^2_A + \frac{1}{2}mv^2_B = \frac{1}{2}(2.0)\left(15^2 + 30^2 + (-10)^2 + 5^2\right) = 1.3 \times 10^3 \text{ J}.$$

The change kinetic energy is then $\Delta K = -5.0 \times 10^2$ J (that is, 500 J of the initial kinetic energy is lost).

73. We apply the conservation of linear momentum to the x and y axes respectively.

$$m_1 v_{1i} = m_1 v_{1f} \cos\theta_1 + m_2 v_{2f} \cos\theta_2$$
$$0 = m_1 v_{1f} \sin\theta_1 - m_2 v_{2f} \sin\theta_2$$

We are given $v_{2f} = 1.20 \times 10^5$ m/s, $\theta_1 = 64.0°$ and $\theta_2 = 51.0°$. Thus, we are left with two unknowns and two equations, which can be readily solved.

(a) We solve for the final alpha particle speed using the y-momentum equation:

$$v_{1f} = \frac{m_2 v_{2f} \sin\theta_2}{m_1 \sin\theta_1} = \frac{(16.0)(1.20 \times 10^5)\sin(51.0°)}{(4.00)\sin(64.0°)} = 4.15 \times 10^5 \text{ m/s}.$$

(b) Plugging our result from part (a) into the x-momentum equation produces the initial alpha particle speed:

$$v_{1i} = \frac{m_1 v_{1f} \cos\theta_1 + m_2 v_{2f} \cos\theta_2}{m_{1i}}$$

$$= \frac{(4.00)(4.15 \times 10^5) \cos(64.0°) + (16.0)(1.2 \times 10^5) \cos(51.0°)}{4.00}$$

$$= 4.84 \times 10^5 \text{ m/s}.$$

74. We orient our $+x$ axis along the initial direction of motion, and specify angles in the "standard" way — so $\theta = -90°$ for the particle B which is assumed to scatter "downward" and $\phi > 0$ for particle A which presumably goes into the first quadrant. We apply the conservation of linear momentum to the x and y axes respectively.

$$m_B v_B = m_B v'_B \cos\theta + m_A v'_A \cos\phi$$
$$0 = m_B v'_B \sin\theta + m_A v'_A \sin\phi$$

(a) Setting $v_B = v$ and $v'_B = v/2$, the y-momentum equation yields

$$m_A v'_A \sin\phi = m_B \frac{v}{2}$$

and the x-momentum equation yields $m_A v'_A \cos\phi = m_B v$.

Dividing these two equations, we find $\tan\phi = \frac{1}{2}$ which yields $\phi = 27°$.

(b) We can *formally* solve for v'_A (using the y-momentum equation and the fact that $\phi = 1/\sqrt{5}$)

$$v'_A = \frac{\sqrt{5}}{2} \frac{m_B}{m_A} v$$

but lacking numerical values for v and the mass ratio, we cannot fully determine the final speed of A. Note: substituting $\cos\phi = 2/\sqrt{5}$, into the x-momentum equation leads to exactly this same relation (that is, no new information is obtained which might help us determine an answer).

75. Suppose the objects enter the collision along lines that make the angles $\theta > 0$ and $\phi > 0$ with the x axis, as shown in the diagram that follows. Both have the same mass m and the same initial speed v. We suppose that after the collision the combined object moves in the positive x direction with speed V. Since the y component of the total momentum of the two-object system is conserved,

$$mv \sin\theta - mv \sin\phi = 0.$$

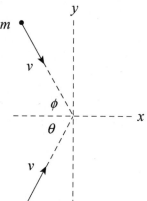

This means $\phi = \theta$. Since the x component is conserved,

$$2mv \cos\theta = 2mV.$$

We now use $V = v/2$ to find that $\cos\theta = 1/2$. This means $\theta = 60°$. The angle between the initial velocities is 120°.

76. We use Eq. 9-88 and simplify with $v_i = 0$, $v_f = v$, and $v_{rel} = u$.

$$v_f - v_i = v_{rel} \ln \frac{M_i}{M_f} \Rightarrow \frac{M_i}{M_f} = e^{v/u}$$

(a) If $v = u$ we obtain $\dfrac{M_i}{M_f} = e^1 \approx 2.7$.

(b) If $v = 2u$ we obtain $\dfrac{M_i}{M_f} = e^2 \approx 7.4$.

77. (a) The thrust of the rocket is given by $T = Rv_{rel}$ where R is the rate of fuel consumption and v_{rel} is the speed of the exhaust gas relative to the rocket. For this problem $R = 480$ kg/s and $v_{rel} = 3.27 \times 10^3$ m/s, so

$$T = (480 \text{ kg/s})(3.27 \times 10^3 \text{ m/s}) = 1.57 \times 10^6 \text{ N}.$$

(b) The mass of fuel ejected is given by $M_{fuel} = R\Delta t$, where Δt is the time interval of the burn. Thus, $M_{fuel} = (480 \text{ kg/s})(250 \text{ s}) = 1.20 \times 10^5$ kg. The mass of the rocket after the burn is

$$M_f = M_i - M_{fuel} = (2.55 \times 10^5 \text{ kg}) - (1.20 \times 10^5 \text{ kg}) = 1.35 \times 10^5 \text{ kg}.$$

(c) Since the initial speed is zero, the final speed is given by

$$v_f = v_{rel} \ln \frac{M_i}{M_f} = (3.27 \times 10^3) \ln \left(\frac{2.55 \times 10^5}{1.35 \times 10^5} \right) = 2.08 \times 10^3 \text{ m/s}.$$

78. We use Eq. 9-88. Then

$$v_f = v_i + v_{rel} \ln \left(\frac{M_i}{M_f} \right) = 105 \text{ m/s} + (253 \text{ m/s}) \ln \left(\frac{6090 \text{ kg}}{6010 \text{ kg}} \right) = 108 \text{ m/s}.$$

79. (a) We consider what must happen to the coal that lands on the faster barge during one minute ($\Delta t = 60$s). In that time, a total of $m = 1000$ kg of coal must experience a change of velocity

$$\Delta v = 20 \text{ km/h} - 10 \text{ km/h} = 10 \text{ km/h} = 2.8 \text{ m/s},$$

where rightwards is considered the positive direction. The rate of change in momentum for the coal is therefore

$$\frac{\Delta \vec{p}}{\Delta t} = \frac{m \Delta \vec{v}}{\Delta t} = \frac{(1000 \text{ kg})(2.8 \text{ m/s})}{60 \text{ s}} = 46 \text{ N}$$

which, by Eq. 9-23, must equal the force exerted by the (faster) barge on the coal. The processes (the shoveling, the barge motions) are constant, so there is no ambiguity in equating $\frac{\Delta p}{\Delta t}$ with $\frac{dp}{dt}$.

(b) The problem states that the frictional forces acting on the barges does not depend on mass, so the loss of mass from the slower barge does not affect its motion (so no extra force is required as a result of the shoveling).

80. (a) We use Eq. 9-68 twice:

$$v_2 = \frac{2m_1}{m_1 + m_2} v_{1i} = \frac{2m_1}{1.5m_1} (4.00 \text{ m/s}) = \frac{16}{3} \text{ m/s}$$

$$v_3 = \frac{2m_2}{m_2 + m_3} v_2 = \frac{2m_2}{1.5m_2} (16/3 \text{ m/s}) = \frac{64}{9} \text{ m/s} = 7.11 \text{ m/s}.$$

(b) Clearly, the speed of block 3 is greater than the (initial) speed of block 1.

(c) The kinetic energy of block 3 is

$$K_{3f} = \frac{1}{2} m_3 v_3^2 = \left(\frac{1}{2}\right)^3 m_1 \left(\frac{16}{9}\right)^2 v_{1i}^2 = \frac{64}{81} K_{1i}.$$

We see the kinetic energy of block 3 is less than the (initial) K of block 1. In the final situation, the initial K is being shared among the three blocks (which are all in motion), so this is not a surprising conclusion.

(d) The momentum of block 3 is

$$p_{3f} = m_3 v_3 = \left(\frac{1}{2}\right)^2 m_1 \left(\frac{16}{9}\right) v_{1i} = \frac{4}{9} p_{1i}$$

and is therefore less than the initial momentum (both of these being considered in magnitude, so questions about $\pm$ sign do not enter the discussion).

81. Using Eq. 9-67 and Eq. 9-68, we have after the first collision

$$v_{1f} = \frac{m_1 - m_2}{m_1 + m_2} v_{1i} = \frac{-m_1}{3m_1} v_{1i} = -\frac{1}{3} v_{1i}$$

$$v_{2f} = \frac{2m_1}{m_1 + m_2} v_{1i} = \frac{2m_1}{3m_1} v_{1i} = \frac{2}{3} v_{1i}.$$

After the second collision, the velocities are

$$v_{2ff} = \frac{m_2-m_3}{m_2+m_3}v_{2f} = \frac{-m_2}{3m_2}\frac{2}{3}v_{1i} = -\frac{2}{9}v_{1i}$$

$$v_{3ff} = \frac{2m_2}{m_2+m_3}v_{2f} = \frac{2m_2}{3m_2}\frac{2}{3}v_{1i} = \frac{4}{9}v_{1i}.$$

(a) Setting $v_{1i} = 4$ m/s, we find $v_{3ff} \approx 1.78$ m/s.

(b) We see that v_{3ff} is less than v_{1i}.

(c) The final kinetic energy of block 3 (expressed in terms of the initial kinetic energy of block 1) is

$$K_{3ff} = \frac{1}{2}m_3 v_3^2 = \frac{1}{2}(4m_1)\left(\frac{16}{9}\right)^2 v_{1i}^2 = \frac{64}{81}K_{1i}.$$

We see that this is less than K_{1i}.

(d) The final momentum of block 3 is $p_{3ff} = m_3 v_{3ff} = (4m_1)\left(\frac{16}{9}\right)v_1 > m_1 v_1$.

82. (a) This is a highly symmetric collision, and when we analyze the y-components of momentum we find their net value is zero. Thus, the stuck-together particles travel along the x axis.

(b) Since it is an elastic collision with identical particles, the final speeds are the same as the initial values. Conservation of momentum along each axis then assures that the angles of approach are the same as the angles of scattering. Therefore, one particle travels along line 2, the other along line 3.

(c) Here the final speeds are less than they were initially. The total x-component cannot be less, however, by momentum conservation, so the loss of speed shows up as a decrease in their y-velocity-components. This leads to smaller angles of scattering. Consequently, one particle travels through region B, the other through region C; the paths are symmetric about the x-axis. We note that this is intermediate between the final states described in parts (b) and (a).

(d) Conservation of momentum along the x-axis leads (because these are identical particles) to the simple observation that the x-component of each particle remains constant:

$$v_{fx} = v\cos\theta = 3.06 \text{ m/s}.$$

(e) As noted above, in this case the speeds are unchanged; both particles are moving at 4.00 m/s in the final state.

83. (a) Momentum conservation gives

$$m_R v_R + m_L v_L = 0 \Rightarrow (0.500 \text{ kg}) v_R + (1.00 \text{ kg})(-1.20 \text{ m/s}) = 0$$

which yields $v_R = 2.40$ m/s. Thus, $\Delta x = v_R t = (2.40 \text{ m/s})(0.800 \text{ s}) = 1.92$ m.

(b) Now we have $m_R v_R + m_L (v_R - 1.20 \text{ m/s}) = 0$, which yields

$$v_R = \frac{(1.2 \text{ m/s}) m_L}{m_L + m_R} = \frac{(1.20 \text{ m/s})(1.00 \text{ kg})}{1.00 \text{ kg} + 0.500 \text{ kg}} = 0.800 \text{ m/s}.$$

Consequently, $\Delta x = v_R t = 0.640$ m.

84. Let m be the mass of the higher floors. By energy conservation, the speed of the higher floors just before impact is

$$mgd = \frac{1}{2} mv^2 \Rightarrow v = \sqrt{2gd}.$$

The magnitude of the impulse during the impact is

$$J = |\Delta p| = m |\Delta v| = mv = m\sqrt{2gd} = mg\sqrt{\frac{2d}{g}} = W\sqrt{\frac{2d}{g}}$$

where $W = mg$ represents the weight of the higher floors. Thus, the average force exerted on the lower floor is

$$F_{avg} = \frac{J}{\Delta t} = \frac{W}{\Delta t} \sqrt{\frac{2d}{g}}$$

With $F_{avg} = sW$, where s is the safety factor, we have

$$s = \frac{1}{\Delta t} \sqrt{\frac{2d}{g}} = \frac{1}{1.5 \times 10^{-3} \text{ s}} \sqrt{\frac{2(4.0 \text{ m})}{9.8 \text{ m/s}^2}} = 6.0 \times 10^2.$$

85. We convert mass rate to SI units: $R = (540 \text{ kg/min})/(60 \text{ s/min}) = 9.00$ kg/s. In the absence of the asked-for additional force, the car would decelerate with a magnitude given by Eq. 9-87:

$$R v_{rel} = M |a|$$

so that if $a = 0$ is desired then the additional force must have a magnitude equal to $R v_{rel}$ (so as to cancel that effect).

$$F = R v_{rel} = (9.00 \text{ kg/s})(3.20 \text{ m/s}) = 28.8 \text{ N}.$$

86. From mechanical energy conservation (or simply using Eq. 2-16 with $\vec{a} = g$ downward) we obtain

$$v = \sqrt{2gh} = \sqrt{2(9.8 \text{ m/s}^2)(1.5 \text{ m})} = 5.4 \text{ m/s}$$

for the speed just as the body makes contact with the ground.

(a) During the compression of the body, the center of mass must decelerate over a distance $d = 0.30$ m. Choosing $+y$ downward, the deceleration a is found using Eq. 2-16.

$$0 = v^2 + 2ad \Rightarrow a = -\frac{v^2}{2d} = -\frac{5.4^2}{2(0.30)}$$

which yields $a = -49 \text{ m/s}^2$. Thus, the magnitude of the net (vertical) force is $m|a| = 49m$ in SI units, which (since $49 \text{ m/s}^2 = 5(9.8 \text{ m/s}^2) = 5g$) can be expressed as $5mg$.

(b) During the deceleration process, the forces on the dinosaur are (in the vertical direction) $\vec{F}_N$ and $m\vec{g}$. If we choose $+y$ upward, and use the final result from part (a), we therefore have

$$F_N - mg = 5mg \Rightarrow F_N = 6mg.$$

In the horizontal direction, there is also a deceleration (from $v_0 = 19$ m/s to zero), in this case due to kinetic friction $f_k = \mu_k F_N = \mu_k(6mg)$. Thus, the net force exerted by the ground on the dinosaur is

$$F_{\text{ground}} = \sqrt{f_k^2 + F_N^2} \approx 7mg.$$

(c) We can applying Newton's second law in the horizontal direction (with the sliding distance denoted as Δx) and then use Eq. 2-16, or we can apply the general notions of energy conservation. The latter approach is shown:

$$\frac{1}{2}mv_0^2 = \mu_k(6mg)\Delta x \Rightarrow \Delta x = \frac{(19 \text{ m/s})^2}{2(6)(0.6)(9.8 \text{ m/s}^2)} \approx 5 \text{ m}.$$

87. Denoting the new speed of the car as v, then the new speed of the man relative to the ground is $v - v_{\text{rel}}$. Conservation of momentum requires

$$\left(\frac{W}{g} + \frac{w}{g}\right)v_0 = \left(\frac{W}{g}\right)v + \left(\frac{w}{g}\right)(v - v_{\text{rel}}).$$

Consequently, the change of velocity is

$$\Delta \vec{v} = v - v_0 = \frac{w\, v_{rel}}{W+w} = \frac{(915\text{ N})(4.00\text{ m/s})}{(2415\text{ N})+(915\text{ N})} = 1.10 \text{ m/s}.$$

88. First, we imagine that the small square piece (of mass m) that was cut from the large plate is returned to it so that the large plate is again a complete 6 m × 6 m ($d = 1.0$ m) square plate (which has its center of mass at the origin). Then we "add" a square piece of "negative mass" ($-m$) at the appropriate location to obtain what is shown in Fig. 9-75. If the mass of the whole plate is M, then the mass of the small square piece cut from it is obtained from a simple ratio of areas:

$$m = \left(\frac{2.0\text{ m}}{6.0\text{ m}}\right)^2 M \Rightarrow M = 9m.$$

(a) The x coordinate of the small square piece is $x = 2.0$ m (the middle of that square "gap" in the figure). Thus the x coordinate of the center of mass of the remaining piece is

$$x_{com} = \frac{(-m)x}{M+(-m)} = \frac{-m(2.0\text{ m})}{9m-m} = -0.25 \text{ m}.$$

(b) Since the y coordinate of the small square piece is zero, we have $y_{com} = 0$.

89. We assume no external forces act on the system composed of the two parts of the last stage. Hence, the total momentum of the system is conserved. Let m_c be the mass of the rocket case and m_p the mass of the payload. At first they are traveling together with velocity v. After the clamp is released m_c has velocity v_c and m_p has velocity v_p. Conservation of momentum yields

$$(m_c + m_p)v = m_c v_c + m_p v_p.$$

(a) After the clamp is released the payload, having the lesser mass, will be traveling at the greater speed. We write $v_p = v_c + v_{rel}$, where v_{rel} is the relative velocity. When this expression is substituted into the conservation of momentum condition, the result is

$$(m_c + m_p)v = m_c v_c + m_p v_c + m_p v_{rel}.$$

Therefore,

$$v_c = \frac{(m_c + m_p)v - m_p v_{rel}}{m_c + m_p} = \frac{(290.0 \text{ kg}+150.0 \text{ kg})(7600 \text{ m/s})-(150.0 \text{ kg})(910.0 \text{ m/s})}{290.0 \text{ kg}+150.0 \text{ kg}}$$
$$= 7290 \text{ m/s}.$$

(b) The final speed of the payload is $v_p = v_c + v_{rel} = 7290$ m/s $+ 910.0$ m/s $= 8200$ m/s.

(c) The total kinetic energy before the clamp is released is

$$K_i = \frac{1}{2}(m_c + m_p)v^2 = \frac{1}{2}(290.0 \text{ kg} + 150.0 \text{ kg})(7600 \text{ m/s})^2 = 1.271 \times 10^{10} \text{ J}.$$

(d) The total kinetic energy after the clamp is released is

$$K_f = \frac{1}{2}m_c v_c^2 + \frac{1}{2}m_p v_p^2 = \frac{1}{2}(290.0 \text{ kg})(7290 \text{ m/s})^2 + \frac{1}{2}(150.0 \text{ kg})(8200 \text{ m/s})^2$$
$$= 1.275 \times 10^{10} \text{ J}.$$

The total kinetic energy increased slightly. Energy originally stored in the spring is converted to kinetic energy of the rocket parts.

90. The velocity of the object is

$$\vec{v} = \frac{d\vec{r}}{dt} = \frac{d}{dt}\left((3500 - 160t)\hat{i} + 2700\hat{j} + 300\hat{k}\right) = -(160 \text{ m/s})\hat{i}.$$

(a) The linear momentum is $\vec{p} = m\vec{v} = (250 \text{ kg})(-160 \text{ m/s}\hat{i}) = (-4.0 \times 10^4 \text{ kg} \cdot \text{m/s})\hat{i}$.

(b) The object is moving west (our $-\hat{i}$ direction).

(c) Since the value of $\vec{p}$ does not change with time, the net force exerted on the object is zero, by Eq. 9-23.

91. (a) If m is the mass of a pellet and v is its velocity as it hits the wall, then its momentum is $p = mv = (2.0 \times 10^{-3} \text{ kg})(500 \text{ m/s}) = 1.0$ kg · m/s, toward the wall.

(b) The kinetic energy of a pellet is

$$K = \frac{1}{2}mv^2 = \frac{1}{2}(2.0 \times 10^{-3} \text{ kg})(500 \text{ m/s})^2 = 2.5 \times 10^2 \text{ J}.$$

(c) The force on the wall is given by the rate at which momentum is transferred from the pellets to the wall. Since the pellets do not rebound, each pellet that hits transfers $p = 1.0$ kg · m/s. If ΔN pellets hit in time Δt, then the average rate at which momentum is transferred is

$$F_{\text{avg}} = \frac{p\Delta N}{\Delta t} = (1.0 \text{ kg} \cdot \text{m/s})(10 \text{ s}^{-1}) = 10 \text{ N}.$$

The force on the wall is in the direction of the initial velocity of the pellets.

(d) If Δt is the time interval for a pellet to be brought to rest by the wall, then the average force exerted on the wall by a pellet is

$$F_{\text{avg}} = \frac{p}{\Delta t} = \frac{1.0 \,\text{kg} \cdot \text{m/s}}{0.6 \times 10^{-3}\,\text{s}} = 1.7 \times 10^3 \,\text{N}.$$

The force is in the direction of the initial velocity of the pellet.

(e) In part (d) the force is averaged over the time a pellet is in contact with the wall, while in part (c) it is averaged over the time for many pellets to hit the wall. During the majority of this time, no pellet is in contact with the wall, so the average force in part (c) is much less than the average force in part (d).

92. One approach is to choose a *moving* coordinate system which travels the center of mass of the body, and another is to do a little extra algebra analyzing it in the original coordinate system (in which the speed of the $m = 8.0$ kg mass is $v_0 = 2$ m/s, as given). Our solution is in terms of the latter approach since we are assuming that this is the approach most students would take. Conservation of linear momentum (along the direction of motion) requires

$$mv_0 = m_1 v_1 + m_2 v_2 \quad \Rightarrow \quad (8.0)(2.0) = (4.0)v_1 + (4.0)v_2$$

which leads to $v_2 = 4 - v_1$ in SI units (m/s). We require

$$\Delta K = \left(\frac{1}{2}m_1 v_1^2 + \frac{1}{2}m_2 v_2^2\right) - \frac{1}{2}mv_0^2 \quad \Rightarrow \quad 16 = \left(\frac{1}{2}(4.0)v_1^2 + \frac{1}{2}(4.0)v_2^2\right) - \frac{1}{2}(8.0)(2.0)^2$$

which simplifies to $v_2^2 = 16 - v_1^2$ in SI units. If we substitute for v_2 from above, we find

$$(4 - v_1)^2 = 16 - v_1^2$$

which simplifies to $2v_1^2 - 8v_1 = 0$, and yields either $v_1 = 0$ or $v_1 = 4$ m/s. If $v_1 = 0$ then $v_2 = 4 - v_1 = 4$ m/s, and if $v_1 = 4$ m/s then $v_2 = 0$.

(a) Since the forward part continues to move in the original direction of motion, the speed of the rear part must be zero.

(b) The forward part has a velocity of 4.0 m/s along the original direction of motion.

93. (a) The initial momentum of the car is

$$\vec{p}_i = m\vec{v}_i = (1400\,\text{kg})(5.3\,\text{m/s})\hat{j} = (7400\,\text{kg} \cdot \text{m/s})\hat{j}$$

and the final momentum is $\vec{p}_f = (7400 \,\text{kg}\cdot\text{m/s})\hat{i}$. The impulse on it equals the change in its momentum:

$$\vec{J} = \vec{p}_f - \vec{p}_i = (7.4\times10^3 \,\text{N}\cdot\text{s})(\hat{i}-\hat{j}).$$

(b) The initial momentum of the car is $\vec{p}_i = (7400 \,\text{kg}\cdot\text{m/s})\hat{i}$ and the final momentum is $\vec{p}_f = 0$. The impulse acting on it is $\vec{J} = \vec{p}_f - \vec{p}_i = (-7.4\times10^3 \,\text{N}\cdot\text{s})\hat{i}$.

(c) The average force on the car is

$$\vec{F}_{avg} = \frac{\Delta\vec{p}}{\Delta t} = \frac{\vec{J}}{\Delta t} = \frac{(7400\,\text{kg}\cdot\text{m/s})(\hat{i}-\hat{j})}{4.6\,\text{s}} = (1600\,\text{N})(\hat{i}-\hat{j})$$

and its magnitude is $F_{avg} = (1600\,\text{N})\sqrt{2} = 2.3\times10^3 \,\text{N}$.

(d) The average force is

$$\vec{F}_{avg} = \frac{\vec{J}}{\Delta t} = \frac{(-7400\,\text{kg}\cdot\text{m/s})\hat{i}}{350\times10^{-3}\,\text{s}} = (-2.1\times10^4 \,\text{N})\hat{i}$$

and its magnitude is $F_{avg} = 2.1\times10^4 \,\text{N}$.

(e) The average force is given above in unit vector notation. Its x and y components have equal magnitudes. The x component is positive and the y component is negative, so the force is 45° below the positive x axis.

94. We first consider the 1200 kg part. The impulse has magnitude J and is (by our choice of coordinates) in the positive direction. Let m_1 be the mass of the part and v_1 be its velocity after the bolts are exploded. We assume both parts are at rest before the explosion. Then $J = m_1 v_1$, so

$$v_1 = \frac{J}{m_1} = \frac{300 \,\text{N}\cdot\text{s}}{1200 \,\text{kg}} = 0.25 \,\text{m/s}.$$

The impulse on the 1800 kg part has the same magnitude but is in the opposite direction, so $-J = m_2 v_2$, where m_2 is the mass and v_2 is the velocity of the part. Therefore,

$$v_2 = -\frac{J}{m_2} = -\frac{300\,\text{N}\cdot\text{s}}{1800\,\text{kg}} = -0.167 \,\text{m/s}.$$

Consequently, the relative speed of the parts after the explosion is

$$u = 0.25 \text{ m/s} - (-0.167 \text{ m/s}) = 0.417 \text{ m/s}.$$

95. We choose our positive direction in the direction of the rebound (so the ball's initial velocity is negative-valued $\vec{v}_i = -5.2 \text{ m/s}$).

(a) The speed of the ball right after the collision is

$$v_f = \sqrt{\frac{2K_f}{m}} = \sqrt{\frac{2(\frac{1}{2}K_i)}{m}} = \sqrt{\frac{\frac{1}{2}mv_i^2}{m}} = \frac{v_i}{\sqrt{2}} \approx 3.7 \text{ m/s}.$$

(b) With $m = 0.15$ kg, the impulse-momentum theorem (Eq. 9-31) yields

$$\vec{J} = m\vec{v}_f - m\vec{v}_i = (0.15 \text{ kg})(3.7 \text{ m/s}) - (0.15 \text{ kg})(-5.2 \text{ m/s}) = 1.3 \text{ N} \cdot \text{s}.$$

(c) Eq. 9-35 leads to $F_{avg} = J/\Delta t = 1.3/0.0076 = 1.8 \times 10^2$ N.

96. Let m_c be the mass of the Chrysler and v_c be its velocity. Let m_f be the mass of the Ford and v_f be its velocity. Then the velocity of the center of mass is

$$v_{com} = \frac{m_c v_c + m_f v_f}{m_c + m_f} = \frac{(2400 \text{ kg})(80 \text{ km/h}) + (1600 \text{ kg})(60 \text{ km/h})}{2400 \text{ kg} + 1600 \text{ kg}} = 72 \text{ km/h}.$$

We note that the two velocities are in the same direction, so the two terms in the numerator have the same sign.

97. Let m_F be the mass of the freight car and v_F be its initial velocity. Let m_C be the mass of the caboose and v be the common final velocity of the two when they are coupled. Conservation of the total momentum of the two-car system leads to

$$m_F v_F = (m_F + m_C)v \Rightarrow v = v_F m_F / (m_F + m_C).$$

The initial kinetic energy of the system is

$$K_i = \frac{1}{2} m_F v_F^2$$

and the final kinetic energy is

$$K_f = \frac{1}{2}(m_F + m_C)v^2 = \frac{1}{2}(m_F + m_C)\frac{m_F^2 v_F^2}{(m_F + m_C)^2} = \frac{1}{2}\frac{m_F^2 v_F^2}{(m_F + m_C)}.$$

Since 27% of the original kinetic energy is lost, we have $K_f = 0.73 K_i$. Thus,

$$\frac{1}{2}\frac{m_F^2 v_F^2}{(m_F+m_C)} = (0.73)\left(\frac{1}{2}m_F v_F^2\right).$$

Simplifying, we obtain $m_F/(m_F+m_C) = 0.73$, which we use in solving for the mass of the caboose:

$$m_C = \frac{0.27}{0.73}m_F = 0.37 m_F = (0.37)(3.18\times 10^4 \text{ kg}) = 1.18\times 10^4 \text{ kg}.$$

98. The fact that they are connected by a spring is not used in the solution. We use Eq. 9-17 for $\vec{v}_{com}$:

$$M\vec{v}_{com} = m_1\vec{v}_1 + m_2\vec{v}_2 = (1.0\text{ kg})(1.7\text{ m/s}) + (3.0\text{ kg})\vec{v}_2$$

which yields $|\vec{v}_2| = 0.57$ m/s. The direction of $\vec{v}_2$ is opposite that of $\vec{v}_1$ (that is, they are both headed towards the center of mass, but from opposite directions).

99. No external forces with horizontal components act on the cart-man system and the vertical forces sum to zero, so the total momentum of the system is conserved. Let m_c be the mass of the cart, v be its initial velocity, and v_c be its final velocity (after the man jumps off). Let m_m be the mass of the man. His initial velocity is the same as that of the cart and his final velocity is zero. Conservation of momentum yields $(m_m + m_c)v = m_c v_c$. Consequently, the final speed of the cart is

$$v_c = \frac{v(m_m+m_c)}{m_c} = \frac{(2.3\text{ m/s})(75\text{ kg}+39\text{ kg})}{39\text{ kg}} = 6.7\text{ m/s}.$$

The cart speeds up by 6.7 m/s − 2.3 m/s = + 4.4 m/s. In order to slow himself, the man gets the cart to push backward on him by pushing forward on it, so the cart speeds up.

100. (a) We find the momentum $\vec{p}_{nr}$ of the residual nucleus from momentum conservation.

$$\vec{p}_{ni} = \vec{p}_e + \vec{p}_v + \vec{p}_{nr} \Rightarrow 0 = (-1.2\times 10^{-22}\text{ kg}\cdot\text{m/s})\hat{i} + (-6.4\times 10^{-23}\text{ kg}\cdot\text{m/s})\hat{j} + \vec{p}_{nr}$$

Thus, $\vec{p}_{nr} = (1.2\times 10^{-22}\text{ kg}\cdot\text{m/s})\hat{i} + (6.4\times 10^{-23}\text{ kg}\cdot\text{m/s})\hat{j}$. Its magnitude is

$$|\vec{p}_{nr}| = \sqrt{(1.2\times 10^{-22}\text{ kg}\cdot\text{m/s})^2 + (6.4\times 10^{-23}\text{ kg}\cdot\text{m/s})^2} = 1.4\times 10^{-22}\text{ kg}\cdot\text{m/s}.$$

(b) The angle measured from the +x axis to $\vec{p}_{nr}$ is

$$\theta = \tan^{-1}\left(\frac{6.4 \times 10^{-23} \text{ kg} \cdot \text{m/s}}{1.2 \times 10^{-22} \text{ kg} \cdot \text{m/s}}\right) = 28°.$$

(c) Combining the two equations $p = mv$ and $K = \frac{1}{2}mv^2$, we obtain (with $p = p_{nr}$ and $m = m_{nr}$)

$$K = \frac{p^2}{2m} = \frac{(1.4 \times 10^{-22} \text{ kg} \cdot \text{m/s})^2}{2(5.8 \times 10^{-26} \text{ kg})} = 1.6 \times 10^{-19} \text{ J}.$$

101. The mass of each ball is m, and the initial speed of one of the balls is $v_{1i} = 2.2 \text{ m/s}$. We apply the conservation of linear momentum to the x and y axes respectively.

$$mv_{1i} = mv_{1f} \cos\theta_1 + mv_{2f} \cos\theta_2$$
$$0 = mv_{1f} \sin\theta_1 - mv_{2f} \sin\theta_2$$

The mass m cancels out of these equations, and we are left with two unknowns and two equations, which is sufficient to solve.

(a) The y-momentum equation can be rewritten as, using $\theta_2 = 60°$ and $v_{2f} = 1.1 \text{ m/s}$,

$$v_{1f} \sin\theta_1 = (1.1 \text{ m/s}) \sin 60° = 0.95 \text{ m/s}.$$

and the x-momentum equation yields

$$v_{1f} \cos\theta_1 = (2.2 \text{ m/s}) - (1.1 \text{ m/s}) \cos 60° = 1.65 \text{ m/s}.$$

Dividing these two equations, we find $\tan\theta_1 = 0.576$ which yields $\theta_1 = 30°$. We plug the value into either equation and find $v_{1f} \approx 1.9 \text{ m/s}$.

(b) From the above, we have $\theta_1 = 30°$, measured *clockwise* from the $+x$-axis, or equivalently, $-30°$, measured *counterclockwise* from the $+x$-axis.

(c) One can check to see if this an elastic collision by computing

$$\frac{2K_i}{m} = v_{1i}^2 \quad \text{and} \quad \frac{2K_f}{m} = v_{1f}^2 + v_{2f}^2$$

and seeing if they are equal (they are), but one must be careful not to use rounded-off values. Thus, it is useful to note that the answer in part (a) can be expressed "exactly" as $v_{1f} = \frac{1}{2}v_{1i}\sqrt{3}$ (and of course $v_{2f} = \frac{1}{2}v_{1i}$ "exactly" — which makes it clear that these two kinetic energy expressions are indeed equal).

102. (a) We use Eq. 9-87. The thrust is

$$Rv_{rel} = Ma = (4.0 \times 10^4 \text{ kg})(2.0 \text{ m/s}^2) = 8.0 \times 10^4 \text{ N}.$$

(b) Since $v_{rel} = 3000$ m/s, we see from part (a) that $R \approx 27$ kg/s.

103. The diagram below shows the situation as the incident ball (the left-most ball) makes contact with the other two.

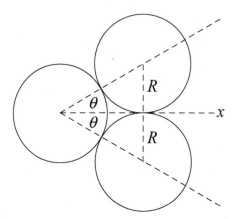

It exerts an impulse of the same magnitude on each ball, along the line that joins the centers of the incident ball and the target ball. The target balls leave the collision along those lines, while the incident ball leaves the collision along the x axis. The three dotted lines that join the centers of the balls in contact form an equilateral triangle, so both of the angles marked θ are 30°. Let v_0 be the velocity of the incident ball before the collision and V be its velocity afterward. The two target balls leave the collision with the same speed. Let v represent that speed. Each ball has mass m. Since the x component of the total momentum of the three-ball system is conserved,

$$mv_0 = mV + 2mv\cos\theta$$

and since the total kinetic energy is conserved,

$$\frac{1}{2}mv_0^2 = \frac{1}{2}mV^2 + 2\left(\frac{1}{2}mv^2\right).$$

We know the directions in which the target balls leave the collision so we first eliminate V and solve for v. The momentum equation gives $V = v_0 - 2v\cos\theta$, so

$$V^2 = v_0^2 - 4v_0v\cos\theta + 4v^2\cos^2\theta$$

and the energy equation becomes $v_0^2 = v_0^2 - 4v_0v\cos\theta + 4v^2\cos^2\theta + 2v^2$. Therefore,

$$v = \frac{2v_0 \cos\theta}{1+2\cos^2\theta} = \frac{2(10 \text{ m/s})\cos 30°}{1+2\cos^2 30°} = 6.93 \text{ m/s}.$$

(a) The discussion and computation above determines the final speed of ball 2 (as labeled in Fig. 9-83) to be 6.9 m/s.

(b) The direction of ball 2 is at 30° counterclockwise from the +x axis.

(c) Similarly, the final speed of ball 3 is 6.9 m/s.

(d) The direction of ball 3 is at −30° counterclockwise from the +x axis.

(e) Now we use the momentum equation to find the final velocity of ball 1:

$$V = v_0 - 2v\cos\theta = 10 \text{ m/s} - 2(6.93 \text{ m/s})\cos 30° = -2.0 \text{ m/s}.$$

So the speed of ball 1 is $|V| = 2.0$ m/s.

(f) The minus sign indicates that it bounces back in the −x direction. The angle is −180°.

104. (a) We use Fig. 9-22 of the text (which treats both angles as positive-valued, even though one of them is in the fourth quadrant; this is why there is an explicit minus sign in Eq. 9-80 as opposed to it being implicitly in the angle). We take the cue ball to be body 1 and the other ball to be body 2. Conservation of the x and the components of the total momentum of the two-ball system leads to:

$$mv_{1i} = mv_{1f}\cos\theta_1 + mv_{2f}\cos\theta_2$$
$$0 = -mv_{1f}\sin\theta_1 + mv_{2f}\sin\theta_2.$$

The masses are the same and cancel from the equations. We solve the second equation for $\sin\theta_2$:

$$\sin\theta_2 = \frac{v_{1f}}{v_{2f}}\sin\theta_1 = \left(\frac{3.50 \text{ m/s}}{2.00 \text{ m/s}}\right)\sin 22.0° = 0.656.$$

Consequently, the angle between the second ball and the initial direction of the first is $\theta_2 = 41.0°$.

(b) We solve the first momentum conservation equation for the initial speed of the cue ball.

$$v_{1i} = v_{1f}\cos\theta_1 + v_{2f}\cos\theta_2 = (3.50 \text{ m/s})\cos 22.0° + (2.00 \text{ m/s})\cos 41.0° = 4.75 \text{ m/s}.$$

(c) With SI units understood, the initial kinetic energy is

$$K_i = \frac{1}{2}mv_i^2 = \frac{1}{2}m(4.75)^2 = 11.3m$$

and the final kinetic energy is

$$K_f = \frac{1}{2}mv_{1f}^2 + \frac{1}{2}mv_{2f}^2 = \frac{1}{2}m((3.50)^2 + (2.00)^2) = 8.1m.$$

Kinetic energy is not conserved.

105. (a) We place the origin of a coordinate system at the center of the pulley, with the x axis horizontal and to the right and with the y axis downward. The center of mass is halfway between the containers, at $x = 0$ and $y = \ell$, where ℓ is the vertical distance from the pulley center to either of the containers. Since the diameter of the pulley is 50 mm, the center of mass is at a horizontal distance of 25 mm from each container.

(b) Suppose 20 g is transferred from the container on the left to the container on the right. The container on the left has mass $m_1 = 480$ g and is at $x_1 = -25$ mm. The container on the right has mass $m_2 = 520$ g and is at $x_2 = +25$ mm. The x coordinate of the center of mass is then

$$x_{com} = \frac{m_1 x_1 + m_2 x_2}{m_1 + m_2} = \frac{(480 \text{ g})(-25 \text{ mm}) + (520 \text{ g})(25 \text{ mm})}{480 \text{ g} + 520 \text{ g}} = 1.0 \text{ mm}.$$

The y coordinate is still ℓ. The center of mass is 26 mm from the lighter container, along the line that joins the bodies.

(c) When they are released the heavier container moves downward and the lighter container moves upward, so the center of mass, which must remain closer to the heavier container, moves downward.

(d) Because the containers are connected by the string, which runs over the pulley, their accelerations have the same magnitude but are in opposite directions. If a is the acceleration of m_2, then $-a$ is the acceleration of m_1. The acceleration of the center of mass is

$$a_{com} = \frac{m_1(-a) + m_2 a}{m_1 + m_2} = a \frac{m_2 - m_1}{m_1 + m_2}.$$

We must resort to Newton's second law to find the acceleration of each container. The force of gravity $m_1 g$, down, and the tension force of the string T, up, act on the lighter container. The second law for it is $m_1 g - T = -m_1 a$. The negative sign appears because a is the acceleration of the heavier container. The same forces act on the heavier container and for it the second law is $m_2 g - T = m_2 a$. The first equation gives $T = m_1 g + m_1 a$. This is substituted into the second equation to obtain $m_2 g - m_1 g - m_1 a = m_2 a$, so

$$a = (m_2 - m_1)g/(m_1 + m_2).$$

Thus,

$$a_{com} = \frac{g(m_2 - m_1)^2}{(m_1 + m_2)^2} = \frac{(9.8 \text{ m/s}^2)(520 \text{ g} - 480 \text{ g})^2}{(480 \text{ g} + 520 \text{ g})^2} = 1.6 \times 10^{-2} \text{ m/s}^2.$$

The acceleration is downward.

106. (a) The momentum change for the 0.15 kg object is

$$\Delta \vec{p} = (0.15)[2\hat{i} + 3.5\hat{j} - 3.2\hat{k} - (5\hat{i} + 6.5\hat{j} + 4\hat{k})] = (-0.450\hat{i} - 0.450\hat{j} - 1.08\hat{k}) \text{ kg·m/s}.$$

(b) By the impulse-momentum theorem (Eq. 9-31), $\vec{J} = \Delta \vec{p}$, we have

$$\vec{J} = (-0.450\hat{i} - 0.450\hat{j} - 1.08\hat{k}) \text{ N·s}.$$

(c) Newton's third law implies $\vec{J}_{wall} = -\vec{J}_{ball}$ (where $\vec{J}_{ball}$ is the result of part (b)), so

$$\vec{J}_{wall} = (0.450\hat{i} + 0.450\hat{j} + 1.08\hat{k}) \text{ N·s}.$$

107. (a) Noting that the initial velocity of the system is zero, we use Eq. 9-19 and Eq. 2-15 (adapted to two dimensions) to obtain

$$\vec{d} = \frac{1}{2}\left(\frac{\vec{F}_1 + \vec{F}_2}{m_1 + m_2}\right)t^2 = \frac{1}{2}\left(\frac{-2\hat{i} + \hat{j}}{0.006}\right)(0.002)^2$$

which has a magnitude of 0.745 mm.

(b) The angle of $\vec{d}$ is 153° counterclockwise from +x-axis.

(c) A similar calculation using Eq. 2-11 (adapted to two dimensions) leads to a center of mass velocity of $\vec{v} = 0.7453$ m/s at 153°. Thus, the center of mass kinetic energy is

$$K_{com} = \frac{1}{2}(m_1 + m_2)v^2 = 0.00167 \text{ J}.$$

108. (a) The change in momentum (taking upwards to be the positive direction) is

$$\Delta \vec{p} = (0.550 \text{ kg})[(3 \text{ m/s})\hat{j} - (-12 \text{ m/s})\hat{j}] = (+8.25 \text{ kg·m/s})\hat{j}.$$

(b) By the impulse-momentum theorem (Eq. 9-31) $\vec{J} = \Delta \vec{p} = (+8.25 \text{ N·s})\hat{j}$.

(c) By Newton's third law, $\vec{J}_c = -\vec{J}_b = (-8.25 \text{ N·s})\hat{j}$.

109. Using Eq. 9-67 and Eq. 9-68, we have after the collision

$$v_1 = \frac{m_1 - m_2}{m_1 + m_2} v_{1i} = \frac{0.6 m_1}{1.4 m_1} v_{1i} = -\tfrac{3}{7}(4 \text{ m/s})$$

$$v_2 = \frac{2 m_1}{m_1 + m_2} v_{1i} = \frac{2 m_1}{1.4 m_1} v_{1i} = \tfrac{1}{7}(4 \text{ m/s}) \ .$$

(a) During the (subsequent) sliding, the kinetic energy of block 1 $K_{1f} = \tfrac{1}{2} m_1 v_1^2$ is converted into thermal form ($\Delta E_{\text{th}} = \mu_k m_1 g\, d_1$). Solving for the sliding distance d_1 we obtain $d_1 = 0.2999$ m ≈ 30 cm.

(b) A very similar computation (but with subscript 2 replacing subscript 1) leads to block 2's sliding distance $d_2 = 3.332$ m ≈ 3.3 m.

110. (a) Since the initial momentum is zero, then the final momenta must add (in the vector sense) to 0. Therefore, with SI units understood, we have

$$\begin{aligned}\vec{p}_3 &= -\vec{p}_1 - \vec{p}_2 = -m_1 \vec{v}_1 - m_2 \vec{v}_2 \\ &= -(16.7 \times 10^{-27})(6.00 \times 10^6\, \hat{\text{i}}) - (8.35 \times 10^{-27})(-8.00 \times 10^6\, \hat{\text{j}}) \\ &= (-1.00 \times 10^{-19}\, \hat{\text{i}} + 0.67 \times 10^{-19}\, \hat{\text{j}})\ \text{kg}\cdot\text{m/s}.\end{aligned}$$

(b) Dividing by $m_3 = 11.7 \times 10^{-27}$ kg and using the Pythagorean theorem we find the speed of the third particle to be $v_3 = 1.03 \times 10^7$ m/s. The total amount of kinetic energy is

$$\tfrac{1}{2} m_1 v_1^2 + \tfrac{1}{2} m_2 v_2^2 + \tfrac{1}{2} m_3 v_3^2 = 1.19 \times 10^{-12} \text{ J}.$$

111. We use m_1 for the mass of the electron and $m_2 = 1840 m_1$ for the mass of the hydrogen atom. Using Eq. 9-68,

$$v_{2f} = \frac{2 m_1}{m_1 + 1840 m_1} v_{1i} = \frac{2}{1841} v_{1i}$$

we compute the final kinetic energy of the hydrogen atom:

$$K_{2f} = \tfrac{1}{2}(1840 m_1)\left(\frac{2 v_{1i}}{1841}\right)^2 = \frac{(1840)(4)}{1841^2}\left(\tfrac{1}{2}(1840 m_1) v_{1i}^2\right)$$

so we find the fraction to be $(1840)(4)/1841^2 \approx 2.2 \times 10^{-3}$, or 0.22%.

112. We treat the car (of mass m_1) as a "point-mass" (which is initially 1.5 m from the right end of the boat). The left end of the boat (of mass m_2) is initially at $x = 0$ (where the dock is), and its left end is at $x = 14$ m. The boat's center of mass (in the absence of the car) is initially at $x = 7.0$ m. We use Eq. 9-5 to calculate the center of mass of the system:

$$x_{com} = \frac{m_1 x_1 + m_2 x_2}{m_1 + m_2} = \frac{(1500 \text{ kg})(14 \text{ m} - 1.5 \text{ m}) + (4000 \text{ kg})(7 \text{ m})}{1500 \text{ kg} + 4000 \text{ kg}} = 8.5 \text{ m}.$$

In the absence of *external* forces, the center of mass of the system does not change. Later, when the car (about to make the jump) is near the left end of the boat (which has moved from the shore an amount δx), the value of the system center of mass is still 8.5 m. The car (at this moment) is thought of as a "point-mass" 1.5 m from the left end, so we must have

$$x_{com} = \frac{m_1 x_1 + m_2 x_2}{m_1 + m_2} = \frac{(1500 \text{ kg})(\delta x + 1.5 \text{ m}) + (4000 \text{ kg})(7 \text{ m} + \delta x)}{1500 \text{ kg} + 4000 \text{ kg}} = 8.5 \text{ m}.$$

Solving this for δx, we find $\delta x = 3.0$ m.

113. By conservation of momentum, the final speed v of the sled satisfies

$$(2900 \text{ kg})(250 \text{ m/s}) = (2900 \text{ kg} + 920 \text{ kg})v$$

which gives $v = 190$ m/s.

114. (a) The magnitude of the impulse is equal to the change in momentum:

$$J = mv - m(-v) = 2mv = 2(0.140 \text{ kg})(7.80 \text{ m/s}) = 2.18 \text{ kg} \cdot \text{m/s}$$

(b) Since in the calculus sense the average of a function is the integral of it divided by the corresponding interval, then the average force is the impulse divided by the time Δt. Thus, our result for the magnitude of the average force is $2mv/\Delta t$. With the given values, we obtain

$$F_{avg} = \frac{2(0.140 \text{ kg})(7.80 \text{ m/s})}{0.00380 \text{ s}} = 575 \text{ N}.$$

115. (a) We locate the coordinate origin at the center of Earth. Then the distance r_{com} of the center of mass of the Earth-Moon system is given by

$$r_{com} = \frac{m_M r_M}{m_M + m_E}$$

where m_M is the mass of the Moon, m_E is the mass of Earth, and r_M is their separation. These values are given in Appendix C. The numerical result is

$$r_{com} = \frac{(7.36 \times 10^{22} \text{ kg})(3.82 \times 10^8 \text{ m})}{7.36 \times 10^{22} \text{ kg} + 5.98 \times 10^{24} \text{ kg}} = 4.64 \times 10^6 \text{ m} \approx 4.6 \times 10^3 \text{ km}.$$

(b) The radius of Earth is $R_E = 6.37 \times 10^6$ m, so $r_{com}/R_E = 0.73 = 73\%$.

116. Conservation of momentum leads to

$$(900 \text{ kg})(1000 \text{ m/s}) = (500 \text{ kg})(v_{shuttle} - 100 \text{ m/s}) + (400 \text{ kg})(v_{shuttle})$$

which yields $v_{shuttle}$ = 1055.6 m/s for the shuttle speed and $v_{shuttle}$ − 100 m/s = 955.6 m/s for the module speed (all measured in the frame of reference of the stationary main spaceship). The fractional increase in the kinetic energy is

$$\frac{\Delta K}{K_i} = \frac{K_f}{K_i} - 1 = \frac{(500 \text{ kg})(955.6 \text{ m/s})^2/2 + (400 \text{ kg})(1055.6 \text{ m/s})^2/2}{(900 \text{ kg})(1000 \text{ m/s})^2/2} = 2.5 \times 10^{-3}.$$

117. (a) The thrust is Rv_{rel} where v_{rel} = 1200 m/s. For this to equal the weight Mg where M = 6100 kg, we must have R = (6100)(9.8)/1200 ≈ 50 kg/s.

(b) Using Eq. 9-42 with the additional effect due to gravity, we have

$$Rv_{rel} - Mg = Ma$$

so that requiring a = 21 m/s² leads to R = (6100)(9.8 + 21)/1200 = 1.6 × 10² kg/s.

118. We denote the mass of the car as M and that of the sumo wrestler as m. Let the initial velocity of the sumo wrestler be $v_0 > 0$ and the final velocity of the car be v. We apply the momentum conservation law.

(a) From $mv_0 = (M + m)v$ we get

$$v = \frac{mv_0}{M + m} = \frac{(242 \text{ kg})(5.3 \text{ m/s})}{2140 \text{ kg} + 242 \text{ kg}} = 0.54 \text{ m/s}.$$

(b) Since $v_{rel} = v_0$, we have

$$mv_0 = Mv + m(v + v_{rel}) = mv_0 + (M + m)v,$$

and obtain $v = 0$ for the final speed of the flatcar.

(c) Now $mv_0 = Mv + m(v − v_{rel})$, which leads to

$$v = \frac{m(v_0 + v_{rel})}{m+M} = \frac{(242 \text{ kg})(5.3 \text{ m/s} + 5.3 \text{ m/s})}{242 \text{ kg} + 2140 \text{ kg}} = 1.1 \text{ m/s}.$$

119. (a) Each block is assumed to have uniform density, so that the center of mass of each block is at its geometric center (the positions of which are given in the table [see problem statement] at $t = 0$). Plugging these positions (and the block masses) into Eq. 9-29 readily gives $x_{com} = -0.50$ m (at $t = 0$).

(b) Note that the left edge of block 2 (the middle of which is still at $x = 0$) is at $x = -2.5$ cm, so that at the moment they touch the right edge of block 1 is at $x = -2.5$ cm and thus the middle of block 1 is at $x = -5.5$ cm. Putting these positions (for the middles) and the block masses into Eq. 9-29 leads to $x_{com} = -1.83$ cm or -0.018 m (at $t = (1.445$ m$)/(0.75$ m/s$) = 1.93$ s).

(c) We could figure where the blocks are at $t = 4.0$ s and use Eq. 9-29 again, but it is easier (and provides more insight) to note that in the absence of *external* forces on the system the center of mass should move at constant velocity:

$$\vec{v}_{com} = \frac{m_1 \vec{v}_1 + m_2 \vec{v}_2}{m_1 + m_2} = 0.25 \text{ m/s } \hat{i}$$

as can be easily verified by putting in the values at $t = 0$. Thus,

$$x_{com} = x_{com\ initial} + \vec{v}_{com}\, t = (-0.50 \text{ m}) + (0.25 \text{ m/s})(4.0 \text{ s}) = +0.50 \text{ m}.$$

120. (a) Since the center of mass of the man-balloon system does not move, the balloon will move downward with a certain speed u relative to the ground as the man climbs up the ladder.

(b) The speed of the man relative to the ground is $v_g = v - u$. Thus, the speed of the center of mass of the system is

$$v_{com} = \frac{mv_g - Mu}{M+m} = \frac{m(v-u) - Mu}{M+m} = 0.$$

This yields

$$u = \frac{mv}{M+m} = \frac{(80 \text{ kg})(2.5 \text{ m/s})}{320 \text{ kg} + 80 \text{ kg}} = 0.50 \text{ m/s}.$$

(c) Now that there is no relative motion within the system, the speed of both the balloon and the man is equal to v_{com}, which is zero. So the balloon will again be stationary.

121. Using Eq. 9-67, we have after the elastic collision

$$v_{1f} = \frac{m_1 - m_2}{m_1 + m_2} v_{1i} = \frac{-200 \text{ g}}{600 \text{ g}} v_{1i} = -\frac{1}{3}(3.00 \text{ m/s}) = -1.00 \text{ m/s}.$$

(a) The impulse is therefore

$$J = m_1 v_{1f} - m_1 v_{1i} = (0.200 \text{ kg})(-1.00 \text{ m/s}) - (0.200 \text{ kg})(3.00 \text{ m/s}) = -0.800 \text{ N·s}$$
$$= -0.800 \text{ kg·m/s},$$

or $|J| = -0.800$ kg·m/s.

(b) For the completely inelastic collision Eq. 9-75 applies

$$v_{1f} = V = \frac{m_1}{m_1 + m_2} v_{1i} = +1.00 \text{ m/s}.$$

Now the impulse is

$$J = m_1 v_{1f} - m_1 v_{1i} = (0.200 \text{ kg})(1.00 \text{ m/s}) - (0.200 \text{ kg})(3.00 \text{ m/s}) = 0.400 \text{ N·s}$$
$$= 0.400 \text{ kg·m/s}.$$

122. We use Eq. 9-88 and simplify with $v_f - v_i = \Delta v$, and $v_{rel} = u$.

$$v_f - v_i = v_{rel} \ln\left(\frac{M_i}{M_f}\right) \Rightarrow \frac{M_f}{M_i} = e^{-\Delta v/u}$$

If $\Delta v = 2.2$ m/s and $u = 1000$ m/s, we obtain $\dfrac{M_i - M_f}{M_i} = 1 - e^{-0.0022} \approx 0.0022.$

123. This is a completely inelastic collision, followed by projectile motion. In the collision, we use momentum conservation.

$$\vec{p}_{shoes} = \vec{p}_{together} \Rightarrow (3.2 \text{ kg})(3.0 \text{ m/s}) = (5.2 \text{ kg})\vec{v}$$

Therefore, $\vec{v} = 1.8$ m/s toward the right as the combined system is projected from the edge of the table. Next, we can use the projectile motion material from Ch. 4 or the energy techniques of Ch. 8; we choose the latter.

$$K_{edge} + U_{edge} = K_{floor} + U_{floor}$$

$$\frac{1}{2}(5.2 \text{ kg})(1.8 \text{ m/s})^2 + (5.2 \text{ kg})(9.8 \text{ m/s}^2)(0.40 \text{ m}) = K_{floor} + 0$$

Therefore, the kinetic energy of the system right before hitting the floor is $K_{floor} = 29$ J.

124. We refer to the discussion in the textbook (Sample Problem 9-10, which uses the same notation that we use here) for some important details in the reasoning. We choose

rightward in Fig. 9-21 as our $+x$ direction. We use the notation $\vec{v}$ when we refer to velocities and v when we refer to speeds (which are necessarily positive). Since the algebra is fairly involved, we find it convenient to introduce the notation $\Delta m = m_2 - m_1$ (which, we note for later reference, is a positive-valued quantity).

(a) Since $\vec{v}_{1i} = +\sqrt{2gh_1}$ where $h_1 = 9.0$ cm, we have

$$\vec{v}_{1f} = \frac{m_1 - m_2}{m_1 + m_2} v_{1i} = -\frac{\Delta m}{m_1 + m_2}\sqrt{2gh_1}$$

which is to say that the *speed* of sphere 1 immediately after the collision is $v_{1f} = (\Delta m/(m_1 + m_2))\sqrt{2gh_1}$ and that $\vec{v}_{1f}$ points in the $-x$ direction. This leads (by energy conservation $m_1 g h_{1f} = \frac{1}{2} m_1 v_{1f}^2$) to

$$h_{1f} = \frac{v_{1f}^2}{2g} = \left(\frac{\Delta m}{m_1 + m_2}\right)^2 h_1 \;.$$

With $m_1 = 50$ g and $m_2 = 85$ g, this becomes $h_{1f} \approx 0.60$ cm.

(b) Eq. 9-68 gives

$$v_{2f} = \frac{2m_1}{m_1 + m_2} v_{1i} = \frac{2m_1}{m_1 + m_2}\sqrt{2gh_1}$$

which leads (by energy conservation $m_2 g h_{2f} = \frac{1}{2} m_2 v_{2f}^2$) to

$$h_{2f} = \frac{v_{2f}^2}{2g} = \left(\frac{2m_1}{m_1 + m_2}\right)^2 h_1 \;.$$

With $m_1 = 50$ g and $m_2 = 85$ g, this becomes $h_{2f} \approx 4.9$ cm.

(c) Fortunately, they hit again at the lowest point (as long as their amplitude of swing was "small" – this is further discussed in Chapter 16). At the risk of using cumbersome notation, we refer to the *next* set of heights as h_{1ff} and h_{2ff}. At the lowest point (before this second collision) sphere 1 has velocity $+\sqrt{2gh_{1f}}$ (rightward in Fig. 9-21) and sphere 2 has velocity $-\sqrt{2gh_{1f}}$ (that is, it points in the $-x$ direction). Thus, the velocity of sphere 1 immediately after the second collision is, using Eq. 9-75,

$$\vec{v}_{1f\!f} = \frac{m_1 - m_2}{m_1 + m_2}\sqrt{2gh_{1f}} + \frac{2m_2}{m_1 + m_2}\left(-\sqrt{2gh_{2f}}\right)$$

$$= \frac{-\Delta m}{m_1 + m_2}\left(\frac{\Delta m}{m_1 + m_2}\sqrt{2gh_1}\right) - \frac{2m_2}{m_1 + m_2}\left(\frac{2m_1}{m_1 + m_2}\sqrt{2gh_1}\right)$$

$$= -\frac{(\Delta m)^2 + 4m_1 m_2}{(m_1 + m_2)^2}\sqrt{2gh_1}.$$

This can be greatly simplified (by expanding $(\Delta m)^2$ and $(m_1 + m_2)^2$) to arrive at the conclusion that the speed of sphere 1 immediately after the second collision is simply $v_{1f\!f} = \sqrt{2gh_1}$ and that $\vec{v}_{1f\!f}$ points in the $-x$ direction. Energy conservation $(m_1 g h_{1f\!f} = \frac{1}{2} m_1 v_{1f\!f}^2)$ leads to

$$h_{1f\!f} = \frac{v_{1f\!f}^2}{2g} = h_1 = 9.0 \text{ cm}.$$

(d) One can reason (energy-wise) that $h_{1f\!f} = 0$ simply based on what we found in part (c). Still, it might be useful to see how this shakes out of the algebra. Eq. 9-76 gives the velocity of sphere 2 immediately after the second collision:

$$v_{2f\!f} = \frac{2m_1}{m_1 + m_2}\sqrt{2gh_{1f}} + \frac{m_2 - m_1}{m_1 + m_2}\left(-\sqrt{2gh_{2f}}\right)$$

$$= \frac{2m_1}{m_1 + m_2}\left(\frac{\Delta m}{m_1 + m_2}\sqrt{2gh_1}\right) + \frac{\Delta m}{m_1 + m_2}\left(\frac{-2m_1}{m_1 + m_2}\sqrt{2gh_1}\right)$$

which vanishes since $(2m_1)(\Delta m) - (\Delta m)(2m_1) = 0$. Thus, the second sphere (after the second collision) stays at the lowest point, which basically recreates the conditions at the start of the problem (so all subsequent swings-and-impacts, neglecting friction, can be easily predicted – as they are just replays of the first two collisions).

125. From mechanical energy conservation (or simply using Eq. 2-16 with $\vec{a} = g$ downward) we obtain

$$v = \sqrt{2gh} = \sqrt{2(9.8 \text{ m/s}^2)(6.0 \text{ m})} = 10.8 \text{ m/s}$$

for the speed just as the $m = 3000$-kg block makes contact with the pile. At the moment of "joining," they are a system of mass $M = 3500$ kg and speed V. With downward positive, momentum conservation leads to

$$mv = MV \Rightarrow V = \frac{(3000)(10.8)}{3500} = 9.3 \text{ m/s}.$$

Now this block-pile "object" must be rapidly decelerated over the small distance $d = 0.030$ m. Using Eq. 2-16 and choosing $+y$ downward, we have

$$0 = V^2 + 2ad \Rightarrow a = -\frac{9.3^2}{2(0.030)} = -1440$$

in SI units (m/s^2). Thus, the net force during the decelerating process has magnitude

$$M|a| = 5.0 \times 10^6 \text{ N}.$$

126. The momentum before the collision (with +x rightward) is

$$(6.0 \text{ kg})(8.0 \text{ m/s}) + (4.0 \text{ kg})(2.0 \text{ m/s}) = 56 \text{ kg} \cdot \text{m/s}.$$

(a) The total momentum at this instant is $(6.0 \text{ kg})(6.4 \text{ m/s}) + (4.0 \text{ kg})\vec{v}$. Since this must equal the initial total momentum (56, using SI units), then we find $\vec{v} = 4.4$ m/s.

(b) The initial kinetic energy was

$$\frac{1}{2}(6.0 \text{ kg})(8.0 \text{ m/s})^2 + \frac{1}{2}(4.0 \text{ kg})(2.0 \text{ m/s})^2 = 200 \text{ J}.$$

The kinetic energy at the instant described in part (a) is

$$\frac{1}{2}(6.0 \text{ kg})(6.4 \text{ m/s})^2 + \frac{1}{2}(4.0 \text{ kg})(4.4 \text{ m/s})^2 = 162 \text{ J}.$$

The "missing" 38 J is not dissipated since there is no friction; it is the energy stored in the spring at this instant when it is compressed. Thus, $U_e = 38$ J.

127. (a) The initial momentum of the system is zero, and it remains so as the electron and proton move toward each other. If p_e is the magnitude of the electron momentum at some instant (during their motion) and p_p is the magnitude of the proton momentum, then these must be equal (and their directions must be opposite) in order to maintain the zero total momentum requirement. Thus, the ratio of their momentum magnitudes is +1.

(b) With v_e and v_p being their respective speeds, we obtain (from the $p_e = p_p$ requirement)

$$m_e v_e = m_p v_p \Rightarrow v_e / v_p = m_p / m_e \approx 1830 \approx 1.83 \times 10^3.$$

(c) We can rewrite $K = \frac{1}{2}mv^2$ as $K = \frac{1}{2}p^2/m$ which immediately leads to

$$K_e / K_p = m_p / m_e \approx 1830 \approx 1.83 \times 10^3.$$

(d) Although the speeds (and kinetic energies) increase, they do so in the proportions indicated above. The answers stay the same.

432 CHAPTER 9

128. In the momentum relationships, we could as easily work with weights as with masses, but because part (b) of this problem asks for kinetic energy—we will find the masses at the outset: $m_1 = 280 \times 10^3/9.8 = 2.86 \times 10^4$ kg and $m_2 = 210 \times 10^3/9.8 = 2.14 \times 10^4$ kg. Both cars are moving in the $+x$ direction: $v_{1i} = 1.52$ m/s and $v_{2i} = 0.914$ m/s.

(a) If the collision is completely elastic, momentum conservation leads to a final speed of

$$V = \frac{m_1 v_{1i} + m_2 v_{2i}}{m_1 + m_2} = 1.26 \text{ m/s}.$$

(b) We compute the total initial kinetic energy and subtract from it the final kinetic energy.

$$K_i - K_f = \frac{1}{2}m_1 v_{1i}^2 + \frac{1}{2}m_2 v_{2i}^2 - \frac{1}{2}(m_1 + m_2)V^2 = 2.25 \times 10^3 \text{ J}.$$

(c) Using Eq. 9-76, we find

$$v_{2f} = \frac{2m_1}{m_1 + m_2} v_{1i} + \frac{m_2 - m_1}{m_1 + m_2} v_{2i} = 1.61 \text{ m/s}$$

(d) Using Eq. 9-75, we find

$$v_{1f} = \frac{m_1 - m_2}{m_1 + m_2} v_{1i} + \frac{2m_2}{m_1 + m_2} v_{2i} = 1.00 \text{ m/s}.$$

129. Using Eq. 9-68 with $m_1 = 3.0$ kg, $v_{1i} = 8.0$ m/s and $v_{2f} = 6.0$ m/s, then

$$v_{2f} = \frac{2m_1}{m_1 + m_2} v_{1i} \Rightarrow m_2 = m_1 \left(\frac{2v_{1i}}{v_{2f}} - 1 \right)$$

leads to $m_2 = M = 5.0$ kg.

130. (a) The center of mass does not move in the absence of external forces (since it was initially at rest).

(b) They collide at their center of mass. If the initial coordinate of P is $x = 0$ and the initial coordinate of Q is $x = 1.0$ m, then Eq. 9-5 gives

$$x_{com} = \frac{m_1 x_1 + m_2 x_2}{m_1 + m_2} = \frac{0 + (0.30 \text{ kg})(1.0 \text{ m})}{0.1 \text{ kg} + 0.3 \text{ kg}} = 0.75 \text{ m}.$$

Thus, they collide at a point 0.75 m from P's original position.

131. The velocities of m_1 and m_2 just after the collision with each other are given by Eq. 9-75 and Eq. 9-76 (setting $v_{1i} = 0$):

$$v_{1f} = \frac{2m_2}{m_1 + m_2} v_{2i}$$

$$v_{2f} = \frac{m_2 - m_1}{m_1 + m_2} v_{2i}$$

After bouncing off the wall, the velocity of m_2 becomes $-v_{2f}$. In these terms, the problem requires

$$v_{1f} = -v_{2f}$$

$$\frac{2m_2}{m_1 + m_2} v_{2i} = -\frac{m_2 - m_1}{m_1 + m_2} v_{2i}$$

which simplifies to

$$2m_2 = -(m_2 - m_1) \Rightarrow m_2 = \frac{m_1}{3}.$$

With $m_1 = 6.6$ kg, we have $m_2 = 2.2$ kg.

132. Momentum conservation (with SI units understood) gives

$$m_1(v_f - 20) + (M - m_1)v_f = Mv_i$$

which yields

$$v_f = \frac{Mv_i + 20\, m_1}{M} = v_i + 20\,\frac{m_1}{M} = 40 + 20\,(m_1/M).$$

(a) The minimum value of v_f is 40 m/s,

(b) The final speed v_f reaches a minimum as m_1 approaches zero.

(c) The maximum value of v_f is 60 m/s.

(d) The final speed v_f reaches a maximum as m_1 approaches M.

133. By the principle of momentum conservation, we must have

$$m_1 \vec{v}_1 + m_2 \vec{v}_2 + m_3 \vec{v}_3 = 0,$$

which implies

$$\vec{v}_3 = -\frac{m_1 \vec{v}_1 + m_2 \vec{v}_2}{m_3}.$$

With

$$m_1 \vec{v}_1 = (0.500)(10.0\hat{i} + 12.0\hat{j}) = 5.00\hat{i} + 6.00\hat{j}$$
$$m_2 \vec{v}_2 = (0.750)(14.0)(\cos 110°\hat{i} + \sin 110°\hat{j}) = -3.59\hat{i} + 9.87\hat{j}$$

(in SI units) and $m_3 = m - m_1 - m_2 = (2.65 - 0.500 - 0.750)\text{kg} = 1.40 \text{ kg}$, we solve for $\vec{v}_3$ and obtain $\vec{v}_3 = (-1.01 \text{ m/s})\hat{i} + (-11.3 \text{ m/s})\hat{j}$.

(a) The magnitude of $\vec{v}_3$ is $|\vec{v}_3| = 11.4$ m/s.

(b) Its angle is 264.9°, which means it is 95.1° clockwise from the +x axis.

134. Using Eq. 9-75 and Eq. 9-76, we find after the collision

(a) $v_{1f} = \dfrac{m_1 - m_2}{m_1 + m_2} v_{1i} + \dfrac{2m_2}{m_1 + m_2} v_{2i} = (-3.8 \text{ m/s})\hat{i}$, and

(b) $v_{2f} = \dfrac{2m_1}{m_1 + m_2} v_{1i} + \dfrac{m_2 - m_1}{m_1 + m_2} v_{2i} = (7.2 \text{ m/s})\hat{i}$.

135. We use Eq. 9-5.

(a) The x coordinate of the center of mass is

$$x_{com} = \dfrac{m_1 x_1 + m_2 x_2 + m_3 x_3 + m_4 x_4}{m_1 + m_2 + m_3 + m_4} = \dfrac{0 + (4)(3) + 0 + (12)(-1)}{m_1 + m_2 + m_3 + m_4} = 0.$$

(b) The y coordinate of the center of mass is

$$y_{com} = \dfrac{m_1 y_1 + m_2 y_2 + m_3 y_3 + m_4 y_4}{m_1 + m_2 + m_3 + m_4} = \dfrac{(2)(3) + 0 + (3)(-2) + 0}{m_1 + m_2 + m_3 + m_4} = 0.$$

(c) We now use Eq. 9-17:

$$\vec{v}_{com} = \dfrac{m_1 \vec{v}_1 + m_2 \vec{v}_2 + m_3 \vec{v}_3 + m_4 \vec{v}_4}{m_1 + m_2 + m_3 + m_4}$$
$$= \dfrac{(2)(-9\hat{j}) + (4)(6\hat{i}) + (3)(6\hat{j}) + (12)(-2\hat{i})}{m_1 + m_2 + m_3 + m_4} = 0.$$

136. Let $M = 22.7$ kg and $m = 3.63$ be the mass of the sled and the cat, respectively. Using the principle of momentum conservation, the speed of the first sled after the cat's first jump with a speed of $v_i = 3.05$ m/s is

$$v_{1f} = \dfrac{mv_i}{M} = 0.488 \text{ m/s}.$$

On the other hand, as the cat lands on the second sled, it sticks to it and the system (sled plus cat) moves forward with a speed

$$v_{2f} = \frac{mv_i}{M+m} = 0.4205 \text{ m/s}.$$

When the cat makes the second jump back to the first sled with a speed v_i, momentum conservation implies

$$Mv_{2ff} = mv_i + (M+m)v_{2f} = mv_i + mv_i = 2mv_i$$

which yields

$$v_{2ff} = \frac{2mv_i}{M} = 0.975 \text{ m/s}.$$

After the cat lands on the first sled, the entire system (cat and the sled) again moves together. By momentum conservation, we have

$$(M+m)v_{1ff} = mv_i + Mv_{1f} = mv_i + mv_i = 2mv_i$$

or

$$v_{1ff} = \frac{2mv_i}{M+m} = 0.841 \text{ m/s}.$$

(a) From the above, we conclude that the first sled moves with a speed $v_{1ff} = 0.841$ m/s a after the cat's two jumps.

(b) Similarly, the speed of the second sled is $v_{2ff} = 0.975$ m/s.

Chapter 10

1. The problem asks us to assume v_{com} and ω are constant. For consistency of units, we write

$$v_{com} = (85\,\text{mi/h})\left(\frac{5280\,\text{ft/mi}}{60\,\text{min/h}}\right) = 7480\,\text{ft/min}.$$

Thus, with $\Delta x = 60\,\text{ft}$, the time of flight is

$$t = \Delta x / v_{com} = (60\,\text{ft})/(7480\,\text{ft/min}) = 0.00802\,\text{min}.$$

During that time, the angular displacement of a point on the ball's surface is

$$\theta = \omega t = (1800\,\text{rev/min})(0.00802\,\text{min}) \approx 14\,\text{rev}.$$

2. (a) The second hand of the smoothly running watch turns through 2π radians during 60 s. Thus,

$$\omega = \frac{2\pi}{60} = 0.105\,\text{rad/s}.$$

(b) The minute hand of the smoothly running watch turns through 2π radians during 3600 s. Thus,

$$\omega = \frac{2\pi}{3600} = 1.75\times 10^{-3}\,\text{rad/s}.$$

(c) The hour hand of the smoothly running 12-hour watch turns through 2π radians during 43200 s. Thus,

$$\omega = \frac{2\pi}{43200} = 1.45\times 10^{-4}\,\text{rad/s}.$$

3. Applying Eq. 2-15 to the vertical axis (with $+y$ downward) we obtain the free-fall time:

$$\Delta y = v_{0y}t + \frac{1}{2}gt^2 \Rightarrow t = \sqrt{\frac{2(10\,\text{m})}{9.8\,\text{m/s}^2}} = 1.4\,\text{s}.$$

Thus, by Eq. 10-5, the magnitude of the average angular velocity is

$$\omega_{avg} = \frac{(2.5\,\text{rev})(2\pi\,\text{rad/rev})}{1.4\,\text{s}} = 11\,\text{rad/s}.$$

4. If we make the units explicit, the function is

$$\theta = (4.0 \text{ rad}/\text{s})t - (3.0 \text{ rad}/\text{s}^2)t^2 + (1.0 \text{ rad}/\text{s}^3)t^3$$

but generally we will proceed as shown in the problem—letting these units be understood. Also, in our manipulations we will generally not display the coefficients with their proper number of significant figures.

(a) Eq. 10-6 leads to

$$\omega = \frac{d}{dt}(4t - 3t^2 + t^3) = 4 - 6t + 3t^2.$$

Evaluating this at $t = 2$ s yields $\omega_2 = 4.0$ rad/s.

(b) Evaluating the expression in part (a) at $t = 4$ s gives $\omega_4 = 28$ rad/s.

(c) Consequently, Eq. 10-7 gives

$$\alpha_{avg} = \frac{\omega_4 - \omega_2}{4 - 2} = 12 \text{ rad}/\text{s}^2.$$

(d) And Eq. 10-8 gives

$$\alpha = \frac{d\omega}{dt} = \frac{d}{dt}(4 - 6t + 3t^2) = -6 + 6t.$$

Evaluating this at $t = 2$ s produces $\alpha_2 = 6.0$ rad/s^2.

(e) Evaluating the expression in part (d) at $t = 4$ s yields $\alpha_4 = 18$ rad/s^2. We note that our answer for α_{avg} does turn out to be the arithmetic average of α_2 and α_4 but point out that this will not always be the case.

5. The falling is the type of constant-acceleration motion you had in Chapter 2. The time it takes for the buttered toast to hit the floor is

$$\Delta t = \sqrt{\frac{2h}{g}} = \sqrt{\frac{2(0.76 \text{ m})}{9.8 \text{ m/s}^2}} = 0.394 \text{ s}.$$

(a) The smallest angle turned for the toast to land butter-side down is $\Delta\theta_{min} = 0.25$ rev $= \pi/2$ rad. This corresponds to an angular speed of

$$\omega_{min} = \frac{\Delta\theta_{min}}{\Delta t} = \frac{\pi/2 \text{ rad}}{0.394 \text{ s}} = 4.0 \text{ rad/s}.$$

(b) The largest angle (less than 1 revolution) turned for the toast to land butter-side down is $\Delta\theta_{max} = 0.75$ rev $= 3\pi/2$ rad. This corresponds to an angular speed of

$$\omega_{max} = \frac{\Delta\theta_{max}}{\Delta t} = \frac{3\pi/2 \text{ rad}}{0.394 \text{ s}} = 12.0 \text{ rad/s}.$$

6. If we make the units explicit, the function is

$$\theta = 2.0 \text{ rad} + \left(4.0 \text{ rad/s}^2\right)t^2 + \left(2.0 \text{ rad/s}^3\right)t^3$$

but in some places we will proceed as indicated in the problem—by letting these units be understood.

(a) We evaluate the function θ at $t = 0$ to obtain $\theta_0 = 2.0$ rad.

(b) The angular velocity as a function of time is given by Eq. 10-6:

$$\omega = \frac{d\theta}{dt} = \left(8.0 \text{ rad/s}^2\right)t + \left(6.0 \text{ rad/s}^3\right)t^2$$

which we evaluate at $t = 0$ to obtain $\omega_0 = 0$.

(c) For $t = 4.0$ s, the function found in the previous part is

$$\omega_4 = (8.0)(4.0) + (6.0)(4.0)^2 = 128 \text{ rad/s}.$$

If we round this to two figures, we obtain $\omega_4 \approx 1.3 \times 10^2$ rad/s.

(d) The angular acceleration as a function of time is given by Eq. 10-8:

$$\alpha = \frac{d\omega}{dt} = 8.0 \text{ rad/s}^2 + \left(12 \text{ rad/s}^3\right)t$$

which yields $\alpha_2 = 8.0 + (12)(2.0) = 32$ rad/s^2 at $t = 2.0$ s.

(e) The angular acceleration, given by the function obtained in the previous part, depends on time; it is not constant.

7. (a) To avoid touching the spokes, the arrow must go through the wheel in not more than

$$\Delta t = \frac{1/8 \text{ rev}}{2.5 \text{ rev}/\text{s}} = 0.050 \text{ s}.$$

The minimum speed of the arrow is then $v_{min} = \dfrac{20 \text{ cm}}{0.050 \text{ s}} = 400 \text{ cm/s} = 4.0 \text{ m/s}$.

(b) No—there is no dependence on radial position in the above computation.

8. (a) We integrate (with respect to time) the $\alpha = 6.0 t^4 - 4.0 t^2$ expression, taking into account that the initial angular velocity is 2.0 rad/s. The result is

$$\omega = 1.2\, t^5 - 1.33\, t^3 + 2.0.$$

(b) Integrating again (and keeping in mind that $\theta_o = 1$) we get

$$\theta = 0.20 t^6 - 0.33\, t^4 + 2.0\, t + 1.0\,.$$

9. We assume the sense of initial rotation is positive. Then, with $\omega_0 = +120$ rad/s and $\omega = 0$ (since it stops at time t), our angular acceleration ("deceleration") will be negative-valued: $\alpha = -4.0$ rad/s^2.

(a) We apply Eq. 10-12 to obtain t.

$$\omega = \omega_0 + \alpha t \quad \Rightarrow \quad t = \dfrac{0 - 120 \text{ rad/s}}{-4.0 \text{ rad/s}^2} = 30 \text{ s}.$$

(b) And Eq. 10-15 gives

$$\theta = \dfrac{1}{2}(\omega_0 + \omega) t = \dfrac{1}{2}(120 \text{ rad/s} + 0)(30 \text{ s}) = 1.8 \times 10^3 \text{ rad}.$$

Alternatively, Eq. 10-14 could be used if it is desired to only use the given information (as opposed to using the result from part (a)) in obtaining θ. If using the result of part (a) is acceptable, then any angular equation in Table 10-1 (except Eq. 10-12) can be used to find θ.

10. (a) We assume the sense of rotation is positive. Applying Eq. 10-12, we obtain

$$\omega = \omega_0 + \alpha t \quad \Rightarrow \quad \alpha = \dfrac{(3000 - 1200) \text{ rev/min}}{(12/60) \text{ min}} = 9.0 \times 10^3 \text{ rev/min}^2.$$

(b) And Eq. 10-15 gives

$$\theta = \dfrac{1}{2}(\omega_0 + \omega) t = \dfrac{1}{2}(1200 \text{ rev/min} + 3000 \text{ rev/min})\left(\dfrac{12}{60} \text{ min}\right) = 4.2 \times 10^2 \text{ rev}.$$

11. (a) With $\omega = 0$ and $\alpha = -4.2$ rad/s^2, Eq. 10-12 yields $t = -\omega_0/\alpha = 3.00$ s.

(b) Eq. 10-4 gives $\theta - \theta_0 = -\omega_0^2 / 2\alpha = 18.9$ rad.

12. We assume the sense of rotation is positive, which (since it starts from rest) means all quantities (angular displacements, accelerations, etc.) are positive-valued.

(a) The angular acceleration satisfies Eq. 10-13:

$$25 \text{ rad} = \frac{1}{2}\alpha(5.0 \text{ s})^2 \Rightarrow \alpha = 2.0 \text{ rad/s}^2.$$

(b) The average angular velocity is given by Eq. 10-5:

$$\omega_{avg} = \frac{\Delta\theta}{\Delta t} = \frac{25 \text{ rad}}{5.0 \text{ s}} = 5.0 \text{ rad/s}.$$

(c) Using Eq. 10-12, the instantaneous angular velocity at $t = 5.0$ s is

$$\omega = (2.0 \text{ rad/s}^2)(5.0 \text{ s}) = 10 \text{ rad/s}.$$

(d) According to Eq. 10-13, the angular displacement at $t = 10$ s is

$$\theta = \omega_0 + \frac{1}{2}\alpha t^2 = 0 + \frac{1}{2}(2.0 \text{ rad/s}^2)(10 \text{ s})^2 = 100 \text{ rad}.$$

Thus, the displacement between $t = 5$ s and $t = 10$ s is $\Delta\theta = 100$ rad $-$ 25 rad $=$ 75 rad.

13. We take $t = 0$ at the start of the interval and take the sense of rotation as positive. Then at the end of the $t = 4.0$ s interval, the angular displacement is $\theta = \omega_0 t + \frac{1}{2}\alpha t^2$. We solve for the angular velocity at the start of the interval:

$$\omega_0 = \frac{\theta - \frac{1}{2}\alpha t^2}{t} = \frac{120 \text{ rad} - \frac{1}{2}(3.0 \text{ rad/s}^2)(4.0 \text{ s})^2}{4.0 \text{ s}} = 24 \text{ rad/s}.$$

We now use $\omega = \omega_0 + \alpha t$ (Eq. 10-12) to find the time when the wheel is at rest:

$$t = -\frac{\omega_0}{\alpha} = -\frac{24 \text{ rad/s}}{3.0 \text{ rad/s}^2} = -8.0 \text{ s}.$$

That is, the wheel started from rest 8.0 s before the start of the described 4.0 s interval.

14. (a) Eq. 10-13 gives

$$\theta - \theta_0 = \omega_0 t + \frac{1}{2}\alpha t^2 = 0 + \frac{1}{2}(1.5 \text{ rad/s}^2)t_1^2$$

where $\theta - \theta_0 = (2 \text{ rev})(2\pi \text{ rad/rev})$. Therefore, $t_1 = 4.09$ s.

(b) We can find the time to go through a full 4 rev (using the same equation to solve for a new time t_2) and then subtract the result of part (a) for t_1 in order to find this answer.

$$(4 \text{ rev})(2\pi \text{ rad/rev}) = 0 + \frac{1}{2}(1.5 \text{ rad/s}^2)t_2^2 \quad \Rightarrow \quad t_2 = 5.789 \text{ s}.$$

Thus, the answer is $5.789 \text{ s} - 4.093 \text{ s} \approx 1.70 \text{ s}$.

15. The problem has (implicitly) specified the positive sense of rotation. The angular acceleration of magnitude 0.25 rad/s² in the negative direction is assumed to be constant over a large time interval, including negative values (for t).

(a) We specify θ_{max} with the condition $\omega = 0$ (this is when the wheel reverses from positive rotation to rotation in the negative direction). We obtain θ_{max} using Eq. 10-14:

$$\theta_{max} = -\frac{\omega_0^2}{2\alpha} = -\frac{(4.7 \text{ rad/s})^2}{2(-0.25 \text{ rad/s}^2)} = 44 \text{ rad}.$$

(b) We find values for t_1 when the angular displacement (relative to its orientation at $t = 0$) is $\theta_1 = 22$ rad (or 22.09 rad if we wish to keep track of accurate values in all intermediate steps and only round off on the final answers). Using Eq. 10-13 and the quadratic formula, we have

$$\theta_1 = \omega_0 t_1 + \frac{1}{2}\alpha t_1^2 \Rightarrow t_1 = \frac{-\omega_0 \pm \sqrt{\omega_0^2 + 2\theta_1 \alpha}}{\alpha}$$

which yields the two roots 5.5 s and 32 s. Thus, the first time the reference line will be at $\theta_1 = 22$ rad is $t = 5.5$ s.

(c) The second time the reference line will be at $\theta_1 = 22$ rad is $t = 32$ s.

(d) We find values for t_2 when the angular displacement (relative to its orientation at $t = 0$) is $\theta_2 = -10.5$ rad. Using Eq. 10-13 and the quadratic formula, we have

$$\theta_2 = \omega_0 t_2 + \frac{1}{2}\alpha t_2^2 \Rightarrow t_2 = \frac{-\omega_0 \pm \sqrt{\omega_0^2 + 2\theta_2 \alpha}}{\alpha}$$

which yields the two roots -2.1 s and 40 s. Thus, at $t = -2.1$ s the reference line will be at $\theta_2 = -10.5$ rad.

(e) At $t = 40$ s the reference line will be at $\theta_2 = -10.5$ rad.

(f) With radians and seconds understood, the graph of θ versus t is shown below (with the points found in the previous parts indicated as small circles).

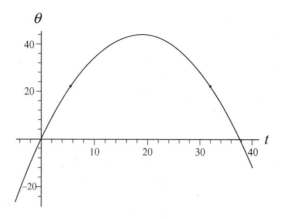

16. The wheel starts turning from rest ($\omega_0 = 0$) at $t = 0$, and accelerates uniformly at $\alpha > 0$, which makes our choice for positive sense of rotation. At t_1 its angular velocity is $\omega_1 = +10$ rev/s, and at t_2 its angular velocity is $\omega_2 = +15$ rev/s. Between t_1 and t_2 it turns through $\Delta\theta = 60$ rev, where $t_2 - t_1 = \Delta t$.

(a) We find α using Eq. 10-14:

$$\omega_2^2 = \omega_1^2 + 2\alpha\Delta\theta \implies \alpha = \frac{(15 \text{ rev/s})^2 - (10 \text{ rev/s})^2}{2(60 \text{ rev})} = 1.04 \text{ rev/s}^2$$

which we round off to 1.0 rev/s^2.

(b) We find Δt using Eq. 10-15:

$$\Delta\theta = \frac{1}{2}(\omega_1 + \omega_2)\Delta t \implies \Delta t = \frac{2(60 \text{ rev})}{10 \text{ rev/s} + 15 \text{ rev/s}} = 4.8 \text{ s}.$$

(c) We obtain t_1 using Eq. 10-12: $\omega_1 = \omega_0 + \alpha t_1 \implies t_1 = \dfrac{10 \text{ rev/s}}{1.04 \text{ rev/s}^2} = 9.6$ s.

(d) Any equation in Table 10-1 involving θ can be used to find θ_1 (the angular displacement during $0 \le t \le t_1$); we select Eq. 10-14.

$$\omega_1^2 = \omega_0^2 + 2\alpha\theta_1 \implies \theta_1 = \frac{(10 \text{ rev/s})^2}{2(1.04 \text{ rev/s}^2)} = 48 \text{ rev.}$$

17. The wheel has angular velocity $\omega_0 = +1.5$ rad/s $= +0.239$ rev/s at $t = 0$, and has constant value of angular acceleration $\alpha < 0$, which indicates our choice for positive sense of rotation. At t_1 its angular displacement (relative to its orientation at $t = 0$) is $\theta_1 =$

+20 rev, and at t_2 its angular displacement is $\theta_2 = +40$ rev and its angular velocity is $\omega_2 = 0$.

(a) We obtain t_2 using Eq. 10-15:

$$\theta_2 = \frac{1}{2}(\omega_0 + \omega_2)t_2 \Rightarrow t_2 = \frac{2(40 \text{ rev})}{0.239 \text{ rev/s}} = 335 \text{ s}$$

which we round off to $t_2 \approx 3.4 \times 10^2$ s.

(b) Any equation in Table 10-1 involving α can be used to find the angular acceleration; we select Eq. 10-16.

$$\theta_2 = \omega_2 t_2 - \frac{1}{2}\alpha t_2^2 \Rightarrow \alpha = -\frac{2(40 \text{ rev})}{(335 \text{ s})^2} = -7.12 \times 10^{-4} \text{ rev/s}^2$$

which we convert to $\alpha = -4.5 \times 10^{-3}$ rad/s².

(c) Using $\theta_1 = \omega_0 t_1 + \frac{1}{2}\alpha t_1^2$ (Eq. 10-13) and the quadratic formula, we have

$$t_1 = \frac{-\omega_0 \pm \sqrt{\omega_0^2 + 2\theta_1 \alpha}}{\alpha} = \frac{-(0.239 \text{ rev/s}) \pm \sqrt{(0.239 \text{ rev/s})^2 + 2(20 \text{ rev})(-7.12 \times 10^{-4} \text{ rev/s}^2)}}{-7.12 \times 10^{-4} \text{ rev/s}^2}$$

which yields two positive roots: 98 s and 572 s. Since the question makes sense only if $t_1 < t_2$ we conclude the correct result is $t_1 = 98$ s.

18. Converting $33\frac{1}{3}$ rev/min to radians-per-second, we get $\omega = 3.49$ rad/s. Combining $v = \omega r$ (Eq. 10-18) with $\Delta t = d/v$ where Δt is the time between bumps (a distance d apart), we arrive at the rate of striking bumps:

$$\frac{1}{\Delta t} = \frac{\omega r}{d} \approx 199/\text{s}.$$

19. We assume the given rate of 1.2×10^{-3} m/y is the linear speed of the top; it is also possible to interpret it as just the horizontal component of the linear speed but the difference between these interpretations is arguably negligible. Thus, Eq. 10-18 leads to

$$\omega = \frac{1.2 \times 10^{-3} \text{ m/y}}{55 \text{ m}} = 2.18 \times 10^{-5} \text{ rad/y}$$

which we convert (since there are about 3.16×10^7 s in a year) to $\omega = 6.9 \times 10^{-13}$ rad/s.

20. (a) Using Eq. 10-6, the angular velocity at $t = 5.0$s is

$$\omega = \left.\frac{d\theta}{dt}\right|_{t=5.0} = \left.\frac{d}{dt}(0.30t^2)\right|_{t=5.0} = 2(0.30)(5.0) = 3.0 \text{ rad / s}.$$

(b) Eq. 10-18 gives the linear speed at $t = 5.0$s: $v = \omega r = (3.0 \text{ rad/s})(10 \text{ m}) = 30$ m/s.

(c) The angular acceleration is, from Eq. 10-8,

$$\alpha = \frac{d\omega}{dt} = \frac{d}{dt}(0.60t) = 0.60 \text{ rad / s}^2.$$

Then, the tangential acceleration at $t = 5.0$s is, using Eq. 10-22,

$$a_t = r\alpha = (10 \text{ m})(0.60 \text{ rad / s}^2) = 6.0 \text{ m / s}^2.$$

(d) The radial (centripetal) acceleration is given by Eq. 10-23:

$$a_r = \omega^2 r = (3.0 \text{ rad / s})^2 (10 \text{ m}) = 90 \text{ m / s}^2.$$

21. (a) We obtain

$$\omega = \frac{(200 \text{ rev / min})(2\pi \text{ rad / rev})}{60 \text{ s / min}} = 20.9 \text{ rad / s}.$$

(b) With $r = 1.20/2 = 0.60$ m, Eq. 10-18 gives $v = r\omega = (0.60 \text{ m})(20.9 \text{ rad/s}) = 12.5$ m/s.

(c) With $t = 1$ min, $\omega = 1000$ rev/min and $\omega_0 = 200$ rev/min, Eq. 10-12 gives

$$\alpha = \frac{\omega - \omega_0}{t} = 800 \text{ rev / min}^2.$$

(d) With the same values used in part (c), Eq. 10-15 becomes

$$\theta = \frac{1}{2}(\omega_0 + \omega)t = \frac{1}{2}(200 \text{ rev/min} + 1000 \text{ rev/min})(1.0 \text{ min}) = 600 \text{ rev}.$$

22. First, we convert the angular velocity: $\omega = (2000 \text{ rev/min})(2\pi/60) = 209$ rad/s. Also, we convert the plane's speed to SI units: $(480)(1000/3600) = 133$ m/s. We use Eq. 10-18 in part (a) and (implicitly) Eq. 4-39 in part (b).

(a) The speed of the tip as seen by the pilot is $v_t = \omega r = (209 \text{ rad/s})(1.5 \text{ m}) = 314 \text{ m/s}$, which (since the radius is given to only two significant figures) we write as $v_t = 3.1 \times 10^2$ m/s.

(b) The plane's velocity $\vec{v}_p$ and the velocity of the tip $\vec{v}_t$ (found in the plane's frame of reference), in any of the tip's positions, must be perpendicular to each other. Thus, the speed as seen by an observer on the ground is

$$v = \sqrt{v_p^2 + v_t^2} = \sqrt{(133 \text{ m/s})^2 + (314 \text{ m/s})^2} = 3.4 \times 10^2 \text{ m/s}.$$

23. (a) Converting from hours to seconds, we find the angular velocity (assuming it is positive) from Eq. 10-18:

$$\omega = \frac{v}{r} = \frac{(2.90 \times 10^4 \text{ km/h})(1.000 \text{ h}/3600 \text{ s})}{3.22 \times 10^3 \text{ km}} = 2.50 \times 10^{-3} \text{ rad/s}.$$

(b) The radial (or centripetal) acceleration is computed according to Eq. 10-23:

$$a_r = \omega^2 r = (2.50 \times 10^{-3} \text{ rad/s})^2 (3.22 \times 10^6 \text{ m}) = 20.2 \text{ m/s}^2.$$

(c) Assuming the angular velocity is constant, then the angular acceleration and the tangential acceleration vanish, since

$$\alpha = \frac{d\omega}{dt} = 0 \text{ and } a_t = r\alpha = 0.$$

24. The function $\theta = \xi e^{\beta t}$ where $\xi = 0.40$ rad and $\beta = 2$ s^{-1} is describing the angular coordinate of a line (which is marked in such a way that all points on it have the same value of angle at a given time) on the object. Taking derivatives with respect to time leads to $\frac{d\theta}{dt} = \xi \beta e^{\beta t}$ and $\frac{d^2\theta}{dt^2} = \xi \beta^2 e^{\beta t}$.

(a) Using Eq. 10-22, we have $a_t = \alpha r = \frac{d^2\theta}{dt^2} r = 6.4$ cm/s^2.

(b) Using Eq. 10-23, we get $a_r = \omega^2 r = \left(\frac{d\theta}{dt}\right)^2 r = 2.6$ cm/s^2.

25. (a) The upper limit for centripetal acceleration (same as the radial acceleration – see Eq. 10-23) places an upper limit of the rate of spin (the angular velocity ω) by considering a point at the rim ($r = 0.25$ m). Thus, $\omega_{max} = \sqrt{a/r} = 40$ rad/s. Now we apply Eq. 10-15 to first half of the motion (where $\omega_0 = 0$):

$$\theta - \theta_0 = \tfrac{1}{2}(\omega_0 + \omega)t \Rightarrow 400 \text{ rad} = \tfrac{1}{2}(0 + 40 \text{ rad/s})t$$

which leads to $t = 20$ s. The second half of the motion takes the same amount of time (the process is essentially the reverse of the first); the total time is therefore 40 s.

(b) Considering the first half of the motion again, Eq. 10-11 leads to

$$\omega = \omega_0 + \alpha t \quad \Rightarrow \quad \alpha = \frac{40 \text{ rad/s}}{20 \text{ s}} = 2.0 \text{ rad/s}^2.$$

26. (a) The tangential acceleration, using Eq. 10-22, is

$$a_t = \alpha r = (14.2 \text{ rad/s}^2)(2.83 \text{ cm}) = 40.2 \text{ cm/s}^2.$$

(b) In rad/s, the angular velocity is $\omega = (2760)(2\pi/60) = 289$ rad/s, so

$$a_r = \omega^2 r = (289 \text{ rad/s})^2 (0.0283 \text{ m}) = 2.36 \times 10^3 \text{ m/s}^2.$$

(c) The angular displacement is, using Eq. 10-14,

$$\theta = \frac{\omega^2}{2\alpha} = \frac{(289 \text{ rad/s})^2}{2(14.2 \text{ rad/s}^2)} = 2.94 \times 10^3 \text{ rad}.$$

Then, using Eq. 10-1, the distance traveled is

$$s = r\theta = (0.0283 \text{ m})(2.94 \times 10^3 \text{ rad}) = 83.2 \text{ m}.$$

27. (a) In the time light takes to go from the wheel to the mirror and back again, the wheel turns through an angle of $\theta = 2\pi/500 = 1.26 \times 10^{-2}$ rad. That time is

$$t = \frac{2\ell}{c} = \frac{2(500 \text{ m})}{2.998 \times 10^8 \text{ m/s}} = 3.34 \times 10^{-6} \text{ s}$$

so the angular velocity of the wheel is

$$\omega = \frac{\theta}{t} = \frac{1.26 \times 10^{-2} \text{ rad}}{3.34 \times 10^{-6} \text{ s}} = 3.8 \times 10^3 \text{ rad/s}.$$

(b) If r is the radius of the wheel, the linear speed of a point on its rim is

$$v = \omega r = (3.8 \times 10^3 \text{ rad/s})(0.050 \text{ m}) = 1.9 \times 10^2 \text{ m/s}.$$

28. (a) The angular acceleration is

$$\alpha = \frac{\Delta\omega}{\Delta t} = \frac{0-150 \text{ rev/min}}{(2.2 \text{ h})(60 \text{ min}/1\text{h})} = -1.14 \text{ rev/min}^2.$$

(b) Using Eq. 10-13 with $t = (2.2)(60) = 132$ min, the number of revolutions is

$$\theta = \omega_0 t + \frac{1}{2}\alpha t^2 = (150 \text{ rev/min})(132 \text{ min}) + \frac{1}{2}(-1.14 \text{ rev/min}^2)(132 \text{ min})^2 = 9.9 \times 10^3 \text{ rev}.$$

(c) With $r = 500$ mm, the tangential acceleration is

$$a_t = \alpha r = (-1.14 \text{ rev/min}^2)\left(\frac{2\pi \text{ rad}}{1 \text{ rev}}\right)\left(\frac{1 \text{ min}}{60 \text{ s}}\right)^2 (500 \text{ mm})$$

which yields $a_t = -0.99$ mm/s^2.

(d) The angular speed of the flywheel is

$$\omega = (75 \text{ rev/min})(2\pi \text{ rad/rev})(1 \text{ min}/60 \text{ s}) = 7.85 \text{ rad/s}.$$

With $r = 0.50$ m, the radial (or centripetal) acceleration is given by Eq. 10-23:

$$a_r = \omega^2 r = (7.85 \text{ rad/s})^2 (0.50 \text{ m}) \approx 31 \text{ m/s}^2$$

which is much bigger than a_t. Consequently, the magnitude of the acceleration is

$$|\vec{a}| = \sqrt{a_r^2 + a_t^2} \approx a_r = 31 \text{ m/s}^2.$$

29. (a) Earth makes one rotation per day and 1 d is (24 h)(3600 s/h) = 8.64×10^4 s, so the angular speed of Earth is

$$\omega = \frac{2\pi \text{ rad}}{8.64 \times 10^4 \text{ s}} = 7.3 \times 10^{-5} \text{ rad/s}.$$

(b) We use $v = \omega r$, where r is the radius of its orbit. A point on Earth at a latitude of 40° moves along a circular path of radius $r = R \cos 40°$, where R is the radius of Earth (6.4×10^6 m). Therefore, its speed is

$$v = \omega(R \cos 40°) = (7.3 \times 10^{-5} \text{ rad/s})(6.4 \times 10^6 \text{ m})\cos 40° = 3.5 \times 10^2 \text{ m/s}.$$

(c) At the equator (and all other points on Earth) the value of ω is the same (7.3×10^{-5} rad/s).

(d) The latitude is 0° and the speed is

$$v = \omega R = (7.3 \times 10^{-5} \text{ rad/s})(6.4 \times 10^6 \text{ m}) = 4.6 \times 10^2 \text{ m/s}.$$

30. Since the belt does not slip, a point on the rim of wheel C has the same tangential acceleration as a point on the rim of wheel A. This means that $\alpha_A r_A = \alpha_C r_C$, where α_A is the angular acceleration of wheel A and α_C is the angular acceleration of wheel C. Thus,

$$\alpha_C = \left(\frac{r_A}{r_C}\right)\alpha_C = \left(\frac{10 \text{ cm}}{25 \text{ cm}}\right)(1.6 \text{ rad/s}^2) = 0.64 \text{ rad/s}^2.$$

Since the angular speed of wheel C is given by $\omega_C = \alpha_C t$, the time for it to reach an angular speed of $\omega = 100$ rev/min = 10.5 rad/s starting from rest is

$$t = \frac{\omega_C}{\alpha_C} = \frac{10.5 \text{ rad/s}}{0.64 \text{ rad/s}^2} = 16 \text{ s}.$$

31. (a) The angular speed in rad/s is

$$\omega = \left(33\frac{1}{3} \text{ rev/min}\right)\left(\frac{2\pi \text{ rad/rev}}{60 \text{ s/min}}\right) = 3.49 \text{ rad/s}.$$

Consequently, the radial (centripetal) acceleration is (using Eq. 10-23)

$$a = \omega^2 r = (3.49 \text{ rad/s})^2 (6.0 \times 10^{-2} \text{ m}) = 0.73 \text{ m/s}^2.$$

(b) Using Ch. 6 methods, we have $ma = f_s \leq f_{s,\max} = \mu_s mg$, which is used to obtain the (minimum allowable) coefficient of friction:

$$\mu_{s,\min} = \frac{a}{g} = \frac{0.73}{9.8} = 0.075.$$

(c) The radial acceleration of the object is $a_r = \omega^2 r$, while the tangential acceleration is $a_t = \alpha r$. Thus,

$$|\vec{a}| = \sqrt{a_r^2 + a_t^2} = \sqrt{(\omega^2 r)^2 + (\alpha r)^2} = r\sqrt{\omega^4 + \alpha^2}.$$

If the object is not to slip at any time, we require

$$f_{s,\max} = \mu_s mg = ma_{\max} = mr\sqrt{\omega_{\max}^4 + \alpha^2}.$$

Thus, since $\alpha = \omega/t$ (from Eq. 10-12), we find

$$\mu_{s,\min} = \frac{r\sqrt{\omega_{\max}^4 + \alpha^2}}{g} = \frac{r\sqrt{\omega_{\max}^4 + (\omega_{\max}/t)^2}}{g} = \frac{(0.060)\sqrt{3.49^4 + (3.4/0.25)^2}}{9.8} = 0.11.$$

32. (a) A complete revolution is an angular displacement of $\Delta\theta = 2\pi$ rad, so the angular velocity in rad/s is given by $\omega = \Delta\theta/T = 2\pi/T$. The angular acceleration is given by

$$\alpha = \frac{d\omega}{dt} = -\frac{2\pi}{T^2}\frac{dT}{dt}.$$

For the pulsar described in the problem, we have

$$\frac{dT}{dt} = \frac{1.26 \times 10^{-5} \text{ s/y}}{3.16 \times 10^7 \text{ s/y}} = 4.00 \times 10^{-13}.$$

Therefore,

$$\alpha = -\left(\frac{2\pi}{(0.033 \text{ s})^2}\right)(4.00 \times 10^{-13}) = -2.3 \times 10^{-9} \text{ rad/s}^2.$$

The negative sign indicates that the angular acceleration is opposite the angular velocity and the pulsar is slowing down.

(b) We solve $\omega = \omega_0 + \alpha t$ for the time t when $\omega = 0$:

$$t = -\frac{\omega_0}{\alpha} = -\frac{2\pi}{\alpha T} = -\frac{2\pi}{(-2.3 \times 10^{-9} \text{ rad/s}^2)(0.033 \text{ s})} = 8.3 \times 10^{10} \text{ s} \approx 2.6 \times 10^3 \text{ years}$$

(c) The pulsar was born 1992–1054 = 938 years ago. This is equivalent to $(938 \text{ y})(3.16 \times 10^7 \text{ s/y}) = 2.96 \times 10^{10}$ s. Its angular velocity at that time was

$$\omega = \omega_0 + \alpha t + \frac{2\pi}{T} + \alpha t = \frac{2\pi}{0.033 \text{ s}} + (-2.3 \times 10^{-9} \text{ rad/s}^2)(-2.96 \times 10^{10} \text{ s}) = 258 \text{ rad/s}.$$

Its period was

$$T = \frac{2\pi}{\omega} = \frac{2\pi}{258 \text{ rad/s}} = 2.4 \times 10^{-2} \text{ s}.$$

33. The kinetic energy (in J) is given by $K = \frac{1}{2}I\omega^2$, where I is the rotational inertia (in kg·m^2) and ω is the angular velocity (in rad/s). We have

$$\omega = \frac{(602 \text{ rev/min})(2\pi \text{ rad/rev})}{60 \text{ s/min}} = 63.0 \text{ rad/s}.$$

Consequently, the rotational inertia is

$$I = \frac{2K}{\omega^2} = \frac{2(24400 \text{ J})}{(63.0 \text{ rad/s})^2} = 12.3 \text{ kg} \cdot \text{m}^2.$$

34. (a) Eq. 10-12 implies that the angular acceleration α should be the slope of the ω vs t graph. Thus, $\alpha = 9/6 = 1.5 \text{ rad/s}^2$.

(b) By Eq. 10-34, K is proportional to ω^2. Since the angular velocity at $t = 0$ is -2 rad/s (and this value squared is 4) and the angular velocity at $t = 4$ s is 4 rad/s (and this value squared is 16), then the ratio of the corresponding kinetic energies must be

$$\frac{K_o}{K_4} = \frac{4}{16} \quad \Rightarrow \quad K_o = \tfrac{1}{4} K_4 = 0.40 \text{ J}.$$

35. We use the parallel axis theorem: $I = I_{com} + Mh^2$, where I_{com} is the rotational inertia about the center of mass (see Table 10-2(d)), M is the mass, and h is the distance between the center of mass and the chosen rotation axis. The center of mass is at the center of the meter stick, which implies $h = 0.50$ m $- 0.20$ m $= 0.30$ m. We find

$$I_{com} = \frac{1}{12} ML^2 = \frac{1}{12}(0.56 \text{ kg})(1.0 \text{ m})^2 = 4.67 \times 10^{-2} \text{ kg} \cdot \text{m}^2.$$

Consequently, the parallel axis theorem yields

$$I = 4.67 \times 10^{-2} \text{ kg} \cdot \text{m}^2 + (0.56 \text{ kg})(0.30 \text{ m})^2 = 9.7 \times 10^{-2} \text{ kg} \cdot \text{m}^2.$$

36. (a) Eq. 10-33 gives

$$I_{total} = md^2 + m(2d)^2 + m(3d)^2 = 14 \, md^2.$$

If the innermost one is removed then we would only obtain $m(2d)^2 + m(3d)^2 = 13 \, md^2$. The percentage difference between these is $(13 - 14)/14 = 0.0714 \approx 7.1\%$.

(b) If, instead, the outermost particle is removed, we would have $md^2 + m(2d)^2 = 5 \, md^2$. The percentage difference in this case is $0.643 \approx 64\%$.

37. Since the rotational inertia of a cylinder is $I = \tfrac{1}{2} MR^2$ (Table 10-2(c)), its rotational kinetic energy is

$$K = \frac{1}{2} I \omega^2 = \frac{1}{4} MR^2 \omega^2.$$

(a) For the smaller cylinder, we have $K = \frac{1}{4}(1.25)(0.25)^2(235)^2 = 1.1\times 10^3$ J.

(b) For the larger cylinder, we obtain $K = \frac{1}{4}(1.25)(0.75)^2(235)^2 = 9.7\times 10^3$ J.

38. The parallel axis theorem (Eq. 10-36) shows that I increases with h. The phrase "out to the edge of the disk" (in the problem statement) implies that the maximum h in the graph is, in fact, the radius R of the disk. Thus, $R = 0.20$ m. Now we can examine, say, the $h = 0$ datum and use the formula for I_{com} (see Table 10-2(c)) for a solid disk, or (which might be a little better, since this is independent of whether it is really a solid disk) we can the difference between the $h = 0$ datum and the $h = h_{max} = R$ datum and relate that difference to the parallel axis theorem (thus the difference is $M(h_{max})^2 = 0.10$ kg·m^2). In either case, we arrive at $M = 2.5$ kg.

39. The particles are treated "point-like" in the sense that Eq. 10-33 yields their rotational inertia, and the rotational inertia for the rods is figured using Table 10-2(e) and the parallel-axis theorem (Eq. 10-36).

(a) With subscript 1 standing for the rod nearest the axis and 4 for the particle farthest from it, we have

$$I = I_1 + I_2 + I_3 + I_4 = \left(\frac{1}{12}Md^2 + M\left(\frac{1}{2}d\right)^2\right) + md^2 + \left(\frac{1}{12}Md^2 + M\left(\frac{3}{2}d\right)^2\right) + m(2d)^2$$

$$= \frac{8}{3}Md^2 + 5md^2 = \frac{8}{3}(1.2 \text{ kg})(0.056 \text{ m})^2 + 5(0.85 \text{ kg})(0.056 \text{ m})^2$$

$$= 0.023 \text{ kg}\cdot\text{m}^2.$$

(b) Using Eq. 10-34, we have

$$K = \frac{1}{2}I\omega^2 = \left(\frac{4}{3}M + \frac{5}{2}m\right)d^2\omega^2 = \left[\frac{4}{3}(1.2 \text{ kg}) + \frac{5}{2}(0.85 \text{ kg})\right](0.056 \text{ m})^2(0.30 \text{ rad/s})^2$$

$$= 1.1\times 10^{-3} \text{ J}.$$

40. (a) We show the figure with its axis of rotation (the thin horizontal line).

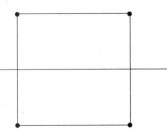

We note that each mass is $r = 1.0$ m from the axis. Therefore, using Eq. 10-26, we obtain

$$I = \sum m_i r_i^2 = 4\,(0.50 \text{ kg})\,(1.0 \text{ m})^2 = 2.0 \text{ kg}\cdot\text{m}^2.$$

(b) In this case, the two masses nearest the axis are $r = 1.0$ m away from it, but the two furthest from the axis are $r = \sqrt{(1.0 \text{ m})^2 + (2.0 \text{ m})^2}$ from it. Here, then, Eq. 10-33 leads to

$$I = \sum m_i r_i^2 = 2(0.50 \text{ kg})(1.0 \text{ m}^2) + 2(0.50 \text{ kg})(5.0 \text{ m}^2) = 6.0 \text{ kg} \cdot \text{m}^2.$$

(c) Now, two masses are on the axis (with $r = 0$) and the other two are a distance $r = \sqrt{(1.0 \text{ m})^2 + (1.0 \text{ m})^2}$ away. Now we obtain $I = 2.0 \text{ kg} \cdot \text{m}^2$.

41. We use the parallel-axis theorem. According to Table 10-2(i), the rotational inertia of a uniform slab about an axis through the center and perpendicular to the large faces is given by

$$I_{com} = \frac{M}{12}(a^2 + b^2).$$

A parallel axis through the corner is a distance $h = \sqrt{(a/2)^2 + (b/2)^2}$ from the center. Therefore,

$$I = I_{com} + Mh^2 = \frac{M}{12}(a^2 + b^2) + \frac{M}{4}(a^2 + b^2) = \frac{M}{3}(a^2 + b^2)$$

$$= \frac{0.172 \text{ kg}}{3}[(0.035 \text{ m})^2 + (0.084 \text{ m})^2]$$

$$= 4.7 \times 10^{-4} \text{ kg} \cdot \text{m}^2.$$

42. (a) Consider three of the disks (starting with the one at point O): ⊕OO. The first one (the one at point O – shown here with the plus sign inside) has rotational inertial (see item (c) in Table 10-2) $I = \frac{1}{2}mR^2$. The next one (using the parallel-axis theorem) has

$$I = \frac{1}{2}mR^2 + mh^2$$

where $h = 2R$. The third one has $I = \frac{1}{2}mR^2 + m(4R)^2$. If we had considered five of the disks OO⊕OO with the one at O in the middle, then the total rotational inertia is

$$I = 5(\tfrac{1}{2}mR^2) + 2(m(2R)^2 + m(4R)^2).$$

The pattern is now clear and we can write down the total I for the collection of fifteen disks:

$$I = 15(\tfrac{1}{2}mR^2) + 2(m(2R)^2 + m(4R)^2 + m(6R)^2 + \ldots + m(14R)^2) = \tfrac{2255}{2}mR^2.$$

The generalization to N disks (where N is assumed to be an odd number) is

$$I = \tfrac{1}{6}(2N^2 + 1)NmR^2.$$

In terms of the total mass ($m = M/15$) and the total length ($R = L/30$), we obtain

$$I = 0.083519 ML^2 \approx (0.08352)(0.1000 \text{ kg})(1.0000 \text{ m})^2 = 8.352 \times 10^{-3} \text{ kg} \cdot \text{m}^2.$$

(b) Comparing to the formula (e) in Table 10-2 (which gives roughly $I = 0.08333\ ML^2$), we find our answer to part (a) is 0.22% lower.

43. (a) Using Table 10-2(c) and Eq. 10-34, the rotational kinetic energy is

$$K = \frac{1}{2} I \omega^2 = \frac{1}{2}\left(\frac{1}{2} MR^2\right) \omega^2 = \frac{1}{4}(500 \text{ kg})(200 \pi \text{ rad/s})^2 (1.0 \text{ m})^2 = 4.9 \times 10^7 \text{ J}.$$

(b) We solve $P = K/t$ (where P is the average power) for the operating time t.

$$t = \frac{K}{P} = \frac{4.9 \times 10^7 \text{ J}}{8.0 \times 10^3 \text{ W}} = 6.2 \times 10^3 \text{ s}$$

which we rewrite as $t \approx 1.0 \times 10^2$ min.

44. (a) We apply Eq. 10-33:

$$I_x = \sum_{i=1}^{4} m_i y_i^2 = \left[50(2.0)^2 + (25)(4.0)^2 + 25(-3.0)^2 + 30(4.0)^2\right] \text{g} \cdot \text{cm}^2 = 1.3 \times 10^3 \text{ g} \cdot \text{cm}^2.$$

(b) For rotation about the y axis we obtain

$$I_y = \sum_{i=1}^{4} m_i x_i^2 = 50(2.0)^2 + (25)(0)^2 + 25(3.0)^2 + 30(2.0)^2 = 5.5 \times 10^2 \text{ g} \cdot \text{cm}^2.$$

(c) And about the z axis, we find (using the fact that the distance from the z axis is $\sqrt{x^2 + y^2}$)

$$I_z = \sum_{i=1}^{4} m_i (x_i^2 + y_i^2) = I_x + I_y = 1.3 \times 10^3 + 5.5 \times 10^2 = 1.9 \times 10^2 \text{ g} \cdot \text{cm}^2.$$

(d) Clearly, the answer to part (c) is $A + B$.

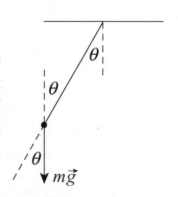

45. Two forces act on the ball, the force of the rod and the force of gravity. No torque about the pivot point is associated with the force of the rod since that force is along the line from the pivot point to the ball. As can be seen from the diagram, the component of the force of gravity that is perpendicular to the rod is $mg \sin \theta$. If ℓ is the length of the rod, then the torque associated with this force has magnitude

$$\tau = mg\ell \sin\theta = (0.75)(9.8)(1.25)\sin 30° = 4.6 \text{ N}\cdot\text{m}.$$

For the position shown, the torque is counter-clockwise.

46. We compute the torques using $\tau = rF \sin\phi$.

(a) For $\phi = 30°$, $\tau_a = (0.152 \text{ m})(111 \text{ N})\sin 30° = 8.4 \text{ N}\cdot\text{m}$.

(b) For $\phi = 90°$, $\tau_b = (0.152 \text{ m})(111 \text{ N})\sin 90° = 17 \text{ N}\cdot\text{m}$.

(c) For $\phi = 180°$, $\tau_c = (0.152 \text{ m})(111 \text{ N})\sin 180° = 0$.

47. We take a torque that tends to cause a counterclockwise rotation from rest to be positive and a torque tending to cause a clockwise rotation to be negative. Thus, a positive torque of magnitude $r_1 F_1 \sin\theta_1$ is associated with $\vec{F}_1$ and a negative torque of magnitude $r_2 F_2 \sin\theta_2$ is associated with $\vec{F}_2$. The net torque is consequently

$$\tau = r_1 F_1 \sin\theta_1 - r_2 F_2 \sin\theta_2.$$

Substituting the given values, we obtain

$$\tau = (1.30 \text{ m})(4.20 \text{ N})\sin 75° - (2.15 \text{ m})(4.90 \text{ N})\sin 60° = -3.85 \text{ N}\cdot\text{m}.$$

48. The net torque is

$$\begin{aligned}\tau &= \tau_A + \tau_B + \tau_C = F_A r_A \sin\phi_A - F_B r_B \sin\phi_B + F_C r_C \sin\phi_C \\ &= (10)(8.0)\sin 135° - (16)(4.0)\sin 90° + (19)(3.0)\sin 160° \\ &= 12 \text{ N}\cdot\text{m}.\end{aligned}$$

49. (a) We use the kinematic equation $\omega = \omega_0 + \alpha t$, where ω_0 is the initial angular velocity, ω is the final angular velocity, α is the angular acceleration, and t is the time. This gives

$$\alpha = \frac{\omega - \omega_0}{t} = \frac{6.20 \text{ rad/s}}{220 \times 10^{-3} \text{ s}} = 28.2 \text{ rad/s}^2.$$

(b) If I is the rotational inertia of the diver, then the magnitude of the torque acting on her is

$$\tau = I\alpha = (12.0 \text{ kg}\cdot\text{m}^2)(28.2 \text{ rad/s}^2) = 3.38 \times 10^2 \text{ N}\cdot\text{m}.$$

50. The rotational inertia is found from Eq. 10-45.

$$I = \frac{\tau}{\alpha} = \frac{32.0}{25.0} = 1.28 \text{ kg} \cdot \text{m}^2$$

51. Combining Eq. 10-45 ($\tau_{net} = I\alpha$) with Eq. 10-38 gives $RF_2 - RF_1 = I\alpha$, where $\alpha = \omega/t$ by Eq. 10-12 (with $\omega_0 = 0$). Using item (c) in Table 10-2 and solving for F_2 we find

$$F_2 = \frac{MR\omega}{2t} + F_1 = \frac{(0.02)(0.02)(250)}{2(1.25)} + 0.1 = 0.140 \text{ N}.$$

52. With counterclockwise positive, the angular acceleration α for both masses satisfies $\tau = mgL_1 - mgL_2 = I\alpha = (mL_1^2 + mL_2^2)\alpha$, by combining Eq. 10-45 with Eq. 10-39 and Eq. 10-33. Therefore, using SI units,

$$\alpha = \frac{g(L_1 - L_2)}{L_1^2 + L_2^2} = \frac{(9.8 \text{ m/s}^2)(0.20 \text{ m} - 0.80 \text{ m})}{(0.20 \text{ m})^2 + (0.80 \text{ m})^2} = -8.65 \text{ rad/s}^2$$

where the negative sign indicates the system starts turning in the clockwise sense. The magnitude of the acceleration vector involves no radial component (yet) since it is evaluated at $t = 0$ when the instantaneous velocity is zero. Thus, for the two masses, we apply Eq. 10-22:

(a) $|\vec{a}_1| = |\alpha|L_1 = (8.65 \text{ rad/s}^2)(0.20 \text{ m}) = 1.7 \text{ m/s}$.

(b) $|\vec{a}_2| = |\alpha|L_2 = (8.65 \text{ rad/s}^2)(0.80 \text{ m}) = 6.9 \text{ m/s}^2$.

53. Combining Eq. 10-34 and Eq. 10-45, we have $RF = I\alpha$, where α is given by ω/t (according to Eq. 10-12, since $\omega_0 = 0$ in this case). We also use the fact that

$$I = I_{plate} + I_{disk}$$

where $I_{disk} = \frac{1}{2}MR^2$ (item (c) in Table 10-2). Therefore,

$$I_{plate} = \frac{RFt}{\omega} - \frac{1}{2}MR^2 = 2.51 \times 10^{-4} \text{ kg} \cdot \text{m}^2.$$

54. According to the sign conventions used in the book, the magnitude of the net torque exerted on the cylinder of mass m and radius R is

$$\tau_{net} = F_1R - F_2R - F_3r = (6.0 \text{ N})(0.12 \text{ m}) - (4.0 \text{ N})(0.12 \text{ m}) - (2.0 \text{ N})(0.050 \text{ m}) = 71 \text{ N} \cdot \text{m}.$$

(a) The resulting angular acceleration of the cylinder (with $I = \frac{1}{2}MR^2$ according to Table 10-2(c)) is

$$\alpha = \frac{\tau_{net}}{I} = \frac{71\,\text{N}\cdot\text{m}}{\frac{1}{2}(2.0\,\text{kg})(0.12\,\text{m})^2} = 9.7\,\text{rad/s}^2.$$

(b) The direction is counterclockwise (which is the positive sense of rotation).

55. (a) We use constant acceleration kinematics. If down is taken to be positive and a is the acceleration of the heavier block m_2, then its coordinate is given by $y = \frac{1}{2}at^2$, so

$$a = \frac{2y}{t^2} = \frac{2(0.750\,\text{m})}{(5.00\,\text{s})^2} = 6.00\times 10^{-2}\,\text{m/s}^2.$$

Block 1 has an acceleration of $6.00\times 10^{-2}\,\text{m/s}^2$ upward.

(b) Newton's second law for block 2 is $m_2 g - T_2 = m_2 a$, where m_2 is its mass and T_2 is the tension force on the block. Thus,

$$T_2 = m_2(g-a) = (0.500\,\text{kg})(9.8\,\text{m/s}^2 - 6.00\times 10^{-2}\,\text{m/s}^2) = 4.87\,\text{N}.$$

(c) Newton's second law for block 1 is $m_1 g - T_1 = -m_1 a$, where T_1 is the tension force on the block. Thus,

$$T_1 = m_1(g+a) = (0.460\,\text{kg})(9.8\,\text{m/s}^2 + 6.00\times 10^{-2}\,\text{m/s}^2) = 4.54\,\text{N}.$$

(d) Since the cord does not slip on the pulley, the tangential acceleration of a point on the rim of the pulley must be the same as the acceleration of the blocks, so

$$\alpha = \frac{a}{R} = \frac{6.00\times 10^{-2}\,\text{m/s}^2}{5.00\times 10^{-2}\,\text{m}} = 1.20\,\text{rad/s}^2.$$

(e) The net torque acting on the pulley is $\tau = (T_2 - T_1)R$. Equating this to $I\alpha$ we solve for the rotational inertia:

$$I = \frac{(T_2 - T_1)R}{\alpha} = \frac{(4.87\,\text{N} - 4.54\,\text{N})(5.00\times 10^{-2}\,\text{m})}{1.20\,\text{rad/s}^2} = 1.38\times 10^{-2}\,\text{kg}\cdot\text{m}^2.$$

56. (a) In this case, the force is mg = (70 kg)(9.8 m/s^2), and the "lever arm" (the perpendicular distance from point O to the line of action of the force) is 0.28 m. Thus, the torque (in absolute value) is (70 kg)(9.8 m/s^2)(0.28 m). Since the moment-of-inertia is $I = 65$ kg·m^2, then Eq. 10-45 gives $|\alpha| = 2.955 \approx 3.0$ rad/s^2.

(b) Now we have another contribution (1.4 m × 300 N) to the net torque, so

$$|\tau_{net}| = (70 \text{ kg})(9.8 \text{ m/s}^2)(0.28 \text{ m}) + (1.4 \text{ m})(300 \text{ N}) = (65 \text{ kg} \cdot \text{m}^2) |\alpha|$$

which leads to $|\alpha| = 9.4 \text{ rad/s}^2$.

57. Since the force acts tangentially at $r = 0.10$ m, the angular acceleration (presumed positive) is

$$\alpha = \frac{\tau}{I} = \frac{Fr}{I} = \frac{(0.5t + 0.3t^2)(0.10)}{1.0 \times 10^{-3}} = 50t + 30t^2$$

in SI units (rad/s^2).

(a) At $t = 3$ s, the above expression becomes $\alpha = 4.2 \times 10^2 \text{ rad/s}^2$.

(b) We integrate the above expression, noting that $\omega_0 = 0$, to obtain the angular speed at $t = 3$ s:

$$\omega = \int_0^3 \alpha \, dt = (25t^2 + 10t^3)\Big|_0^3 = 5.0 \times 10^2 \text{ rad/s}.$$

58. (a) We apply Eq. 10-34:

$$K = \frac{1}{2} I \omega^2 = \frac{1}{2} \left(\frac{1}{3} mL^2\right) \omega^2 = \frac{1}{6} mL^2 \omega^2 = \frac{1}{6}(0.42 \text{ kg})(0.75 \text{ m})^2 (4.0 \text{ rad/s})^2 = 0.63 \text{ J}.$$

(b) Simple conservation of mechanical energy leads to $K = mgh$. Consequently, the center of mass rises by

$$h = \frac{K}{mg} = \frac{mL^2 \omega^2}{6mg} = \frac{L^2 \omega^2}{6g} = \frac{(0.75 \text{ m})^2 (4.0 \text{ rad/s})^2}{6(9.8 \text{ m/s}^2)} = 0.153 \text{ m} \approx 0.15 \text{ m}.$$

59. The initial angular speed is $\omega = (280 \text{ rev/min})(2\pi/60) = 29.3 \text{ rad/s}$.

(a) Since the rotational inertia is (Table 10-2(a)) $I = (32 \text{ kg})(1.2 \text{ m})^2 = 46.1 \text{ kg} \cdot \text{m}^2$, the work done is

$$W = \Delta K = 0 - \frac{1}{2} I \omega^2 = -\frac{1}{2}(46.1 \text{ kg} \cdot \text{m}^2)(29.3 \text{ rad/s})^2$$

which yields $|W| = 1.98 \times 10^4$ J.

(b) The average power (in absolute value) is therefore

$$|P| = \frac{|W|}{\Delta t} = \frac{19.8 \times 10^3}{15} = 1.32 \times 10^3 \text{ W}.$$

60. (a) The speed of v of the mass m after it has descended $d = 50$ cm is given by $v^2 = 2ad$ (Eq. 2-16). Thus, using $g = 980$ cm/s^2, we have

$$v = \sqrt{2ad} = \sqrt{\frac{2(2mg)d}{M+2m}} = \sqrt{\frac{4(50)(980)(50)}{400+2(50)}} = 1.4\times10^2 \text{ cm/s}.$$

(b) The answer is still 1.4×10^2 cm/s = 1.4 m/s, since it is independent of R.

61. With $\omega = (1800)(2\pi/60) = 188.5$ rad/s, we apply Eq. 10-55:

$$P = \tau\omega \Rightarrow \tau = \frac{74600 \text{ W}}{188.5 \text{ rad/s}} = 396 \text{ N}\cdot\text{m}.$$

62. (a) We use the parallel-axis theorem to find the rotational inertia:

$$I = I_{com} + Mh^2 = \frac{1}{2}MR^2 + Mh^2 = \frac{1}{2}(20 \text{ kg})(0.10 \text{ m})^2 + (20 \text{ kg})(0.50 \text{ m})^2 = 0.15 \text{ kg}\cdot\text{m}^2.$$

(b) Conservation of energy requires that $Mgh = \frac{1}{2}I\omega^2$, where ω is the angular speed of the cylinder as it passes through the lowest position. Therefore,

$$\omega = \sqrt{\frac{2Mgh}{I}} = \sqrt{\frac{2(20 \text{ kg})(9.8 \text{ m/s}^2)(0.050 \text{ m})}{0.15 \text{ kg}\cdot\text{m}^2}} = 11 \text{ rad/s}.$$

63. We use ℓ to denote the length of the stick. Since its center of mass is $\ell/2$ from either end, its initial potential energy is $\frac{1}{2}mg\ell$, where m is its mass. Its initial kinetic energy is zero. Its final potential energy is zero, and its final kinetic energy is $\frac{1}{2}I\omega^2$, where I is its rotational inertia about an axis passing through one end of the stick and ω is the angular velocity just before it hits the floor. Conservation of energy yields

$$\frac{1}{2}mg\ell = \frac{1}{2}I\omega^2 \Rightarrow \omega = \sqrt{\frac{mg\ell}{I}}.$$

The free end of the stick is a distance ℓ from the rotation axis, so its speed as it hits the floor is (from Eq. 10-18)

$$v = \omega\ell = \sqrt{\frac{mg\ell^3}{I}}.$$

Using Table 10-2 and the parallel-axis theorem, the rotational inertial is $I = \frac{1}{3}m\ell^2$, so

$$v = \sqrt{3g\ell} = \sqrt{3(9.8 \text{ m/s}^2)(1.00 \text{ m})} = 5.42 \text{ m/s}.$$

64. (a) Eq. 10-33 gives

$$I_{total} = md^2 + m(2d)^2 + m(3d)^2 = 14\, md^2,$$

where $d = 0.020$ m and $m = 0.010$ kg. The work done is $W = \Delta K = \frac{1}{2}I\omega_f^2 - \frac{1}{2}I\omega_i^2$, where $\omega_f = 20$ rad/s and $\omega_i = 0$. This gives $W = 11.2$ mJ.

(b) Now, $\omega_f = 40$ rad/s and $\omega_i = 20$ rad/s, and we get $W = 33.6$ mJ.

(c) In this case, $\omega_f = 60$ rad/s and $\omega_i = 40$ rad/s. This gives $W = 56.0$ mJ.

(d) Eq. 10-34 indicates that the slope should be $\frac{1}{2}I$. Therefore, it should be

$$7md^2 = 2.80 \times 10^{-5}\text{ J·s}^2/\text{rad}^2.$$

65. Using the parallel axis theorem and items (e) and (h) in Table 10-2, the rotational inertia is

$$I = \tfrac{1}{12}mL^2 + m(L/2)^2 + \tfrac{1}{2}mR^2 + m(R+L)^2 = 10.83mR^2,$$

where $L = 2R$ has been used. If we take the base of the rod to be at the coordinate origin ($x = 0$, $y = 0$) then the center of mass is at

$$y = \frac{mL/2 + m(L+R)}{m+m} = 2R.$$

Comparing the position shown in the textbook figure to its upside down (inverted) position shows that the change in center of mass position (in absolute value) is $|\Delta y| = 4R$. The corresponding loss in gravitational potential energy is converted into kinetic energy. Thus,

$$K = (2m)g(4R) \quad \Rightarrow \quad \omega = 9.82 \text{ rad/s}.$$

where Eq. 10-34 has been used.

66. From Table 10-2, the rotational inertia of the spherical shell is $2MR^2/3$, so the kinetic energy (after the object has descended distance h) is

$$K = \frac{1}{2}\left(\frac{2}{3}MR^2\right)\omega_{sphere}^2 + \frac{1}{2}I\omega_{pulley}^2 + \frac{1}{2}mv^2.$$

Since it started from rest, then this energy must be equal (in the absence of friction) to the potential energy mgh with which the system started. We substitute v/r for the pulley's angular speed and v/R for that of the sphere and solve for v.

$$v = \sqrt{\frac{mgh}{\frac{1}{2}m + \frac{1}{2}\frac{I}{r^2} + \frac{M}{3}}} = \sqrt{\frac{2gh}{1 + (I/mr^2) + (2M/3m)}}$$

$$= \sqrt{\frac{2(9.8)(0.82)}{1 + 3.0 \times 10^{-3}/((0.60)(0.050)^2) + 2(4.5)/3(0.60)}} = 1.4 \text{ m/s}$$

67. (a) We use conservation of mechanical energy to find an expression for ω^2 as a function of the angle θ that the chimney makes with the vertical. The potential energy of the chimney is given by $U = Mgh$, where M is its mass and h is the altitude of its center of mass above the ground. When the chimney makes the angle θ with the vertical, $h = (H/2)\cos\theta$. Initially the potential energy is $U_i = Mg(H/2)$ and the kinetic energy is zero. The kinetic energy is $\frac{1}{2}I\omega^2$ when the chimney makes the angle θ with the vertical, where I is its rotational inertia about its bottom edge. Conservation of energy then leads to

$$MgH/2 = Mg(H/2)\cos\theta + \frac{1}{2}I\omega^2 \Rightarrow \omega^2 = (MgH/I)(1 - \cos\theta).$$

The rotational inertia of the chimney about its base is $I = MH^2/3$ (found using Table 10-2(e) with the parallel axis theorem). Thus

$$\omega = \sqrt{\frac{3g}{H}(1 - \cos\theta)} = \sqrt{\frac{3(9.80 \text{ m/s}^2)}{55.0 \text{ m}}(1 - \cos 35.0°)} = 0.311 \text{ rad/s}.$$

(b) The radial component of the acceleration of the chimney top is given by $a_r = H\omega^2$, so

$$a_r = 3g(1 - \cos\theta) = 3(9.80 \text{ m/s}^2)(1 - \cos 35.0°) = 5.32 \text{ m/s}^2.$$

(c) The tangential component of the acceleration of the chimney top is given by $a_t = H\alpha$, where α is the angular acceleration. We are unable to use Table 10-1 since the acceleration is not uniform. Hence, we differentiate

$$\omega^2 = (3g/H)(1 - \cos\theta)$$

with respect to time, replacing $d\omega/dt$ with α, and $d\theta/dt$ with ω, and obtain

$$\frac{d\omega^2}{dt} = 2\omega\alpha = (3g/H)\omega\sin\theta \Rightarrow \alpha = (3g/2H)\sin\theta.$$

Consequently,

$$a_t = H\alpha = \frac{3g}{2}\sin\theta = \frac{3(9.80 \text{ m/s}^2)}{2}\sin 35.0° = 8.43 \text{ m/s}^2.$$

(d) The angle θ at which $a_t = g$ is the solution to $\frac{3g}{2} \sin\theta = g$. Thus, $\sin\theta = 2/3$ and we obtain $\theta = 41.8°$.

68. The rotational inertia of the passengers is (to a good approximation) given by Eq. 10-53: $I = \sum mR^2 = NmR^2$ where N is the number of people and m is the (estimated) mass per person. We apply Eq. 10-52:

$$W = \frac{1}{2}I\omega^2 = \frac{1}{2}NmR^2\omega^2$$

where $R = 38$ m and $N = 36 \times 60 = 2160$ persons. The rotation rate is constant so that $\omega = \theta/t$ which leads to $\omega = 2\pi/120 = 0.052$ rad/s. The mass (in kg) of the average person is probably in the range $50 \leq m \leq 100$, so the work should be in the range

$$\frac{1}{2}(2160)(50)(38)^2(0.052)^2 \leq W \leq \frac{1}{2}(2160)(100)(38)^2(0.052)^2$$

$$2\times 10^5 \text{ J} \leq W \leq 4\times 10^5 \text{ J}.$$

69. We choose positive coordinate directions (different choices for each item) so that each is accelerating positively, which will allow us to set $a_2 = a_1 = R\alpha$ (for simplicity, we denote this as a). Thus, we choose rightward positive for $m_2 = M$ (the block on the table), downward positive for $m_1 = M$ (the block at the end of the string) and (somewhat unconventionally) clockwise for positive sense of disk rotation. This means that we interpret θ given in the problem as a positive-valued quantity. Applying Newton's second law to m_1, m_2 and (in the form of Eq. 10-45) to M, respectively, we arrive at the following three equations (where we allow for the possibility of friction f_2 acting on m_2).

$$m_1 g - T_1 = m_1 a_1$$
$$T_2 - f_2 = m_2 a_2$$
$$T_1 R - T_2 R = I\alpha$$

(a) From Eq. 10-13 (with $\omega_0 = 0$) we find

$$\theta = \omega_0 t + \frac{1}{2}\alpha t^2 \Rightarrow \alpha = \frac{2\theta}{t^2} = \frac{2(1.30 \text{ rad})}{(0.0910 \text{ s})^2} = 314 \text{ rad/s}^2.$$

(b) From the fact that $a = R\alpha$ (noted above), we obtain

$$a = \frac{2R\theta}{t^2} = \frac{2(0.024 \text{ m})(1.30 \text{ rad})}{(0.0910 \text{ s})^2} = 7.54 \text{ m/s}^2.$$

(c) From the first of the above equations, we find

$$T_1 = m_1(g-a_1) = M\left(g - \frac{2R\theta}{t^2}\right) = (6.20 \text{ kg})\left(9.80 \text{ m/s}^2 - \frac{2(0.024 \text{ m})(1.30 \text{ rad})}{(0.0910 \text{ s})^2}\right)$$
$$= 14.0 \text{ N}.$$

(d) From the last of the above equations, we obtain the second tension:

$$T_2 = T_1 - \frac{I\alpha}{R} = M\left(g - \frac{2R\theta}{t^2}\right) - \frac{2I\theta}{Rt^2} = 14.0 \text{ N} - \frac{(7.40\times 10^{-4} \text{ kg}\cdot\text{m}^2)(314 \text{ rad/s}^2)}{0.024 \text{ m}}$$
$$= 4.36 \text{ N}.$$

70. In the calculation below, M_1 and M_2 are the ring masses, R_{1i} and R_{2i} are their inner radii, and R_{1o} and R_{2o} are their outer radii. Referring to item (b) in Table 10-2, we compute

$$I = \tfrac{1}{2}M_1(R_{1i}^2 + R_{1o}^2) + \tfrac{1}{2}M_2(R_{2i}^2 + R_{2o}^2) = 0.00346 \text{ kg}\cdot\text{m}^2.$$

Thus, with Eq. 10-38 ($\tau = rF$ where $r = R_{2o}$) and $\tau = I\alpha$ (Eq. 10-45), we find

$$\alpha = \frac{(0.140)(12.0)}{0.00346} = 485 \text{ rad/s}^2.$$

Then Eq. 10-12 gives $\omega = \alpha t = 146$ rad/s.

71. The volume of each disk is $\pi r^2 h$ where we are using h to denote the thickness (which equals 0.00500 m). If we use R (which equals 0.0400 m) for the radius of the larger disk and r (which equals 0.0200 m) for the radius of the smaller one, then the mass of each is $m = \rho\pi r^2 h$ and $M = \rho\pi R^2 h$ where $\rho = 1400$ kg/m^3 is the given density. We now use the parallel axis theorem as well as item (c) in Table 10-2 to obtain the rotation inertia of the two-disk assembly:

$$I = \tfrac{1}{2}MR^2 + \tfrac{1}{2}mr^2 + m(r+R)^2 = \rho\pi h[\tfrac{1}{2}R^4 + \tfrac{1}{2}r^4 + r^2(r+R)^2] = 6.16\times 10^{-5} \text{ kg}\cdot\text{m}^2.$$

72. (a) The longitudinal separation between Helsinki and the explosion site is $\Delta\theta = 102° - 25° = 77°$. The spin of the earth is constant at

$$\omega = \frac{1 \text{ rev}}{1 \text{ day}} = \frac{360°}{24 \text{ h}}$$

so that an angular displacement of $\Delta\theta$ corresponds to a time interval of

$$\Delta t = (77°)\left(\frac{24 \text{ h}}{360°}\right) = 5.1 \text{ h}.$$

(b) Now $\Delta\theta = 102° - (-20°) = 122°$ so the required time shift would be

$$\Delta t = (122°)\left(\frac{24\text{ h}}{360°}\right) = 8.1\text{ h}.$$

73. We choose positive coordinate directions (different choices for each item) so that each is accelerating positively, which will allow us to set $a_1 = a_2 = R\alpha$ (for simplicity, we denote this as a). Thus, we choose upward positive for m_1, downward positive for m_2 and (somewhat unconventionally) clockwise for positive sense of disk rotation. Applying Newton's second law to $m_1 m_2$ and (in the form of Eq. 10-45) to M, respectively, we arrive at the following three equations.

$$T_1 - m_1 g = m_1 a_1$$
$$m_2 g - T_2 = m_2 a_2$$
$$T_2 R - T_1 R = I\alpha$$

(a) The rotational inertia of the disk is $I = \frac{1}{2} MR^2$ (Table 10-2(c)), so we divide the third equation (above) by R, add them all, and use the earlier equality among accelerations — to obtain:

$$m_2 g - m_1 g = \left(m_1 + m_2 + \frac{1}{2} M\right) a$$

which yields $a = \frac{4}{25} g = 1.57\text{ m/s}^2$.

(b) Plugging back in to the first equation, we find

$$T_1 = \frac{29}{25} m_1 g = 4.55\text{ N}$$

where it is important in this step to have the mass in SI units: $m_1 = 0.40$ kg.

(c) Similarly, with $m_2 = 0.60$ kg, we find

$$T_2 = \frac{5}{6} m_2 g = 4.94\text{ N}.$$

74. (a) Constant angular acceleration kinematics can be used to compute the angular acceleration α. If ω_0 is the initial angular velocity and t is the time to come to rest, then

$$0 = \omega_0 + \alpha t \Rightarrow \alpha = -\frac{\omega_0}{t}$$

which yields $-39/32 = -1.2$ rev/s or (multiplying by 2π) -7.66 rad/s^2 for the value of α.

(b) We use $\tau = I\alpha$, where τ is the torque and I is the rotational inertia. The contribution of the rod to I is $M\ell^2/12$ (Table 10-2(e)), where M is its mass and ℓ is its length. The contribution of each ball is $m(\ell/2)^2$, where m is the mass of a ball. The total rotational inertia is

$$I = \frac{M\ell^2}{12} + 2\frac{m\ell^2}{4} = \frac{(6.40 \text{ kg})(1.20 \text{ m})^2}{12} + \frac{(1.06 \text{ kg})(1.20 \text{ m})^2}{2}$$

which yields $I = 1.53$ kg·m². The torque, therefore, is

$$\tau = (1.53 \text{ kg·m}^2)(-7.66 \text{ rad/s}^2) = -11.7 \text{ N·m}.$$

(c) Since the system comes to rest the mechanical energy that is converted to thermal energy is simply the initial kinetic energy

$$K_i = \frac{1}{2}I\omega_0^2 = \frac{1}{2}(1.53 \text{ kg·m}^2)((2\pi)(39) \text{ rad/s})^2 = 4.59 \times 10^4 \text{ J}.$$

(d) We apply Eq. 10-13:

$$\theta = \omega_0 t + \frac{1}{2}\alpha t^2 = ((2\pi)(39) \text{ rad/s})(32.0 \text{ s}) + \frac{1}{2}(-7.66 \text{ rad/s}^2)(32.0 \text{ s})^2$$

which yields 3920 rad or (dividing by 2π) 624 rev for the value of angular displacement θ.

(e) Only the mechanical energy that is converted to thermal energy can still be computed without additional information. It is 4.59×10^4 J no matter how τ varies with time, as long as the system comes to rest.

75. The *Hint* given in the problem would make the computation in part (a) very straightforward (without doing the integration as we show here), but we present this further level of detail in case that hint is not obvious or — simply — in case one wishes to see how the calculus supports our intuition.

(a) The (centripetal) force exerted on an infinitesimal portion of the blade with mass dm located a distance r from the rotational axis is (Newton's second law) $dF = (dm)\omega^2 r$, where dm can be written as $(M/L)dr$ and the angular speed is

$$\omega = (320)(2\pi/60) = 33.5 \text{ rad/s}.$$

Thus for the entire blade of mass M and length L the total force is given by

$$F = \int dF = \int \omega^2 r\, dm = \frac{M}{L}\int_0^L \omega^2 r\, dr = \frac{M\omega^2 L}{2} = \frac{(110\,\text{kg})(33.5\,\text{rad/s})^2(7.80\,\text{m})}{2}$$
$$= 4.81\times 10^5\,\text{N}.$$

(b) About its center of mass, the blade has $I = ML^2/12$ according to Table 10-2(e), and using the parallel-axis theorem to "move" the axis of rotation to its end-point, we find the rotational inertia becomes $I = ML^2/3$. Using Eq. 10-45, the torque (assumed constant) is

$$\tau = I\alpha = \left(\frac{1}{3}ML^2\right)\left(\frac{\Delta\omega}{\Delta t}\right) = \frac{1}{3}(110\,\text{kg})(7.8\,\text{m})^2\left(\frac{33.5\,\text{rad/s}}{6.7\,\text{s}}\right) = 1.12\times 10^4\,\text{N}\cdot\text{m}.$$

(c) Using Eq. 10-52, the work done is

$$W = \Delta K = \frac{1}{2}I\omega^2 - 0 = \frac{1}{2}\left(\frac{1}{3}ML^2\right)\omega^2 = \frac{1}{6}(110\,\text{kg})(7.80\,\text{m})^2(33.5\,\text{rad/s})^2 = 1.25\times 10^6\,\text{J}.$$

76. The wheel starts turning from rest ($\omega_0 = 0$) at $t = 0$, and accelerates uniformly at $\alpha = 2.00\,\text{rad/s}^2$. Between t_1 and t_2 the wheel turns through $\Delta\theta = 90.0$ rad, where $t_2 - t_1 = \Delta t = 3.00$ s. We solve (b) first.

(b) We use Eq. 10-13 (with a slight change in notation) to describe the motion for $t_1 \le t \le t_2$:

$$\Delta\theta = \omega_1 \Delta t + \frac{1}{2}\alpha(\Delta t)^2 \Rightarrow \omega_1 = \frac{\Delta\theta}{\Delta t} - \frac{\alpha\Delta t}{2}$$

which we plug into Eq. 10-12, set up to describe the motion during $0 \le t \le t_1$:

$$\omega_1 = \omega_0 + \alpha t_1 \Rightarrow \frac{\Delta\theta}{\Delta t} - \frac{\alpha\Delta t}{2} = \alpha t_1 \Rightarrow \frac{90.0}{3.00} - \frac{(2.00)(3.00)}{2} = (2.00)t_1$$

yielding $t_1 = 13.5$ s.

(a) Plugging into our expression for ω_1 (in previous part) we obtain

$$\omega_1 = \frac{\Delta\theta}{\Delta t} - \frac{\alpha\Delta t}{2} = \frac{90.0}{3.00} - \frac{(2.00)(3.00)}{2} = 27.0\,\text{rad/s}.$$

77. To get the time to reach the maximum height, we use Eq. 4-23, setting the left-hand side to zero. Thus, we find

$$t = \frac{(60\,\text{m/s})\sin(20°)}{9.8\,\text{m/s}^2} = 2.094\,\text{s}.$$

Then (assuming $\alpha = 0$) Eq. 10-13 gives

$$\theta - \theta_0 = \omega_0 t = (90 \text{ rad/s})(2.094 \text{ s}) = 188 \text{ rad},$$

which is equivalent to roughly 30 rev.

78. We choose ± directions such that the initial angular velocity is $\omega_0 = -317$ rad/s and the values for α, τ and F are positive.

(a) Combining Eq. 10-12 with Eq. 10-45 and Table 10-2(f) (and using the fact that $\omega = 0$) we arrive at the expression

$$\tau = \left(\frac{2}{5} MR^2\right)\left(-\frac{\omega_0}{t}\right) = -\frac{2}{5}\frac{MR^2 \omega_0}{t}.$$

With $t = 15.5$ s, $R = 0.226$ m and $M = 1.65$ kg, we obtain $\tau = 0.689$ N·m.

(b) From Eq. 10-40, we find $F = \tau/R = 3.05$ N.

(c) Using again the expression found in part (a), but this time with $R = 0.854$ m, we get $\tau = 9.84$ N·m.

(d) Now, $F = \tau/R = 11.5$ N.

79. The center of mass is initially at height $h = \frac{L}{2}\sin 40°$ when the system is released (where $L = 2.0$ m). The corresponding potential energy Mgh (where $M = 1.5$ kg) becomes rotational kinetic energy $\frac{1}{2} I\omega^2$ as it passes the horizontal position (where I is the rotational inertia about the pin). Using Table 10-2 (e) and the parallel axis theorem, we find

$$I = \tfrac{1}{12} ML^2 + M(L/2)^2 = \tfrac{1}{3} ML^2.$$

Therefore,

$$Mg\frac{L}{2}\sin 40° = \frac{1}{2}\left(\frac{1}{3} ML^2\right)\omega^2 \Rightarrow \omega = \sqrt{\frac{3g \sin 40°}{L}} = 3.1 \text{ rad/s}.$$

80. (a) Eq. 10-12 leads to $\alpha = -\omega_0/t = -(25.0 \text{ rad/s})/(20.0 \text{ s}) = -1.25 \text{ rad/s}^2$.

(b) Eq. 10-15 leads to $\theta = \frac{1}{2}\omega_0 t = \frac{1}{2}(25.0 \text{ rad/s})(20.0 \text{ s}) = 250$ rad.

(c) Dividing the previous result by 2π we obtain $\theta = 39.8$ rev.

81. (a) With $r = 0.780$ m, the rotational inertia is

$$I = Mr^2 = (1.30 \text{ kg})(0.780 \text{ m})^2 = 0.791 \text{ kg} \cdot \text{m}^2.$$

(b) The torque that must be applied to counteract the effect of the drag is

$$\tau = rf = (0.780 \text{ m})(2.30 \times 10^{-2} \text{ N}) = 1.79 \times 10^{-2} \text{ N} \cdot \text{m}.$$

82. The motion consists of two stages. The first, the interval $0 \le t \le 20$ s, consists of constant angular acceleration given by

$$\alpha = \frac{5.0 \text{ rad/s}}{2.0 \text{ s}} = 2.5 \text{ rad/s}^2.$$

The second stage, $20 < t \le 40$ s, consists of constant angular velocity $\omega = \Delta\theta / \Delta t$. Analyzing the first stage, we find

$$\theta_1 = \frac{1}{2}\alpha t^2 \bigg|_{t=20} = 500 \text{ rad}, \quad \omega = \alpha t \big|_{t=20} = 50 \text{ rad/s}.$$

Analyzing the second stage, we obtain

$$\theta_2 = \theta_1 + \omega \Delta t = 500 \text{ rad} + (50 \text{ rad/s})(20 \text{ s}) = 1.5 \times 10^3 \text{ rad}.$$

83. The magnitude of torque is the product of the force magnitude and the distance from the pivot to the line of action of the force. In our case, it is the gravitational force that passes through the walker's center of mass. Thus,

$$\tau = I\alpha = rF = rmg.$$

(a) Without the pole, with $I = 15$ kg·m², the angular acceleration is

$$\alpha = \frac{rF}{I} = \frac{rmg}{I} = \frac{(0.050 \text{ m})(70 \text{ kg})(9.8 \text{ m/s}^2)}{15 \text{ kg} \cdot \text{m}^2} = 2.3 \text{ rad/s}^2.$$

(b) When the walker carries a pole, the torque due to the gravitational force through the pole's center of mass opposes the torque due to the gravitational force that passes through the walker's center of mass. Therefore,

$$\tau_{net} = \sum_i r_i F_i = (0.050 \text{ m})(70 \text{ kg})(9.8 \text{ m/s}^2) - (0.10 \text{ m})(14 \text{ kg})(9.8 \text{ m/s}^2) = 20.58 \text{ N} \cdot \text{m},$$

and the resulting angular acceleration is

$$\alpha = \frac{\tau_{net}}{I} = \frac{20.58 \text{ N} \cdot \text{m}}{15 \text{ kg} \cdot \text{m}^2} \approx 1.4 \text{ rad/s}^2.$$

84. The angular displacements of disks A and B can be written as:

$$\theta_A = \omega_A t, \quad \theta_B = \frac{1}{2}\alpha_B t^2.$$

(a) The time when $\theta_A = \theta_B$ is given by

$$\omega_A t = \frac{1}{2}\alpha_B t^2 \implies t = \frac{2\omega_A}{\alpha_B} = \frac{2(9.5 \text{ rad/s})}{(2.2 \text{ rad/s}^2)} = 8.6 \text{ s}.$$

(b) The difference in the angular displacement is

$$\Delta\theta = \theta_A - \theta_B = \omega_A t - \frac{1}{2}\alpha_B t^2 = 9.5t - 1.1t^2.$$

For their reference lines to align momentarily, we only require $\Delta\theta = 2\pi N$, where N is an integer. The quadratic equation can be readily solve to yield

$$t_N = \frac{9.5 \pm \sqrt{(9.5)^2 - 4(1.1)(2\pi N)}}{2(1.1)} = \frac{9.5 \pm \sqrt{90.25 - 27.6N}}{2.2}.$$

The solution $t_0 = 8.63$ s (taking the positive root) coincides with the result obtained in (a), while $t_0 = 0$ (taking the negative root) is the moment when both disks begin to rotate. In fact, two solutions exist for $N = 0, 1, 2,$ and 3.

85. Eq. 10-40 leads to $\tau = mgr = (70 \text{ kg})(9.8 \text{ m/s}^2)(0.20 \text{ m}) = 1.4 \times 10^2$ N·m.

86. (a) Using Eq. 10-15, we have $60.0 \text{ rad} = \frac{1}{2}(\omega_1 + \omega_2)(6.00 \text{ s})$. With $\omega_2 = 15.0$ rad/s, then $\omega_1 = 5.00$ rad/s.

(b) Eq. 10-12 gives $\alpha = (15.0 \text{ rad/s} - 5.0 \text{ rad/s})/(6.00 \text{ s}) = 1.67 \text{ rad/s}^2$.

(c) Interpreting ω now as ω_1 and θ as $\theta_1 = 10.0$ rad (and $\omega_0 = 0$) Eq. 10-14 leads to

$$\theta_0 = -\frac{\omega_1^2}{2\alpha} + \theta_1 = 2.50 \text{ rad}.$$

87. With rightward positive for the block and clockwise negative for the wheel (as is conventional), then we note that the tangential acceleration of the wheel is of opposite

sign from the block's acceleration (which we simply denote as a); that is, $a_t = -a$. Applying Newton's second law to the block leads to $P - T = ma$, where $m = 2.0$ kg. Applying Newton's second law (for rotation) to the wheel leads to $-TR = I\alpha$, where $I = 0.050$ kg·m².

Noting that $R\alpha = a_t = -a$, we multiply this equation by R and obtain

$$-TR^2 = -Ia \Rightarrow T = a\frac{I}{R^2}.$$

Adding this to the above equation (for the block) leads to $P = (m + I/R^2)a$.

Thus, $a = 0.92$ m/s² and therefore $\alpha = -4.6$ rad/s² (or $|\alpha| = 4.6$ rad/s²), where the negative sign in α should not be mistaken for a deceleration (it simply indicates the clockwise sense to the motion).

88. (a) The time for one revolution is the circumference of the orbit divided by the speed v of the Sun: $T = 2\pi R/v$, where R is the radius of the orbit. We convert the radius:

$$R = (2.3 \times 10^4 \text{ ly})(9.46 \times 10^{12} \text{ km/ly}) = 2.18 \times 10^{17} \text{ km}$$

where the ly $\leftrightarrow$ km conversion can be found in Appendix D or figured "from basics" (knowing the speed of light). Therefore, we obtain

$$T = \frac{2\pi(2.18 \times 10^{17} \text{ km})}{250 \text{ km/s}} = 5.5 \times 10^{15} \text{ s}.$$

(b) The number of revolutions N is the total time t divided by the time T for one revolution; that is, $N = t/T$. We convert the total time from years to seconds and obtain

$$N = \frac{(4.5 \times 10^9 \text{ y})(3.16 \times 10^7 \text{ s/y})}{5.5 \times 10^{15} \text{ s}} = 26.$$

89. We assume the sense of initial rotation is positive. Then, with $\omega_0 > 0$ and $\omega = 0$ (since it stops at time t), our angular acceleration is negative-valued.

(a) The angular acceleration is constant, so we can apply Eq. 10-12 ($\omega = \omega_0 + \alpha t$). To obtain the requested units, we have $t = 30/60 = 0.50$ min. Thus,

$$\alpha = -\frac{33.33 \text{ rev/min}}{0.50 \text{ min}} = -66.7 \text{ rev/min}^2 \approx -67 \text{ rev/min}^2.$$

(b) We use Eq. 10-13:

$$\theta = \omega_0 t + \frac{1}{2}\alpha t^2 = (33.33 \text{ rev/min})(0.50 \text{ min}) + \frac{1}{2}(-66.7 \text{ rev/min}^2)(0.50 \text{ min})^2 = 8.3 \text{ rev}.$$

90. We use conservation of mechanical energy. The center of mass is at the midpoint of the cross bar of the **H** and it drops by $L/2$, where L is the length of any one of the rods. The gravitational potential energy decreases by $MgL/2$, where M is the mass of the body. The initial kinetic energy is zero and the final kinetic energy may be written $\frac{1}{2}I\omega^2$, where I is the rotational inertia of the body and ω is its angular velocity when it is vertical. Thus,

$$0 = -MgL/2 + \frac{1}{2}I\omega^2 \Rightarrow \omega = \sqrt{MgL/I}.$$

Since the rods are thin the one along the axis of rotation does not contribute to the rotational inertia. All points on the other leg are the same distance from the axis of rotation, so that leg contributes $(M/3)L^2$, where $M/3$ is its mass. The cross bar is a rod that rotates around one end, so its contribution is $(M/3)L^2/3 = ML^2/9$. The total rotational inertia is

$$I = (ML^2/3) + (ML^2/9) = 4ML^2/9.$$

Consequently, the angular velocity is

$$\omega = \sqrt{\frac{MgL}{I}} = \sqrt{\frac{MgL}{4ML^2/9}} = \sqrt{\frac{9g}{4L}} = \sqrt{\frac{9(9.800 \text{ m/s}^2)}{4(0.600 \text{ m})}} = 6.06 \text{ rad/s}.$$

91. (a) According to Table 10-2, the rotational inertia formulas for the cylinder (radius R) and the hoop (radius r) are given by

$$I_C = \frac{1}{2}MR^2 \text{ and } I_H = Mr^2.$$

Since the two bodies have the same mass, then they will have the same rotational inertia if

$$R^2/2 = R_H^2 \rightarrow R_H = R/\sqrt{2}.$$

(b) We require the rotational inertia to be written as $I = Mk^2$, where M is the mass of the given body and k is the radius of the "equivalent hoop." It follows directly that $k = \sqrt{I/M}$.

92. (a) We use $\tau = I\alpha$, where τ is the net torque acting on the shell, I is the rotational inertia of the shell, and α is its angular acceleration. Therefore,

$$I = \frac{\tau}{\alpha} = \frac{960 \text{ N} \cdot \text{m}}{6.20 \text{ rad/s}^2} = 155 \text{ kg} \cdot \text{m}^2.$$

(b) The rotational inertia of the shell is given by $I = (2/3) MR^2$ (see Table 10-2 of the text). This implies

$$M = \frac{3I}{2R^2} = \frac{3(155 \text{ kg} \cdot \text{m}^2)}{2(1.90 \text{ m})^2} = 64.4 \text{ kg}.$$

93. We choose positive coordinate directions so that each is accelerating positively, which will allow us to set $a_{\text{box}} = R\alpha$ (for simplicity, we denote this as a). Thus, we choose downhill positive for the $m = 2.0$ kg box and (as is conventional) counterclockwise for positive sense of wheel rotation. Applying Newton's second law to the box and (in the form of Eq. 10-45) to the wheel, respectively, we arrive at the following two equations (using θ as the incline angle 20°, not as the angular displacement of the wheel).

$$mg \sin \theta - T = ma$$
$$TR = I\alpha$$

Since the problem gives $a = 2.0$ m/s^2, the first equation gives the tension $T = m(g \sin \theta - a) = 2.7$ N. Plugging this and $R = 0.20$ m into the second equation (along with the fact that $\alpha = a/R$) we find the rotational inertia

$$I = TR^2/a = 0.054 \text{ kg} \cdot \text{m}^2.$$

94. Analyzing the forces tending to drag the $M = 5124$ kg stone down the oak beam, we find

$$F = Mg(\sin \theta + \mu_s \cos \theta)$$

where $\mu_s = 0.22$ (static friction is assumed to be at its maximum value) and the incline angle θ for the oak beam is $\sin^{-1}(3.9/10) = 23°$ (but the incline angle for the spruce log is the complement of that). We note that the component of the weight of the workers (N of them) which is perpendicular to the spruce log is $Nmg \cos(90° - \theta) = Nmg \sin \theta$, where $m = 85$ kg. The corresponding torque is therefore $Nmg\ell \sin \theta$ where $\ell = 4.5 - 0.7 = 3.8$ m. This must (at least) equal the magnitude of torque due to F, so with $r = 0.7$ m, we have

$$Mgr(\sin \theta + \mu_s \cos \theta) = Ngm\ell \sin \theta.$$

This expression yields $N \approx 17$ for the number of workers.

95. The centripetal acceleration at a point P which is r away from the axis of rotation is given by Eq. 10-23: $a = v^2/r = \omega^2 r$, where $v = \omega r$, with $\omega = 2000$ rev/min ≈ 209.4 rad/s.

(a) If points A and P are at a radial distance $r_A = 1.50$ m and $r = 0.150$ m from the axis, the difference in their acceleration is

$$\Delta a = a_A - a = \omega^2 (r_A - r) = (209.4 \text{ rad/s})^2 (1.50 \text{ m} - 0.150 \text{ m}) \approx 5.92 \times 10^4 \text{ m/s}^2$$

(b) The slope is given by $a/r = \omega^2 = 4.39 \times 10^4 / \text{s}^2$.

96. Let T be the tension on the rope. From Newton's second law, we have

$$T - mg = ma \Rightarrow T = m(g + a).$$

Since the box has an upward acceleration $a = 0.80 \text{ m/s}^2$, the tension is given by

$$T = (30 \text{ kg})(9.8 \text{ m/s}^2 + 0.8 \text{ m/s}^2) = 318 \text{ N}.$$

The rotation of the device is described by $F_{\text{app}} R - Tr = I\alpha = Ia/r$. The moment of inertia can then be obtained as

$$I = \frac{r(F_{\text{app}} R - Tr)}{a} = \frac{(0.20 \text{ m})[(140 \text{ N})(0.50 \text{ m}) - (318 \text{ N})(0.20 \text{ m})]}{0.80 \text{ m/s}^2} = 1.6 \text{ kg} \cdot \text{m}^2.$$

97. The distances from P to the particles are as follows:

$$r_1 = a \text{ for } m_1 = 2M (\text{lower left})$$
$$r_2 = \sqrt{b^2 - a^2} \text{ for } m_2 = M (\text{top})$$
$$r_3 = a \text{ for } m_1 = 2M (\text{lower right})$$

The rotational inertia of the system about P is

$$I = \sum_{i=1}^{3} m_i r_i^2 = (3a^2 + b^2) M$$

which yields $I = 0.208 \text{ kg} \cdot \text{m}^2$ for $M = 0.40$ kg, $a = 0.30$ m and $b = 0.50$ m. Applying Eq. 10-52, we find

$$W = \frac{1}{2} I \omega^2 = \frac{1}{2} (0.208 \text{ kg} \cdot \text{m}^2)(5.0 \text{ rad/s})^2 = 2.6 \text{ J}.$$

98. In the figure below, we show a pull tab of a beverage can. Since the tab is pivoted, when pulling on one end upward with a force $\vec{F}_1$, a force $\vec{F}_2$ will be exerted on the other

end. The torque produced by $\vec{F}_1$ must be balanced by the torque produced by $\vec{F}_2$ so that the tab does not rotate.

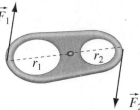

The two forces are related by

$$r_1 F_1 = r_2 F_2$$

where $r_1 \approx 1.8$ cm and $r_2 \approx 0.73$ cm. Thus, if $F_1 = 10$ N,

$$F_2 = \left(\frac{r_1}{r_2}\right) F_1 \approx \left(\frac{1.8 \text{ cm}}{0.73 \text{ cm}}\right)(10 \text{ N}) \approx 25 \text{ N}.$$

99. (a) We apply Eq. 10-18, using the subscript J for the Jeep.

$$\omega = \frac{v_J}{r_J} = \frac{114 \text{ km/h}}{0.100 \text{ km}}$$

which yields 1140 rad/h or (dividing by 3600) 0.32 rad/s for the value of the angular speed ω.

(b) Since the cheetah has the same angular speed, we again apply Eq. 10-18, using the subscript c for the cheetah.

$$v_c = r_c \omega = (92 \text{ m})(1140 \text{ rad/h}) = 1.048 \times 10^5 \text{ m/h} \approx 1.0 \times 10^2 \text{ km/h}$$

for the cheetah's speed.

100. Using Eq. 10-7 and Eq. 10-18, the average angular acceleration is

$$\alpha_{avg} = \frac{\Delta \omega}{\Delta t} = \frac{\Delta v}{r \Delta t} = \frac{25-12}{(0.75/2)(6.2)} = 5.6 \text{ rad/s}^2 .$$

101. We make use of Table 10-2(e) and the parallel-axis theorem in Eq. 10-36.

(a) The moment of inertia is

$$I = \frac{1}{12} ML^2 + Mh^2 = \frac{1}{12}(3.0 \text{ kg})(4.0 \text{ m})^2 + (3.0 \text{ kg})(1.0 \text{ m})^2 = 7.0 \text{ kg} \cdot \text{m}^2 .$$

(b) The rotational kinetic energy is

$$K_{rot} = \frac{1}{2}I\omega^2 \Rightarrow \omega = \sqrt{\frac{2K_{rot}}{I}} = \sqrt{\frac{2(20\text{ J})}{7\text{ kg}\cdot\text{m}^2}} = 2.4\text{ rad/s}$$

The linear speed of the end B is given by $v_B = \omega r_{AB} = (2.4\text{ rad/s})(3.00\text{ m}) = 7.2\text{ m/s}$, where r_{AB} is the distance between A and B.

(c) The maximum angle θ is attained when all the rotational kinetic energy is transformed into potential energy. Moving from the vertical position ($\theta = 0$) to the maximum angle θ, the center of mass is elevated by $\Delta y = d_{AC}(1-\cos\theta)$, where $d_{AC} = 1.00$ m is the distance between A and the center of mass of the rod. Thus, the change in potential energy is

$$\Delta U = mg\Delta y = mgd_{AC}(1-\cos\theta) \Rightarrow 20\text{ J} = (3.0\text{ kg})(9.8\text{ m/s}^2)(1.0\text{ m})(1-\cos\theta)$$

which yields $\cos\theta = 0.32$, or $\theta \approx 71°$.

102. (a) The linear speed at $t = 15.0$ s is

$$v = a_t t = (0.500\text{ m/s}^2)(15.0\text{ s}) = 7.50\text{ m/s}.$$

The radial (centripetal) acceleration at that moment is

$$a_r = \frac{v^2}{r} = \frac{(7.50\text{ m/s})^2}{30.0\text{ m}} = 1.875\text{ m/s}^2.$$

Thus, the net acceleration has magnitude:

$$a = \sqrt{a_t^2 + a_r^2} = \sqrt{(0.500\text{ m/s}^2)^2 + (1.875\text{ m/s}^2)^2} = 1.94\text{ m/s}^2.$$

(b) We note that $\vec{a}_t \| \vec{v}$. Therefore, the angle between $\vec{v}$ and $\vec{a}$ is

$$\tan^{-1}\left(\frac{a_r}{a_t}\right) = \tan^{-1}\left(\frac{1.875}{0.5}\right) = 75.1°$$

so that the vector is pointing more toward the center of the track than in the direction of motion.

103. (a) Using Eq. 10-1, the angular displacement is

$$\theta = \frac{5.6\,\text{m}}{8.0 \times 10^{-2}\,\text{m}} = 1.4 \times 10^2 \text{ rad}.$$

(b) We use $\theta = \tfrac{1}{2}\alpha t^2$ (Eq. 10-13) to obtain t:

$$t = \sqrt{\frac{2\theta}{\alpha}} = \sqrt{\frac{2(1.4 \times 10^2 \text{ rad})}{1.5\,\text{rad/s}^2}} = 14\,\text{s}.$$

104. We apply Eq. 10-12 twice, assuming the sense of rotation is positive. We have $\omega > 0$ and $\alpha < 0$. Since the angular velocity at $t = 1$ min is $\omega_1 = (0.90)(250) = 225$ rev/min, we have

$$\omega_1 = \omega_0 + \alpha t \Rightarrow \alpha = \frac{225 - 250}{1} = -25 \text{ rev/min}^2.$$

Next, between $t = 1$ min and $t = 2$ min we have the interval $\Delta t = 1$ min. Consequently, the angular velocity at $t = 2$ min is

$$\omega_2 = \omega_1 + \alpha \Delta t = 225 + (-25)(1) = 200 \text{ rev/min}.$$

105. (a) Using Table 10-2(c), the rotational inertia is

$$I = \frac{1}{2}mR^2 = \frac{1}{2}(1210\,\text{kg})\left(\frac{1.21\,\text{m}}{2}\right)^2 = 221\,\text{kg}\cdot\text{m}^2.$$

(b) The rotational kinetic energy is, by Eq. 10-34,

$$K = \frac{1}{2}I\omega^2 = \frac{1}{2}(2.21\times 10^2\,\text{kg}\cdot\text{m}^2)[(1.52\,\text{rev/s})(2\pi\,\text{rad/rev})]^2 = 1.10\times 10^4\,\text{J}.$$

106. (a) We obtain

$$\omega = \frac{(33.33 \text{ rev/min})(2\pi \text{ rad/rev})}{60 \text{ s/min}} = 3.5 \text{ rad/s}.$$

(b) Using Eq. 10-18, we have $v = r\omega = (15)(3.49) = 52$ cm/s.

(c) Similarly, when $r = 7.4$ cm we find $v = r\omega = 26$ cm/s. The goal of this exercise is to observe what is and is not the same at different locations on a body in rotational motion (ω is the same, v is not), as well as to emphasize the importance of radians when working with equations such as Eq. 10-18.

107. With $v = 50(1000/3600) = 13.9$ m/s, Eq. 10-18 leads to

$$\omega = \frac{v}{r} = \frac{13.9}{110} = 0.13 \text{ rad/s}.$$

108. (a) The angular speed ω associated with Earth's spin is $\omega = 2\pi/T$, where $T = 86400$ s (one day). Thus,

$$\omega = \frac{2\pi}{86400 \text{ s}} = 7.3 \times 10^{-5} \text{ rad/s}$$

and the angular acceleration α required to accelerate the Earth from rest to ω in one day is $\alpha = \omega/T$. The torque needed is then

$$\tau = I\alpha = \frac{I\omega}{T} = \frac{(9.7 \times 10^{37} \text{ kg} \cdot \text{m}^2)(7.3 \times 10^{-5} \text{ rad/s})}{86400 \text{ s}} = 8.2 \times 10^{28} \text{ N} \cdot \text{m}$$

where we used

$$I = \frac{2}{5}MR^2 = \frac{2}{5}(5.98 \times 10^{24} \text{ kg})(6.37 \times 10^6 \text{ m})^2 = 9.7 \times 10^{37} \text{ kg} \cdot \text{m}^2$$

for Earth's rotational inertia.

(b) Using the values from part (a), the kinetic energy of the Earth associated with its rotation about its own axis is $K = \frac{1}{2}I\omega^2 = 2.6 \times 10^{29}$ J. This is how much energy would need to be supplied to bring it (starting from rest) to the current angular speed.

(c) The associated power is

$$P = \frac{K}{T} = \frac{2.57 \times 10^{29} \text{ J}}{86400 \text{ s}} = 3.0 \times 10^{24} \text{ W}.$$

109. The translational kinetic energy of the molecule is

$$K_t = \frac{1}{2}mv^2 = \frac{1}{2}(5.30 \times 10^{-26} \text{ kg})(500 \text{ m/s})^2 = 6.63 \times 10^{-21} \text{ J}.$$

With $I = 1.94 \times 10^{-46}$ kg·m^2, we employ Eq. 10-34:

$$K_r = \frac{2}{3}K_t \quad \Rightarrow \quad \frac{1}{2}I\omega^2 = \frac{2}{3}(6.63 \times 10^{-21} \text{ J})$$

which leads to $\omega = 6.75 \times 10^{12}$ rad/s.

110. (a) The rotational inertia relative to the specified axis is

$$I = \sum m_i r_i^2 = (2M)L^2 + (2M)L^2 + M(2L)^2$$

which is found to be $I = 4.6$ kg·m². Then, with $\omega = 1.2$ rad/s, we obtain the kinetic energy from Eq. 10-34:

$$K = \frac{1}{2}I\omega^2 = 3.3 \text{ J}.$$

(b) In this case the axis of rotation would appear as a standard y axis with origin at P. Each of the $2M$ balls are a distance of $r = L \cos 30°$ from that axis. Thus, the rotational inertia in this case is

$$I = \sum m_i r_i^2 = (2M)r^2 + (2M)r^2 + M(2L)^2$$

which is found to be $I = 4.0$ kg·m². Again, from Eq. 10-34 we obtain the kinetic energy

$$K = \frac{1}{2}I\omega^2 = 2.9 \text{ J}.$$

111. (a) The linear speed of a point on belt 1 is

$$v_1 = r_A \omega_A = (15 \text{ cm})(10 \text{ rad/s}) = 1.5 \times 10^2 \text{ cm/s}.$$

(b) The angular speed of pulley B is

$$r_B \omega_B = r_A \omega_A \quad \Rightarrow \quad \omega_B = \frac{r_A \omega_A}{r_B} = \left(\frac{15 \text{ cm}}{10 \text{ cm}}\right)(10 \text{ rad/s}) = 15 \text{ rad/s}.$$

(c) Since the two pulleys are rigidly attached to each other, the angular speed of pulley B' is the same as that of pulley B, i.e., $\omega_B' = 15$ rad/s.

(d) The linear speed of a point on belt 2 is

$$v_2 = r_{B'} \omega_B' = (5 \text{ cm})(15 \text{ rad/s}) = 75 \text{ cm/s}.$$

(e) The angular speed of pulley C is

$$r_C \omega_C = r_{B'} \omega_B' \quad \Rightarrow \quad \omega_C = \frac{r_{B'} \omega_B'}{r_C} = \left(\frac{5 \text{ cm}}{25 \text{ cm}}\right)(15 \text{ rad/s}) = 3.0 \text{ rad/s}$$

112. (a) The particle at A has $r = 0$ with respect to the axis of rotation. The particle at B is $r = L = 0.50$ m from the axis; similarly for the particle directly above A in the figure. The particle diagonally opposite A is a distance $r = \sqrt{2}L = 0.71$ m from the axis. Therefore,

$$I = \sum m_i r_i^2 = 2mL^2 + m\left(\sqrt{2}L\right)^2 = 0.20 \text{ kg} \cdot \text{m}^2.$$

(b) One imagines rotating the figure (about point A) clockwise by 90° and noting that the center of mass has fallen a distance equal to L as a result. If we let our reference position for gravitational potential be the height of the center of mass at the instant AB swings through vertical orientation, then

$$K_0 + U_0 = K + U \quad \Rightarrow \quad 0 + (4m)gh_0 = K + 0.$$

Since $h_0 = L = 0.50$ m, we find $K = 3.9$ J. Then, using Eq. 10-34, we obtain

$$K = \frac{1}{2} I_A \omega^2 \Rightarrow \omega = 6.3 \text{ rad/s}.$$

113. Using Eq. 10-12, we have

$$\omega = \omega_0 + \alpha t \Rightarrow \alpha = \frac{2.6 \text{ rad/s} - 8.0 \text{ rad/s}}{3.0 \text{ s}} = -1.8 \text{ rad/s}^2.$$

Using this value in Eq. 10-14 leads to

$$\omega^2 = \omega_0^2 + 2\alpha\theta \Rightarrow \theta = \frac{0 - (8.0 \text{ rad/s})^2}{2(-1.8 \text{ rad/s}^2)} = 18 \text{ rad}.$$

114. We make use of Table 10-2(e) as well as the parallel-axis theorem, Eq. 10-34, where needed. We use ℓ (as a subscript) to refer to the long rod and s to refer to the short rod.

(a) The rotational inertia is

$$I = I_s + I_\ell = \frac{1}{12} m_s L_s^2 + \frac{1}{3} m_\ell L_\ell^2 = 0.019 \text{ kg} \cdot \text{m}^2.$$

(b) We note that the center of the short rod is a distance of $h = 0.25$ m from the axis. The rotational inertia is

$$I = I_s + I_\ell = \frac{1}{12} m_s L_s^2 + m_s h^2 + \frac{1}{12} m_\ell L_\ell^2$$

which again yields $I = 0.019$ kg$\cdot$m^2.

115. We employ energy methods in this solution; thus, considerations of positive versus negative sense (regarding the rotation of the wheel) are not relevant.

(a) The speed of the box is related to the angular speed of the wheel by $v = R\omega$, so that

$$K_{box} = \frac{1}{2}m_{box}v^2 \Rightarrow v = \sqrt{\frac{2K_{box}}{m_{box}}} = 1.41 \text{ m/s}$$

implies that the angular speed is $\omega = 1.41/0.20 = 0.71$ rad/s. Thus, the kinetic energy of rotation is $\frac{1}{2}I\omega^2 = 10.0$ J.

(b) Since it was released from rest at what we will consider to be the reference position for gravitational potential, then (with SI units understood) energy conservation requires

$$K_0 + U_0 = K + U \quad \Rightarrow \quad 0 + 0 = (6.0 + 10.0) + m_{box}g(-h).$$

Therefore, $h = 16.0/58.8 = 0.27$ m.

116. (a) One particle is on the axis, so $r = 0$ for it. For each of the others, the distance from the axis is

$$r = (0.60 \text{ m}) \sin 60° = 0.52 \text{ m}.$$

Therefore, the rotational inertia is $I = \sum m_i r_i^2 = 0.27 \text{ kg} \cdot \text{m}^2$.

(b) The two particles that are nearest the axis are each a distance of $r = 0.30$ m from it. The particle "opposite" from that side is a distance $r = (0.60 \text{ m}) \sin 60° = 0.52$ m from the axis. Thus, the rotational inertia is

$$I = \sum m_i r_i^2 = 0.22 \text{ kg} \cdot \text{m}^2.$$

(c) The distance from the axis for each of the particles is $r = \frac{1}{2}(0.60 \text{ m}) \sin 60°$. The rotational inertia is

$$I = 3(0.50 \text{ kg})(0.26 \text{ m})^2 = 0.10 \text{ kg} \cdot \text{m}^2.$$

Chapter 11

1. The velocity of the car is a constant

$$\vec{v} = +(80 \text{ km/h})(1000 \text{ m/km})(1 \text{ h}/3600 \text{ s})\,\hat{i} = (+22 \text{ m/s})\hat{i},$$

and the radius of the wheel is $r = 0.66/2 = 0.33$ m.

(a) In the car's reference frame (where the lady perceives herself to be at rest) the road is moving towards the rear at $\vec{v}_{\text{road}} = -v = -22 \text{ m/s}$, and the motion of the tire is purely rotational. In this frame, the center of the tire is "fixed" so $v_{\text{center}} = 0$.

(b) Since the tire's motion is only rotational (not translational) in this frame, Eq. 10-18 gives $\vec{v}_{\text{top}} = (+22 \text{ m/s})\hat{i}$.

(c) The bottom-most point of the tire is (momentarily) in firm contact with the road (not skidding) and has the same velocity as the road: $\vec{v}_{\text{bottom}} = (-22 \text{ m/s})\hat{i}$. This also follows from Eq. 10-18.

(d) This frame of reference is not accelerating, so "fixed" points within it have zero acceleration; thus, $a_{\text{center}} = 0$.

(e) Not only is the motion purely rotational in this frame, but we also have $\omega =$ constant, which means the only acceleration for points on the rim is radial (centripetal). Therefore, the magnitude of the acceleration is

$$a_{\text{top}} = \frac{v^2}{r} = \frac{(22 \text{ m/s})^2}{0.33 \text{ m}} = 1.5 \times 10^3 \text{ m/s}^2.$$

(f) The magnitude of the acceleration is the same as in part (d): $a_{\text{bottom}} = 1.5 \times 10^3$ m/s^2.

(g) Now we examine the situation in the road's frame of reference (where the road is "fixed" and it is the car that appears to be moving). The center of the tire undergoes purely translational motion while points at the rim undergo a combination of translational and rotational motions. The velocity of the center of the tire is $\vec{v} = (+22 \text{ m/s})\hat{i}$.

(h) In part (b), we found $\vec{v}_{\text{top,car}} = +v$ and we use Eq. 4-39:

$$\vec{v}_{\text{top,ground}} = \vec{v}_{\text{top,car}} + \vec{v}_{\text{car,ground}} = v\hat{i} + v\hat{i} = 2v\hat{i}$$

which yields $2v = +44$ m/s. This is consistent with Fig. 11-3(c).

(i) We can proceed as in part (h) or simply recall that the bottom-most point is in firm contact with the (zero-velocity) road. Either way – the answer is zero.

(j) The translational motion of the center is constant; it does not accelerate.

(k) Since we are transforming between constant-velocity frames of reference, the accelerations are unaffected. The answer is as it was in part (e): 1.5×10^3 m/s^2.

(l) As explained in part (k), $a = 1.5 \times 10^3$ m/s^2.

2. The initial speed of the car is

$$v = (80 \text{ km/h})(1000 \text{ m/km})(1 \text{ h}/3600 \text{ s}) = 22.2 \text{ m/s}.$$

The tire radius is $R = 0.750/2 = 0.375$ m.

(a) The initial speed of the car is the initial speed of the center of mass of the tire, so Eq. 11-2 leads to

$$\omega_0 = \frac{v_{com0}}{R} = \frac{22.2 \text{ m/s}}{0.375 \text{ m}} = 59.3 \text{ rad/s}.$$

(b) With $\theta = (30.0)(2\pi) = 188$ rad and $\omega = 0$, Eq. 10-14 leads to

$$\omega^2 = \omega_0^2 + 2\alpha\theta \quad \Rightarrow \quad |\alpha| = \frac{(59.3 \text{ rad/s})^2}{2(188 \text{ rad})} = 9.31 \text{ rad/s}^2.$$

(c) Eq. 11-1 gives $R\theta = 70.7$ m for the distance traveled.

3. Let M be the mass of the car (presumably including the mass of the wheels) and v be its speed. Let I be the rotational inertia of one wheel and ω be the angular speed of each wheel. The kinetic energy of rotation is

$$K_{rot} = 4\left(\frac{1}{2}I\omega^2\right),$$

where the factor 4 appears because there are four wheels. The total kinetic energy is given by $K = \frac{1}{2}Mv^2 + 4(\frac{1}{2}I\omega^2)$. The fraction of the total energy that is due to rotation is

$$\text{fraction} = \frac{K_{rot}}{K} = \frac{4I\omega^2}{Mv^2 + 4I\omega^2}.$$

For a uniform disk (relative to its center of mass) $I = \frac{1}{2}mR^2$ (Table 10-2(c)). Since the wheels roll without sliding $\omega = v/R$ (Eq. 11-2). Thus the numerator of our fraction is

$$4I\omega^2 = 4\left(\frac{1}{2}mR^2\right)\left(\frac{v}{R}\right)^2 = 2mv^2$$

and the fraction itself becomes

$$\text{fraction} = \frac{2mv^2}{Mv^2 + 2mv^2} = \frac{2m}{M+2m} = \frac{2(10)}{1000} = \frac{1}{50} = 0.020.$$

The wheel radius cancels from the equations and is not needed in the computation.

4. We use the results from section 11.3.

(a) We substitute $I = \frac{2}{5}MR^2$ (Table 10-2(f)) and $a = -0.10g$ into Eq. 11-10:

$$-0.10g = -\frac{g\sin\theta}{1+\left(\frac{2}{5}MR^2\right)/MR^2} = -\frac{g\sin\theta}{7/5}$$

which yields $\theta = \sin^{-1}(0.14) = 8.0°$.

(b) The acceleration would be more. We can look at this in terms of forces or in terms of energy. In terms of forces, the uphill static friction would then be absent so the downhill acceleration would be due only to the downhill gravitational pull. In terms of energy, the rotational term in Eq. 11-5 would be absent so that the potential energy it started with would simply become $\frac{1}{2}mv^2$ (without it being "shared" with another term) resulting in a greater speed (and, because of Eq. 2-16, greater acceleration).

5. By Eq. 10-52, the work required to stop the hoop is the negative of the initial kinetic energy of the hoop. The initial kinetic energy is $K = \frac{1}{2}I\omega^2 + \frac{1}{2}mv^2$ (Eq. 11-5), where $I = mR^2$ is its rotational inertia about the center of mass, $m = 140$ kg, and $v = 0.150$ m/s is the speed of its center of mass. Eq. 11-2 relates the angular speed to the speed of the center of mass: $\omega = v/R$. Thus,

$$K = \frac{1}{2}mR^2\left(\frac{v^2}{R^2}\right) + \frac{1}{2}mv^2 = mv^2 = (140\text{ kg})(0.150\text{ m/s})^2$$

which implies that the work required is -3.15 J.

6. From $I = \frac{2}{3}MR^2$ (Table 10-2(g)) we find

$$M = \frac{3I}{2R^2} = \frac{3(0.040 \text{ kg} \cdot \text{m}^2)}{2(0.15 \text{ m})^2} = 2.7 \text{ kg}.$$

It also follows from the rotational inertia expression that $\frac{1}{2}I\omega^2 = \frac{1}{3}MR^2\omega^2$. Furthermore, it rolls without slipping, $v_{\text{com}} = R\omega$, and we find

$$\frac{K_{\text{rot}}}{K_{\text{com}} + K_{\text{rot}}} = \frac{\frac{1}{3}MR^2\omega^2}{\frac{1}{2}mR^2\omega^2 + \frac{1}{3}MR^2\omega^2}.$$

(a) Simplifying the above ratio, we find $K_{\text{rot}}/K = 0.4$. Thus, 40% of the kinetic energy is rotational, or

$$K_{\text{rot}} = (0.4)(20 \text{ J}) = 8.0 \text{ J}.$$

(b) From $K_{\text{rot}} = \frac{1}{3}MR^2\omega^2 = 8.0 \text{ J}$ (and using the above result for M) we find

$$\omega = \frac{1}{0.15 \text{ m}}\sqrt{\frac{3(8.0 \text{ J})}{2.7 \text{ kg}}} = 20 \text{ rad/s}$$

which leads to $v_{\text{com}} = (0.15 \text{ m})(20 \text{ rad/s}) = 3.0 \text{ m/s}$.

(c) We note that the inclined distance of 1.0 m corresponds to a height $h = 1.0 \sin 30° = 0.50$ m. Mechanical energy conservation leads to

$$K_i = K_f + U_f \Rightarrow 20 \text{ J} = K_f + Mgh$$

which yields (using the values of M and h found above) $K_f = 6.9$ J.

(d) We found in part (a) that 40% of this must be rotational, so

$$\frac{1}{3}MR^2\omega_f^2 = (0.40)K_f \Rightarrow \omega_f = \frac{1}{0.15 \text{ m}}\sqrt{\frac{3(0.40)(6.9 \text{ J})}{2.7 \text{ kg}}}$$

which yields $\omega_f = 12$ rad/s and leads to

$$v_{\text{com}f} = R\omega_f = (0.15 \text{ m})(12 \text{ rad/s}) = 1.8 \text{ m/s}.$$

7. With $\vec{F}_{\text{app}} = (10 \text{ N})\hat{i}$, we solve the problem by applying Eq. 9-14 and Eq. 11-37.

(a) Newton's second law in the x direction leads to

$$F_{app} - f_s = ma \Rightarrow f_s = 10\,\text{N} - (10\,\text{kg})(0.60\,\text{m/s}^2) = 4.0\,\text{N}.$$

In unit vector notation, we have $\vec{f}_s = (-4.0\,\text{N})\hat{i}$ which points leftward.

(b) With $R = 0.30$ m, we find the magnitude of the angular acceleration to be

$$|\alpha| = |a_{com}|/R = 2.0\,\text{rad/s}^2,$$

from Eq. 11-6. The only force not directed towards (or away from) the center of mass is $\vec{f}_s$, and the torque it produces is clockwise:

$$|\tau| = I|\alpha| \Rightarrow (0.30\,\text{m})(4.0\,\text{N}) = I(2.0\,\text{rad/s}^2)$$

which yields the wheel's rotational inertia about its center of mass: $I = 0.60\,\text{kg}\cdot\text{m}^2$.

8. Using the floor as the reference position for computing potential energy, mechanical energy conservation leads to

$$U_{release} = K_{top} + U_{top} \Rightarrow mgh = \frac{1}{2}mv_{com}^2 + \frac{1}{2}I\omega^2 + mg(2R).$$

Substituting $I = \frac{2}{5}mr^2$ (Table 10-2(f)) and $\omega = v_{com}/r$ (Eq. 11-2), we obtain

$$mgh = \frac{1}{2}mv_{com}^2 + \frac{1}{2}\left(\frac{2}{5}mr^2\right)\left(\frac{v_{com}}{r}\right)^2 + 2mgR \Rightarrow gh = \frac{7}{10}v_{com}^2 + 2gR$$

where we have canceled out mass m in that last step.

(a) To be on the verge of losing contact with the loop (at the top) means the normal force is vanishingly small. In this case, Newton's second law along the vertical direction ($+y$ downward) leads to

$$mg = ma_r \Rightarrow g = \frac{v_{com}^2}{R - r}$$

where we have used Eq. 10-23 for the radial (centripetal) acceleration (of the center of mass, which at this moment is a distance $R - r$ from the center of the loop). Plugging the result $v_{com}^2 = g(R - r)$ into the previous expression stemming from energy considerations gives

$$gh = \frac{7}{10}(g)(R - r) + 2gR$$

which leads to $h = 2.7R - 0.7r \approx 2.7R$. With $R = 14.0$ cm, we have $h = (2.7)(14.0 \text{ cm}) = 37.8$ cm.

(b) The energy considerations shown above (now with $h = 6R$) can be applied to point Q (which, however, is only at a height of R) yielding the condition

$$g(6R) = \frac{7}{10} v_{com}^2 + gR$$

which gives us $v_{com}^2 = 50g R/7$. Recalling previous remarks about the radial acceleration, Newton's second law applied to the horizontal axis at Q leads to

$$N = m \frac{v_{com}^2}{R-r} = m \frac{50gR}{7(R-r)}$$

which (for $R \gg r$) gives

$$N \approx \frac{50mg}{7} = \frac{50(2.80 \times 10^{-4} \text{ kg})(9.80 \text{ m/s}^2)}{7} = 1.96 \times 10^{-2} \text{ N}.$$

(b) The direction is toward the center of the loop.

9. (a) We find its angular speed as it leaves the roof using conservation of energy. Its initial kinetic energy is $K_i = 0$ and its initial potential energy is $U_i = Mgh$ where $h = 6.0 \sin 30° = 3.0$ m (we are using the edge of the roof as our reference level for computing U). Its final kinetic energy (as it leaves the roof) is (Eq. 11-5)

$$K_f = \tfrac{1}{2} Mv^2 + \tfrac{1}{2} I\omega^2.$$

Here we use v to denote the speed of its center of mass and ω is its angular speed — at the moment it leaves the roof. Since (up to that moment) the ball rolls without sliding we can set $v = R\omega = v$ where $R = 0.10$ m. Using $I = \tfrac{1}{2} MR^2$ (Table 10-2(c)), conservation of energy leads to

$$Mgh = \frac{1}{2} Mv^2 + \frac{1}{2} I\omega^2 = \frac{1}{2} MR^2 \omega^2 + \frac{1}{4} MR^2 \omega^2 = \frac{3}{4} MR^2 \omega^2.$$

The mass M cancels from the equation, and we obtain

$$\omega = \frac{1}{R} \sqrt{\frac{4}{3} gh} = \frac{1}{0.10 \text{ m}} \sqrt{\frac{4}{3} (9.8 \text{ m/s}^2)(3.0 \text{ m})} = 63 \text{ rad/s}.$$

(b) Now this becomes a projectile motion of the type examined in Chapter 4. We put the origin at the position of the center of mass when the ball leaves the track (the "initial" position for this part of the problem) and take $+x$ leftward and $+y$ downward. The result of part (a) implies $v_0 = R\omega = 6.3$ m/s, and we see from the figure that (with these positive direction choices) its components are

$$v_{0x} = v_0 \cos 30° = 5.4 \text{ m/s}$$
$$v_{0y} = v_0 \sin 30° = 3.1 \text{ m/s}.$$

The projectile motion equations become

$$x = v_{0x}t \quad \text{and} \quad y = v_{0y}t + \frac{1}{2}gt^2.$$

We first find the time when $y = H = 5.0$ m from the second equation (using the quadratic formula, choosing the positive root):

$$t = \frac{-v_{0y} + \sqrt{v_{0y}^2 + 2gH}}{g} = 0.74 \text{ s}.$$

Then we substitute this into the x equation and obtain $x = (5.4 \text{ m/s})(0.74 \text{ s}) = 4.0$ m.

10. We plug $a = -3.5$ m/s^2 (where the magnitude of this number was estimated from the "rise over run" in the graph), $\theta = 30°$, $M = 0.50$ kg and $R = 0.060$ m into Eq. 11-10 and solve for the rotational inertia. We find $I = 7.2 \times 10^{-4}$ kg·m^2.

11. To find where the ball lands, we need to know its speed as it leaves the track (using conservation of energy). Its initial kinetic energy is $K_i = 0$ and its initial potential energy is $U_i = MgH$. Its final kinetic energy (as it leaves the track) is $K_f = \frac{1}{2}Mv^2 + \frac{1}{2}I\omega^2$ (Eq. 11-5) and its final potential energy is Mgh. Here we use v to denote the speed of its center of mass and ω is its angular speed — at the moment it leaves the track. Since (up to that moment) the ball rolls without sliding we can set $\omega = v/R$. Using $I = \frac{2}{5}MR^2$ (Table 10-2(f)), conservation of energy leads to

$$MgH = \frac{1}{2}Mv^2 + \frac{1}{2}I\omega^2 + Mgh = \frac{1}{2}Mv^2 + \frac{2}{10}Mv^2 + Mgh = \frac{7}{10}Mv^2 + Mgh.$$

The mass M cancels from the equation, and we obtain

$$v = \sqrt{\frac{10}{7}g(H-h)} = \sqrt{\frac{10}{7}(9.8 \text{ m/s}^2)(6.0 \text{ m} - 2.0 \text{ m})} = 7.48 \text{ m/s}.$$

Now this becomes a projectile motion of the type examined in Chapter 4. We put the origin at the position of the center of mass when the ball leaves the track (the "initial" position for this part of the problem) and take $+x$ rightward and $+y$ downward. Then (since the initial velocity is purely horizontal) the projectile motion equations become

$$x = vt \text{ and } y = -\frac{1}{2}gt^2.$$

Solving for x at the time when $y = h$, the second equation gives $t = \sqrt{2h/g}$. Then, substituting this into the first equation, we find

$$x = v\sqrt{\frac{2h}{g}} = (7.48 \text{ m/s})\sqrt{\frac{2(2.0 \text{ m})}{9.8 \text{ m/s}^2}} = 4.8 \text{ m}.$$

12. (a) Let the turning point be designated P. By energy conservation, the mechanical energy at $x = 7.0$ m is equal to the mechanical energy at P. Thus, with Eq. 11-5, we have

$$75 \text{ J} = \tfrac{1}{2}mv_p^2 + \tfrac{1}{2}I_{com}\omega_p^2 + U_p$$

Using item (f) of Table 10-2 and Eq. 11-2 (which means, if this is to be a turning point, that $\omega_p = v_p = 0$), we find $U_p = 75$ J. On the graph, this seems to correspond to $x = 2.0$ m, and we conclude that there is a turning point (and this is it). The ball, therefore, does not reach the origin.

(b) We note that there is no point (on the graph, to the right of $x = 7.0$ m) which is shown "higher" than 75 J, so we suspect that there is no turning point in this direction, and we seek the velocity v_p at $x = 13$ m. If we obtain a real, nonzero answer, then our suspicion is correct (that it does reach this point P at $x = 13$ m). By energy conservation, the mechanical energy at $x = 7.0$ m is equal to the mechanical energy at P. Therefore,

$$75 \text{ J} = \tfrac{1}{2}mv_p^2 + \tfrac{1}{2}I_{com}\omega_p^2 + U_p$$

Again, using item (f) of Table 11-2, Eq. 11-2 (less trivially this time) and $U_p = 60$ J (from the graph), as well as the numerical data given in the problem, we find $v_p = 7.3$ m/s.

13. (a) We choose clockwise as the negative rotational sense and rightwards as the positive translational direction. Thus, since this is the moment when it begins to roll smoothly, Eq. 11-2 becomes $v_{com} = -R\omega = (-0.11 \text{ m})\omega$.

This velocity is positive-valued (rightward) since ω is negative-valued (clockwise) as shown in Fig. 11-57.

(b) The force of friction exerted on the ball of mass m is $-\mu_k mg$ (negative since it points left), and setting this equal to ma_{com} leads to

$$a_{com} = -\mu g = -(0.21)(9.8 \text{ m/s}^2) = -2.1 \text{ m/s}^2$$

where the minus sign indicates that the center of mass acceleration points left, opposite to its velocity, so that the ball is decelerating.

(c) Measured about the center of mass, the torque exerted on the ball due to the frictional force is given by $\tau = -\mu mgR$. Using Table 10-2(f) for the rotational inertia, the angular acceleration becomes (using Eq. 10-45)

$$\alpha = \frac{\tau}{I} = \frac{-\mu mgR}{2mR^2/5} = \frac{-5\mu g}{2R} = \frac{-5(0.21)(9.8 \text{ m/s}^2)}{2(0.11 \text{ m})} = -47 \text{ rad/s}^2$$

where the minus sign indicates that the angular acceleration is clockwise, the same direction as ω (so its angular motion is "speeding up").

(d) The center-of-mass of the sliding ball decelerates from $v_{com,0}$ to v_{com} during time t according to Eq. 2-11: $v_{com} = v_{com,0} - \mu gt$. During this time, the angular speed of the ball increases (in magnitude) from zero to $|\omega|$ according to Eq. 10-12:

$$|\omega| = |\alpha|t = \frac{5\mu gt}{2R} = \frac{v_{com}}{R}$$

where we have made use of our part (a) result in the last equality. We have two equations involving v_{com}, so we eliminate that variable and find

$$t = \frac{2v_{com,0}}{7\mu g} = \frac{2(8.5 \text{ m/s})}{7(0.21)(9.8 \text{ m/s}^2)} = 1.2 \text{ s}.$$

(e) The skid length of the ball is (using Eq. 2-15)

$$\Delta x = v_{com,0} t - \frac{1}{2}(\mu g)t^2 = (8.5 \text{ m/s})(1.2 \text{ s}) - \frac{1}{2}(0.21)(9.8 \text{ m/s}^2)(1.2 \text{ s})^2 = 8.6 \text{ m}.$$

(f) The center of mass velocity at the time found in part (d) is

$$v_{com} = v_{com,0} - \mu gt = 8.5 \text{ m/s} - (0.21)(9.8 \text{ m/s}^2)(1.2 \text{ s}) = 6.1 \text{ m/s}.$$

14. To find the center of mass speed v on the plateau, we use the projectile motion equations of Chapter 4. With $v_{oy} = 0$ (and using "h" for h_2) Eq. 4-22 gives the time-of-flight as $t = \sqrt{2h/g}$. Then Eq. 4-21 (squared, and using d for the horizontal displacement) gives $v^2 = gd^2/2h$. Now, to find the speed v_p at point P, we apply energy conservation, i.e.,

mechanical energy on the plateau is equal to the mechanical energy at P. With Eq. 11-5, we obtain

$$\tfrac{1}{2}mv^2 + \tfrac{1}{2}I_{com}\omega^2 + mgh_1 = \tfrac{1}{2}mv_p^2 + \tfrac{1}{2}I_{com}\omega_p^2$$

Using item (f) of Table 10-2, Eq. 11-2, and our expression (above) $v^2 = gd^2/2h$, we obtain

$$gd^2/2h + 10gh_1/7 = v_p^2$$

which yields (using the values stated in the problem) $v_p = 1.34$ m/s.

15. The physics of a rolling object usually requires a separate and very careful discussion (above and beyond the basics of rotation discussed in chapter 10); this is done in the first three sections of chapter 11. Also, the normal force on something (which is here the center of mass of the ball) following a circular trajectory is discussed in section 6-6 (see particularly sample problem 6-7). Adapting Eq. 6-19 to the consideration of forces at the *bottom* of an arc, we have

$$F_N - Mg = Mv^2/r$$

which tells us (since we are given $F_N = 2Mg$) that the center of mass speed (squared) is $v^2 = gr$, where r is the arc radius (0.48 m) Thus, the ball's angular speed (squared) is

$$\omega^2 = v^2/R^2 = gr/R^2,$$

where R is the ball's radius. Plugging this into Eq. 10-5 and solving for the rotational inertia (about the center of mass), we find

$$I_{com} = 2MhR^2/r - MR^2 = MR^2[2(0.36/0.48) - 1].$$

Thus, using the β notation suggested in the problem, we find

$$\beta = 2(0.36/0.48) - 1 = 0.50.$$

16. The physics of a rolling object usually requires a separate and very careful discussion (above and beyond the basics of rotation discussed in chapter 11); this is done in the first three sections of Chapter 11. Using energy conservation with Eq. 11-5 and solving for the rotational inertia (about the center of mass), we find

$$I_{com} = 2MhR^2/r - MR^2 = MR^2[2g(H-h)/v^2 - 1].$$

Thus, using the β notation suggested in the problem, we find

$$\beta = 2g(H-h)/v^2 - 1.$$

To proceed further, we need to find the center of mass speed v, which we do using the projectile motion equations of Chapter 4. With $v_{oy} = 0$, Eq. 4-22 gives the time-of-flight as $t = \sqrt{2h/g}$. Then Eq. 4-21 (squared, and using d for the horizontal displacement) gives $v^2 = gd^2/2h$. Plugging this into our expression for β gives

$$2g(H-h)/v^2 - 1 = 4h(H-h)/d^2 - 1$$

Therefore, with the values given in the problem, we find $\beta = 0.25$.

17. (a) The derivation of the acceleration is found in §11-4; Eq. 11-13 gives

$$a_{com} = -\frac{g}{1 + I_{com}/MR_0^2}$$

where the positive direction is upward. We use $I_{com} = 950$ g·cm², $M = 120$g, $R_0 = 0.320$ cm and $g = 980$ cm/s² and obtain

$$|a_{com}| = \frac{980 \text{ cm/s}^2}{1 + (950 \text{ g·cm}^2)/(120 \text{ g})(0.32 \text{ cm})^2} = 12.5 \text{ cm/s}^2 \approx 13 \text{ cm/s}^2.$$

(b) Taking the coordinate origin at the initial position, Eq. 2-15 leads to $y_{com} = \frac{1}{2}a_{com}t^2$. Thus, we set $y_{com} = -120$ cm, and find

$$t = \sqrt{\frac{2y_{com}}{a_{com}}} = \sqrt{\frac{2(-120 \text{ cm})}{-12.5 \text{ cm/s}^2}} = 4.38 \text{ s} \approx 4.4 \text{ s}.$$

(c) As it reaches the end of the string, its center of mass velocity is given by Eq. 2-11:

$$v_{com} = a_{com}t = (-12.5 \text{ cm/s}^2)(4.38 \text{ s}) = -54.8 \text{ cm/s},$$

so its linear speed then is approximately $|v_{com}| = 55$ cm/s.

(d) The translational kinetic energy is

$$\tfrac{1}{2}mv_{com}^2 = \tfrac{1}{2}(0.120 \text{ kg})(0.548 \text{ m/s})^2 = 1.8 \times 10^{-2} \text{ J}.$$

(e) The angular velocity is given by $\omega = -v_{com}/R_0$ and the rotational kinetic energy is

$$\frac{1}{2}I_{com}\omega^2 = \frac{1}{2}I_{com}\frac{v_{com}^2}{R_0^2} = \frac{1}{2}\frac{(9.50 \times 10^{-5} \text{ kg·m}^2)(0.548 \text{ m/s})^2}{(3.2 \times 10^{-3} \text{ m})^2}$$

which yields $K_{rot} = 1.4$ J.

(f) The angular speed is

$$\omega = \frac{|v_{com}|}{R_0} = \frac{0.548 \text{ m/s}}{3.2 \times 10^{-3} \text{ m}} = 1.7 \times 10^2 \text{ rad/s} = 27 \text{ rev/s}.$$

18. (a) The derivation of the acceleration is found in § 11-4; Eq. 11-13 gives

$$a_{com} = -\frac{g}{1 + I_{com}/MR_0^2}$$

where the positive direction is upward. We use $I_{com} = MR^2/2$ where the radius is $R = 0.32$ m and $M = 116$ kg is the *total* mass (thus including the fact that there are two disks) and obtain

$$a = -\frac{g}{1 + (MR^2/2)/MR_0^2} = \frac{g}{1 + (R/R_0)^2/2}$$

which yields $a = -g/51$ upon plugging in $R_0 = R/10 = 0.032$ m. Thus, the magnitude of the center of mass acceleration is 0.19 m/s^2.

(b) As observed in §11-4, our result in part (a) applies to both the descending and the rising yoyo motions.

(c) The external forces on the center of mass consist of the cord tension (upward) and the pull of gravity (downward). Newton's second law leads to

$$T - Mg = ma \Rightarrow T = M\left(g - \frac{g}{51}\right) = 1.1 \times 10^3 \text{ N}.$$

(d) Our result in part (c) indicates that the tension is well below the ultimate limit for the cord.

(e) As we saw in our acceleration computation, all that mattered was the ratio R/R_0 (and, of course, g). So if it's a scaled-up version, then such ratios are unchanged and we obtain the same result.

(f) Since the tension also depends on mass, then the larger yoyo will involve a larger cord tension.

19. If we write $\vec{r} = x\hat{i} + y\hat{j} + z\hat{k}$, then (using Eq. 3-30) we find $\vec{r} \times \vec{F}$ is equal to

$$(yF_z - zF_y)\hat{i} + (zF_x - xF_z)\hat{j} + (xF_y - yF_x)\hat{k}.$$

(a) In the above expression, we set (with SI units understood) $x = 0$, $y = -4.0$, $z = 3.0$, $F_x = 2.0$, $F_y = 0$ and $F_z = 0$. Then we obtain

$$\vec{\tau} = \vec{r} \times \vec{F} = \left(6.0\hat{j} + 8.0\hat{k}\right) \text{N} \cdot \text{m}.$$

This has magnitude $\sqrt{(6.0 \text{ N} \cdot \text{m})^2 + (8.0 \text{ N} \cdot \text{m})^2} = 10 \text{ N} \cdot \text{m}$ and is seen to be parallel to the yz plane. Its angle (measured counterclockwise from the $+y$ direction) is $\tan^{-1}(8/6) = 53°$.

(b) In the above expression, we set $x = 0$, $y = -4.0$, $z = 3.0$, $F_x = 0$, $F_y = 2.0$ and $F_z = 4.0$. Then we obtain $\vec{\tau} = \vec{r} \times \vec{F} = (-22 \text{ N} \cdot \text{m})\hat{i}$. This has magnitude $22 \text{ N} \cdot \text{m}$ and points in the $-x$ direction.

20. If we write $\vec{r} = x\hat{i} + y\hat{j} + z\hat{k}$, then (using Eq. 3-30) we find $\vec{r} \times \vec{F}$ is equal to

$$\left(yF_z - zF_y\right)\hat{i} + \left(zF_x - xF_z\right)\hat{j} + \left(xF_y - yF_x\right)\hat{k}.$$

(a) In the above expression, we set (with SI units understood) $x = -2.0$, $y = 0$, $z = 4.0$, $F_x = 6.0$, $F_y = 0$ and $F_z = 0$. Then we obtain $\vec{\tau} = \vec{r} \times \vec{F} = (24 \text{ N} \cdot \text{m})\hat{j}$.

(b) The values are just as in part (a) with the exception that now $F_x = -6.0$. We find $\vec{\tau} = \vec{r} \times \vec{F} = (-24 \text{ N} \cdot \text{m})\hat{j}$.

(c) In the above expression, we set $x = -2.0$, $y = 0$, $z = 4.0$, $F_x = 0$, $F_y = 0$ and $F_z = 6.0$. We get $\vec{\tau} = \vec{r} \times \vec{F} = (12 \text{ N} \cdot \text{m})\hat{j}$.

(d) The values are just as in part (c) with the exception that now $F_z = -6.0$. We find $\vec{\tau} = \vec{r} \times \vec{F} = (-12 \text{ N} \cdot \text{m})\hat{j}$.

21. If we write $\vec{r} = x\hat{i} + y\hat{j} + z\hat{k}$, then (using Eq. 3-30) we find $\vec{r} \times \vec{F}$ is equal to

$$\left(yF_z - zF_y\right)\hat{i} + \left(zF_x - xF_z\right)\hat{j} + \left(xF_y - yF_x\right)\hat{k}.$$

With (using SI units) $x = 0$, $y = -4.0$, $z = 5.0$, $F_x = 0$, $F_y = -2.0$ and $F_z = 3.0$ (these latter terms being the individual forces that contribute to the net force), the expression above yields

$$\vec{\tau} = \vec{r} \times \vec{F} = (-2.0 \text{ N} \cdot \text{m})\hat{i}.$$

22. If we write $\vec{r}' = x'\hat{i} + y'\hat{j} + z'\hat{k}$, then (using Eq. 3-30) we find $\vec{r}' \times \vec{F}$ is equal to

$$\left(y'F_z - z'F_y\right)\hat{i} + \left(z'F_x - x'F_z\right)\hat{j} + \left(x'F_y - y'F_x\right)\hat{k}.$$

(a) Here, $\vec{r}' = \vec{r}$ where $\vec{r} = 3.0\hat{i} - 2.0\hat{j} + 4.0\hat{k}$, and $\vec{F} = \vec{F}_1$. Thus, dropping the prime in the above expression, we set (with SI units understood) $x = 3.0$, $y = -2.0$, $z = 4.0$, $F_x = 3.0$, $F_y = -4.0$ and $F_z = 5.0$. Then we obtain

$$\vec{\tau} = \vec{r} \times \vec{F}_1 = \left(6.0\hat{i} - 3.0\hat{j} - 6.0\hat{k}\right) \text{N} \cdot \text{m}.$$

(b) This is like part (a) but with $\vec{F} = \vec{F}_2$. We plug in $F_x = -3.0$, $F_y = -4.0$ and $F_z = -5.0$ and obtain

$$\vec{\tau} = \vec{r} \times \vec{F}_2 = \left(26\hat{i} + 3.0\hat{j} - 18\hat{k}\right) \text{N} \cdot \text{m}.$$

(c) We can proceed in either of two ways. We can add (vectorially) the answers from parts (a) and (b), or we can first add the two force vectors and then compute $\vec{\tau} = \vec{r} \times (\vec{F}_1 + \vec{F}_2)$ (these total force components are computed in the next part). The result is

$$\vec{\tau} = \vec{r} \times (\vec{F}_1 + \vec{F}_2) = \left(32\hat{i} - 24\hat{k}\right) \text{N} \cdot \text{m}.$$

(d) Now $\vec{r}' = \vec{r} - \vec{r}_o$ where $\vec{r}_o = 3.0\hat{i} + 2.0\hat{j} + 4.0\hat{k}$. Therefore, in the above expression, we set $x' = 0$, $y' = -4.0$, $z' = 0$, and

$$F_x = 3.0 - 3.0 = 0$$
$$F_y = -4.0 - 4.0 = -8.0$$
$$F_z = 5.0 - 5.0 = 0.$$

We get $\vec{\tau} = \vec{r}' \times (\vec{F}_1 + \vec{F}_2) = 0$.

23. If we write $\vec{r} = x\hat{i} + y\hat{j} + z\hat{k}$, then (using Eq. 3-30) we find $\vec{r} \times \vec{F}$ is equal to

$$\left(yF_z - zF_y\right)\hat{i} + \left(zF_x - xF_z\right)\hat{j} + \left(xF_y - yF_x\right)\hat{k}.$$

(a) Plugging in, we find $\vec{\tau} = \left[(3.0\,\text{m})(6.0\,\text{N}) - (4.0\,\text{m})(-8.0\,\text{N})\right]\hat{k} = (50\,\text{N} \cdot \text{m})\,\hat{k}$.

(b) We use Eq. 3-27, $|\vec{r} \times \vec{F}| = rF \sin\phi$, where ϕ is the angle between $\vec{r}$ and $\vec{F}$. Now $r = \sqrt{x^2 + y^2} = 5.0$ m and $F = \sqrt{F_x^2 + F_y^2} = 10$ N. Thus,

$$rF = (5.0\,\text{m})(10\,\text{N}) = 50\,\text{N} \cdot \text{m},$$

the same as the magnitude of the vector product calculated in part (a). This implies $\sin \phi = 1$ and $\phi = 90°$.

24. Eq. 11-14 (along with Eq. 3-30) gives

$$\vec{\tau} = \vec{r} \times \vec{F} = 4.00\hat{i} + (12.0 + 2.00F_x)\hat{j} + (14.0 + 3.00F_x)\hat{k}$$

with SI units understood. Comparing this with the known expression for the torque (given in the problem statement), we see that F_x must satisfy two conditions:

$$12.0 + 2.00F_x = 2.00 \quad \text{and} \quad 14.0 + 3.00F_x = -1.00.$$

The answer ($F_x = -5.00$ N) satisfies both conditions.

25. We use the notation $\vec{r}'$ to indicate the vector pointing from the axis of rotation directly to the position of the particle. If we write $\vec{r}' = x'\hat{i} + y'\hat{j} + z'\hat{k}$, then (using Eq. 3-30) we find $\vec{r}' \times \vec{F}$ is equal to

$$\left(y'F_z - z'F_y\right)\hat{i} + \left(z'F_x - x'F_z\right)\hat{j} + \left(x'F_y - y'F_x\right)\hat{k}.$$

(a) Here, $\vec{r}' = \vec{r}$. Dropping the primes in the above expression, we set (with SI units understood) $x = 0$, $y = 0.5$, $z = -2.0$, $F_x = 2.0$, $F_y = 0$ and $F_z = -3.0$. Then we obtain

$$\vec{\tau} = \vec{r} \times \vec{F} = \left(-1.5\hat{i} - 4.0\hat{j} - 1.0\hat{k}\right) \text{N·m}.$$

(b) Now $\vec{r}' = \vec{r} - \vec{r}_o$ where $\vec{r}_o = 2.0\hat{i} - 3.0\hat{k}$. Therefore, in the above expression, we set $x' = -2.0$, $y' = 0.5$, $z' = 1.0$, $F_x = 2.0$, $F_y = 0$ and $F_z = -3.0$. Thus, we obtain

$$\vec{\tau} = \vec{r}' \times \vec{F} = \left(-1.5\hat{i} - 4.0\hat{j} - 1.0\hat{k}\right) \text{N·m}.$$

26. If we write $\vec{r}' = x'\hat{i} + y'\hat{j} + z'\hat{k}$, then (using Eq. 3-30) we find $\vec{r}' \times \vec{v}$ is equal to

$$\left(y'v_z - z'v_y\right)\hat{i} + \left(z'v_x - x'v_z\right)\hat{j} + \left(x'v_y - y'v_x\right)\hat{k}.$$

(a) Here, $\vec{r}' = \vec{r}$ where $\vec{r} = 3.0\hat{i} - 4.0\hat{j}$. Thus, dropping the primes in the above expression, we set (with SI units understood) $x = 3.0$, $y = -4.0$, $z = 0$, $v_x = 30$, $v_y = 60$ and $v_z = 0$. Then (with $m = 2.0$ kg) we obtain

$$\vec{\ell} = m(\vec{r} \times \vec{v}) = (6.0 \times 10^2 \text{ kg·m}^2/\text{s})\hat{k}.$$

(b) Now $\vec{r}' = \vec{r} - \vec{r}_o$ where $\vec{r}_o = -2.0\hat{i} - 2.0\hat{j}$. Therefore, in the above expression, we set $x' = 5.0$, $y' = -2.0$, $z' = 0$, $v_x = 30$, $v_y = 60$ and $v_z = 0$. We get

$$\vec{\ell} = m(\vec{r}' \times \vec{v}) = (7.2 \times 10^2 \ \text{kg} \cdot \text{m}^2/\text{s})\hat{k}.$$

27. For the 3.1 kg particle, Eq. 11-21 yields

$$\ell_1 = r_{\perp 1} m v_1 = (2.8 \ \text{m})(3.1 \ \text{kg})(3.6 \ \text{m/s}) = 31.2 \ \text{kg} \cdot \text{m}^2/\text{s}.$$

Using the right-hand rule for vector products, we find this $(\vec{r}_1 \times \vec{p}_1)$ is out of the page, or along the $+z$ axis, perpendicular to the plane of Fig. 11-40. And for the 6.5 kg particle, we find

$$\ell_2 = r_{\perp 2} m v_2 = (1.5 \ \text{m})(6.5 \ \text{kg})(2.2 \ \text{m/s}) = 21.4 \ \text{kg} \cdot \text{m}^2/\text{s}.$$

And we use the right-hand rule again, finding that this $(\vec{r}_2 \times \vec{p}_2)$ is into the page, or in the $-z$ direction.

(a) The two angular momentum vectors are in opposite directions, so their vector sum is the *difference* of their magnitudes: $L = \ell_1 - \ell_2 = 9.8 \ \text{kg} \cdot \text{m}^2/\text{s}$.

(b) The direction of the net angular momentum is along the $+z$ axis.

28. We note that the component of $\vec{v}$ perpendicular to $\vec{r}$ has magnitude $v \sin \theta_2$ where $\theta_2 = 30°$. A similar observation applies to $\vec{F}$.

(a) Eq. 11-20 leads to $\ell = r m v_\perp = (3.0 \ \text{m})(2.0 \ \text{kg})(4.0 \ \text{m/s}) \sin 30° = 12 \ \text{kg} \cdot \text{m}^2/\text{s}.$

(b) Using the right-hand rule for vector products, we find $\vec{r} \times \vec{p}$ points out of the page, or along the $+z$ axis, perpendicular to the plane of the figure.

(c) Eq. 10-38 leads to $\tau = r F \sin \theta_2 = (3.0 \ \text{m})(2.0 \ \text{N}) \sin 30° = 3.0 \ \text{N} \cdot \text{m}.$

(d) Using the right-hand rule for vector products, we find $\vec{r} \times \vec{F}$ is also out of the page, or along the $+z$ axis, perpendicular to the plane of the figure.

29. (a) We use $\vec{\ell} = m\vec{r} \times \vec{v}$, where $\vec{r}$ is the position vector of the object, $\vec{v}$ is its velocity vector, and m is its mass. Only the x and z components of the position and velocity vectors are nonzero, so Eq. 3-30 leads to $\vec{r} \times \vec{v} = (-x v_z + z v_x)\hat{j}$. Therefore,

$$\vec{\ell} = m(-x v_z + z v_x)\hat{j} = (0.25 \ \text{kg})(-(2.0 \ \text{m})(5.0 \ \text{m/s}) + (-2.0 \ \text{m})(-5.0 \ \text{m/s}))\hat{j} = 0.$$

(b) If we write $\vec{r} = x\hat{i} + y\hat{j} + z\hat{k}$, then (using Eq. 3-30) we find $\vec{r} \times \vec{F}$ is equal to

$$(yF_z - zF_y)\hat{i} + (zF_x - xF_z)\hat{j} + (xF_y - yF_x)\hat{k}.$$

With $x = 2.0$, $z = -2.0$, $F_y = 4.0$ and all other components zero (and SI units understood) the expression above yields

$$\vec{\tau} = \vec{r} \times \vec{F} = \left(8.0\hat{i} + 8.0\hat{k}\right) \text{ N} \cdot \text{m}.$$

30. (a) The acceleration vector is obtained by dividing the force vector by the (scalar) mass:

$$\vec{a} = \vec{F}/m = (3.00 \text{ m/s}^2)\hat{i} - (4.00 \text{ m/s}^2)\hat{j} + (2.00 \text{ m/s}^2)\hat{k}.$$

(b) Use of Eq. 11-18 leads directly to

$$\vec{L} = (42.0 \text{ kg} \cdot \text{m}^2/\text{s})\hat{i} + (24.0 \text{ kg} \cdot \text{m}^2/\text{s})\hat{j} + (60.0 \text{ kg} \cdot \text{m}^2/\text{s})\hat{k}.$$

(c) Similarly, the torque is

$$\vec{\tau} = \vec{r} \times \vec{F} = (-8.00 \text{ N} \cdot \text{m})\hat{i} - (26.0 \text{ N} \cdot \text{m})\hat{j} - (40.0 \text{ N} \cdot \text{m})\hat{k}.$$

(d) We note (using the Pythagorean theorem) that the magnitude of the velocity vector is 7.35 m/s and that of the force is 10.8 N. The dot product of these two vectors is $\vec{v} \cdot \vec{F} = -48$ (in SI units). Thus, Eq. 3-20 yields

$$\theta = \cos^{-1}[-48.0/(7.35 \times 10.8)] = 127°.$$

31. (a) Since the speed is (momentarily) zero when it reaches maximum height, the angular momentum is zero then.

(b) With the convention (used in several places in the book) that clockwise sense is to be associated with the negative sign, we have $L = -r_\perp mv$ where $r_\perp = 2.00$ m, $m = 0.400$ kg, and v is given by free-fall considerations (as in chapter 2). Specifically, y_{max} is determined by Eq. 2-16 with the speed at max height set to zero; we find $y_{max} = v_o^2/2g$ where $v_o = 40.0$ m/s. Then with $y = \frac{1}{2}y_{max}$, Eq. 2-16 can be used to give $v = v_o/\sqrt{2}$. In this way we arrive at $L = -22.6$ kg·m^2/s.

(c) As mentioned in the previous part, we use the minus sign in writing $\tau = -r_\perp F$ with the force F being equal (in magnitude) to mg. Thus, $\tau = -7.84$ N·m.

(d) Due to the way $r_\perp$ is defined it does not matter how far up the ball is. The answer is the same as in part (c), $\tau = -7.84$ N·m.

32. We use a right-handed coordinate system with $\hat{k}$ directed out of the xy plane so as to be consistent with counterclockwise rotation (and the right-hand rule). Thus, all the angular momenta being considered are along the $-\hat{k}$ direction; for example, in part (b) $\vec{\ell} = -4.0 t^2\, \hat{k}$ in SI units. We use Eq. 11-23.

(a) The angular momentum is constant so its derivative is zero. There is no torque in this instance.

(b) Taking the derivative with respect to time, we obtain the torque:

$$\vec{\tau} = \frac{d\vec{\ell}}{dt} = \left(-4.0\hat{k}\right)\frac{dt^2}{dt} = (-8.0 t\ \text{N} \cdot \text{m})\hat{k}.$$

This vector points in the $-\hat{k}$ direction (causing the clockwise motion to speed up) for all $t > 0$.

(c) With $\vec{\ell} = (-4.0\sqrt{t})\hat{k}$ in SI units, the torque is

$$\vec{\tau} = \left(-4.0\hat{k}\right)\frac{d\sqrt{t}}{dt} = \left(-4.0\hat{k}\right)\left(\frac{1}{2\sqrt{t}}\right) = \left(-\frac{2.0}{\sqrt{t}}\hat{k}\right) \text{N} \cdot \text{m}.$$

This vector points in the $-\hat{k}$ direction (causing the clockwise motion to speed up) for all $t > 0$ (and it is undefined for $t < 0$).

(d) Finally, we have

$$\vec{\tau} = \left(-4.0\hat{k}\right)\frac{dt^{-2}}{dt} = \left(-4.0\hat{k}\right)\left(\frac{-2}{t^3}\right) = \left(\frac{8.0}{t^3}\hat{k}\right) \text{N} \cdot \text{m}.$$

This vector points in the $+\hat{k}$ direction (causing the initially clockwise motion to slow down) for all $t > 0$.

33. If we write (for the general case) $\vec{r} = x\hat{i} + y\hat{j} + z\hat{k}$, then (using Eq. 3-30) we find $\vec{r} \times \vec{v}$ is equal to

$$(yv_z - zv_y)\hat{i} + (zv_x - xv_z)\hat{j} + (xv_y - yv_x)\hat{k}.$$

(a) The angular momentum is given by the vector product $\vec{\ell} = m\vec{r} \times \vec{v}$, where $\vec{r}$ is the position vector of the particle, $\vec{v}$ is its velocity, and $m = 3.0$ kg is its mass. Substituting (with SI units understood) $x = 3$, $y = 8$, $z = 0$, $v_x = 5$, $v_y = -6$ and $v_z = 0$ into the above expression, we obtain

$$\vec{\ell} = (3.0)[(3.0)(-6.0)-(8.0)(5.0)]\hat{k} = (-1.7\times 10^2 \text{ kg}\cdot\text{m}^2/\text{s})\hat{k}.$$

(b) The torque is given by Eq. 11-14, $\vec{\tau} = \vec{r}\times\vec{F}$. We write $\vec{r} = x\hat{i}+y\hat{j}$ and $\vec{F} = F_x\hat{i}$ and obtain

$$\vec{\tau} = \left(x\hat{i}+y\hat{j}\right)\times\left(F_x\hat{i}\right) = -yF_x\hat{k}$$

since $\hat{i}\times\hat{i} = 0$ and $\hat{j}\times\hat{i} = -\hat{k}$. Thus, we find

$$\vec{\tau} = -(8.0\text{m})(-7.0\text{N})\hat{k} = (56\text{ N}\cdot\text{m})\hat{k}.$$

(c) According to Newton's second law $\vec{\tau} = d\vec{\ell}/dt$, so the rate of change of the angular momentum is 56 kg·m²/s², in the positive z direction.

34. The rate of change of the angular momentum is

$$\frac{d\vec{\ell}}{dt} = \vec{\tau}_1+\vec{\tau}_2 = (2.0\text{ N}\cdot\text{m})\hat{i} - (4.0\text{ N}\cdot\text{m})\hat{j}.$$

Consequently, the vector $d\vec{\ell}/dt$ has a magnitude $\sqrt{(2.0\text{ N}\cdot\text{m})^2+(-4.0\text{ N}\cdot\text{m})^2} = 4.5\text{ N}\cdot\text{m}$ and is at an angle θ (in the xy plane, or a plane parallel to it) measured from the positive x axis, where

$$\theta = \tan^{-1}\left(\frac{-4.0\text{ N}\cdot\text{m}}{2.0\text{ N}\cdot\text{m}}\right) = -63°,$$

the negative sign indicating that the angle is measured clockwise as viewed "from above" (by a person on the $+z$ axis).

35. (a) We note that

$$\vec{v} = \frac{d\vec{r}}{dt} = 8.0t\,\hat{i} - (2.0+12t)\hat{j}$$

with SI units understood. From Eq. 11-18 (for the angular momentum) and Eq. 3-30, we find the particle's angular momentum is $8t^2\hat{k}$. Using Eq. 11-23 (relating its time-derivative to the (single) torque) then yields $\vec{\tau} = (48t\,\hat{k})\text{ N}\cdot\text{m}$.

(b) From our (intermediate) result in part (a), we see the angular momentum increases in proportion to t^2.

36. (a) Eq. 10-34 gives $\alpha = \tau/I$ and Eq. 10-12 leads to $\omega = \alpha t = \tau t/I$. Therefore, the angular momentum at $t = 0.033$ s is

$$I\omega = \tau t = (16\,\text{N}\cdot\text{m})(0.033\,\text{s}) = 0.53\,\text{kg}\cdot\text{m}^2/\text{s}$$

where this is essentially a derivation of the angular version of the impulse-momentum theorem.

(b) We find

$$\omega = \frac{\tau t}{I} = \frac{(16\,\text{N}\cdot\text{m})(0.033\,\text{s})}{1.2\times 10^{-3}\,\text{kg}\cdot\text{m}^2} = 440\,\text{rad/s}$$

which we convert as follows: $\omega = (440\,\text{rad/s})(60\,\text{s/min})(1\,\text{rev}/2\pi\,\text{rad}) \approx 4.2\times 10^3\,\text{rev/min}$.

37. (a) Since $\tau = dL/dt$, the average torque acting during any interval Δt is given by $\tau_{\text{avg}} = (L_f - L_i)/\Delta t$, where L_i is the initial angular momentum and L_f is the final angular momentum. Thus,

$$\tau_{\text{avg}} = \frac{0.800\,\text{kg}\cdot\text{m}^2/\text{s} - 3.00\,\text{kg}\cdot\text{m}^2/\text{s}}{1.50\,\text{s}} = -1.47\,\text{N}\cdot\text{m},$$

or $|\tau_{\text{avg}}| = 1.47\,\text{N}\cdot\text{m}$. In this case the negative sign indicates that the direction of the torque is opposite the direction of the initial angular momentum, implicitly taken to be positive.

(b) The angle turned is $\theta = \omega_0 t + \alpha t^2/2$. If the angular acceleration α is uniform, then so is the torque and $\alpha = \tau/I$. Furthermore, $\omega_0 = L_i/I$, and we obtain

$$\theta = \frac{L_i t + \tau t^2/2}{I} = \frac{(3.00\,\text{kg}\cdot\text{m}^2/\text{s})(1.50\,\text{s}) + (-1.467\,\text{N}\cdot\text{m})(1.50\,\text{s})^2/2}{0.140\,\text{kg}\cdot\text{m}^2} = 20.4\,\text{rad}.$$

(c) The work done on the wheel is

$$W = \tau\theta = (-1.47\,\text{N}\cdot\text{m})(20.4\,\text{rad}) = -29.9\,\text{J}$$

where more precise values are used in the calculation than what is shown here. An equally good method for finding W is Eq. 10-52, which, if desired, can be rewritten as $W = (L_f^2 - L_i^2)/2I$.

(d) The average power is the work done by the flywheel (the negative of the work done on the flywheel) divided by the time interval:

$$P_{\text{avg}} = -\frac{W}{\Delta t} = -\frac{-29.8\,\text{J}}{1.50\,\text{s}} = 19.9\,\text{W}.$$

38. We relate the motions of the various disks by examining their linear speeds (using Eq. 10-18). The fact that the linear speed at the rim of disk A must equal the linear speed at the rim of disk C leads to $\omega_A = 2\omega_C$. The fact that the linear speed at the hub of disk A must equal the linear speed at the rim of disk B leads to $\omega_A = \frac{1}{2}\omega_B$. Thus, $\omega_B = 4\omega_C$. The ratio of their angular momenta depend on these angular velocities as well as their rotational inertias (see item (c) in Table 11-2), which themselves depend on their masses. If h is the thickness and ρ is the density of each disk, then each mass is $\rho\pi R^2 h$. Therefore,

$$\frac{L_C}{L_B} = \frac{(\frac{1}{2})\rho\pi R_C^2 h R_C^2 \omega_C}{(\frac{1}{2})\rho\pi R_B^2 h R_B^2 \omega_B} = 1024.$$

39. (a) A particle contributes mr_2 to the rotational inertia. Here r is the distance from the origin O to the particle. The total rotational inertia is

$$I = m(3d)^2 + m(2d)^2 + m(d)^2 = 14md^2 = 14(2.3\times10^{-2}\text{kg})(0.12\text{ m})^2$$
$$= 4.6\times10^{-3}\text{ kg}\cdot\text{m}^2.$$

(b) The angular momentum of the middle particle is given by $L_m = I_m\omega$, where $I_m = 4md^2$ is its rotational inertia. Thus

$$L_m = 4md^2\omega = 4(2.3\times10^{-2}\text{kg})(0.12\text{ m})^2(0.85\text{ rad/s}) = 1.1\times10^{-3}\text{ kg}\cdot\text{m}^2/\text{s}.$$

(c) The total angular momentum is

$$I\omega = 14md^2\omega = 14(2.3\times10^{-2}\text{kg})(0.12\text{ m})^2(0.85\text{ rad/s}) = 3.9\times10^{-3}\text{ kg}\cdot\text{m}^2/\text{s}.$$

40. The results may be found by integrating Eq. 11-29 with respect to time, keeping in mind that $\vec{L}_i = 0$ and that the integration may be thought of as "adding the areas" under the line-segments (in the plot of the torque versus time – with "areas" under the time axis contributing negatively). It is helpful to keep in mind, also, that the area of a triangle is $\frac{1}{2}$ (base)(height).

(a) We find that $\vec{L} = 24$ kg·m²/s at $t = 7.0$ s.

(b) Similarly, $\vec{L} = 1.5$ kg·m²/s at $t = 20$ s.

41. (a) For the hoop, we use Table 10-2(h) and the parallel-axis theorem to obtain

$$I_1 = I_{com} + mh^2 = \frac{1}{2}mR^2 + mR^2 = \frac{3}{2}mR^2.$$

Of the thin bars (in the form of a square), the member along the rotation axis has (approximately) no rotational inertia about that axis (since it is thin), and the member farthest from it is very much like it (by being parallel to it) except that it is displaced by a distance h; it has rotational inertia given by the parallel axis theorem:

$$I_2 = I_{com} + mh^2 = 0 + mR^2 = mR^2.$$

Now the two members of the square perpendicular to the axis have the same rotational inertia (that is $I_3 = I_4$). We find I_3 using Table 10-2(e) and the parallel-axis theorem:

$$I_3 = I_{com} + mh^2 = \frac{1}{12}mR^2 + m\left(\frac{R}{2}\right)^2 = \frac{1}{3}mR^2.$$

Therefore, the total rotational inertia is

$$I_1 + I_2 + I_3 + I_4 = \frac{19}{6}mR^2 = 1.6\,\text{kg}\cdot\text{m}^2.$$

(b) The angular speed is constant:

$$\omega = \frac{\Delta\theta}{\Delta t} = \frac{2\pi}{2.5} = 2.5\,\text{rad/s}.$$

Thus, $L = I_{total}\omega = 4.0\,\text{kg}\cdot\text{m}^2/\text{s}$.

42. Torque is the time derivative of the angular momentum. Thus, the change in the angular momentum is equal to the time integral of the torque. With $\tau = (5.00 + 2.00t)\,\text{N}\cdot\text{m}$, the angular momentum as a function of time is (in units $\text{kg}\cdot\text{m}^2/\text{s}$)

$$L(t) = \int \tau\,dt = \int (5.00 + 2.00t)\,dt = L_0 + 5.00t + 1.00t^2$$

Since $L = 5.00\,\text{kg}\cdot\text{m}^2/\text{s}$ when $t = 1.00\,\text{s}$, the integration constant is $L_0 = -1$. Thus, the complete expression of the angular momentum is

$$L(t) = -1 + 5.00t + 1.00t^2.$$

At $t = 3.00\,\text{s}$, we have $L(t = 3.00) = -1 + 5.00(3.00) + 1.00(3.00)^2 = 23.0\,\text{kg}\cdot\text{m}^2/\text{s}$.

43. (a) No external torques act on the system consisting of the man, bricks, and platform, so the total angular momentum of the system is conserved. Let I_i be the initial rotational inertia of the system and let I_f be the final rotational inertia. Then $I_i\omega_i = I_f\omega_f$ and

$$\omega_f = \left(\frac{I_i}{I_f}\right)\omega_i = \left(\frac{6.0\,\text{kg}\cdot\text{m}^2}{2.0\,\text{kg}\cdot\text{m}^2}\right)(1.2\,\text{rev/s}) = 3.6\,\text{rev/s}.$$

(b) The initial kinetic energy is $K_i = \frac{1}{2}I_i\omega_i^2$, the final kinetic energy is $K_f = \frac{1}{2}I_f\omega_f^2$, and their ratio is

$$\frac{K_f}{K_i} = \frac{I_f\omega_f^2/2}{I_i\omega_i^2/2} = \frac{(2.0\,\text{kg}\cdot\text{m}^2)(3.6\,\text{rev/s})^2/2}{(6.0\,\text{kg}\cdot\text{m}^2)(1.2\,\text{rev/s})^2/2} = 3.0.$$

(c) The man did work in decreasing the rotational inertia by pulling the bricks closer to his body. This energy came from the man's store of internal energy.

44. We use conservation of angular momentum:

$$I_m\omega_m = I_p\omega_p.$$

The respective angles θ_m and θ_p by which the motor and probe rotate are therefore related by

$$\int I_m\omega_m\,dt = I_m\theta_m = \int I_p\omega_p\,dt = I_p\theta_p$$

which gives

$$\theta_m = \frac{I_p\theta_p}{I_m} = \frac{(12\,\text{kg}\cdot\text{m}^2)(30°)}{2.0\times 10^{-3}\,\text{kg}\cdot\text{m}^2} = 180000°.$$

The number of revolutions for the rotor is then $(1.8\times 10^5)°/(360°/\text{rev}) = 5.0\times 10^2$ rev.

45. (a) No external torques act on the system consisting of the two wheels, so its total angular momentum is conserved. Let I_1 be the rotational inertia of the wheel that is originally spinning (at ω_i) and I_2 be the rotational inertia of the wheel that is initially at rest. Then $I_1\omega_i = (I_1 + I_2)\omega_f$ and

$$\omega_f = \frac{I_1}{I_1 + I_2}\omega_i$$

where ω_f is the common final angular velocity of the wheels. Substituting $I_2 = 2I_1$ and $\omega_i = 800$ rev/min, we obtain $\omega_f = 267$ rev/min.

(b) The initial kinetic energy is $K_i = \frac{1}{2}I_1\omega_i^2$ and the final kinetic energy is $K_f = \frac{1}{2}(I_1 + I_2)\omega_f^2$. We rewrite this as

$$K_f = \frac{1}{2}(I_1 + 2I_1)\left(\frac{I_1\omega_i}{I_1 + 2I_1}\right)^2 = \frac{1}{6}I\omega_i^2.$$

Therefore, the fraction lost, $(K_i - K_f)/K_i$ is

$$1 - \frac{K_f}{K_i} = 1 - \frac{I\omega_i^2/6}{I\omega_i^2/2} = \frac{2}{3} = 0.667.$$

46. Using Eq. 11-31 with angular momentum conservation, $\vec{L}_i = \vec{L}_f$ (Eq. 11-33) leads to the ratio of rotational inertias being inversely proportional to the ratio of angular velocities. Thus, $I_f/I_i = 6/5 = 1.0 + 0.2$. We interpret the "1.0" as the ratio of disk rotational inertias (which does not change in this problem) and the "0.2" as the ratio of the roach rotational inertial to that of the disk. Thus, the answer is 0.20.

47. (a) We apply conservation of angular momentum: $I_1\omega_1 + I_2\omega_2 = (I_1 + I_2)\omega$. The angular speed after coupling is therefore

$$\omega = \frac{I_1\omega_1 + I_2\omega_2}{I_1 + I_2} = \frac{(3.3\,\text{kg}\cdot\text{m}^2)(450\,\text{rev/min}) + (6.6\,\text{kg}\cdot\text{m}^2)(900\,\text{rev/min})}{3.3\,\text{kg}\cdot\text{m}^2 + 6.6\,\text{kg}\cdot\text{m}^2}$$
$$= 750\,\text{rev/min}.$$

(b) In this case, we obtain

$$\omega = \frac{I_1\omega_1 + I_2\omega_2}{I_1 + I_2} = \frac{(3.3\,\text{kg}\cdot\text{m}^2)(450\,\text{rev/min}) + (6.6\,\text{kg}\cdot\text{m}^2)(-900\,\text{rev/min})}{3.3\,\text{kg}\cdot\text{m}^2 + 6.6\,\text{kg}\cdot\text{m}^2}$$
$$= -450\,\text{rev/min}$$

or $|\omega| = 450\,\text{rev/min}$.

(c) The minus sign indicates that $\vec{\omega}$ is in the direction of the second disk's initial angular velocity - clockwise.

48. Angular momentum conservation $I_i\omega_i = I_f\omega_f$ leads to

$$\frac{\omega_f}{\omega_i} = \frac{I_i}{I_f}\omega_i = 3$$

which implies

$$\frac{K_f}{K_i} = \frac{I_f\omega_f^2/2}{I_i\omega_i^2/2} = \frac{I_f}{I_i}\left(\frac{\omega_f}{\omega_i}\right)^2 = 3.$$

49. No external torques act on the system consisting of the train and wheel, so the total angular momentum of the system (which is initially zero) remains zero. Let $I = MR^2$ be the rotational inertia of the wheel. Its final angular momentum is

$$\vec{L}_f = I\omega \hat{k} = -MR^2|\omega|\hat{k},$$

where $\hat{k}$ is *up* in Fig. 11-47 and that last step (with the minus sign) is done in recognition that the wheel's clockwise rotation implies a negative value for ω. The linear speed of a point on the track is ωR and the speed of the train (going counterclockwise in Fig. 11-47 with speed v' relative to an outside observer) is therefore $v' = v - |\omega|R$ where v is its speed relative to the tracks. Consequently, the angular momentum of the train is $m(v - |\omega|R)R\hat{k}$. Conservation of angular momentum yields

$$0 = -MR^2|\omega|\hat{k} + m(v - |\omega|R)R\hat{k}.$$

When this equation is solved for the angular speed, the result is

$$|\omega| = \frac{mvR}{(M+m)R^2} = \frac{v}{(M/m+1)R} = \frac{(0.15 \text{ m/s})}{(1.1+1)(0.43 \text{ m})} = 0.17 \text{ rad/s}.$$

50. So that we don't get confused about $\pm$ signs, we write the angular *speed* to the lazy Susan as $|\omega|$ and reserve the ω symbol for the angular velocity (which, using a common convention, is negative-valued when the rotation is clockwise). When the roach "stops" we recognize that it comes to rest relative to the lazy Susan (not relative to the ground).

(a) Angular momentum conservation leads to

$$mvR + I\omega_0 = (mR^2 + I)\omega_f$$

which we can write (recalling our discussion about angular speed versus angular velocity) as

$$mvR - I|\omega_0| = -(mR^2 + I)|\omega_f|.$$

We solve for the final angular speed of the system:

$$|\omega_f| = \frac{mvR - I|\omega_0|}{mR^2 + I} = \frac{(0.17 \text{ kg})(2.0 \text{ m/s})(0.15 \text{ m}) - (5.0 \times 10^{-3} \text{ kg} \cdot \text{m}^2)(2.8 \text{ rad/s})}{(5.0 \times 10^{-3} \text{ kg} \cdot \text{m}^2) + (0.17 \text{ kg})(0.15 \text{ m})^2}$$
$$= 4.2 \text{ rad/s}.$$

(b) No, $K_f \neq K_i$ and — if desired — we can solve for the difference:

$$K_i - K_f = \frac{mI}{2} \frac{v^2 + \omega_0^2 R^2 + 2Rv|\omega_0|}{mR^2 + I}$$

which is clearly positive. Thus, some of the initial kinetic energy is "lost" — that is, transferred to another form. And the culprit is the roach, who must find it difficult to stop (and "internalize" that energy).

51. We assume that from the moment of grabbing the stick onward, they maintain rigid postures so that the system can be analyzed as a symmetrical rigid body with center of mass midway between the skaters.

(a) The total linear momentum is zero (the skaters have the same mass and equal-and-opposite velocities). Thus, their center of mass (the middle of the 3.0 m long stick) remains fixed and they execute circular motion (of radius $r = 1.5$ m) about it.

(b) Using Eq. 10-18, their angular velocity (counterclockwise as seen in Fig. 11-48) is

$$\omega = \frac{v}{r} = \frac{1.4 \text{ m/s}}{1.5 \text{ m}} = 0.93 \text{ rad/s}.$$

(c) Their rotational inertia is that of two particles in circular motion at $r = 1.5$ m, so Eq. 10-33 yields

$$I = \sum mr^2 = 2(50 \text{ kg})(1.5 \text{ m})^2 = 225 \text{ kg} \cdot \text{m}^2.$$

Therefore, Eq. 10-34 leads to

$$K = \frac{1}{2} I \omega^2 = \frac{1}{2}(225 \text{ kg} \cdot \text{m}^2)(0.93 \text{ rad/s})^2 = 98 \text{ J}.$$

(d) Angular momentum is conserved in this process. If we label the angular velocity found in part (a) ω_i and the rotational inertia of part (b) as I_i, we have

$$I_i \omega_i = (225 \text{ kg} \cdot \text{m}^2)(0.93 \text{ rad/s}) = I_f \omega_f.$$

The final rotational inertia is $\sum mr_f^2$ where $r_f = 0.5$ m so $I_f = 25$ kg·m². Using this value, the above expression gives $\omega_f = 8.4$ rad/s.

(e) We find

$$K_f = \frac{1}{2} I_f \omega_f^2 = \frac{1}{2}(25 \text{ kg} \cdot \text{m}^2)(8.4 \text{ rad/s})^2 = 8.8 \times 10^2 \text{ J}.$$

(f) We account for the large increase in kinetic energy (part (e) minus part (c)) by noting that the skaters do a great deal of work (converting their internal energy into mechanical energy) as they pull themselves closer — "fighting" what appears to them to be large "centrifugal forces" trying to keep them apart.

52. The gravitational force acts at the center of mass and cannot provide a torque to change the bola's angular momentum during the flight. So, the angular momentum before and after the configuration change must be equal. We treat both configurations as a rigid object rotating around a fixed point. The initial and final rotational inertias are

$$I_i = m(2\ell)^2 + m(2\ell)^2 + m(0)^2 = 8m\ell^2$$
$$I_f = m\ell^2 + m\ell^2 + m\ell^2 = 3m\ell^2.$$

(a) Since angular momentum is conserved, $L_i = L_f$, or $I_i\omega_i = I_f\omega_f$. Thus,

$$\frac{\omega_f}{\omega_i} = \frac{I_i}{I_f} = \frac{8m\ell^2}{3m\ell^2} = \frac{8}{3} = 2.7.$$

(b) The initial and final kinetic energies are $K_i = I_i\omega_i^2/2$ and $K_f = I_f\omega_f^2/2$, respectively. Thus, we find the ratio to be

$$\frac{K_f}{K_i} = \frac{I_f\omega_f^2/2}{I_i\omega_i^2/2} = \frac{I_f}{I_i}\left(\frac{\omega_f}{\omega_i}\right)^2 = \frac{I_f}{I_i}\left(\frac{I_i}{I_f}\right)^2 = \frac{I_i}{I_f} = \frac{8}{3} = 2.7.$$

53. For simplicity, we assume the record is turning freely, without any work being done by its motor (and without any friction at the bearings or at the stylus trying to slow it down). Before the collision, the angular momentum of the system (presumed positive) is $I_i\omega_i$ where $I_i = 5.0\times10^{-4}$ kg·m² and $\omega_i = 4.7$ rad/s. The rotational inertia afterwards is

$$I_f = I_i + mR^2$$

where $m = 0.020$ kg and $R = 0.10$ m. The mass of the record (0.10 kg), although given in the problem, is not used in the solution. Angular momentum conservation leads to

$$I_i\omega_i = I_f\omega_f \Rightarrow \omega_f = \frac{I_i\omega_i}{I_i + mR^2} = 3.4 \text{ rad/s}.$$

54. Table 10-2 gives the rotational inertia of a thin rod rotating about a perpendicular axis through its center. The angular speeds of the two arms are, respectively,

$$\omega_1 = \frac{(0.500 \text{ rev})(2\pi \text{ rad/rev})}{0.700 \text{ s}} = 4.49 \text{ rad/s}$$

$$\omega_2 = \frac{(1.00 \text{ rev})(2\pi \text{ rad/rev})}{0.700 \text{ s}} = 8.98 \text{ rad/s}.$$

Treating each arm as a thin rod of mass 4.0 kg and length 0.60 m, the angular momenta of the two arms are

$$L_1 = I\omega_1 = mr^2\omega_1 = (4.0 \text{ kg})(0.60 \text{ m})^2(4.49 \text{ rad/s}) = 6.46 \text{ kg} \cdot \text{m}^2/\text{s}$$
$$L_2 = I\omega_2 = mr^2\omega_2 = (4.0 \text{ kg})(0.60 \text{ m})^2(8.98 \text{ rad/s}) = 12.92 \text{ kg} \cdot \text{m}^2/\text{s}.$$

From the athlete's reference frame, one arm rotates clockwise, while the other rotates counterclockwise. Thus, the total angular momentum about the common rotation axis though the shoulders is

$$L = L_2 - L_1 = 12.92 \text{ kg} \cdot \text{m}^2/\text{s} - 6.46 \text{ kg} \cdot \text{m}^2/\text{s} = 6.46 \text{ kg} \cdot \text{m}^2/\text{s}.$$

55. The axis of rotation is in the middle of the rod, with $r = 0.25$ m from either end. By Eq. 11-19, the initial angular momentum of the system (which is just that of the bullet, before impact) is $rmv \sin\theta$ where $m = 0.003$ kg and $\theta = 60°$. Relative to the axis, this is counterclockwise and thus (by the common convention) positive. After the collision, the moment of inertia of the system is

$$I = I_{\text{rod}} + mr^2$$

where $I_{\text{rod}} = ML^2/12$ by Table 10-2(e), with $M = 4.0$ kg and $L = 0.5$ m. Angular momentum conservation leads to

$$rmv \sin\theta = \left(\frac{1}{12}ML^2 + mr^2\right)\omega.$$

Thus, with $\omega = 10$ rad/s, we obtain

$$v = \frac{\left(\frac{1}{12}(4.0 \text{ kg})(0.5 \text{ m})^2 + (0.003 \text{ kg})(0.25 \text{ m})^2\right)(10 \text{ rad/s})}{(0.25 \text{ m})(0.003 \text{ kg})\sin 60°} = 1.3 \times 10^3 \text{ m/s}.$$

56. We denote the cockroach with subscript 1 and the disk with subscript 2. The cockroach has a mass $m_1 = m$, while the mass of the disk is $m_2 = 4.00 \, m$.

(a) Initially the angular momentum of the system consisting of the cockroach and the disk is

$$L_i = m_1 v_{1i} r_{1i} + I_2 \omega_{2i} = m_1 \omega_0 R^2 + \frac{1}{2} m_2 \omega_0 R^2.$$

After the cockroach has completed its walk, its position (relative to the axis) is $r_{1f} = R/2$ so the final angular momentum of the system is

$$L_f = m_1 \omega_f \left(\frac{R}{2}\right)^2 + \frac{1}{2} m_2 \omega_f R^2.$$

Then from $L_f = L_i$ we obtain

$$\omega_f \left(\frac{1}{4} m_1 R^2 + \frac{1}{2} m_2 R\right) = \omega_0 \left(m_1 R^2 + \frac{1}{2} m_2 R^2\right).$$

Thus,

$$\omega_f = \left(\frac{m_1 R^2 + m_2 R^2/2}{m_1 R^2/4 + m_2 R^2/2}\right)\omega_0 = \left(\frac{1+(m_2/m_1)/2}{1/4+(m_2/m_1)/2}\right)\omega_0 = \left(\frac{1+2}{1/4 + 2}\right)\omega_0 = 1.33\omega_0.$$

With $\omega_0 = 0.260$ rad/s, we have $\omega_f = 0.347$ rad/s.

(b) We substitute $I = L/\omega$ into $K = \frac{1}{2} I \omega^2$ and obtain $K = \frac{1}{2} L \omega$. Since we have $L_i = L_f$, the kinetic energy ratio becomes

$$\frac{K}{K_0} = \frac{L_f \omega_f / 2}{L_i \omega_i / 2} = \frac{\omega_f}{\omega_0} = 1.33.$$

(c) The cockroach does positive work while walking toward the center of the disk, increasing the total kinetic energy of the system.

57. By angular momentum conservation (Eq. 11-33), the total angular momentum after the explosion must be equal to before the explosion:

$$L'_p + L'_r = L_p + L_r$$

$$\left(\tfrac{L}{2}\right) m v_p + \tfrac{1}{12} M L^2 \omega' = I_p \omega + \tfrac{1}{12} M L^2 \omega$$

where one must be careful to avoid confusing the length of the rod ($L = 0.800$ m) with the angular momentum symbol. Note that $I_p = m(L/2)^2$ by Eq.10-33, and

$$\omega' = v_{\text{end}}/r = (v_p - 6)/(L/2),$$

510   CHAPTER 11

where the latter relation follows from the penultimate sentence in the problem (and "6" stands for "6.00 m/s" here). Since $M = 3m$ and $\omega = 20$ rad/s, we end up with enough information to solve for the particle speed: $v_p = 11.0$ m/s.

58. (a) With $r = 0.60$ m, we obtain $I = 0.060 + (0.501)r^2 = 0.24$ kg · m².

(b) Invoking angular momentum conservation, with SI units understood,

$$\ell_0 = L_f \;\Rightarrow\; mv_0 r = I\omega \;\Rightarrow\; (0.001)v_0(0.60) = (0.24)(4.5)$$

which leads to $v_0 = 1.8 \times 10^3$ m/s.

59. Their angular velocities, when they are stuck to each other, are equal, regardless of whether they share the same central axis. The initial rotational inertia of the system is

$$I_0 = I_{\text{big disk}} + I_{\text{small disk}} \quad \text{where} \quad I_{\text{big disk}} = \frac{1}{2} MR^2$$

using Table 10-2(c). Similarly, since the small disk is initially concentric with the big one, $I_{\text{small disk}} = \frac{1}{2} mr^2$. After it slides, the rotational inertia of the small disk is found from the parallel axis theorem (using $h = R - r$). Thus, the new rotational inertia of the system is

$$I = \frac{1}{2} MR^2 + \frac{1}{2} mr^2 + m(R-r)^2.$$

(a) Angular momentum conservation, $I_0 \omega_0 = I\omega$, leads to the new angular velocity:

$$\omega = \omega_0 \frac{(MR^2/2) + (mr^2/2)}{(MR^2/2) + (mr^2/2) + m(R-r)^2}.$$

Substituting $M = 10m$ and $R = 3r$, this becomes $\omega = \omega_0(91/99)$. Thus, with $\omega_0 = 20$ rad/s, we find $\omega = 18$ rad/s.

(b) From the previous part, we know that

$$\frac{I_0}{I} = \frac{91}{99} \quad \text{and} \quad \frac{\omega}{\omega_0} = \frac{91}{99}.$$

Plugging these into the ratio of kinetic energies, we have

$$\frac{K}{K_0} = \frac{I\omega^2/2}{I_0 \omega_0^2/2} = \frac{I}{I_0}\left(\frac{\omega}{\omega_0}\right)^2 = \frac{99}{91}\left(\frac{91}{99}\right)^2 = 0.92.$$

60. The initial rotational inertia of the system is $I_i = I_{disk} + I_{student}$, where $I_{disk} = 300$ kg·m² (which, incidentally, does agree with Table 10-2(c)) and $I_{student} = mR^2$ where $m = 60$ kg and $R = 2.0$ m.

The rotational inertia when the student reaches $r = 0.5$ m is $I_f = I_{disk} + mr^2$. Angular momentum conservation leads to

$$I_i\omega_i = I_f\omega_f \Rightarrow \omega_f = \omega_i \frac{I_{disk} + mR^2}{I_{disk} + mr^2}$$

which yields, for $\omega_i = 1.5$ rad/s, a final angular velocity of $\omega_f = 2.6$ rad/s.

61. We make the unconventional choice of *clockwise* sense as positive, so that the angular velocities in this problem are positive. With $r = 0.60$ m and $I_0 = 0.12$ kg·m², the rotational inertia of the putty-rod system (after the collision) is

$$I = I_0 + (0.20)r^2 = 0.19 \text{ kg·m}^2.$$

Invoking angular momentum conservation $L_0 = L_f$ or $I_0\omega_0 = I\omega$, we have

$$\omega = \frac{I_0}{I}\omega_0 = \frac{0.12 \text{ kg·m}^2}{0.19 \text{ kg·m}^2}(2.4 \text{ rad/s}) = 1.5 \text{ rad/s}.$$

62. We treat the ballerina as a rigid object rotating around a fixed axis, initially and then again near maximum height. Her initial rotational inertia (trunk and one leg extending outward at a 90° angle) is

$$I_i = I_{trunk} + I_{leg} = 0.660 \text{ kg·m}^2 + 1.44 \text{ kg·m}^2 = 2.10 \text{ kg·m}^2.$$

Similarly, her final rotational inertia (trunk and *both* legs extending outward at a $\theta = 30°$ angle) is

$$I_f = I_{trunk} + 2I_{leg}\sin^2\theta = 0.660 \text{ kg·m}^2 + 2(1.44 \text{ kg·m}^2)\sin^2 30° = 1.38 \text{ kg·m}^2,$$

where we have used the fact that the effective length of the extended leg at an angle θ is $L_\perp = L\sin\theta$ and $I \sim L_\perp^2$. Once air-borne, there is no external torque about the ballerina's center of mass and her angular momentum cannot change. Therefore, $L_i = L_f$ or $I_i\omega_i = I_f\omega_f$, and the ratio of the angular speeds is

$$\frac{\omega_f}{\omega_i} = \frac{I_i}{I_f} = \frac{2.10 \text{ kg·m}^2}{1.38 \text{ kg·m}^2} = 1.52.$$

63. (a) We consider conservation of angular momentum (Eq. 11-33) about the center of the rod:

$$L_i = L_f \Rightarrow -dmv + \frac{1}{12}ML^2\omega = 0$$

where negative is used for "clockwise." Item (e) in Table 11-2 and Eq. 11-21 (with $r_\perp = d$) have also been used. This leads to

$$d = \frac{ML^2\omega}{12\,mv} = \frac{M(0.60\text{ m})^2(80\text{ rad/s})}{12(M/3)(40\text{ m/s})} = 0.180\text{ m}.$$

(b) Increasing d causes the magnitude of the negative (clockwise) term in the above equation to increase. This would make the total angular momentum negative before the collision, and (by Eq. 11-33) also negative afterwards. Thus, the system would rotate clockwise if d were greater.

64. The aerialist is in extended position with $I_1 = 19.9\text{ kg}\cdot\text{m}^2$ during the first and last quarter of the turn, so the total angle rotated in t_1 is $\theta_1 = 0.500$ rev. In t_2 he is in a tuck position with $I_2 = 3.93\text{ kg}\cdot\text{m}^2$, and the total angle rotated is $\theta_2 = 3.500$ rev. Since there is no external torque about his center of mass, angular momentum is conserved, $I_1\omega_1 = I_2\omega_2$. Therefore, the total flight time can be written as

$$t = t_1 + t_2 = \frac{\theta_1}{\omega_1} + \frac{\theta_2}{\omega_2} = \frac{\theta_1}{I_2\omega_2/I_1} + \frac{\theta_2}{\omega_2} = \frac{1}{\omega_2}\left(\frac{I_1}{I_2}\theta_1 + \theta_2\right).$$

Substituting the values given, we find ω_2 to be

$$\omega_2 = \frac{1}{t}\left(\frac{I_1}{I_2}\theta_1 + \theta_2\right) = \frac{1}{1.87\text{ s}}\left(\frac{19.9\text{ kg}\cdot\text{m}^2}{3.93\text{ kg}\cdot\text{m}^2}(0.500\text{ rev}) + 3.50\text{ rev}\right) = 3.23\text{ rev/s}.$$

65. This is a completely inelastic collision which we analyze using angular momentum conservation. Let m and v_0 be the mass and initial speed of the ball and R the radius of the merry-go-round. The initial angular momentum is

$$\vec{\ell}_0 = \vec{r}_0 \times \vec{p}_0 \Rightarrow \ell_0 = R(mv_0)\cos 37°$$

where $\phi = 37°$ is the angle between $\vec{v}_0$ and the line tangent to the outer edge of the merry-go-around. Thus, $\ell_0 = 19\text{ kg}\cdot\text{m}^2/\text{s}$. Now, with SI units understood,

$$\ell_0 = L_f \Rightarrow 19\text{ kg}\cdot\text{m}^2 = I\omega = \left(150 + (30)R^2 + (1.0)R^2\right)\omega$$

so that $\omega = 0.070$ rad/s.

66. We make the unconventional choice of *clockwise* sense as positive, so that the angular velocities (and angles) in this problem are positive. Mechanical energy conservation applied to the particle (before impact) leads to

$$mgh = \frac{1}{2}mv^2 \Rightarrow v = \sqrt{2gh}$$

for its speed right before undergoing the completely inelastic collision with the rod. The collision is described by angular momentum conservation:

$$mvd = \left(I_{rod} + md^2\right)\omega$$

where I_{rod} is found using Table 10-2(e) and the parallel axis theorem:

$$I_{rod} = \frac{1}{12}Md^2 + M\left(\frac{d}{2}\right)^2 = \frac{1}{3}Md^2.$$

Thus, we obtain the angular velocity of the system immediately after the collision:

$$\omega = \frac{md\sqrt{2gh}}{(Md^2/3) + md^2}$$

which means the system has kinetic energy $\left(I_{rod} + md^2\right)\omega^2/2$ which will turn into potential energy in the final position, where the block has reached a height H (relative to the lowest point) and the center of mass of the stick has increased its height by $H/2$. From trigonometric considerations, we note that $H = d(1 - \cos\theta)$, so we have

$$\frac{1}{2}\left(I_{rod} + md^2\right)\omega^2 = mgH + Mg\frac{H}{2} \Rightarrow \frac{1}{2}\frac{m^2d^2(2gh)}{(Md^2/3) + md^2} = \left(m + \frac{M}{2}\right)gd(1-\cos\theta)$$

from which we obtain

$$\theta = \cos^{-1}\left(1 - \frac{m^2h}{(m+M/2)(m+M/3)}\right) = \cos^{-1}\left(1 - \frac{h/d}{(1+M/2m)(1+M/3m)}\right)$$

$$= \cos^{-1}\left(1 - \frac{(20\text{ cm}/40\text{ cm})}{(1+1)(1+2/3)}\right) = \cos^{-1}(0.85)$$

$$= 32°.$$

67. (a) If we consider a short time interval from just before the wad hits to just after it hits and sticks, we may use the principle of conservation of angular momentum. The initial angular momentum is the angular momentum of the falling putty wad. The wad initially moves along a line that is $d/2$ distant from the axis of rotation, where $d = 0.500$ m is the length of the rod. The angular momentum of the wad is $mvd/2$ where $m = 0.0500$ kg and $v = 3.00$ m/s are the mass and initial speed of the wad. After the wad sticks, the rod has angular velocity ω and angular momentum $I\omega$, where I is the rotational inertia of the system consisting of the rod with the two balls and the wad at its end. Conservation of angular momentum yields $mvd/2 = I\omega$ where

$$I = (2M + m)(d/2)^2$$

and $M = 2.00$ kg is the mass of each of the balls. We solve

$$mvd/2 = (2M + m)(d/2)^2 \omega$$

for the angular speed:

$$\omega = \frac{2mv}{(2M+m)d} = \frac{2(0.0500 \text{ kg})(3.00 \text{ m/s})}{(2(2.00 \text{ kg}) + 0.0500 \text{ kg})(0.500 \text{ m})} = 0.148 \text{ rad/s}.$$

(b) The initial kinetic energy is $K_i = \frac{1}{2}mv^2$, the final kinetic energy is $K_f = \frac{1}{2}I\omega^2$, and their ratio is $K_f/K_i = I\omega^2/mv^2$. When $I = (2M+m)d^2/4$ and $\omega = 2mv/(2M+m)d$ are substituted, this becomes

$$\frac{K_f}{K_i} = \frac{m}{2M+m} = \frac{0.0500 \text{ kg}}{2(2.00 \text{ kg}) + 0.0500 \text{ kg}} = 0.0123.$$

(c) As the rod rotates, the sum of its kinetic and potential energies is conserved. If one of the balls is lowered a distance h, the other is raised the same distance and the sum of the potential energies of the balls does not change. We need consider only the potential energy of the putty wad. It moves through a 90° arc to reach the lowest point on its path, gaining kinetic energy and losing gravitational potential energy as it goes. It then swings up through an angle θ, losing kinetic energy and gaining potential energy, until it momentarily comes to rest. Take the lowest point on the path to be the zero of potential energy. It starts a distance $d/2$ above this point, so its initial potential energy is $U_i = mgd/2$. If it swings up to the angular position θ, as measured from its lowest point, then its final height is $(d/2)(1 - \cos\theta)$ above the lowest point and its final potential energy is

$$U_f = mg(d/2)(1 - \cos\theta).$$

The initial kinetic energy is the sum of that of the balls and wad:

$$K_i = \frac{1}{2}I\omega^2 = \frac{1}{2}(2M+m)(d/2)^2\omega^2.$$

At its final position, we have $K_f = 0$. Conservation of energy provides the relation:

$$mg\frac{d}{2} + \frac{1}{2}(2M+m)\left(\frac{d}{2}\right)^2\omega^2 = mg\frac{d}{2}(1-\cos\theta).$$

When this equation is solved for $\cos\theta$, the result is

$$\cos\theta = -\frac{1}{2}\left(\frac{2M+m}{mg}\right)\left(\frac{d}{2}\right)\omega^2 = -\frac{1}{2}\left(\frac{2(2.00\text{ kg})+0.0500\text{ kg}}{(0.0500\text{ kg})(9.8\text{ m/s}^2)}\right)\left(\frac{0.500\text{ m}}{2}\right)(0.148\text{ rad/s})^2$$
$$= -0.0226.$$

Consequently, the result for θ is 91.3°. The total angle through which it has swung is 90° + 91.3° = 181°.

68. (a) The angular speed of the top is $\omega = 30$ rev/s $= 30(2\pi)$ rad/s. The precession rate of the top can be obtained by using Eq. 11-46:

$$\Omega = \frac{Mgr}{I\omega} = \frac{(0.50\text{ kg})(9.8\text{ m/s}^2)(0.040\text{ m})}{(5.0\times10^{-4}\text{ kg}\cdot\text{m}^2)(60\pi\text{ rad/s})} = 2.08\text{ rad/s} \approx 0.33\text{ rev/s}.$$

(b) The direction of the precession is clockwise as viewed from overhead.

69. The precession rate can be obtained by using Eq. 11-46 with $r = (11/2)$ cm = 0.055 m. Noting that $I_{\text{disk}} = MR^2/2$ and its angular speed is

$$\omega = 1000\text{ rev/min} = \frac{2\pi(1000)}{60}\text{ rad/s} \approx 1.0\times10^2\text{ rad/s},$$

we have

$$\Omega = \frac{Mgr}{(MR^2/2)\omega} = \frac{2gr}{R^2\omega} = \frac{2(9.8\text{ m/s}^2)(0.055\text{ m})}{(0.50\text{ m})^2(1.0\times10^2\text{ rad/s})} \approx 0.041\text{ rad/s}.$$

70. Item (*i*) in Table 10-2 gives the moment of inertia about the center of mass in terms of width a (0.15 m) and length b (0.20 m). In using the parallel axis theorem, the distance from the center to the point about which it spins (as described in the problem) is $\sqrt{(a/4)^2 + (b/4)^2}$. If we denote the thickness as h (0.012 m) then the volume is abh, which means the mass is ρabh (where $\rho = 2640$ kg/m^3 is the density). We can write the kinetic energy in terms of the angular momentum by substituting $\omega = L/I$ into Eq. 10-34:

$$K = \frac{1}{2}\frac{L^2}{I} = \frac{1}{2}\frac{(0.104)^2}{\rho abh((a^2+b^2)/12 + (a/4)^2 + (b/4)^2)} = 0.62 \text{ J}.$$

71. We denote the cat with subscript 1 and the ring with subscript 2. The cat has a mass $m_1 = M/4$, while the mass of the ring is $m_2 = M = 8.00$ kg. The moment of inertia of the ring is $I_2 = m_2(R_1^2 + R_2^2)/2$ (Table 10-2), and $I_1 = m_1 r^2$ for the cat, where r is the perpendicular distance from the axis of rotation.

Initially the angular momentum of the system consisting of the cat (at $r = R_2$) and the ring is

$$L_i = m_1 v_{1i} r_{1i} + I_2 \omega_{2i} = m_1 \omega_0 R_2^2 + \frac{1}{2} m_2 (R_1^2 + R_2^2) \omega_0 = m_1 R_2^2 \omega_0 \left[1 + \frac{1}{2}\frac{m_2}{m_1}\left(\frac{R_1^2}{R_2^2} + 1\right)\right].$$

After the cat has crawled to the inner edge at $r = R_1$ the final angular momentum of the system is

$$L_f = m_1 \omega_f R_1^2 + \frac{1}{2} m_2 (R_1^2 + R_2^2) \omega_f = m_1 R_1^2 \omega_f \left[1 + \frac{1}{2}\frac{m_2}{m_1}\left(1 + \frac{R_2^2}{R_1^2}\right)\right].$$

Then from $L_f = L_i$ we obtain

$$\frac{\omega_f}{\omega_0} = \left(\frac{R_2}{R_1}\right)^2 \frac{1 + \frac{1}{2}\frac{m_2}{m_1}\left(\frac{R_1^2}{R_2^2} + 1\right)}{1 + \frac{1}{2}\frac{m_2}{m_1}\left(1 + \frac{R_2^2}{R_1^2}\right)} = (2.0)^2 \frac{1 + 2(0.25+1)}{1 + 2(1+4)} = 1.273.$$

Thus, $\omega_f = 1.273 \omega_0$. Using $\omega_0 = 8.00$ rad/s, we have $\omega_f = 10.2$ rad/s. By substituting $I = L/\omega$ into $K = I\omega^2/2$, we obtain $K = L\omega/2$. Since $L_i = L_f$, the kinetic energy ratio becomes

$$\frac{K_f}{K_i} = \frac{L_f \omega_f / 2}{L_i \omega_i / 2} = \frac{\omega_f}{\omega_0} = 1.273.$$

which implies $\Delta K = K_f - K_i = 0.273 K_i$. The cat does positive work while walking toward the center of the ring, increasing the total kinetic energy of the system.

Since the initial kinetic energy is given by

$$K_i = \frac{1}{2}\left[m_1 R_2^2 + \frac{1}{2}m_2(R_1^2 + R_2^2)\right]\omega_0^2 = \frac{1}{2}m_1 R_2^2 \omega_0^2 \left[1 + \frac{1}{2}\frac{m_2}{m_1}\left(\frac{R_1^2}{R_2^2}+1\right)\right]$$

$$= \frac{1}{2}(2.00 \text{ kg})(0.800 \text{ m})^2 (8.00 \text{ rad/s})^2 [1 + (1/2)(4)(0.5^2 + 1)]$$

$$= 143.36 \text{ J},$$

the increase in kinetic energy is

$$\Delta K = (0.273)(143.36 \text{ J}) = 39.1 \text{ J}.$$

72. The total angular momentum (about the origin) before the collision (using Eq. 11-18 and Eq. 3-30 for each particle and then adding the terms) is

$$\vec{L}_i = [(0.5 \text{ m})(2.5 \text{ kg})(3.0 \text{ m/s}) + (0.1 \text{ m})(4.0 \text{ kg})(4.5 \text{ m/s})]\hat{k}.$$

The final angular momentum of the stuck-together particles (after the collision) measured relative to the origin is (using Eq. 11-33)

$$\vec{L}_f = \vec{L}_i = (5.55 \text{ kg·m}^2/\text{s})\hat{k}.$$

73. (a) The diagram below shows the particles and their lines of motion. The origin is marked O and may be anywhere. The angular momentum of particle 1 has magnitude

$$\ell_1 = mvr_1 \sin\theta_1 = mv(d+h)$$

and it is into the page. The angular momentum of particle 2 has magnitude

$$\ell_2 = mvr_2 \sin\theta_2 = mvh$$

and it is out of the page. The net angular momentum has magnitude

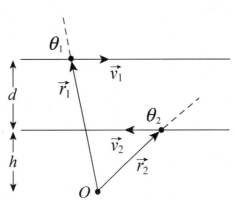

$$L = mv(d+h) - mvh = mvd$$
$$= (2.90 \times 10^{-4} \text{ kg})(5.46 \text{ m/s})(0.042 \text{ m})$$
$$= 6.65 \times 10^{-5} \text{ kg·m}^2/\text{s}.$$

and is into the page. This result is independent of the location of the origin.

(b) As indicated above, the expression does not change.

(c) Suppose particle 2 is traveling to the right. Then

$$L = mv(d+h) + mvh = mv(d+2h).$$

This result depends on h, the distance from the origin to one of the lines of motion. If the origin is midway between the lines of motion, then $h = -d/2$ and $L = 0$.

(d) As we have seen in part (c), the result depends on the choice of origin.

74. (a) We use Table 10-2(e) and the parallel-axis theorem to obtain the rod's rotational inertia about an axis through one end:

$$I = I_{com} + Mh^2 = \frac{1}{12} ML^2 + M\left(\frac{L}{2}\right)^2 = \frac{1}{3} ML^2$$

where $L = 6.00$ m and $M = 10.0/9.8 = 1.02$ kg. Thus, the inertia is $I = 12.2$ kg·m².

(b) Using $\omega = (240)(2\pi/60) = 25.1$ rad/s, Eq. 11-31 gives the magnitude of the angular momentum as

$$I\omega = (12.2 \text{ kg·m}^2)(25.1 \text{ rad/s}) = 308 \text{ kg·m}^2/\text{s}.$$

Since it is rotating clockwise as viewed from above, then the right-hand rule indicates that its direction is down.

75. We use $L = I\omega$ and $K = \frac{1}{2}I\omega^2$ and observe that the speed of points on the rim (corresponding to the speed of points on the belt) of wheels A and B must be the same (so $\omega_A R_A = \omega_B R_B$).

(a) If $L_A = L_B$ (call it L) then the ratio of rotational inertias is

$$\frac{I_A}{I_B} = \frac{L/\omega_A}{L/\omega_B} = \frac{\omega_B}{\omega_A} = \frac{R_A}{R_B} = \frac{1}{3} = 0.333.$$

(b) If we have $K_A = K_B$ (call it K) then the ratio of rotational inertias becomes

$$\frac{I_A}{I_B} = \frac{2K/\omega_A^2}{2K/\omega_B^2} = \left(\frac{\omega_B}{\omega_A}\right)^2 = \left(\frac{R_A}{R_B}\right)^2 = \frac{1}{9} = 0.111.$$

76. Both $\vec{r}$ and $\vec{v}$ lie in the xy plane. The position vector $\vec{r}$ has an x component that is a function of time (being the integral of the x component of velocity, which is itself time-dependent) and a y component that is constant ($y = -2.0$ m). In the cross product $\vec{r} \times \vec{v}$, all that matters is the y component of $\vec{r}$ since $v_x \neq 0$ but $v_y = 0$:

$$\vec{r} \times \vec{v} = -yv_x \hat{k}.$$

(a) The angular momentum is $\vec{\ell} = m(\vec{r} \times \vec{v})$ where the mass is $m = 2.0$ kg in this case. With SI units understood and using the above cross-product expression, we have

$$\vec{\ell} = (2.0)\left(-(-2.0)(-6.0t^2)\right)\hat{k} = -24t^2 \hat{k}$$

in kg·m²/s. This implies the particle is moving clockwise (as observed by someone on the +z axis) for $t > 0$.

(b) The torque is caused by the (net) force $\vec{F} = m\vec{a}$ where

$$\vec{a} = \frac{d\vec{v}}{dt} = (-12t\,\hat{i})\,\text{m/s}^2.$$

The remark above that only the y component of $\vec{r}$ still applies, since $a_y = 0$. We use $\vec{\tau} = \vec{r} \times \vec{F} = m(\vec{r} \times \vec{a})$ and obtain

$$\vec{\tau} = (2.0)(-(-2.0)(-12t))\hat{k} = (-48t\,\hat{k})\,\text{N}\cdot\text{m}.$$

The torque on the particle (as observed by someone on the +z axis) is clockwise, causing the particle motion (which was clockwise to begin with) to increase.

(c) We replace $\vec{r}$ with $\vec{r}\,'$ (measured relative to the new reference point) and note (again) that only its y component matters in these calculations. Thus, with $y' = -2.0 - (-3.0) = 1.0$ m, we find

$$\vec{\ell}\,' = (2.0)\left(-(1.0)(-6.0t^2)\right)\hat{k} = (12t^2\,\hat{k})\,\text{kg}\cdot\text{m}^2/\text{s}..$$

The fact that this is positive implies that the particle is moving counterclockwise relative to the new reference point.

(d) Using $\vec{\tau}\,' = \vec{r}\,' \times \vec{F} = m(\vec{r}\,' \times \vec{a})$, we obtain

$$\vec{\tau} = (2.0)(-(1.0)(-12t))\hat{k} = (24t\,\hat{k})\,\text{N}\cdot\text{m}..$$

The torque on the particle (as observed by someone on the +z axis) is counterclockwise, relative to the new reference point.

77. As the wheel-axel system rolls down the inclined plane by a distance d, the decrease in potential energy is $\Delta U = mgd\sin\theta$. This must be equal to the total kinetic energy gained:

$$mgd\sin\theta = \frac{1}{2}mv^2 + \frac{1}{2}I\omega^2.$$

Since the axel rolls without slipping, the angular speed is given by $\omega = v/r$, where r is the radius of the axel. The above equation then becomes

$$mgd\sin\theta = \frac{1}{2}I\omega^2\left(\frac{mr^2}{I}+1\right) = K_{rot}\left(\frac{mr^2}{I}+1\right)$$

(a) With $m=10.0$ kg, $d = 2.00$ m, $r = 0.200$ m, and $I = 0.600$ kg·m², $mr^2/I = 2/3$, the rotational kinetic energy may be obtained as $98\text{ J} = K_{rot}(5/3)$, or $K_{rot} = 58.8$ J.

(b) The translational kinetic energy is $K_{trans} = (98 - 58.8)\text{J} = 39.2$ J.

78. (a) The acceleration is given by Eq. 11-13:

$$a_{com} = \frac{g}{1+I_{com}/MR_0^2}$$

where upward is the positive translational direction. Taking the coordinate origin at the initial position, Eq. 2-15 leads to

$$y_{com} = v_{com,0}t + \frac{1}{2}a_{com}t^2 = v_{com,0}t - \frac{\frac{1}{2}gt^2}{1+I_{com}/MR_0^2}$$

where $y_{com} = -1.2$ m and $v_{com,0} = -1.3$ m/s. Substituting $I_{com} = 0.000095$ kg·m², $M = 0.12$ kg, $R_0 = 0.0032$ m and $g = 9.8$ m/s², we use the quadratic formula and find

$$t = \frac{\left(1+\frac{I_{com}}{MR_0^2}\right)\left(v_{com,0} \mp \sqrt{v_{com,0}^2 - \frac{2gy_{com}}{1+I_{com}/MR_0^2}}\right)}{g}$$

$$= \frac{\left(1+\frac{0.000095}{(0.12)(0.0032)^2}\right)\left(-1.3 \mp \sqrt{(1.3)^2 - \frac{2(9.8)(-1.2)}{1+0.000095/(0.12)(0.0032)^2}}\right)}{9.8}$$

$$= -21.7 \text{ or } 0.885$$

where we choose $t = 0.89$ s as the answer.

(b) We note that the initial potential energy is $U_i = Mgh$ and $h = 1.2$ m (using the bottom as the reference level for computing U). The initial kinetic energy is as shown in Eq. 11-5, where the initial angular and linear speeds are related by Eq. 11-2. Energy conservation leads to

$$K_f = K_i + U_i = \frac{1}{2}mv_{com,0}^2 + \frac{1}{2}I\left(\frac{v_{com,0}}{R_0}\right)^2 + Mgh$$

$$= \frac{1}{2}(0.12 \text{ kg})(1.3 \text{ m/s})^2 + \frac{1}{2}(9.5\times10^{-5} \text{ kg}\cdot\text{m}^2)\left(\frac{1.3 \text{ m/s}}{0.0032 \text{ m}}\right)^2 + (0.12 \text{ kg})(9.8 \text{ m/s}^2)(1.2 \text{ m})$$

$$= 9.4 \text{ J}.$$

(c) As it reaches the end of the string, its center of mass velocity is given by Eq. 2-11:

$$v_{com} = v_{com,0} + a_{com}t = v_{com,0} - \frac{gt}{1 + I_{com}/MR_0^2}.$$

Thus, we obtain

$$v_{com} = -1.3 \text{ m/s} - \frac{(9.8 \text{ m/s}^2)(0.885 \text{ s})}{1 + \frac{0.000095 \text{ kg}\cdot\text{m}^2}{(0.12 \text{ kg})(0.0032 \text{ m})^2}} = -1.41 \text{ m/s}$$

so its linear speed at that moment is approximately 1.4 m/s.

(d) The translational kinetic energy is $\frac{1}{2}mv_{com}^2 = \frac{1}{2}(0.12 \text{ kg})(-1.41 \text{ m/s})^2 = 0.12 \text{ J}$.

(e) The angular velocity at that moment is given by

$$\omega = -\frac{v_{com}}{R_0} = -\frac{-1.41 \text{ m/s}}{0.0032 \text{ m}} = 441 \text{ rad/s} \approx 4.4\times10^2 \text{ rad/s}.$$

(f) And the rotational kinetic energy is

$$\frac{1}{2}I_{com}\omega^2 = \frac{1}{2}(9.50\times10^{-5} \text{ kg}\cdot\text{m}^2)(441 \text{ rad/s})^2 = 9.2 \text{ J}.$$

79. (a) When the small sphere is released at the edge of the large "bowl" (the hemisphere of radius R), its center of mass is at the same height at that edge, but when it is at the bottom of the "bowl" its center of mass is a distance r above the bottom surface of the hemisphere. Since the small sphere descends by $R - r$, its loss in gravitational potential energy is $mg(R-r)$, which, by conservation of mechanical energy, is equal to its kinetic energy at the bottom of the track. Thus,

$$K = mg(R-r) = (5.6\times10^{-4} \text{ kg})(9.8 \text{ m/s}^2)(0.15 \text{ m} - 0.0025 \text{ m}) = 8.1\times10^{-4} \text{ J}.$$

(b) Using Eq. 11-5 for K, the asked-for fraction becomes

$$\frac{K_{rot}}{K} = \frac{\frac{1}{2}I\omega^2}{\frac{1}{2}I\omega^2 + \frac{1}{2}Mv_{com}^2} = \frac{1}{1+\left(\frac{M}{I}\right)\left(\frac{v_{com}}{\omega}\right)^2}.$$

Substituting $v_{com} = R\omega$ (Eq. 11-2) and $I = \frac{2}{5}MR^2$ (Table 10-2(f)), we obtain

$$\frac{K_{rot}}{K} = \frac{1}{1+\left(\frac{5}{2R^2}\right)R^2} = \frac{2}{7} \approx 0.29.$$

(c) The small sphere is executing circular motion so that when it reaches the bottom, it experiences a radial acceleration upward (in the direction of the normal force which the "bowl" exerts on it). From Newton's second law along the vertical axis, the normal force F_N satisfies $F_N - mg = ma_{com}$ where

$$a_{com} = v_{com}^2 /(R-r).$$

Therefore,

$$F_N = mg + \frac{mv_{com}^2}{R-r} = \frac{mg(R-r) + mv_{com}^2}{R-r}.$$

But from part (a), $mg(R-r) = K$, and from Eq. 11-5, $\frac{1}{2}mv_{com}^2 = K - K_{rot}$. Thus,

$$F_N = \frac{K + 2(K - K_{rot})}{R-r} = 3\left(\frac{K}{R-r}\right) - 2\left(\frac{K_{rot}}{R-r}\right).$$

We now plug in $R - r = K/mg$ and use the result of part (b):

$$F_N = 3mg - 2mg\left(\frac{2}{7}\right) = \frac{17}{7}mg = \frac{17}{7}(5.6 \times 10^{-4} \text{ kg})(9.8 \text{ m/s}^2) = 1.3 \times 10^{-2} \text{ N}.$$

80. Conservation of energy implies that mechanical energy at maximum height up the ramp is equal to the mechanical energy on the floor. Thus, using Eq. 11-5, we have

$$\frac{1}{2}mv_f^2 + \frac{1}{2}I_{com}\omega_f^2 + mgh = \frac{1}{2}mv^2 + \frac{1}{2}I_{com}\omega^2$$

where $v_f = \omega_f = 0$ at the point on the ramp where it (momentarily) stops. We note that the height h relates to the distance traveled along the ramp d by $h = d\sin(15°)$. Using item (f) in Table 10-2 and Eq. 11-2, we obtain

$$mgd\sin 15° = \frac{1}{2}mv^2 + \frac{1}{2}\left(\frac{2}{5}mR^2\right)\left(\frac{v}{R}\right)^2 = \frac{1}{2}mv^2 + \frac{1}{5}mv^2 = \frac{7}{10}mv^2.$$

After canceling m and plugging in $d = 1.5$ m, we find $v = 2.33$ m/s.

81. (a) Interpreting h as the height increase for the center of mass of the body, then (using Eq. 11-5) mechanical energy conservation, $K_i = U_f$, leads to

$$\frac{1}{2}mv_{com}^2 + \frac{1}{2}I\omega^2 = mgh \Rightarrow \frac{1}{2}mv^2 + \frac{1}{2}I\left(\frac{v}{R}\right)^2 = mg\left(\frac{3v^2}{4g}\right)$$

from which v cancels and we obtain $I = \frac{1}{2}mR^2$.

(b) From Table 10-2(c), we see that the body could be a solid cylinder.

82. (a) Using Eq. 2-16 for the translational (center-of-mass) motion, we find

$$v^2 = v_0^2 + 2a\Delta x \Rightarrow a = -\frac{v_0^2}{2\Delta x}$$

which yields $a = -4.11$ for $v_0 = 43$ and $\Delta x = 225$ (SI units understood). The magnitude of the linear acceleration of the center of mass is therefore 4.11 m/s^2.

(b) With $R = 0.250$ m, Eq. 11-6 gives

$$|\alpha| = |a|/R = 16.4 \text{ rad/s}^2.$$

If the wheel is going rightward, it is rotating in a clockwise sense. Since it is slowing down, this angular acceleration is counterclockwise (opposite to ω) so (with the usual convention that counterclockwise is positive) there is no need for the absolute value signs for α.

(c) Eq. 11-8 applies with Rf_s representing the magnitude of the frictional torque. Thus,

$$Rf_s = I\alpha = (0.155 \text{ kg·m}^2)(16.4 \text{ rad/s}^2) = 2.55 \text{ N·m}.$$

83. If the polar cap melts, the resulting body of water will effectively increase the equatorial radius of the Earth from R_e to $R_e' = R_e + \Delta R$, thereby increasing the moment of inertia of the Earth and slowing its rotation (by conservation of angular momentum), causing the duration T of a day to increase by ΔT. We note that (in rad/s) $\omega = 2\pi/T$ so

$$\frac{\omega'}{\omega} = \frac{2\pi/T'}{2\pi/T} = \frac{T}{T'}$$

from which it follows that

$$\frac{\Delta \omega}{\omega} = \frac{\omega'}{\omega} - 1 = \frac{T}{T'} - 1 = -\frac{\Delta T}{T'}.$$

We can approximate that last denominator as T so that we end up with the simple relationship $|\Delta\omega|/\omega = \Delta T/T$. Now, conservation of angular momentum gives us

$$\Delta L = 0 = \Delta(I\omega) \approx I(\Delta\omega) + \omega(\Delta I)$$

so that $|\Delta\omega|/\omega = \Delta I/I$. Thus, using our expectation that rotational inertia is proportional to the equatorial radius squared (supported by Table 10-2(f) for a perfect uniform sphere, but then this isn't a perfect uniform sphere) we have

$$\frac{\Delta T}{T} = \frac{\Delta I}{I} = \frac{\Delta(R_e^2)}{R_e^2} \approx \frac{2\Delta R_e}{R_e} = \frac{2(30\,\text{m})}{6.37 \times 10^6\,\text{m}}$$

so with $T = 86400$ s we find (approximately) that $\Delta T = 0.8$ s. The radius of the earth can be found in Appendix C or on the inside front cover of the textbook.

84. With $r_\perp = 1300$ m, Eq. 11-21 gives

$$\ell = r_\perp mv = (1300\,\text{m})(1200\,\text{kg})(80\,\text{m/s}) = 1.2 \times 10^8\,\text{kg} \cdot \text{m}^2/\text{s}.$$

85. (a) In terms of the radius of gyration k, the rotational inertia of the merry-go-round is $I = Mk^2$. We obtain

$$I = (180\,\text{kg})(0.910\,\text{m})^2 = 149\,\text{kg} \cdot \text{m}^2.$$

(b) An object moving along a straight line has angular momentum about any point that is not on the line. The magnitude of the angular momentum of the child about the center of the merry-go-round is given by Eq. 11-21, mvR, where R is the radius of the merry-go-round. Therefore,

$$|\vec{L}_{\text{child}}| = (44.0\,\text{kg})(3.00\,\text{m/s})(1.20\,\text{m}) = 158\,\text{kg} \cdot \text{m}^2/\text{s}.$$

(c) No external torques act on the system consisting of the child and the merry-go-round, so the total angular momentum of the system is conserved. The initial angular momentum is given by mvR; the final angular momentum is given by $(I + mR^2)\omega$, where ω is the final common angular velocity of the merry-go-round and child. Thus $mvR = (I + mR^2)\omega$ and

$$\omega = \frac{mvR}{I + mR^2} = \frac{158\,\text{kg} \cdot \text{m}^2/\text{s}}{149\,\text{kg} \cdot \text{m}^2 + (44.0\,\text{kg})(1.20\,\text{m})^2} = 0.744\,\text{rad/s}.$$

86. For a constant (single) torque, Eq. 11-29 becomes $\vec{\tau} = \frac{d\vec{L}}{dt} = \frac{\Delta \vec{L}}{\Delta t}$. Thus, we obtain $\Delta t = 600/50 = 12$ s.

87. This problem involves the vector cross product of vectors lying in the *xy* plane. For such vectors, if we write $\vec{r}' = x'\hat{i} + y'\hat{j}$, then (using Eq. 3-30) we find

$$\vec{r}' \times \vec{v} = (x'v_y - y'v_x)\hat{k}.$$

(a) Here, $\vec{r}'$ points in either the $+\hat{i}$ or the $-\hat{i}$ direction (since the particle moves along the *x* axis). It has no y' or z' components, and neither does $\vec{v}$, so it is clear from the above expression (or, more simply, from the fact that $\hat{i} \times \hat{i} = 0$) that $\vec{\ell} = m(\vec{r}' \times \vec{v}) = 0$ in this case.

(b) The net force is in the $-\hat{i}$ direction (as one finds from differentiating the velocity expression, yielding the acceleration), so, similar to what we found in part (a), we obtain $\vec{\tau} = \vec{r}' \times \vec{F} = 0$.

(c) Now, $\vec{r}' = \vec{r} - \vec{r}_o$ where $\vec{r}_o = 2.0\hat{i} + 5.0\hat{j}$ (with SI units understood) and points from (2.0, 5.0, 0) to the instantaneous position of the car (indicated by $\vec{r}$ which points in either the $+x$ or $-x$ directions, or nowhere (if the car is passing through the origin)). Since $\vec{r} \times \vec{v} = 0$ we have (plugging into our general expression above)

$$\vec{\ell} = m(\vec{r}' \times \vec{v}) = -m(\vec{r}_o \times \vec{v}) = -(3.0)\left((2.0)(0) - (5.0)(-2.0t^3)\right)\hat{k}$$

which yields $\vec{\ell} = (-30t^3\hat{k})$ kg·m/s^2.

(d) The acceleration vector is given by $\vec{a} = \frac{d\vec{v}}{dt} = -6.0t^2\hat{i}$ in SI units, and the net force on the car is $m\vec{a}$. In a similar argument to that given in the previous part, we have

$$\vec{\tau} = m(\vec{r}' \times \vec{a}) = -m(\vec{r}_o \times \vec{a}) = -(3.0)\left((2.0)(0) - (5.0)(-6.0t^2)\right)\hat{k}$$

which yields $\vec{\tau} = (-90t^2\hat{k})$ N·m.

(e) In this situation, $\vec{r}' = \vec{r} - \vec{r}_o$ where $\vec{r}_o = 2.0\hat{i} - 5.0\hat{j}$ (with SI units understood) and points from (2.0, −5.0, 0) to the instantaneous position of the car (indicated by $\vec{r}$ which points in either the $+x$ or $-x$ directions, or nowhere (if the car is passing through the origin)). Since $\vec{r} \times \vec{v} = 0$ we have (plugging into our general expression above)

$$\vec{\ell} = m(\vec{r}' \times \vec{v}) = -m(\vec{r}_o \times \vec{v}) = -(3.0)\left((2.0)(0) - (-5.0)(-2.0t^3)\right)\hat{k}$$

which yields $\vec{\ell} = (30t^3 \hat{k})$ kg·m²/s.

(f) Again, the acceleration vector is given by $\vec{a} = -6.0t^2 \hat{i}$ in SI units, and the net force on the car is $m\vec{a}$. In a similar argument to that given in the previous part, we have

$$\vec{\tau} = m(\vec{r}' \times \vec{a}) = -m(\vec{r}_o \times \vec{a}) = -(3.0)\big((2.0)(0) - (-5.0)(-6.0t^2)\big)\hat{k}$$

which yields $\vec{\tau} = (90t^2 \hat{k})$ N·m.

88. The rotational kinetic energy is $K = \tfrac{1}{2}I\omega^2$, where $I = mR^2$ is its rotational inertia about the center of mass (Table 10-2(a)), $m = 140$ kg, and $\omega = v_{com}/R$ (Eq. 11-2). The ratio is

$$\frac{K_{transl}}{K_{rot}} = \frac{\tfrac{1}{2}mv_{com}^2}{\tfrac{1}{2}(mR^2)(v_{com}/R)^2} = 1.00.$$

89. We note that its mass is $M = 36/9.8 = 3.67$ kg and its rotational inertia is $I_{com} = \tfrac{2}{5}MR^2$ (Table 10-2(f)).

(a) Using Eq. 11-2, Eq. 11-5 becomes

$$K = \frac{1}{2}I_{com}\omega^2 + \frac{1}{2}Mv_{com}^2 = \frac{1}{2}\left(\frac{2}{5}MR^2\right)\left(\frac{v_{com}}{R}\right)^2 + \frac{1}{2}Mv_{com}^2 = \frac{7}{10}Mv_{com}^2$$

which yields $K = 61.7$ J for $v_{com} = 4.9$ m/s.

(b) This kinetic energy turns into potential energy Mgh at some height $h = d\sin\theta$ where the sphere comes to rest. Therefore, we find the distance traveled up the $\theta = 30°$ incline from energy conservation:

$$\frac{7}{10}Mv_{com}^2 = Mgd\sin\theta \Rightarrow d = \frac{7v_{com}^2}{10g\sin\theta} = 3.43\,\text{m}.$$

(c) As shown in the previous part, M cancels in the calculation for d. Since the answer is independent of mass, then, it is also independent of the sphere's weight.

90. The speed of the center of mass of the car is $v = (40)(1000/3600) = 11$ m/s. The angular speed of the wheels is given by Eq. 11-2: $\omega = v/R$ where the wheel radius R is not given (but will be seen to cancel in these calculations).

(a) For one wheel of mass $M = 32$ kg, Eq. 10-34 gives (using Table 10-2(c))

$$K_{rot} = \frac{1}{2}I\omega^2 = \frac{1}{2}\left(\frac{1}{2}MR^2\right)\left(\frac{v}{R}\right)^2 = \frac{1}{4}Mv^2$$

which yields $K_{rot} = 9.9 \times 10^2$ J. The time given in the problem (10 s) is not used in the solution.

(b) Adding the above to the wheel's translational kinetic energy, $\frac{1}{2}Mv^2$, leads to

$$K_{wheel} = \frac{1}{2}Mv^2 + \frac{1}{4}Mv^2 = \frac{3}{4}(32 \text{ kg})(11 \text{ m/s})^2 = 3.0 \times 10^3 \text{ J}.$$

(c) With $M_{car} = 1700$ kg and the fact that there are four wheels, we have

$$\frac{1}{2}M_{car}v^2 + 4\left(\frac{3}{4}Mv^2\right) = 1.2 \times 10^5 \text{ J}.$$

91. We denote the wheel with subscript 1 and the whole system with subscript 2. We take clockwise as the negative sense for rotation (as is the usual convention).

(a) Conservation of angular momentum gives $L = I_1\omega_1 = I_2\omega_2$, where $I_1 = m_1R_1^2$. Thus

$$\omega_2 = \omega_1 \frac{I_1}{I_2} = (-57.7 \text{ rad/s})\frac{(37 \text{ N}/9.8 \text{ m/s}^2)(0.35 \text{ m})^2}{2.1 \text{ kg}\cdot\text{m}^2} = -12.7 \text{ rad/s},$$

or $|\omega_2| = 12.7$ rad/s.

(b) The system rotates clockwise (as seen from above) at the rate of 12.7 rad/s.

92. Information relevant to this calculation can be found in Appendix C or on the inside front cover of the textbook. The angular speed is constant so

$$\omega = \frac{2\pi}{T} = \frac{2\pi}{86400} = 7.3 \times 10^{-5} \text{ rad/s}.$$

Thus, with $m = 84$ kg and $R = 6.37 \times 10^6$ m, we find $\ell = mR^2\omega = 2.5 \times 10^{11}$ kg·m^2/s.

93. The initial angular momentum of the system is zero. The final angular momentum of the girl-plus-merry-go-round is $(I + MR^2)\omega$ which we will take to be positive. The final angular momentum we associate with the thrown rock is negative: $-mRv$, where v is the speed (positive, by definition) of the rock relative to the ground.

(a) Angular momentum conservation leads to

$$0 = (I + MR^2)\omega - mRv \Rightarrow \omega = \frac{mRv}{I + MR^2}.$$

(b) The girl's linear speed is given by Eq. 10-18:

$$R\omega = \frac{mvR^2}{I + MR^2}.$$

94. (a) With $\vec{p} = m\vec{v} = -16\hat{j}$ kg·m/s, we take the vector cross product (using either Eq. 3-30 or, more simply, Eq. 11-20 and the right-hand rule): $\vec{\ell} = \vec{r} \times \vec{p} = (-32\,\text{kg}\cdot\text{m}^2/\text{s})\hat{k}$.

(b) Now the axis passes through the point $\vec{R} = 4.0\hat{j}$ m, parallel with the z axis. With $\vec{r}' = \vec{r} - \vec{R} = 2.0\hat{i}$ m, we again take the cross product and arrive at the same result as before:

$$\vec{\ell}' = \vec{r}' \times \vec{p} = (-32\,\text{kg}\cdot\text{m}^2/\text{s})\hat{k}.$$

(c) Torque is defined in Eq. 11-14: $\vec{\tau} = \vec{r} \times \vec{F} = (12\,\text{N}\cdot\text{m})\hat{k}$.

(d) Using the notation from part (b), $\vec{\tau}' = \vec{r}' \times \vec{F} = 0$.

95. We make the unconventional choice of *clockwise* sense as positive, so that the angular acceleration is positive (as is the linear acceleration of the center of mass, since we take rightwards as positive).

(a) We approach this in the manner of Eq. 11-3 (*pure rotation* about point P) but use torques instead of energy. The torque (relative to point P) is $\tau = I_P\alpha$, where

$$I_P = \frac{1}{2}MR^2 + MR^2 = \frac{3}{2}MR^2$$

with the use of the parallel-axis theorem and Table 10-2(c). The torque is due to the $F_{app} = 12$ N force and can be written as $\tau = F_{app}(2R)$. In this way, we find

$$\tau = I_P\alpha = \left(\frac{3}{2}MR^2\right)\alpha = 2RF_{app}$$

which leads to

$$\alpha = \frac{2RF_{app}}{3MR^2/2} = \frac{4F_{app}}{3MR} = \frac{4(12\,\text{N})}{3(10\,\text{kg})(0.10\,\text{m})} = 16\,\text{rad/s}^2.$$

Hence, $a_{com} = R\alpha = 1.6$ m/s^2.

(b) As shown above, $\alpha = 16$ rad/s^2.

(c) Applying Newton's second law in its linear form yields $(12\,\text{N}) - f = Ma_{com}$. Therefore, $f = -4.0$ N. Contradicting what we assumed in setting up our force equation, the friction force is found to point *rightward* with magnitude 4.0 N, i.e., $\vec{f} = (4.0\,\text{N})\hat{i}$.

96. (a) Sample Problem 10-8 gives $I = 19.64$ kg·m^2 and $\omega = 1466$ rad/s. Thus, the angular momentum is

$$L = I\omega = (19.64\,\text{kg}\cdot\text{m}^2)(1466\,\text{rad/s}) \approx 2.9\times 10^4\,\text{kg}\cdot\text{m}^2/\text{s}.$$

(b) We rewrite Eq. 11-29 as $|\vec{\tau}_{avg}| = |\Delta\vec{L}|/\Delta t$ and plug in $|\Delta\vec{L}| = 2.9\times 10^4$ kg·m^2/s and $\Delta t = 0.025$ s, which leads to $|\vec{\tau}_{avg}| = 1.2\times 10^6$ N·m.

97. Since we will be taking the vector cross product in the course of our calculations, below, we note first that when the two vectors in a cross product $\vec{A}\times\vec{B}$ are in the xy plane, we have $\vec{A} = A_x\hat{i} + A_y\hat{j}$ and $\vec{B} = B_x\hat{i} + B_y\hat{j}$, and Eq. 3-30 leads to

$$\vec{A}\times\vec{B} = (A_xB_y - A_yB_x)\hat{k}.$$

Now, we choose coordinates centered on point O, with $+x$ rightwards and $+y$ upwards. In unit-vector notation, the initial position of the particle, then, is $\vec{r}_0 = s\hat{i}$ and its later position (halfway to the ground) is $\vec{r} = s\hat{i} - \tfrac{1}{2}h\hat{j}$. Using either the free-fall equations of Ch. 2 or the energy techniques of Ch. 8, we find the speed at its later position to be $v = \sqrt{2g|\Delta y|} = \sqrt{gh}$. Its momentum there is $\vec{p} = -M\sqrt{gh}\hat{j}$. We find the angular momentum using Eq. 11-18 and our observation, above, about the cross product of two vectors in the xy plane:

$$\vec{\ell} = \vec{r}\times\vec{p} = -sM\sqrt{gh}\,\hat{k}$$

Therefore, its magnitude is

$$|\vec{\ell}| = sM\sqrt{gh} = (0.45\,\text{m})(0.25\,\text{kg})\sqrt{(9.8\,\text{m/s}^2)(1.8\,\text{m})} = 0.47\,\text{kg}\cdot\text{m}^2/\text{s}.$$

98. This problem involves the vector cross product of vectors lying in the xy plane. For such vectors, if we write $\vec{r} = x\hat{i} + y\hat{j}$, then (using Eq. 3-30) we find

$$\vec{r}\times\vec{p} = (\Delta x p_y - \Delta y p_x)\hat{k}.$$

The momentum components are

$$p_x = p \cos \theta$$
$$p_y = p \sin \theta$$

where $p = 2.4$ (SI units understood) and $\theta = 115°$. The mass (0.80 kg) given in the problem is not used in the solution. Thus, with $x = 2.0$, $y = 3.0$ and the momentum components described above, we obtain

$$\vec{\ell} = \vec{r} \times \vec{p} = (7.4 \,\mathrm{kg} \cdot \mathrm{m}^2/\mathrm{s}) \hat{k}.$$

Chapter 12

1. (a) The center of mass is given by

$$x_{com} = \frac{0+0+0+(m)(2.00 \text{ m})+(m)(2.00 \text{ m})+(m)(2.00 \text{ m})}{6m} = 1.00 \text{ m}.$$

(b) Similarly, we have

$$y_{com} = \frac{0+(m)(2.00 \text{ m})+(m)(4.00 \text{ m})+(m)(4.00 \text{ m})+(m)(2.00 \text{ m})+0}{6m} = 2.00 \text{ m}.$$

(c) Using Eq. 12-14 and noting that the gravitational effects are different at the different locations in this problem, we have

$$x_{cog} = \frac{\sum_{i=1}^{6} x_i m_i g_i}{\sum_{i=1}^{6} m_i g_i} = \frac{x_1 m_1 g_1 + x_2 m_2 g_2 + x_3 m_3 g_3 + x_4 m_4 g_4 + x_5 m_5 g_5 + x_6 m_6 g_6}{m_1 g_1 + m_2 g_2 + m_3 g_3 + m_4 g_4 + m_5 g_5 + m_6 g_6} = 0.987 \text{ m}.$$

(d) Similarly, y_{cog} = [0 + (2.00)(m)(7.80) + (4.00)(m)(7.60) + (4.00)(m)(7.40) + (2.00)(m)(7.60) + 0]/(8.00m + 7.80m + 7.60m + 7.40m + 7.60m + 7.80m) = 1.97 m.

2. The situation is somewhat similar to that depicted for problem 10 (see the figure that accompanies that problem). By analyzing the forces at the "kink" where $\vec{F}$ is exerted, we find (since the acceleration is zero) $2T \sin\theta = F$, where θ is the angle (taken positive) between each segment of the string and its "relaxed" position (when the two segments are collinear). Setting $T = F$ therefore yields $\theta = 30°$. Since $\alpha = 180° - 2\theta$ is the angle between the two segments, then we find $\alpha = 120°$.

3. The object exerts a downward force of magnitude $F = 3160$ N at the midpoint of the rope, causing a "kink" similar to that shown for problem 10 (see the figure that accompanies that problem). By analyzing the forces at the "kink" where $\vec{F}$ is exerted, we find (since the acceleration is zero) $2T \sin\theta = F$, where θ is the angle (taken positive) between each segment of the string and its "relaxed" position (when the two segments are colinear). In this problem, we have

$$\theta = \tan^{-1}\left(\frac{0.35 \text{ m}}{1.72 \text{ m}}\right) = 11.5°.$$

Therefore, $T = F/(2\sin\theta) = 7.92 \times 10^3$ N.

4. From $\vec{\tau} = \vec{r} \times \vec{F}$, we note that persons 1 through 4 exert torques pointing out of the page (relative to the fulcrum), and persons 5 through 8 exert torques pointing into the page.

(a) Among persons 1 through 4, the largest magnitude of torque is (330 N)(3 m) = 990 N·m, due to the weight of person 2.

(b) Among persons 5 through 8, the largest magnitude of torque is (330 N)(3 m) = 990 N·m, due to the weight of person 7.

5. Three forces act on the sphere: the tension force $\vec{T}$ of the rope (acting along the rope), the force of the wall $\vec{F}_N$ (acting horizontally away from the wall), and the force of gravity $m\vec{g}$ (acting downward). Since the sphere is in equilibrium they sum to zero. Let θ be the angle between the rope and the vertical. Then Newton's second law gives

$$\begin{aligned}\text{vertical component:} & \quad T\cos\theta - mg = 0 \\ \text{horizontal component:} & \quad F_N - T\sin\theta = 0.\end{aligned}$$

(a) We solve the first equation for the tension: $T = mg/\cos\theta$. We substitute $\cos\theta = L/\sqrt{L^2 + r^2}$ to obtain

$$T = \frac{mg\sqrt{L^2 + r^2}}{L} = \frac{(0.85 \text{ kg})(9.8 \text{ m/s}^2)\sqrt{(0.080 \text{ m})^2 + (0.042 \text{ m})^2}}{0.080 \text{ m}} = 9.4 \text{ N}.$$

(b) We solve the second equation for the normal force: $F_N = T\sin\theta$. Using $\sin\theta = r/\sqrt{L^2 + r^2}$, we obtain

$$F_N = \frac{Tr}{\sqrt{L^2 + r^2}} = \frac{mg\sqrt{L^2 + r^2}}{L} \cdot \frac{r}{\sqrt{L^2 + r^2}} = \frac{mgr}{L} = \frac{(0.85 \text{ kg})(9.8 \text{ m/s}^2)(0.042 \text{ m})}{(0.080 \text{ m})} = 4.4 \text{ N}.$$

6. Our notation is as follows: $M = 1360$ kg is the mass of the automobile; $L = 3.05$ m is the horizontal distance between the axles; $\ell = (3.05 - 1.78)$ m = 1.27 m is the horizontal distance from the rear axle to the center of mass; F_1 is the force exerted on each front wheel; and, F_2 is the force exerted on each back wheel.

(a) Taking torques about the rear axle, we find

$$F_1 = \frac{Mg\ell}{2L} = \frac{(1360\,\text{kg})(9.80\,\text{m/s}^2)(1.27\,\text{m})}{2(3.05\,\text{m})} = 2.77 \times 10^3\,\text{N}.$$

(b) Equilibrium of forces leads to $2F_1 + 2F_2 = Mg$, from which we obtain $F_2 = 3.89 \times 10^3\,\text{N}$.

7. We take the force of the left pedestal to be F_1 at $x = 0$, where the x axis is along the diving board. We take the force of the right pedestal to be F_2 and denote its position as $x = d$. W is the weight of the diver, located at $x = L$. The following two equations result from setting the sum of forces equal to zero (with upwards positive), and the sum of torques (about x_2) equal to zero:

$$F_1 + F_2 - W = 0$$
$$F_1 d + W(L - d) = 0$$

(a) The second equation gives

$$F_1 = -\frac{L-d}{d}W = -\left(\frac{3.0\,\text{m}}{1.5\,\text{m}}\right)(580\,\text{N}) = -1160\,\text{N}$$

which should be rounded off to $F_1 = -1.2 \times 10^3\,\text{N}$. Thus, $|F_1| = 1.2 \times 10^3\,\text{N}$.

(b) Since F_1 is negative, indicating that this force is downward.

(c) The first equation gives $F_2 = W - F_1 = 580\,\text{N} + 1160\,\text{N} = 1740\,\text{N}$

which should be rounded off to $F_2 = 1.7 \times 10^3\,\text{N}$. Thus, $|F_2| = 1.7 \times 10^3\,\text{N}$.

(d) The result is positive, indicating that this force is upward.

(e) The force of the diving board on the left pedestal is upward (opposite to the force of the pedestal on the diving board), so this pedestal is being stretched.

(f) The force of the diving board on the right pedestal is downward, so this pedestal is being compressed.

8. Let $\ell_1 = 1.5\,\text{m}$ and $\ell_2 = (5.0 - 1.5)\,\text{m} = 3.5\,\text{m}$. We denote tension in the cable closer to the window as F_1 and that in the other cable as F_2. The force of gravity on the scaffold itself (of magnitude $m_s g$) is at its midpoint, $\ell_3 = 2.5\,\text{m}$ from either end.

(a) Taking torques about the end of the plank farthest from the window washer, we find

$$F_1 = \frac{m_w g \ell_2 + m_s g \ell_3}{\ell_1 + \ell_2} = \frac{(80\,\text{kg})(9.8\,\text{m/s}^2)(3.5\,\text{m}) + (60\,\text{kg})(9.8\,\text{m/s}^2)(2.5\,\text{m})}{5.0\,\text{m}}$$
$$= 8.4 \times 10^2\,\text{N}.$$

(b) Equilibrium of forces leads to

$$F_1 + F_2 = m_s g + m_w g = (60\,\text{kg} + 80\,\text{kg})(9.8\,\text{m/s}^2) = 1.4 \times 10^3\,\text{N}$$

which (using our result from part (a)) yields $F_2 = 5.3 \times 10^2\,\text{N}$.

9. The forces on the ladder are shown in the diagram on the right. F_1 is the force of the window, horizontal because the window is frictionless. F_2 and F_3 are components of the force of the ground on the ladder. M is the mass of the window cleaner and m is the mass of the ladder.

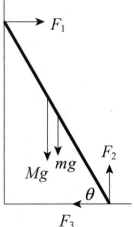

The force of gravity on the man acts at a point 3.0 m up the ladder and the force of gravity on the ladder acts at the center of the ladder. Let θ be the angle between the ladder and the ground. We use $\cos\theta = d/L$ or $\sin\theta = \sqrt{L^2 - d^2}/L$ to find $\theta = 60°$. Here L is the length of the ladder (5.0 m) and d is the distance from the wall to the foot of the ladder (2.5 m).

(a) Since the ladder is in equilibrium the sum of the torques about its foot (or any other point) vanishes. Let ℓ be the distance from the foot of the ladder to the position of the window cleaner. Then,

$$Mg\ell\cos\theta + mg(L/2)\cos\theta - F_1 L \sin\theta = 0,$$

and

$$F_1 = \frac{(M\ell + mL/2)g\cos\theta}{L\sin\theta} = \frac{[(75\,\text{kg})(3.0\,\text{m}) + (10\,\text{kg})(2.5\,\text{m})](9.8\,\text{m/s}^2)\cos 60°}{(5.0\,\text{m})\sin 60°}$$
$$= 2.8 \times 10^2\,\text{N}.$$

This force is outward, away from the wall. The force of the ladder on the window has the same magnitude but is in the opposite direction: it is approximately 280 N, inward.

(b) The sum of the horizontal forces and the sum of the vertical forces also vanish:

$$F_1 - F_3 = 0$$
$$F_2 - Mg - mg = 0$$

The first of these equations gives $F_3 = F_1 = 2.8 \times 10^2\,\text{N}$ and the second gives

$$F_2 = (M+m)g = (75\,\text{kg}+10\,\text{kg})(9.8\,\text{m/s}^2) = 8.3\times10^2\,\text{N}$$

The magnitude of the force of the ground on the ladder is given by the square root of the sum of the squares of its components:

$$F = \sqrt{F_2^2 + F_3^2} = \sqrt{(2.8\times10^2\,\text{N})^2 + (8.3\times10^2\,\text{N})^2} = 8.8\times10^2\,\text{N}.$$

(c) The angle ϕ between the force and the horizontal is given by

$$\tan\phi = F_3/F_2 = 830/280 = 2.94,$$

so $\phi = 71°$. The force points to the left and upward, 71° above the horizontal. We note that this force is not directed along the ladder.

10. The angle of each half of the rope, measured from the dashed line, is

$$\theta = \tan^{-1}\left(\frac{0.30\,\text{m}}{9.0\,\text{m}}\right) = 1.9°.$$

Analyzing forces at the "kink" (where $\vec{F}$ is exerted) we find

$$T = \frac{F}{2\sin\theta} = \frac{550\,\text{N}}{2\sin 1.9°} = 8.3\times 10^3\,\text{N}.$$

11. The x axis is along the meter stick, with the origin at the zero position on the scale. The forces acting on it are shown on the diagram below. The nickels are at $x = x_1 = 0.120$ m, and m is their total mass. The knife edge is at $x = x_2 = 0.455$ m and exerts force $\vec{F}$. The mass of the meter stick is M, and the force of gravity acts at the center of the stick, $x = x_3 = 0.500$ m. Since the meter stick is in equilibrium, the sum of the torques about x_2 must vanish:

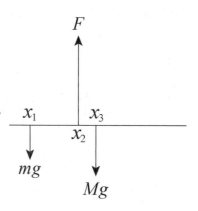

$$Mg(x_3 - x_2) - mg(x_2 - x_1) = 0.$$

Thus,

$$M = \frac{x_2 - x_1}{x_3 - x_2}m = \left(\frac{0.455\,\text{m}-0.120\,\text{m}}{0.500\,\text{m}-0.455\,\text{m}}\right)(10.0\,\text{g}) = 74.4\text{ g}.$$

12. (a) Analyzing vertical forces where string 1 and string 2 meet, we find

$$T_1 = \frac{w_A}{\cos\phi} = \frac{40\text{N}}{\cos 35°} = 49\text{N}.$$

(b) Looking at the horizontal forces at that point leads to

$$T_2 = T_1 \sin 35° = (49\,\text{N})\sin 35° = 28\,\text{N}.$$

(c) We denote the components of T_3 as T_x (rightward) and T_y (upward). Analyzing horizontal forces where string 2 and string 3 meet, we find $T_x = T_2 = 28$ N. From the vertical forces there, we conclude $T_y = w_B = 50$ N. Therefore,

$$T_3 = \sqrt{T_x^2 + T_y^2} = 57\,\text{N}.$$

(d) The angle of string 3 (measured from vertical) is

$$\theta = \tan^{-1}\left(\frac{T_x}{T_y}\right) = \tan^{-1}\left(\frac{28}{50}\right) = 29°.$$

13. (a) Analyzing the horizontal forces (which add to zero) we find $F_h = F_3 = 5.0$ N.

(b) Equilibrium of vertical forces leads to $F_v = F_1 + F_2 = 30$ N.

(c) Computing torques about point O, we obtain

$$F_v d = F_2 b + F_3 a \Rightarrow d = \frac{(10\,\text{N})(3.0\,\text{m}) + (5.0\,\text{N})(2.0\,\text{m})}{30\,\text{N}} = 1.3\,\text{m}.$$

14. The forces exerted horizontally by the obstruction and vertically (upward) by the floor are applied at the bottom front corner C of the crate, as it verges on tipping. The center of the crate, which is where we locate the gravity force of magnitude $mg = 500$ N, is a horizontal distance $\ell = 0.375$ m from C. The applied force of magnitude $F = 350$ N is a vertical distance h from C. Taking torques about C, we obtain

$$h = \frac{mg\ell}{F} = \frac{(500\,\text{N})(0.375\,\text{m})}{350\,\text{N}} = 0.536\,\text{m}.$$

15. Setting up equilibrium of torques leads to a simple "level principle" ratio:

$$F_\perp = (40\,\text{N})\frac{d}{L} = (40\,\text{N})\frac{2.6\,\text{cm}}{12\,\text{cm}} = 8.7\,\text{N}.$$

16. With pivot at the left end, Eq. 12-9 leads to

$$-m_s g \frac{L}{2} - Mgx + T_R L = 0$$

where m_s is the scaffold's mass (50 kg) and M is the total mass of the paint cans (75 kg). The variable x indicates the center of mass of the paint can collection (as measured from the left end), and T_R is the tension in the right cable (722 N). Thus we obtain $x = 0.702$ m.

17. The (vertical) forces at points A, B and P are F_A, F_B and F_P, respectively. We note that $F_P = W$ and is upward. Equilibrium of forces and torques (about point B) lead to

$$F_A + F_B + W = 0$$
$$bW - aF_A = 0.$$

(a) From the second equation, we find

$$F_A = bW/a = (15/5)W = 3W = 3(900 \text{ N}) = 2.7 \times 10^3 \text{ N}.$$

(b) The direction is upward since $F_A > 0$.

(c) Using this result in the first equation above, we obtain

$$F_B = W - F_A = -4W = -4(900 \text{ N}) = -3.6 \times 10^3 \text{ N},$$

or $|F_B| = 3.6 \times 10^3 \text{ N}$.

(d) F_B points downward, as indicated by the minus sign.

18. Our system consists of the lower arm holding a bowling ball. As shown in the free-body diagram, the forces on the lower arm consist of $\vec{T}$ from the biceps muscle, $\vec{F}$ from the bone of the upper arm, and the gravitational forces, $m\vec{g}$ and $M\vec{g}$. Since the system is in static equilibrium, the net force acting on the system is zero:

$$0 = \sum F_{\text{net},y} = T - F - (m+M)g$$

In addition, the net torque about O must also vanish:

$$0 = \sum_O \tau_{\text{net}} = (d)(T) + (0)F - (D)(mg) - L(Mg).$$

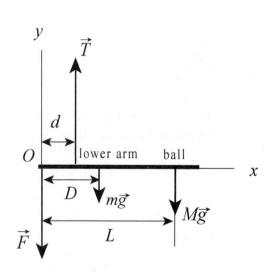

(a) From the torque equation, we find the force on the lower arms by the biceps muscle to be

$$T = \frac{(mD + ML)g}{d} = \frac{[(1.8 \text{ kg})(0.15 \text{ m}) + (7.2 \text{ kg})(0.33 \text{ m})](9.8 \text{ m/s}^2)}{0.040 \text{ m}}$$
$$= 648 \text{ N} \approx 6.5 \times 10^2 \text{ N}.$$

(b) Substituting the above result into the force equation, we find F to be

$$F = T - (M+m)g = 648 \text{ N} - (7.2 \text{ kg} + 1.8 \text{ kg})(9.8 \text{ m/s}^2) = 560 \text{ N} = 5.6 \times 10^2 \text{ N}.$$

19. (a) With the pivot at the hinge, Eq. 12-9 gives $TL\cos\theta - mg\frac{L}{2} = 0$. This leads to $\theta = 78°$. Then the geometric relation $\tan\theta = L/D$ gives $D = 0.64$ m.

(b) A higher (steeper) slope for the cable results in a smaller tension. Thus, making D greater than the value of part (a) should prevent rupture.

20. With pivot at the left end of the lower scaffold, Eq. 12-9 leads to

$$-m_2 g \frac{L_2}{2} - mgd + T_R L_2 = 0$$

where m_2 is the lower scaffold's mass (30 kg) and L_2 is the lower scaffold's length (2.00 m). The mass of the package ($m = 20$ kg) is a distance $d = 0.50$ m from the pivot, and T_R is the tension in the rope connecting the right end of the lower scaffold to the larger scaffold above it. This equation yields $T_R = 196$ N. Then Eq. 12-8 determines T_L (the tension in the cable connecting the right end of the lower scaffold to the larger scaffold above it): $T_L = 294$ N. Next, we analyze the larger scaffold (of length $L_1 = L_2 + 2d$ and mass m_1, given in the problem statement) placing our pivot at its left end and using Eq. 12-9:

$$-m_1 g \frac{L_1}{2} - T_L d - T_R (L_1 - d) + T L_1 = 0.$$

This yields $T = 457$ N.

21. We consider the wheel as it leaves the lower floor. The floor no longer exerts a force on the wheel, and the only forces acting are the force F applied horizontally at the axle, the force of gravity mg acting vertically at the center of the wheel, and the force of the step corner, shown as the two components f_h and f_v. If the minimum force is applied the wheel does not accelerate, so both the total force and the total torque acting on it are zero.

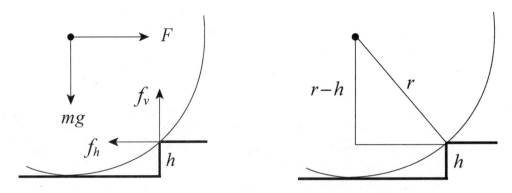

We calculate the torque around the step corner. The second diagram indicates that the distance from the line of F to the corner is $r - h$, where r is the radius of the wheel and h is the height of the step.

The distance from the line of mg to the corner is $\sqrt{r^2 + (r-h)^2} = \sqrt{2rh - h^2}$. Thus,

$$F(r-h) - mg\sqrt{2rh - h^2} = 0.$$

The solution for F is

$$F = \frac{\sqrt{2rh - h^2}}{r - h} mg = \frac{\sqrt{2(6.00 \times 10^{-2}\,\text{m})(3.00 \times 10^{-2}\,\text{m}) - (3.00 \times 10^{-2}\,\text{m})^2}}{(6.00 \times 10^{-2}\,\text{m}) - (3.00 \times 10^{-2}\,\text{m})}(0.800\,\text{kg})(9.80\,\text{m/s}^2)$$
$$= 13.6\,\text{N}.$$

22. As shown in the free-body diagram, the forces on the climber consist of $\vec{T}$ from the rope, normal force $\vec{F}_N$ on her feet, upward static frictional force $\vec{f}_s$ and downward gravitational force $m\vec{g}$. Since the climber is in static equilibrium, the net force acting on her is zero. Applying Newton's second law to the vertical and horizontal directions, we have

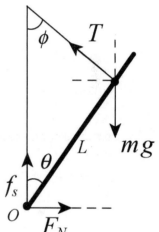

$$0 = \sum F_{\text{net},x} = F_N - T \sin\phi$$
$$0 = \sum F_{\text{net},y} = T \cos\phi + f_s - mg.$$

In addition, the net torque about O (contact point between her feet and the wall) must also vanish:

$$0 = \sum_O \tau_{\text{net}} = mgL \sin\theta - TL \sin(180° - \theta - \phi).$$

From the torque equation, we obtain $T = mg \sin\theta / \sin(180° - \theta - \phi)$. Substituting the expression into the force equations, and noting that $f_s = \mu_s F_N$, we find the coefficient of static friction to be

$$\mu_s = \frac{f_s}{F_N} = \frac{mg - T\cos\phi}{T\sin\phi} = \frac{mg - mg\sin\theta\cos\phi/\sin(180° - \theta - \phi)}{mg\sin\theta\sin\phi/\sin(180° - \theta - \phi)}$$
$$= \frac{1 - \sin\theta\cos\phi/\sin(180° - \theta - \phi)}{\sin\theta\sin\phi/\sin(180° - \theta - \phi)}.$$

With $\theta = 40°$ and $\phi = 30°$, the result is

$$\mu_s = \frac{1-\sin\theta\cos\phi/\sin(180°-\theta-\phi)}{\sin\theta\sin\phi/\sin(180°-\theta-\phi)} = \frac{1-\sin 40°\cos 30°/\sin(180°-40°-30°)}{\sin 40°\sin 30°/\sin(180°-40°-30°)} = 1.19.$$

23. (a) All forces are vertical and all distances are measured along an axis inclined at $\theta = 30°$. Thus, any trigonometric factor cancels out and the application of torques about the contact point (referred to in the problem) leads to

$$F_{tripcep} = \frac{(15\,\text{kg})(9.8\,\text{m/s}^2)(35\,\text{cm}) - (2.0\,\text{kg})(9.8\,\text{m/s}^2)(15\,\text{cm})}{2.5\,\text{cm}} = 1.9 \times 10^3 \text{ N}.$$

(b) The direction is upward since $F_{tricep} > 0$

(c) Equilibrium of forces (with upwards positive) leads to

$$F_{tripcep} + F_{humer} + (15\,\text{kg})(9.8\,\text{m/s}^2) - (2.0\,\text{kg})(9.8\,\text{m/s}^2) = 0$$

and thus to $F_{humer} = -2.1 \times 10^3 \text{ N}$, or $|F_{humer}| = 2.1 \times 10^3 \text{ N}$.

(d) The minus sign implies that F_{humer} points downward.

24. As shown in the free-body diagram, the forces on the climber consist of the normal forces F_{N1} on his hands from the ground and F_{N2} on his feet from the wall, static frictional force f_s and downward gravitational force mg. Since the climber is in static equilibrium, the net force acting on him is zero. Applying Newton's second law to the vertical and horizontal directions, we have

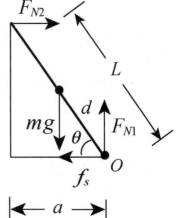

$$0 = \sum F_{net,x} = F_{N2} - f_s$$
$$0 = \sum F_{net,y} = F_{N1} - mg.$$

In addition, the net torque about O (contact point between his feet and the wall) must also vanish:

$$0 = \sum_O \tau_{net} = mgd\cos\theta - F_{N2}L\sin\theta.$$

The torque equation gives $F_{N2} = mgd\cos\theta/L\sin\theta = mgd\cot\theta/L$. On the other hand, from the force equation we have $F_{N2} = f_s$ and $F_{N1} = mg$. These expressions can be combined to yield

$$f_s = F_{N2} = F_{N1}\cot\theta\frac{d}{L}.$$

On the other hand, the frictional force can also be written as $f_s = \mu_s F_{N1}$, where μ_s is the coefficient of static friction between his feet and the ground. From the above equation and the values given in the problem statement, we find μ_s to be

$$\mu_s = \cot\theta \frac{d}{L} = \frac{a}{\sqrt{L^2 - a^2}} \frac{d}{L} = \frac{0.914 \text{ m}}{\sqrt{(2.10 \text{ m})^2 - (0.914 \text{ m})^2}} \frac{0.940 \text{ m}}{2.10 \text{ m}} = 0.216.$$

25. The beam is in equilibrium: the sum of the forces and the sum of the torques acting on it each vanish. As shown in the figure, the beam makes an angle of 60° with the vertical and the wire makes an angle of 30° with the vertical.

(a) We calculate the torques around the hinge. Their sum is

$$TL \sin 30° - W(L/2) \sin 60° = 0.$$

Here W is the force of gravity acting at the center of the beam, and T is the tension force of the wire. We solve for the tension:

$$T = \frac{W \sin 60°}{2 \sin 30°} = \frac{(222 \text{N}) \sin 60°}{2 \sin 30°} = 192 \text{ N}.$$

(b) Let F_h be the horizontal component of the force exerted by the hinge and take it to be positive if the force is outward from the wall. Then, the vanishing of the horizontal component of the net force on the beam yields $F_h - T \sin 30° = 0$ or

$$F_h = T \sin 30° = (192.3 \text{ N}) \sin 30° = 96.1 \text{ N}.$$

(c) Let F_v be the vertical component of the force exerted by the hinge and take it to be positive if it is upward. Then, the vanishing of the vertical component of the net force on the beam yields $F_v + T \cos 30° - W = 0$ or

$$F_v = W - T \cos 30° = 222 \text{ N} - (192.3 \text{ N}) \cos 30° = 55.5 \text{ N}.$$

26. (a) The problem asks for the person's pull (his force exerted on the rock) but since we are examining forces and torques *on the person*, we solve for the reaction force F_{N1} (exerted leftward on the hands by the rock). At that point, there is also an upward force of static friction on his hands f_1 which we will take to be at its maximum value $\mu_1 F_{N1}$. We note that equilibrium of horizontal forces requires $F_{N1} = F_{N2}$ (the force exerted leftward on his feet); on this feet there is also an upward static friction force of magnitude $\mu_2 F_{N2}$. Equilibrium of vertical forces gives

$$f_1 + f_2 - mg = 0 \Rightarrow F_{N1} = \frac{mg}{\mu_1 + \mu_2} = 3.4 \times 10^2 \text{ N}.$$

(b) Computing torques about the point where his feet come in contact with the rock, we find

$$mg(d+w) - f_1 w - F_{N1} h = 0 \implies h = \frac{mg(d+w) - \mu_1 F_{N1} w}{F_{N1}} = 0.88 \text{ m}.$$

(c) Both intuitively and mathematically (since both coefficients are in the denominator) we see from part (a) that F_{N1} would increase in such a case.

(d) As for part (b), it helps to plug part (a) into part (b) and simplify:

$$h = (d+w)\mu_2 + d\mu_1$$

from which it becomes apparent that h should decrease if the coefficients decrease.

27. (a) We note that the angle between the cable and the strut is

$$\alpha = \theta - \phi = 45° - 30° = 15°.$$

The angle between the strut and any vertical force (like the weights in the problem) is $\beta = 90° - 45° = 45°$. Denoting $M = 225$ kg and $m = 45.0$ kg, and ℓ as the length of the boom, we compute torques about the hinge and find

$$T = \frac{M g \ell \sin\beta + mg\left(\frac{\ell}{2}\right)\sin\beta}{\ell \sin\alpha} = \frac{Mg\sin\beta + mg\sin\beta/2}{\sin\alpha}.$$

The unknown length ℓ cancels out and we obtain $T = 6.63 \times 10^3$ N.

(b) Since the cable is at 30° from horizontal, then horizontal equilibrium of forces requires that the horizontal hinge force be

$$F_x = T\cos 30° = 5.74 \times 10^3 \text{ N}.$$

(c) And vertical equilibrium of forces gives the vertical hinge force component:

$$F_y = Mg + mg + T\sin 30° = 5.96 \times 10^3 \text{ N}.$$

28. (a) The sign is attached in two places: at $x_1 = 1.00$ m (measured rightward from the hinge) and at $x_2 = 3.00$ m. We assume the downward force due to the sign's weight is equal at these two attachment points: each being *half* the sign's weight of mg. The angle where the cable comes into contact (also at x_2) is

$$\theta = \tan^{-1}(d_v/d_h) = \tan^{-1}(4.00 \text{ m}/3.00 \text{ m})$$

and the force exerted there is the tension T. Computing torques about the hinge, we find

$$T = \frac{\frac{1}{2}mgx_1 + \frac{1}{2}mgx_2}{x_2 \sin\theta} = \frac{\frac{1}{2}(50.0 \text{ kg})(9.8 \text{ m/s}^2)(1.00 \text{ m}) + \frac{1}{2}(50.0 \text{ kg})(9.8 \text{m/s}^2)(3.00 \text{ m})}{(3.00 \text{ m})(0.800)}$$

$= 408$ N.

(b) Equilibrium of horizontal forces requires the horizontal hinge force be

$$F_x = T \cos\theta = 245 \text{ N}.$$

(c) The direction of the horizontal force is rightward.

(d) Equilibrium of vertical forces requires the vertical hinge force be

$$F_y = mg - T \sin\theta = 163 \text{ N}.$$

(e) The direction of the vertical force is upward.

29. The bar is in equilibrium, so the forces and the torques acting on it each sum to zero. Let T_l be the tension force of the left–hand cord, T_r be the tension force of the right–hand cord, and m be the mass of the bar. The equations for equilibrium are:

vertical force components $\quad T_l \cos\theta + T_r \cos\phi - mg = 0$
horizontal force components $\quad -T_l \sin\theta + T_r \sin\phi = 0$
torques $\quad mgx - T_r L \cos\phi = 0.$

The origin was chosen to be at the left end of the bar for purposes of calculating the torque. The unknown quantities are T_l, T_r, and x. We want to eliminate T_l and T_r, then solve for x. The second equation yields $T_l = T_r \sin\phi / \sin\theta$ and when this is substituted into the first and solved for T_r the result is

$$T_r = \frac{mg \sin\theta}{\sin\phi \cos\theta + \cos\phi \sin\theta}.$$

This expression is substituted into the third equation and the result is solved for x:

$$x = L \frac{\sin\theta \cos\phi}{\sin\phi \cos\theta + \cos\phi \sin\theta} = L \frac{\sin\theta \cos\phi}{\sin(\theta + \phi)}.$$

The last form was obtained using the trigonometric identity $\sin(A + B) = \sin A \cos B + \cos A \sin B$. For the special case of this problem $\theta + \phi = 90°$ and $\sin(\theta + \phi) = 1$. Thus,

$$x = L \sin\theta \cos\phi = (6.10 \text{ m}) \sin 36.9° \cos 53.1° = 2.20 \text{ m}.$$

30. (a) Computing torques about point A, we find

$$T_{max} L \sin\theta = W x_{max} + W_b \left(\frac{L}{2}\right).$$

We solve for the maximum distance:

$$x_{max} = \left(\frac{T_{max}\sin\theta - W_b/2}{W}\right) L = \left(\frac{(500\text{ N})\sin 30.0° - (200\text{ N})/2}{300\text{ N}}\right)(3.00\text{ m}) = 1.50\text{ m}.$$

(b) Equilibrium of horizontal forces gives $F_x = T_{max}\cos\theta = 433\text{ N}.$

(c) And equilibrium of vertical forces gives $F_y = W + W_b - T_{max}\sin\theta = 250\text{ N}.$

31. The problem states that each hinge supports half the door's weight, so each vertical hinge force component is $F_y = mg/2 = 1.3 \times 10^2$ N. Computing torques about the top hinge, we find the horizontal hinge force component (at the bottom hinge) is

$$F_h = \frac{(27\text{ kg})(9.8\text{ m/s}^2)(0.91\text{ m}/2)}{2.1\text{ m} - 2(0.30\text{ m})} = 80\text{ N}.$$

Equilibrium of horizontal forces demands that the horizontal component of the top hinge force has the same magnitude (though opposite direction).

(a) In unit-vector notation, the force on the door at the top hinge is

$$F_{top} = (-80\text{ N})\hat{i} + (1.3\times 10^2\text{ N})\hat{j}.$$

(b) Similarly, the force on the door at the bottom hinge is

$$F_{bottom} = (+80\text{ N})\hat{i} + (1.3\times 10^2\text{ N})\hat{j}$$

32. (a) Computing torques about the hinge, we find the tension in the wire:

$$TL\sin\theta - Wx = 0 \Rightarrow T = \frac{Wx}{L\sin\theta}.$$

(b) The horizontal component of the tension is $T\cos\theta$, so equilibrium of horizontal forces requires that the horizontal component of the hinge force is

$$F_x = \left(\frac{Wx}{L\sin\theta}\right)\cos\theta = \frac{Wx}{L\tan\theta}.$$

(c) The vertical component of the tension is $T\sin\theta$, so equilibrium of vertical forces requires that the vertical component of the hinge force is

$$F_y = W - \left(\frac{Wx}{L\sin\theta}\right)\sin\theta = W\left(1 - \frac{x}{L}\right).$$

33. We examine the box when it is about to tip. Since it will rotate about the lower right edge, that is where the normal force of the floor is exerted. This force is labeled F_N on the diagram below. The force of friction is denoted by f, the applied force by F, and the force of gravity by W. Note that the force of gravity is applied at the center of the box. When the minimum force is applied the box does not accelerate, so the sum of the horizontal force components vanishes: $F - f = 0$, the sum of the vertical force components vanishes: $F_N - W = 0$, and the sum of the torques vanishes:

$$FL - WL/2 = 0.$$

Here L is the length of a side of the box and the origin was chosen to be at the lower right edge.

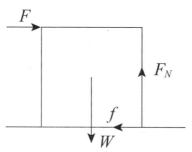

(a) From the torque equation, we find

$$F = \frac{W}{2} = \frac{890\,\text{N}}{2} = 445\,\text{N}.$$

(b) The coefficient of static friction must be large enough that the box does not slip. The box is on the verge of slipping if $\mu_s = f/F_N$. According to the equations of equilibrium

$$F_N = W = 890\,\text{N and } f = F = 445\,\text{N},$$

so

$$\mu_s = \frac{445\,\text{N}}{890\,\text{N}} = 0.50.$$

(c) The box can be rolled with a smaller applied force if the force points upward as well as to the right. Let θ be the angle the force makes with the horizontal. The torque equation then becomes

$$FL \cos \theta + FL \sin \theta - WL/2 = 0,$$

with the solution

$$F = \frac{W}{2(\cos \theta + \sin \theta)}.$$

We want $\cos \theta + \sin \theta$ to have the largest possible value. This occurs if $\theta = 45°$, a result we can prove by setting the derivative of $\cos \theta + \sin \theta$ equal to zero and solving for θ. The minimum force needed is

$$F = \frac{W}{4 \cos 45°} = \frac{890 \text{ N}}{4 \cos 45°} = 315 \text{ N}.$$

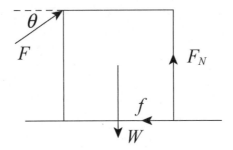

34. As shown in the free-body diagram, the forces on the climber consist of the normal force from the wall, the vertical component F_v and the horizontal component F_h of the force acting on her four fingertips, and the downward gravitational force mg. Since the climber is in static equilibrium, the net force acting on her is zero. Applying Newton's second law to the vertical and horizontal directions, we have

$$0 = \sum F_{net,x} = 4F_h - F_N$$
$$0 = \sum F_{net,y} = 4F_v - mg.$$

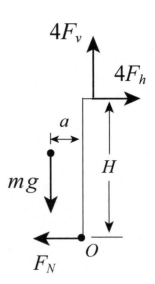

In addition, the net torque about O (contact point between her feet and the wall) must also vanish:

$$0 = \sum_O \tau_{net} = (mg)a - (4F_h)H.$$

(a) From the torque equation, we find the horizontal component of the force on her fingertip to be

$$F_h = \frac{mga}{4H} = \frac{(70 \text{ kg})(9.8 \text{ m/s}^2)(0.20 \text{ m})}{4(2.0 \text{ m})} \approx 17 \text{ N}.$$

(b) From the y-component of the force equation, we obtain

$$F_v = \frac{mg}{4} = \frac{(70 \text{ kg})(9.8 \text{ m/s}^2)}{4} \approx 1.7 \times 10^2 \text{ N}.$$

35. (a) With the pivot at the hinge, Eq. 12-9 yields

$$TL\cos\theta - F_a y = 0.$$

This leads to $T = (F_a/\cos\theta)(y/L)$ so that we can interpret $F_a/\cos\theta$ as the slope on the tension graph (which we estimate to be 600 in SI units). Regarding the F_h graph, we use Eq. 12–7 to get

$$F_h = T\cos\theta - F_a = (-F_a)(y/L) - F_a$$

after substituting our previous expression. The result implies that the slope on the F_h graph (which we estimate to be −300) is equal to $-F_a$, or $F_a = 300$ N and (plugging back in) $\theta = 60.0°$.

(b) As mentioned in the previous part, $F_a = 300$ N.

36. (a) With $F = ma = -\mu_k mg$ the magnitude of the deceleration is

$$|a| = \mu_k g = (0.40)(9.8 \text{ m/s}^2) = 3.92 \text{ m/s}^2.$$

(b) As hinted in the problem statement, we can use Eq. 12-9, evaluating the torques about the car's center of mass, and bearing in mind that the friction forces are acting horizontally at the bottom of the wheels; the total friction force there is $f_k = \mu_k gm = 3.92m$ (with SI units understood – and m is the car's mass), a vertical distance of 0.75 meter below the center of mass. Thus, torque equilibrium leads to

$$(3.92m)(0.75) + F_{Nr}(2.4) - F_{Nf}(1.8) = 0.$$

Eq. 12-8 also holds (the acceleration is horizontal, not vertical), so we have $F_{Nr} + F_{Nf} = mg$, which we can solve simultaneously with the above torque equation. The mass is obtained from the car's weight: $m = 11000/9.8$, and we obtain $F_{Nr} = 3929 \approx 4000$ N. Since each involves two wheels then we have (roughly) 2.0×10^3 N on each rear wheel.

(c) From the above equation, we also have $F_{Nf} = 7071 \approx 7000$ N, or 3.5×10^3 N on each front wheel, as the values of the individual normal forces.

(d) Eq. 6-2 directly yields (approximately) 7.9×10^2 N of friction on each rear wheel,

(e) Similarly, Eq. 6-2 yields 1.4×10^3 N on each front wheel.

37. The free-body diagram on the right shows the forces acting on the plank. Since the roller is frictionless the force it exerts is normal to the plank and makes the angle θ with the vertical. Its magnitude is designated F. W is the force of gravity; this force acts at the center of the plank, a distance $L/2$ from the point where the plank touches the floor. F_N is the normal force of the floor and f is the force of friction. The distance from the foot of the plank to the wall is denoted by d. This quantity is not given directly but it can be computed using $d = h/\tan\theta$.

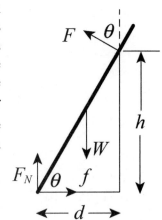

The equations of equilibrium are:

horizontal force components $\qquad F\sin\theta - f = 0$

vertical force components $\qquad F\cos\theta - W + F_N = 0$

torques $\qquad F_N d - fh - W\left(d - \tfrac{L}{2}\cos\theta\right) = 0.$

The point of contact between the plank and the roller was used as the origin for writing the torque equation.

When $\theta = 70°$ the plank just begins to slip and $f = \mu_s F_N$, where μ_s is the coefficient of static friction. We want to use the equations of equilibrium to compute F_N and f for $\theta = 70°$, then use $\mu_s = f/F_N$ to compute the coefficient of friction.

The second equation gives $F = (W - F_N)/\cos\theta$ and this is substituted into the first to obtain

$$f = (W - F_N)\sin\theta/\cos\theta = (W - F_N)\tan\theta.$$

This is substituted into the third equation and the result is solved for F_N:

$$F_N = \frac{d - (L/2)\cos\theta + h\tan\theta}{d + h\tan\theta} W = \frac{h(1+\tan^2\theta) - (L/2)\sin\theta}{h(1+\tan^2\theta)} W,$$

where we have use $d = h/\tan\theta$ and multiplied both numerator and denominator by $\tan\theta$. We use the trigonometric identity $1 + \tan^2\theta = 1/\cos^2\theta$ and multiply both numerator and denominator by $\cos^2\theta$ to obtain

$$F_N = W\left(1 - \frac{L}{2h}\cos^2\theta\sin\theta\right).$$

Now we use this expression for F_N in $f = (W - F_N)\tan\theta$ to find the friction:

$$f = \frac{WL}{2h}\sin^2\theta\cos\theta.$$

We substitute these expressions for f and F_N into $\mu_s = f/F_N$ and obtain

$$\mu_s = \frac{L\sin^2\theta\cos\theta}{2h - L\sin\theta\cos^2\theta}.$$

Evaluating this expression for $\theta = 70°$, we obtain

$$\mu_s = \frac{(6.1\,\text{m})\sin^2 70°\cos 70°}{2(3.05\,\text{m}) - (6.1\,\text{m})\sin 70°\cos^2 70°} = 0.34.$$

38. The phrase "loosely bolted" means that there is no torque exerted by the bolt at that point (where A connects with B). The force exerted on A at the hinge has x and y components F_x and F_y. The force exerted on A at the bolt has components G_x and G_y and those exerted on B are simply $-G_x$ and $-G_y$ by Newton's third law. The force exerted on B at its hinge has components H_x and H_y. If a horizontal force is positive, it points rightward, and if a vertical force is positive it points upward.

(a) We consider the combined $A\cup B$ system, which has a total weight of Mg where $M = 122$ kg and the line of action of that downward force of gravity is $x = 1.20$ m from the wall. The vertical distance between the hinges is $y = 1.80$ m. We compute torques about the bottom hinge and find

$$F_x = -\frac{Mgx}{y} = -797\,\text{N}.$$

If we examine the forces on A alone and compute torques about the bolt, we instead find

$$F_y = \frac{m_A g x}{\ell} = 265\,\text{N}$$

where $m_A = 54.0$ kg and $\ell = 2.40$ m (the length of beam A). Thus, in unit-vector notation, we have

$$\vec{F} = F_x\hat{i} + F_y\hat{j} = (-797\,\text{N})\hat{i} + (265\,\text{N})\hat{j}.$$

(b) Equilibrium of horizontal and vertical forces on beam A readily yields $G_x = -F_x = 797$ N and $G_y = m_A g - F_y = 265$ N. In unit-vector notation, we have

$$\vec{G} = G_x\hat{i} + G_y\hat{j} = (+797\,\text{N})\hat{i} + (265\,\text{N})\hat{j}$$

(c) Considering again the combined A∪B system, equilibrium of horizontal and vertical forces readily yields $H_x = -F_x = 797$ N and $H_y = Mg - F_y = 931$ N. In unit-vector notation, we have

$$\vec{H} = H_x\hat{i} + H_y\hat{j} = (+797 \text{ N})\hat{i} + (931 \text{ N})\hat{j}$$

(d) As mentioned above, Newton's third law (and the results from part (b)) immediately provide $-G_x = -797$ N and $-G_y = -265$ N for the force components acting on B at the bolt. In unit-vector notation, we have

$$-\vec{G} = -G_x\hat{i} - G_y\hat{j} = (-797 \text{ N})\hat{i} - (265 \text{ N})\hat{j}$$

39. The force diagram shown below depicts the situation just before the crate tips, when the normal force acts at the front edge. However, it may also be used to calculate the angle for which the crate begins to slide. W is the force of gravity on the crate, F_N is the normal force of the plane on the crate, and f is the force of friction. We take the x axis to be down the plane and the y axis to be in the direction of the normal force. We assume the acceleration is zero but the crate is on the verge of sliding.

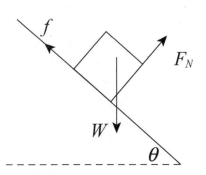

(a) The x and y components of Newton's second law are

$$W\sin\theta - f = 0 \text{ and } F_N - W\cos\theta = 0$$

respectively. The y equation gives $F_N = W\cos\theta$. Since the crate is about to slide

$$f = \mu_s F_N = \mu_s W \cos\theta,$$

where μ_s is the coefficient of static friction. We substitute into the x equation and find

$$W\sin\theta - \mu_s W \cos\theta = 0 \Rightarrow \tan\theta = \mu_s.$$

This leads to $\theta = \tan^{-1}\mu_s = \tan^{-1} 0.60 = 31.0°$.

In developing an expression for the total torque about the center of mass when the crate is about to tip, we find that the normal force and the force of friction act at the front edge. The torque associated with the force of friction tends to turn the crate clockwise and has magnitude fh, where h is the perpendicular distance from the bottom of the crate to the center of gravity. The torque associated with the normal force tends to turn the crate counterclockwise and has magnitude $F_N\ell/2$, where ℓ is the length of an edge. Since the total torque vanishes, $fh = F_N\ell/2$. When the crate is about to tip, the acceleration of the center of gravity vanishes, so $f = W\sin\theta$ and $F_N = W\cos\theta$. Substituting these expressions into the torque equation, we obtain

$$\theta = \tan^{-1}\frac{\ell}{2h} = \tan^{-1}\frac{1.2\,\text{m}}{2(0.90\,\text{m})} = 33.7°.$$

As θ is increased from zero the crate slides before it tips.

(b) It starts to slide when $\theta = 31°$.

(c) The crate begins to slide when $\theta = \tan^{-1}\mu_s = \tan^{-1} 0.70 = 35.0°$ and begins to tip when $\theta = 33.7°$. Thus, it tips first as the angle is increased.

(d) Tipping begins at $\theta = 33.7° \approx 34°$.

40. Let x be the horizontal distance between the firefighter and the origin O (see figure) that makes the ladder on the verge of sliding. The forces on the firefighter + ladder system consist of the horizontal force F_w from the wall, the vertical component F_{py} and the horizontal component F_{px} of the force $\vec{F}_p$ on the ladder from the pavement, and the downward gravitational forces Mg and mg, where M and m are the masses of the firefighter and the ladder, respectively. Since the system is in static equilibrium, the net force acting on the system is zero. Applying Newton's second law to the vertical and horizontal directions, we have

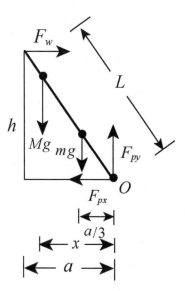

$$0 = \sum F_{\text{net},x} = F_w - F_{px}$$
$$0 = \sum F_{\text{net},y} = F_{py} - (M+m)g.$$

Since the ladder is on the verge of sliding, $F_{px} = \mu_s F_{py}$. Therefore, we have

$$F_w = F_{px} = \mu_s F_{py} = \mu_s(M+m)g.$$

In addition, the net torque about O (contact point between the ladder and the wall) must also vanish:

$$0 = \sum_O \tau_{\text{net}} = -h(F_w) + x(Mg) + \frac{a}{3}(mg) = 0.$$

Solving for x, we obtain

$$x = \frac{hF_w - (a/3)mg}{Mg} = \frac{h\mu_s(M+m)g - (a/3)mg}{Mg} = \frac{h\mu_s(M+m) - (a/3)m}{M}$$

Substituting the values given in the problem statement (with $a = \sqrt{L^2 - h^2} = 7.58$ m), the fraction of ladder climbed is

$$\frac{x}{a} = \frac{h\mu_s(M+m) - (a/3)m}{Ma} = \frac{(9.3 \text{ m})(0.53)(72 \text{ kg} + 45 \text{ kg}) - (7.58 \text{ m}/3)(45 \text{ kg})}{(72 \text{ kg})(7.58 \text{ m})}$$

$$= 0.848 \approx 85\%.$$

41. The diagrams below show the forces on the two sides of the ladder, separated. F_A and F_E are the forces of the floor on the two feet, T is the tension force of the tie rod, W is the force of the man (equal to his weight), F_h is the horizontal component of the force exerted by one side of the ladder on the other, and F_v is the vertical component of that force. Note that the forces exerted by the floor are normal to the floor since the floor is frictionless. Also note that the force of the left side on the right and the force of the right side on the left are equal in magnitude and opposite in direction.

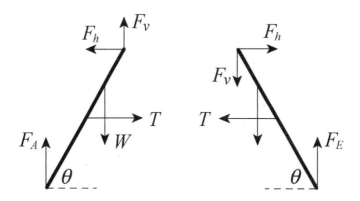

Since the ladder is in equilibrium, the vertical components of the forces on the left side of the ladder must sum to zero: $F_v + F_A - W = 0$. The horizontal components must sum to zero: $T - F_h = 0$. The torques must also sum to zero. We take the origin to be at the hinge and let L be the length of a ladder side. Then

$$F_A L \cos\theta - W(L/4)\cos\theta - T(L/2)\sin\theta = 0.$$

Here we recognize that the man is one–fourth the length of the ladder side from the top and the tie rod is at the midpoint of the side.

The analogous equations for the right side are $F_E - F_v = 0$, $F_h - T = 0$, and $F_E L \cos\theta - T(L/2)\sin\theta = 0$.

There are 5 different equations:

$$F_v + F_A - W = 0,$$
$$T - F_h = 0$$
$$F_A L \cos\theta - W(L/4)\cos\theta - T(L/2)\sin\theta = 0$$
$$F_E - F_v = 0$$
$$F_E L \cos\theta - T(L/2)\sin\theta = 0.$$

The unknown quantities are F_A, F_E, F_v, F_h, and T.

(a) First we solve for T by systematically eliminating the other unknowns. The first equation gives $F_A = W - F_v$ and the fourth gives $F_v = F_E$. We use these to substitute into the remaining three equations to obtain

$$T - F_h = 0$$
$$WL\cos\theta - F_E L\cos\theta - W(L/4)\cos\theta - T(L/2)\sin\theta = 0$$
$$F_E L\cos\theta - T(L/2)\sin\theta = 0.$$

The last of these gives $F_E = T\sin\theta/2\cos\theta = (T/2)\tan\theta$. We substitute this expression into the second equation and solve for T. The result is

$$T = \frac{3W}{4\tan\theta}.$$

To find $\tan\theta$, we consider the right triangle formed by the upper half of one side of the ladder, half the tie rod, and the vertical line from the hinge to the tie rod. The lower side of the triangle has a length of 0.381 m, the hypotenuse has a length of 1.22 m, and the vertical side has a length of $\sqrt{(1.22\,\text{m})^2 - (0.381\,\text{m})^2} = 1.16\,\text{m}$. This means

$$\tan\theta = (1.16\text{m})/(0.381\text{m}) = 3.04.$$

Thus,

$$T = \frac{3(854\,\text{N})}{4(3.04)} = 211\,\text{N}.$$

(b) We now solve for F_A. Since $F_v = F_E$ and $F_E = T\sin\theta/2\cos\theta$, $F_v = 3W/8$. We substitute this into $F_v + F_A - W = 0$ and solve for F_A. We find

$$F_A = W - F_v = W - 3W/8 = 5W/8 = 5(884\,\text{N})/8 = 534\,\text{N}.$$

(c) We have already obtained an expression for F_E: $F_E = 3W/8$. Evaluating it, we get $F_E = 320$ N.

42. (a) Eq. 12-9 leads to

$$TL\sin\theta - m_p g x - m_b g\left(\frac{L}{2}\right) = 0 \ .$$

This can be written in the form of a straight line (in the graph) with

$$T = (\text{"slope"})\frac{x}{L} + \text{"y-intercept"},$$

where "slope" = $m_p g/\sin\theta$ and "y-intercept" = $m_b g/2\sin\theta$. The graph suggests that the slope (in SI units) is 200 and the y-intercept is 500. These facts, combined with the given $m_p + m_b = 61.2$ kg datum, lead to the conclusion:

$$\sin\theta = 61.22g/1200 \Rightarrow \theta = 30.0°.$$

(b) It also follows that $m_p = 51.0$ kg.

(c) Similarly, $m_b = 10.2$ kg.

43. (a) The shear stress is given by F/A, where F is the magnitude of the force applied parallel to one face of the aluminum rod and A is the cross-sectional area of the rod. In this case F is the weight of the object hung on the end: $F = mg$, where m is the mass of the object. If r is the radius of the rod then $A = \pi r^2$. Thus, the shear stress is

$$\frac{F}{A} = \frac{mg}{\pi r^2} = \frac{(1200\,\text{kg})(9.8\,\text{m/s}^2)}{\pi(0.024\,\text{m})^2} = 6.5\times 10^6\,\text{N/m}^2.$$

(b) The shear modulus G is given by

$$G = \frac{F/A}{\Delta x/L}$$

where L is the protrusion of the rod and Δx is its vertical deflection at its end. Thus,

$$\Delta x = \frac{(F/A)L}{G} = \frac{(6.5\times 10^6\,\text{N/m}^2)(0.053\,\text{m})}{3.0\times 10^{10}\,\text{N/m}^2} = 1.1\times 10^{-5}\,\text{m}.$$

44. (a) The Young's modulus is given by

$$E = \frac{\text{stress}}{\text{strain}} = \text{slope of the stress-strain curve} = \frac{150\times 10^6\,\text{N/m}^2}{0.002} = 7.5\times 10^{10}\,\text{N/m}^2.$$

(b) Since the linear range of the curve extends to about 2.9×10^8 N/m², this is approximately the yield strength for the material.

45. (a) Let F_A and F_B be the forces exerted by the wires on the log and let m be the mass of the log. Since the log is in equilibrium $F_A + F_B - mg = 0$. Information given about the

stretching of the wires allows us to find a relationship between F_A and F_B. If wire A originally had a length L_A and stretches by ΔL_A, then $\Delta L_A = F_A L_A / AE$, where A is the cross–sectional area of the wire and E is Young's modulus for steel (200×10^9 N/m^2). Similarly, $\Delta L_B = F_B L_B / AE$. If ℓ is the amount by which B was originally longer than A then, since they have the same length after the log is attached, $\Delta L_A = \Delta L_B + \ell$. This means

$$\frac{F_A L_A}{AE} = \frac{F_B L_B}{AE} + \ell.$$

We solve for F_B:

$$F_B = \frac{F_A L_A}{L_B} - \frac{AE\ell}{L_B}.$$

We substitute into $F_A + F_B - mg = 0$ and obtain

$$F_A = \frac{mgL_B + AE\ell}{L_A + L_B}.$$

The cross–sectional area of a wire is

$$A = \pi r^2 = \pi (1.20 \times 10^{-3} \text{m})^2 = 4.52 \times 10^{-6} \text{m}^2.$$

Both L_A and L_B may be taken to be 2.50 m without loss of significance. Thus

$$F_A = \frac{(103 \text{ kg})(9.8 \text{ m/s}^2)(2.50 \text{ m}) + (4.52 \times 10^{-6} \text{ m}^2)(200 \times 10^9 \text{ N/m}^2)(2.0 \times 10^{-3} \text{ m})}{2.50 \text{ m} + 2.50 \text{ m}}$$

$$= 866 \text{ N}.$$

(b) From the condition $F_A + F_B - mg = 0$, we obtain

$$F_B = mg - F_A = (103 \text{ kg})(9.8 \text{ m/s}^2) - 866 \text{ N} = 143 \text{ N}.$$

(c) The net torque must also vanish. We place the origin on the surface of the log at a point directly above the center of mass. The force of gravity does not exert a torque about this point. Then, the torque equation becomes $F_A d_A - F_B d_B = 0$, which leads to

$$\frac{d_A}{d_B} = \frac{F_B}{F_A} = \frac{143 \text{ N}}{866 \text{ N}} = 0.165.$$

46. Since the force is (stress × area) and the displacement is (strain × length), we can write the work integral (eq. 7-32) as

$$W = \int F dx = \int (\text{stress}) A \, (\text{differential strain}) L = AL \int (\text{stress}) (\text{differential strain})$$

which means the work is (wire-area) × (wire-length) × (graph-area-under-curve). Since the area of a triangle (see the graph in the problem statement) is $\frac{1}{2}$(base)(height) then we determine the work done to be

$$W = (2.00 \times 10^{-6} \text{ m}^2)(0.800 \text{ m})(\tfrac{1}{2})(1.0 \times 10^{-3})(7.0 \times 10^7 \text{ N/m}^2) = 0.0560 \text{ J}.$$

47. (a) Since the brick is now horizontal and the cylinders were initially the same length ℓ, then both have been compressed an equal amount $\Delta \ell$. Thus,

$$\frac{\Delta \ell}{\ell} = \frac{F_A}{A_A E_A} \quad \text{and} \quad \frac{\Delta \ell}{\ell} = \frac{F_B}{A_B E_B}$$

which leads to

$$\frac{F_A}{F_B} = \frac{A_A E_A}{A_B E_B} = \frac{(2A_B)(2E_B)}{A_B E_B} = 4.$$

When we combine this ratio with the equation $F_A + F_B = W$, we find $F_A/W = 4/5 = 0.80$.

(b) This also leads to the result $F_B/W = 1/5 = 0.20$.

(c) Computing torques about the center of mass, we find $F_A d_A = F_B d_B$ which leads to

$$\frac{d_A}{d_B} = \frac{F_B}{F_A} = \frac{1}{4} = 0.25.$$

48. 46. Since the force is (stress × area) and the displacement is (strain × length), we can write the work integral (eq. 7-32) as

$$W = \int F dx = \int (\text{stress}) A \, (\text{differential strain}) L = AL \int (\text{stress})(\text{differential strain})$$

which means the work is (thread cross-sectional area) × (thread length) × (graph-area-under-curve). The area under the curve is

$$\text{graph area} = \frac{1}{2} a s_1 + \frac{1}{2}(a+b)(s_2 - s_1) + \frac{1}{2}(b+c)(s_3 - s_2) = \frac{1}{2}\left[a s_2 + b(s_3 - s_1) + c(s_3 - s_2)\right]$$

$$= \frac{1}{2}\left[(0.12 \times 10^9 \text{ N/m}^2)(1.4) + (0.30 \times 10^9 \text{ N/m}^2)(1.0) + (0.80 \times 10^9 \text{ N/m}^2)(0.60)\right]$$

$$= 4.74 \times 10^8 \text{ N/m}^2.$$

(a) The kinetic energy that would put the thread on the verge of breaking is simply equal to W:

$$K = W = AL(\text{graph area}) = (8.0\times10^{-12} \text{ m}^2)(8.0\times10^{-3} \text{ m})(4.74\times10^8 \text{ N/m}^2) = 3.03\times10^{-5} \text{ J}.$$

(b) The kinetic energy of the fruit fly of mass 6.00 mg and speed 1.70 m/s is

$$K_f = \frac{1}{2}m_f v_f^2 = \frac{1}{2}(6.00\times10^{-6} \text{ kg})(1.70 \text{ m/s})^2 = 8.67\times10^{-6} \text{ J}.$$

(c) Since $K_f < W$, the fruit fly will not be able to break the thread.

(d) The kinetic energy of a bumble bee of mass 0.388 g and speed 0.420 m/s is

$$K_b = \frac{1}{2}m_b v_b^2 = \frac{1}{2}(3.99\times10^{-4} \text{ kg})0.420 \text{ m/s})^2 = 3.42\times10^{-5} \text{ J}.$$

(e) On the other hand, since $K_b > W$, the bumble bee will be able to break the thread.

49. The flat roof (as seen from the air) has area A = 150 m × 5.8 m = 870 m². The volume of material directly above the tunnel (which is at depth d = 60 m) is therefore

$$V = A \times d = (870 \text{ m}^2) \times (60\text{m}) = 52200 \text{ m}^3.$$

Since the density is ρ = 2.8 g/cm³ = 2800 kg/m³, we find the mass of material supported by the steel columns to be $m = \rho V = 1.46 \times 10^8$ m³.

(a) The weight of the material supported by the columns is $mg = 1.4 \times 10^9$ N.

(b) The number of columns needed is

$$n = \frac{1.43\times10^9 \text{ N}}{\frac{1}{2}(400\times10^6 \text{ N}/\text{m}^2)(960\times10^{-4}\text{m}^2)} = 75.$$

50. On the verge of breaking, the length of the thread is

$$L = L_0 + \Delta L = L_0(1 + \Delta L/L_0) = L_0(1+2) = 3L_0,$$

where $L_0 = 0.020$ m is the original length, and strain $= \Delta L/L_0 = 2$, as given in the problem. The free-body diagram of the system is shown on the right. The condition for equilibrium is

$$mg = 2T\sin\theta$$

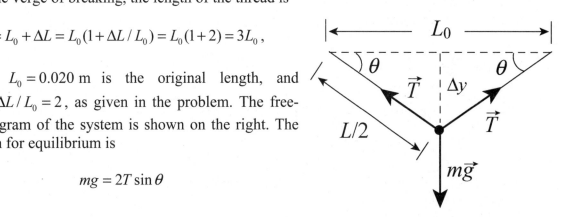

where m is the mass of the insect and $T = A(\text{stress})$. Since the volume of the thread remains constant is it is being stretched, we have $V = A_0 L_0 = AL$, or $A = A_0(L_0/L) = A_0/3$. The vertical distance Δy is

$$\Delta y = \sqrt{(L/2)^2 - (L0/2)^2} = \sqrt{\frac{9L_0^2}{4} - \frac{L_0^2}{4}} = \sqrt{2}L_0.$$

Thus, the mass of the insect is

$$m = \frac{2T\sin\theta}{g} = \frac{2(A_0/3)(\text{stress})\sin\theta}{g} = \frac{2A_0(\text{stress})}{3g}\frac{\Delta y}{3L_0/2} = \frac{4\sqrt{2}A_0(\text{stress})}{9g}$$

$$= \frac{4\sqrt{2}(8.00\times10^{-12}\text{ m}^2)(8.20\times10^8\text{ N/m}^2)}{9(9.8\text{ m/s}^2)}$$

$$= 4.21\times10^{-4}\text{ kg}$$

or 0.421 g.

51. Let the forces that compress stoppers A and B be F_A and F_B, respectively. Then equilibrium of torques about the axle requires $FR = r_A F_A + r_B F_B$. If the stoppers are compressed by amounts $|\Delta y_A|$ and $|\Delta y_B|$ respectively, when the rod rotates a (presumably small) angle θ (in radians), then $|\Delta y_A| = r_A \theta$ and $|\Delta y_B| = r_B \theta$.

Furthermore, if their "spring constants" k are identical, then $k = |F/\Delta y|$ leads to the condition $F_A/r_A = F_B/r_B$ which provides us with enough information to solve.

(a) Simultaneous solution of the two conditions leads to

$$F_A = \frac{Rr_A}{r_A^2 + r_B^2}F = \frac{(5.0\text{ cm})(7.0\text{ cm})}{(7.0\text{ cm})^2 + (4.0\text{ cm})^2}(220\text{ N}) = 118\text{ N} \approx 1.2\times10^2\text{ N}.$$

(b) It also yields

$$F_B = \frac{Rr_B}{r_A^2 + r_B^2}F = \frac{(5.0\text{ cm})(4.0\text{ cm})}{(7.0\text{ cm})^2 + (4.0\text{ cm})^2}(220\text{ N}) = 68\text{ N}.$$

52. (a) With pivot at the hinge (at the left end), Eq. 12-9 gives

$$-mgx - Mg\frac{L}{2} + F_h h = 0$$

where m is the man's mass and M is that of the ramp; F_h is the leftward push of the right wall onto the right edge of the ramp. This equation can be written to be of the form (for a straight line in a graph)

$$F_h = (\text{"slope"})x + (\text{"y-intercept"}),$$

where the "slope" is mg/h and the "y-intercept" is $MgD/2h$. Since $h = 0.480$ m and $D = 4.00$ m, and the graph seems to intercept the vertical axis at 20 kN, then we find $M = 500$ kg.

(b) Since the "slope" (estimated from the graph) is $(5000 \text{ N})/(4 \text{ m})$, then the man's mass must be $m = 62.5$ kg.

53. With the x axis parallel to the incline (positive uphill), then

$$\Sigma F_x = 0 \Rightarrow T\cos 25° - mg\sin 45° = 0.$$

Therefore, $T = 76$ N.

54. The beam has a mass $M = 40.0$ kg and a length $L = 0.800$ m. The mass of the package of tamale is $m = 10.0$ kg.

(a) Since the system is in static equilibrium, the normal force on the beam from roller A is equal to half of the weight of the beam:

$$F_A = Mg/2 = (40.0 \text{ kg})(9.80 \text{ m/s}^2)/2 = 196 \text{ N}.$$

(b) The normal force on the beam from roller B is equal to half of the weight of the beam plus the weight of the tamale:

$$F_B = Mg/2 + mg = (40.0 \text{ kg})(9.80 \text{ m/s}^2)/2 + (10.0 \text{ kg})(9.80 \text{ m/s}^2) = 294 \text{ N}.$$

(c) When the right-hand end of the beam is centered over roller B, the normal force on the beam from roller A is equal to the weight of the beam plus half of the weight of the tamale:

$$F_A = Mg + mg/2 = (40.0 \text{ kg})(9.8 \text{ m/s}^2) + (10.0 \text{ kg})(9.80 \text{ m/s}^2)/2 = 441 \text{ N}.$$

(d) Similarly, the normal force on the beam from roller B is equal to half of the weight of the tamale:

$$F_B = mg/2 = (10.0 \text{ kg})(9.80 \text{ m/s}^2)/2 = 49.0 \text{ N}.$$

(e) We choose the rotational axis to pass through roller B. When the beam is on the verge of losing contact with roller A, the net torque is zero. The balancing equation may be written as

$$mgx = Mg(L/4 - x) \Rightarrow x = \frac{L}{4}\frac{M}{M+m}.$$

Substituting the values given, we obtain $x = 0.160$ m.

55. (a) The forces acting on bucket are the force of gravity, down, and the tension force of cable A, up. Since the bucket is in equilibrium and its weight is

$$W_B = m_B g = (817 \text{kg})(9.80 \text{m/s}^2) = 8.01 \times 10^3 \text{N},$$

the tension force of cable A is $T_A = 8.01 \times 10^3 \text{N}$.

(b) We use the coordinates axes defined in the diagram. Cable A makes an angle of $\theta_2 = 66.0°$ with the negative y axis, cable B makes an angle of $27.0°$ with the positive y axis, and cable C is along the x axis. The y components of the forces must sum to zero since the knot is in equilibrium. This means $T_B \cos 27.0° - T_A \cos 66.0° = 0$ and

$$T_B = \frac{\cos 66.0°}{\cos 27.0°} T_A = \left(\frac{\cos 66.0°}{\cos 27.0°}\right)(8.01 \times 10^3 \text{N}) = 3.65 \times 10^3 \text{N}.$$

(c) The x components must also sum to zero. This means

$$T_C + T_B \sin 27.0° - T_A \sin 66.0° = 0$$

Which yields

$$T_C = T_A \sin 66.0° - T_B \sin 27.0° = (8.01 \times 10^3 \text{N}) \sin 66.0° - (3.65 \times 10^3 \text{N}) \sin 27.0°$$
$$= 5.66 \times 10^3 \text{ N}.$$

56. (a) Eq. 12-8 leads to $T_1 \sin 40° + T_2 \sin\theta = mg$. Also, Eq. 12-7 leads to

$$T_1 \cos 40° - T_2 \cos\theta = 0.$$

Combining these gives the expression

$$T_2 = \frac{mg}{\cos\theta \tan 40° + \sin\theta}.$$

To minimize this, we can plot it or set its derivative equal to zero. In either case, we find that it is at its minimum at $\theta = 50°$.

(b) At $\theta = 50°$, we find $T_2 = 0.77 mg$.

57. The cable that goes around the lowest pulley is cable 1 and has tension $T_1 = F$. That pulley is supported by the cable 2 (so $T_2 = 2T_1 = 2F$) and goes around the middle pulley. The middle pulley is supported by cable 3 (so $T_3 = 2T_2 = 4F$) and goes around the top pulley. The top pulley is supported by the upper cable with tension T, so $T = 2T_3 = 8F$. Three cables are supporting the block (which has mass $m = 6.40$ kg):

$$T_1 + T_2 + T_3 = mg \Rightarrow F = \frac{mg}{7} = 8.96 \text{ N}.$$

Therefore, $T = 8(8.96 \text{ N}) = 71.7 \text{ N}.$

58. Since all surfaces are frictionless, the contact force $\vec{F}$ exerted by the lower sphere on the upper one is along that 45° line, and the forces exerted by walls and floors are "normal" (perpendicular to the wall and floor surfaces, respectively). Equilibrium of forces on the top sphere leads to the two conditions

$$F_{\text{wall}} = F \cos 45° \quad \text{and} \quad F \sin 45° = mg.$$

And (using Newton's third law) equilibrium of forces on the bottom sphere leads to the two conditions

$$F'_{\text{wall}} = F \cos 45° \quad \text{and} \quad F'_{\text{floor}} = F \sin 45° + mg.$$

(a) Solving the above equations, we find $F'_{\text{floor}} = 2mg$.

(b) We obtain for the left side of the container, $F'_{\text{wall}} = mg$.

(c) We obtain for the right side of the container, $F_{\text{wall}} = mg$.

(d) We get $F = mg / \sin 45° = \sqrt{2} mg$.

59. (a) The center of mass of the top brick cannot be further (to the right) with respect to the brick below it (brick 2) than $L/2$; otherwise, its center of gravity is past any point of support and it will fall. So $a_1 = L/2$ in the maximum case.

(b) With brick 1 (the top brick) in the maximum situation, then the combined center of mass of brick 1 and brick 2 is halfway between the middle of brick 2 and its right edge. That point (the combined com) must be supported, so in the maximum case, it is just above the right edge of brick 3. Thus, $a_2 = L/4$.

(c) Now the total center of mass of bricks 1, 2 and 3 is one-third of the way between the middle of brick 3 and its right edge, as shown by this calculation:

$$x_{\text{com}} = \frac{2m(0) + m(-L/2)}{3m} = -\frac{L}{6}$$

where the origin is at the right edge of brick 3. This point is above the right edge of brick 4 in the maximum case, so $a_3 = L/6$.

(d) A similar calculation

$$x'_{com} = \frac{3m(0) + m(-L/2)}{4m} = -\frac{L}{8}$$

shows that $a_4 = L/8$.

(e) We find $h = \sum_{i=1}^{4} a_i = 25L/24$.

60. (a) If L (= 1500 cm) is the unstretched length of the rope and $\Delta L = 2.8$ cm is the amount it stretches then the strain is

$$\Delta L / L = (2.8\,\text{cm})/(1500\,\text{cm}) = 1.9 \times 10^{-3}.$$

(b) The stress is given by F/A where F is the stretching force applied to one end of the rope and A is the cross–sectional area of the rope. Here F is the force of gravity on the rock climber. If m is the mass of the rock climber then $F = mg$. If r is the radius of the rope then $A = \pi r^2$. Thus the stress is

$$\frac{F}{A} = \frac{mg}{\pi r^2} = \frac{(95\,\text{kg})(9.8\,\text{m/s}^2)}{\pi(4.8 \times 10^{-3}\,\text{m})^2} = 1.3 \times 10^7 \text{ N/m}^2.$$

(c) Young's modulus is the stress divided by the strain:

$$E = (1.3 \times 10^7 \text{ N/m}^2) / (1.9 \times 10^{-3}) = 6.9 \times 10^9 \text{ N/m}^2.$$

61. We denote the mass of the slab as m, its density as ρ, and volume as $V = LTW$. The angle of inclination is $\theta = 26°$.

(a) The component of the weight of the slab along the incline is

$$F_1 = mg \sin\theta = \rho V g \sin\theta$$
$$= (3.2 \times 10^3 \text{ kg/m}^3)(43\,\text{m})(2.5\,\text{m})(12\,\text{m})(9.8\,\text{m/s}^2) \sin 26° \approx 1.8 \times 10^7 \text{ N}.$$

(b) The static force of friction is

$$f_s = \mu_s F_N = \mu_s mg \cos\theta = \mu_s \rho V g \cos\theta$$
$$= (0.39)(3.2 \times 10^3 \text{ kg/m}^3)(43\,\text{m})(2.5\,\text{m})(12\,\text{m})(9.8\,\text{m/s}^2) \cos 26° \approx 1.4 \times 10^7 \text{ N}.$$

(c) The minimum force needed from the bolts to stabilize the slab is

$$F_2 = F_1 - f_s = 1.77 \times 10^7 \text{ N} - 1.42 \times 10^7 \text{ N} = 3.5 \times 10^6 \text{ N}.$$

If the minimum number of bolts needed is n, then $F_2 / nA \leq 3.6 \times 10^8 \text{ N/m}^2$, or

$$n \geq \frac{3.5 \times 10^6 \text{ N}}{(3.6 \times 10^8 \text{ N/m}^2)(6.4 \times 10^{-4} \text{ m}^2)} = 15.2$$

Thus 16 bolts are needed.

62. The notation and coordinates are as shown in Fig. 12-6 in the textbook. Here, the ladder's center of mass is halfway up the ladder (unlike in the textbook figure). Also, we label the x and y forces at the ground f_s and F_N, respectively. Now, balancing forces, we have

$$\Sigma F_x = 0 \Rightarrow f_s = F_w$$
$$\Sigma F_y = 0 \Rightarrow F_N = mg$$

Since $f_s = f_{s,\max}$, we divide the equations to obtain

$$\frac{f_{s,\max}}{F_N} = \mu_s = \frac{F_w}{mg} \,.$$

Now, from $\Sigma \tau_z = 0$ (with axis at the ground) we have $mg(a/2) - F_w h = 0$. But from the Pythagorean theorem, $h = \sqrt{L^2 - a^2}$, where L = length of ladder. Therefore,

$$\frac{F_w}{mg} = \frac{a/2}{h} = \frac{a}{2\sqrt{L^2 - a^2}} \,.$$

In this way, we find

$$\mu_s = \frac{a}{2\sqrt{L^2 - a^2}} \Rightarrow a = \frac{2\mu_s L}{\sqrt{1 + 4\mu_s^2}} = 3.4 \text{ m}.$$

63. Analyzing forces at the knot (particularly helpful is a graphical view of the vector right–triangle with horizontal "side" equal to the static friction force f_s and vertical "side" equal to the weight $m_B g$ of block B), we find $f_s = m_B g \tan \theta$ where $\theta = 30°$. For f_s to be at its maximum value, then it must equal $\mu_s m_A g$ where the weight of block A is $m_A g = (10 \text{ kg})(9.8 \text{ m/s}^2)$. Therefore,

$$\mu_s m_A g = m_B g \tan \theta \Rightarrow \mu_s = \frac{5.0}{10} \tan 30° = 0.29.$$

64. To support a load of $W = mg = (670 \text{ kg})(9.8 \text{ m/s}^2) = 6566$ N, the steel cable must stretch an amount proportional to its "free" length:

$$\Delta L = \left(\frac{W}{AY}\right) L \quad \text{where } A = \pi r^2$$

and $r = 0.0125$ m.

(a) If $L = 12$ m, then $\Delta L = \left(\dfrac{6566 \text{ N}}{\pi(0.0125 \text{ m})^2 (2.0\times10^{11} \text{ N/m}^2)}\right)(12 \text{ m}) = 8.0\times10^{-4}$ m.

(b) Similarly, when $L = 350$ m, we find $\Delta L = 0.023$ m.

65. With the pivot at the hinge, Eq. 12-9 leads to

$$-mg\sin\theta_1 \frac{L}{2} + TL\sin(180° - \theta_1 - \theta_2) = 0.$$

where $\theta_1 = 60°$ and $T = mg/2$. This yields $\theta_2 = 60°$.

66. (a) Setting up equilibrium of torques leads to

$$F_{\text{far end}} L = (73\text{ kg})(9.8\text{ m/s}^2)\frac{L}{4} + (2700\text{ N})\frac{L}{2}$$

which yields $F_{\text{far end}} = 1.5 \times 10^3$ N.

(b) Then, equilibrium of vertical forces provides

$$F_{\text{near end}} = (73)(9.8) + 2700 - F_{\text{far end}} = 1.9\times10^3 \text{ N}.$$

67. (a) and (b) With $+x$ rightward and $+y$ upward (we assume the adult is pulling with force $\vec{P}$ to the right), we have

$$\Sigma F_y = 0 \Rightarrow W = T\cos\theta = 270 \text{ N}$$
$$\Sigma F_x = 0 \Rightarrow P = T\sin\theta = 72 \text{ N}$$

where $\theta = 15°$.

(c) Dividing the above equations leads to

$$\frac{P}{W} = \tan\theta.$$

Thus, with $W = 270$ N and $P = 93$ N, we find $\theta = 19°$.

68. We denote the tension in the upper left string (bc) as T' and the tension in the lower right string (ab) as T. The supported weight is $Mg = 19.6$ N. The force equilibrium conditions lead to

$$T'\cos 60° = T\cos 20° \qquad \text{horizontal forces}$$
$$T'\sin 60° = W + T\sin 20° \qquad \text{vertical forces.}$$

(a) We solve the above simultaneous equations and find

$$T = \frac{W}{\tan 60°\cos 20° - \sin 20°} = 15\text{N}.$$

(b) Also, we obtain $T' = T\cos 20° / \cos 60° = 29$ N.

69. (a) Because of Eq. 12-3, we can write

$$\vec{T} + (m_B g \angle -90°) + (m_A g \angle -150°) = 0.$$

Solving the equation, we obtain $\vec{T} = (106.34 \angle 63.963°)$. Thus, the magnitude of the tension in the upper cord is 106 N,

(b) and its angle (measured ccw from the +x axis) is 64.0°.

70. (a) The angle between the beam and the floor is

$$\sin^{-1}(d/L) = \sin^{-1}(1.5/2.5) = 37°,$$

so that the angle between the beam and the weight vector $\vec{W}$ of the beam is 53°. With $L = 2.5$ m being the length of beam, and choosing the axis of rotation to be at the base,

$$\Sigma \tau_z = 0 \Rightarrow PL - W\left(\frac{L}{2}\right)\sin 53° = 0$$

Thus, $P = \frac{1}{2} W \sin 53° = 200$ N.

(b) Note that

$$\vec{P} + \vec{W} = (200 \angle 90°) + (500 \angle -127°) = (360 \angle -146°)$$

using magnitude-angle notation (with angles measured relative to the beam, where "uphill" along the beam would correspond to 0°) with the unit Newton understood. The "net force of the floor" $\vec{F_f}$ is equal and opposite to this (so that the total net force on the beam is zero), so that $|\vec{F_f}| = 360$ N and is directed 34° counterclockwise from the beam.

(c) Converting that angle to one measured from true horizontal, we have $\theta = 34° + 37° = 71°$. Thus, $f_s = F_f \cos\theta$ and $F_N = F_f \sin\theta$. Since $f_s = f_{s,\max}$, we divide the equations to obtain

$$\frac{F_N}{f_{s,\max}} = \frac{1}{\mu_s} = \tan\theta \ .$$

Therefore, $\mu_s = 0.35$.

71. The cube has side length l and volume $V = l^3$. We use $p = B\Delta V / V$ for the pressure p. We note that

$$\frac{\Delta V}{V} = \frac{\Delta l^3}{l^3} = \frac{(l+\Delta l)^3 - l^3}{l^3} \approx \frac{3l^2 \Delta l}{l^3} = 3\frac{\Delta l}{l}.$$

Thus, the pressure required is

$$p = \frac{3B\Delta l}{l} = \frac{3(1.4\times 10^{11}\ \text{N/m}^2)(85.5\,\text{cm} - 85.0\,\text{cm})}{85.5\,\text{cm}} = 2.4\times 10^9\ \text{N/m}^2.$$

72. Adopting the usual convention that torques that would produce counterclockwise rotation are positive, we have (with axis at the hinge)

$$\sum \tau_z = 0 \Rightarrow TL\sin 60^\circ - Mg\left(\frac{L}{2}\right) = 0$$

where $L = 5.0$ m and $M = 53$ kg. Thus, $T = 300$ N. Now (with F_p for the force of the hinge)

$$\sum F_x = 0 \Rightarrow F_{px} = -T\cos\theta = -150\,\text{N}$$
$$\sum F_y = 0 \Rightarrow F_{py} = Mg - T\sin\theta = 260\,\text{N}$$

where $\theta = 60^\circ$. Therefore, $\vec{F}_p = (-1.5\times 10^2\ \text{N})\hat{i} + (2.6\times 10^2\ \text{N})\hat{j}$.

73. (a) Choosing an axis through the hinge, perpendicular to the plane of the figure and taking torques that would cause counterclockwise rotation as positive, we require the net torque to vanish:

$$FL\sin 90^\circ - Th\sin 65^\circ = 0$$

where the length of the beam is $L = 3.2$ m and the height at which the cable attaches is $h = 2.0$ m. Note that the weight of the beam does not enter this equation since its line of action is directed towards the hinge. With $F = 50$ N, the above equation yields $T = 88$ N.

(b) To find the components of $\vec{F}_p$ we balance the forces:

$$\sum F_x = 0 \Rightarrow F_{px} = T\cos 25^\circ - F$$
$$\sum F_y = 0 \Rightarrow F_{py} = T\sin 25^\circ + W$$

where W is the weight of the beam (60 N). Thus, we find that the hinge force components are $F_{px} = 30$ N rightward and $F_{py} = 97$ N upward. In unit-vector notation, $\vec{F}_p = (30 \text{ N})\hat{i} + (97 \text{ N})\hat{j}$.

74. (a) Computing the torques about the hinge, we have $TL \sin 40° = W \dfrac{L}{2} \sin 50°$ where the length of the beam is $L = 12$ m and the tension is $T = 400$ N. Therefore, the weight is $W = 671$ N, which means that the gravitational force on the beam is $\vec{F}_w = (-671 \text{ N})\hat{j}$.

(b) Equilibrium of horizontal and vertical forces yields, respectively,

$$F_{\text{hinge } x} = T = 400 \text{ N}$$
$$F_{\text{hinge } y} = W = 671 \text{ N}$$

where the hinge force components are rightward (for x) and upward (for y). In unit-vector notation, we have $\vec{F}_{\text{hinge}} = (400 \text{ N})\hat{i} + (671 \text{ N})\hat{j}$

75. We locate the origin of the x axis at the edge of the table and choose rightwards positive. The criterion (in part (a)) is that the center of mass of the block above another must be no further than the edge of the one below; the criterion in part (b) is more subtle and is discussed below. Since the edge of the table corresponds to $x = 0$ then the total center of mass of the blocks must be zero.

(a) We treat this as three items: one on the upper left (composed of two bricks, one directly on top of the other) of mass $2m$ whose center is above the left edge of the bottom brick; a single brick at the upper right of mass m which necessarily has its center over the right edge of the bottom brick (so $a_1 = L/2$ trivially); and, the bottom brick of mass m. The total center of mass is

$$\frac{(2m)(a_2 - L) + ma_2 + m(a_2 - L/2)}{4m} = 0$$

which leads to $a_2 = 5L/8$. Consequently, $h = a_2 + a_1 = 9L/8$.

(b) We have four bricks (each of mass m) where the center of mass of the top and the center of mass of the bottom one have the same value $x_{cm} = b_2 - L/2$. The middle layer consists of two bricks, and we note that it is possible for each of their centers of mass to be beyond the respective edges of the bottom one! This is due to the fact that the top brick is exerting downward forces (each equal to half its weight) on the middle blocks — and in the extreme case, this may be thought of as a pair of concentrated forces exerted at the innermost edges of the middle bricks. Also, in the extreme case, the support force (upward) exerted on a middle block (by the bottom one) may be thought of as a concentrated force located at the edge of the bottom block (which is the point about which we compute torques, in the following).

If (as indicated in our sketch, where $\vec{F}_{top}$ has magnitude $mg/2$) we consider equilibrium of torques on the rightmost brick, we obtain

$$mg\left(b_1 - \frac{1}{2}L\right) = \frac{mg}{2}(L - b_1)$$

which leads to $b_1 = 2L/3$. Once we conclude from symmetry that $b_2 = L/2$ then we also arrive at $h = b_2 + b_1 = 7L/6$.

76. One arm of the balance has length ℓ_1 and the other has length ℓ_2. The two cases described in the problem are expressed (in terms of torque equilibrium) as

$$m_1 \ell_1 = m\ell_2 \quad \text{and} \quad m\ell_1 = m_2 \ell_2.$$

We divide equations and solve for the unknown mass: $m = \sqrt{m_1 m_2}$.

77. Since GA exerts a leftward force T at the corner A, then (by equilibrium of horizontal forces at that point) the force F_{diag} in CA must be pulling with magnitude

$$F_{diag} = \frac{T}{\sin 45°} = T\sqrt{2}.$$

This analysis applies equally well to the force in DB. And these diagonal bars are pulling on the bottom horizontal bar exactly as they do to the top bar, so the bottom bar CD is the "mirror image" of the top one (it is also under tension T). Since the figure is symmetrical (except for the presence of the turnbuckle) under 90° rotations, we conclude that the side bars (DA and BC) also are under tension T (a conclusion that also follows from considering the vertical components of the pull exerted at the corners by the diagonal bars).

(a) Bars that are in tension are BC, CD and DA.

(b) The magnitude of the forces causing tension is $T = 535$ N.

(c) The magnitude of the forces causing compression on CA and DB is

$$F_{diag} = \sqrt{2}T = (1.41)535 \text{ N} = 757 \text{ N}.$$

78. (a) For computing torques, we choose the axis to be at support 2 and consider torques which encourage counterclockwise rotation to be positive. Let m = mass of gymnast and M = mass of beam. Thus, equilibrium of torques leads to

$$Mg(1.96\,\text{m}) - mg(0.54\,\text{m}) - F_1(3.92\,\text{m}) = 0.$$

Therefore, the upward force at support 1 is $F_1 = 1163$ N (quoting more figures than are significant — but with an eye toward using this result in the remaining calculation). In unit-vector notation, we have $\vec{F}_1 \approx (1.16 \times 10^3 \text{ N})\hat{j}$.

(b) Balancing forces in the vertical direction, we have $F_1 + F_2 - Mg - mg = 0$, so that the upward force at support 2 is $F_2 = 1.74 \times 10^3$ N. In unit-vector notation, we have $\vec{F}_2 \approx (1.74 \times 10^3 \text{ N})\hat{j}$.

79. (a) Let $d = 0.00600$ m. In order to achieve the same final lengths, wires 1 and 3 must stretch an amount d more than wire 2 stretches:

$$\Delta L_1 = \Delta L_3 = \Delta L_2 + d.$$

Combining this with Eq. 12-23 we obtain

$$F_1 = F_3 = F_2 + \frac{dAE}{L}.$$

Now, Eq. 12-8 produces $F_1 + F_3 + F_2 - mg = 0$. Combining this with the previous relation (and using Table 12-1) leads to $F_1 = 1380$ N $\approx 1.38 \times 10^3$ N.

(b) Similarly, $F_2 = 180$ N.

80. Our system is the second finger bone. Since the system is in static equilibrium, the net force acting on it is zero. In addition, the torque about any point must be zero. We set up the torque equation about point O where $\vec{F}_c$ act:

$$0 = \sum_O \tau_{net} = -\left(\frac{d}{3}\right)F_t \sin\alpha + (d)F_v \sin\theta + (d)F_h \sin\phi.$$

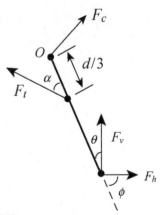

Solving for F_t and substituting the values given, we obtain

$$F_t = \frac{3(F_v \sin\theta + F_h \sin\phi)}{\sin\alpha} = \frac{3[(162.4 \text{ N})\sin 10° + (13.4 \text{ N})\sin 80°]}{\sin 45°} = 175.6 \text{ N}$$
$$\approx 1.8 \times 10^2 \text{ N}.$$

81. When it is about to move, we are still able to apply the equilibrium conditions, but (to obtain the critical condition) we set static friction equal to its maximum value and picture the normal force $\vec{F}_N$ as a concentrated force (upward) at the bottom corner of the cube, directly below the point O where P is being applied. Thus, the line of action of $\vec{F}_N$ passes through point O and exerts no torque about O (of course, a similar observation applied to the pull P). Since $F_N = mg$ in this problem, we have $f_{smax} = \mu mg$ applied a distance h away from O. And the line of action of force of gravity (of magnitude mg), which is best pictured as a concentrated force at the center of the cube, is a distance $L/2$ away from O. Therefore, equilibrium of torques about O produces

$$\mu mgh = mg\left(\frac{L}{2}\right) \Rightarrow \mu = \frac{L}{2h} = \frac{(8.0 \text{ cm})}{2(7.0 \text{ cm})} = 0.57$$

for the critical condition we have been considering. We now interpret this in terms of a range of values for μ.

(a) For it to slide but not tip, a value of μ *less* than that derived above is needed, since then — static friction will be exceeded for a smaller value of P, before the pull is strong enough to cause it to tip. Thus, $\mu < L/2h = 0.57$ is required.

(b) And for it to tip but not slide, we need μ *greater* than that derived above is needed, since now — static friction will not be exceeded even for the value of P which makes the cube rotate about its front lower corner. That is, we need to have $\mu > L/2h = 0.57$ in this case.

82. The assumption stated in the problem (that the density does not change) is not meant to be realistic; those who are familiar with Poisson's ratio (and other topics related to the strengths of materials) might wish to think of this problem as treating a fictitious material (which happens to have the same value of E as aluminum, given in Table 12-1) whose density does not significantly change during stretching. Since the mass does not change, either, then the constant-density assumption implies the volume (which is the circular area times its length) stays the same:

$$(\pi r^2 L)_{new} = (\pi r^2 L)_{old} \quad \Rightarrow \quad \Delta L = L[(1000/999.9)^2 - 1].$$

Now, Eq. 12-23 gives

$$F = \pi r^2 E \Delta L/L = \pi r^2 (7.0 \times 10^9 \text{ N/m}^2)[(1000/999.9)^2 - 1].$$

Using either the new or old value for r gives the answer $F = 44$ N.

83. Where the crosspiece comes into contact with the beam, there is an upward force of $2F$ (where F is the upward force exerted by each man). By equilibrium of vertical forces, $W = 3F$ where W is the weight of the beam. If the beam is uniform, its center of gravity is

a distance $L/2$ from the man in front, so that computing torques about the front end leads to

$$W\frac{L}{2} = 2Fx = 2\left(\frac{W}{3}\right)x$$

which yields $x = 3L/4$ for the distance from the crosspiece to the front end. It is therefore a distance $L/4$ from the rear end (the "free" end).

84. (a) Setting up equilibrium of torques leads to a simple "level principle" ratio:

$$F_{\text{catch}} = (11\text{kg})(9.8\text{m/s}^2)\frac{(91/2-10)\text{cm}}{91\text{cm}} = 42\text{ N}.$$

(b) Then, equilibrium of vertical forces provides

$$F_{\text{hinge}} = (11\text{kg})(9.8\text{m/s}^2) - F_{\text{catch}} = 66\text{ N}.$$

85. We choose an axis through the top (where the ladder comes into contact with the wall), perpendicular to the plane of the figure and take torques that would cause counterclockwise rotation as positive. Note that the line of action of the applied force $\vec{F}$ intersects the wall at a height of $(8.0\text{ m})/5 = 1.6\text{ m}$; in other words, the *moment arm* for the applied force (in terms of where we have chosen the axis) is $r_\perp = (4/5)(8.0\text{ m}) = 6.4\text{ m}$. The moment arm for the weight is half the horizontal distance from the wall to the base of the ladder; this works out to be $\sqrt{(10\text{ m})^2 - (8\text{ m})^2}/2 = 3.0\text{ m}$. Similarly, the moment arms for the x and y components of the force at the ground $(\vec{F}_g)$ are 8.0 m and 6.0 m, respectively. Thus, with lengths in meters, we have

$$\Sigma\tau_z = F(6.4\text{ m}) + W(3.0\text{ m}) + F_{gx}(8.0\text{ m}) - F_{gy}(6.0\text{ m}) = 0.$$

In addition, from balancing the vertical forces we find that $W = F_{gy}$ (keeping in mind that the wall has no friction). Therefore, the above equation can be written as

$$\Sigma\tau_z = F(6.4\text{ m}) + W(3.0\text{ m}) + F_{gx}(8.0\text{ m}) - W(6.0\text{ m}) = 0.$$

(a) With $F = 50$ N and $W = 200$ N, the above equation yields $F_{gx} = 35$ N. Thus, in unit vector notation we obtain

$$\vec{F}_g = (35\text{ N})\hat{i} + (200\text{ N})\hat{j}.$$

(b) With $F = 150$ N and $W = 200$ N, the above equation yields $F_{gx} = -45$ N. Therefore, in unit vector notation we obtain

$$\vec{F}_g = (-45 \text{ N})\hat{i} + (200 \text{ N})\hat{j}.$$

(c) Note that the phrase "start to move towards the wall" implies that the friction force is pointed away from the wall (in the $-\hat{i}$ direction). Now, if $f = -F_{gx}$ and $F_N = F_{gy} = 200$ N are related by the (maximum) static friction relation ($f = f_{s,\max} = \mu_s F_N$) with $\mu_s = 0.38$, then we find $F_{gx} = -76$ N. Returning this to the above equation, we obtain

$$F = \frac{(200 \text{ N})(3.0 \text{ m}) + (76 \text{ N})(8.0 \text{ m})}{6.4 \text{ m}} = 1.9 \times 10^2 \text{ N}.$$

86. The force F exerted on the beam is $F = 7900$ N, as computed in the Sample Problem. Let $F/A = S_u/6$, where $S_u = 50 \times 10^6$ N/m^2 is the ultimate strength (see Table 12-1), then

$$A = \frac{6F}{S_u} = \frac{6(7900 \text{ N})}{50 \times 10^6 \text{ N/m}^2} = 9.5 \times 10^{-4} \text{ m}^2.$$

Thus the thickness is $\sqrt{A} = \sqrt{9.5 \times 10^{-4} \text{ m}^2} = 0.031 \text{ m}$.

Chapter 13

1. The magnitude of the force of one particle on the other is given by $F = Gm_1m_2/r^2$, where m_1 and m_2 are the masses, r is their separation, and G is the universal gravitational constant. We solve for r:

$$r = \sqrt{\frac{Gm_1m_2}{F}} = \sqrt{\frac{(6.67 \times 10^{-11}\,\text{N}\cdot\text{m}^2/\text{kg}^2)(5.2\,\text{kg})(2.4\,\text{kg})}{2.3 \times 10^{-12}\,\text{N}}} = 19\,\text{m}.$$

2. We use subscripts s, e, and m for the Sun, Earth and Moon, respectively. Plugging in the numerical values (say, from Appendix C) we find

$$\frac{F_{sm}}{F_{em}} = \frac{Gm_sm_m/r_{sm}^2}{Gm_em_m/r_{em}^2} = \frac{m_s}{m_e}\left(\frac{r_{em}}{r_{sm}}\right)^2 = \frac{1.99 \times 10^{30}\,\text{kg}}{5.98 \times 10^{24}\,\text{kg}}\left(\frac{3.82 \times 10^8\,\text{m}}{1.50 \times 10^{11}\,\text{m}}\right)^2 = 2.16.$$

3. The gravitational force between the two parts is

$$F = \frac{Gm(M-m)}{r^2} = \frac{G}{r^2}(mM - m^2)$$

which we differentiate with respect to m and set equal to zero:

$$\frac{dF}{dm} = 0 = \frac{G}{r^2}(M - 2m) \Rightarrow M = 2m.$$

This leads to the result $m/M = 1/2$.

4. The gravitational force between you and the moon at its initial position (directly opposite of Earth from you) is

$$F_0 = \frac{GM_m m}{(R_{ME} + R_E)^2}$$

where M_m is the mass of the moon, R_{ME} is the distance between the moon and the Earth, and R_E is the radius of the Earth. At its final position (directly above you), the gravitational force between you and the moon is

$$F_1 = \frac{GM_m m}{(R_{ME} - R_E)^2}.$$

(a) The ratio of the moon's gravitational pulls at the two different positions is

$$\frac{F_1}{F_0} = \frac{GM_m m/(R_{ME} - R_E)^2}{GM_m m/(R_{ME} + R_E)^2} = \left(\frac{R_{ME} + R_E}{R_{ME} - R_E}\right)^2 = \left(\frac{3.82 \times 10^8 \text{ m} + 6.37 \times 10^6 \text{ m}}{3.82 \times 10^8 \text{ m} - 6.37 \times 10^6 \text{ m}}\right)^2 = 1.06898.$$

Therefore, the increase is 0.06898, or approximately, 6.9%.

(b) The change of the gravitational pull may be approximated as

$$F_1 - F_0 = \frac{GM_m m}{(R_{ME} - R_E)^2} - \frac{GM_m m}{(R_{ME} + R_E)^2} \approx \frac{GM_m m}{R_{ME}^2}\left(1 + 2\frac{R_E}{R_{ME}}\right) - \frac{GM_m m}{R_{ME}^2}\left(1 - 2\frac{R_E}{R_{ME}}\right) = \frac{4GM_m m R_E}{R_{ME}^3}.$$

On the other hand, your weight, as measured on a scale on Earth is

$$F_g = mg_E = \frac{GM_E m}{R_E^2}.$$

Since the moon pulls you "up," the percentage decrease of weight is

$$\frac{F_1 - F_0}{F_g} = 4\left(\frac{M_m}{M_E}\right)\left(\frac{R_E}{R_{ME}}\right)^3 = 4\left(\frac{7.36 \times 10^{22} \text{ kg}}{5.98 \times 10^{24} \text{ kg}}\right)\left(\frac{6.37 \times 10^6 \text{ m}}{3.82 \times 10^8 \text{ m}}\right)^3 = 2.27 \times 10^{-7} \approx (2.3 \times 10^{-5})\%.$$

5. We require the magnitude of force (given by Eq. 13-1) exerted by particle C on A be equal to that exerted by B on A. Thus,

$$\frac{Gm_A m_C}{r^2} = \frac{Gm_A m_B}{d^2}.$$

We substitute in $m_B = 3m_A$ and $m_B = 3m_A$, and (after canceling "m_A") solve for r. We find $r = 5d$. Thus, particle C is placed on the x axis, to left of particle A (so it is at a negative value of x), at $x = -5.00d$.

6. Using $F = GmM/r^2$, we find that the topmost mass pulls upward on the one at the origin with 1.9×10^{-8} N, and the rightmost mass pulls rightward on the one at the origin with 1.0×10^{-8} N. Thus, the (x, y) components of the net force, which can be converted to polar components (here we use magnitude-angle notation), are

$$\vec{F}_{net} = (1.04 \times 10^{-8}, 1.85 \times 10^{-8}) \Rightarrow (2.13 \times 10^{-8} \angle 60.6°).$$

(a) The magnitude of the force is 2.13×10^{-8} N.

(b) The direction of the force relative to the $+x$ axis is $60.6°$.

7. At the point where the forces balance $GM_e m/r_1^2 = GM_s m/r_2^2$, where M_e is the mass of Earth, M_s is the mass of the Sun, m is the mass of the space probe, r_1 is the distance from the center of Earth to the probe, and r_2 is the distance from the center of the Sun to the probe. We substitute $r_2 = d - r_1$, where d is the distance from the center of Earth to the center of the Sun, to find

$$\frac{M_e}{r_1^2} = \frac{M_s}{(d-r_1)^2}.$$

Taking the positive square root of both sides, we solve for r_1. A little algebra yields

$$r_1 = \frac{d\sqrt{M_e}}{\sqrt{M_s} + \sqrt{M_e}} = \frac{(150\times 10^9 \text{ m})\sqrt{5.98\times 10^{24} \text{ kg}}}{\sqrt{1.99\times 10^{30} \text{ kg}} + \sqrt{5.98\times 10^{24} \text{ kg}}} = 2.60\times 10^8 \text{ m}.$$

Values for M_e, M_s, and d can be found in Appendix C.

8. The gravitational forces on m_5 from the two 5.00g masses m_1 and m_4 cancel each other. Contributions to the net force on m_5 come from the remaining two masses:

$$F_{net} = \frac{(6.67\times 10^{-11} \text{ N}\cdot\text{m}^2/\text{kg}^2)(2.50\times 10^{-3} \text{ kg})(3.00\times 10^{-3} \text{ kg} - 1.00\times 10^{-3} \text{ kg})}{(\sqrt{2}\times 10^{-1} \text{ m})^2}$$

$$= 1.67\times 10^{-14} \text{ N}.$$

The force is directed along the diagonal between m_2 and m_3, towards m_2. In unit-vector notation, we have

$$\vec{F}_{net} = F_{net}(\cos 45°\hat{i} + \sin 45°\hat{j}) = (1.18\times 10^{-14} \text{ N})\hat{i} + (1.18\times 10^{-14} \text{ N})\hat{j}$$

9. The gravitational force from Earth on you (with mass m) is

$$F_g = \frac{GM_E m}{R_E^2} = mg$$

where $g = GM_E/R_E^2 = 9.8 \text{ m/s}^2$. If r is the distance between you and a tiny black hole of mass $M_b = 1\times 10^{11}$ kg that has the same gravitational pull on you as the Earth, then

$$F_g = \frac{GM_b m}{r^2} = mg.$$

Combining the two equations, we obtain

$$mg = \frac{GM_E m}{R_E^2} = \frac{GM_b m}{r^2} \Rightarrow r = \sqrt{\frac{GM_b}{g}} = \sqrt{\frac{(6.67\times 10^{-11}\text{ m}^3/\text{kg}\cdot\text{s}^2)(1\times 10^{11}\text{ kg})}{9.8\text{ m/s}^2}} \approx 0.8\text{ m}.$$

10. (a) We are told the value of the force when particle C is removed (that is, as its position x goes to infinity), which is a situation in which any force caused by C vanishes (because Eq. 13-1 has r^2 in the denominator). Thus, this situation only involves the force exerted by A on B:

$$\frac{Gm_A m_B}{(0.20\text{ m})^2} = 4.17\times 10^{-10}\text{ N}.$$

Since $m_B = 1.0$ kg, then this yields $m_A = 0.25$ kg.

(b) We note (from the graph) that the net force on B is zero when $x = 0.40$ m. Thus, at that point, the force exerted by C must have the same magnitude (but opposite direction) as the force exerted by A (which is the one discussed in part (a)). Therefore

$$\frac{Gm_C m_B}{(0.40\text{ m})^2} = 4.17\times 10^{-10}\text{ N} \Rightarrow m_C = 1.00\text{ kg}.$$

11. (a) The distance between any of the spheres at the corners and the sphere at the center is

$$r = \ell/2\cos 30° = \ell/\sqrt{3}$$

where ℓ is the length of one side of the equilateral triangle. The net (downward) contribution caused by the two bottom-most spheres (each of mass m) to the total force on m_4 has magnitude

$$2F_y = 2\left(\frac{Gm_4 m}{r^2}\right)\sin 30° = 3\frac{Gm_4 m}{\ell^2}.$$

This must equal the magnitude of the pull from M, so

$$3\frac{Gm_4 m}{\ell^2} = \frac{Gm_4 m}{\left(\ell/\sqrt{3}\right)^2}$$

which readily yields $m = M$.

(b) Since m_4 cancels in that last step, then the amount of mass in the center sphere is not relevant to the problem. The net force is still zero.

12. All the forces are being evaluated at the origin (since particle A is there), and all forces (except the net force) are along the location-vectors $\vec{r}$ which point to particles B and C. We note that the angle for the location-vector pointing to particle B is 180° –

30.0° = 150° (measured ccw from the +x axis). The component along, say, the x axis of one of the force-vectors $\vec{F}$ is simply Fx/r in this situation (where F is the magnitude of $\vec{F}$). Since the force itself (see Eq. 13-1) is inversely proportional to r^2 then the aforementioned x component would have the form $GmMx/r^3$; similarly for the other components. With $m_A = 0.0060$ kg, $m_B = 0.0120$ kg, and $m_C = 0.0080$ kg, we therefore have

$$F_{net\,x} = \frac{Gm_A m_B x_B}{r_B^3} + \frac{Gm_A m_C x_C}{r_C^3} = (2.77 \times 10^{-14}\,\text{N})\cos(-163.8°)$$

and

$$F_{net\,y} = \frac{Gm_A m_B y_B}{r_B^3} + \frac{Gm_A m_C y_C}{r_C^3} = (2.77 \times 10^{-14}\,\text{N})\sin(-163.8°)$$

where $r_B = d_{AB} = 0.50$ m, and $(x_B, y_B) = (r_B\cos(150°), r_B\sin(150°))$ (with SI units understood). A fairly quick way to solve for r_C is to consider the vector difference between the net force and the force exerted by A, and then employ the Pythagorean theorem. This yields $r_C = 0.40$ m.

(a) By solving the above equations, the x coordinate of particle C is $x_C = -0.20$ m.

(b) Similarly, the y coordinate of particle C is $y_C = -0.35$ m.

13. If the lead sphere were not hollowed the magnitude of the force it exerts on m would be $F_1 = GMm/d^2$. Part of this force is due to material that is removed. We calculate the force exerted on m by a sphere that just fills the cavity, at the position of the cavity, and subtract it from the force of the solid sphere.

The cavity has a radius $r = R/2$. The material that fills it has the same density (mass to volume ratio) as the solid sphere. That is $M_c/r^3 = M/R^3$, where M_c is the mass that fills the cavity. The common factor $4\pi/3$ has been canceled. Thus,

$$M_c = \left(\frac{r^3}{R^3}\right)M = \left(\frac{R^3}{8R^3}\right)M = \frac{M}{8}.$$

The center of the cavity is $d - r = d - R/2$ from m, so the force it exerts on m is

$$F_2 = \frac{G(M/8)m}{(d - R/2)^2}.$$

The force of the hollowed sphere on m is

$$F = F_1 - F_2 = GMm\left(\frac{1}{d^2} - \frac{1}{8(d-R/2)^2}\right) = \frac{GMm}{d^2}\left(1 - \frac{1}{8(1-R/2d)^2}\right)$$

$$= \frac{(6.67\times 10^{-11} \text{ m}^3/\text{s}^2\cdot\text{kg})(2.95 \text{ kg})(0.431 \text{ kg})}{(9.00\times 10^{-2}\text{m})^2}\left(1 - \frac{1}{8[1-(4\times 10^{-2}\text{m})/(2\cdot 9\times 10^{-2}\text{m})]^2}\right)$$

$$= 8.31\times 10^{-9} \text{ N}.$$

14. Using Eq. 13-1, we find

$$\vec{F}_{AB} = \frac{2Gm_A^2}{d^2}\hat{j} \quad \text{and} \quad \vec{F}_{AC} = -\frac{4Gm_A^2}{3d^2}\hat{i}.$$

Since the vector sum of all three forces must be zero, we find the third force (using magnitude-angle notation) is

$$\vec{F}_{AD} = \frac{Gm_A^2}{d^2}(2.404 \angle -56.3°).$$

This tells us immediately the direction of the vector $\vec{r}$ (pointing from the origin to particle D), but to find its magnitude we must solve (with $m_D = 4m_A$) the following equation:

$$2.404\left(\frac{Gm_A^2}{d^2}\right) = \frac{Gm_A m_D}{r^2}.$$

This yields $r = 1.29d$. In magnitude-angle notation, then, $\vec{r} = (1.29 \angle -56.3°)$, with SI units understood. The "exact" answer without regard to significant figure considerations is

$$\vec{r} = \left(2\sqrt{\frac{6}{13\sqrt{13}}}, -3\sqrt{\frac{6}{13\sqrt{13}}}\right).$$

(a) In (x, y) notation, the x coordinate is $x = 0.716d$.

(b) Similarly, the y coordinate is $y = -1.07d$.

15. All the forces are being evaluated at the origin (since particle A is there), and all forces are along the location-vectors $\vec{r}$ which point to particles B, C and D. In three dimensions, the Pythagorean theorem becomes $r = \sqrt{x^2 + y^2 + z^2}$. The component along, say, the x axis of one of the force-vectors $\vec{F}$ is simply Fx/r in this situation (where F is the magnitude of $\vec{F}$). Since the force itself (see Eq. 13-1) is inversely proportional to r^2 then the aforementioned x component would have the form $GmMx/r^3$; similarly for the other components. For example, the z component of the force exerted on particle A by particle B is

$$\frac{Gm_A m_B z_B}{r_B^3} = \frac{Gm_A(2m_A)(2d)}{((2d)^2 + d^2 + (2d)^2)^3} = \frac{4Gm_A^2}{27\,d^2}.$$

In this way, each component can be written as some multiple of Gm_A^2/d^2. For the z component of the force exerted on particle A by particle C, that multiple is $-9\sqrt{14}/196$. For the x components of the forces exerted on particle A by particles B and C, those multiples are $4/27$ and $-3\sqrt{14}/196$, respectively. And for the y components of the forces exerted on particle A by particles B and C, those multiples are $2/27$ and $3\sqrt{14}/98$, respectively. To find the distance r to particle D one method is to solve (using the fact that the vector add to zero)

$$\left(\frac{Gm_A m_D}{r^2}\right)^2 = [(4/27 - 3\sqrt{14}/196)^2 + (2/27 + 3\sqrt{14}/98)^2 + (4/27 - 9\sqrt{14}/196)^2]\left(\frac{Gm_A^2}{d^2}\right)^2$$

(where $m_D = 4m_A$) for r. This gives $r = 4.357d$. The individual values of x, y and z (locating the particle D) can then be found by considering each component of the $Gm_A m_D/r^2$ force separately.

(a) The x component of $\vec{r}$ would be

$$Gm_A\, m_D\, x/r^3 = -(4/27 - 3\sqrt{14}/196)Gm_A^2/d^2,$$

which yields $x = -1.88d$.

(b) Similarly, $y = -3.90d$,

(c) and $z = 0.489d$.

In this way we are able to deduce that $(x, y, z) = (1.88d, 3.90d, 0.49d)$.

16. Since the rod is an extended object, we cannot apply Equation 13-1 directly to find the force. Instead, we consider a small differential element of the rod, of mass dm of thickness dr at a distance r from m_1. The gravitational force between dm and m_1 is

$$dF = \frac{Gm_1 dm}{r^2} = \frac{Gm_1(M/L)dr}{r^2},$$

where we have substituted $dm = (M/L)dr$ since mass is uniformly distributed. The direction of $d\vec{F}$ is to the right (see figure). The total force can be found by integrating over the entire length of the rod:

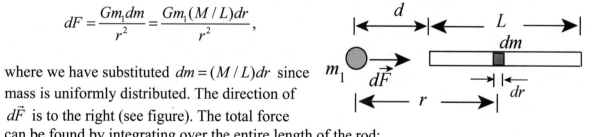

$$F = \int dF = \frac{Gm_1 M}{L}\int_d^{L+d}\frac{dr}{r^2} = -\frac{Gm_1 M}{L}\left(\frac{1}{L+d} - \frac{1}{d}\right) = \frac{Gm_1 M}{d(L+d)}.$$

Substituting the values given in the problem statement, we obtain

$$F = \frac{Gm_1M}{d(L+d)} = \frac{(6.67 \times 10^{-11} \text{ m}^3/\text{kg} \cdot \text{s}^2)(0.67 \text{ kg})(5.0 \text{ kg})}{(0.23 \text{ m})(3.0 \text{ m} + 0.23 \text{ m})} = 3.0 \times 10^{-10} \text{ N}.$$

17. The acceleration due to gravity is given by $a_g = GM/r^2$, where M is the mass of Earth and r is the distance from Earth's center. We substitute $r = R + h$, where R is the radius of Earth and h is the altitude, to obtain $a_g = GM/(R+h)^2$. We solve for h and obtain $h = \sqrt{GM/a_g} - R$. According to Appendix C, $R = 6.37 \times 10^6$ m and $M = 5.98 \times 10^{24}$ kg, so

$$h = \sqrt{\frac{(6.67 \times 10^{-11} \text{ m}^3/\text{s}^2 \cdot \text{kg})(5.98 \times 10^{24} \text{ kg})}{(4.9 \text{m/s}^2)}} - 6.37 \times 10^6 \text{ m} = 2.6 \times 10^6 \text{ m}.$$

18. We follow the method shown in Sample Problem 13-3. Thus,

$$a_g = \frac{GM_E}{r^2} \Rightarrow da_g = -2\frac{GM_E}{r^3} dr$$

which implies that the change in weight is

$$W_{\text{top}} - W_{\text{bottom}} \approx m(da_g).$$

But since $W_{\text{bottom}} = GmM_E/R^2$ (where R is Earth's mean radius), we have

$$mda_g = -2\frac{GmM_E}{R^3} dr = -2W_{\text{bottom}}\frac{dr}{R} = -2(600 \text{ N})\frac{1.61 \times 10^3 \text{ m}}{6.37 \times 10^6 \text{ m}} = -0.303 \text{ N}$$

for the weight change (the minus sign indicating that it is a decrease in W). We are not including any effects due to the Earth's rotation (as treated in Eq. 13-13).

19. (a) The gravitational acceleration at the surface of the Moon is $g_{\text{moon}} = 1.67$ m/s^2 (see Appendix C). The ratio of weights (for a given mass) is the ratio of g-values, so

$$W_{\text{moon}} = (100 \text{ N})(1.67/9.8) = 17 \text{ N}.$$

(b) For the force on that object caused by Earth's gravity to equal 17 N, then the free-fall acceleration at its location must be $a_g = 1.67$ m/s^2. Thus,

$$a_g = \frac{Gm_E}{r^2} \Rightarrow r = \sqrt{\frac{Gm_E}{a_g}} = 1.5 \times 10^7 \text{ m}$$

so the object would need to be a distance of $r/R_E = 2.4$ "radii" from Earth's center.

20. The free-body diagram of the force acting on the plumb line is shown on the right. The mass of the sphere is

$$M = \rho V = \rho\left(\frac{4\pi}{3}R^3\right) = \frac{4\pi}{3}(2.6\times 10^3 \text{ kg/m}^3)(2.00\times 10^3 \text{ m})^3$$
$$= 8.71\times 10^{13} \text{ kg.}$$

The force between the "spherical" mountain and the plumb line is $F = GMm/r^2$. Suppose at equilibrium the line makes an angle θ with the vertical and the net force acting on the line is zero. Therefore,

$$0 = \sum F_{\text{net}, x} = T\sin\theta - F = T\sin\theta - \frac{GMm}{r^2}$$

$$0 = \sum F_{\text{net}, y} = T\cos - mg$$

The two equations can be combined to give $\tan\theta = \dfrac{F}{mg} = \dfrac{GM}{gr^2}$. The distance the lower end moves toward the sphere is

$$x = l\tan\theta = l\frac{GM}{gr^2} = (0.50 \text{ m})\frac{(6.67\times 10^{-11} \text{ m}^3/\text{kg}\cdot\text{s}^2)(8.71\times 10^{13} \text{ kg})}{(9.8)(3\times 2.00\times 10^3 \text{ m})^2}$$
$$= 8.2\times 10^{-6} \text{ m.}$$

21. (a) The gravitational acceleration is $a_g = \dfrac{GM}{R^2} = 7.6 \text{ m/s}^2$.

(b) Note that the total mass is $5M$. Thus, $a_g = \dfrac{G(5M)}{(3R)^2} = 4.2 \text{ m/s}^2$.

22. (a) Plugging $R_h = 2GM_h/c^2$ into the indicated expression, we find

$$a_g = \frac{GM_h}{(1.001R_h)^2} = \frac{GM_h}{(1.001)^2(2GM_h/c^2)^2} = \frac{c^4}{(2.002)^2 G}\frac{1}{M_h}$$

which yields $a_g = (3.02\times 10^{43} \text{ kg}\cdot\text{m/s}^2)/M_h$.

(b) Since M_h is in the denominator of the above result, a_g decreases as M_h increases.

(c) With $M_h = (1.55\times 10^{12})(1.99\times 10^{30} \text{ kg})$, we obtain $a_g = 9.82 \text{ m/s}^2$.

(d) This part refers specifically to the very large black hole treated in the previous part. With that mass for M in Eq. 13–16, and $r = 2.002GM/c^2$, we obtain

$$da_g = -2\frac{GM}{(2.002GM/c^2)^3}dr = -\frac{2c^6}{(2.002)^3(GM)^2}dr$$

where $dr \to 1.70$ m as in Sample Problem 13-3. This yields (in absolute value) an acceleration difference of 7.30×10^{-15} m/s^2.

(e) The miniscule result of the previous part implies that, in this case, any effects due to the differences of gravitational forces on the body are negligible.

23. From Eq. 13-14, we see the extreme case is when "g" becomes zero, and plugging in Eq. 13-15 leads to

$$0 = \frac{GM}{R^2} - R\omega^2 \Rightarrow M = \frac{R^3\omega^2}{G}.$$

Thus, with $R = 20000$ m and $\omega = 2\pi$ rad/s, we find $M = 4.7 \times 10^{24}$ kg $\approx 5 \times 10^{24}$ kg.

24. (a) What contributes to the GmM/r^2 force on m is the (spherically distributed) mass M contained within r (where r is measured from the center of M). At point A we see that $M_1 + M_2$ is at a smaller radius than $r = a$ and thus contributes to the force:

$$|F_{on\,m}| = \frac{G(M_1 + M_2)m}{a^2}.$$

(b) In the case $r = b$, only M_1 is contained within that radius, so the force on m becomes GM_1m/b^2.

(c) If the particle is at C, then no other mass is at smaller radius and the gravitational force on it is zero.

25. (a) The magnitude of the force on a particle with mass m at the surface of Earth is given by $F = GMm/R^2$, where M is the total mass of Earth and R is Earth's radius. The acceleration due to gravity is

$$a_g = \frac{F}{m} = \frac{GM}{R^2} = \frac{(6.67\times 10^{-11}\text{ m}^3/\text{s}^2\cdot\text{kg})(5.98\times 10^{24}\text{ kg})}{(6.37\times 10^6\text{ m})^2} = 9.83\text{ m/s}^2.$$

(b) Now $a_g = GM/R^2$, where M is the total mass contained in the core and mantle together and R is the outer radius of the mantle (6.345×10^6 m, according to Fig. 13-43). The total mass is

$$M = (1.93 \times 10^{24} \text{ kg} + 4.01 \times 10^{24} \text{ kg}) = 5.94 \times 10^{24} \text{ kg}.$$

The first term is the mass of the core and the second is the mass of the mantle. Thus,

$$a_g = \frac{(6.67 \times 10^{-11} \text{ m}^3/\text{s}^2 \cdot \text{kg})(5.94 \times 10^{24} \text{ kg})}{(6.345 \times 10^6 \text{ m})^2} = 9.84 \text{ m/s}^2.$$

(c) A point 25 km below the surface is at the mantle-crust interface and is on the surface of a sphere with a radius of $R = 6.345 \times 10^6$ m. Since the mass is now assumed to be uniformly distributed the mass within this sphere can be found by multiplying the mass per unit volume by the volume of the sphere: $M = (R^3/R_e^3)M_e$, where M_e is the total mass of Earth and R_e is the radius of Earth. Thus,

$$M = \left(\frac{6.345 \times 10^6 \text{ m}}{6.37 \times 10^6 \text{ m}}\right)^3 (5.98 \times 10^{24} \text{ kg}) = 5.91 \times 10^{24} \text{ kg}.$$

The acceleration due to gravity is

$$a_g = \frac{GM}{R^2} = \frac{(6.67 \times 10^{-11} \text{ m}^3/\text{s}^2 \cdot \text{kg})(5.91 \times 10^{24} \text{ kg})}{(6.345 \times 10^6 \text{ m})^2} = 9.79 \text{ m/s}^2.$$

26. (a) Using Eq. 13-1, we set GmM/r^2 equal to $\frac{1}{2} GmM/R^2$, and we find $r = R\sqrt{2}$. Thus, the distance from the surface is $(\sqrt{2} - 1)R = 0.414R$.

(b) Setting the density ρ equal to M/V where $V = \frac{4}{3}\pi R^3$, we use Eq. 13-19:

$$F = \frac{4\pi Gmr\rho}{3} = \frac{4\pi Gmr}{3}\left(\frac{M}{4\pi R^3/3}\right) = \frac{GMmr}{R^3} = \frac{1}{2}\frac{GMm}{R^2} \Rightarrow r = R/2.$$

27. Using the fact that the volume of a sphere is $4\pi R^3/3$, we find the density of the sphere:

$$\rho = \frac{M_{total}}{\frac{4}{3}\pi R^3} = \frac{1.0 \times 10^4 \text{ kg}}{\frac{4}{3}\pi(1.0 \text{ m})^3} = 2.4 \times 10^3 \text{ kg/m}^3.$$

When the particle of mass m (upon which the sphere, or parts of it, are exerting a gravitational force) is at radius r (measured from the center of the sphere), then whatever mass M is at a radius less than r must contribute to the magnitude of that force (GMm/r^2).

(a) At $r = 1.5$ m, all of M_{total} is at a smaller radius and thus all contributes to the force:

$$\left|F_{\text{on } m}\right| = \frac{GmM_{\text{total}}}{r^2} = m\left(3.0\times 10^{-7}\,\text{N/kg}\right).$$

(b) At $r = 0.50$ m, the portion of the sphere at radius smaller than that is

$$M = \rho\left(\frac{4}{3}\pi r^3\right) = 1.3\times 10^3\,\text{kg}.$$

Thus, the force on m has magnitude $GMm/r^2 = m\,(3.3\times 10^{-7}\,\text{N/kg})$.

(c) Pursuing the calculation of part (b) algebraically, we find

$$\left|F_{\text{on } m}\right| = \frac{Gm\rho\left(\frac{4}{3}\pi r^3\right)}{r^2} = mr\left(6.7\times 10^{-7}\,\frac{\text{N}}{\text{kg}\cdot\text{m}}\right).$$

28. The difference between free-fall acceleration g and the gravitational acceleration a_g at the equator of the star is (see Equation 13.14):

$$a_g - g = \omega^2 R$$

where

$$\omega = \frac{2\pi}{T} = \frac{2\pi}{0.041\,\text{s}} = 153\,\text{rad/s}$$

is the angular speed of the star. The gravitational acceleration at the equator is

$$a_g = \frac{GM}{R^2} = \frac{(6.67\times 10^{-11}\,\text{m}^3/\text{kg}\cdot\text{s}^2)(1.98\times 10^{30}\,\text{kg})}{(1.2\times 10^4\,\text{m})^2} = 9.17\times 10^{11}\,\text{m/s}^2.$$

Therefore, the percentage difference is

$$\frac{a_g - g}{a_g} = \frac{\omega^2 R}{a_g} = \frac{(153\,\text{rad/s})^2(1.2\times 10^4\,\text{m})}{9.17\times 10^{11}\,\text{m/s}^2} = 3.06\times 10^{-4} \approx 0.031\%.$$

29. (a) The density of a uniform sphere is given by $\rho = 3M/4\pi R^3$, where M is its mass and R is its radius. The ratio of the density of Mars to the density of Earth is

$$\frac{\rho_M}{\rho_E} = \frac{M_M}{M_E}\frac{R_E^3}{R_M^3} = 0.11\left(\frac{0.65\times 10^4\,\text{km}}{3.45\times 10^3\,\text{km}}\right)^3 = 0.74.$$

(b) The value of a_g at the surface of a planet is given by $a_g = GM/R^2$, so the value for Mars is

$$a_g M = \frac{M_M}{M_E} \frac{R_E^2}{R_M^2} a_{gE} = 0.11 \left(\frac{0.65 \times 10^4 \text{ km}}{3.45 \times 10^3 \text{ km}} \right)^2 (9.8 \text{ m/s}^2) = 3.8 \text{ m/s}^2.$$

(c) If v is the escape speed, then, for a particle of mass m

$$\frac{1}{2} mv^2 = G \frac{mM}{R} \Rightarrow v = \sqrt{\frac{2GM}{R}}.$$

For Mars, the escape speed is

$$v = \sqrt{\frac{2(6.67 \times 10^{-11} \text{ m}^3/\text{s}^2 \cdot \text{kg})(0.11)(5.98 \times 10^{24} \text{ kg})}{3.45 \times 10^6 \text{ m}}} = 5.0 \times 10^3 \text{ m/s}.$$

30. (a) The gravitational potential energy is

$$U = -\frac{GMm}{r} = -\frac{(6.67 \times 10^{-11} \text{ m}^3/\text{s}^2 \cdot \text{kg})(5.2 \text{ kg})(2.4 \text{ kg})}{19 \text{ m}} = -4.4 \times 10^{-11} \text{ J}.$$

(b) Since the change in potential energy is

$$\Delta U = -\frac{GMm}{3r} - \left(-\frac{GMm}{r} \right) = -\frac{2}{3} (-4.4 \times 10^{-11} \text{ J}) = 2.9 \times 10^{-11} \text{ J},$$

the work done by the gravitational force is $W = -\Delta U = -2.9 \times 10^{-11}$ J.

(c) The work done by you is $W' = \Delta U = 2.9 \times 10^{-11}$ J.

31. The amount of (kinetic) energy needed to escape is the same as the (absolute value of the) gravitational potential energy at its original position. Thus, an object of mass m on a planet of mass M and radius R needs $K = GmM/R$ in order to (barely) escape.
(a) Setting up the ratio, we find

$$\frac{K_m}{K_E} = \frac{M_m}{M_E} \frac{R_E}{R_m} = 0.0451$$

using the values found in Appendix C.

(b) Similarly, for the Jupiter escape energy (divided by that for Earth) we obtain

$$\frac{K_J}{K_E} = \frac{M_J}{M_E}\frac{R_E}{R_J} = 28.5.$$

32. (a) The potential energy at the surface is (according to the graph) -5.0×10^9 J, so (since U is inversely proportional to r – see Eq. 13-21) at an r-value a factor of 5/4 times what it was at the surface then U must be a factor of 4/5 what it was. Thus, at $r = 1.25R_s$ $U = -4.0 \times 10^9$ J. Since mechanical energy is assumed to be conserved in this problem, we have $K + U = -2.0 \times 10^9$ J at this point. Since $U = -4.0 \times 10^9$ J here, then $K = 2.0 \times 10^9$ J at this point.

(b) To reach the point where the mechanical energy equals the potential energy (that is, where $U = -2.0 \times 10^9$ J) means that U must reduce (from its value at $r = 1.25R_s$) by a factor of 2 – which means the r value must increase (relative to $r = 1.25R_s$) by a corresponding factor of 2. Thus, the turning point must be at $r = 2.5R_s$.

33. The equation immediately preceding Eq. 13-28 shows that $K = -U$ (with U evaluated at the planet's surface: -5.0×10^9 J) is required to "escape." Thus, $K = 5.0 \times 10^9$ J.

34. The gravitational potential energy is

$$U = -\frac{Gm(M-m)}{r} = -\frac{G}{r}(Mm - m^2)$$

which we differentiate with respect to m and set equal to zero (in order to minimize). Thus, we find $M - 2m = 0$ which leads to the ratio $m/M = 1/2$ to obtain the least potential energy.

Note that a second derivative of U with respect to m would lead to a positive result regardless of the value of m – which means its graph is everywhere concave upward and thus its extremum is indeed a minimum.

35. (a) The work done by you in moving the sphere of mass m_B equals the change in the potential energy of the three-sphere system. The initial potential energy is

$$U_i = -\frac{Gm_A m_B}{d} - \frac{Gm_A m_C}{L} - \frac{Gm_B m_C}{L-d}$$

and the final potential energy is

$$U_f = -\frac{Gm_A m_B}{L-d} - \frac{Gm_A m_C}{L} - \frac{Gm_B m_C}{d}.$$

The work done is

$$W = U_f - U_i = Gm_B \left[m_A \left(\frac{1}{d} - \frac{1}{L-d} \right) + m_C \left(\frac{1}{L-d} - \frac{1}{d} \right) \right]$$

$$= Gm_B \left[m_A \frac{L-2d}{d(L-d)} + m_C \frac{2d-L}{d(L-d)} \right] = Gm_B (m_A - m_C) \frac{L-2d}{d(L-d)}$$

$$= (6.67 \times 10^{-11} \text{ m}^3/\text{s}^2 \cdot \text{kg})(0.010 \text{ kg})(0.080 \text{ kg} - 0.020 \text{ kg}) \frac{0.12 \text{ m} - 2(0.040 \text{ m})}{(0.040 \text{ m})(0.12 - 0.040 \text{ m})}$$

$$= +5.0 \times 10^{-13} \text{ J}.$$

(b) The work done by the force of gravity is $-(U_f - U_i) = -5.0 \times 10^{-13}$ J.

36. (a) From Eq. 13-28, we see that $v_0 = \sqrt{GM/2R_E}$ in this problem. Using energy conservation, we have

$$\frac{1}{2} mv_0^2 - GMm/R_E = -GMm/r$$

which yields $r = 4R_E/3$. So the multiple of R_E is 4/3 or 1.33.

(b) Using the equation in the textbook immediately preceding Eq. 13-28, we see that in this problem we have $K_i = GMm/2R_E$, and the above manipulation (using energy conservation) in this case leads to $r = 2R_E$. So the multiple of R_E is 2.00.

(c) Again referring to the equation in the textbook immediately preceding Eq. 13-28, we see that the mechanical energy = 0 for the "escape condition."

37. (a) We use the principle of conservation of energy. Initially the particle is at the surface of the asteroid and has potential energy $U_i = -GMm/R$, where M is the mass of the asteroid, R is its radius, and m is the mass of the particle being fired upward. The initial kinetic energy is $\frac{1}{2}mv^2$. The particle just escapes if its kinetic energy is zero when it is infinitely far from the asteroid. The final potential and kinetic energies are both zero. Conservation of energy yields

$$-GMm/R + \tfrac{1}{2}mv^2 = 0.$$

We replace GM/R with $a_g R$, where a_g is the acceleration due to gravity at the surface. Then, the energy equation becomes $-a_g R + \tfrac{1}{2}v^2 = 0$. We solve for v:

$$v = \sqrt{2a_g R} = \sqrt{2(3.0 \text{ m/s}^2)(500 \times 10^3 \text{ m})} = 1.7 \times 10^3 \text{ m/s}.$$

(b) Initially the particle is at the surface; the potential energy is $U_i = -GMm/R$ and the kinetic energy is $K_i = \tfrac{1}{2}mv^2$. Suppose the particle is a distance h above the surface when it momentarily comes to rest. The final potential energy is $U_f = -GMm/(R+h)$ and the final kinetic energy is $K_f = 0$. Conservation of energy yields

$$-\frac{GMm}{R} + \frac{1}{2}mv^2 = -\frac{GMm}{R+h}.$$

We replace GM with $a_g R^2$ and cancel m in the energy equation to obtain

$$-a_g R + \frac{1}{2}v^2 = -\frac{a_g R^2}{(R+h)}.$$

The solution for h is

$$h = \frac{2a_g R^2}{2a_g R - v^2} - R = \frac{2(3.0 \text{ m/s}^2)(500 \times 10^3 \text{ m})^2}{2(3.0 \text{ m/s}^2)(500 \times 10^3 \text{ m}) - (1000 \text{ m/s})^2} - (500 \times 10^3 \text{ m})$$
$$= 2.5 \times 10^5 \text{ m}.$$

(c) Initially the particle is a distance h above the surface and is at rest. Its potential energy is $U_i = -GMm/(R+h)$ and its initial kinetic energy is $K_i = 0$. Just before it hits the asteroid its potential energy is $U_f = -GMm/R$. Write $\frac{1}{2}mv_f^2$ for the final kinetic energy. Conservation of energy yields

$$-\frac{GMm}{R+h} = -\frac{GMm}{R} + \frac{1}{2}mv^2.$$

We substitute $a_g R^2$ for GM and cancel m, obtaining

$$-\frac{a_g R^2}{R+h} = -a_g R + \frac{1}{2}v^2.$$

The solution for v is

$$v = \sqrt{2a_g R - \frac{2a_g R^2}{R+h}} = \sqrt{2(3.0 \text{ m/s}^2)(500 \times 10^3 \text{ m}) - \frac{2(3.0 \text{ m/s}^2)(500 \times 10^3 \text{ m})^2}{(500 \times 10^3 \text{ m}) + (1000 \times 10^3 \text{ m})}}$$
$$= 1.4 \times 10^3 \text{ m/s}.$$

38. Energy conservation for this situation may be expressed as follows:

$$K_1 + U_1 = K_2 + U_2 \Rightarrow K_1 - \frac{GmM}{r_1} = K_2 - \frac{GmM}{r_2}.$$

where $M = 5.0 \times 10^{23}$ kg, $r_1 = R = 3.0 \times 10^6$ m and $m = 10$ kg.

(a) If $K_1 = 5.0 \times 10^7$ J and $r_2 = 4.0 \times 10^6$ m, then the above equation leads to

$$K_2 = K_1 + GmM\left(\frac{1}{r_2} - \frac{1}{r_1}\right) = 2.2 \times 10^7 \text{ J}.$$

(b) In this case, we require $K_2 = 0$ and $r_2 = 8.0 \times 10^6$ m, and solve for K_1:

$$K_1 = K_2 + GmM\left(\frac{1}{r_1} - \frac{1}{r_2}\right) = 6.9 \times 10^7 \text{ J}.$$

39. (a) The momentum of the two-star system is conserved, and since the stars have the same mass, their speeds and kinetic energies are the same. We use the principle of conservation of energy. The initial potential energy is $U_i = -GM^2/r_i$, where M is the mass of either star and r_i is their initial center-to-center separation. The initial kinetic energy is zero since the stars are at rest. The final potential energy is $U_f = -2GM^2/r_i$ since the final separation is $r_i/2$. We write Mv^2 for the final kinetic energy of the system. This is the sum of two terms, each of which is $\frac{1}{2}Mv^2$. Conservation of energy yields

$$-\frac{GM^2}{r_i} = -\frac{2GM^2}{r_i} + Mv^2.$$

The solution for v is

$$v = \sqrt{\frac{GM}{r_i}} = \sqrt{\frac{(6.67 \times 10^{-11} \text{ m}^3/\text{s}^2 \cdot \text{kg})(10^{30} \text{ kg})}{10^{10} \text{ m}}} = 8.2 \times 10^4 \text{ m/s}.$$

(b) Now the final separation of the centers is $r_f = 2R = 2 \times 10^5$ m, where R is the radius of either of the stars. The final potential energy is given by $U_f = -GM^2/r_f$ and the energy equation becomes $-GM^2/r_i = -GM^2/r_f + Mv^2$. The solution for v is

$$v = \sqrt{GM\left(\frac{1}{r_f} - \frac{1}{r_i}\right)} = \sqrt{(6.67 \times 10^{-11} \text{ m}^3/\text{s}^2 \cdot \text{kg})(10^{30} \text{ kg})\left(\frac{1}{2 \times 10^5 \text{ m}} - \frac{1}{10^{10} \text{ m}}\right)}$$
$$= 1.8 \times 10^7 \text{ m/s}.$$

40. (a) The initial gravitational potential energy is

$$U_i = -\frac{GM_A M_B}{r_i} = -\frac{(6.67 \times 10^{-11} \text{ m}^3/\text{s}^2 \cdot \text{kg})(20 \text{ kg})(10 \text{ kg})}{0.80 \text{ m}}$$
$$= -1.67 \times 10^{-8} \text{ J} \approx -1.7 \times 10^{-8} \text{ J}.$$

(b) We use conservation of energy (with $K_i = 0$):

$$U_i = K + U \Rightarrow -1.7 \times 10^{-8} = K - \frac{(6.67 \times 10^{-11} \text{ m}^3/\text{s}^2 \cdot \text{kg})(20 \text{ kg})(10 \text{ kg})}{0.60 \text{ m}}$$

which yields $K = 5.6 \times 10^{-9}$ J. Note that the value of r is the difference between 0.80 m and 0.20 m.

41. Let $m = 0.020$ kg and $d = 0.600$ m (the original edge-length, in terms of which the final edge-length is $d/3$). The total initial gravitational potential energy (using Eq. 13-21 and some elementary trigonometry) is

$$U_i = -\frac{4Gm^2}{d} - \frac{2Gm^2}{\sqrt{2}\, d} .$$

Since U is inversely proportional to r then reducing the size by $1/3$ means increasing the magnitude of the potential energy by a factor of 3, so

$$U_f = 3U_i \Rightarrow \Delta U = 2U_i = 2(4 + \sqrt{2})\left(-\frac{Gm^2}{d}\right) = -4.82 \times 10^{-13} \text{ J} .$$

42. (a) Applying Eq. 13-21 and the Pythagorean theorem leads to

$$U = -\left(\frac{GM^2}{2D} + \frac{2GmM}{\sqrt{y^2 + D^2}}\right)$$

where M is the mass of particle B (also that of particle C) and m is the mass of particle A. The value given in the problem statement (for infinitely large y, for which the second term above vanishes) determines M, since D is given. Thus $M = 0.50$ kg.

(b) We estimate (from the graph) the $y = 0$ value to be $U_o = -3.5 \times 10^{-10}$ J. Using this, our expression above determines m. We obtain $m = 1.5$ kg.

43. The period T and orbit radius r are related by the law of periods: $T^2 = (4\pi^2/GM)r^3$, where M is the mass of Mars. The period is 7 h 39 min, which is 2.754×10^4 s. We solve for M:

$$M = \frac{4\pi^2 r^3}{GT^2} = \frac{4\pi^2 (9.4 \times 10^6 \text{ m})^3}{(6.67 \times 10^{-11} \text{m}^3/\text{s}^2 \cdot \text{kg})(2.754 \times 10^4 \text{ s})^2} = 6.5 \times 10^{23} \text{ kg}.$$

44. From Eq. 13-37, we obtain $v = \sqrt{GM/r}$ for the speed of an object in circular orbit (of radius r) around a planet of mass M. In this case, $M = 5.98 \times 10^{24}$ kg and

$$r = (700 + 6370)\text{m} = 7070 \text{ km} = 7.07 \times 10^6 \text{ m}.$$

The speed is found to be $v = 7.51 \times 10^3$ m/s. After multiplying by 3600 s/h and dividing by 1000 m/km this becomes $v = 2.7 \times 10^4$ km/h.

(a) For a head-on collision, the relative speed of the two objects must be $2v = 5.4 \times 10^4$ km/h.

(b) A perpendicular collision is possible if one satellite is, say, orbiting above the equator and the other is following a longitudinal line. In this case, the relative speed is given by the Pythagorean theorem: $\sqrt{v^2 + v^2} = 3.8 \times 10^4$ km/h.

45. Let N be the number of stars in the galaxy, M be the mass of the Sun, and r be the radius of the galaxy. The total mass in the galaxy is NM and the magnitude of the gravitational force acting on the Sun is $F = GNM^2/r^2$. The force points toward the galactic center. The magnitude of the Sun's acceleration is $a = v^2/R$, where v is its speed. If T is the period of the Sun's motion around the galactic center then $v = 2\pi R/T$ and $a = 4\pi^2 R/T^2$. Newton's second law yields $GNM^2/R^2 = 4\pi^2 MR/T^2$. The solution for N is

$$N = \frac{4\pi^2 R^3}{GT^2 M}.$$

The period is 2.5×10^8 y, which is 7.88×10^{15} s, so

$$N = \frac{4\pi^2 (2.2 \times 10^{20} \text{ m})^3}{(6.67 \times 10^{-11} \text{ m}^3/\text{s}^2 \cdot \text{kg})(7.88 \times 10^{15} \text{ s})^2 (2.0 \times 10^{30} \text{ kg})} = 5.1 \times 10^{10}.$$

46. Kepler's law of periods, expressed as a ratio, is

$$\left(\frac{a_M}{a_E}\right)^3 = \left(\frac{T_M}{T_E}\right)^2 \Rightarrow (1.52)^3 = \left(\frac{T_M}{1\text{y}}\right)^2$$

where we have substituted the mean-distance (from Sun) ratio for the semi-major axis ratio. This yields $T_M = 1.87$ y. The value in Appendix C (1.88 y) is quite close, and the small apparent discrepancy is not significant, since a more precise value for the semi-major axis ratio is $a_M/a_E = 1.523$ which does lead to $T_M = 1.88$ y using Kepler's law. A question can be raised regarding the use of a ratio of mean distances for the ratio of semi-major axes, but this requires a more lengthy discussion of what is meant by a "mean distance" than is appropriate here.

47. (a) The greatest distance between the satellite and Earth's center (the apogee distance) and the least distance (perigee distance) are, respectively,

$$R_a = (6.37 \times 10^6 \text{ m} + 360 \times 10^3 \text{ m}) = 6.73 \times 10^6 \text{ m}$$
$$R_p = (6.37 \times 10^6 \text{ m} + 180 \times 10^3 \text{ m}) = 6.55 \times 10^6 \text{ m}.$$

Here 6.37×10^6 m is the radius of Earth. From Fig. 13-13, we see that the semi-major axis is

$$a = \frac{R_a + R_p}{2} = \frac{6.73 \times 10^6 \text{ m} + 6.55 \times 10^6 \text{ m}}{2} = 6.64 \times 10^6 \text{ m}.$$

(b) The apogee and perigee distances are related to the eccentricity e by $R_a = a(1 + e)$ and $R_p = a(1 - e)$. Add to obtain $R_a + R_p = 2a$ and $a = (R_a + R_p)/2$. Subtract to obtain $R_a - R_p = 2ae$. Thus,

$$e = \frac{R_a - R_p}{2a} = \frac{R_a - R_p}{R_a + R_p} = \frac{6.73 \times 10^6 \text{ m} - 6.55 \times 10^6 \text{ m}}{6.73 \times 10^6 \text{ m} + 6.55 \times 10^6 \text{ m}} = 0.0136.$$

48. Kepler's law of periods, expressed as a ratio, is

$$\left(\frac{r_s}{r_m}\right)^3 = \left(\frac{T_s}{T_m}\right)^2 \Rightarrow \left(\frac{1}{2}\right)^3 = \left(\frac{T_s}{1 \text{ lunar month}}\right)^2$$

which yields $T_s = 0.35$ lunar month for the period of the satellite.

49. (a) If r is the radius of the orbit then the magnitude of the gravitational force acting on the satellite is given by GMm/r^2, where M is the mass of Earth and m is the mass of the satellite. The magnitude of the acceleration of the satellite is given by v^2/r, where v is its speed. Newton's second law yields $GMm/r^2 = mv^2/r$. Since the radius of Earth is 6.37×10^6 m the orbit radius is $r = (6.37 \times 10^6 \text{ m} + 160 \times 10^3 \text{ m}) = 6.53 \times 10^6$ m. The solution for v is

$$v = \sqrt{\frac{GM}{r}} = \sqrt{\frac{(6.67 \times 10^{-11} \text{ m}^3/\text{s}^2 \cdot \text{kg})(5.98 \times 10^{24} \text{ kg})}{6.53 \times 10^6 \text{ m}}} = 7.82 \times 10^3 \text{ m/s}.$$

(b) Since the circumference of the circular orbit is $2\pi r$, the period is

$$T = \frac{2\pi r}{v} = \frac{2\pi(6.53 \times 10^6 \text{ m})}{7.82 \times 10^3 \text{ m/s}} = 5.25 \times 10^3 \text{ s}.$$

This is equivalent to 87.5 min.

50. (a) The distance from the center of an ellipse to a focus is ae where a is the semimajor axis and e is the eccentricity. Thus, the separation of the foci (in the case of Earth's orbit) is

$$2ae = 2(1.50 \times 10^{11} \text{ m})(0.0167) = 5.01 \times 10^9 \text{ m}.$$

(b) To express this in terms of solar radii (see Appendix C), we set up a ratio:

$$\frac{5.01 \times 10^9 \text{ m}}{6.96 \times 10^8 \text{ m}} = 7.20.$$

51. (a) The period of the comet is 1420 years (and one month), which we convert to $T = 4.48 \times 10^{10}$ s. Since the mass of the Sun is 1.99×10^{30} kg, then Kepler's law of periods gives

$$(4.48 \times 10^{10} \text{ s})^2 = \left(\frac{4\pi^2}{(6.67 \times 10^{-11} \text{ m}^3/\text{kg} \cdot \text{s}^2)(1.99 \times 10^{30} \text{ kg})}\right) a^3 \Rightarrow a = 1.89 \times 10^{13} \text{ m}.$$

(b) Since the distance from the focus (of an ellipse) to its center is ea and the distance from center to the aphelion is a, then the comet is at a distance of

$$ea + a = (0.11 + 1)(1.89 \times 10^{13} \text{ m}) = 2.1 \times 10^{13} \text{ m}$$

when it is farthest from the Sun. To express this in terms of Pluto's orbital radius (found in Appendix C), we set up a ratio:

$$\left(\frac{2.1 \times 10^{13}}{5.9 \times 10^{12}}\right) R_P = 3.6 R_P.$$

52. To "hover" above Earth ($M_E = 5.98 \times 10^{24}$ kg) means that it has a period of 24 hours (86400 s). By Kepler's law of periods,

$$(86400)^2 = \left(\frac{4\pi^2}{GM_E}\right) r^3 \Rightarrow r = 4.225 \times 10^7 \text{ m}.$$

Its altitude is therefore $r - R_E$ (where $R_E = 6.37 \times 10^6$ m) which yields 3.58×10^7 m.

53. (a) If we take the logarithm of Kepler's law of periods, we obtain

$$2 \log(T) = \log(4\pi^2/GM) + 3 \log(a) \Rightarrow \log(a) = \frac{2}{3} \log(T) - \frac{1}{3} \log(4\pi^2/GM)$$

where we are ignoring an important subtlety about units (the arguments of logarithms cannot have units, since they are transcendental functions). Although the problem can be continued in this way, we prefer to set it up without units, which requires taking a ratio. If we divide Kepler's law (applied to the Jupiter-moon system, where M is mass of Jupiter) by the law applied to Earth orbiting the Sun (of mass M_o), we obtain

$$(T/T_E)^2 = \left(\frac{M_o}{M}\right)\left(\frac{a}{r_E}\right)^3$$

where $T_E = 365.25$ days is Earth's orbital period and $r_E = 1.50 \times 10^{11}$ m is its mean distance from the Sun. In this case, it is perfectly legitimate to take logarithms and obtain

$$\log\left(\frac{r_E}{a}\right) = \frac{2}{3}\log\left(\frac{T_E}{T}\right) + \frac{1}{3}\log\left(\frac{M_o}{M}\right)$$

(written to make each term positive) which is the way we plot the data (log (r_E/a) on the vertical axis and log (T_E/T) on the horizontal axis).

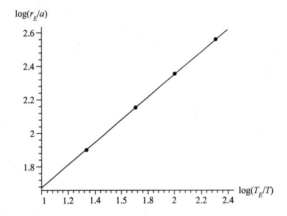

(b) When we perform a least-squares fit to the data, we obtain

$$\log\,(r_E/a) = 0.666 \log\,(T_E/T) + 1.01,$$

which confirms the expectation of slope = 2/3 based on the above equation.

(c) And the 1.01 intercept corresponds to the term $1/3 \log (M_o/M)$ which implies

$$\frac{M_o}{M} = 10^{3.03} \Rightarrow M = \frac{M_o}{1.07 \times 10^3}.$$

Plugging in $M_o = 1.99 \times 10^{30}$ kg (see Appendix C), we obtain $M = 1.86 \times 10^{27}$ kg for Jupiter's mass. This is reasonably consistent with the value 1.90×10^{27} kg found in Appendix C.

54. (a) The period is $T = 27(3600) = 97200$ s, and we are asked to assume that the orbit is circular (of radius $r = 100000$ m). Kepler's law of periods provides us with an approximation to the asteroid's mass:

$$(97200)^2 = \left(\frac{4\pi^2}{GM}\right)(100000)^3 \Rightarrow M = 6.3 \times 10^{16} \text{ kg}.$$

(b) Dividing the mass M by the given volume yields an average density equal to

$$\rho = 6.3 \times 10^{16} / 1.41 \times 10^{13} = 4.4 \times 10^3 \text{ kg/m}^3,$$

which is about 20% less dense than Earth.

55. In our system, we have $m_1 = m_2 = M$ (the mass of our Sun, 1.99×10^{30} kg). With $r = 2r_1$ in this system (so r_1 is one-half the Earth-to-Sun distance r), and $v = \pi r/T$ for the speed, we have

$$\frac{Gm_1 m_2}{r^2} = m_1 \frac{(\pi r/T)^2}{r/2} \Rightarrow T = \sqrt{\frac{2\pi^2 r^3}{GM}}.$$

With $r = 1.5 \times 10^{11}$ m, we obtain $T = 2.2 \times 10^7$ s. We can express this in terms of Earth-years, by setting up a ratio:

$$T = \left(\frac{T}{1 \text{ y}}\right)(1 \text{ y}) = \left(\frac{2.2 \times 10^7 \text{ s}}{3.156 \times 10^7 \text{ s}}\right)(1 \text{ y}) = 0.71 \text{ y}.$$

56. The two stars are in circular orbits, not about each other, but about the two-star system's center of mass (denoted as O), which lies along the line connecting the centers of the two stars. The gravitational force between the stars provides the centripetal force necessary to keep their orbits circular. Thus, for the visible, Newton's second law gives

$$F = \frac{Gm_1 m_2}{r^2} = \frac{m_1 v^2}{r_1}$$

where r is the distance between the centers of the stars. To find the relation between r and r_1, we locate the center of mass relative to m_1. Using Equation 9-1, we obtain

$$r_1 = \frac{m_1(0) + m_2 r}{m_1 + m_2} = \frac{m_2 r}{m_1 + m_2} \Rightarrow r = \frac{m_1 + m_2}{m_2} r_1.$$

On the other hand, since the orbital speed of m_1 is $v = 2\pi r_1/T$, then $r_1 = vT/2\pi$ and the expression for r can be rewritten as

$$r = \frac{m_1 + m_2}{m_2} \frac{vT}{2\pi}.$$

Substituting r and r_1 into the force equation, we obtain

$$F = \frac{4\pi^2 G m_1 m_2^3}{(m_1 + m_2)^2 v^2 T^2} = \frac{2\pi m_1 v}{T}$$

or

$$\frac{m_2^3}{(m_1+m_2)^2} = \frac{v^3 T}{2\pi G} = \frac{(2.7\times 10^5 \text{ m/s})^3 (1.70 \text{ days})(86400 \text{ s/day})}{2\pi (6.67\times 10^{-11} \text{ m}^3/\text{kg}\cdot\text{s}^2)} = 6.90\times 10^{30} \text{ kg}$$
$$= 3.467 M_s,$$

where $M_s = 1.99\times 10^{30}$ kg is the mass of the sun. With $m_1 = 6M_s$, we write $m_2 = \alpha M_s$ and solve the following cubic equation for α:

$$\frac{\alpha^3}{(6+\alpha)^2} - 3.467 = 0.$$

The equation has one real solution: $\alpha = 9.3$, which implies $m_2 / M_s \approx 9$.

57. From Kepler's law of periods (where $T = 2.4(3600) = 8640$ s), we find the planet's mass M:

$$(8640 \text{ s})^2 = \left(\frac{4\pi^2}{GM}\right)(8.0\times 10^6 \text{ m})^3 \Rightarrow M = 4.06\times 10^{24} \text{ kg}.$$

But we also know $a_g = GM/R^2 = 8.0$ m/s² so that we are able to solve for the planet's radius:

$$R = \sqrt{\frac{GM}{a_g}} = 5.8\times 10^6 \text{ m}.$$

58. (a) We make use of

$$\frac{m_2^3}{(m_1+m_2)^2} = \frac{v^3 T}{2\pi G}$$

where $m_1 = 0.9 M_{\text{Sun}}$ is the estimated mass of the star. With $v = 70$ m/s and $T = 1500$ days (or $1500 \times 86400 = 1.3 \times 10^8$ s), we find

$$\frac{m_2^3}{(0.9 M_{\text{Sun}} + m_2)^2} = 1.06 \times 10^{23} \text{ kg}.$$

Since $M_{\text{Sun}} \approx 2.0 \times 10^{30}$ kg, we find $m_2 \approx 7.0 \times 10^{27}$ kg. Dividing by the mass of Jupiter (see Appendix C), we obtain $m \approx 3.7 m_J$.

(b) Since $v = 2\pi r_1/T$ is the speed of the star, we find

$$r_1 = \frac{vT}{2\pi} = \frac{(70 \text{ m/s})(1.3\times 10^8 \text{ s})}{2\pi} = 1.4\times 10^9 \text{ m}$$

for the star's orbital radius. If r is the distance between the star and the planet, then $r_2 = r - r_1$ is the orbital radius of the planet, and is given by

$$r_2 = r_1\left(\frac{m_1+m_2}{m_2}-1\right) = r_1\frac{m_1}{m_2} = 3.7\times 10^{11} \text{ m}.$$

Dividing this by 1.5×10^{11} m (Earth's orbital radius, r_E) gives $r_2 = 2.5 r_E$.

59. Each star is attracted toward each of the other two by a force of magnitude GM^2/L^2, along the line that joins the stars. The net force on each star has magnitude $2(GM^2/L^2)\cos 30°$ and is directed toward the center of the triangle. This is a centripetal force and keeps the stars on the same circular orbit if their speeds are appropriate. If R is the radius of the orbit, Newton's second law yields $(GM^2/L^2)\cos 30° = Mv^2/R$.

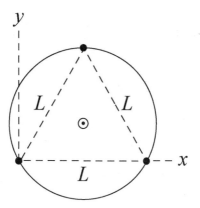

The stars rotate about their center of mass (marked by a circled dot on the diagram above) at the intersection of the perpendicular bisectors of the triangle sides, and the radius of the orbit is the distance from a star to the center of mass of the three-star system. We take the coordinate system to be as shown in the diagram, with its origin at the left-most star. The altitude of an equilateral triangle is $\left(\sqrt{3}/2\right)L$, so the stars are located at $x = 0$, $y = 0$; $x = L$, $y = 0$; and $x = L/2$, $y = \sqrt{3}L/2$. The x coordinate of the center of mass is $x_c = (L + L/2)/3 = L/2$ and the y coordinate is $y_c = \left(\sqrt{3}L/2\right)/3 = L/2\sqrt{3}$. The distance from a star to the center of mass is

$$R = \sqrt{x_c^2 + y_c^2} = \sqrt{(L^2/4)+(L^2/12)} = L/\sqrt{3}.$$

Once the substitution for R is made Newton's second law becomes $\left(2GM^2/L^2\right)\cos 30° = \sqrt{3}Mv^2/L$. This can be simplified somewhat by recognizing that $\cos 30° = \sqrt{3}/2$, and we divide the equation by M. Then, $GM/L^2 = v^2/L$ and $v = \sqrt{GM/L}$.

60. Although altitudes are given, it is the orbital radii which enter the equations. Thus, $r_A = (6370 + 6370)$ km $= 12740$ km, and $r_B = (19110 + 6370)$ km $= 25480$ km

(a) The ratio of potential energies is

$$\frac{U_B}{U_A} = \frac{-GmM/r_B}{-GmM/r_A} = \frac{r_A}{r_B} = \frac{1}{2}.$$

(b) Using Eq. 13-38, the ratio of kinetic energies is

$$\frac{K_B}{K_A} = \frac{GmM/2r_B}{GmM/2r_A} = \frac{r_A}{r_B} = \frac{1}{2}.$$

(c) From Eq. 13-40, it is clear that the satellite with the largest value of r has the smallest value of $|E|$ (since r is in the denominator). And since the values of E are negative, then the smallest value of $|E|$ corresponds to the largest energy E. Thus, satellite B has the largest energy.

(d) The difference is

$$\Delta E = E_B - E_A = -\frac{GmM}{2}\left(\frac{1}{r_B} - \frac{1}{r_A}\right).$$

Being careful to convert the r values to meters, we obtain $\Delta E = 1.1 \times 10^8$ J. The mass M of Earth is found in Appendix C.

61. (a) We use the law of periods: $T^2 = (4\pi^2/GM)r^3$, where M is the mass of the Sun (1.99×10^{30} kg) and r is the radius of the orbit. The radius of the orbit is twice the radius of Earth's orbit: $r = 2r_e = 2(150 \times 10^9 \text{ m}) = 300 \times 10^9$ m. Thus,

$$T = \sqrt{\frac{4\pi^2 r^3}{GM}} = \sqrt{\frac{4\pi^2 (300 \times 10^9 \text{ m})^3}{(6.67 \times 10^{-11} \text{ m}^3/\text{s}^2 \cdot \text{kg})(1.99 \times 10^{30} \text{kg})}} = 8.96 \times 10^7 \text{ s}.$$

Dividing by (365 d/y) (24 h/d) (60 min/h) (60 s/min), we obtain $T = 2.8$ y.

(b) The kinetic energy of any asteroid or planet in a circular orbit of radius r is given by $K = GMm/2r$, where m is the mass of the asteroid or planet. We note that it is proportional to m and inversely proportional to r. The ratio of the kinetic energy of the asteroid to the kinetic energy of Earth is $K/K_e = (m/m_e)(r_e/r)$. We substitute $m = 2.0 \times 10^{-4} m_e$ and $r = 2r_e$ to obtain $K/K_e = 1.0 \times 10^{-4}$.

62. (a) Circular motion requires that the force in Newton's second law provide the necessary centripetal acceleration:

$$\frac{GmM}{r^2} = m\frac{v^2}{r}.$$

Since the left-hand side of this equation is the force given as 80 N, then we can solve for the combination mv^2 by multiplying both sides by $r = 2.0 \times 10^7$ m. Thus, $mv^2 = (2.0 \times 10^7 \text{ m})(80 \text{ N}) = 1.6 \times 10^9$ J. Therefore,

$$K = \frac{1}{2}mv^2 = \frac{1}{2}\left(1.6 \times 10^9 \text{ J}\right) = 8.0 \times 10^8 \text{ J}.$$

(b) Since the gravitational force is inversely proportional to the square of the radius, then

$$\frac{F'}{F} = \left(\frac{r}{r'}\right)^2.$$

Thus, $F' = (80 \text{ N})(2/3)^2 = 36$ N.

63. The energy required to raise a satellite of mass m to an altitude h (at rest) is given by

$$E_1 = \Delta U = GM_E m \left(\frac{1}{R_E} - \frac{1}{R_E + h}\right),$$

and the energy required to put it in circular orbit once it is there is

$$E_2 = \frac{1}{2}mv_{\text{orb}}^2 = \frac{GM_E m}{2(R_E + h)}.$$

Consequently, the energy difference is

$$\Delta E = E_1 - E_2 = GM_E m \left[\frac{1}{R_E} - \frac{3}{2(R_E + h)}\right].$$

(a) Solving the above equation, the height h_0 at which $\Delta E = 0$ is given by

$$\frac{1}{R_E} - \frac{3}{2(R_E + h_0)} = 0 \Rightarrow h_0 = \frac{R_E}{2} = 3.19 \times 10^6 \text{ m}.$$

(b) For greater height $h > h_0$, $\Delta E > 0$ implying $E_1 > E_2$. Thus, the energy of lifting is greater.

64. (a) From Eq. 13-40, we see that the energy of each satellite is $-GM_E m/2r$. The total energy of the two satellites is twice that result:

$$E = E_A + E_B = -\frac{GM_E m}{r} = -\frac{(6.67 \times 10^{-11} \text{ m}^3/\text{kg} \cdot \text{s}^2)(5.98 \times 10^{24} \text{kg})(125 \text{ kg})}{7.87 \times 10^6 \text{ m}}$$
$$= -6.33 \times 10^9 \text{ J}.$$

(b) We note that the speed of the wreckage will be zero (immediately after the collision), so it has no kinetic energy at that moment. Replacing m with $2m$ in the potential energy expression, we therefore find the total energy of the wreckage at that instant is

$$E = -\frac{GM_E(2m)}{2r} = -\frac{(6.67 \times 10^{-11} \text{ m}^3/\text{kg} \cdot \text{s}^2)(5.98 \times 10^{24} \text{kg})2(125 \text{ kg})}{2(7.87 \times 10^6 \text{ m})} = -6.33 \times 10^9 \text{ J}.$$

(c) An object with zero speed at that distance from Earth will simply fall towards the Earth, its trajectory being toward the center of the planet.

65. (a) From Kepler's law of periods, we see that T is proportional to $r^{3/2}$.

(b) Eq. 13-38 shows that K is inversely proportional to r.

(c) and (d) From the previous part, knowing that K is proportional to v^2, we find that v is proportional to $1/\sqrt{r}$. Thus, by Eq. 13-31, the angular momentum (which depends on the product rv) is proportional to $r/\sqrt{r} = \sqrt{r}$.

66. (a) The pellets will have the same speed v but opposite direction of motion, so the *relative speed* between the pellets and satellite is $2v$. Replacing v with $2v$ in Eq. 13-38 is equivalent to multiplying it by a factor of 4. Thus,

$$K_{rel} = 4\left(\frac{GM_E m}{2r}\right) = \frac{2(6.67 \times 10^{-11} \text{ m}^3/\text{kg} \cdot \text{s}^2)(5.98 \times 10^{24} \text{ kg})(0.0040 \text{ kg})}{(6370 + 500) \times 10^3 \text{ m}} = 4.6 \times 10^5 \text{ J}.$$

(b) We set up the ratio of kinetic energies:

$$\frac{K_{rel}}{K_{bullet}} = \frac{4.6 \times 10^5 \text{ J}}{\frac{1}{2}(0.0040 \text{ kg})(950 \text{ m/s})^2} = 2.6 \times 10^2.$$

67. (a) The force acting on the satellite has magnitude GMm/r^2, where M is the mass of Earth, m is the mass of the satellite, and r is the radius of the orbit. The force points toward the center of the orbit. Since the acceleration of the satellite is v^2/r, where v is its speed, Newton's second law yields $GMm/r^2 = mv^2/r$ and the speed is given by $v = \sqrt{GM/r}$. The radius of the orbit is the sum of Earth's radius and the altitude of the satellite: $r = (6.37 \times 10^6 + 640 \times 10^3)$ m $= 7.01 \times 10^6$ m. Thus,

$$v = \sqrt{\frac{(6.67 \times 10^{-11}\, \text{m}^3/\text{s}^2 \cdot \text{kg})(5.98 \times 10^{24}\, \text{kg})}{7.01 \times 10^6\, \text{m}}} = 7.54 \times 10^3\, \text{m/s}.$$

(b) The period is

$$T = 2\pi r/v = 2\pi(7.01 \times 10^6\, \text{m})/(7.54 \times 10^3\, \text{m/s}) = 5.84 \times 10^3\, \text{s} \approx 97\, \text{min}.$$

(c) If E_0 is the initial energy then the energy after n orbits is $E = E_0 - nC$, where $C = 1.4 \times 10^5$ J/orbit. For a circular orbit the energy and orbit radius are related by $E = -GMm/2r$, so the radius after n orbits is given by $r = -GMm/2E$.
The initial energy is

$$E_0 = -\frac{(6.67 \times 10^{-11}\, \text{m}^3/\text{s}^2 \cdot \text{kg})(5.98 \times 10^{24}\, \text{kg})(220\, \text{kg})}{2(7.01 \times 10^6\, \text{m})} = -6.26 \times 10^9\, \text{J},$$

the energy after 1500 orbits is

$$E = E_0 - nC = -6.26 \times 10^9\, \text{J} - (1500\, \text{orbit})(1.4 \times 10^5\, \text{J/orbit}) = -6.47 \times 10^9\, \text{J},$$

and the orbit radius after 1500 orbits is

$$r = -\frac{(6.67 \times 10^{-11}\, \text{m}^3/\text{s}^2 \cdot \text{kg})(5.98 \times 10^{24}\, \text{kg})(220\, \text{kg})}{2(-6.47 \times 10^9\, \text{J})} = 6.78 \times 10^6\, \text{m}.$$

The altitude is $h = r - R = (6.78 \times 10^6\, \text{m} - 6.37 \times 10^6\, \text{m}) = 4.1 \times 10^5\, \text{m}$. Here R is the radius of Earth. This torque is internal to the satellite-Earth system, so the angular momentum of that system is conserved.

(d) The speed is

$$v = \sqrt{\frac{GM}{r}} = \sqrt{\frac{(6.67 \times 10^{-11}\, \text{m}^3/\text{s}^2 \cdot \text{kg})(5.98 \times 10^{24}\, \text{kg})}{6.78 \times 10^6\, \text{m}}} = 7.67 \times 10^3\, \text{m/s} \approx 7.7\, \text{km/s}.$$

(e) The period is

$$T = \frac{2\pi r}{v} = \frac{2\pi(6.78 \times 10^6\, \text{m})}{7.67 \times 10^3\, \text{m/s}} = 5.6 \times 10^3\, \text{s} \approx 93\, \text{min}.$$

(f) Let F be the magnitude of the average force and s be the distance traveled by the satellite. Then, the work done by the force is $W = -Fs$. This is the change in energy: $-Fs = \Delta E$. Thus, $F = -\Delta E/s$. We evaluate this expression for the first orbit. For a complete orbit $s = 2\pi r = 2\pi(7.01 \times 10^6\, \text{m}) = 4.40 \times 10^7\, \text{m}$, and $\Delta E = -1.4 \times 10^5\, \text{J}$. Thus,

$$F = -\frac{\Delta E}{s} = \frac{1.4 \times 10^5 \text{ J}}{4.40 \times 10^7 \text{ m}} = 3.2 \times 10^{-3} \text{ N}.$$

(g) The resistive force exerts a torque on the satellite, so its angular momentum is not conserved.

(h) The satellite-Earth system is essentially isolated, so its momentum is very nearly conserved.

68. The orbital radius is $r = R_E + h = 6370 \text{ km} + 400 \text{ km} = 6770 \text{ km} = 6.77 \times 10^6$ m.

(a) Using Kepler's law given in Eq. 13-34, we find the period of the ships to be

$$T_0 = \sqrt{\frac{4\pi^2 r^3}{GM}} = \sqrt{\frac{4\pi^2 (6.77 \times 10^6 \text{ m})^3}{(6.67 \times 10^{-11} \text{m}^3/\text{s}^2 \cdot \text{kg})(5.98 \times 10^{24} \text{kg})}} = 5.54 \times 10^3 \text{ s} \approx 92.3 \text{ min}.$$

(b) The speed of the ships is

$$v_0 = \frac{2\pi r}{T_0} = \frac{2\pi(6.77 \times 10^6 \text{ m})}{5.54 \times 10^3 \text{ s}} = 7.68 \times 10^3 \text{ m/s}^2.$$

(c) The new kinetic energy is

$$K = \frac{1}{2}mv^2 = \frac{1}{2}m(0.99v_0)^2 = \frac{1}{2}(2000 \text{ kg})(0.99)^2(7.68 \times 10^3 \text{ m/s})^2 = 5.78 \times 10^{10} \text{ J}.$$

(d) Immediately after the burst, the potential energy is the same as it was before the burst. Therefore,

$$U = -\frac{GMm}{r} = -\frac{(6.67 \times 10^{-11} \text{m}^3/\text{s}^2 \cdot \text{kg})(5.98 \times 10^{24} \text{kg})(2000 \text{ kg})}{6.77 \times 10^6 \text{ m}} = -1.18 \times 10^{11} \text{ J}.$$

(e) In the new elliptical orbit, the total energy is

$$E = K + U = 5.78 \times 10^{10} \text{ J} + (-1.18 \times 10^{11} \text{ J}) = -6.02 \times 10^{10} \text{ J}.$$

(f) For elliptical orbit, the total energy can be written as (see Eq. 13-42) $E = -GMm/2a$, where a is the semi-major axis. Thus,

$$a = -\frac{GMm}{2E} = -\frac{(6.67 \times 10^{-11} \text{m}^3/\text{s}^2 \cdot \text{kg})(5.98 \times 10^{24} \text{kg})(2000 \text{ kg})}{2(-6.02 \times 10^{10} \text{ J})} = 6.63 \times 10^6 \text{ m}.$$

(g) To find the period, we use Eq. 13-34 but replace r with a. The result is

$$T = \sqrt{\frac{4\pi^2 a^3}{GM}} = \sqrt{\frac{4\pi^2 (6.63 \times 10^6 \text{ m})^3}{(6.67 \times 10^{-11} \text{m}^3/\text{s}^2 \cdot \text{kg})(5.98 \times 10^{24} \text{kg})}} = 5.37 \times 10^3 \text{ s} \approx 89.5 \text{ min}.$$

(h) The orbital period T for Picard's elliptical orbit is shorter than Igor's by

$$\Delta T = T_0 - T = 5540 \text{ s} - 5370 \text{ s} = 170 \text{ s}.$$

Thus, Picard will arrive back at point P ahead of Igor by 170 s – 90 s = 80 s.

69. We define the "effective gravity" in his environment as $g_{\text{eff}} = 220/60 = 3.67 \text{ m/s}^2$. Thus, using equations from Chapter 2 (and selecting downwards as the positive direction), we find the "fall-time" to be

$$\Delta y = v_0 t + \frac{1}{2} g_{\text{eff}} t^2 \Rightarrow t = \sqrt{\frac{2(2.1 \text{ m})}{3.67 \text{ m/s}^2}} = 1.1 \text{ s}.$$

70. We estimate the planet to have radius $r = 10$ m. To estimate the mass m of the planet, we require its density equal that of Earth (and use the fact that the volume of a sphere is $4\pi r^3/3$):

$$\frac{m}{4\pi r^3/3} = \frac{M_E}{4\pi R_E^3/3} \Rightarrow m = M_E \left(\frac{r}{R_E}\right)^3$$

which yields (with $M_E \approx 6 \times 10^{24}$ kg and $R_E \approx 6.4 \times 10^6$ m) $m = 2.3 \times 10^7$ kg.

(a) With the above assumptions, the acceleration due to gravity is

$$a_g = \frac{Gm}{r^2} = \frac{(6.7 \times 10^{-11} \text{ m}^3/\text{s}^2 \cdot \text{kg})(2.3 \times 10^7 \text{ kg})}{(10 \text{ m})^2} = 1.5 \times 10^{-5} \text{ m/s}^2 \approx 2 \times 10^{-5} \text{ m/s}^2.$$

(b) Eq. 13-28 gives the escape speed:

$$v = \sqrt{\frac{2Gm}{r}} \approx 0.02 \text{ m/s}.$$

71. Using energy conservation (and Eq. 13-21) we have

$$K_1 - \frac{GMm}{r_1} = K_2 - \frac{GMm}{r_2}.$$

Plugging in two pairs of values (for (K_1, r_1) and (K_2, r_2)) from the graph and using the value of G and M (for earth) given in the book, we find

(a) $m \approx 1.0 \times 10^3$ kg.

(b) Similarly, $v = (2K/m)^{1/2} \approx 1.5 \times 10^3$ m/s (at $r = 1.945 \times 10^7$ m).

72. (a) The gravitational acceleration a_g is defined in Eq. 13-11. The problem is concerned with the difference between a_g evaluated at $r = 50R_h$ and a_g evaluated at $r = 50R_h + h$ (where h is the estimate of your height). Assuming h is much smaller than $50R_h$ then we can approximate h as the dr which is present when we consider the differential of Eq. 13-11:

$$|da_g| = \frac{2GM}{r^3} dr \approx \frac{2GM}{50^3 R_h^3} h = \frac{2GM}{50^3 (2GM/c^2)^3} h.$$

If we approximate $|da_g| = 10$ m/s^2 and $h \approx 1.5$ m, we can solve this for M. Giving our results in terms of the Sun's mass means dividing our result for M by 2×10^{30} kg. Thus, admitting some tolerance into our estimate of h we find the "critical" black hole mass should in the range of 105 to 125 solar masses.

(b) Interestingly, this turns out to be lower limit (which will surprise many students) since the above expression shows $|da_g|$ is inversely proportional to M. It should perhaps be emphasized that a distance of $50R_h$ from a small black hole is much smaller than a distance of $50R_h$ from a large black hole.

73. The magnitudes of the individual forces (acting on m_C, exerted by m_A and m_B respectively) are

$$F_{AC} = \frac{Gm_A m_C}{r_{AC}^2} = 2.7 \times 10^{-8} \text{ N} \quad \text{and} \quad F_{BC} = \frac{Gm_B m_C}{r_{BC}^2} = 3.6 \times 10^{-8} \text{ N}$$

where $r_{AC} = 0.20$ m and $r_{BC} = 0.15$ m. With $r_{AB} = 0.25$ m, the angle $\vec{F}_A$ makes with the x axis can be obtained as

$$\theta_A = \pi + \cos^{-1}\left(\frac{r_{AC}^2 + r_{AB}^2 - r_{BC}^2}{2r_{AC}r_{AB}}\right) = \pi + \cos^{-1}(0.80) = 217°.$$

Similarly, the angle $\vec{F}_B$ makes with the x axis can be obtained as

$$\theta_B = -\cos^{-1}\left(\frac{r_{AB}^2 + r_{BC}^2 - r_{AC}^2}{2r_{AB}r_{BC}}\right) = -\cos^{-1}(0.60) = -53°.$$

The net force acting on m_C then becomes

$$\vec{F}_C = F_{AC}(\cos\theta_A \hat{i} + \sin\theta_A \hat{j}) + F_{BC}(\cos\theta_B \hat{i} + \sin\theta_B \hat{j})$$
$$= (F_{AC}\cos\theta_A + F_{BC}\cos\theta_B)\hat{i} + (F_{AC}\sin\theta_A + F_{BC}\sin\theta_B)\hat{j}$$
$$= (-4.4\times 10^{-8}\text{ N})\hat{j}$$

74. The key point here is that angular momentum is conserved:

$$I_p \omega_p = I_a \omega_a$$

which leads to $\omega_p = (r_a/r_p)^2 \omega_a$, but $r_p = 2a - r_a$ where a is determined by Eq. 13-34 (particularly, see the paragraph after that equation in the textbook). Therefore,

$$\omega_p = \frac{r_a^2 \omega_a}{(2(GMT^2/4\pi^2)^{1/3} - r_a)^2} = 9.24 \times 10^{-5} \text{ rad/s}.$$

75. (a) Using Kepler's law of periods, we obtain

$$T = \sqrt{\left(\frac{4\pi^2}{GM}\right) r^3} = 2.15 \times 10^4 \text{ s}.$$

(b) The speed is constant (before she fires the thrusters), so $v_o = 2\pi r/T = 1.23 \times 10^4$ m/s.

(c) A two percent reduction in the previous value gives $v = 0.98 v_o = 1.20 \times 10^4$ m/s.

(d) The kinetic energy is $K = \frac{1}{2}mv^2 = 2.17 \times 10^{11}$ J.

(e) The potential energy is $U = -GmM/r = -4.53 \times 10^{11}$ J.

(f) Adding these two results gives $E = K + U = -2.35 \times 10^{11}$ J.

(g) Using Eq. 13-42, we find the semi-major axis to be

$$a = \frac{-GMm}{2E} = 4.04 \times 10^7 \text{ m}.$$

(h) Using Kepler's law of periods for elliptical orbits (using a instead of r) we find the new period is

$$T' = \sqrt{\left(\frac{4\pi^2}{GM}\right) a^3} = 2.03 \times 10^4 \text{ s}.$$

This is smaller than our result for part (a) by $T - T' = 1.22 \times 10^3$ s.

(i) Elliptical orbit has a smaller period.

76. (a) With $M = 2.0 \times 10^{30}$ kg and $r = 10000$ m, we find

$$a_g = \frac{GM}{r^2} = 1.3 \times 10^{12} \text{ m/s}^2 \ .$$

(b) Although a close answer may be gotten by using the constant acceleration equations of Chapter 2, we show the more general approach (using energy conservation):

$$K_o + U_o = K + U$$

where $K_o = 0$, $K = \frac{1}{2}mv^2$ and U given by Eq. 13-21. Thus, with $r_o = 10001$ m, we find

$$v = \sqrt{2GM\left(\frac{1}{r} - \frac{1}{r_o}\right)} = 1.6 \times 10^6 \text{ m/s} \ .$$

77. We note that r_A (the distance from the origin to sphere A, which is the same as the separation between A and B) is 0.5, $r_C = 0.8$, and $r_D = 0.4$ (with SI units understood). The force $\vec{F}_k$ that the kth sphere exerts on m_B has magnitude $Gm_k m_B / r_k^2$ and is directed from the origin towards m_k so that it is conveniently written as

$$\vec{F}_k = \frac{Gm_k m_B}{r_k^2}\left(\frac{x_k}{r_k}\hat{i} + \frac{y_k}{r_k}\hat{j}\right) = \frac{Gm_k m_B}{r_k^3}\left(x_k\hat{i} + y_k\hat{j}\right).$$

Consequently, the vector addition (where k equals A, B and D) to obtain the net force on m_B becomes

$$\vec{F}_{\text{net}} = \sum_k \vec{F}_k = Gm_B\left(\left(\sum_k \frac{m_k x_k}{r_k^3}\right)\hat{i} + \left(\sum_k \frac{m_k y_k}{r_k^3}\right)\hat{j}\right) = (3.7 \times 10^{-5} \text{ N})\hat{j}.$$

78. (a) We note that r_C (the distance from the origin to sphere C, which is the same as the separation between C and B) is 0.8, $r_D = 0.4$, and the separation between spheres C and D is $r_{CD} = 1.2$ (with SI units understood). The total potential energy is therefore

$$-\frac{GM_B M_C}{r_C^2} - \frac{GM_B M_D}{r_D^2} - \frac{GM_C M_D}{r_{CD}^2} = -1.3 \times 10^{-4} \text{ J}$$

using the mass-values given in the previous problem.

(b) Since any gravitational potential energy term (of the sort considered in this chapter) is necessarily negative ($-GmM/r^2$ where all variables are positive) then having another mass to include in the computation can only lower the result (that is, make the result more negative).

(c) The observation in the previous part implies that the work I do in removing sphere A (to obtain the case considered in part (a)) must lead to an increase in the system energy; thus, I do positive work.

(d) To put sphere A back in, I do negative work, since I am causing the system energy to become more negative.

79. We use $F = Gm_s m_m/r^2$, where m_s is the mass of the satellite, m_m is the mass of the meteor, and r is the distance between their centers. The distance between centers is $r = R + d = 15$ m $+ 3$ m $= 18$ m. Here R is the radius of the satellite and d is the distance from its surface to the center of the meteor. Thus,

$$F = \frac{(6.67 \times 10^{-11} \text{ N} \cdot \text{m}^2/\text{kg}^2)(20 \text{kg})(7.0 \text{kg})}{(18 \text{m})^2} = 2.9 \times 10^{-11} \text{ N}.$$

80. (a) Since the volume of a sphere is $4\pi R^3/3$, the density is

$$\rho = \frac{M_{\text{total}}}{\frac{4}{3}\pi R^3} = \frac{3 M_{\text{total}}}{4\pi R^3}.$$

When we test for gravitational acceleration (caused by the sphere, or by parts of it) at radius r (measured from the center of the sphere), the mass M which is at radius less than r is what contributes to the reading (GM/r^2). Since $M = \rho(4\pi r^3/3)$ for $r \leq R$ then we can write this result as

$$\frac{G\left(\frac{3M_{\text{total}}}{4\pi R^3}\right)\left(\frac{4\pi r^3}{3}\right)}{r^2} = \frac{GM_{\text{total}} r}{R^3}$$

when we are considering points on or inside the sphere. Thus, the value a_g referred to in the problem is the case where $r = R$:

$$a_g = \frac{GM_{\text{total}}}{R^2},$$

and we solve for the case where the acceleration equals $a_g/3$:

$$\frac{GM_{\text{total}}}{3R^2} = \frac{GM_{\text{total}} r}{R^3} \Rightarrow r = \frac{R}{3}.$$

(b) Now we treat the case of an external test point. For points with $r > R$ the acceleration is GM_{total}/r^2, so the requirement that it equal $a_g/3$ leads to

$$\frac{GM_{total}}{3R^2} = \frac{GM_{total}}{r^2} \quad \Rightarrow \quad r = \sqrt{3}R.$$

81. Energy conservation for this situation may be expressed as follows:

$$K_1 + U_1 = K_2 + U_2 \quad \Rightarrow \quad \frac{1}{2}mv_1^2 - \frac{GmM}{r_1} = \frac{1}{2}mv_2^2 - \frac{GmM}{r_2}$$

where $M = 5.98 \times 10^{24}$ kg, $r_1 = R = 6.37 \times 10^6$ m and $v_1 = 10000$ m/s. Setting $v_2 = 0$ to find the maximum of its trajectory, we solve the above equation (noting that m cancels in the process) and obtain $r_2 = 3.2 \times 10^7$ m. This implies that its *altitude* is $r_2 - R = 2.5 \times 10^7$ m.

82. (a) Because it is moving in a circular orbit, F/m must equal the centripetal acceleration:

$$\frac{80 \text{ N}}{50 \text{ kg}} = \frac{v^2}{r}.$$

But $v = 2\pi r/T$, where $T = 21600$ s, so we are led to

$$1.6 \text{ m/s}^2 = \frac{4\pi^2}{T^2}r$$

which yields $r = 1.9 \times 10^7$ m.

(b) From the above calculation, we infer $v^2 = (1.6 \text{ m/s}^2)r$ which leads to $v^2 = 3.0 \times 10^7$ m²/s². Thus, $K = \frac{1}{2}mv^2 = 7.6 \times 10^8$ J.

(c) As discussed in § 13-4, F/m also tells us the gravitational acceleration:

$$a_g = 1.6 \text{ m/s}^2 = \frac{GM}{r^2}.$$

We therefore find $M = 8.6 \times 10^{24}$ kg.

83. (a) We write the centripetal acceleration (which is the same for each, since they have identical mass) as $r\omega^2$ where ω is the unknown angular speed. Thus,

$$\frac{G(M)(M)}{(2r)^2} = \frac{GM^2}{4r^2} = Mr\omega^2$$

which gives $\omega = \frac{1}{2}\sqrt{MG/r^3} = 2.2 \times 10^{-7}$ rad/s.

(b) To barely escape means to have total energy equal to zero (see discussion prior to Eq. 13-28). If m is the mass of the meteoroid, then

$$\frac{1}{2}mv^2 - \frac{GmM}{r} - \frac{GmM}{r} = 0 \Rightarrow v = \sqrt{\frac{4GM}{r}} = 8.9 \times 10^4 \text{ m/s}.$$

84. See Appendix C. We note that, since $v = 2\pi r/T$, the centripetal acceleration may be written as $a = 4\pi^2 r/T^2$. To express the result in terms of g, we divide by 9.8 m/s^2.

(a) The acceleration associated with Earth's spin ($T = 24$ h $= 86400$ s) is

$$a = g \frac{4\pi^2 (6.37 \times 10^6 \text{ m})}{(86400 \text{ s})^2 (9.8 \text{ m/s}^2)} = 3.4 \times 10^{-3} g.$$

(b) The acceleration associated with Earth's motion around the Sun ($T = 1$ y $= 3.156 \times 10^7$ s) is

$$a = g \frac{4\pi^2 (1.5 \times 10^{11} \text{ m})}{(3.156 \times 10^7 \text{ s})^2 (9.8 \text{ m/s}^2)} = 6.1 \times 10^{-4} g.$$

(c) The acceleration associated with the Solar System's motion around the galactic center ($T = 2.5 \times 10^8$ y $= 7.9 \times 10^{15}$ s) is

$$a = g \frac{4\pi^2 (2.2 \times 10^{20} \text{ m})}{(7.9 \times 10^{15} \text{ s})^2 (9.8 \text{ m/s}^2)} = 1.4 \times 10^{-11} g.$$

85. We use m_1 for the 20 kg of the sphere at $(x_1, y_1) = (0.5, 1.0)$ (SI units understood), m_2 for the 40 kg of the sphere at $(x_2, y_2) = (-1.0, -1.0)$, and m_3 for the 60 kg of the sphere at $(x_3, y_3) = (0, -0.5)$. The mass of the 20 kg object at the origin is simply denoted m. We note that $r_1 = \sqrt{1.25}$, $r_2 = \sqrt{2}$, and $r_3 = 0.5$ (again, with SI units understood). The force $\vec{F}_n$ that the nth sphere exerts on m has magnitude $Gm_n m / r_n^2$ and is directed from the origin towards m_n, so that it is conveniently written as

$$\vec{F}_n = \frac{Gm_n m}{r_n^2} \left(\frac{x_n}{r_n} \hat{i} + \frac{y_n}{r_n} \hat{j} \right) = \frac{Gm_n m}{r_n^3} \left(x_n \hat{i} + y_n \hat{j} \right).$$

Consequently, the vector addition to obtain the net force on m becomes

$$\vec{F}_{\text{net}} = \sum_{n=1}^{3} \vec{F}_n = Gm \left(\left(\sum_{n=1}^{3} \frac{m_n x_n}{r_n^3} \right) \hat{i} + \left(\sum_{n=1}^{3} \frac{m_n y_n}{r_n^3} \right) \hat{j} \right) = -9.3 \times 10^{-9} \hat{i} - 3.2 \times 10^{-7} \hat{j}$$

in SI units. Therefore, we find the net force magnitude is $|\vec{F}_{\text{net}}| = 3.2 \times 10^{-7}$ N.

86. We apply the work-energy theorem to the object in question. It starts from a point at the surface of the Earth with zero initial speed and arrives at the center of the Earth with final speed v_f. The corresponding increase in its kinetic energy, $\frac{1}{2}mv_f^2$, is equal to the work done on it by Earth's gravity: $\int F\, dr = \int (-Kr)dr$ (using the notation of that Sample Problem referred to in the problem statement). Thus,

$$\frac{1}{2}mv_f^2 = \int_R^0 F\, dr = \int_R^0 (-Kr)\, dr = \frac{1}{2}KR^2$$

where R is the radius of Earth. Solving for the final speed, we obtain $v_f = R\sqrt{K/m}$. We note that the acceleration of gravity $a_g = g = 9.8$ m/s^2 on the surface of Earth is given by

$$a_g = GM/R^2 = G(4\pi R^3/3)\rho/R^2,$$

where ρ is Earth's average density. This permits us to write $K/m = 4\pi G\rho/3 = g/R$. Consequently,

$$v_f = R\sqrt{\frac{K}{m}} = R\sqrt{\frac{g}{R}} = \sqrt{gR} = \sqrt{(9.8 \text{ m/s}^2)(6.37 \times 10^6 \text{ m})} = 7.9 \times 10^3 \text{ m/s}.$$

87. (a) The total energy is conserved, so there is no difference between its values at aphelion and perihelion.

(b) Since the change is small, we use differentials:

$$dU = \left(\frac{GM_E M_S}{r^2}\right) dr \approx \left[\frac{(6.67 \times 10^{-11})(1.99 \times 10^{30})(5.98 \times 10^{24})}{(1.5 \times 10^{11})^2}\right](5 \times 10^9)$$

which yields $\Delta U \approx 1.8 \times 10^{32}$ J. A more direct subtraction of the values of the potential energies leads to the same result.

(c) From the previous two parts, we see that the variation in the kinetic energy ΔK must also equal 1.8×10^{32} J.

(d) With $\Delta K \approx dK = mv\, dv$, where $v \approx 2\pi R/T$, we have

$$1.8 \times 10^{32} \approx (5.98 \times 10^{24})\left(\frac{2\pi (1.5 \times 10^{11})}{3.156 \times 10^7}\right)\Delta v$$

which yields a difference of $\Delta v \approx 0.99$ km/s in Earth's speed (relative to the Sun) between aphelion and perihelion.

88. Let the distance from Earth to the spaceship be r. $R_{em} = 3.82 \times 10^8$ m is the distance from Earth to the moon. Thus,

$$F_m = \frac{GM_m m}{(R_{em} - r)^2} = F_E = \frac{GM_e m}{r^2},$$

where m is the mass of the spaceship. Solving for r, we obtain

$$r = \frac{R_{em}}{\sqrt{M_m/M_e} + 1} = \frac{3.82 \times 10^8 \text{ m}}{\sqrt{(7.36 \times 10^{22} \text{ kg})/(5.98 \times 10^{24} \text{ kg})} + 1} = 3.44 \times 10^8 \text{ m}.$$

89. We integrate Eq. 13-1 with respect to r from $3R_E$ to $4R_E$ and obtain the work equal to $-GM_E m(1/(4R_E) - 1/(3R_E)) = GM_E m/12R_E$.

90. If the angular velocity were any greater, loose objects on the surface would not go around with the planet but would travel out into space.

(a) The magnitude of the gravitational force exerted by the planet on an object of mass m at its surface is given by $F = GmM/R^2$, where M is the mass of the planet and R is its radius. According to Newton's second law this must equal mv^2/R, where v is the speed of the object. Thus,

$$\frac{GM}{R^2} = \frac{v^2}{R}.$$

Replacing M with $(4\pi/3)\rho R^3$ (where ρ is the density of the planet) and v with $2\pi R/T$ (where T is the period of revolution), we find

$$\frac{4\pi}{3} G\rho R = \frac{4\pi^2 R}{T^2}.$$

We solve for T and obtain

$$T = \sqrt{\frac{3\pi}{G\rho}}.$$

(b) The density is 3.0×10^3 kg/m^3. We evaluate the equation for T:

$$T = \sqrt{\frac{3\pi}{(6.67 \times 10^{-11} \text{ m}^3/\text{s}^2 \cdot \text{kg})(3.0 \times 10^3 \text{ kg/m}^3)}} = 6.86 \times 10^3 \text{ s} = 1.9 \text{ h}.$$

91. (a) It is possible to use $v^2 = v_0^2 + 2a\Delta y$ as we did for free-fall problems in Chapter 2 because the acceleration can be considered approximately constant over this interval.

However, our approach will not assume constant acceleration; we use energy conservation:

$$\frac{1}{2}mv_0^2 - \frac{GMm}{r_0} = \frac{1}{2}mv^2 - \frac{GMm}{r} \Rightarrow v = \sqrt{\frac{2GM(r_0-r)}{r_0 r}}$$

which yields $v = 1.4 \times 10^6$ m/s.

(b) We estimate the height of the apple to be $h = 7$ cm $= 0.07$ m. We may find the answer by evaluating Eq. 13-11 at the surface (radius r in part (a)) and at radius $r + h$, being careful not to round off, and then taking the difference of the two values, or we may take the differential of that equation — setting dr equal to h. We illustrate the latter procedure:

$$|da_g| = \left|-2\frac{GM}{r^3}dr\right| \approx 2\frac{GM}{r^3}h = 3 \times 10^6 \text{ m/s}^2.$$

92. (a) The gravitational force exerted on the baby (denoted with subscript b) by the obstetrician (denoted with subscript o) is given by

$$F_{bo} = \sqrt{\frac{Gm_o m_b}{r_{bo}^2}} = \sqrt{\frac{(6.67 \times 10^{-11} \text{N} \cdot \text{m}^2/\text{kg}^2)(70 \text{kg})(3 \text{kg})}{(1\text{m})^2}} = 1 \times 10^{-8} \text{N}.$$

(b) The maximum (minimum) forces exerted by Jupiter on the baby occur when it is separated from the Earth by the shortest (longest) distance r_{min} (r_{max}), respectively. Thus

$$F_{bJ}^{max} = \sqrt{\frac{Gm_J m_b}{r_{min}^2}} = \sqrt{\frac{(6.67 \times 10^{-11} \text{N} \cdot \text{m}^2/\text{kg}^2)(2 \times 10^{27} \text{kg})(3 \text{kg})}{(6 \times 10^{11} \text{m})^2}} = 1 \times 10^{-6} \text{N}.$$

(c) And we obtain

$$F_{bJ}^{min} = \sqrt{\frac{Gm_J m_b}{r_{max}^2}} = \sqrt{\frac{(6.67 \times 10^{-11} \text{N} \cdot \text{m}^2/\text{kg}^2)(2 \times 10^{27} \text{kg})(3 \text{kg})}{(9 \times 10^{11} \text{m})^2}} = 5 \times 10^{-7} \text{N}.$$

(d) No. The gravitational force exerted by Jupiter on the baby is greater than that by the obstetrician by a factor of up to 1×10^{-6} N/1×10^{-8} N $= 100$.

93. The magnitude of the net gravitational force on one of the smaller stars (of mass m) is

$$\frac{GMm}{r^2} + \frac{Gmm}{(2r)^2} = \frac{Gm}{r^2}\left(M + \frac{m}{4}\right).$$

This supplies the centripetal force needed for the motion of the star:

$$\frac{Gm}{r^2}\left(M + \frac{m}{4}\right) = m\frac{v^2}{r} \quad \text{where } v = \frac{2pr}{T}.$$

Plugging in for speed v, we arrive at an equation for period T:

$$T = \frac{2\pi r^{3/2}}{\sqrt{G(M + m/4)}}.$$

94. (a) We note that $height = R - R_{\text{Earth}}$ where $R_{\text{Earth}} = 6.37 \times 10^6$ m. With $M = 5.98 \times 10^{24}$ kg, $R_0 = 6.57 \times 10^6$ m and $R = 7.37 \times 10^6$ m, we have

$$K_i + U_i = K + U \Rightarrow \frac{1}{2}m(3.70 \times 10^3)^2 - \frac{GmM}{R_0} = K - \frac{GmM}{R},$$

which yields $K = 3.83 \times 10^7$ J.

(b) Again, we use energy conservation.

$$K_i + U_i = K_f + U_f \Rightarrow \frac{1}{2}m(3.70 \times 10^3)^2 - \frac{GmM}{R_0} = 0 - \frac{GmM}{R_f}$$

Therefore, we find $R_f = 7.40 \times 10^6$ m. This corresponds to a distance of 1034.9 km ≈ 1.03×10^3 km above the Earth's surface.

95. Energy conservation for this situation may be expressed as follows:

$$K_1 + U_1 = K_2 + U_2 \Rightarrow \frac{1}{2}mv_1^2 - \frac{GmM}{r_1} = \frac{1}{2}mv_2^2 - \frac{GmM}{r_2}$$

where $M = 7.0 \times 10^{24}$ kg, $r_2 = R = 1.6 \times 10^6$ m and $r_1 = \infty$ (which means that $U_1 = 0$). We are told to assume the meteor starts at rest, so $v_1 = 0$. Thus, $K_1 + U_1 = 0$ and the above equation is rewritten as

$$\frac{1}{2}mv_2^2 - \frac{GmM}{r_2} \Rightarrow v_2 = \sqrt{\frac{2GM}{R}} = 2.4 \times 10^4 \text{ m/s}.$$

96. The initial distance from each fixed sphere to the ball is $r_0 = \infty$, which implies the initial gravitational potential energy is zero. The distance from each fixed sphere to the ball when it is at $x = 0.30$ m is $r = 0.50$ m, by the Pythagorean theorem.

(a) With $M = 20$ kg and $m = 10$ kg, energy conservation leads to

$$K_i + U_i = K + U \Rightarrow 0 + 0 = K - 2\frac{GmM}{r}$$

which yields $K = 2GmM/r = 5.3 \times 10^{-8}$ J.

(b) Since the y-component of each force will cancel, the net force points in the $-x$ direction, with a magnitude $2F_x = 2(GmM/r^2)\cos\theta$, where $\theta = \tan^{-1}(4/3) = 53°$. Thus, the result is $\vec{F}_{net} = (-6.4 \times 10^{-8}\text{ N})\hat{i}$.

97. The kinetic energy in its circular orbit is $\frac{1}{2}mv^2$ where $v = 2\pi r/T$. Using the values stated in the problem and using Eq. 13-41, we directly find $E = -1.87 \times 10^9$ J.

98. (a) From Ch. 2, we have $v^2 = v_0^2 + 2a\Delta x$, where a may be interpreted as an average acceleration in cases where the acceleration is not uniform. With $v_0 = 0$, $v = 11000$ m/s and $\Delta x = 220$ m, we find $a = 2.75 \times 10^5$ m/s². Therefore,

$$a = \left(\frac{2.75 \times 10^5 \text{ m/s}^2}{9.8 \text{ m/s}^2}\right)g = 2.8 \times 10^4\, g.$$

(b) The acceleration is certainly deadly enough to kill the passengers.

(c) Again using $v^2 = v_0^2 + 2a\Delta x$, we find

$$a = \frac{(7000 \text{ m/s})^2}{2(3500 \text{ m})} = 7000 \text{ m/s}^2 = 714g.$$

(d) Energy conservation gives the craft's speed v (in the absence of friction and other dissipative effects) at altitude $h = 700$ km after being launched from $R = 6.37 \times 10^6$ m (the surface of Earth) with speed $v_0 = 7000$ m/s. That altitude corresponds to a distance from Earth's center of $r = R + h = 7.07 \times 10^6$ m.

$$\frac{1}{2}mv_0^2 - \frac{GMm}{R} = \frac{1}{2}mv^2 - \frac{GMm}{r}.$$

With $M = 5.98 \times 10^{24}$ kg (the mass of Earth) we find $v = 6.05 \times 10^3$ m/s. But to orbit at that radius requires (by Eq. 13-37)

$$v' = \sqrt{GM/r} = 7.51 \times 10^3 \text{ m/s}.$$

The difference between these is $v' - v = 1.46 \times 10^3$ m/s $\approx 1.5 \times 10^3$ m/s, which presumably is accounted for by the action of the rocket engine.

99. (a) All points on the ring are the same distance ($r = \sqrt{x^2 + R^2}$) from the particle, so the gravitational potential energy is simply $U = -GMm/\sqrt{x^2 + R^2}$, from Eq. 13-21. The corresponding force (by symmetry) is expected to be along the x axis, so we take a (negative) derivative of U (with respect to x) to obtain it (see Eq. 8-20). The result for the magnitude of the force is $GMmx(x^2 + R^2)^{-3/2}$.

(b) Using our expression for U, then the magnitude of the loss in potential energy as the particle falls to the center is $GMm(1/R - 1/\sqrt{x^2 + R^2})$. This must "turn into" kinetic energy ($\frac{1}{2}mv^2$), so we solve for the speed and obtain

$$v = [2GM(R^{-1} - (R^2 + x^2)^{-1/2})]^{1/2}.$$

100. Consider that we are examining the forces on the mass in the lower left-hand corner of the square. Note that the mass in the upper right-hand corner is $20\sqrt{2} = 28$ cm $= 0.28$ m away. Now, the *nearest* masses each pull with a force of $GmM/r^2 = 3.8 \times 10^{-9}$ N, one upward and the other rightward. The net force caused by these two forces is (3.8×10^{-9}, 3.8×10^{-9}) $\rightarrow$ ($5.3 \times 10^{-9} \angle 45°$), where the rectangular components are shown first -- and then the polar components (magnitude-angle notation). Now, the mass in the upper right-hand corner also pulls at 45°, so its force-magnitude (1.9×10^{-9}) will simply add to the magnitude just calculated. Thus, the final result is 7.2×10^{-9} N.

101. (a) Their initial potential energy is $-Gm^2/R_i$ and they started from rest, so energy conservation leads to

$$-\frac{Gm^2}{R_i} = K_{total} - \frac{Gm^2}{0.5R_i} \Rightarrow K_{total} = \frac{Gm^2}{R_i}.$$

(b) They have equal mass, and this is being viewed in the center-of-mass frame, so their speeds are identical and their kinetic energies are the same. Thus,

$$K = \frac{1}{2}K_{total} = \frac{Gm^2}{2R_i}.$$

(c) With $K = \frac{1}{2}mv^2$, we solve the above equation and find $v = \sqrt{Gm/R_i}$.

(d) Their relative speed is $2v = 2\sqrt{Gm/R_i}$. This is the (instantaneous) rate at which the gap between them is closing.

(e) The premise of this part is that we assume we are not moving (that is, that body A acquires no kinetic energy in the process). Thus, $K_{total} = K_B$ and the logic of part (a) leads to $K_B = Gm^2/R_i$.

(f) And $\frac{1}{2}mv_B^2 = K_B$ yields $v_B = \sqrt{2Gm/R_i}$.

(g) The answer to part (f) is incorrect, due to having ignored the accelerated motion of "our" frame (that of body A). Our computations were therefore carried out in a noninertial frame of reference, for which the energy equations of Chapter 8 are not directly applicable.

102. Gravitational acceleration is defined in Eq. 13-11 (which we are treating as a positive quantity). The problem, then, is asking for the magnitude difference of $a_{g\,net}$ when the contributions from the Moon and the Sun are in the same direction ($a_{g\,net} = a_{gSun} + a_{gMoon}$) as opposed to when they are in opposite directions ($a_{g\,net} = a_{gSun} - a_{gMoon}$). The difference (in absolute value) is clearly $2a_{gMoon}$. In specifically wanting the *percentage* change, the problem is requesting us to divide this difference by the average of the two $a_{g\,net}$ values being considered (that average is easily seen to be equal to a_{gSun}), and finally multiply by 100% in order to quote the result in the right format. Thus,

$$\frac{2a_{gMoon}}{a_{gSun}} = 2\left(\frac{M_{Moon}}{M_{Sun}}\right)\left(\frac{r_{Sun\ to\ Eearth}}{r_{Moon\ to\ Earth}}\right)^2 = 2\left(\frac{7.36 \times 10^{22}}{1.99 \times 10^{30}}\right)\left(\frac{1.50 \times 10^{11}}{3.82 \times 10^8}\right)^2 = 0.011 = 1.1\%.$$

103. (a) Kepler's law of periods is

$$T^2 = \left(\frac{4\pi^2}{GM}\right)r^3.$$

With $M = 6.0 \times 10^{30}$ kg and $T = 300(86400) = 2.6 \times 10^7$ s, we obtain $r = 1.9 \times 10^{11}$ m.

(b) That its orbit is circular suggests that its speed is constant, so

$$v = \frac{2\pi r}{T} = 4.6 \times 10^4 \text{ m/s}.$$

104. Using Eq. 13-21, the potential energy of the dust particle is

$$U = -GmM_E/R - GmM_m/r = -Gm(M_E/R + M_m/r).$$

Chapter 14

1. The pressure increase is the applied force divided by the area: $\Delta p = F/A = F/\pi r^2$, where r is the radius of the piston. Thus

$$\Delta p = (42 \text{ N})/\pi(0.011 \text{ m})^2 = 1.1 \times 10^5 \text{ Pa}.$$

This is equivalent to 1.1 atm.

2. We note that the container is cylindrical, the important aspect of this being that it has a uniform cross-section (as viewed from above); this allows us to relate the pressure at the bottom simply to the total weight of the liquids. Using the fact that 1L = 1000 cm^3, we find the weight of the first liquid to be

$$W_1 = m_1 g = \rho_1 V_1 g = (2.6 \text{ g/cm}^3)(0.50 \text{ L})(1000 \text{ cm}^3/\text{L})(980 \text{ cm/s}^2) = 1.27 \times 10^6 \text{ g} \cdot \text{cm/s}^2$$
$$= 12.7 \text{ N}.$$

In the last step, we have converted grams to kilograms and centimeters to meters. Similarly, for the second and the third liquids, we have

$$W_2 = m_2 g = \rho_2 V_2 g = (1.0 \text{ g/cm}^3)(0.25 \text{ L})(1000 \text{ cm}^3/\text{L})(980 \text{ cm/s}^2) = 2.5 \text{ N}$$

and

$$W_3 = m_3 g = \rho_3 V_3 g = (0.80 \text{ g/cm}^3)(0.40 \text{ L})(1000 \text{ cm}^3/\text{L})(980 \text{ cm/s}^2) = 3.1 \text{ N}.$$

The total force on the bottom of the container is therefore $F = W_1 + W_2 + W_3 = 18$ N.

3. The air inside pushes outward with a force given by $p_i A$, where p_i is the pressure inside the room and A is the area of the window. Similarly, the air on the outside pushes inward with a force given by $p_o A$, where p_o is the pressure outside. The magnitude of the net force is $F = (p_i - p_o)A$. Since 1 atm = 1.013×10^5 Pa,

$$F = (1.0 \text{ atm} - 0.96 \text{ atm})(1.013 \times 10^5 \text{ Pa/atm})(3.4 \text{ m})(2.1 \text{ m}) = 2.9 \times 10^4 \text{ N}.$$

4. Knowing the standard air pressure value in several units allows us to set up a variety of conversion factors:

(a) $P = (28 \text{ lb/in.}^2) \left(\dfrac{1.01 \times 10^5 \text{ Pa}}{14.7 \text{ lb/in}^2} \right) = 190 \text{ kPa}.$

(b) $(120 \text{ mmHg}) \left(\dfrac{1.01 \times 10^5 \text{ Pa}}{760 \text{ mmHg}} \right) = 15.9 \text{ kPa}$, $(80 \text{ mmHg}) \left(\dfrac{1.01 \times 10^5 \text{ Pa}}{760 \text{ mmHg}} \right) = 10.6 \text{ kPa}$.

5. Let the volume of the expanded air sacs be V_a and that of the fish with its air sacs collapsed be V. Then

$$\rho_{\text{fish}} = \dfrac{m_{\text{fish}}}{V} = 1.08 \text{ g/cm}^3 \quad \text{and} \quad \rho_w = \dfrac{m_{\text{fish}}}{V + V_a} = 1.00 \text{ g/cm}^3$$

where ρ_w is the density of the water. This implies

$$\rho_{\text{fish}} V = \rho_w (V + V_a) \text{ or } (V + V_a)/V = 1.08/1.00,$$

which gives $V_a/(V + V_a) = 0.074 = 7.4\%$.

6. The magnitude F of the force required to pull the lid off is $F = (p_o - p_i)A$, where p_o is the pressure outside the box, p_i is the pressure inside, and A is the area of the lid. Recalling that $1 \text{N/m}^2 = 1$ Pa, we obtain

$$p_i = p_o - \dfrac{F}{A} = 1.0 \times 10^5 \text{ Pa} - \dfrac{480 \text{ N}}{77 \times 10^{-4} \text{ m}^2} = 3.8 \times 10^4 \text{ Pa}.$$

7. (a) The pressure difference results in forces applied as shown in the figure. We consider a team of horses pulling to the right. To pull the sphere apart, the team must exert a force at least as great as the horizontal component of the total force determined by "summing" (actually, integrating) these force vectors.

We consider a force vector at angle θ. Its leftward component is $\Delta p \cos \theta dA$, where dA is the area element for where the force is applied. We make use of the symmetry of the problem and let dA be that of a ring of constant θ on the surface. The radius of the ring is $r = R \sin \theta$, where R is the radius of the sphere. If the angular width of the ring is $d\theta$, in radians, then its width is $R d\theta$ and its area is $dA = 2\pi R^2 \sin \theta\, d\theta$. Thus the net horizontal component of the force of the air is given by

$$F_h = 2\pi R^2 \Delta p \int_0^{\pi/2} \sin \theta \cos \theta\, d\theta = \pi R^2 \Delta p \sin^2 \theta \Big|_0^{\pi/2} = \pi R^2 \Delta p.$$

(b) We use 1 atm = 1.01×10^5 Pa to show that $\Delta p = 0.90$ atm = 9.09×10^4 Pa. The sphere radius is $R = 0.30$ m, so

$$F_h = \pi (0.30 \text{ m})^2 (9.09 \times 10^4 \text{ Pa}) = 2.6 \times 10^4 \text{ N}.$$

(c) One team of horses could be used if one half of the sphere is attached to a sturdy wall. The force of the wall on the sphere would balance the force of the horses.

8. We estimate the pressure difference (specifically due to hydrostatic effects) as follows:

$$\Delta p = \rho g h = (1.06 \times 10^3 \text{ kg/m}^3)(9.8 \text{ m/s}^2)(1.83 \text{ m}) = 1.90 \times 10^4 \text{ Pa}.$$

9. Recalling that 1 atm = 1.01×10^5 Pa, Eq. 14-8 leads to

$$\rho g h = (1024 \text{ kg/m}^3)(9.80 \text{ m/s}^2)(10.9 \times 10^3 \text{ m})\left(\frac{1 \text{ atm}}{1.01 \times 10^5 \text{ Pa}}\right) \approx 1.08 \times 10^3 \text{ atm}.$$

10. Note that 0.05 atm equals 5065 Pa. Application of Eq. 14-7 with the notation in this problem leads to

$$d_{max} = \frac{p}{\rho_{liquid} g} = \frac{0.05 \text{ atm}}{\rho_{liquid} g} = \frac{5065 \text{ Pa}}{\rho_{liquid} g}.$$

Thus the difference of this quantity between fresh water (998 kg/m^3) and Dead Sea water (1500 kg/m^3) is

$$\Delta d_{max} = \frac{5065 \text{ Pa}}{g}\left(\frac{1}{\rho_{fw}} - \frac{1}{\rho_{sw}}\right) = \frac{5065 \text{ Pa}}{9.8 \text{ m/s}^2}\left(\frac{1}{998 \text{ kg/m}^3} - \frac{1}{1500 \text{ kg/m}^3}\right) = 0.17 \text{ m}.$$

11. The pressure p at the depth d of the hatch cover is $p_0 + \rho g d$, where ρ is the density of ocean water and p_0 is atmospheric pressure. The downward force of the water on the hatch cover is $(p_0 + \rho g d)A$, where A is the area of the cover. If the air in the submarine is at atmospheric pressure then it exerts an upward force of $p_0 A$. The minimum force that must be applied by the crew to open the cover has magnitude

$$F = (p_0 + \rho g d)A - p_0 A = \rho g d A = (1024 \text{ kg/m}^3)(9.8 \text{ m/s}^2)(100 \text{ m})(1.2 \text{ m})(0.60 \text{ m})$$
$$= 7.2 \times 10^5 \text{ N}.$$

12. With $A = 0.000500$ m^2 and $F = pA$ (with p given by Eq. 14-9), then we have $\rho g h A = 9.80$ N. This gives $h \approx 2.0$ m, which means $d + h = 2.80$ m.

13. In this case, Bernoulli's equation reduces to Eq. 14-10. Thus,

$$p_g = \rho g(-h) = -(1800 \text{ kg/m}^3)(9.8 \text{ m/s}^2)(1.5 \text{ m}) = -2.6 \times 10^4 \text{ Pa}.$$

14. Using Eq. 14-7, we find the gauge pressure to be $p_{gauge} = \rho g h$, where ρ is the density of the fluid medium, and h is the vertical distance to the point where the pressure is equal to the atmospheric pressure.

The gauge pressure at a depth of 20 m in seawater is

$$p_1 = \rho_{sw} g d = (1024 \text{ kg/m}^3)(9.8 \text{ m/s}^2)(20 \text{ m}) = 2.00 \times 10^5 \text{ Pa}.$$

On the other hand, the gauge pressure at an altitude of 7.6 km is

$$p_2 = \rho_{air} gh = (0.87 \text{ kg/m}^3)(9.8 \text{ m/s}^2)(7600 \text{ m}) = 6.48 \times 10^4 \text{ Pa}.$$

Therefore, the change in pressure is

$$\Delta p = p_1 - p_2 = 2.00 \times 10^5 \text{ Pa} - 6.48 \times 10^4 \text{ Pa} \approx 1.4 \times 10^5 \text{ Pa}.$$

15. The hydrostatic blood pressure is the gauge pressure in the column of blood between feet and brain. We calculate the gauge pressure using Eq. 14-7.

(a) The gauge pressure at the brain of the giraffe is

$$p_{brain} = p_{heart} - \rho gh = 250 \text{ torr} - (1.06 \times 10^3 \text{ kg/m}^3)(9.8 \text{ m/s}^2)(2.0 \text{ m}) \frac{1 \text{ torr}}{133.33 \text{ Pa}} = 94 \text{ torr}.$$

(b) The gauge pressure at the feet of the giraffe is

$$p_{feet} = p_{heart} + \rho gh = 250 \text{ torr} + (1.06 \times 10^3 \text{ kg/m}^3)(9.8 \text{ m/s}^2)(2.0 \text{ m}) \frac{1 \text{ torr}}{133.33 \text{ Pa}} = 406 \text{ torr}$$
$$\approx 4.1 \times 10^2 \text{ torr}.$$

(c) The increase in the blood pressure at the brain as the giraffe lower is head to the level of its feet is

$$\Delta p = p_{feet} - p_{brain} = 406 \text{ torr} - 94 \text{ torr} = 312 \text{ torr} \approx 3.1 \times 10^2 \text{ torr}.$$

16. Since the pressure (caused by liquid) at the bottom of the barrel is doubled due to the presence of the narrow tube, so is the hydrostatic force. The ratio is therefore equal to 2.0. The difference between the hydrostatic force and the weight is accounted for by the additional upward force exerted by water on the top of the barrel due to the increased pressure introduced by the water in the tube.

17. The hydrostatic blood pressure is the gauge pressure in the column of blood between feet and brain. We calculate the gauge pressure using Eq. 14-7.

(a) The gauge pressure at the heart of the *Argentinosaurus* is

$$p_{heart} = p_{brain} + \rho gh = 80 \text{ torr} + (1.06 \times 10^3 \text{ kg/m}^3)(9.8 \text{ m/s}^2)(21 \text{ m} - 9.0 \text{ m}) \frac{1 \text{ torr}}{133.33 \text{ Pa}}.$$
$$= 1.0 \times 10^3 \text{ torr}.$$

(b) The gauge pressure at the feet of the *Argentinosaurus* is

$$p_{feet} = p_{brain} + \rho g h' = 80 \text{ torr} + (1.06\times 10^3 \text{ kg/m}^3)(9.8 \text{ m/s}^2)(21 \text{ m})\frac{1 \text{ torr}}{133.33 \text{ Pa}}$$
$$= 80 \text{ torr} + 1642 \text{ torr} = 1722 \text{ torr} \approx 1.7\times 10^3 \text{ torr}.$$

18. At a depth h without the snorkel tube, the external pressure on the diver is

$$p = p_0 + \rho g h$$

where p_0 is the atmospheric pressure. Thus, with a snorkel tube of length h, the pressure difference between the internal air pressure and the water pressure against the body is

$$\Delta p = p = p_0 = \rho g h.$$

(a) If $h = 0.20$ m, then

$$\Delta p = \rho g h = (998 \text{ kg/m}^3)(9.8 \text{ m/s}^2)(0.20 \text{ m})\frac{1 \text{ atm}}{1.01\times 10^5 \text{ Pa}} = 0.019 \text{ atm}.$$

(b) Similarly, if $h = 4.0$ m, then

$$\Delta p = \rho g h = (998 \text{ kg/m}^3)(9.8 \text{ m/s}^2)(4.0 \text{ m})\frac{1 \text{ atm}}{1.01\times 10^5 \text{ Pa}} \approx 0.39 \text{ atm}.$$

19. When the levels are the same the height of the liquid is $h = (h_1 + h_2)/2$, where h_1 and h_2 are the original heights. Suppose h_1 is greater than h_2. The final situation can then be achieved by taking liquid with volume $A(h_1 - h)$ and mass $\rho A(h_1 - h)$, in the first vessel, and lowering it a distance $h - h_2$. The work done by the force of gravity is

$$W = \rho A(h_1 - h)g(h - h_2).$$

We substitute $h = (h_1 + h_2)/2$ to obtain

$$W = \frac{1}{4}\rho g A(h_1 - h_2)^2 = \frac{1}{4}(1.30\times 10^3 \text{ kg/m}^3)(9.80 \text{ m/s}^2)(4.00\times 10^{-4} \text{ m}^2)(1.56 \text{ m} - 0.854 \text{ m})^2$$
$$= 0.635 \text{ J}$$

20. To find the pressure at the brain of the pilot, we note that the inward acceleration can be treated from the pilot's reference frame as though it is an outward gravitational acceleration against which the heart must push the blood. Thus, with $a = 4g$, we have

$$p_{brain} = p_{heart} - \rho a r = 120 \text{ torr} - (1.06\times 10^3 \text{ kg/m}^3)(4\times 9.8 \text{ m/s}^2)(0.30 \text{ m})\frac{1 \text{ torr}}{133 \text{ Pa}}$$
$$= 120 \text{ torr} - 94 \text{ torr} = 26 \text{ torr}.$$

21. Letting $p_a = p_b$, we find

$$\rho_c g(6.0 \text{ km} + 32 \text{ km} + D) + \rho_m(y - D) = \rho_c g(32 \text{ km}) + \rho_m y$$

and obtain

$$D = \frac{(6.0\text{km})\rho_c}{\rho_m - \rho_c} = \frac{(6.0\text{km})(2.9\text{g/cm}^3)}{3.3\text{g/cm}^3 - 2.9\text{g/cm}^3} = 44\text{ km}.$$

22. (a) The force on face A of area A_A due to the water pressure alone is

$$F_A = p_A A_A = \rho_w g h_A A_A = \rho_w g(2d)d^2 = 2(1.0\times10^3 \text{ kg/m}^3)(9.8\text{m/s}^2)(5.0\text{m})^3$$
$$= 2.5\times10^6 \text{ N}.$$

Adding the contribution from the atmospheric pressure,

$$F_0 = (1.0 \times 10^5 \text{ Pa})(5.0 \text{ m})^2 = 2.5 \times 10^6 \text{ N},$$

we have

$$F_A' = F_0 + F_A = 2.5\times10^6 \text{ N} + 2.5\times10^6 \text{ N} = 5.0\times10^6 \text{ N}.$$

(b) The force on face B due to water pressure alone is

$$F_B = p_{avgB} A_B = \rho_\omega g \left(\frac{5d}{2}\right) d^2 = \frac{5}{2}\rho_w g d^3 = \frac{5}{2}(1.0\times10^3 \text{ kg/m}^3)(9.8\text{m/s}^2)(5.0\text{m})^3$$
$$= 3.1\times10^6 \text{ N}.$$

Adding the contribution from the atmospheric pressure,

$$F_0 = (1.0 \times 10^5 \text{ Pa})(5.0 \text{ m})^2 = 2.5 \times 10^6 \text{ N},$$

we obtain

$$F_B' = F_0 + F_B = 2.5\times10^6 \text{ N} + 3.1\times10^6 \text{ N} = 5.6\times10^6 \text{ N}.$$

23. We can integrate the pressure (which varies linearly with depth according to Eq. 14-7) over the area of the wall to find out the net force on it, and the result turns out fairly intuitive (because of that linear dependence): the force is the "average" water pressure multiplied by the area of the wall (or at least the part of the wall that is exposed to the water), where "average" pressure is taken to mean $\frac{1}{2}$(pressure at surface + pressure at bottom). Assuming the pressure at the surface can be taken to be zero (in the gauge pressure sense explained in section 14-4), then this means the force on the wall is $\frac{1}{2}\rho g h$ multiplied by the appropriate area. In this problem the area is hw (where w is the 8.00 m width), so the force is $\frac{1}{2}\rho g h^2 w$, and the change in force (as h is changed) is

$$\tfrac{1}{2}\rho g w\,(h_f^{\,2}-h_i^{\,2}) = \tfrac{1}{2}(998\text{ kg/m}^3)(9.80\text{ m/s}^2)(8.00\text{ m})(4.00^2-2.00^2)\text{m}^2 = 4.69\times 10^5\text{ N}.$$

24. (a) At depth y the gauge pressure of the water is $p = \rho g y$, where ρ is the density of the water. We consider a horizontal strip of width W at depth y, with (vertical) thickness dy, across the dam. Its area is $dA = W\,dy$ and the force it exerts on the dam is $dF = p\,dA = \rho g y W\,dy$. The total force of the water on the dam is

$$F = \int_0^D \rho g y W\,dy = \tfrac{1}{2}\rho g W D^2 = \tfrac{1}{2}(1.00\times 10^3\text{ kg/m}^3)(9.80\text{ m/s}^2)(314\text{ m})(35.0\text{ m})^2$$
$$= 1.88\times 10^9\text{ N}.$$

(b) Again we consider the strip of water at depth y. Its moment arm for the torque it exerts about O is $D - y$ so the torque it exerts is

$$d\tau = dF(D-y) = \rho g y W(D-y)\,dy$$

and the total torque of the water is

$$\tau = \int_0^D \rho g y W(D-y)\,dy = \rho g W\left(\tfrac{1}{2}D^3 - \tfrac{1}{3}D^3\right) = \tfrac{1}{6}\rho g W D^3$$
$$= \tfrac{1}{6}(1.00\times 10^3\text{ kg/m}^3)(9.80\text{ m/s}^2)(314\text{ m})(35.0\text{ m})^3 = 2.20\times 10^{10}\text{ N}\cdot\text{m}.$$

(c) We write $\tau = rF$, where r is the effective moment arm. Then,

$$r = \frac{\tau}{F} = \frac{\tfrac{1}{6}\rho g W D^3}{\tfrac{1}{2}\rho g W D^2} = \frac{D}{3} = \frac{35.0\text{ m}}{3} = 11.7\text{ m}.$$

25. As shown in Eq. 14-9, the atmospheric pressure p_0 bearing down on the barometer's mercury pool is equal to the pressure $\rho g h$ at the base of the mercury column: $p_0 = \rho g h$. Substituting the values given in the problem statement, we find the atmospheric pressure to be

$$p_0 = \rho g h = (1.3608\times 10^4\text{ kg/m}^3)(9.7835\text{ m/s}^2)(0.74035\text{ m})\frac{1\text{ torr}}{133.33\text{ Pa}} = 739.26\text{ torr}.$$

26. The gauge pressure you can produce is

$$p = -\rho g h = -\frac{(1000\text{ kg/m}^3)(9.8\text{ m/s}^2)(4.0\times 10^{-2}\text{ m})}{1.01\times 10^5\text{ Pa/atm}} = -3.9\times 10^{-3}\text{ atm}.$$

where the minus sign indicates that the pressure inside your lung is less than the outside pressure.

27. (a) We use the expression for the variation of pressure with height in an incompressible fluid: $p_2 = p_1 - \rho g(y_2 - y_1)$. We take y_1 to be at the surface of Earth, where the pressure is $p_1 = 1.01 \times 10^5$ Pa, and y_2 to be at the top of the atmosphere, where the pressure is $p_2 = 0$. For this calculation, we take the density to be uniformly 1.3 kg/m^3. Then,

$$y_2 - y_1 = \frac{p_1}{\rho g} = \frac{1.01 \times 10^5 \text{ Pa}}{(1.3 \text{ kg/m}^3)(9.8 \text{ m/s}^2)} = 7.9 \times 10^3 \text{ m} = 7.9 \text{ km}.$$

(b) Let h be the height of the atmosphere. Now, since the density varies with altitude, we integrate

$$p_2 = p_1 - \int_0^h \rho g \, dy.$$

Assuming $\rho = \rho_0 (1 - y/h)$, where ρ_0 is the density at Earth's surface and $g = 9.8$ m/s^2 for $0 \leq y \leq h$, the integral becomes

$$p_2 = p_1 - \int_0^h \rho_0 g \left(1 - \frac{y}{h}\right) dy = p_1 - \frac{1}{2}\rho_0 g h.$$

Since $p_2 = 0$, this implies

$$h = \frac{2p_1}{\rho_0 g} = \frac{2(1.01 \times 10^5 \text{ Pa})}{(1.3 \text{ kg/m}^3)(9.8 \text{ m/s}^2)} = 16 \times 10^3 \text{ m} = 16 \text{ km}.$$

28. (a) According to Pascal's principle $F/A = f/a \rightarrow F = (A/a)f$.

(b) We obtain

$$f = \frac{a}{A} F = \frac{(3.80 \text{ cm})^2}{(53.0 \text{ cm})^2} (20.0 \times 10^3 \text{ N}) = 103 \text{ N}.$$

The ratio of the squares of diameters is equivalent to the ratio of the areas. We also note that the area units cancel.

29. Eq. 14-13 combined with Eq. 5-8 and Eq. 7-21 (in absolute value) gives

$$mg = kx \frac{A_1}{A_2}.$$

With $A_2 = 18 A_1$ (and the other values given in the problem) we find $m = 8.50$ kg.

30. (a) The pressure (including the contribution from the atmosphere) at a depth of $h_{top} = L/2$ (corresponding to the top of the block) is

$$p_{top} = p_{atm} + \rho g h_{top} = 1.01 \times 10^5 \text{ Pa} + (1030 \text{ kg/m}^3)(9.8 \text{ m/s}^2)(0.300 \text{ m}) = 1.04 \times 10^5 \text{ Pa}$$

where the unit Pa (Pascal) is equivalent to N/m². The force on the top surface (of area $A = L^2 = 0.36 \text{ m}^2$) is

$$F_{top} = p_{top} A = 3.75 \times 10^4 \text{ N}.$$

(b) The pressure at a depth of $h_{bot} = 3L/2$ (that of the bottom of the block) is

$$p_{bot} = p_{atm} + \rho g h_{bot} = 1.01 \times 10^5 \text{ Pa} + (1030 \text{ kg/m}^3)(9.8 \text{ m/s}^2)(0.900 \text{ m}) = 1.10 \times 10^5 \text{ Pa}$$

where we recall that the unit Pa (Pascal) is equivalent to N/m². The force on the bottom surface is

$$F_{bot} = p_{bot} A = 3.96 \times 10^4 \text{ N}.$$

(c) Taking the difference $F_{bot} - F_{top}$ cancels the contribution from the atmosphere (including any numerical uncertainties associated with that value) and leads to

$$F_{bot} - F_{top} = \rho g (h_{bot} - h_{top}) A = \rho g L^3 = 2.18 \times 10^3 \text{ N}$$

which is to be expected on the basis of Archimedes' principle. Two other forces act on the block: an upward tension T and a downward pull of gravity mg. To remain stationary, the tension must be

$$T = mg - (F_{bot} - F_{top}) = (450 \text{ kg})(9.80 \text{ m/s}^2) - 2.18 \times 10^3 \text{ N} = 2.23 \times 10^3 \text{ N}.$$

(d) This has already been noted in the previous part: $F_b = 2.18 \times 10^3 \text{ N}$, and $T + F_b = mg$.

31. (a) The anchor is completely submerged in water of density ρ_w. Its effective weight is $W_{eff} = W - \rho_w g V$, where W is its actual weight (mg). Thus,

$$V = \frac{W - W_{eff}}{\rho_w g} = \frac{200 \text{ N}}{(1000 \text{ kg/m}^3)(9.8 \text{ m/s}^2)} = 2.04 \times 10^{-2} \text{ m}^3.$$

(b) The mass of the anchor is $m = \rho V$, where ρ is the density of iron (found in Table 14-1). Its weight in air is

$$W = mg = \rho V g = (7870 \text{ kg/m}^3)(2.04 \times 10^{-2} \text{ m}^3)(9.80 \text{ m/s}^2) = 1.57 \times 10^3 \text{ N}.$$

32. (a) Archimedes' principle makes it clear that a body, in order to float, displaces an amount of the liquid which corresponds to the weight of the body. The problem (indirectly) tells us that the weight of the boat is $W = 35.6$ kN. In salt water of density $\rho' = 1100$ kg/m^3, it must displace an amount of liquid having weight equal to 35.6 kN.

(b) The displaced volume of salt water is equal to

$$V' = \frac{W}{\rho' g} = \frac{3.56 \times 10^3 \text{ N}}{(1.10 \times 10^3 \text{ kg/m}^3)(9.80 \text{ m/s}^2)} = 3.30 \text{ m}^3.$$

In freshwater, it displaces a volume of $V = W/\rho g = 3.63$ m^3, where $\rho = 1000$ kg/m^3. The difference is $V - V' = 0.330$ m^3.

33. The problem intends for the children to be completely above water. The total downward pull of gravity on the system is

$$3(356 \text{ N}) + N \rho_{wood} g V$$

where N is the (minimum) number of logs needed to keep them afloat and V is the volume of each log: $V = \pi (0.15 \text{ m})^2 (1.80 \text{ m}) = 0.13$ m^3. The buoyant force is $F_b = \rho_{water} g V_{submerged}$ where we require $V_{submerged} \leq NV$. The density of water is 1000 kg/m^3. To obtain the minimum value of N we set $V_{submerged} = NV$ and then round our "answer" for N up to the nearest integer:

$$3(356 \text{ N}) + N \rho_{wood} g V = \rho_{water} g N V \quad \Rightarrow \quad N = \frac{3(356 \text{ N})}{gV(\rho_{water} - \rho_{wood})}$$

which yields $N = 4.28 \rightarrow 5$ logs.

34. Taking "down" as the positive direction, then using Eq. 14-16 in Newton's second law, we have $5g - 3g = 5a$ (where "5" = 5.00 kg, and "3" = 3.00 kg and $g = 9.8$ m/s^2). This gives $a = \frac{2}{5}g$. Then (see Eq. 2-15) $\frac{1}{2}at^2 = 0.0784$ m (in the downward direction).

35. (a) Let V be the volume of the block. Then, the submerged volume is $V_s = 2V/3$. Since the block is floating, the weight of the displaced water is equal to the weight of the block, so $\rho_w V_s = \rho_b V$, where ρ_w is the density of water, and ρ_b is the density of the block. We substitute $V_s = 2V/3$ to obtain

$$\rho_b = 2\rho_w/3 = 2(1000 \text{ kg/m}^3)/3 \approx 6.7 \times 10^2 \text{ kg/m}^3.$$

(b) If ρ_o is the density of the oil, then Archimedes' principle yields $\rho_o V_s = \rho_b V$. We substitute $V_s = 0.90V$ to obtain $\rho_o = \rho_b/0.90 = 7.4 \times 10^2$ kg/m^3.

36. Work is the integral of the force (over distance – see Eq. 7-32), and referring to the equation immediately preceding Eq. 14-7, we see the work can be written as

$$W = \int \rho_{water} \, gA(-y) \, dy$$

where we are using $y = 0$ to refer to the water surface (and the $+y$ direction is upward). Let $h = 0.500$ m. Then, the integral has a lower limit of $-h$ and an upper limit of y_f, with $y_f/h = -\rho_{cylinder}/\rho_{water} = -0.400$. The integral leads to

$$W = \tfrac{1}{2}\rho_{water}gAh^2(1 - 0.4^2) = 4.11 \text{ kJ}.$$

37. (a) The downward force of gravity mg is balanced by the upward buoyant force of the liquid: $mg = \rho g V_s$. Here m is the mass of the sphere, ρ is the density of the liquid, and V_s is the submerged volume. Thus $m = \rho V_s$. The submerged volume is half the total volume of the sphere, so $V_s = \tfrac{1}{2}(4\pi/3)r_o^3$, where r_o is the outer radius. Therefore,

$$m = \frac{2\pi}{3}\rho r_o^3 = \left(\frac{2\pi}{3}\right)(800 \text{ kg/m}^3)(0.090 \text{ m})^3 = 1.22 \text{ kg}.$$

(b) The density ρ_m of the material, assumed to be uniform, is given by $\rho_m = m/V$, where m is the mass of the sphere and V is its volume. If r_i is the inner radius, the volume is

$$V = \frac{4\pi}{3}(r_o^3 - r_i^3) = \frac{4\pi}{3}\left((0.090 \text{ m})^3 - (0.080 \text{ m})^3\right) = 9.09 \times 10^{-4} \text{ m}^3.$$

The density is

$$\rho_m = \frac{1.22 \text{ kg}}{9.09 \times 10^{-4} \text{ m}^3} = 1.3 \times 10^3 \text{ kg/m}^3.$$

38. If the alligator floats, by Archimedes' principle the buoyancy force is equal to the alligator's weight (see Eq. 14-17). Therefore,

$$F_b = F_g = m_{H_2O}g = (\rho_{H_2O}Ah)g.$$

If the mass is to increase by a small amount $m \rightarrow m' = m + \Delta m$, then

$$F_b \rightarrow F_b' = \rho_{H_2O}A(h + \Delta h)g.$$

With $\Delta F_b = F_b' - F_b = 0.010mg$, the alligator sinks by

$$\Delta h = \frac{\Delta F_b}{\rho_{H_2O}Ag} = \frac{0.01mg}{\rho_{H_2O}Ag} = \frac{0.010(130 \text{ kg})}{(998 \text{ kg/m}^3)(0.20 \text{ m}^2)} = 6.5 \times 10^{-3} \text{ m} = 6.5 \text{ mm}.$$

39. Let V_i be the total volume of the iceberg. The non-visible portion is below water, and thus the volume of this portion is equal to the volume V_f of the fluid displaced by the iceberg. The fraction of the iceberg that is visible is

$$\text{frac} = \frac{V_i - V_f}{V_i} = 1 - \frac{V_f}{V_i}.$$

Since iceberg is floating, Eq. 14-18 applies:

$$F_g = m_i g = m_f g \Rightarrow m_i = m_f.$$

Since $m = \rho V$, the above equation implies

$$\rho_i V_i = \rho_f V_f \Rightarrow \frac{V_f}{V_i} = \frac{\rho_i}{\rho_f}.$$

Thus, the visible fraction is

$$\text{frac} = 1 - \frac{V_f}{V_i} = 1 - \frac{\rho_i}{\rho_f}$$

(a) If the iceberg ($\rho_i = 917$ kg/m^3) floats in saltwater with $\rho_f = 1024$ kg/m^3, then the fraction would be

$$\text{frac} = 1 - \frac{\rho_i}{\rho_f} = 1 - \frac{917 \text{ kg/m}^3}{1024 \text{ kg/m}^3} = 0.10 = 10\%.$$

(b) On the other hand, if the iceberg floats in fresh water ($\rho_f = 1000$ kg/m^3), then the fraction would be

$$\text{frac} = 1 - \frac{\rho_i}{\rho_f} = 1 - \frac{917 \text{ kg/m}^3}{1000 \text{ kg/m}^3} = 0.083 = 8.3\%.$$

40. (a) An object of the same density as the surrounding liquid (in which case the "object" could just be a packet of the liquid itself) is not going to accelerate up or down (and thus won't gain any kinetic energy). Thus, the point corresponding to zero K in the graph must correspond to the case where the density of the object equals ρ_{liquid}. Therefore, $\rho_{\text{ball}} = 1.5$ g/cm^3 (or 1500 kg/m^3).

(b) Consider the $\rho_{\text{liquid}} = 0$ point (where $K_{\text{gained}} = 1.6$ J). In this case, the ball is falling through perfect vacuum, so that $v^2 = 2gh$ (see Eq. 2-16) which means that $K = \frac{1}{2}mv^2 = 1.6$ J can be used to solve for the mass. We obtain $m_{\text{ball}} = 4.082$ kg. The volume of the ball is then given by $m_{\text{ball}}/\rho_{\text{ball}} = 2.72 \times 10^{-3}$ m^3.

41. For our estimate of $V_{submerged}$ we interpret "almost completely submerged" to mean

$$V_{submerged} \approx \frac{4}{3}\pi r_o^3 \quad \text{where } r_o = 60 \text{ cm}.$$

Thus, equilibrium of forces (on the iron sphere) leads to

$$F_b = m_{iron} g \quad \Rightarrow \quad \rho_{water} g V_{submerged} = \rho_{iron} g \left(\frac{4}{3}\pi r_o^3 - \frac{4}{3}\pi r_i^3 \right)$$

where r_i is the inner radius (half the inner diameter). Plugging in our estimate for $V_{submerged}$ as well as the densities of water (1.0 g/cm^3) and iron (7.87 g/cm^3), we obtain the inner diameter:

$$2r_i = 2r_o \left(1 - \frac{1.0 \text{ g/cm}^3}{7.87 \text{ g/cm}^3} \right)^{1/3} = 57.3 \text{ cm}.$$

42. From the "kink" in the graph it is clear that $d = 1.5$ cm. Also, the $h = 0$ point makes it clear that the (true) weight is 0.25 N. We now use Eq. 14-19 at $h = d = 1.5$ cm to obtain

$$F_b = (0.25 \text{ N} - 0.10 \text{ N}) = 0.15 \text{ N}.$$

Thus, $\rho_{liquid} g V = 0.15$, where $V = (1.5 \text{ cm})(5.67 \text{ cm}^2) = 8.5 \times 10^{-6}$ m^3. Thus, $\rho_{liquid} = 1800$ kg/m^3 = 1.8 g/cm^3.

43. The volume V_{cav} of the cavities is the difference between the volume V_{cast} of the casting as a whole and the volume V_{iron} contained: $V_{cav} = V_{cast} - V_{iron}$. The volume of the iron is given by $V_{iron} = W/g\rho_{iron}$, where W is the weight of the casting and ρ_{iron} is the density of iron. The effective weight in water (of density ρ_w) is $W_{eff} = W - g\rho_w V_{cast}$. Thus, $V_{cast} = (W - W_{eff})/g\rho_w$ and

$$V_{cav} = \frac{W - W_{eff}}{g\rho_w} - \frac{W}{g\rho_{iron}} = \frac{6000 \text{ N} - 4000 \text{ N}}{(9.8 \text{ m/s}^2)(1000 \text{ kg/m}^3)} - \frac{6000 \text{ N}}{(9.8 \text{ m/s}^2)(7.87 \times 10^3 \text{ kg/m}^3)}$$
$$= 0.126 \text{ m}^3.$$

44. Due to the buoyant force, the ball accelerates upward (while in the water) at rate a given by Newton's second law:

$$\rho_{water} V g - \rho_{ball} V g = \rho_{ball} V a \quad \Rightarrow \quad \rho_{ball} = \rho_{water}(1 + \text{``}a\text{''})$$

where – for simplicity – we are using in that last expression an acceleration "a" measured in "gees" (so that "a" = 2, for example, means that $a = 2(9.80 \text{ m/s}^2) = 19.6$ m/s^2). In this problem, with $\rho_{ball} = 0.300\, \rho_{water}$, we find therefore that "a" = 7/3. Using Eq. 2-16, then the speed of the ball as it emerges from the water is

$$v = \sqrt{2a\Delta y},$$

were $a = (7/3)g$ and $\Delta y = 0.600$ m. This causes the ball to reach a maximum height h_{max} (measured above the water surface) given by $h_{max} = v^2/2g$ (see Eq. 2-16 again). Thus, $h_{max} = (7/3)\Delta y = 1.40$ m.

45. (a) If the volume of the car below water is V_1 then $F_b = \rho_w V_1 g = W_{car}$, which leads to

$$V_1 = \frac{W_{car}}{\rho_w g} = \frac{(1800\,\text{kg})(9.8\,\text{m/s}^2)}{(1000\,\text{kg/m}^3)(9.8\,\text{m/s}^2)} = 1.80\,\text{m}^3.$$

(b) We denote the total volume of the car as V and that of the water in it as V_2. Then

$$F_b = \rho_w V g = W_{car} + \rho_w V_2 g$$

which gives

$$V_2 = V - \frac{W_{car}}{\rho_w g} = (0.750\,\text{m}^3 + 5.00\,\text{m}^3 + 0.800\,\text{m}^3) - \frac{1800\,\text{kg}}{1000\,\text{kg/m}^3} = 4.75\,\text{m}^3.$$

46. (a) Since the lead is not displacing any water (of density ρ_w), the lead's volume is not contributing to the buoyant force F_b. If the immersed volume of wood is V_i, then

$$F_b = \rho_w V_i g = 0.900\,\rho_w V_{wood}\, g = 0.900\,\rho_w g \left(\frac{m_{wood}}{\rho_{wood}}\right),$$

which, when floating, equals the weights of the wood and lead:

$$F_b = 0.900\,\rho_w g \left(\frac{m_{wood}}{\rho_{wood}}\right) = (m_{wood} + m_{lead})g.$$

Thus,

$$m_{lead} = 0.900\,\rho_w \left(\frac{m_{wood}}{\rho_{wood}}\right) - m_{wood} = \frac{(0.900)(1000\,\text{kg/m}^3)(3.67\,\text{kg})}{600\,\text{kg/m}^3} - 3.67\,\text{kg} = 1.84\,\text{kg}.$$

(b) In this case, the volume $V_{lead} = m_{lead}/\rho_{lead}$ also contributes to F_b. Consequently,

$$F_b = 0.900\,\rho_w g \left(\frac{m_{wood}}{\rho_{wood}}\right) + \left(\frac{\rho_w}{\rho_{lead}}\right) m_{lead}\, g = (m_{wood} + m_{lead})g,$$

which leads to

$$m_{\text{lead}} = \frac{0.900(\rho_w/\rho_{\text{wood}})m_{\text{wood}} - m_{\text{wood}}}{1 - \rho_w/\rho_{\text{lead}}} = \frac{1.84 \text{ kg}}{1 - (1.00 \times 10^3 \text{ kg/m}^3 / 1.13 \times 10^4 \text{ kg/m}^3)}$$

$$= 2.01 \text{ kg}.$$

47. (a) When the model is suspended (in air) the reading is F_g (its true weight, neglecting any buoyant effects caused by the air). When the model is submerged in water, the reading is lessened because of the buoyant force: $F_g - F_b$. We denote the difference in readings as Δm. Thus,

$$F_g - (F_g - F_b) = \Delta mg$$

which leads to $F_b = \Delta mg$. Since $F_b = \rho_w g V_m$ (the weight of water displaced by the model) we obtain

$$V_m = \frac{\Delta m}{\rho_w} = \frac{0.63776 \text{ kg}}{1000 \text{ kg/m}} \approx 6.378 \times 10^{-4} \text{ m}^3.$$

(b) The $\frac{1}{20}$ scaling factor is discussed in the problem (and for purposes of significant figures is treated as exact). The actual volume of the dinosaur is

$$V_{\text{dino}} = 20^3 \, V_m = 5.102 \text{ m}^3.$$

(c) Using $\rho \approx \rho_w = 1000 \text{ kg/m}^3$, we find

$$\rho = \frac{m_{\text{dino}}}{V_{\text{dino}}} \Rightarrow m_{\text{dino}} = (1000 \text{ kg/m}^3)(5.102 \text{ m}^3)$$

which yields 5.102×10^3 kg for the *T. rex* mass.

48. Let ρ be the density of the cylinder (0.30 g/cm³ or 300 kg/m³) and ρ_{Fe} be the density of the iron (7.9 g/cm³ or 7900 kg/m³). The volume of the cylinder is

$$V_c = (6 \times 12) \text{ cm}^3 = 72 \text{ cm}^3 = 0.000072 \text{ m}^3,$$

and that of the ball is denoted V_b. The part of the cylinder that is submerged has volume

$$V_s = (4 \times 12) \text{ cm}^3 = 48 \text{ cm}^3 = 0.000048 \text{ m}^3.$$

Using the ideas of section 14-7, we write the equilibrium of forces as

$$\rho g V_c + \rho_{\text{Fe}} g V_b = \rho_w g V_s + \rho_w g V_b \Rightarrow V_b = 3.8 \text{ cm}^3$$

where we have used $\rho_w = 998$ kg/m^3 (for water, see Table 14-1). Using $V_b = \frac{4}{3}\pi r^3$ we find $r = 9.7$ mm.

49. We use the equation of continuity. Let v_1 be the speed of the water in the hose and v_2 be its speed as it leaves one of the holes. $A_1 = \pi R^2$ is the cross-sectional area of the hose. If there are N holes and A_2 is the area of a single hole, then the equation of continuity becomes

$$v_1 A_1 = v_2 (N A_2) \Rightarrow v_2 = \frac{A_1}{NA_2} v_1 = \frac{R^2}{Nr^2} v_1$$

where R is the radius of the hose and r is the radius of a hole. Noting that $R/r = D/d$ (the ratio of diameters) we find

$$v_2 = \frac{D^2}{Nd^2} v_1 = \frac{(1.9 \text{ cm})^2}{24 (0.13 \text{ cm})^2} (0.91 \text{ m/s}) = 8.1 \text{ m/s}.$$

50. We use the equation of continuity and denote the depth of the river as h. Then,

$$(8.2 \text{ m})(3.4 \text{ m})(2.3 \text{ m/s}) + (6.8 \text{ m})(3.2 \text{ m})(2.6 \text{ m/s}) = h(10.5 \text{ m})(2.9 \text{ m/s})$$

which leads to $h = 4.0$ m.

51. This problem involves use of continuity equation (Eq. 14-23): $A_1 v_1 = A_2 v_2$.

(a) Initially the flow speed is $v_i = 1.5$ m/s and the cross-sectional area is $A_i = HD$. At point a, as can be seen from Fig. 14-47, the cross-sectional area is

$$A_a = (H-h)D - (b-h)d.$$

Thus, by continuity equation, the speed at point a is

$$v_a = \frac{A_i v_i}{A_a} = \frac{HD v_i}{(H-h)D - (b-h)d} = \frac{(14 \text{ m})(55 \text{ m})(1.5 \text{ m/s})}{(14 \text{ m} - 0.80 \text{ m})(55 \text{ m}) - (12 \text{ m} - 0.80 \text{ m})(30 \text{ m})} = 2.96 \text{ m/s}$$
$$\approx 3.0 \text{ m/s}.$$

(b) Similarly, at point b, the cross-sectional area is $A_b = HD - bd$, and therefore, by continuity equation, the speed at point b is

$$v_b = \frac{A_i v_i}{A_b} = \frac{HD v_i}{HD - bd} = \frac{(14 \text{ m})(55 \text{ m})(1.5 \text{ m/s})}{(14 \text{ m})(55 \text{ m}) - (12 \text{ m})(30 \text{ m})} = 2.8 \text{ m/s}.$$

52. The left and right sections have a total length of 60.0 m, so (with a speed of 2.50 m/s) it takes $60.0/2.50 = 24.0$ seconds to travel through those sections. Thus it takes $(88.8 - 24.0)$ s $= 64.8$ s to travel through the middle section. This implies that the speed in the middle section is $v_{mid} = (110 \text{ m})/(64.8 \text{ s}) = 0.772$ m/s. Now Eq. 14-23 (plus that fact that $A = \pi r^2$) implies $r_{mid} = r_A\sqrt{(2.5 \text{ m/s})/(0.772 \text{ m/s})}$ where $r_A = 2.00$ cm. Therefore, $r_{mid} = 3.60$ cm.

53. Suppose that a mass Δm of water is pumped in time Δt. The pump increases the potential energy of the water by Δmgh, where h is the vertical distance through which it is lifted, and increases its kinetic energy by $\frac{1}{2}\Delta mv^2$, where v is its final speed. The work it does is $\Delta W = \Delta mgh + \frac{1}{2}\Delta mv^2$ and its power is

$$P = \frac{\Delta W}{\Delta t} = \frac{\Delta m}{\Delta t}\left(gh + \frac{1}{2}v^2\right).$$

Now the rate of mass flow is $\Delta m/\Delta t = \rho_w Av$, where ρ_w is the density of water and A is the area of the hose. The area of the hose is $A = \pi r^2 = \pi(0.010 \text{ m})^2 = 3.14 \times 10^{-4}$ m^2 and

$$\rho_w Av = (1000 \text{ kg/m}^3)(3.14 \times 10^{-4} \text{ m}^2)(5.00 \text{ m/s}) = 1.57 \text{ kg/s}.$$

Thus,

$$P = \rho Av\left(gh + \frac{1}{2}v^2\right) = (1.57 \text{ kg/s})\left((9.8 \text{ m/s}^2)(3.0 \text{ m}) + \frac{(5.0 \text{ m/s})^2}{2}\right) = 66 \text{ W}.$$

54. (a) The equation of continuity provides $(26 + 19 + 11)$ L/min = 56 L/min for the flow rate in the main (1.9 cm diameter) pipe.

(b) Using $v = R/A$ and $A = \pi d^2/4$, we set up ratios:

$$\frac{v_{56}}{v_{26}} = \frac{56/\pi(1.9)^2/4}{26/\pi(1.3)^2/4} \approx 1.0.$$

55. (a) We use the equation of continuity: $A_1v_1 = A_2v_2$. Here A_1 is the area of the pipe at the top and v_1 is the speed of the water there; A_2 is the area of the pipe at the bottom and v_2 is the speed of the water there. Thus

$$v_2 = (A_1/A_2)v_1 = [(4.0 \text{ cm}^2)/(8.0 \text{ cm}^2)](5.0 \text{ m/s}) = 2.5 \text{ m/s}.$$

(b) We use the Bernoulli equation:

$$p_1 + \tfrac{1}{2}\rho v_1^2 + \rho gh_1 = p_2 + \tfrac{1}{2}\rho v_2^2 + \rho gh_2,$$

where ρ is the density of water, h_1 is its initial altitude, and h_2 is its final altitude. Thus

$$p_2 = p_1 + \frac{1}{2}\rho(v_1^2 - v_2^2) + \rho g(h_1 - h_2)$$

$$= 1.5 \times 10^5 \text{ Pa} + \frac{1}{2}(1000 \text{ kg/m}^3)\left[(5.0 \text{ m/s})^2 - (2.5 \text{ m/s})^2\right] + (1000 \text{ kg/m}^3)(9.8 \text{ m/s}^2)(10 \text{ m})$$

$$= 2.6 \times 10^5 \text{ Pa}.$$

56. We use Bernoulli's equation:

$$p_2 - p_i = \rho g D + \frac{1}{2}\rho(v_1^2 - v_2^2)$$

where $\rho = 1000$ kg/m^3, $D = 180$ m, $v_1 = 0.40$ m/s and $v_2 = 9.5$ m/s. Therefore, we find $\Delta p = 1.7 \times 10^6$ Pa, or 1.7 MPa. The SI unit for pressure is the Pascal (Pa) and is equivalent to N/m^2.

57. (a) The equation of continuity leads to

$$v_2 A_2 = v_1 A_1 \implies v_2 = v_1\left(\frac{r_1^2}{r_2^2}\right)$$

which gives $v_2 = 3.9$ m/s.

(b) With $h = 7.6$ m and $p_1 = 1.7 \times 10^5$ Pa, Bernoulli's equation reduces to

$$p_2 = p_1 - \rho g h + \frac{1}{2}\rho(v_1^2 - v_2^2) = 8.8 \times 10^4 \text{ Pa}.$$

58. (a) We use $Av = $ const. The speed of water is

$$v = \frac{(25.0 \text{ cm})^2 - (5.00 \text{ cm})^2}{(25.0 \text{ cm})^2}(2.50 \text{ m/s}) = 2.40 \text{ m/s}.$$

(b) Since $p + \frac{1}{2}\rho v^2 = $ const., the pressure difference is

$$\Delta p = \frac{1}{2}\rho \Delta v^2 = \frac{1}{2}(1000 \text{ kg/m}^3)\left[(2.50 \text{ m/s})^2 - (2.40 \text{ m/s})^2\right] = 245 \text{ Pa}.$$

59. (a) We use the Bernoulli equation:

$$p_1 + \frac{1}{2}\rho v_1^2 + \rho g h_1 = p_2 + \frac{1}{2}\rho v_2^2 + \rho g h_2,$$

where h_1 is the height of the water in the tank, p_1 is the pressure there, and v_1 is the speed of the water there; h_2 is the altitude of the hole, p_2 is the pressure there, and v_2 is the speed of the water there. ρ is the density of water. The pressure at the top of the tank and at the hole is atmospheric, so $p_1 = p_2$. Since the tank is large we may neglect the water speed at the top; it is much smaller than the speed at the hole. The Bernoulli equation then becomes $\rho g h_1 = \frac{1}{2}\rho v_2^2 + \rho g h_2$ and

$$v_2 = \sqrt{2g(h_1 - h_2)} = \sqrt{2(9.8\,\text{m/s}^2)(0.30\,\text{m})} = 2.42\,\text{m/s}.$$

The flow rate is $A_2 v_2 = (6.5 \times 10^{-4}\,\text{m}^2)(2.42\,\text{m/s}) = 1.6 \times 10^{-3}\,\text{m}^3/\text{s}$.

(b) We use the equation of continuity: $A_2 v_2 = A_3 v_3$, where $A_3 = \frac{1}{2} A_2$ and v_3 is the water speed where the area of the stream is half its area at the hole. Thus

$$v_3 = (A_2/A_3)v_2 = 2v_2 = 4.84\,\text{m/s}.$$

The water is in free fall and we wish to know how far it has fallen when its speed is doubled to 4.84 m/s. Since the pressure is the same throughout the fall, $\frac{1}{2}\rho v_2^2 + \rho g h_2 = \frac{1}{2}\rho v_3^2 + \rho g h_3$. Thus

$$h_2 - h_3 = \frac{v_3^2 - v_2^2}{2g} = \frac{(4.84\,\text{m/s})^2 - (2.42\,\text{m/s})^2}{2(9.8\,\text{m/s}^2)} = 0.90\,\text{m}.$$

60. (a) The speed v of the fluid flowing out of the hole satisfies $\frac{1}{2}\rho v^2 = \rho g h$ or $v = \sqrt{2gh}$. Thus, $\rho_1 v_1 A_1 = \rho_2 v_2 A_2$, which leads to

$$\rho_1 \sqrt{2gh}\, A_1 = \rho_2 \sqrt{2gh}\, A_2 \quad \Rightarrow \quad \frac{\rho_1}{\rho_2} = \frac{A_2}{A_1} = 2.$$

(b) The ratio of volume flow is

$$\frac{R_1}{R_2} = \frac{v_1 A_1}{v_2 A_2} = \frac{A_1}{A_2} = \frac{1}{2}$$

(c) Letting $R_1/R_2 = 1$, we obtain $v_1/v_2 = A_2/A_1 = 2 = \sqrt{h_1/h_2}$. Thus

$$h_2 = h_1/4 = (12.0\,\text{cm})/4 = 3.00\,\text{cm}.$$

61. We rewrite the formula for work W (when the force is constant in a direction parallel to the displacement d) in terms of pressure:

$$W = Fd = \left(\frac{F}{A}\right)(Ad) = pV$$

where V is the volume of the water being forced through, and p is to be interpreted as the pressure difference between the two ends of the pipe. Thus,

$$W = (1.0 \times 10^5 \text{ Pa})(1.4 \text{ m}^3) = 1.4 \times 10^5 \text{ J}.$$

62. (a) The volume of water (during 10 minutes) is

$$V = (v_1 t) A_1 = (15 \text{ m/s})(10 \text{ min})(60 \text{ s/min})\left(\frac{\pi}{4}\right)(0.03 \text{ m})^2 = 6.4 \text{ m}^3.$$

(b) The speed in the left section of pipe is

$$v_2 = v_1\left(\frac{A_1}{A_2}\right) = v_1\left(\frac{d_1}{d_2}\right)^2 = (15 \text{ m/s})\left(\frac{3.0 \text{ cm}}{5.0 \text{ cm}}\right)^2 = 5.4 \text{ m/s}.$$

(c) Since $p_1 + \frac{1}{2}\rho v_1^2 + \rho g h_1 = p_2 + \frac{1}{2}\rho v_2^2 + \rho g h_2$ and $h_1 = h_2$, $p_1 = p_0$, which is the atmospheric pressure,

$$p_2 = p_0 + \frac{1}{2}\rho(v_1^2 - v_2^2) = 1.01 \times 10^5 \text{ Pa} + \frac{1}{2}(1.0 \times 10^3 \text{ kg/m}^3)\left[(15 \text{ m/s})^2 - (5.4 \text{ m/s})^2\right]$$
$$= 1.99 \times 10^5 \text{ Pa} = 1.97 \text{ atm}.$$

Thus, the gauge pressure is $(1.97 \text{ atm} - 1.00 \text{ atm}) = 0.97 \text{ atm} = 9.8 \times 10^4$ Pa.

63. (a) The friction force is

$$f = A\Delta p = \rho_\omega g d A = (1.0 \times 10^3 \text{ kg/m}^3)(9.8 \text{ m/s}^2)(6.0 \text{ m})\left(\frac{\pi}{4}\right)(0.040 \text{ m})^2 = 74 \text{ N}.$$

(b) The speed of water flowing out of the hole is $v = \sqrt{2gd}$. Thus, the volume of water flowing out of the pipe in $t = 3.0$ h is

$$V = Avt = \frac{\pi}{4}(0.040 \text{ m})^2 \sqrt{2(9.8 \text{ m/s}^2)(6.0 \text{ m})}\,(3.0 \text{ h})(3600 \text{ s/h}) = 1.5 \times 10^2 \text{ m}^3.$$

64. (a) We note (from the graph) that the pressures are equal when the value of inverse-area-squared is 16 (in SI units). This is the point at which the areas of the two pipe sections are equal. Thus, if $A_1 = 1/\sqrt{16}$ when the pressure difference is zero, then A_2 is 0.25 m^2.

(b) Using Bernoulli's equation (in the form Eq. 14-30) we find the pressure difference may be written in the form a straight line: $mx + b$ where x is inverse-area-squared (the horizontal axis in the graph), m is the slope, and b is the intercept (seen to be -300 kN/m^2). Specifically, Eq. 14-30 predicts that b should be $-\frac{1}{2}\rho v_2^2$. Thus, with $\rho = 1000$ kg/m^3 we obtain $v_2 = \sqrt{600}$ m/s. Then the volume flow rate (see Eq. 14-24) is

$$R = A_2 v_2 = (0.25 \text{ m}^2)(\sqrt{600} \text{ m/s}) = 6.12 \text{ m}^3/\text{s}.$$

If the more accurate value (see Table 14-1) $\rho = 998$ kg/m^3 is used, then the answer is 6.13 m^3/s.

65. (a) Since Sample Problem 14-8 deals with a similar situation, we use the final equation (labeled "Answer") from it:

$$v = \sqrt{2gh} \implies v = v_0 \text{ for the projectile motion.}$$

The stream of water emerges horizontally ($\theta_0 = 0°$ in the notation of Chapter 4), and setting $y - y_0 = -(H - h)$ in Eq. 4-22, we obtain the "time-of-flight"

$$t = \sqrt{\frac{-2(H-h)}{-g}} = \sqrt{\frac{2}{g}(H-h)}.$$

Using this in Eq. 4-21, where $x_0 = 0$ by choice of coordinate origin, we find

$$x = v_0 t = \sqrt{2gh}\sqrt{\frac{2(H-h)}{g}} = 2\sqrt{h(H-h)} = 2\sqrt{(10 \text{ cm})(40 \text{ cm} - 10 \text{ cm})} = 35 \text{ cm}.$$

(b) The result of part (a) (which, when squared, reads $x^2 = 4h(H - h)$) is a quadratic equation for h once x and H are specified. Two solutions for h are therefore mathematically possible, but are they both physically possible? For instance, are both solutions positive and less than H? We employ the quadratic formula:

$$h^2 - Hh + \frac{x^2}{4} = 0 \implies h = \frac{H \pm \sqrt{H^2 - x^2}}{2}$$

which permits us to see that both roots are physically possible, so long as $x < H$. Labeling the larger root h_1 (where the plus sign is chosen) and the smaller root as h_2 (where the minus sign is chosen), then we note that their sum is simply

$$h_1 + h_2 = \frac{H + \sqrt{H^2 - x^2}}{2} + \frac{H - \sqrt{H^2 - x^2}}{2} = H.$$

Thus, one root is related to the other (generically labeled h' and h) by $h' = H - h$. Its numerical value is $h' = 40$ cm $- 10$ cm $= 30$ cm.

(c) We wish to maximize the function $f = x^2 = 4h(H - h)$. We differentiate with respect to h and set equal to zero to obtain

$$\frac{df}{dh} = 4H - 8h = 0 \Rightarrow h = \frac{H}{2}$$

or $h = (40$ cm$)/2 = 20$ cm, as the depth from which an emerging stream of water will travel the maximum horizontal distance.

66. By Eq. 14-23, we note that the speeds in the left and right sections are $\frac{1}{4} v_{mid}$ and $\frac{1}{9} v_{mid}$, respectively, where $v_{mid} = 0.500$ m/s. We also note that 0.400 m^3 of water has a mass of 399 kg (see Table 14-1). Then Eq. 14-31 (and the equation below it) gives

$$W = \frac{1}{2} m \, v_{mid}^2 \left(\frac{1}{9^2} - \frac{1}{4^2}\right) = -2.50 \text{ J}.$$

67. (a) The continuity equation yields $Av = aV$, and Bernoulli's equation yields $\Delta p + \frac{1}{2}\rho v^2 = \frac{1}{2}\rho V^2$, where $\Delta p = p_1 - p_2$. The first equation gives $V = (A/a)v$. We use this to substitute for V in the second equation, and obtain $\Delta p + \frac{1}{2}\rho v^2 = \frac{1}{2}\rho(A/a)^2 v^2$. We solve for v. The result is

$$v = \sqrt{\frac{2\Delta p}{\rho((A/a)^2 - 1)}} = \sqrt{\frac{2a^2 \Delta p}{\rho(A^2 - a^2)}}.$$

(b) We substitute values to obtain

$$v = \sqrt{\frac{2(32 \times 10^{-4} \text{m}^2)^2 (55 \times 10^3 \text{Pa} - 41 \times 10^3 \text{Pa})}{(1000 \text{kg/m}^3)((64 \times 10^{-4} \text{m}^2)^2 - (32 \times 10^{-4} \text{m}^2)^2)}} = 3.06 \text{ m/s}.$$

Consequently, the flow rate is

$$Av = (64 \times 10^{-4} \text{m}^2)(3.06 \text{ m/s}) = 2.0 \times 10^{-2} \text{m}^3/\text{s}.$$

68. We use the result of part (a) in the previous problem.

(a) In this case, we have $\Delta p = p_1 = 2.0$ atm. Consequently,

$$v = \sqrt{\frac{2\Delta p}{\rho((A/a)^2 - 1)}} = \sqrt{\frac{4(1.01 \times 10^5 \text{ Pa})}{(1000 \text{ kg/m}^3)[(5a/a)^2 - 1]}} = 4.1 \text{ m/s}.$$

(b) And the equation of continuity yields $V = (A/a)v = (5a/a)v = 5v = 21$ m/s.

(c) The flow rate is given by

$$Av = \frac{\pi}{4}(5.0 \times 10^{-4} \text{ m}^2)(4.1 \text{ m/s}) = 8.0 \times 10^{-3} \text{ m}^3/\text{s}.$$

69. (a) This is similar to the situation treated in Sample Problem 14-7, and we refer to some of its steps (and notation). Combining Eq. 14-35 and Eq. 14-36 in a manner very similar to that shown in the textbook, we find

$$R = A_1 A_2 \sqrt{\frac{2\Delta p}{\rho(A_1^2 - A_2^2)}}$$

for the flow rate expressed in terms of the pressure difference and the cross-sectional areas. Note that this reduces to Eq. 14-38 for the case $A_2 = A_1/2$ treated in the Sample Problem. Note that $\Delta p = p_1 - p_2 = -7.2 \times 10^3$ Pa and $A_1^2 - A_2^2 = -8.66 \times 10^{-3}$ m^4, so that the square root is well defined. Therefore, we obtain $R = 0.0776$ m^3/s.

(b) The mass rate of flow is $\rho R = 69.8$ kg/s.

70. (a) Bernoulli's equation gives $p_A = p_B + \frac{1}{2}\rho_{air}v^2$. But $\Delta p = p_A - p_B = \rho g h$ in order to balance the pressure in the two arms of the U-tube. Thus $\rho g h = \frac{1}{2}\rho_{air}v^2$, or

$$v = \sqrt{\frac{2\rho g h}{\rho_{air}}}.$$

(b) The plane's speed relative to the air is

$$v = \sqrt{\frac{2\rho g h}{\rho_{air}}} = \sqrt{\frac{2(810 \text{ kg/m}^3)(9.8 \text{ m/s}^2)(0.260 \text{ m})}{1.03 \text{ kg/m}^3}} = 63.3 \text{ m/s}.$$

71. We use the formula for v obtained in the previous problem:

$$v = \sqrt{\frac{2\Delta p}{\rho_{air}}} = \sqrt{\frac{2(180 \text{ Pa})}{0.031 \text{ kg/m}^3}} = 1.1 \times 10^2 \text{ m/s}.$$

72. We use Bernoulli's equation $p_1 + \frac{1}{2}\rho v_1^2 + \rho g h_1 = p_2 + \frac{1}{2}\rho v_2^2 + \rho g h_2$.

When the water level rises to height h_2, just on the verge of flooding, v_2, the speed of water in pipe M, is given by

$$\rho g(h_1 - h_2) = \frac{1}{2}\rho v_2^2 \Rightarrow v_2 = \sqrt{2g(h_1 - h_2)} = 13.86 \text{ m/s}.$$

By continuity equation, the corresponding rainfall rate is

$$v_1 = \left(\frac{A_2}{A_1}\right)v_2 = \frac{\pi(0.030 \text{ m})^2}{(30 \text{ m})(60 \text{ m})}(13.86 \text{ m/s}) = 2.177 \times 10^{-5} \text{ m/s} \approx 7.8 \text{ cm/h}.$$

73. The normal force $\vec{F}_N$ exerted (upward) on the glass ball of mass m has magnitude 0.0948 N. The buoyant force exerted by the milk (upward) on the ball has magnitude

$$F_b = \rho_{milk}\, g\, V$$

where $V = \frac{4}{3}\pi r^3$ is the volume of the ball. Its radius is $r = 0.0200$ m. The milk density is $\rho_{milk} = 1030$ kg/m^3. The (actual) weight of the ball is, of course, downward, and has magnitude $F_g = m_{glass}\, g$. Application of Newton's second law (in the case of zero acceleration) yields

$$F_N + \rho_{milk}\, g\, V - m_{glass}\, g = 0$$

which leads to $m_{glass} = 0.0442$ kg. We note the above equation is equivalent to Eq.14-19 in the textbook.

74. The volume rate of flow is $R = vA$ where $A = \pi r^2$ and $r = d/2$. Solving for speed, we obtain

$$v = \frac{R}{A} = \frac{R}{\pi(d/2)^2} = \frac{4R}{\pi d^2}.$$

(a) With $R = 7.0 \times 10^{-3}$ m^3/s and $d = 14 \times 10^{-3}$ m, our formula yields $v = 45$ m/s, which is about 13% of the speed of sound (which we establish by setting up a ratio: v/v_s where $v_s = 343$ m/s).

(b) With the contracted trachea ($d = 5.2 \times 10^{-3}$ m) we obtain $v = 330$ m/s, or 96% of the speed of sound.

75. If we examine both sides of the U-tube at the level where the low-density liquid (with $\rho = 0.800$ g/cm^3 = 800 kg/m^3) meets the water (with $\rho_w = 0.998$ g/cm^3 = 998 kg/m^3), then the pressures there on either side of the tube must agree:

$$\rho g h = \rho_w g h_w$$

where $h = 8.00$ cm $= 0.0800$ m, and Eq. 14-9 has been used. Thus, the height of the water column (as measured from that level) is $h_w = (800/998)(8.00 \text{ cm}) = 6.41$ cm. The volume of water in that column is therefore

$$V = \pi r^2 h_w = \pi (1.50 \text{ cm})^2 (6.41 \text{ cm}) = 45.3 \text{ cm}^3.$$

76. Since (using Eq. 5-8) $F_g = mg = \rho_{skier} g V$ and (Eq. 14-16) the buoyant force is $F_b = \rho_{snow} g V$, then their ratio is

$$\frac{F_b}{F_g} = \frac{\rho_{snow} g V}{\rho_{skier} g V} = \frac{\rho_{snow}}{\rho_{skier}} = \frac{96}{1020} = 0.094 \text{ (or 9.4\%)}.$$

77. (a) We consider a point D on the surface of the liquid in the container, in the same tube of flow with points A, B and C. Applying Bernoulli's equation to points D and C, we obtain

$$p_D + \frac{1}{2}\rho v_D^2 + \rho g h_D = p_C + \frac{1}{2}\rho v_C^2 + \rho g h_C$$

which leads to

$$v_C = \sqrt{\frac{2(p_D - p_C)}{\rho} + 2g(h_D - h_C) + v_D^2} \approx \sqrt{2g(d + h_2)}$$

where in the last step we set $p_D = p_C = p_{air}$ and $v_D/v_C \approx 0$. Plugging in the values, we obtain

$$v_c = \sqrt{2(9.8 \text{ m/s}^2)(0.40 \text{ m} + 0.12 \text{ m})} = 3.2 \text{ m/s}.$$

(b) We now consider points B and C:

$$p_B + \frac{1}{2}\rho v_B^2 + \rho g h_B = p_C + \frac{1}{2}\rho v_C^2 + \rho g h_C .$$

Since $v_B = v_C$ by equation of continuity, and $p_C = p_{air}$, Bernoulli's equation becomes

$$p_B = p_C + \rho g (h_C - h_B) = p_{air} - \rho g (h_1 + h_2 + d)$$
$$= 1.0 \times 10^5 \text{ Pa} - (1.0 \times 10^3 \text{ kg/m}^3)(9.8 \text{ m/s}^2)(0.25 \text{ m} + 0.40 \text{ m} + 0.12 \text{ m})$$
$$= 9.2 \times 10^4 \text{ Pa}.$$

(c) Since $p_B \geq 0$, we must let $p_{air} - \rho g (h_1 + d + h_2) \geq 0$, which yields

$$h_1 \leq h_{1,\max} = \frac{p_{air}}{\rho} - d - h_2 \leq \frac{p_{air}}{\rho} = 10.3 \text{ m}.$$

78. To be as general as possible, we denote the ratio of body density to water density as f (so that $f = \rho/\rho_w = 0.95$ in this problem). Floating involves equilibrium of vertical forces acting on the body (Earth's gravity pulls down and the buoyant force pushes up). Thus,

$$F_b = F_g \Rightarrow \rho_w g V_w = \rho g V$$

where V is the total volume of the body and V_w is the portion of it which is submerged.

(a) We rearrange the above equation to yield

$$\frac{V_w}{V} = \frac{\rho}{\rho_w} = f$$

which means that 95% of the body is submerged and therefore 5.0% is above the water surface.

(b) We replace ρ_w with $1.6\rho_w$ in the above equilibrium of forces relationship, and find

$$\frac{V_w}{V} = \frac{\rho}{1.6\rho_w} = \frac{f}{1.6}$$

which means that 59% of the body is submerged and thus 41% is above the quicksand surface.

(c) The answer to part (b) suggests that a person in that situation is able to breathe.

79. We note that in "gees" (where acceleration is expressed as a multiple of g) the given acceleration is $0.225/9.8 = 0.02296$. Using $m = \rho V$, Newton's second law becomes

$$\rho_{wat} V g - \rho_{bub} V g = \rho_{bub} V a \quad \Rightarrow \quad \rho_{bub} = \rho_{wat}(1 + \text{"}a\text{"})$$

where in the final expression "a" is to be understood to be in "gees." Using $\rho_{wat} = 998$ kg/m³ (see Table 14-1) we find $\rho_{bub} = 975.6$ kg/m³. Using volume $V = \frac{4}{3}\pi r^3$ for the bubble, we then find its mass: $m_{bub} = 5.11 \times 10^{-7}$ kg.

80. The downward force on the balloon is mg and the upward force is $F_b = \rho_{out} V g$. Newton's second law (with $m = \rho_{in} V$) leads to

$$\rho_{out} V g - \rho_{in} V g = \rho_{in} V a \Rightarrow \left(\frac{\rho_{out}}{\rho_{in}} - 1\right) g = a.$$

The problem specifies $\rho_{out}/\rho_{in} = 1.39$ (the outside air is cooler and thus more dense than the hot air inside the balloon). Thus, the upward acceleration is $(1.39 - 1.00)(9.80 \text{ m/s}^2) = 3.82 \text{ m/s}^2$.

81. We consider the can with nearly its total volume submerged, and just the rim above water. For calculation purposes, we take its submerged volume to be $V = 1200 \text{ cm}^3$. To float, the total downward force of gravity (acting on the tin mass m_t and the lead mass m_ℓ) must be equal to the buoyant force upward:

$$(m_t + m_\ell)g = \rho_w V g \Rightarrow m_\ell = (1 \text{ g/cm}^3)(1200 \text{ cm}^3) - 130 \text{ g}$$

which yields 1.07×10^3 g for the (maximum) mass of the lead (for which the can still floats). The given density of lead is not used in the solution.

82. If the mercury level in one arm of the tube is lowered by an amount x, it will rise by x in the other arm. Thus, the net difference in mercury level between the two arms is $2x$, causing a pressure difference of $\Delta p = 2\rho_{Hg} g x$, which should be compensated for by the water pressure $p_w = \rho_w g h$, where $h = 11.2$ cm. In these units, $\rho_w = 1.00$ g/cm^3 and $\rho_{Hg} = 13.6$ g/cm^3 (see Table 14-1). We obtain

$$x = \frac{\rho_w g h}{2\rho_{Hg} g} = \frac{(1.00 \text{ g/cm}^3)(11.2 \text{ cm})}{2(13.6 \text{ g/cm}^3)} = 0.412 \text{ cm}.$$

83. Neglecting the buoyant force caused by air, then the 30 N value is interpreted as the true weight W of the object. The buoyant force of the water on the object is therefore $(30 - 20) \text{ N} = 10$ N, which means

$$F_b = \rho_w V g \Rightarrow V = \frac{10 \text{ N}}{(1000 \text{ kg/m}^3)(9.8 \text{ m/s}^2)} = 1.02 \times 10^{-3} \text{ m}^3$$

is the volume of the object. When the object is in the second liquid, the buoyant force is $(30 - 24) \text{ N} = 6.0$ N, which implies

$$\rho_2 = \frac{6.0 \text{ N}}{(9.8 \text{ m/s}^2)(1.02 \times 10^{-3} \text{ m}^3)} = 6.0 \times 10^2 \text{ kg/m}^3.$$

84. An object of mass $m = \rho V$ floating in a liquid of density ρ_{liquid} is able to float if the downward pull of gravity mg is equal to the upward buoyant force $F_b = \rho_{liquid} g V_{sub}$ where V_{sub} is the portion of the object which is submerged. This readily leads to the relation:

$$\frac{\rho}{\rho_{liquid}} = \frac{V_{sub}}{V}$$

for the fraction of volume submerged of a floating object. When the liquid is water, as described in this problem, this relation leads to

$$\frac{\rho}{\rho_w} = 1$$

since the object "floats fully submerged" in water (thus, the object has the same density as water). We assume the block maintains an "upright" orientation in each case (which is not necessarily realistic).

(a) For liquid A,

$$\frac{\rho}{\rho_A} = \frac{1}{2}$$

so that, in view of the fact that $\rho = \rho_w$, we obtain $\rho_A/\rho_w = 2$.

(b) For liquid B, noting that two-thirds *above* means one-third *below*,

$$\frac{\rho}{\rho_B} = \frac{1}{3}$$

so that $\rho_B/\rho_w = 3$.

(c) For liquid C, noting that one-fourth *above* means three-fourths *below*,

$$\frac{\rho}{\rho_C} = \frac{3}{4}$$

so that $\rho_C/\rho_w = 4/3$.

85. Equilibrium of forces (on the floating body) is expressed as

$$F_b = m_{body}\, g \Rightarrow \rho_{liquid}\, gV_{submerged} = \rho_{body}\, gV_{total}$$

which leads to

$$\frac{V_{submerged}}{V_{total}} = \frac{\rho_{body}}{\rho_{liquid}}.$$

We are told (indirectly) that two-thirds of the body is below the surface, so the fraction above is 2/3. Thus, with $\rho_{body} = 0.98$ g/cm^3, we find $\rho_{liquid} \approx 1.5$ g/cm^3 — certainly much more dense than normal seawater (the Dead Sea is about seven times saltier than the ocean due to the high evaporation rate and low rainfall in that region).

Chapter 15

1. The textbook notes (in the discussion immediately after Eq. 15-7) that the acceleration amplitude is $a_m = \omega^2 x_m$, where ω is the angular frequency ($\omega = 2\pi f$ since there are 2π radians in one cycle). Therefore, in this circumstance, we obtain

$$a_m = \omega^2 x_m = (2\pi f)^2 x_m = (2\pi(6.60 \text{ Hz}))^2 (0.0220 \text{ m}) = 37.8 \text{ m/s}^2.$$

2. (a) The angular frequency ω is given by $\omega = 2\pi f = 2\pi/T$, where f is the frequency and T is the period. The relationship $f = 1/T$ was used to obtain the last form. Thus

$$\omega = 2\pi/(1.00 \times 10^{-5} \text{ s}) = 6.28 \times 10^5 \text{ rad/s}.$$

(b) The maximum speed v_m and maximum displacement x_m are related by $v_m = \omega x_m$, so

$$x_m = \frac{v_m}{\omega} = \frac{1.00 \times 10^3 \text{ m/s}}{6.28 \times 10^5 \text{ rad/s}} = 1.59 \times 10^{-3} \text{ m}.$$

3. (a) The amplitude is half the range of the displacement, or $x_m = 1.0$ mm.

(b) The maximum speed v_m is related to the amplitude x_m by $v_m = \omega x_m$, where ω is the angular frequency. Since $\omega = 2\pi f$, where f is the frequency,

$$v_m = 2\pi f x_m = 2\pi(120 \text{ Hz})(1.0 \times 10^{-3} \text{ m}) = 0.75 \text{ m/s}.$$

(c) The maximum acceleration is

$$a_m = \omega^2 x_m = (2\pi f)^2 x_m = (2\pi(120 \text{ Hz}))^2 (1.0 \times 10^{-3} \text{ m}) = 5.7 \times 10^2 \text{ m/s}^2.$$

4. (a) The acceleration amplitude is related to the maximum force by Newton's second law: $F_{\max} = m a_m$. The textbook notes (in the discussion immediately after Eq. 15-7) that the acceleration amplitude is $a_m = \omega^2 x_m$, where ω is the angular frequency ($\omega = 2\pi f$ since there are 2π radians in one cycle). The frequency is the reciprocal of the period: $f = 1/T = 1/0.20 = 5.0$ Hz, so the angular frequency is $\omega = 10\pi$ (understood to be valid to two significant figures). Therefore,

$$F_{\max} = m\omega^2 x_m = (0.12 \text{ kg})(10\pi \text{ rad/s})^2 (0.085 \text{ m}) = 10 \text{ N}.$$

(b) Using Eq. 15-12, we obtain

$$\omega = \sqrt{\frac{k}{m}} \Rightarrow k = m\omega^2 = (0.12\text{kg})(10\pi \text{ rad/s})^2 = 1.2 \times 10^2 \text{ N/m}.$$

5. (a) During simple harmonic motion, the speed is (momentarily) zero when the object is at a "turning point" (that is, when $x = +x_m$ or $x = -x_m$). Consider that it starts at $x = +x_m$ and we are told that $t = 0.25$ second elapses until the object reaches $x = -x_m$. To execute a full cycle of the motion (which takes a period T to complete), the object which started at $x = +x_m$ must return to $x = +x_m$ (which, by symmetry, will occur 0.25 second *after* it was at $x = -x_m$). Thus, $T = 2t = 0.50$ s.

(b) Frequency is simply the reciprocal of the period: $f = 1/T = 2.0$ Hz.

(c) The 36 cm distance between $x = +x_m$ and $x = -x_m$ is $2x_m$. Thus, $x_m = 36/2 = 18$ cm.

6. (a) Since the problem gives the frequency $f = 3.00$ Hz, we have $\omega = 2\pi f = 6\pi$ rad/s (understood to be valid to three significant figures). Each spring is considered to support one fourth of the mass m_{car} so that Eq. 15-12 leads to

$$\omega = \sqrt{\frac{k}{m_{car}/4}} \Rightarrow k = \frac{1}{4}(1450\text{kg})(6\pi \text{ rad/s})^2 = 1.29 \times 10^5 \text{ N/m}.$$

(b) If the new mass being supported by the four springs is $m_{total} = [1450 + 5(73)]$ kg = 1815 kg, then Eq. 15-12 leads to

$$\omega_{new} = \sqrt{\frac{k}{m_{total}/4}} \Rightarrow f_{new} = \frac{1}{2\pi}\sqrt{\frac{1.29 \times 10^5 \text{ N/m}}{(1815/4) \text{ kg}}} = 2.68 \text{ Hz}.$$

7. (a) The motion repeats every 0.500 s so the period must be $T = 0.500$ s.

(b) The frequency is the reciprocal of the period: $f = 1/T = 1/(0.500 \text{ s}) = 2.00$ Hz.

(c) The angular frequency ω is $\omega = 2\pi f = 2\pi(2.00 \text{ Hz}) = 12.6$ rad/s.

(d) The angular frequency is related to the spring constant k and the mass m by $\omega = \sqrt{k/m}$. We solve for k and obtain

$$k = m\omega^2 = (0.500 \text{ kg})(12.6 \text{ rad/s})^2 = 79.0 \text{ N/m}.$$

(e) Let x_m be the amplitude. The maximum speed is

$$v_m = \omega x_m = (12.6 \text{ rad/s})(0.350 \text{ m}) = 4.40 \text{ m/s}.$$

(f) The maximum force is exerted when the displacement is a maximum and its magnitude is given by $F_m = kx_m = (79.0 \text{ N/m})(0.350 \text{ m}) = 27.6 \text{ N}$.

8. (a) The problem describes the time taken to execute one cycle of the motion. The period is $T = 0.75$ s.

(b) Frequency is simply the reciprocal of the period: $f = 1/T \approx 1.3$ Hz, where the SI unit abbreviation Hz stands for Hertz, which means a cycle-per-second.

(c) Since 2π radians are equivalent to a cycle, the angular frequency ω (in radians-per-second) is related to frequency f by $\omega = 2\pi f$ so that $\omega \approx 8.4$ rad/s.

9. The magnitude of the maximum acceleration is given by $a_m = \omega^2 x_m$, where ω is the angular frequency and x_m is the amplitude.

(a) The angular frequency for which the maximum acceleration is g is given by $\omega = \sqrt{g/x_m}$, and the corresponding frequency is given by

$$f = \frac{\omega}{2\pi} = \frac{1}{2\pi}\sqrt{\frac{g}{x_m}} = \frac{1}{2\pi}\sqrt{\frac{9.8 \text{ m/s}^2}{1.0\times 10^{-6} \text{ m}}} = 498 \text{ Hz}.$$

(b) For frequencies greater than 498 Hz, the acceleration exceeds g for some part of the motion.

10. We note (from the graph) that $x_m = 6.00$ cm. Also the value at $t = 0$ is $x_o = -2.00$ cm. Then Eq. 15-3 leads to

$$\phi = \cos^{-1}(-2.00/6.00) = +1.91 \text{ rad or } -4.37 \text{ rad}.$$

The other "root" (+4.37 rad) can be rejected on the grounds that it would lead to a positive slope at $t = 0$.

11. (a) Making sure our calculator is in radians mode, we find

$$x = 6.0 \cos\left(3\pi(2.0) + \frac{\pi}{3}\right) = 3.0 \text{ m}.$$

(b) Differentiating with respect to time and evaluating at $t = 2.0$ s, we find

$$v = \frac{dx}{dt} = -3\pi(6.0)\sin\left(3\pi(2.0) + \frac{\pi}{3}\right) = -49 \text{ m/s}.$$

(c) Differentiating again, we obtain

$$a = \frac{dv}{dt} = -(3\pi)^2(6.0)\cos\left(3\pi(2.0)+\frac{\pi}{3}\right) = -2.7\times 10^2 \text{ m/s}^2.$$

(d) In the second paragraph after Eq. 15-3, the textbook defines the phase of the motion. In this case (with $t = 2.0$ s) the phase is $3\pi(2.0) + \pi/3 \approx 20$ rad.

(e) Comparing with Eq. 15-3, we see that $\omega = 3\pi$ rad/s. Therefore, $f = \omega/2\pi = 1.5$ Hz.

(f) The period is the reciprocal of the frequency: $T = 1/f \approx 0.67$ s.

12. We note (from the graph) that $v_m = \omega x_m = 5.00$ cm/s. Also the value at $t = 0$ is $v_o = 4.00$ cm/s. Then Eq. 15-6 leads to

$$\phi = \sin^{-1}(-4.00/5.00) = -0.927 \text{ rad or } +5.36 \text{ rad}.$$

The other "root" (+4.07 rad) can be rejected on the grounds that it would lead to a positive slope at $t = 0$.

13. When displaced from equilibrium, the net force exerted by the springs is $-2kx$ acting in a direction so as to return the block to its equilibrium position ($x = 0$). Since the acceleration $a = d^2x/dt^2$, Newton's second law yields

$$m\frac{d^2x}{dt^2} = -2kx.$$

Substituting $x = x_m \cos(\omega t + \phi)$ and simplifying, we find

$$\omega^2 = \frac{2k}{m}$$

where ω is in radians per unit time. Since there are 2π radians in a cycle, and frequency f measures cycles per second, we obtain

$$f = \frac{\omega}{2\pi} = \frac{1}{2\pi}\sqrt{\frac{2k}{m}} = \frac{1}{2\pi}\sqrt{\frac{2(7580 \text{ N/m})}{0.245 \text{ kg}}} = 39.6 \text{ Hz}.$$

14. The statement that "the spring does not affect the collision" justifies the use of elastic collision formulas in section 10-5. We are told the period of SHM so that we can find the mass of block 2:

$$T = 2\pi\sqrt{\frac{m_2}{k}} \Rightarrow m_2 = \frac{kT^2}{4\pi^2} = 0.600 \text{ kg}.$$

At this point, the rebound speed of block 1 can be found from Eq. 10-30:

$$|v_{1f}| = \left|\frac{0.200 \text{ kg} - 0.600 \text{ kg}}{0.200 \text{ kg} + 0.600 \text{ kg}}\right|(8.00 \text{ m/s}) = 4.00 \text{ m/s}.$$

This becomes the initial speed v_0 of the projectile motion of block 1. A variety of choices for the positive axis directions are possible, and we choose left as the $+x$ direction and down as the $+y$ direction, in this instance. With the "launch" angle being zero, Eq. 4-21 and Eq. 4-22 (with $-g$ replaced with $+g$) lead to

$$x - x_0 = v_0 t = v_0 \sqrt{\frac{2h}{g}} = (4.00 \text{ m/s})\sqrt{\frac{2(4.90 \text{ m})}{9.8 \text{ m/s}^2}}.$$

Since $x - x_0 = d$, we arrive at $d = 4.00$ m.

15. (a) Eq. 15-8 leads to

$$a = -\omega^2 x \Rightarrow \omega = \sqrt{\frac{-a}{x}} = \sqrt{\frac{123 \text{ m/s}^2}{0.100 \text{ m}}} = 35.07 \text{ rad/s}.$$

Therefore, $f = \omega/2\pi = 5.58$ Hz.

(b) Eq. 15-12 provides a relation between ω (found in the previous part) and the mass:

$$\omega = \sqrt{\frac{k}{m}} \Rightarrow m = \frac{400 \text{ N/m}}{(35.07 \text{ rad/s})^2} = 0.325 \text{ kg}.$$

(c) By energy conservation, $\frac{1}{2}kx_m^2$ (the energy of the system at a turning point) is equal to the sum of kinetic and potential energies at the time t described in the problem.

$$\frac{1}{2}kx_m^2 = \frac{1}{2}mv^2 + \frac{1}{2}kx^2 \Rightarrow x_m = \sqrt{\frac{m}{k}v^2 + x^2}.$$

Consequently, $x_m = \sqrt{(0.325 \text{ kg}/400 \text{ N/m})(13.6 \text{ m/s})^2 + (0.100 \text{ m})^2} = 0.400$ m.

16. From highest level to lowest level is twice the amplitude x_m of the motion. The period is related to the angular frequency by Eq. 15-5. Thus, $x_m = \frac{1}{2}d$ and $\omega = 0.503$ rad/h. The phase constant ϕ in Eq. 15-3 is zero since we start our clock when $x_0 = x_m$ (at the highest point). We solve for t when x is one-fourth of the total distance from highest to lowest level, or (which is the same) half the distance from highest level to middle level (where we locate the origin of coordinates). Thus, we seek t when the ocean surface is at $x = \frac{1}{2}x_m = \frac{1}{4}d$. With $x = x_m \cos(\omega t + \phi)$, we obtain

$$\frac{1}{4}d = \left(\frac{1}{2}d\right)\cos(0.503t + 0) \Rightarrow \frac{1}{2} = \cos(0.503t)$$

which has $t = 2.08$ h as the smallest positive root. The calculator is in radians mode during this calculation.

17. The maximum force that can be exerted by the surface must be less than $\mu_s F_N$ or else the block will not follow the surface in its motion. Here, μ_s is the coefficient of static friction and F_N is the normal force exerted by the surface on the block. Since the block does not accelerate vertically, we know that $F_N = mg$, where m is the mass of the block. If the block follows the table and moves in simple harmonic motion, the magnitude of the maximum force exerted on it is given by

$$F = ma_m = m\omega^2 x_m = m(2\pi f)^2 x_m,$$

where a_m is the magnitude of the maximum acceleration, ω is the angular frequency, and f is the frequency. The relationship $\omega = 2\pi f$ was used to obtain the last form. We substitute $F = m(2\pi f)^2 x_m$ and $F_N = mg$ into $F < \mu_s F_N$ to obtain $m(2\pi f)^2 x_m < \mu_s mg$. The largest amplitude for which the block does not slip is

$$x_m = \frac{\mu_s g}{(2\pi f)^2} = \frac{(0.50)(9.8 \text{ m/s}^2)}{(2\pi \times 2.0 \text{ Hz})^2} = 0.031 \text{ m}.$$

A larger amplitude requires a larger force at the end points of the motion. The surface cannot supply the larger force and the block slips.

18. They pass each other at time t, at $x_1 = x_2 = \frac{1}{2}x_m$ where

$$x_1 = x_m \cos(\omega t + \phi_1) \quad \text{and} \quad x_2 = x_m \cos(\omega t + \phi_2).$$

From this, we conclude that $\cos(\omega t + \phi_1) = \cos(\omega t + \phi_2) = \frac{1}{2}$, and therefore that the phases (the arguments of the cosines) are either both equal to $\pi/3$ or one is $\pi/3$ while the other is $-\pi/3$. Also at this instant, we have $v_1 = -v_2 \neq 0$ where

$$v_1 = -x_m \omega \sin(\omega t + \phi_1) \quad \text{and} \quad v_2 = -x_m \omega \sin(\omega t + \phi_2).$$

This leads to $\sin(\omega t + \phi_1) = -\sin(\omega t + \phi_2)$. This leads us to conclude that the phases have opposite sign. Thus, one phase is $\pi/3$ and the other phase is $-\pi/3$; the ωt term cancels if we take the phase difference, which is seen to be $\pi/3 - (-\pi/3) = 2\pi/3$.

19. (a) Let

$$x_1 = \frac{A}{2}\cos\left(\frac{2\pi t}{T}\right)$$

be the coordinate as a function of time for particle 1 and

$$x_2 = \frac{A}{2}\cos\left(\frac{2\pi t}{T} + \frac{\pi}{6}\right)$$

be the coordinate as a function of time for particle 2. Here T is the period. Note that since the range of the motion is A, the amplitudes are both $A/2$. The arguments of the cosine functions are in radians. Particle 1 is at one end of its path ($x_1 = A/2$) when $t = 0$. Particle 2 is at $A/2$ when $2\pi t/T + \pi/6 = 0$ or $t = -T/12$. That is, particle 1 lags particle 2 by one-twelfth a period. We want the coordinates of the particles 0.50 s later; that is, at $t = 0.50$ s,

$$x_1 = \frac{A}{2}\cos\left(\frac{2\pi \times 0.50 \text{ s}}{1.5 \text{ s}}\right) = -0.25A$$

and

$$x_2 = \frac{A}{2}\cos\left(\frac{2\pi \times 0.50 \text{ s}}{1.5 \text{ s}} + \frac{\pi}{6}\right) = -0.43A.$$

Their separation at that time is $x_1 - x_2 = -0.25A + 0.43A = 0.18A$.

(b) The velocities of the particles are given by

$$v_1 = \frac{dx_1}{dt} = \frac{\pi A}{T}\sin\left(\frac{2\pi t}{T}\right)$$

and

$$v_2 = \frac{dx_2}{dt} = \frac{\pi A}{T}\sin\left(\frac{2\pi t}{T} + \frac{\pi}{6}\right).$$

We evaluate these expressions for $t = 0.50$ s and find they are both negative-valued, indicating that the particles are moving in the same direction.

20. We note that the ratio of Eq. 15-6 and Eq. 15-3 is $v/x = -\omega\tan(\omega t + \phi)$ where $\omega = 1.20$ rad/s in this problem. Evaluating this at $t = 0$ and using the values from the graphs shown in the problem, we find

$$\phi = \tan^{-1}(-v_0/x_0\omega) = \tan^{-1}(+4.00/(2 \times 1.20)) = 1.03 \text{ rad (or } -5.25 \text{ rad)}.$$

One can check that the other "root" (4.17 rad) is unacceptable since it would give the wrong signs for the individual values of v_0 and x_0.

21. Both parts of this problem deal with the critical case when the maximum acceleration becomes equal to that of free fall. The textbook notes (in the discussion immediately after Eq. 15-7) that the acceleration amplitude is $a_m = \omega^2 x_m$, where ω is the angular frequency; this is the expression we set equal to $g = 9.8$ m/s^2.

(a) Using Eq. 15-5 and $T = 1.0$ s, we have

$$\left(\frac{2\pi}{T}\right)^2 x_m = g \Rightarrow x_m = \frac{gT^2}{4\pi^2} = 0.25 \text{ m}.$$

(b) Since $\omega = 2\pi f$, and $x_m = 0.050$ m is given, we find

$$(2\pi f)^2 x_m = g \Rightarrow f = \frac{1}{2\pi}\sqrt{\frac{g}{x_m}} = 2.2 \text{ Hz}.$$

22. Eq. 15-12 gives the angular velocity:

$$\omega = \sqrt{\frac{k}{m}} = \sqrt{\frac{100 \text{ N/m}}{2.00 \text{ kg}}} = 7.07 \text{ rad/s}.$$

Energy methods (discussed in §15-4) provide one method of solution. Here, we use trigonometric techniques based on Eq. 15-3 and Eq. 15-6.

(a) Dividing Eq. 15-6 by Eq. 15-3, we obtain

$$\frac{v}{x} = -\omega \tan(\omega t + \phi)$$

so that the phase $(\omega t + \phi)$ is found from

$$\omega t + \phi = \tan^{-1}\left(\frac{-v}{\omega x}\right) = \tan^{-1}\left(\frac{-3.415 \text{ m/s}}{(7.07 \text{ rad/s})(0.129 \text{ m})}\right).$$

With the calculator in radians mode, this gives the phase equal to -1.31 rad. Plugging this back into Eq. 15-3 leads to $0.129 \text{ m} = x_m \cos(-1.31) \Rightarrow x_m = 0.500 \text{ m}$.

(b) Since $\omega t + \phi = -1.31$ rad at $t = 1.00$ s, we can use the above value of ω to solve for the phase constant ϕ. We obtain $\phi = -8.38$ rad (though this, as well as the previous result, can have 2π or 4π (and so on) added to it without changing the physics of the situation). With this value of ϕ, we find $x_0 = x_m \cos\phi = -0.251$ m.

(c) And we obtain $v_0 = -x_m \omega \sin\phi = 3.06$ m/s.

23. Let the spring constants be k_1 and k_2. When displaced from equilibrium, the magnitude of the net force exerted by the springs is $|k_1 x + k_2 x|$ acting in a direction so as

to return the block to its equilibrium position ($x = 0$). Since the acceleration $a = d^2x/dt^2$, Newton's second law yields

$$m\frac{d^2x}{dt^2} = -k_1 x - k_2 x.$$

Substituting $x = x_m \cos(\omega t + \phi)$ and simplifying, we find

$$\omega^2 = \frac{k_1 + k_2}{m}$$

where ω is in radians per unit time. Since there are 2π radians in a cycle, and frequency f measures cycles per second, we obtain

$$f = \frac{\omega}{2\pi} = \frac{1}{2\pi}\sqrt{\frac{k_1 + k_2}{m}}.$$

The single springs each acting alone would produce simple harmonic motions of frequency

$$f_1 = \frac{1}{2\pi}\sqrt{\frac{k_1}{m}} = 30 \text{ Hz}, \quad f_2 = \frac{1}{2\pi}\sqrt{\frac{k_2}{m}} = 45 \text{ Hz},$$

respectively. Comparing these expressions, it is clear that

$$f = \sqrt{f_1^2 + f_2^2} = \sqrt{(30 \text{ Hz})^2 + (45 \text{ Hz})^2} = 54 \text{ Hz}.$$

24. To be on the verge of slipping means that the force exerted on the smaller block (at the point of maximum acceleration) is $f_{max} = \mu_s mg$. The textbook notes (in the discussion immediately after Eq. 15-7) that the acceleration amplitude is $a_m = \omega^2 x_m$, where $\omega = \sqrt{k/(m+M)}$ is the angular frequency (from Eq. 15-12). Therefore, using Newton's second law, we have

$$ma_m = \mu_s mg \Rightarrow \frac{k}{m+M} x_m = \mu_s g$$

which leads to

$$x_m = \frac{\mu_s g(m+M)}{k} = \frac{(0.40)(9.8 \text{ m/s}^2)(1.8 \text{ kg} + 10 \text{ kg})}{200 \text{ N/m}} = 0.23 \text{ m} = 23 \text{ cm}.$$

25. (a) We interpret the problem as asking for the equilibrium position; that is, the block is gently lowered until forces balance (as opposed to being suddenly released and allowed to oscillate). If the amount the spring is stretched is x, then we examine force-components along the incline surface and find

$$kx = mg\sin\theta \Rightarrow x = \frac{mg\sin\theta}{k} = \frac{(14.0\text{ N})\sin 40.0°}{120\text{ N/m}} = 0.0750\text{ m}$$

at equilibrium. The calculator is in degrees mode in the above calculation. The distance from the top of the incline is therefore $(0.450 + 0.75)\text{ m} = 0.525\text{ m}$.

(b) Just as with a vertical spring, the effect of gravity (or one of its components) is simply to shift the equilibrium position; it does not change the characteristics (such as the period) of simple harmonic motion. Thus, Eq. 15-13 applies, and we obtain

$$T = 2\pi\sqrt{\frac{14.0\text{ N}/9.80\text{ m/s}^2}{120\text{ N/m}}} = 0.686\text{ s}.$$

26. We wish to find the effective spring constant for the combination of springs shown in the figure. We do this by finding the magnitude F of the force exerted on the mass when the total elongation of the springs is Δx. Then $k_{\text{eff}} = F/\Delta x$. Suppose the left-hand spring is elongated by Δx_ℓ and the right-hand spring is elongated by Δx_r. The left-hand spring exerts a force of magnitude $k\Delta x_\ell$ on the right-hand spring and the right-hand spring exerts a force of magnitude $k\Delta x_r$ on the left-hand spring. By Newton's third law these must be equal, so $\Delta x_\ell = \Delta x_r$. The two elongations must be the same and the total elongation is twice the elongation of either spring: $\Delta x = 2\Delta x_\ell$. The left-hand spring exerts a force on the block and its magnitude is $F = k\Delta x_\ell$. Thus $k_{\text{eff}} = k\Delta x_\ell/2\Delta x_\ell = k/2$. The block behaves as if it were subject to the force of a single spring, with spring constant $k/2$. To find the frequency of its motion replace k_{eff} in $f = (1/2\pi)\sqrt{k_{\text{eff}}/m}$ with $k/2$ to obtain

$$f = \frac{1}{2\pi}\sqrt{\frac{k}{2m}}.$$

With $m = 0.245$ kg and $k = 6430$ N/m, the frequency is $f = 18.2$ Hz.

27. When the block is at the end of its path and is momentarily stopped, its displacement is equal to the amplitude and all the energy is potential in nature. If the spring potential energy is taken to be zero when the block is at its equilibrium position, then

$$E = \frac{1}{2}kx_m^2 = \frac{1}{2}(1.3\times 10^2\text{ N/m})(0.024\text{ m})^2 = 3.7\times 10^{-2}\text{ J}.$$

28. (a) The energy at the turning point is all potential energy: $E = \frac{1}{2}kx_m^2$ where $E = 1.00$ J and $x_m = 0.100$ m. Thus,

$$k = \frac{2E}{x_m^2} = 200\text{ N/m}.$$

(b) The energy as the block passes through the equilibrium position (with speed $v_m = 1.20$ m/s) is purely kinetic:

$$E = \frac{1}{2}mv_m^2 \Rightarrow m = \frac{2E}{v_m^2} = 1.39 \text{ kg}.$$

(c) Eq. 15-12 (divided by 2π) yields

$$f = \frac{1}{2\pi}\sqrt{\frac{k}{m}} = 1.91 \text{ Hz}.$$

29. The total energy is given by $E = \frac{1}{2}kx_m^2$, where k is the spring constant and x_m is the amplitude. We use the answer from part (b) to do part (a), so it is best to look at the solution for part (b) first.

(a) The fraction of the energy that is kinetic is

$$\frac{K}{E} = \frac{E-U}{E} = 1 - \frac{U}{E} = 1 - \frac{1}{4} = \frac{3}{4} = 0.75$$

where the result from part (b) has been used.

(b) When $x = \frac{1}{2}x_m$ the potential energy is $U = \frac{1}{2}kx^2 = \frac{1}{8}kx_m^2$. The ratio is

$$\frac{U}{E} = \frac{kx_m^2/8}{kx_m^2/2} = \frac{1}{4} = 0.25.$$

(c) Since $E = \frac{1}{2}kx_m^2$ and $U = \frac{1}{2}kx^2$, $U/E = x^2/x_m^2$. We solve $x^2/x_m^2 = 1/2$ for x. We should get $x = x_m/\sqrt{2}$.

30. The total mechanical energy is equal to the (maximum) kinetic energy as it passes through the equilibrium position ($x = 0$):

$$\tfrac{1}{2}mv^2 = \tfrac{1}{2}(2.0 \text{ kg})(0.85 \text{ m/s})^2 = 0.72 \text{ J}.$$

Looking at the graph in the problem, we see that $U(x=10)=0.5$ J. Since the potential function has the form $U(x) = bx^2$, the constant is $b = 5.0 \times 10^{-3}$ J/cm^2. Thus, $U(x) = 0.72$ J when $x = 12$ cm.

(a) Thus, the mass does turn back before reaching $x = 15$ cm.

(b) It turns back at $x = 12$ cm.

31. (a) Eq. 15-12 (divided by 2π) yields

$$f = \frac{1}{2\pi}\sqrt{\frac{k}{m}} = \frac{1}{2\pi}\sqrt{\frac{1000 \text{ N/m}}{5.00 \text{ kg}}} = 2.25 \text{ Hz}.$$

(b) With $x_o = 0.500$ m, we have $U_0 = \frac{1}{2}kx_0^2 = 125$ J.

(c) With $v_o = 10.0$ m/s, the initial kinetic energy is $K_0 = \frac{1}{2}mv_0^2 = 250$ J.

(d) Since the total energy $E = K_o + U_o = 375$ J is conserved, then consideration of the energy at the turning point leads to

$$E = \frac{1}{2}kx_m^2 \Rightarrow x_m = \sqrt{\frac{2E}{k}} = 0.866 \text{ m}.$$

32. We infer from the graph (since mechanical energy is conserved) that the *total* energy in the system is 6.0 J; we also note that the amplitude is apparently $x_m = 12$ cm $= 0.12$ m. Therefore we can set the maximum *potential* energy equal to 6.0 J and solve for the spring constant k:

$$\tfrac{1}{2}k x_m^2 = 6.0 \text{ J} \quad \Rightarrow \quad k = 8.3 \times 10^2 \text{ N/m}.$$

33. The textbook notes (in the discussion immediately after Eq. 15-7) that the acceleration amplitude is $a_m = \omega^2 x_m$, where ω is the angular frequency and $x_m = 0.0020$ m is the amplitude. Thus, $a_m = 8000$ m/s^2 leads to $\omega = 2000$ rad/s. Using Newton's second law with $m = 0.010$ kg, we have

$$F = ma = m(-a_m \cos(\omega t + \phi)) = -(80 \text{ N})\cos\left(2000t - \frac{\pi}{3}\right)$$

where t is understood to be in seconds.

(a) Eq. 15-5 gives $T = 2\pi/\omega = 3.1 \times 10^{-3}$ s.

(b) The relation $v_m = \omega x_m$ can be used to solve for v_m, or we can pursue the alternate (though related) approach of energy conservation. Here we choose the latter. By Eq. 15-12, the spring constant is $k = \omega^2 m = 40000$ N/m. Then, energy conservation leads to

$$\frac{1}{2}kx_m^2 = \frac{1}{2}mv_m^2 \Rightarrow v_m = x_m\sqrt{\frac{k}{m}} = 4.0 \text{ m/s}.$$

(c) The total energy is $\tfrac{1}{2}kx_m^2 = \tfrac{1}{2}mv_m^2 = 0.080$ J.

(d) At the maximum displacement, the force acting on the particle is

$$F = kx = (4.0 \times 10^4 \text{ N/m})(2.0 \times 10^{-3} \text{ m}) = 80 \text{ N}.$$

(e) At half of the maximum displacement, $x = 1.0$ mm, and the force is

$$F = kx = (4.0 \times 10^4 \text{ N/m})(1.0 \times 10^{-3} \text{ m}) = 40 \text{ N}.$$

34. We note that the ratio of Eq. 15-6 and Eq. 15-3 is $v/x = -\omega \tan(\omega t + \phi)$ where ω is given by Eq. 15-12. Since the kinetic energy is $\frac{1}{2}mv^2$ and the potential energy is $\frac{1}{2}kx^2$ (which may be conveniently written as $\frac{1}{2}m\omega^2 x^2$) then the ratio of kinetic to potential energy is simply

$$(v/x)^2/\omega^2 = \tan^2(\omega t + \phi),$$

which at $t = 0$ is $\tan^2\phi$. Since $\phi = \pi/6$ in this problem, then the ratio of kinetic to potential energy at $t = 0$ is $\tan^2(\pi/6) = 1/3$.

35. The problem consists of two distinct parts: the completely inelastic collision (which is assumed to occur instantaneously, the bullet embedding itself in the block before the block moves through significant distance) followed by simple harmonic motion (of mass $m + M$ attached to a spring of spring constant k).

(a) Momentum conservation readily yields $v' = mv/(m + M)$. With $m = 9.5$ g, $M = 5.4$ kg and $v = 630$ m/s, we obtain $v' = 1.1$ m/s.

(b) Since v' occurs at the equilibrium position, then $v' = v_m$ for the simple harmonic motion. The relation $v_m = \omega x_m$ can be used to solve for x_m, or we can pursue the alternate (though related) approach of energy conservation. Here we choose the latter:

$$\frac{1}{2}(m+M)(v')^2 = \frac{1}{2}kx_m^2 \quad \Rightarrow \quad \frac{1}{2}(m+M)\frac{m^2v^2}{(m+M)^2} = \frac{1}{2}kx_m^2$$

which simplifies to

$$x_m = \frac{mv}{\sqrt{k(m+M)}} = \frac{(9.5 \times 10^{-3} \text{ kg})(630 \text{ m/s})}{\sqrt{(6000 \text{ N/m})(9.5 \times 10^{-3} \text{ kg} + 5.4 \text{ kg})}} = 3.3 \times 10^{-2} \text{ m}.$$

36. We note that the spring constant is

$$k = 4\pi^2 m_1/T^2 = 1.97 \times 10^5 \text{ N/m}.$$

It is important to determine where in its simple harmonic motion (which "phase" of its motion) block 2 is when the impact occurs. Since $\omega = 2\pi/T$ and the given value of t (when the collision takes place) is one-fourth of T, then $\omega t = \pi/2$ and the location then of block 2 is $x = x_m\cos(\omega t + \phi)$ where $\phi = \pi/2$ which gives $x = x_m\cos(\pi/2 + \pi/2) = -x_m$. This means block 2 is at a turning point in its motion (and thus has zero speed right before the impact occurs); this means, too, that the spring is stretched an amount of 1 cm = 0.01 m at this moment. To calculate its after-collision speed (which will be the same as that of block 1 right after the impact, since they stick together in the process) we use momentum conservation and obtain v = (4.0 kg)(6.0 m/s)/(6.0 kg) = 4.0 m/s. Thus, at the end of the impact itself (while block 1 is still at the same position as before the impact) the system (consisting now of a total mass M = 6.0 kg) has kinetic energy

$$K = \tfrac{1}{2}(6.0 \text{ kg})(4.0 \text{ m/s})^2 = 48 \text{ J}$$

and potential energy

$$U = \tfrac{1}{2}kx^2 = \tfrac{1}{2}(1.97 \times 10^5 \text{ N/m})(0.010 \text{ m})^2 \approx 10 \text{ J},$$

meaning the total mechanical energy in the system at this stage is approximately $E = K + U$ = 58 J. When the system reaches its new turning point (at the new amplitude X) then this amount must equal its (maximum) potential energy there: $E = \tfrac{1}{2}(1.97 \times 10^5 \text{ N/m}) X^2$. Therefore, we find

$$X = \sqrt{\frac{2E}{k}} = \sqrt{\frac{2(58 \text{ J})}{1.97 \times 10^5 \text{ N/m}}} = 0.024 \text{ m}.$$

37. (a) The object oscillates about its equilibrium point, where the downward force of gravity is balanced by the upward force of the spring. If ℓ is the elongation of the spring at equilibrium, then $k\ell = mg$, where k is the spring constant and m is the mass of the object. Thus $k/m = g/\ell$ and

$$f = \omega/2\pi = (1/2\pi)\sqrt{k/m} = (1/2\pi)\sqrt{g/\ell}.$$

Now the equilibrium point is halfway between the points where the object is momentarily at rest. One of these points is where the spring is unstretched and the other is the lowest point, 10 cm below. Thus ℓ = 5.0 cm = 0.050 m and

$$f = \frac{1}{2\pi}\sqrt{\frac{9.8 \text{ m/s}^2}{0.050 \text{ m}}} = 2.2 \text{ Hz}.$$

(b) Use conservation of energy. We take the zero of gravitational potential energy to be at the initial position of the object, where the spring is unstretched. Then both the initial potential and kinetic energies are zero. We take the y axis to be positive in the downward direction and let y = 0.080 m. The potential energy when the object is at this point is $U = \tfrac{1}{2}ky^2 - mgy$. The energy equation becomes $0 = \tfrac{1}{2}ky^2 - mgy + \tfrac{1}{2}mv^2$. We solve for the speed:

$$v = \sqrt{2gy - \frac{k}{m}y^2} = \sqrt{2gy - \frac{g}{\ell}y^2} = \sqrt{2(9.8\,\text{m/s}^2)(0.080\,\text{m}) - \left(\frac{9.8\,\text{m/s}^2}{0.050\,\text{m}}\right)(0.080\,\text{m})^2}$$

$$= 0.56\,\text{m/s}$$

(c) Let m be the original mass and Δm be the additional mass. The new angular frequency is $\omega' = \sqrt{k/(m+\Delta m)}$. This should be half the original angular frequency, or $\tfrac{1}{2}\sqrt{k/m}$. We solve $\sqrt{k/(m+\Delta m)} = \tfrac{1}{2}\sqrt{k/m}$ for m. Square both sides of the equation, then take the reciprocal to obtain $m + \Delta m = 4m$. This gives

$$m = \Delta m/3 = (300\text{ g})/3 = 100\text{ g} = 0.100\text{ kg}.$$

(d) The equilibrium position is determined by the balancing of the gravitational and spring forces: $ky = (m + \Delta m)g$. Thus $y = (m + \Delta m)g/k$. We will need to find the value of the spring constant k. Use $k = m\omega^2 = m(2\pi f)^2$. Then

$$y\frac{(m+\Delta m)g}{m(2\pi f)^2} = \frac{(0.100\text{ kg}+0.300\text{ kg})(9.80\text{ m/s}^2)}{(0.100\text{ kg})(2\pi \times 2.24\text{ Hz})^2} = 0.200\text{ m}.$$

This is measured from the initial position.

38. From Eq. 15-23 (in absolute value) we find the torsion constant:

$$\kappa = \left|\frac{\tau}{\theta}\right| = \frac{0.20\text{ N}\cdot\text{m}}{0.85\text{ rad}} = 0.235\text{ N}\cdot\text{m/rad}.$$

With $I = 2mR^2/5$ (the rotational inertia for a solid sphere — from Chapter 11), Eq. 15–23 leads to

$$T = 2\pi\sqrt{\frac{\tfrac{2}{5}mR^2}{\kappa}} = 2\pi\sqrt{\frac{\tfrac{2}{5}(95\text{ kg})(0.15\text{ m})^2}{0.235\text{ N}\cdot\text{m/rad}}} = 12\text{ s}.$$

39. (a) We take the angular displacement of the wheel to be $\theta = \theta_m \cos(2\pi t/T)$, where θ_m is the amplitude and T is the period. We differentiate with respect to time to find the angular velocity: $\Omega = -(2\pi/T)\theta_m \sin(2\pi t/T)$. The symbol Ω is used for the angular velocity of the wheel so it is not confused with the angular frequency. The maximum angular velocity is

$$\Omega_m = \frac{2\pi\theta_m}{T} = \frac{(2\pi)(\pi\text{ rad})}{0.500\text{ s}} = 39.5\text{ rad/s}.$$

(b) When $\theta = \pi/2$, then $\theta/\theta_m = 1/2$, $\cos(2\pi t/T) = 1/2$, and

$$\sin(2\pi t/T) = \sqrt{1-\cos^2(2\pi t/T)} = \sqrt{1-(1/2)^2} = \sqrt{3}/2$$

where the trigonometric identity $\cos^2\theta + \sin^2\theta = 1$ is used. Thus,

$$\Omega = -\frac{2\pi}{T}\theta_m \sin\left(\frac{2\pi t}{T}\right) = -\left(\frac{2\pi}{0.500 \text{ s}}\right)(\pi \text{ rad})\left(\frac{\sqrt{3}}{2}\right) = -34.2 \text{ rad/s}.$$

During another portion of the cycle its angular speed is +34.2 rad/s when its angular displacement is $\pi/2$ rad.

(c) The angular acceleration is

$$\alpha = \frac{d^2\theta}{dt^2} = -\left(\frac{2\pi}{T}\right)^2 \theta_m \cos(2\pi t/T) = -\left(\frac{2\pi}{T}\right)^2 \theta.$$

When $\theta = \pi/4$,

$$\alpha = -\left(\frac{2\pi}{0.500 \text{ s}}\right)^2\left(\frac{\pi}{4}\right) = -124 \text{ rad/s}^2,$$

or $|\alpha| = 124 \text{ rad/s}^2$.

40. (a) Comparing the given expression to Eq. 15-3 (after changing notation $x \to \theta$), we see that $\omega = 4.43$ rad/s. Since $\omega = \sqrt{g/L}$ then we can solve for the length: $L = 0.499$ m.

(b) Since $v_m = \omega x_m = \omega L\theta_m = (4.43 \text{ rad/s})(0.499 \text{ m})(0.0800 \text{ rad})$ and $m = 0.0600$ kg, then we can find the maximum kinetic energy: $\frac{1}{2}mv_m^2 = 9.40 \times 10^{-4}$ J.

41. (a) Referring to Sample Problem 15-5, we see that the distance between P and C is $h = \frac{2}{3}L - \frac{1}{2}L = \frac{1}{6}L$. The parallel axis theorem (see Eq. 15–30) leads to

$$I = \frac{1}{12}mL^2 + mh^2 = \left(\frac{1}{12} + \frac{1}{36}\right)mL^2 = \frac{1}{9}mL^2.$$

Eq. 15-29 then gives

$$T = 2\pi\sqrt{\frac{I}{mgh}} = 2\pi\sqrt{\frac{L^2/9}{gL/6}} = 2\pi\sqrt{\frac{2L}{3g}}$$

which yields $T = 1.64$ s for $L = 1.00$ m.

(b) We note that this T is identical to that computed in Sample Problem 15-5. As far as the characteristics of the periodic motion are concerned, the center of oscillation provides a pivot which is equivalent to that chosen in the Sample Problem (pivot at the edge of the stick).

42. We require

$$T = 2\pi\sqrt{\frac{L_o}{g}} = 2\pi\sqrt{\frac{I}{mgh}}$$

similar to the approach taken in part (b) of Sample Problem 15-5, but treating in our case a more general possibility for I. Canceling 2π, squaring both sides, and canceling g leads directly to the result; $L_o = I/mh$.

43. (a) A uniform disk pivoted at its center has a rotational inertia of $\frac{1}{2}Mr^2$, where M is its mass and r is its radius. The disk of this problem rotates about a point that is displaced from its center by $r + L$, where L is the length of the rod, so, according to the parallel-axis theorem, its rotational inertia is $\frac{1}{2}Mr^2 + \frac{1}{2}M(L+r)^2$. The rod is pivoted at one end and has a rotational inertia of $mL^2/3$, where m is its mass. The total rotational inertia of the disk and rod is

$$I = \frac{1}{2}Mr^2 + M(L+r)^2 + \frac{1}{3}mL^2$$

$$= \frac{1}{2}(0.500\text{kg})(0.100\text{m})^2 + (0.500\text{kg})(0.500\text{m}+0.100\text{m})^2 + \frac{1}{3}(0.270\text{kg})(0.500\text{m})^2$$

$$= 0.205\,\text{kg}\cdot\text{m}^2.$$

(b) We put the origin at the pivot. The center of mass of the disk is

$$\ell_d = L + r = 0.500\text{ m} + 0.100\text{ m} = 0.600\text{ m}$$

away and the center of mass of the rod is $\ell_r = L/2 = (0.500\text{ m})/2 = 0.250$ m away, on the same line. The distance from the pivot point to the center of mass of the disk-rod system is

$$d = \frac{M\ell_d + m\ell_r}{M+m} = \frac{(0.500\text{ kg})(0.600\text{ m}) + (0.270\text{ kg})(0.250\text{ m})}{0.500\text{ kg} + 0.270\text{ kg}} = 0.477\text{ m}.$$

(c) The period of oscillation is

$$T = 2\pi\sqrt{\frac{I}{(M+m)gd}} = 2\pi\sqrt{\frac{0.205\text{ kg}\cdot\text{m}^2}{(0.500\text{ kg}+0.270\text{ kg})(9.80\text{ m/s}^2)(0.447\text{ m})}} = 1.50\text{ s}.$$

44. We use Eq. 15-29 and the parallel-axis theorem $I = I_{cm} + mh^2$ where $h = d$, the unknown. For a meter stick of mass m, the rotational inertia about its center of mass is $I_{cm} = mL^2/12$ where $L = 1.0$ m. Thus, for $T = 2.5$ s, we obtain

$$T = 2\pi\sqrt{\frac{mL^2/12 + md^2}{mgd}} = 2\pi\sqrt{\frac{L^2}{12gd} + \frac{d}{g}}.$$

Squaring both sides and solving for d leads to the quadratic formula:

$$d = \frac{g(T/2\pi)^2 \pm \sqrt{d^2(T/2\pi)^4 - L^2/3}}{2}.$$

Choosing the plus sign leads to an impossible value for d ($d = 1.5 > L$). If we choose the minus sign, we obtain a physically meaningful result: $d = 0.056$ m.

45. We use Eq. 15-29 and the parallel-axis theorem $I = I_{cm} + mh^2$ where $h = d$. For a solid disk of mass m, the rotational inertia about its center of mass is $I_{cm} = mR^2/2$. Therefore,

$$T = 2\pi\sqrt{\frac{mR^2/2 + md^2}{mgd}} = 2\pi\sqrt{\frac{R^2 + 2d^2}{2gd}} = 2\pi\sqrt{\frac{(2.35 \text{ cm})^2 + 2(1.75 \text{ cm})^2}{2(980 \text{ cm/s}^2)(1.75 \text{ cm})}} = 0.366 \text{ s}.$$

46. To use Eq. 15-29 we need to locate the center of mass and we need to compute the rotational inertia about A. The center of mass of the stick shown horizontal in the figure is at A, and the center of mass of the other stick is 0.50 m below A. The two sticks are of equal mass so the center of mass of the system is $h = \frac{1}{2}(0.50 \text{ m}) = 0.25 \text{m}$ below A, as shown in the figure. Now, the rotational inertia of the system is the sum of the rotational inertia I_1 of the stick shown horizontal in the figure and the rotational inertia I_2 of the stick shown vertical. Thus, we have

$$I = I_1 + I_2 = \frac{1}{12}ML^2 + \frac{1}{3}ML^2 = \frac{5}{12}ML^2$$

where $L = 1.00$ m and M is the mass of a meter stick (which cancels in the next step). Now, with $m = 2M$ (the total mass), Eq. 15–29 yields

$$T = 2\pi\sqrt{\frac{\frac{5}{12}ML^2}{2Mgh}} = 2\pi\sqrt{\frac{5L}{6g}}$$

where $h = L/4$ was used. Thus, $T = 1.83$ s.

47. From Eq. 15-28, we find the length of the pendulum when the period is $T = 8.85$ s:

$$L = \frac{gT^2}{4\pi^2}.$$

The new length is $L' = L - d$ where $d = 0.350$ m. The new period is

$$T' = 2\pi\sqrt{\frac{L'}{g}} = 2\pi\sqrt{\frac{L}{g} - \frac{d}{g}} = 2\pi\sqrt{\frac{T^2}{4\pi^2} - \frac{d}{g}}$$

which yields $T' = 8.77$ s.

48. (a) The rotational inertia of a uniform rod with pivot point at its end is $I = mL^2/12 + mL^2 = 1/3\,ML^2$. Therefore, Eq. 15-29 leads to

$$T = 2\pi\sqrt{\frac{\frac{1}{3}ML^2}{Mg(L/2)}} \quad \Rightarrow \quad \frac{3gT^2}{8\pi^2}$$

so that $L = 0.84$ m.

(b) By energy conservation

$$E_{\text{bottom of swing}} = E_{\text{end of swing}} \quad \Rightarrow \quad K_m = U_m$$

where $U = Mg\ell(1 - \cos\theta)$ with ℓ being the distance from the axis of rotation to the center of mass. If we use the small angle approximation ($\cos\theta \approx 1 - \tfrac{1}{2}\theta^2$ with θ in radians (Appendix E)), we obtain

$$U_m = (0.5\ \text{kg})(9.8\ \text{m/s}^2)\left(\frac{L}{2}\right)\left(\frac{1}{2}\theta_m^2\right)$$

where $\theta_m = 0.17$ rad. Thus, $K_m = U_m = 0.031$ J. If we calculate $(1 - \cos\theta)$ straightforwardly (without using the small angle approximation) then we obtain within 0.3% of the same answer.

49. This is similar to the situation treated in Sample Problem 15-5, except that O is no longer at the end of the stick. Referring to the center of mass as C (assumed to be the geometric center of the stick), we see that the distance between O and C is $h = x$. The parallel axis theorem (see Eq. 15-30) leads to

$$I = \frac{1}{12}mL^2 + mh^2 = m\left(\frac{L^2}{12} + x^2\right).$$

Eq. 15-29 gives

$$T = 2\pi\sqrt{\frac{I}{mgh}} = 2\pi\sqrt{\frac{\left(\frac{L^2}{12} + x^2\right)}{gx}} = 2\pi\sqrt{\frac{\left(L^2 + 12x^2\right)}{12gx}}.$$

(a) Minimizing T by graphing (or special calculator functions) is straightforward, but the standard calculus method (setting the derivative equal to zero and solving) is somewhat awkward. We pursue the calculus method but choose to work with $12gT^2/2\pi$ instead of T (it should be clear that $12gT^2/2\pi$ is a minimum whenever T is a minimum). The result is

$$\frac{d\left(\frac{12gT^2}{2\pi}\right)}{dx} = 0 = \frac{d\left(\frac{L^2}{x}+12x\right)}{dx} = -\frac{L^2}{x^2}+12$$

which yields $x = L/\sqrt{12} = (1.85 \text{ m})/\sqrt{12} = 0.53 \text{ m}$ as the value of x which should produce the smallest possible value of T.

(b) With $L = 1.85$ m and $x = 0.53$ m, we obtain $T = 2.1$ s from the expression derived in part (a).

50. Consider that the length of the spring as shown in the figure (with one of the block's corners lying directly above the block's center) is some value L (its rest length). If the (constant) distance between the block's center and the point on the wall where the spring attaches is a distance r, then $r\cos\theta = d/\sqrt{2}$ and $r\cos\theta = L$ defines the angle θ measured from a line on the block drawn from the center to the top corner to the line of r (a straight line from the center of the block to the point of attachment of the spring on the wall). In terms of this angle, then, the problem asks us to consider the dynamics that results from increasing θ from its original value θ_o to $\theta_o + 3°$ and then releasing the system and letting it oscillate. If the new (stretched) length of spring is L' (when $\theta = \theta_o + 3°$), then it is a straightforward trigonometric exercise to show that

$$(L')^2 = r^2 + (d/\sqrt{2})^2 - 2r(d/\sqrt{2})\cos(\theta_o + 3°) = L^2 + d^2 - d^2\cos(3°) + \sqrt{2}\,Ld\sin(3°).$$

since $\theta_o = 45°$. The difference between L' (as determined by this expression) and the original spring length L is the amount the spring has been stretched (denoted here as x_m). If one plots x_m versus L over a range that seems reasonable considering the figure shown in the problem (say, from $L = 0.03$ m to $L = 0.10$ m) one quickly sees that $x_m \approx 0.00222$ m is an excellent approximation (and is very close to what one would get by approximating x_m as the arc length of the path made by that upper block corner as the block is turned through 3°, even though this latter procedure should in principle overestimate x_m). Using this value of x_m with the given spring constant leads to a potential energy of $U = \frac{1}{2}k x_m^2 = 0.00296$ J. Setting this equal to the kinetic energy the block has as it passes back through the initial position, we have

$$K = 0.00296 \text{ J} = \frac{1}{2} I \omega_m^2$$

where ω_m is the maximum angular speed of the block (and is not to be confused with the angular frequency ω of the oscillation, though they are related by $\omega_m = \theta_o \omega$ if θ_o is expressed in radians). The rotational inertia of the block is $I = \frac{1}{6}Md^2 = 0.0018$ kg·m². Thus, we can solve the above relation for the maximum angular speed of the block:

$$\omega_m = \sqrt{\frac{2K}{I}} = \sqrt{\frac{2(0.00296 \text{ J})}{0.0018 \text{ kg} \cdot \text{m}^2}} = 1.81 \text{ rad/s}.$$

Therefore the angular frequency of the oscillation is $\omega = \omega_m/\theta_o = 34.6$ rad/s. Using Eq. 15-5, then, the period is $T = 0.18$ s.

51. If the torque exerted by the spring on the rod is proportional to the angle of rotation of the rod and if the torque tends to pull the rod toward its equilibrium orientation, then the rod will oscillate in simple harmonic motion. If $\tau = -C\theta$, where τ is the torque, θ is the angle of rotation, and C is a constant of proportionality, then the angular frequency of oscillation is $\omega = \sqrt{C/I}$ and the period is

$$T = 2\pi/\omega = 2\pi\sqrt{I/C},$$

where I is the rotational inertia of the rod. The plan is to find the torque as a function of θ and identify the constant C in terms of given quantities. This immediately gives the period in terms of given quantities. Let ℓ_0 be the distance from the pivot point to the wall. This is also the equilibrium length of the spring. Suppose the rod turns through the angle θ, with the left end moving away from the wall. This end is now $(L/2)\sin\theta$ further from the wall and has moved a distance $(L/2)(1 - \cos\theta)$ to the right. The length of the spring is now

$$\ell = \sqrt{(L/2)^2(1-\cos\theta)^2 + [\ell_0 + (L/2)\sin\theta]^2}.$$

If the angle θ is small we may approximate $\cos\theta$ with 1 and $\sin\theta$ with θ in radians. Then the length of the spring is given by $\ell \approx \ell_0 + L\theta/2$ and its elongation is $\Delta x = L\theta/2$. The force it exerts on the rod has magnitude $F = k\Delta x = kL\theta/2$. Since θ is small we may approximate the torque exerted by the spring on the rod by $\tau = -FL/2$, where the pivot point was taken as the origin. Thus $\tau = -(kL^2/4)\theta$. The constant of proportionality C that relates the torque and angle of rotation is $C = kL^2/4$. The rotational inertia for a rod pivoted at its center is $I = mL^2/12$, where m is its mass. See Table 10-2. Thus the period of oscillation is

$$T = 2\pi\sqrt{\frac{I}{C}} = 2\pi\sqrt{\frac{mL^2/12}{kL^2/4}} = 2\pi\sqrt{\frac{m}{3k}}.$$

With $m = 0.600$ kg and $k = 1850$ N/m, we obtain $T = 0.0653$ s.

52. (a) For the "physical pendulum" we have

$$T = 2\pi\sqrt{\frac{I}{mgh}} = 2\pi\sqrt{\frac{I_{com} + mh^2}{mgh}}.$$

If we substitute r for h and use item (i) in Table 10-2, we have

$$T = \frac{2\pi}{\sqrt{g}} \sqrt{\frac{a^2+b^2}{12r} + r}$$

In the figure below, we plot T as a function of r, for $a = 0.35$ m and $b = 0.45$ m.

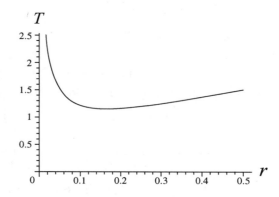

(b) The minimum of T can be located by setting its derivative to zero, $dT/dr = 0$. This yields

$$r = \sqrt{\frac{a^2+b^2}{12}} = \sqrt{\frac{(0.35 \text{ m})^2 + (0.45 \text{ m})^2}{12}} = 0.16 \text{ m}.$$

(c) The direction from the center does not matter, so the locus of points is a circle around the center, of radius $[(a^2 + b^2)/12]^{1/2}$.

53. Replacing x and v in Eq. 15-3 and Eq. 15-6 with θ and $d\theta/dt$, respectively, we identify 4.44 rad/s as the angular frequency ω. Then we evaluate the expressions at $t = 0$ and divide the second by the first:

$$\left(\frac{d\theta/dt}{\theta}\right)_{\text{at } t=0} = -\omega \tan\phi.$$

(a) The value of θ at $t = 0$ is 0.0400 rad, and the value of $d\theta/dt$ then is –0.200 rad/s, so we are able to solve for the phase constant:

$$\phi = \tan^{-1}[0.200/(0.0400 \times 4.44)] = 0.845 \text{ rad}.$$

(b) Once ϕ is determined we can plug back in to $\theta_o = \theta_m \cos\phi$ to solve for the angular amplitude. We find $\theta_m = 0.0602$ rad.

54. We note that the initial angle is $\theta_o = 7° = 0.122$ rad (though it turns out this value will cancel in later calculations). If we approximate the initial stretch of the spring as the arc-length that the corresponding point on the plate has moved through ($x = r\theta_o$ where $r = 0.025$ m) then the initial potential energy is approximately $\frac{1}{2}kx^2 = 0.0093$ J. This should

equal to the kinetic energy of the plate ($\frac{1}{2}I\omega_m^2$ where this ω_m is the maximum angular speed of the plate, not the angular frequency ω). Noting that the maximum angular speed of the plate is $\omega_m = \omega\theta_o$ where $\omega = 2\pi/T$ with $T = 20$ ms $= 0.02$ s as determined from the graph, then we can find the rotational inertial from $\frac{1}{2} I \omega_m^2 = 0.0093$ J. Thus, $I = 1.3 \times 10^{-5}$ kg·m^2.

55. (a) The period of the pendulum is given by $T = 2\pi\sqrt{I/mgd}$, where I is its rotational inertia, $m = 22.1$ g is its mass, and d is the distance from the center of mass to the pivot point. The rotational inertia of a rod pivoted at its center is $mL^2/12$ with $L = 2.20$ m. According to the parallel-axis theorem, its rotational inertia when it is pivoted a distance d from the center is $I = mL^2/12 + md^2$. Thus,

$$T = 2\pi\sqrt{\frac{m(L^2/12 + d^2)}{mgd}} = 2\pi\sqrt{\frac{L^2 + 12d^2}{12gd}}.$$

Minimizing T with respect to d, $dT/d(d)=0$, we obtain $d = L/\sqrt{12}$. Therefore, the minimum period T is

$$T_{min} = 2\pi\sqrt{\frac{L^2 + 12(L/\sqrt{12})^2}{12g(L/\sqrt{12})}} = 2\pi\sqrt{\frac{2L}{\sqrt{12}g}} = 2\pi\sqrt{\frac{2(2.20\text{ m})}{\sqrt{12}(9.80\text{ m/s}^2)}} = 2.26\text{ s}.$$

(b) If d is chosen to minimize the period, then as L is increased the period will increase as well.

(c) The period does not depend on the mass of the pendulum, so T does not change when m increases.

56. The table of moments of inertia in Chapter 11, plus the parallel axis theorem found in that chapter, leads to

$$I_P = \tfrac{1}{2}MR^2 + Mh^2 = \tfrac{1}{2}(2.5\text{ kg})(0.21\text{ m})^2 + (2.5\text{ kg})(0.97\text{ m})^2 = 2.41 \text{ kg·m}^2$$

where P is the hinge pin shown in the figure (the point of support for the physical pendulum), which is a distance $h = 0.21$ m $+ 0.76$ m away from the center of the disk.

(a) Without the torsion spring connected, the period is

$$T = 2\pi\sqrt{\frac{I_P}{Mgh}} = 2.00\text{ s}.$$

(b) Now we have two "restoring torques" acting in tandem to pull the pendulum back to the vertical position when it is displaced. The magnitude of the torque-sum is (Mgh +

$\kappa)\theta = I_P \alpha$, where the small angle approximation ($\sin\theta \approx \theta$ in radians) and Newton's second law (for rotational dynamics) have been used. Making the appropriate adjustment to the period formula, we have

$$T' = 2\pi \sqrt{\frac{I_P}{Mgh + \kappa}}.$$

The problem statement requires $T = T' + 0.50$ s. Thus, $T' = (2.00 - 0.50)$s $= 1.50$ s. Consequently,

$$\kappa = \frac{4\pi^2}{T'^2} I_P - Mgh = 18.5 \text{ N·m/rad}.$$

57. (a) We want to solve $e^{-bt/2m} = 1/3$ for t. We take the natural logarithm of both sides to obtain $-bt/2m = \ln(1/3)$. Therefore, $t = -(2m/b)\ln(1/3) = (2m/b)\ln 3$. Thus,

$$t = \frac{2(1.50 \text{ kg})}{0.230 \text{ kg/s}} \ln 3 = 14.3 \text{ s}.$$

(b) The angular frequency is

$$\omega' = \sqrt{\frac{k}{m} - \frac{b^2}{4m^2}} = \sqrt{\frac{8.00 \text{ N/m}}{1.50 \text{ kg}} - \frac{(0.230 \text{ kg/s})^2}{4(1.50 \text{ kg})^2}} = 2.31 \text{ rad/s}.$$

The period is $T = 2\pi/\omega' = (2\pi)/(2.31 \text{ rad/s}) = 2.72$ s and the number of oscillations is

$$t/T = (14.3 \text{ s})/(2.72 \text{ s}) = 5.27.$$

58. Referring to the numbers in Sample Problem 15-7, we have $m = 0.25$ kg, $b = 0.070$ kg/s and $T = 0.34$ s. Thus, when $t = 20T$, the damping factor becomes

$$e^{-bt/2m} = e^{-(0.070)(20)(0.34)/2(0.25)} = 0.39.$$

59. Since the energy is proportional to the amplitude squared (see Eq. 15-21), we find the fractional change (assumed small) is

$$\frac{E' - E}{E} \approx \frac{dE}{E} = \frac{dx_m^2}{x_m^2} = \frac{2x_m dx_m}{x_m^2} = 2\frac{dx_m}{x_m}.$$

Thus, if we approximate the fractional change in x_m as dx_m/x_m, then the above calculation shows that multiplying this by 2 should give the fractional energy change. Therefore, if x_m decreases by 3%, then E must decrease by 6.0 %.

60. (a) From Hooke's law, we have

$$k = \frac{(500 \text{ kg})(9.8 \text{ m/s}^2)}{10 \text{cm}} = 4.9 \times 10^2 \text{ N/cm}.$$

(b) The amplitude decreasing by 50% during one period of the motion implies

$$e^{-bT/2m} = \frac{1}{2} \quad \text{where} \quad T = \frac{2\pi}{\omega'}.$$

Since the problem asks us to estimate, we let $\omega' \approx \omega = \sqrt{k/m}$. That is, we let

$$\omega' \approx \sqrt{\frac{49000 \text{ N/m}}{500 \text{ kg}}} \approx 9.9 \text{ rad/s},$$

so that $T \approx 0.63$ s. Taking the (natural) log of both sides of the above equation, and rearranging, we find

$$b = \frac{2m}{T} \ln 2 \approx \frac{2(500 \text{ kg})}{0.63 \text{ s}} (0.69) = 1.1 \times 10^3 \text{ kg/s}.$$

Note: if one worries about the $\omega' \approx \omega$ approximation, it is quite possible (though messy) to use Eq. 15-43 in its full form and solve for b. The result would be (quoting more figures than are significant)

$$b = \frac{2 \ln 2 \sqrt{mk}}{\sqrt{(\ln 2)^2 + 4\pi^2}} = 1086 \text{ kg/s}$$

which is in good agreement with the value gotten "the easy way" above.

61. (a) We set $\omega = \omega_d$ and find that the given expression reduces to $x_m = F_m/b\omega$ at resonance.

(b) In the discussion immediately after Eq. 15-6, the book introduces the velocity amplitude $v_m = \omega x_m$. Thus, at resonance, we have $v_m = \omega F_m/b\omega = F_m/b$.

62. With $\omega = 2\pi/T$ then Eq. 15-28 can be used to calculate the angular frequencies for the given pendulums. For the given range of $2.00 < \omega < 4.00$ (in rad/s), we find only two of the given pendulums have appropriate values of ω: pendulum (d) with length of 0.80 m (for which $\omega = 3.5$ rad/s) and pendulum (e) with length of 1.2 m (for which $\omega = 2.86$ rad/s).

63. With $M = 1000$ kg and $m = 82$ kg, we adapt Eq. 15-12 to this situation by writing

$$\omega = \frac{2\pi}{T} = \sqrt{\frac{k}{M+4m}}.$$

If $d = 4.0$ m is the distance traveled (at constant car speed v) between impulses, then we may write $T = v/d$, in which case the above equation may be solved for the spring constant:

$$\frac{2\pi v}{d} = \sqrt{\frac{k}{M+4m}} \Rightarrow k = (M+4m)\left(\frac{2\pi v}{d}\right)^2.$$

Before the people got out, the equilibrium compression is $x_i = (M + 4m)g/k$, and afterward it is $x_f = Mg/k$. Therefore, with $v = 16000/3600 = 4.44$ m/s, we find the rise of the car body on its suspension is

$$x_i - x_f = \frac{4mg}{k} = \frac{4mg}{M+4m}\left(\frac{d}{2\pi v}\right)^2 = 0.050 \text{ m}.$$

64. Its total mechanical energy is equal to its maximum potential energy $\frac{1}{2}kx_m^2$, and its potential energy at $t = 0$ is $\frac{1}{2}kx_o^2$ where $x_o = x_m\cos(\pi/5)$ in this problem. The ratio is therefore $\cos^2(\pi/5) = 0.655 = 65.5\%$.

65. (a) From the graph, we find $x_m = 7.0$ cm $= 0.070$ m, and $T = 40$ ms $= 0.040$ s. Thus, the angular frequency is $\omega = 2\pi/T = 157$ rad/s. Using $m = 0.020$ kg, the maximum kinetic energy is then $\frac{1}{2}mv^2 = \frac{1}{2}m\omega^2 x_m^2 = 1.2$ J.

(b) Using Eq. 15-5, we have $f = \omega/2\pi = 50$ oscillations per second. Of course, Eq. 15-2 can also be used for this.

66. (a) From the graph we see that $x_m = 7.0$ cm $= 0.070$ m and $T = 40$ ms $= 0.040$ s. The maximum speed is $x_m\omega = x_m 2\pi/T = 11$ m/s.

(b) The maximum acceleration is $x_m\omega^2 = x_m(2\pi/T)^2 = 1.7 \times 10^3$ m/s^2.

67. Setting 15 mJ (0.015 J) equal to the maximum kinetic energy leads to $v_{max} = 0.387$ m/s. Then one can use either an "exact" approach using $v_{max} = \sqrt{2gL(1-\cos(\theta_{max}))}$ or the "SHM" approach where

$$v_{max} = L\omega_{max} = L\omega\theta_{max} = L\sqrt{g/L}\,\theta_{max}$$

to find L. Both approaches lead to $L = 1.53$ m.

68. Since $\omega = 2\pi f$ where $f = 2.2$ Hz, we find that the angular frequency is $\omega = 13.8$ rad/s. Thus, with $x = 0.010$ m, the acceleration amplitude is $a_m = x_m\omega^2 = 1.91$ m/s^2. We set up a ratio:

$$a_m = \left(\frac{a_m}{g}\right)g = \left(\frac{1.91}{9.8}\right)g = 0.19g.$$

69. (a) Assume the bullet becomes embedded and moves with the block before the block moves a significant distance. Then the momentum of the bullet-block system is conserved during the collision. Let m be the mass of the bullet, M be the mass of the block, v_0 be the initial speed of the bullet, and v be the final speed of the block and bullet. Conservation of momentum yields $mv_0 = (m + M)v$, so

$$v = \frac{mv_0}{m+M} = \frac{(0.050 \text{ kg})(150 \text{ m/s})}{0.050 \text{ kg} + 4.0 \text{ kg}} = 1.85 \text{ m/s}.$$

When the block is in its initial position the spring and gravitational forces balance, so the spring is elongated by Mg/k. After the collision, however, the block oscillates with simple harmonic motion about the point where the spring and gravitational forces balance with the bullet embedded. At this point the spring is elongated a distance $\ell = (M+m)g/k$, somewhat different from the initial elongation. Mechanical energy is conserved during the oscillation. At the initial position, just after the bullet is embedded, the kinetic energy is $\frac{1}{2}(M+m)v^2$ and the elastic potential energy is $\frac{1}{2}k(Mg/k)^2$. We take the gravitational potential energy to be zero at this point. When the block and bullet reach the highest point in their motion the kinetic energy is zero. The block is then a distance y_m above the position where the spring and gravitational forces balance. Note that y_m is the amplitude of the motion. The spring is compressed by $y_m - \ell$, so the elastic potential energy is $\frac{1}{2}k(y_m - \ell)^2$. The gravitational potential energy is $(M + m)gy_m$. Conservation of mechanical energy yields

$$\frac{1}{2}(M+m)v^2 + \frac{1}{2}k\left(\frac{Mg}{k}\right)^2 = \frac{1}{2}k(y_m - \ell)^2 + (M+m)gy_m.$$

We substitute $\ell = (M+m)g/k$. Algebraic manipulation leads to

$$y_m = \sqrt{\frac{(m+M)v^2}{k} - \frac{mg^2}{k^2}(2M+m)}$$

$$= \sqrt{\frac{(0.050 \text{ kg} + 4.0 \text{ kg})(1.85 \text{ m/s})^2}{500 \text{ N/m}} - \frac{(0.050 \text{ kg})(9.8 \text{ m/s}^2)^2}{(500 \text{ N/m})^2}[2(4.0 \text{ kg}) + 0.050 \text{ kg}]}$$

$$= 0.166 \text{ m}.$$

(b) The original energy of the bullet is $E_0 = \frac{1}{2}mv_0^2 = \frac{1}{2}(0.050 \text{ kg})(150 \text{ m/s})^2 = 563 \text{ J}$. The kinetic energy of the bullet-block system just after the collision is

$$E = \frac{1}{2}(m+M)v^2 = \frac{1}{2}(0.050 \text{ kg} + 4.0 \text{ kg})(1.85 \text{ m/s})^2 = 6.94 \text{ J}.$$

Since the block does not move significantly during the collision, the elastic and gravitational potential energies do not change. Thus, E is the energy that is transferred. The ratio is

$$E/E_0 = (6.94 \text{ J})/(563 \text{ J}) = 0.0123 \text{ or } 1.23\%.$$

70. (a) We note that

$$\omega = \sqrt{k/m} = \sqrt{1500/0.055} = 165.1 \text{ rad/s}.$$

We consider the most direct path in each part of this problem. That is, we consider in part (a) the motion directly from $x_1 = +0.800 x_m$ at time t_1 to $x_2 = +0.600 x_m$ at time t_2 (as opposed to, say, the block moving from $x_1 = +0.800 x_m$ through $x = +0.600 x_m$, through $x = 0$, reaching $x = -x_m$ and after returning back through $x = 0$ then getting to $x_2 = +0.600 x_m$). Eq. 15-3 leads to

$$\omega t_1 + \phi = \cos^{-1}(0.800) = 0.6435 \text{ rad}$$

$$\omega t_2 + \phi = \cos^{-1}(0.600) = 0.9272 \text{ rad}.$$

Subtracting the first of these equations from the second leads to

$$\omega(t_2 - t_1) = 0.9272 - 0.6435 = 0.2838 \text{ rad}.$$

Using the value for ω computed earlier, we find $t_2 - t_1 = 1.72 \times 10^{-3}$ s.

(b) Let t_3 be when the block reaches $x = -0.800 x_m$ in the direct sense discussed above. Then the reasoning used in part (a) leads here to

$$\omega(t_3 - t_1) = (2.4981 - 0.6435) \text{ rad} = 1.8546 \text{ rad}$$

and thus to $t_3 - t_1 = 11.2 \times 10^{-3}$ s.

71. (a) The problem gives the frequency $f = 440$ Hz, where the SI unit abbreviation Hz stands for Hertz, which means a cycle-per-second. The angular frequency ω is similar to frequency except that ω is in radians-per-second. Recalling that 2π radians are equivalent to a cycle, we have $\omega = 2\pi f \approx 2.8 \times 10^3$ rad/s.

(b) In the discussion immediately after Eq. 15-6, the book introduces the velocity amplitude $v_m = \omega x_m$. With $x_m = 0.00075$ m and the above value for ω, this expression yields $v_m = 2.1$ m/s.

(c) In the discussion immediately after Eq. 15-7, the book introduces the acceleration amplitude $a_m = \omega^2 x_m$, which (if the more precise value $\omega = 2765$ rad/s is used) yields $a_m = 5.7$ km/s.

72. (a) The textbook notes (in the discussion immediately after Eq. 15-7) that the acceleration amplitude is $a_m = \omega^2 x_m$, where ω is the angular frequency ($\omega = 2\pi f$ since there are 2π radians in one cycle). Therefore, in this circumstance, we obtain

$$a_m = (2\pi(1000 \text{ Hz}))^2 (0.00040 \text{ m}) = 1.6 \times 10^4 \text{ m/s}^2.$$

(b) Similarly, in the discussion after Eq. 15-6, we find $v_m = \omega x_m$ so that

$$v_m = (2\pi(1000 \text{ Hz}))(0.00040 \text{ m}) = 2.5 \text{ m/s}.$$

(c) From Eq. 15-8, we have (in absolute value)

$$|a| = (2\pi(1000 \text{ Hz}))^2 (0.00020 \text{ m}) = 7.9 \times 10^3 \text{ m/s}^2.$$

(d) This can be approached with the energy methods of §15-4, but here we will use trigonometric relations along with Eq. 15-3 and Eq. 15-6. Thus, allowing for both roots stemming from the square root,

$$\sin(\omega t + \phi) = \pm\sqrt{1 - \cos^2(\omega t + \phi)} \quad \Rightarrow \quad -\frac{v}{\omega x_m} = \pm\sqrt{1 - \frac{x^2}{x_m^2}}.$$

Taking absolute values and simplifying, we obtain

$$|v| = 2\pi f \sqrt{x_m^2 - x^2} = 2\pi(1000)\sqrt{0.00040^2 - 0.00020^2} = 2.2 \text{ m/s}.$$

73. (a) The rotational inertia is $I = \tfrac{1}{2} MR^2 = \tfrac{1}{2}(3.00 \text{ kg})(0.700 \text{ m})^2 = 0.735 \text{ kg} \cdot \text{m}^2$.

(b) Using Eq. 15-22 (in absolute value), we find

$$\kappa = \frac{\tau}{\theta} = \frac{0.0600 \text{ N} \cdot \text{m}}{2.5 \text{ rad}} = 0.0240 \text{ N} \cdot \text{m/rad}.$$

(c) Using Eq. 15-5, Eq. 15-23 leads to

$$\omega = \sqrt{\frac{\kappa}{I}} = \sqrt{\frac{0.024 \text{ N} \cdot \text{m/rad}}{0.735 \text{ kg} \cdot \text{m}^2}} = 0.181 \text{ rad/s}.$$

74. (a) We use Eq. 15-29 and the parallel-axis theorem $I = I_{cm} + mh^2$ where $h = R = 0.126$ m. For a solid disk of mass m, the rotational inertia about its center of mass is $I_{cm} = mR^2/2$. Therefore,

$$T = 2\pi\sqrt{\frac{mR^2/2 + mR^2}{mgR}} = 2\pi\sqrt{\frac{3R}{2g}} = 0.873 \text{ s}.$$

(b) We seek a value of $r \neq R$ such that

$$2\pi\sqrt{\frac{R^2 + 2r^2}{2gr}} = 2\pi\sqrt{\frac{3R}{2g}}$$

and are led to the quadratic formula:

$$r = \frac{3R \pm \sqrt{(3R)^2 - 8R^2}}{4} = R \text{ or } \frac{R}{2}.$$

Thus, our result is $r = 0.126/2 = 0.0630$ m.

75. (a) The frequency for small amplitude oscillations is $f = (1/2\pi)\sqrt{g/L}$, where L is the length of the pendulum. This gives

$$f = (1/2\pi)\sqrt{(9.80 \text{ m}/\text{s}^2)/(2.0 \text{ m})} = 0.35 \text{ Hz}.$$

(b) The forces acting on the pendulum are the tension force $\vec{T}$ of the rod and the force of gravity $m\vec{g}$. Newton's second law yields $\vec{T} + m\vec{g} = m\vec{a}$, where m is the mass and $\vec{a}$ is the acceleration of the pendulum. Let $\vec{a} = \vec{a}_e + \vec{a}'$, where $\vec{a}_e$ is the acceleration of the elevator and $\vec{a}'$ is the acceleration of the pendulum relative to the elevator. Newton's second law can then be written $m(\vec{g} - \vec{a}_e) + \vec{T} = m\vec{a}'$. Relative to the elevator the motion is exactly the same as it would be in an inertial frame where the acceleration due to gravity is $\vec{g} - \vec{a}_e$. Since $\vec{g}$ and $\vec{a}_e$ are along the same line and in opposite directions we can find the frequency for small amplitude oscillations by replacing g with $g + a_e$ in the expression $f = (1/2\pi)\sqrt{g/L}$. Thus

$$f = \frac{1}{2\pi}\sqrt{\frac{g + a_e}{L}} = \frac{1}{2\pi}\sqrt{\frac{9.8 \text{ m}/\text{s}^2 + 2.0 \text{ m}/\text{s}^2}{2.0 \text{ m}}} = 0.39 \text{ Hz}.$$

(c) Now the acceleration due to gravity and the acceleration of the elevator are in the same direction and have the same magnitude. That is, $\vec{g} - \vec{a}_e = 0$. To find the frequency for small amplitude oscillations, replace g with zero in $f = (1/2\pi)\sqrt{g/L}$. The result is zero. The pendulum does not oscillate.

76. Since the particle has zero speed (momentarily) at $x \neq 0$, then it must be at its turning point; thus, $x_o = x_m = 0.37$ cm. It is straightforward to infer from this that the phase

constant ϕ in Eq. 15-2 is zero. Also, $f = 0.25$ Hz is given, so we have $\omega = 2\pi f = \pi/2$ rad/s. The variable t is understood to take values in seconds.

(a) The period is $T = 1/f = 4.0$ s.

(b) As noted above, $\omega = \pi/2$ rad/s.

(c) The amplitude, as observed above, is 0.37 cm.

(d) Eq. 15-3 becomes $x = (0.37$ cm$) \cos(\pi t/2)$.

(e) The derivative of x is $v = -(0.37$ cm/s$)(\pi/2) \sin(\pi t/2) \approx (-0.58$ cm/s$) \sin(\pi t/2)$.

(f) From the previous part, we conclude $v_m = 0.58$ cm/s.

(g) The acceleration-amplitude is $a_m = \omega^2 x_m = 0.91$ cm/s^2.

(h) Making sure our calculator is in radians mode, we find $x = (0.37) \cos(\pi(3.0)/2) = 0$. It is important to avoid rounding off the value of π in order to get precisely zero, here.

(i) With our calculator still in radians mode, we obtain $v = -(0.58$ cm/s$)\sin(\pi(3.0)/2) = 0.58$ cm/s.

77. Since $T = 0.500$ s, we note that $\omega = 2\pi/T = 4\pi$ rad/s. We work with SI units, so $m = 0.0500$ kg and $v_m = 0.150$ m/s.

(a) Since $\omega = \sqrt{k/m}$, the spring constant is

$$k = \omega^2 m = (4\pi \text{ rad/s})^2 (0.0500 \text{ kg}) = 7.90 \text{ N/m}.$$

(b) We use the relation $v_m = x_m \omega$ and obtain

$$x_m = \frac{v_m}{\omega} = \frac{0.150}{4\pi} = 0.0119 \text{ m}.$$

(c) The frequency is $f = \omega/2\pi = 2.00$ Hz (which is equivalent to $f = 1/T$).

78. (a) Hooke's law readily yields $(0.300$ kg$)(9.8$ m/s$^2)/(0.0200$ m$) = 147$ N/m.

(b) With $m = 2.00$ kg, the period is

$$T = 2\pi \sqrt{\frac{m}{k}} = 0.733 \text{ s}.$$

79. Using $\Delta m = 2.0$ kg, $T_1 = 2.0$ s and $T_2 = 3.0$ s, we write

$$T_1 = 2\pi\sqrt{\frac{m}{k}} \quad \text{and} \quad T_2 = 2\pi\sqrt{\frac{m+\Delta m}{k}}.$$

Dividing one relation by the other, we obtain

$$\frac{T_2}{T_1} = \sqrt{\frac{m+\Delta m}{m}}$$

which (after squaring both sides) simplifies to $m = \dfrac{\Delta m}{(T_2/T_1)^2 - 1} = 1.6\,\text{kg}.$

80. (a) Comparing with Eq. 15-3, we see $\omega = 10$ rad/s in this problem. Thus, $f = \omega/2\pi = 1.6$ Hz.

(b) Since $v_m = \omega x_m$ and $x_m = 10$ cm (see Eq. 15-3), then $v_m = (10\,\text{rad/s})(10\,\text{cm}) = 100$ cm/s or 1.0 m/s.

(c) The maximum occurs at $t = 0$.

(d) Since $a_m = \omega^2 x_m$ then $v_m = (10\,\text{rad/s})^2(10\,\text{cm}) = 1000$ cm/s² or 10 m/s².

(e) The acceleration extremes occur at the displacement extremes: $x = \pm x_m$ or $x = \pm 10$ cm.

(f) Using Eq. 15-12, we find

$$\omega = \sqrt{\frac{k}{m}} \Rightarrow k = (0.10\,\text{kg})(10\,\text{rad/s})^2 = 10\,\text{N/m}.$$

Thus, Hooke's law gives $F = -kx = -10x$ in SI units.

81. (a) We require $U = \tfrac{1}{2}E$ at some value of x. Using Eq. 15-21, this becomes

$$\tfrac{1}{2}kx^2 = \tfrac{1}{2}\left(\tfrac{1}{2}kx_m^2\right) \Rightarrow x = \frac{x_m}{\sqrt{2}}.$$

We compare the given expression x as a function of t with Eq. 15-3 and find $x_m = 5.0$ m. Thus, the value of x we seek is $x = 5.0/\sqrt{2} \approx 3.5$ m.

(b) We solve the given expression (with $x = 5.0/\sqrt{2}$), making sure our calculator is in radians mode:

$$t = \frac{\pi}{4} + \frac{3}{\pi}\cos^{-1}\left(\frac{1}{\sqrt{2}}\right) = 1.54\,\text{s}.$$

Since we are asked for the interval $t_{eq} - t$ where t_{eq} specifies the instant the particle passes through the equilibrium position, then we set $x = 0$ and find

$$t_{eq} = \frac{\pi}{4} + \frac{3}{\pi} \cos^{-1}(0) = 2.29 \text{ s}.$$

Consequently, the time interval is $t_{eq} - t = 0.75$ s.

82. The distance from the relaxed position of the bottom end of the spring to its equilibrium position when the body is attached is given by Hooke's law:

$$\Delta x = F/k = (0.20 \text{ kg})(9.8 \text{ m/s}^2)/(19 \text{ N/m}) = 0.103 \text{ m}.$$

(a) The body, once released, will not only fall through the Δx distance but continue through the equilibrium position to a "turning point" equally far on the other side. Thus, the total descent of the body is $2\Delta x = 0.21$ m.

(b) Since $f = \omega/2\pi$, Eq. 15-12 leads to

$$f = \frac{1}{2\pi}\sqrt{\frac{k}{m}} = 1.6 \text{ Hz}.$$

(c) The maximum distance from the equilibrium position is the amplitude: $x_m = \Delta x = 0.10$ m.

83. We use $v_m = \omega x_m = 2\pi f x_m$, where the frequency is $180/(60 \text{ s}) = 3.0$ Hz and the amplitude is half the stroke, or $x_m = 0.38$ m. Thus,

$$v_m = 2\pi(3.0 \text{ Hz})(0.38 \text{ m}) = 7.2 \text{ m/s}.$$

84. (a) The rotational inertia of a hoop is $I = mR^2$, and the energy of the system becomes

$$E = \frac{1}{2}I\omega^2 + \frac{1}{2}kx^2$$

and θ is in radians. We note that $r\omega = v$ (where $v = dx/dt$). Thus, the energy becomes

$$E = \frac{1}{2}\left(\frac{mR^2}{r^2}\right)v^2 + \frac{1}{2}kx^2$$

which looks like the energy of the simple harmonic oscillator discussed in §15-4 *if* we identify the mass m in that section with the term mR^2/r^2 appearing in this problem. Making this identification, Eq. 15-12 yields

$$\omega = \sqrt{\frac{k}{mR^2/r^2}} = \frac{r}{R}\sqrt{\frac{k}{m}}.$$

(b) If $r = R$ the result of part (a) reduces to $\omega = \sqrt{k/m}$.

(c) And if $r = 0$ then $\omega = 0$ (the spring exerts no restoring torque on the wheel so that it is not brought back towards its equilibrium position).

85. (a) Hooke's law readily yields

$$k = (15 \text{ kg})(9.8 \text{ m/s}^2)/(0.12 \text{ m}) = 1225 \text{ N/m}.$$

Rounding to three significant figures, the spring constant is therefore 1.23 kN/m.

(b) We are told $f = 2.00$ Hz $= 2.00$ cycles/sec. Since a cycle is equivalent to 2π radians, we have $\omega = 2\pi(2.00) = 4\pi$ rad/s (understood to be valid to three significant figures). Using Eq. 15-12, we find

$$\omega = \sqrt{\frac{k}{m}} \Rightarrow m = \frac{1225 \text{ N/m}}{(4\pi \text{ rad/s})^2} = 7.76 \text{ kg}.$$

Consequently, the weight of the package is $mg = 76.0$ N.

86. (a) First consider a single spring with spring constant k and unstretched length L. One end is attached to a wall and the other is attached to an object. If it is elongated by Δx the magnitude of the force it exerts on the object is $F = k \Delta x$. Now consider it to be two springs, with spring constants k_1 and k_2, arranged so spring 1 is attached to the object. If spring 1 is elongated by Δx_1 then the magnitude of the force exerted on the object is $F = k_1 \Delta x_1$. This must be the same as the force of the single spring, so $k \Delta x = k_1 \Delta x_1$. We must determine the relationship between Δx and Δx_1. The springs are uniform so equal unstretched lengths are elongated by the same amount and the elongation of any portion of the spring is proportional to its unstretched length. This means spring 1 is elongated by $\Delta x_1 = CL_1$ and spring 2 is elongated by $\Delta x_2 = CL_2$, where C is a constant of proportionality. The total elongation is

$$\Delta x = \Delta x_1 + \Delta x_2 = C(L_1 + L_2) = CL_2(n + 1),$$

where $L_1 = nL_2$ was used to obtain the last form. Since $L_2 = L_1/n$, this can also be written $\Delta x = CL_1(n + 1)/n$. We substitute $\Delta x_1 = CL_1$ and $\Delta x = CL_1(n + 1)/n$ into $k \Delta x = k_1 \Delta x_1$ and solve for k_1. With $k = 8600$ N/m and $n = L_1/L_2 = 0.70$, we obtain

$$k_1 = \left(\frac{n+1}{n}\right)k = \left(\frac{0.70+1.0}{0.70}\right)(8600 \text{ N/m}) = 20886 \text{ N/m} \approx 2.1 \times 10^4 \text{ N/m}$$

(b) Now suppose the object is placed at the other end of the composite spring, so spring 2 exerts a force on it. Now $k\,\Delta x = k_2\,\Delta x_2$. We use $\Delta x_2 = CL_2$ and $\Delta x = CL_2(n+1)$, then solve for k_2. The result is $k_2 = k(n+1)$.

$$k_2 = (n+1)k = (0.70+1.0)(8600 \text{ N/m}) = 14620 \text{ N/m} \approx 1.5\times10^4 \text{ N/m}$$

(c) To find the frequency when spring 1 is attached to mass m, we replace k in $(1/2\pi)\sqrt{k/m}$ with $k(n+1)/n$. With $f = (1/2\pi)\sqrt{k/m}$, we obtain, for $f = 200$ Hz and $n = 0.70$

$$f_1 = \frac{1}{2\pi}\sqrt{\frac{(n+1)k}{nm}} = \sqrt{\frac{n+1}{n}}f = \sqrt{\frac{0.70+1.0}{0.70}}(200 \text{ Hz}) = 3.1\times10^2 \text{ Hz}.$$

(d) To find the frequency when spring 2 is attached to the mass, we replace k with $k(n+1)$ to obtain

$$f_2 = \frac{1}{2\pi}\sqrt{\frac{(n+1)k}{m}} = \sqrt{n+1}f = \sqrt{0.70+1.0}(200 \text{ Hz}) = 2.6\times10^2 \text{ Hz}.$$

87. The magnitude of the downhill component of the gravitational force acting on each ore car is

$$w_x = (10000 \text{ kg})(9.8 \text{ m/s}^2)\sin\theta$$

where $\theta = 30°$ (and it is important to have the calculator in degrees mode during this problem). We are told that a downhill pull of $3w_x$ causes the cable to stretch $x = 0.15$ m. Since the cable is expected to obey Hooke's law, its spring constant is

$$k = \frac{3w_x}{x} = 9.8\times10^5 \text{ N/m}.$$

(a) Noting that the oscillating mass is that of *two* of the cars, we apply Eq. 15-12 (divided by 2π).

$$f = \frac{\omega}{2\pi} = \frac{1}{2\pi}\sqrt{\frac{k}{m}} = \frac{1}{2\pi}\sqrt{\frac{9.8\times10^5 \text{ N/m}}{20000 \text{ kg}}} = 1.1 \text{ Hz}.$$

(b) The difference between the equilibrium positions of the end of the cable when supporting two as opposed to three cars is

$$\Delta x = \frac{3w_x - 2w_x}{k} = 0.050 \text{ m}.$$

88. Since the centripetal acceleration is horizontal and Earth's gravitational $\vec{g}$ is downward, we can define the magnitude of an "effective" gravitational acceleration using the Pythagorean theorem:

$$g_{eff} = \sqrt{g^2 + (v^2/R)^2}.$$

Then, since frequency is the reciprocal of the period, Eq. 15-28 leads to

$$f = \frac{1}{2\pi}\sqrt{\frac{g_{eff}}{L}} = \frac{1}{2\pi}\sqrt{\frac{\sqrt{g^2+v^4/R^2}}{L}}.$$

With $v = 70$ m/s, $R = 50$ m, and $L = 0.20$ m, we have $f \approx 3.5$ s^{-1} = 3.5 Hz.

89. (a) The spring stretches until the magnitude of its upward force on the block equals the magnitude of the downward force of gravity: $ky = mg$, where $y = 0.096$ m is the elongation of the spring at equilibrium, k is the spring constant, and $m = 1.3$ kg is the mass of the block. Thus

$$k = mg/y = (1.3 \text{ kg})(9.8 \text{ m/s}^2)/(0.096 \text{ m}) = 1.33 \times 10^2 \text{ N/m}.$$

(b) The period is given by

$$T = \frac{1}{f} = \frac{2\pi}{\omega} = 2\pi\sqrt{\frac{m}{k}} = 2\pi\sqrt{\frac{1.3 \text{ kg}}{133 \text{ N/m}}} = 0.62 \text{ s}.$$

(c) The frequency is $f = 1/T = 1/0.62$ s = 1.6 Hz.

(d) The block oscillates in simple harmonic motion about the equilibrium point determined by the forces of the spring and gravity. It is started from rest 5.0 cm below the equilibrium point so the amplitude is 5.0 cm.

(e) The block has maximum speed as it passes the equilibrium point. At the initial position, the block is not moving but it has potential energy

$$U_i = -mgy_i + \frac{1}{2}ky_i^2 = -(1.3 \text{ kg})(9.8 \text{ m/s}^2)(0.146 \text{ m}) + \frac{1}{2}(133 \text{ N/m})(0.146 \text{ m})^2 = -0.44 \text{ J}.$$

When the block is at the equilibrium point, the elongation of the spring is $y = 9.6$ cm and the potential energy is

$$U_f = -mgy + \frac{1}{2}ky^2 = -(1.3 \text{ kg})(9.8 \text{ m/s}^2)(0.096 \text{ m}) + \frac{1}{2}(133 \text{ N/m})(0.096 \text{ m})^2 = -0.61 \text{ J}.$$

We write the equation for conservation of energy as $U_i = U_f + \frac{1}{2}mv^2$ and solve for v:

$$v = \sqrt{\frac{2(U_i - U_f)}{m}} = \sqrt{\frac{2(-0.44\,\text{J} + 0.61\,\text{J})}{1.3\,\text{kg}}} = 0.51\,\text{m/s}.$$

90. (a) The Hooke's law force (of magnitude (100)(0.30) = 30 N) is directed upward and the weight (20 N) is downward. Thus, the net force is 10 N upward.

(b) The equilibrium position is where the upward Hooke's law force balances the weight, which corresponds to the spring being stretched (from unstretched length) by 20 N/100 N/m = 0.20 m. Thus, relative to the equilibrium position, the block (at the instant described in part (a)) is at what one might call *the bottom turning point* (since $v = 0$) at $x = -x_m$ where the amplitude is $x_m = 0.30 - 0.20 = 0.10$ m.

(c) Using Eq. 15-13 with $m = W/g \approx 2.0$ kg, we have

$$T = 2\pi\sqrt{\frac{m}{k}} = 0.90\,\text{s}.$$

(d) The maximum kinetic energy is equal to the maximum potential energy $\frac{1}{2}kx_m^2$. Thus,

$$K_m = U_m = \frac{1}{2}(100\,\text{N/m})(0.10\,\text{m})^2 = 0.50\,\text{J}.$$

91. We note that for a horizontal spring, the relaxed position is the equilibrium position (in a regular simple harmonic motion setting); thus, we infer that the given $v = 5.2$ m/s at $x = 0$ is the maximum value v_m (which equals ωx_m where $\omega = \sqrt{k/m} = 20$ rad/s).

(a) Since $\omega = 2\pi f$, we find $f = 3.2$ Hz.

(b) We have $v_m = 5.2$ m/s = (20 rad/s)x_m, which leads to $x_m = 0.26$ m.

(c) With meters, seconds and radians understood,

$$x = (0.26\,\text{m})\cos(20t + \phi)$$
$$v = -(5.2\,\text{m/s})\sin(20t + \phi).$$

The requirement that $x = 0$ at $t = 0$ implies (from the first equation above) that either $\phi = +\pi/2$ or $\phi = -\pi/2$. Only one of these choices meets the further requirement that $v > 0$ when $t = 0$; that choice is $\phi = -\pi/2$. Therefore,

$$x = (0.26\,\text{m})\cos\left(20t - \frac{\pi}{2}\right) = (0.26\,\text{m})\sin(20t).$$

92. (a) Eq. 15-21 leads to

$$E = \frac{1}{2}kx_m^2 \Rightarrow x_m = \sqrt{\frac{2E}{k}} = \sqrt{\frac{2(4.0 \text{ J})}{200 \text{ N/m}}} = 0.20 \text{ m}.$$

(b) Since $T = 2\pi\sqrt{m/k} = 2\pi\sqrt{0.80 \text{ kg}/200 \text{ N/m}} \approx 0.4 \text{ s}$, then the block completes $10/0.4 = 25$ cycles during the specified interval.

(c) The maximum kinetic energy is the total energy, 4.0 J.

(d) This can be approached more than one way; we choose to use energy conservation:

$$E = K + U \Rightarrow 4.0 = \frac{1}{2}mv^2 + \frac{1}{2}kx^2.$$

Therefore, when $x = 0.15$ m, we find $v = 2.1$ m/s.

93. The time for one cycle is $T = (50 \text{ s})/20 = 2.5$ s. Thus, from Eq. 15-23, we find

$$I = \kappa\left(\frac{T}{2\pi}\right)^2 = (0.50)\left(\frac{2.5}{2\pi}\right)^2 = 0.079 \text{ kg} \cdot \text{m}^2.$$

94. The period formula, Eq. 15-29, requires knowing the distance h from the axis of rotation and the center of mass of the system. We also need the rotational inertia I about the axis of rotation. From the figure, we see $h = L + R$ where $R = 0.15$ m. Using the parallel-axis theorem, we find

$$I = \frac{1}{2}MR^2 + M(L+R)^2,$$

where $M = 1.0$ kg. Thus, Eq. 15-29, with $T = 2.0$ s, leads to

$$2.0 = 2\pi\sqrt{\frac{\frac{1}{2}MR^2 + M(L+R)^2}{Mg(L+R)}}$$

which leads to $L = 0.8315$ m.

95. (a) By Eq. 15-13, the mass of the block is

$$m_b = \frac{kT_0^2}{4\pi^2} = 2.43 \text{ kg}.$$

Therefore, with $m_p = 0.50$ kg, the new period is

$$T = 2\pi\sqrt{\frac{m_p + m_b}{k}} = 0.44 \text{ s}.$$

(b) The speed before the collision (since it is at its maximum, passing through equilibrium) is $v_0 = x_m \omega_0$ where $\omega_0 = 2\pi/T_0$; thus, $v_0 = 3.14$ m/s. Using momentum conservation (along the horizontal direction) we find the speed after the collision.

$$V = v_0 \frac{m_b}{m_p + m_b} = 2.61 \text{ m/s}.$$

The equilibrium position has not changed, so (for the new system of greater mass) this represents the maximum speed value for the subsequent harmonic motion: $V = x'_m \omega$ where $\omega = 2\pi/T = 14.3$ rad/s. Therefore, $x'_m = 0.18$ m.

96. (a) Hooke's law provides the spring constant: $k = (20 \text{ N})/(0.20 \text{ m}) = 1.0 \times 10^2$ N/m.

(b) The attached mass is $m = (5.0 \text{ N})/(9.8 \text{ m/s}^2) = 0.51$ kg. Consequently, Eq. 15-13 leads to

$$T = 2\pi\sqrt{\frac{m}{k}} = 2\pi\sqrt{\frac{0.51 \text{ kg}}{100 \text{ N/m}}} = 0.45 \text{ s}.$$

97. (a) Hooke's law provides the spring constant:

$$k = (4.00 \text{ kg})(9.8 \text{ m/s}^2)/(0.160 \text{ m}) = 245 \text{ N/m}.$$

(b) The attached mass is $m = 0.500$ kg. Consequently, Eq. 15-13 leads to

$$T = 2\pi\sqrt{\frac{m}{k}} = 2\pi\sqrt{\frac{0.500}{245}} = 0.284 \text{ s}.$$

98. (a) We are told that when $t = 4T$, with $T = 2\pi/\omega' \approx 2\pi\sqrt{m/k}$ (neglecting the second term in Eq. 15-43),

$$e^{-bt/2m} = \frac{3}{4}.$$

Thus,

$$T \approx 2\pi\sqrt{(2.00 \text{ kg})/(10.0 \text{ N/m})} = 2.81 \text{ s}$$

and we find

$$\frac{b(4T)}{2m} = \ln\left(\frac{4}{3}\right) = 0.288 \quad \Rightarrow \quad b = \frac{2(2.00 \text{ kg})(0.288)}{4(2.81 \text{ s})} = 0.102 \text{ kg/s}.$$

(b) Initially, the energy is $E_o = \frac{1}{2}kx_{mo}^2 = \frac{1}{2}(10.0)(0.250)^2 = 0.313$ J. At $t = 4T$,

$$E = \tfrac{1}{2}k(\tfrac{3}{4}x_{mo})^2 = 0.176 \text{ J}.$$

Therefore, $E_o - E = 0.137$ J.

99. Since d_m is the amplitude of oscillation, then the maximum acceleration being set to $0.2g$ provides the condition: $\omega^2 d_m = 0.2g$. Since d_s is the amount the spring stretched in order to achieve vertical equilibrium of forces, then we have the condition $kd_s = mg$. Since we can write this latter condition as $m\omega^2 d_s = mg$, then $\omega^2 = g/d_s$. Plugging this into our first condition, we obtain

$$d_s = d_m/0.2 = (10 \text{ cm})/0.2 = 50 \text{ cm}.$$

100. We note (from the graph) that $a_m = \omega^2 x_m = 4.00$ cm/s^2. Also the value at $t = 0$ is $a_o = 1.00$ cm/s^2. Then Eq. 15-7 leads to

$$\phi = \cos^{-1}(-1.00/4.00) = +1.82 \text{ rad or } -4.46 \text{ rad}.$$

The other "root" (+4.46 rad) can be rejected on the grounds that it would lead to a negative slope at $t = 0$.

101. (a) The graphs suggest that $T = 0.40$ s and $\kappa = 4/0.2 = 0.02$ N·m/rad. With these values, Eq. 15-23 can be used to determine the rotational inertia:

$$I = \kappa T^2/4\pi^2 = 8.11 \times 10^{-5} \text{ kg·m}^2.$$

(b) We note (from the graph) that $\theta_{max} = 0.20$ rad. Setting the maximum kinetic energy ($\tfrac{1}{2}I\omega_{max}^2$) equal to the maximum potential energy (see the hint in the problem) leads to $\omega_{max} = \theta_{max}\sqrt{\kappa/I} = 3.14$ rad/s.

102. The angular frequency of the simple harmonic oscillation is given by Eq. 15-13:

$$\omega = \sqrt{\frac{k}{m}}.$$

Thus, for two different masses m_1 and m_2, with the same spring constant k, the ratio of the frequencies would be

$$\frac{\omega_1}{\omega_2} = \frac{\sqrt{k/m_1}}{\sqrt{k/m_2}} = \sqrt{\frac{m_2}{m_1}}.$$

In our case, with $m_1 = m$ and $m_2 = 2.5m$, the ratio is $\dfrac{\omega_1}{\omega_2} = \sqrt{\dfrac{m_2}{m_1}} = \sqrt{2.5} = 1.58$.

103. For simple harmonic motion, Eq. 15-24 must reduce to

$$\tau = -L(F_g \sin\theta) \to -L(F_g \theta)$$

where θ is in radians. We take the percent difference (in absolute value)

$$\left|\frac{(-LF_g \sin\theta) - (-LF_g \theta)}{-LF_g \sin\theta}\right| = \left|1 - \frac{\theta}{\sin\theta}\right|$$

and set this equal to 0.010 (corresponding to 1.0%). In order to solve for θ (since this is not possible "in closed form"), several approaches are available. Some calculators have built-in numerical routines to facilitate this, and most math software packages have this capability. Alternatively, we could expand $\sin\theta \approx \theta - \theta^3/6$ (valid for small θ) and thereby find an approximate solution (which, in turn, might provide a seed value for a numerical search). Here we show the latter approach:

$$\left|1 - \frac{\theta}{\theta - \theta^3/6}\right| \approx 0.010 \Rightarrow \frac{1}{1 - \theta^2/6} \approx 1.010$$

which leads to $\theta \approx \sqrt{6(0.01/1.01)} = 0.24$ rad $= 14.0°$. A more accurate value (found numerically) for the θ value which results in a 1.0% deviation is 13.986°.

104. (a) The graph makes it clear that the period is $T = 0.20$ s.

(b) The period of the simple harmonic oscillator is given by Eq. 15-13:

$$T = 2\pi\sqrt{\frac{m}{k}}.$$

Thus, using the result from part (a) with $k = 200$ N/m, we obtain $m = 0.203 \approx 0.20$ kg.

(c) The graph indicates that the speed is (momentarily) zero at $t = 0$, which implies that the block is at $x_0 = \pm x_m$. From the graph we also note that the slope of the velocity curve (hence, the acceleration) is positive at $t = 0$, which implies (from $ma = -kx$) that the value of x is negative. Therefore, with $x_m = 0.20$ m, we obtain $x_0 = -0.20$ m.

(d) We note from the graph that $v = 0$ at $t = 0.10$ s, which implied $a = \pm a_m = \pm \omega^2 x_m$. Since acceleration is the instantaneous slope of the velocity graph, then (looking again at the graph) we choose the negative sign. Recalling $\omega^2 = k/m$ we obtain $a = -197 \approx -2.0 \times 10^2$ m/s^2.

(e) The graph shows $v_m = 6.28$ m/s, so

$$K_m = \frac{1}{2}mv_m^2 = 4.0 \text{ J}.$$

105. (a) From the graph, it is clear that $x_m = 0.30$ m.

(b) With $F = -kx$, we see k is the (negative) slope of the graph — which is $75/0.30 = 250$ N/m. Plugging this into Eq. 15-13 yields

$$T = 2\pi\sqrt{\frac{m}{k}} = 0.28 \text{ s}.$$

(c) As discussed in §15-2, the maximum acceleration is

$$a_m = \omega^2 x_m = \frac{k}{m}x_m = 1.5\times 10^2 \text{ m/s}^2.$$

Alternatively, we could arrive at this result using $a_m = (2\pi/T)^2 x_m$.

(d) Also in §15-2 is $v_m = \omega x_m$ so that the maximum kinetic energy is

$$K_m = \frac{1}{2}mv_m^2 = \frac{1}{2}m\omega^2 x_m^2 = \frac{1}{2}kx_m^2$$

which yields $11.3 \approx 11$ J. We note that the above manipulation reproduces the notion of energy conservation for this system (maximum kinetic energy being equal to the maximum potential energy).

106. (a) The potential energy at the turning point is equal (in the absence of friction) to the total kinetic energy (translational plus rotational) as it passes through the equilibrium position:

$$\frac{1}{2}kx_m^2 = \frac{1}{2}Mv_{cm}^2 + \frac{1}{2}I_{cm}^2\omega^2 = \frac{1}{2}Mv_{cm}^2 + \frac{1}{2}\left(\frac{1}{2}MR^2\right)\left(\frac{v_{cm}}{R}\right)^2$$

$$= \frac{1}{2}Mv_{cm}^2 + \frac{1}{4}Mv_{cm}^2 = \frac{3}{4}Mv_{cm}^2$$

which leads to $Mv_{cm}^2 = 2kx_m^2/3 = 0.125$ J. The translational kinetic energy is therefore $\frac{1}{2}Mv_{cm}^2 = kx_m^2/3 = 0.0625$ J.

(b) And the rotational kinetic energy is $\frac{1}{4}Mv_{cm}^2 = kx_m^2/6 = 0.03125 \text{ J} \approx 3.13\times 10^{-2}$ J.

(c) In this part, we use v_{cm} to denote the speed at any instant (and not just the maximum speed as we had done in the previous parts). Since the energy is constant, then

$$\frac{dE}{dt} = \frac{d}{dt}\left(\frac{3}{4}Mv_{cm}^2\right) + \frac{d}{dt}\left(\frac{1}{2}kx^2\right) = \frac{3}{2}Mv_{cm}a_{cm} + kxv_{cm} = 0$$

which leads to

$$a_{cm} = -\left(\frac{2k}{3M}\right)x.$$

Comparing with Eq. 15-8, we see that $\omega = \sqrt{2k/3M}$ for this system. Since $\omega = 2\pi/T$, we obtain the desired result: $T = 2\pi\sqrt{3M/2k}$.

107. (a) From Eq. 16-12, $T = 2\pi\sqrt{m/k} = 0.45$ s.

(b) For a vertical spring, the distance between the unstretched length and the equilibrium length (with a mass m attached) is mg/k, where in this problem $mg = 10$ N and $k = 200$ N/m (so that the distance is 0.05 m). During simple harmonic motion, the convention is to establish $x = 0$ at the equilibrium length (the middle level for the oscillation) and to write the total energy without any gravity term; i.e.,

$$E = K + U,$$

where $U = kx^2/2$. Thus, as the block passes through the unstretched position, the energy is $E = 2.0 + \frac{1}{2}k(0.05)^2 = 2.25$ J. At its topmost and bottommost points of oscillation, the energy (using this convention) is all elastic potential: $\frac{1}{2}kx_m^2$. Therefore, by energy conservation,

$$2.25 = \frac{1}{2}kx_m^2 \Rightarrow x_m = \pm 0.15 \text{ m}.$$

This gives the amplitude of oscillation as 0.15 m, but how far are these points from the *unstretched* position? We add (or subtract) the 0.05 m value found above and obtain 0.10 m for the top-most position and 0.20 m for the bottom-most position.

(c) As noted in part (b), $x_m = \pm 0.15$ m.

(d) The maximum kinetic energy equals the maximum potential energy (found in part (b)) and is equal to 2.25 J.

108. Using Eq. 15-12, we find $\omega = \sqrt{k/m} = 10$ rad/s. We also use $v_m = x_m\omega$ and $a_m = x_m\omega^2$.

(a) The amplitude (meaning "displacement amplitude") is $x_m = v_m/\omega = 3/10 = 0.30$ m.

(b) The acceleration-amplitude is $a_m = (0.30 \text{ m})(10 \text{ rad/s})^2 = 30 \text{ m/s}^2$.

(c) One interpretation of this question is "what is the most negative value of the acceleration?" in which case the answer is $-a_m = -30$ m/s^2. Another interpretation is "what is the smallest value of the absolute-value of the acceleration?" in which case the answer is zero.

(d) Since the period is $T = 2\pi/\omega = 0.628$ s. Therefore, seven cycles of the motion requires $t = 7T = 4.4$ s.

109. The mass is $m = \dfrac{0.108 \text{ kg}}{6.02 \times 10^{23}} = 1.8 \times 10^{-25}$ kg. Using Eq. 15-12 and the fact that $f = \omega/2\pi$, we have

$$1 \times 10^{13} \text{ Hz} = \frac{1}{2\pi}\sqrt{\frac{k}{m}} \Rightarrow k = \left(2\pi \times 10^{13}\right)^2 \left(1.8 \times 10^{-25}\right) \approx 7 \times 10^2 \text{ N/m}.$$

110. (a) Eq. 15-28 gives

$$T = 2\pi\sqrt{\frac{L}{g}} = 2\pi\sqrt{\frac{17 m}{9.8 \text{ m/s}^2}} = 8.3 \text{ s}.$$

(b) Plugging $I = mL^2$ into Eq. 15-25, we see that the mass m cancels out. Thus, the characteristics (such as the period) of the periodic motion do not depend on the mass.

111. (a) The net horizontal force is F since the batter is assumed to exert no horizontal force on the bat. Thus, the horizontal acceleration (which applies as long as F acts on the bat) is $a = F/m$.

(b) The only torque on the system is that due to F, which is exerted at P, at a distance $L_o - \tfrac{1}{2}L$ from C. Since $L_o = 2L/3$ (see Sample Problem 15-5), then the distance from C to P is $\tfrac{2}{3}L - \tfrac{1}{2}L = \tfrac{1}{6}L$. Since the net torque is equal to the rotational inertia ($I = 1/12 mL^2$ about the center of mass) multiplied by the angular acceleration, we obtain

$$\alpha = \frac{\tau}{I} = \frac{F(\tfrac{1}{6}L)}{\tfrac{1}{12}mL^2} = \frac{2F}{mL}.$$

(c) The distance from C to O is $r = L/2$, so the contribution to the acceleration at O stemming from the angular acceleration (in the counterclockwise direction of Fig. 15-11) is $\alpha r = \tfrac{1}{2}\alpha L$ (leftward in that figure). Also, the contribution to the acceleration at O due to the result of part (a) is F/m (rightward in that figure). Thus, if we choose rightward as positive, then the net acceleration of O is

$$a_O = \frac{F}{m} - \frac{1}{2}\alpha L = \frac{F}{m} - \frac{1}{2}\left(\frac{2F}{mL}\right)L = 0.$$

(d) Point O stays relatively stationary in the batting process, and that might be possible due to a force exerted by the batter or due to a finely tuned cancellation such as we have shown here. We assumed that the batter exerted no force, and our first expectation is that the impulse delivered by the impact would make all points on the bat go into motion, but for this particular choice of impact point, we have seen that the point being held by the batter is naturally stationary and exerts no force on the batter's hands which would otherwise have to "fight" to keep a good hold of it.

112. (a) A plot of x versus t (in SI units) is shown below:

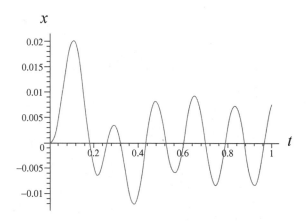

If we expand the plot near the end of that time interval we have

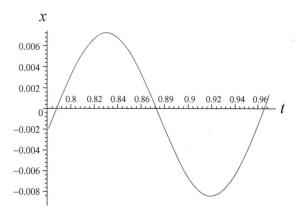

This is close enough to a regular sine wave cycle that we can estimate its period ($T = 0.18$ s, so $\omega = 35$ rad/s) and its amplitude ($y_m = 0.008$ m).

(b) Now, with the new driving frequency ($\omega_d = 13.2$ rad/s), the x versus t graph (for the first one second of motion) is as shown below:

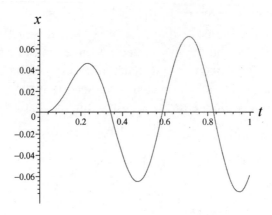

It is a little more difficult in this case to estimate a regular sine-curve-like amplitude and period (for the part of the above graph near the end of that time interval), but we arrive at roughly $y_m = 0.07$ m, $T = 0.48$ s, and $\omega = 13$ rad/s.

(c) Now, with $\omega_d = 20$ rad/s, we obtain (for the behavior of the graph, below, near the end of the interval) the estimates: $y_m = 0.03$ m, $T = 0.31$ s, and $\omega = 20$ rad/s.

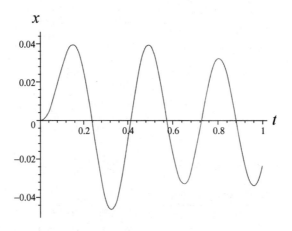

Chapter 16

1. (a) The angular wave number is $k = \dfrac{2\pi}{\lambda} = \dfrac{2\pi}{1.80\,\text{m}} = 3.49\,\text{m}^{-1}$.

(b) The speed of the wave is $v = \lambda f = \dfrac{\lambda \omega}{2\pi} = \dfrac{(1.80\,\text{m})(110\,\text{rad/s})}{2\pi} = 31.5\,\text{m/s}$.

2. The distance d between the beetle and the scorpion is related to the transverse speed v_t and longitudinal speed v_ℓ as

$$d = v_t t_t = v_\ell t_\ell$$

where t_t and t_ℓ are the arrival times of the wave in the transverse and longitudinal directions, respectively. With $v_t = 50$ m/s and $v_\ell = 150$ m/s, we have

$$\dfrac{t_t}{t_\ell} = \dfrac{v_\ell}{v_t} = \dfrac{150\,\text{m/s}}{50\,\text{m/s}} = 3.0.$$

Thus, if

$$\Delta t = t_t - t_\ell = 3.0 t_\ell - t_\ell = 2.0 t_\ell = 4.0 \times 10^{-3}\,\text{s} \;\Rightarrow\; t_\ell = 2.0 \times 10^{-3}\,\text{s},$$

then $d = v_\ell t_\ell = (150\,\text{m/s})(2.0 \times 10^{-3}\,\text{s}) = 0.30\,\text{m} = 30$ cm.

3. (a) The motion from maximum displacement to zero is one-fourth of a cycle so 0.170 s is one-fourth of a period. The period is $T = 4(0.170\,\text{s}) = 0.680$ s.

(b) The frequency is the reciprocal of the period:

$$f = \dfrac{1}{T} = \dfrac{1}{0.680\,\text{s}} = 1.47\,\text{Hz}.$$

(c) A sinusoidal wave travels one wavelength in one period:

$$v = \dfrac{\lambda}{T} = \dfrac{1.40\,\text{m}}{0.680\,\text{s}} = 2.06\,\text{m/s}.$$

4. (a) The speed of the wave is the distance divided by the required time. Thus,

$$v = \frac{853 \text{ seats}}{39 \text{ s}} = 21.87 \text{ seats/s} \approx 22 \text{ seats/s}.$$

(b) The width w is equal to the distance the wave has moved during the average time required by a spectator to stand and then sit. Thus,

$$w = vt = (21.87 \text{ seats/s})(1.8 \text{ s}) \approx 39 \text{ seats}.$$

5. Let $y_1 = 2.0$ mm (corresponding to time t_1) and $y_2 = -2.0$ mm (corresponding to time t_2). Then we find

$$kx + 600t_1 + \phi = \sin^{-1}(2.0/6.0)$$

and

$$kx + 600t_2 + \phi = \sin^{-1}(-2.0/6.0).$$

Subtracting equations gives

$$600(t_1 - t_2) = \sin^{-1}(2.0/6.0) - \sin^{-1}(-2.0/6.0).$$

Thus we find $t_1 - t_2 = 0.011$ s (or 1.1 ms).

6. Setting $x = 0$ in $u = -\omega y_m \cos(kx - \omega t + \phi)$ (see Eq. 16-21 or Eq. 16-28) gives

$$u = -\omega y_m \cos(-\omega t + \phi)$$

as the function being plotted in the graph. We note that it has a positive "slope" (referring to its t-derivative) at $t = 0$:

$$\frac{du}{dt} = \frac{d(-\omega y_m \cos(-\omega t + \phi))}{dt} = -y_m \omega^2 \sin(-\omega t + \phi) > 0 \text{ at } t = 0.$$

This implies that $-\sin\phi > 0$ and consequently that ϕ is in either the third or fourth quadrant. The graph shows (at $t = 0$) $u = -4$ m/s, and (at some later t) $u_{max} = 5$ m/s. We note that $u_{max} = y_m \omega$. Therefore,

$$u = -u_{max} \cos(-\omega t + \phi)\big|_{t=0} \Rightarrow \phi = \cos^{-1}(\tfrac{4}{5}) = \pm 0.6435 \text{ rad}$$

(bear in mind that $\cos\theta = \cos(-\theta)$), and we must choose $\phi = -0.64$ rad (since this is about $-37°$ and is in fourth quadrant). Of course, this answer added to $2n\pi$ is still a valid answer (where n is any integer), so that, for example, $\phi = -0.64 + 2\pi = 5.64$ rad is also an acceptable result.

7. Using $v = f\lambda$, we find the length of one cycle of the wave is

$$\lambda = 350/500 = 0.700 \text{ m} = 700 \text{ mm}.$$

From $f = 1/T$, we find the time for one cycle of oscillation is $T = 1/500 = 2.00 \times 10^{-3}$ s = 2.00 ms.

(a) A cycle is equivalent to 2π radians, so that $\pi/3$ rad corresponds to one-sixth of a cycle. The corresponding length, therefore, is $\lambda/6 = 700/6 = 117$ mm.

(b) The interval 1.00 ms is half of T and thus corresponds to half of one cycle, or half of 2π rad. Thus, the phase difference is $(1/2)2\pi = \pi$ rad.

8. (a) The amplitude is $y_m = 6.0$ cm.

(b) We find λ from $2\pi/\lambda = 0.020\pi$: $\lambda = 1.0 \times 10^2$ cm.

(c) Solving $2\pi f = \omega = 4.0\pi$, we obtain $f = 2.0$ Hz.

(d) The wave speed is $v = \lambda f = (100 \text{ cm})(2.0 \text{ Hz}) = 2.0 \times 10^2$ cm/s.

(e) The wave propagates in the $-x$ direction, since the argument of the trig function is $kx + \omega t$ instead of $kx - \omega t$ (as in Eq. 16-2).

(f) The maximum transverse speed (found from the time derivative of y) is

$$u_{max} = 2\pi f y_m = (4.0\pi \text{ s}^{-1})(6.0 \text{ cm}) = 75 \text{ cm/s}.$$

(g) $y(3.5 \text{ cm}, 0.26 \text{ s}) = (6.0 \text{ cm}) \sin[0.020\pi(3.5) + 4.0\pi(0.26)] = -2.0$ cm.

9. (a) Recalling from Ch. 12 the simple harmonic motion relation $u_m = y_m \omega$, we have

$$\omega = \frac{16}{0.040} = 400 \text{ rad/s}.$$

Since $\omega = 2\pi f$, we obtain $f = 64$ Hz.

(b) Using $v = f\lambda$, we find $\lambda = 80/64 = 1.26$ m ≈ 1.3 m.

(c) The amplitude of the transverse displacement is $y_m = 4.0$ cm $= 4.0 \times 10^{-2}$ m.

(d) The wave number is $k = 2\pi/\lambda = 5.0$ rad/m.

(e) The angular frequency, as obtained in part (a), is $\omega = 16/0.040 = 4.0 \times 10^2$ rad/s.

(f) The function describing the wave can be written as

$$y = 0.040 \sin(5x - 400t + \phi)$$

where distances are in meters and time is in seconds. We adjust the phase constant ϕ to satisfy the condition $y = 0.040$ at $x = t = 0$. Therefore, $\sin \phi = 1$, for which the "simplest" root is $\phi = \pi/2$. Consequently, the answer is

$$y = 0.040 \sin\left(5x - 400t + \frac{\pi}{2}\right).$$

(g) The sign in front of ω is minus.

10. With length in centimeters and time in seconds, we have

$$u = \frac{du}{dt} = 225\pi \sin(\pi x - 15\pi t).$$

Squaring this and adding it to the square of $15\pi y$, we have

$$u^2 + (15\pi y)^2 = (225\pi)^2 [\sin^2(\pi x - 15\pi t) + \cos^2(\pi x - 15\pi t)]$$

so that

$$u = \sqrt{(225\pi)^2 - (15\pi y)^2} = 15\pi \sqrt{15^2 - y^2}.$$

Therefore, where $y = 12$, u must be $\pm 135\pi$. Consequently, the *speed* there is 424 cm/s = 4.24 m/s.

11. (a) The amplitude y_m is half of the 6.00 mm vertical range shown in the figure, i.e., $y_m = 3.0$ mm.

(b) The speed of the wave is $v = d/t = 15$ m/s, where $d = 0.060$ m and $t = 0.0040$ s. The angular wave number is $k = 2\pi/\lambda$ where $\lambda = 0.40$ m. Thus,

$$k = \frac{2\pi}{\lambda} = 16 \text{ rad/m}.$$

(c) The angular frequency is found from

$$\omega = kv = (16 \text{ rad/m})(15 \text{ m/s}) = 2.4 \times 10^2 \text{ rad/s}.$$

(d) We choose the minus sign (between kx and ωt) in the argument of the sine function because the wave is shown traveling to the right [in the $+x$ direction] – see section 16-5). Therefore, with SI units understood, we obtain

$$y = y_m \sin(kx - kvt) \approx 0.0030 \sin(16x - 2.4 \times 10^2 \, t).$$

12. The slope that they are plotting is the physical slope of sinusoidal waveshape (not to be confused with the more abstract "slope" of its time development; the physical slope is an x-derivative whereas the more abstract "slope" would be the t-derivative). Thus, where the figure shows a maximum slope equal to 0.2 (with no unit), it refers to the maximum of the following function:

$$\frac{dy}{dx} = \frac{d y_m \sin(kx - \omega t)}{dx} = y_m k \cos(kx - \omega t).$$

The problem additionally gives $t = 0$, which we can substitute into the above expression if desired. In any case, the maximum of the above expression is $y_m k$, where

$$k = \frac{2\pi}{\lambda} = \frac{2\pi}{0.40 \text{ m}} = 15.7 \text{ rad/m}.$$

Therefore, setting $y_m k$ equal to 0.20 allows us to solve for the amplitude y_m. We find

$$y_m = \frac{0.20}{15.7 \text{ rad/m}} = 0.0127 \text{ m} \approx 1.3 \text{ cm}.$$

13. From Eq. 16-10, a general expression for a sinusoidal wave traveling along the $+x$ direction is

$$y(x, t) = y_m \sin(kx - \omega t + \phi)$$

(a) The figure shows that at $x = 0$, $y(0, t) = y_m \sin(-\omega t + \phi)$ is a positive sine function, i.e., $y(0, t) = +y_m \sin \omega t$. Therefore, the phase constant must be $\phi = \pi$. At $t = 0$, we then have

$$y(x, 0) = y_m \sin(kx + \pi) = -y_m \sin kx$$

which is a negative sine function. A plot of $y(x,0)$ is depicted on the right.

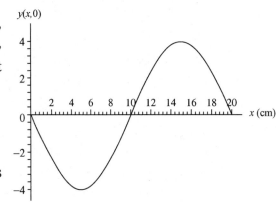

(b) From the figure we see that the amplitude is $y_m = 4.0$ cm.

(c) The angular wave number is given by $k = 2\pi/\lambda = \pi/10 = 0.31$ rad/cm.

(d) The angular frequency is $\omega = 2\pi/T = \pi/5 = 0.63$ rad/s.

(e) As found in part (a), the phase is $\phi = \pi$.

(f) The sign is minus since the wave is traveling in the $+x$ direction.

(g) Since the frequency is $f = 1/T = 0.10$ s, the speed of the wave is $v = f\lambda = 2.0$ cm/s.

(h) From the results above, the wave may be expressed as

$$y(x,t) = 4.0\sin\left(\frac{\pi x}{10} - \frac{\pi t}{5} + \pi\right) = -4.0\sin\left(\frac{\pi x}{10} - \frac{\pi t}{5}\right).$$

Taking the derivative of y with respect to t, we find

$$u(x,t) = \frac{\partial y}{\partial t} = 4.0\left(\frac{\pi}{t}\right)\cos\left(\frac{\pi x}{10} - \frac{\pi t}{5}\right)$$

which yields $u(0, 5.0) = -2.5$ cm/s.

14. From $v = \sqrt{\tau/\mu}$, we have

$$\frac{v_{new}}{v_{old}} = \frac{\sqrt{\tau_{new}/\mu_{new}}}{\sqrt{\tau_{old}/\mu_{old}}} = \sqrt{2}.$$

15. The wave speed v is given by $v = \sqrt{\tau/\mu}$, where τ is the tension in the rope and μ is the linear mass density of the rope. The linear mass density is the mass per unit length of rope:

$$\mu = m/L = (0.0600 \text{ kg})/(2.00 \text{ m}) = 0.0300 \text{ kg/m}.$$

Thus,

$$v = \sqrt{\frac{500 \text{ N}}{0.0300 \text{ kg/m}}} = 129 \text{ m/s}.$$

16. The volume of a cylinder of height ℓ is $V = \pi r^2 \ell = \pi d^2 \ell/4$. The strings are long, narrow cylinders, one of diameter d_1 and the other of diameter d_2 (and corresponding linear densities μ_1 and μ_2). The mass is the (regular) density multiplied by the volume: $m = \rho V$, so that the mass-per-unit length is

$$\mu = \frac{m}{\ell} = \frac{\rho \pi d^2 \ell/4}{\ell} = \frac{\pi \rho d^2}{4}$$

and their ratio is

$$\frac{\mu_1}{\mu_2} = \frac{\pi \rho d_1^2/4}{\pi \rho d_2^2/4} = \left(\frac{d_1}{d_2}\right)^2.$$

Therefore, the ratio of diameters is

$$\frac{d_1}{d_2} = \sqrt{\frac{\mu_1}{\mu_2}} = \sqrt{\frac{3.0}{0.29}} = 3.2.$$

17. (a) The amplitude of the wave is $y_m = 0.120$ mm.

(b) The wave speed is given by $v = \sqrt{\tau/\mu}$, where τ is the tension in the string and μ is the linear mass density of the string, so the wavelength is $\lambda = v/f = \sqrt{\tau/\mu}/f$ and the angular wave number is

$$k = \frac{2\pi}{\lambda} = 2\pi f \sqrt{\frac{\mu}{\tau}} = 2\pi(100\,\text{Hz})\sqrt{\frac{0.50\,\text{kg/m}}{10\,\text{N}}} = 141\,\text{m}^{-1}.$$

(c) The frequency is $f = 100$ Hz, so the angular frequency is

$$\omega = 2\pi f = 2\pi(100\,\text{Hz}) = 628\,\text{rad/s}.$$

(d) We may write the string displacement in the form $y = y_m \sin(kx + \omega t)$. The plus sign is used since the wave is traveling in the negative x direction. In summary, the wave can be expressed as

$$y = (0.120\,\text{mm})\sin\left[(141\,\text{m}^{-1})x + (628\,\text{s}^{-1})t\right].$$

18. We use $v = \sqrt{\tau/\mu} \propto \sqrt{\tau}$ to obtain

$$\tau_2 = \tau_1 \left(\frac{v_2}{v_1}\right)^2 = (120\,\text{N})\left(\frac{180\,\text{m/s}}{170\,\text{m/s}}\right)^2 = 135\,\text{N}.$$

19. (a) The wave speed is given by $v = \lambda/T = \omega/k$, where λ is the wavelength, T is the period, ω is the angular frequency ($2\pi/T$), and k is the angular wave number ($2\pi/\lambda$). The displacement has the form $y = y_m \sin(kx + \omega t)$, so $k = 2.0$ m^{-1} and $\omega = 30$ rad/s. Thus

$$v = (30\,\text{rad/s})/(2.0\,\text{m}^{-1}) = 15\,\text{m/s}.$$

(b) Since the wave speed is given by $v = \sqrt{\tau/\mu}$, where τ is the tension in the string and μ is the linear mass density of the string, the tension is

$$\tau = \mu v^2 = (1.6\times 10^{-4}\,\text{kg/m})(15\,\text{m/s})^2 = 0.036\,\text{N}.$$

20. (a) Comparing with Eq. 16-2, we see that $k = 20$/m and $\omega = 600$/s. Therefore, the speed of the wave is (see Eq. 16-13) $v = \omega/k = 30$ m/s.

(b) From Eq. 16–26, we find

$$\mu = \frac{\tau}{v^2} = \frac{15}{30^2} = 0.017\,\text{kg/m} = 17\,\text{g/m}.$$

21. (a) We read the amplitude from the graph. It is about 5.0 cm.

(b) We read the wavelength from the graph. The curve crosses $y = 0$ at about $x = 15$ cm and again with the same slope at about $x = 55$ cm, so

$$\lambda = (55\,\text{cm} - 15\,\text{cm}) = 40\,\text{cm} = 0.40\,\text{m}.$$

(c) The wave speed is $v = \sqrt{\tau/\mu}$, where τ is the tension in the string and μ is the linear mass density of the string. Thus,

$$v = \sqrt{\frac{3.6\,\text{N}}{25\times 10^{-3}\,\text{kg/m}}} = 12\,\text{m/s}.$$

(d) The frequency is $f = v/\lambda = (12\,\text{m/s})/(0.40\,\text{m}) = 30$ Hz and the period is

$$T = 1/f = 1/(30\,\text{Hz}) = 0.033\,\text{s}.$$

(e) The maximum string speed is

$$u_m = \omega y_m = 2\pi f y_m = 2\pi(30\,\text{Hz})(5.0\,\text{cm}) = 940\,\text{cm/s} = 9.4\,\text{m/s}.$$

(f) The angular wave number is $k = 2\pi/\lambda = 2\pi/(0.40\,\text{m}) = 16\,\text{m}^{-1}$.

(g) The angular frequency is $\omega = 2\pi f = 2\pi(30\,\text{Hz}) = 1.9\times 10^2$ rad/s

(h) According to the graph, the displacement at $x = 0$ and $t = 0$ is 4.0×10^{-2} m. The formula for the displacement gives $y(0, 0) = y_m \sin \phi$. We wish to select ϕ so that

$$5.0 \times 10^{-2} \sin \phi = 4.0 \times 10^{-2}.$$

The solution is either 0.93 rad or 2.21 rad. In the first case the function has a positive slope at $x = 0$ and matches the graph. In the second case it has negative slope and does not match the graph. We select $\phi = 0.93$ rad.

(i) The string displacement has the form $y(x, t) = y_m \sin(kx + \omega t + \phi)$. A plus sign appears in the argument of the trigonometric function because the wave is moving in the negative x direction. Using the results obtained above, the expression for the displacement is

$$y(x,t) = (5.0 \times 10^{-2}\,\text{m}) \sin\left[(16\,\text{m}^{-1})x + (190\,\text{s}^{-1})t + 0.93\right].$$

22. (a) The general expression for $y(x, t)$ for the wave is $y(x, t) = y_m \sin(kx - \omega t)$, which, at $x = 10$ cm, becomes $y(x = 10\,\text{cm}, t) = y_m \sin[k(10\,\text{cm} - \omega t)]$. Comparing this with the expression given, we find $\omega = 4.0$ rad/s, or $f = \omega/2\pi = 0.64$ Hz.

(b) Since $k(10\,\text{cm}) = 1.0$, the wave number is $k = 0.10$/cm. Consequently, the wavelength is $\lambda = 2\pi/k = 63$ cm.

(c) The amplitude is $y_m = 5.0$ cm.

(d) In part (b), we have shown that the angular wave number is $k = 0.10$/cm.

(e) The angular frequency is $\omega = 4.0$ rad/s.

(f) The sign is minus since the wave is traveling in the $+x$ direction.

Summarizing the results obtained above by substituting the values of k and ω into the general expression for $y(x, t)$, with centimeters and seconds understood, we obtain

$$y(x,t) = 5.0 \sin(0.10x - 4.0t).$$

(g) Since $v = \omega/k = \sqrt{\tau/\mu}$, the tension is

$$\tau = \frac{\omega^2 \mu}{k^2} = \frac{(4.0\,\text{g/cm})(4.0\,\text{s}^{-1})^2}{(0.10\,\text{cm}^{-1})^2} = 6400\,\text{g} \cdot \text{cm/s}^2 = 0.064\,\text{N}.$$

23. The pulses have the same speed v. Suppose one pulse starts from the left end of the wire at time $t = 0$. Its coordinate at time t is $x_1 = vt$. The other pulse starts from the right end, at $x = L$, where L is the length of the wire, at time $t = 30$ ms. If this time is denoted by t_0 then the coordinate of this wave at time t is $x_2 = L - v(t - t_0)$. They meet when $x_1 = x_2$, or, what is the same, when $vt = L - v(t - t_0)$. We solve for the time they meet: $t = (L + vt_0)/2v$ and the coordinate of the meeting point is $x = vt = (L + vt_0)/2$. Now, we calculate the wave speed:

$$v = \sqrt{\frac{\tau L}{m}} = \sqrt{\frac{(250\,\text{N})(10.0\,\text{m})}{0.100\,\text{kg}}} = 158\,\text{m/s}.$$

Here τ is the tension in the wire and L/m is the linear mass density of the wire. The coordinate of the meeting point is

$$x = \frac{10.0\,\text{m} + (158\,\text{m/s})(30.0\times 10^{-3}\,\text{s})}{2} = 7.37\,\text{m}.$$

This is the distance from the left end of the wire. The distance from the right end is $L - x = (10.0\,\text{m} - 7.37\,\text{m}) = 2.63\,\text{m}$.

24. (a) The tension in each string is given by $\tau = Mg/2$. Thus, the wave speed in string 1 is

$$v_1 = \sqrt{\frac{\tau}{\mu_1}} = \sqrt{\frac{Mg}{2\mu_1}} = \sqrt{\frac{(500\,\text{g})(9.80\,\text{m/s}^2)}{2(3.00\,\text{g/m})}} = 28.6\,\text{m/s}.$$

(b) And the wave speed in string 2 is

$$v_2 = \sqrt{\frac{Mg}{2\mu_2}} = \sqrt{\frac{(500\,\text{g})(9.80\,\text{m/s}^2)}{2(5.00\,\text{g/m})}} = 22.1\,\text{m/s}.$$

(c) Let $v_1 = \sqrt{M_1 g/(2\mu_1)} = v_2 = \sqrt{M_2 g/(2\mu_2)}$ and $M_1 + M_2 = M$. We solve for M_1 and obtain

$$M_1 = \frac{M}{1 + \mu_2/\mu_1} = \frac{500\,\text{g}}{1 + 5.00/3.00} = 187.5\,\text{g} \approx 188\,\text{g}.$$

(d) And we solve for the second mass: $M_2 = M - M_1 = (500\,\text{g} - 187.5\,\text{g}) \approx 313\,\text{g}$.

25. (a) The wave speed at any point on the rope is given by $v = \sqrt{\tau/\mu}$, where τ is the tension at that point and μ is the linear mass density. Because the rope is hanging the tension varies from point to point. Consider a point on the rope a distance y from the bottom end. The forces acting on it are the weight of the rope below it, pulling down, and the tension, pulling up. Since the rope is in equilibrium, these forces balance. The weight of the rope below is given by $\mu g y$, so the tension is $\tau = \mu g y$. The wave speed is $v = \sqrt{\mu g y / \mu} = \sqrt{g y}$.

(b) The time dt for the wave to move past a length dy, a distance y from the bottom end, is $dt = dy/v = dy/\sqrt{gy}$ and the total time for the wave to move the entire length of the rope is

$$t = \int_0^L \frac{dy}{\sqrt{gy}} = 2\sqrt{\frac{y}{g}}\bigg|_0^L = 2\sqrt{\frac{L}{g}}.$$

26. Using Eq. 16–33 for the average power and Eq. 16–26 for the speed of the wave, we solve for $f = \omega/2\pi$:

$$f = \frac{1}{2\pi y_m}\sqrt{\frac{2P_{avg}}{\mu\sqrt{\tau/\mu}}} = \frac{1}{2\pi(7.70\times 10^{-3}\,\text{m})}\sqrt{\frac{2(85.0\,\text{W})}{\sqrt{(36.0\,\text{N})(0.260\,\text{kg}/2.70\,\text{m})}}} = 198\,\text{Hz}.$$

27. We note from the graph (and from the fact that we are dealing with a cosine-squared, see Eq. 16-30) that the wave frequency is $f = \frac{1}{2\,\text{ms}} = 500$ Hz, and that the wavelength $\lambda = 0.20$ m. We also note from the graph that the maximum value of dK/dt is 10 W. Setting this equal to the maximum value of Eq. 16-29 (where we just set that cosine term equal to 1) we find

$$\tfrac{1}{2}\mu v \omega^2 y_m^2 = 10$$

with SI units understood. Substituting in $\mu = 0.002$ kg/m, $\omega = 2\pi f$ and $v = f\lambda$, we solve for the wave amplitude:

$$y_m = \sqrt{\frac{10}{2\pi^2\mu\lambda f^3}} = 0.0032\,\text{m}.$$

28. Comparing $y(x,t) = (3.00\,\text{mm})\sin[(4.00\,\text{m}^{-1})x - (7.00\,\text{s}^{-1})t]$ to the general expression $y(x,t) = y_m \sin(kx - \omega t)$, we see that $k = 4.00$ m^{-1} and $\omega = 7.00$ rad/s. The speed of the wave is

$$v = \omega/k = (7.00\,\text{rad/s})/(4.00\,\text{m}^{-1}) = 1.75\,\text{m/s}.$$

29. The wave $y(x,t) = (2.00\,\text{mm})[(20\,\text{m}^{-1})x - (4.0\,\text{s}^{-1})t]^{1/2}$ is of the form $h(kx - \omega t)$ with angular wave number $k = 20$ m^{-1} and angular frequency $\omega = 4.0$ rad/s. Thus, the speed of the wave is

$$v = \omega/k = (4.0\,\text{rad/s})/(20\,\text{m}^{-1}) = 0.20\,\text{m/s}.$$

30. The wave $y(x,t) = (4.00\,\text{mm})\,h[(30\,\text{m}^{-1})x + (6.0\,\text{s}^{-1})t]$ is of the form $h(kx - \omega t)$ with angular wave number $k = 30$ m^{-1} and angular frequency $\omega = 6.0$ rad/s. Thus, the speed of the wave is

$$v = \omega/k = (6.0\,\text{rad/s})/(30\,\text{m}^{-1}) = 0.20\,\text{m/s}.$$

31. The displacement of the string is given by

$$y = y_m \sin(kx - \omega t) + y_m \sin(kx - \omega t + \phi) = 2y_m \cos\left(\tfrac{1}{2}\phi\right)\sin\left(kx - \omega t + \tfrac{1}{2}\phi\right),$$

where $\phi = \pi/2$. The amplitude is

$$A = 2y_m \cos\left(\tfrac{1}{2}\phi\right) = 2y_m \cos(\pi/4) = 1.41 y_m.$$

32. (a) Let the phase difference be ϕ. Then from Eq. 16–52, $2y_m \cos(\phi/2) = 1.50 y_m$, which gives

$$\phi = 2\cos^{-1}\left(\frac{1.50 y_m}{2 y_m}\right) = 82.8°.$$

(b) Converting to radians, we have $\phi = 1.45$ rad.

(c) In terms of wavelength (the length of each cycle, where each cycle corresponds to 2π rad), this is equivalent to $1.45 \text{ rad}/2\pi = 0.230$ wavelength.

33. (a) The amplitude of the second wave is $y_m = 9.00$ mm, as stated in the problem.

(b) The figure indicates that $\lambda = 40$ cm $= 0.40$ m, which implies that the angular wave number is $k = 2\pi/0.40 = 16$ rad/m.

(c) The figure (along with information in the problem) indicates that the speed of each wave is $v = dx/t = (56.0 \text{ cm})/(8.0 \text{ ms}) = 70$ m/s. This, in turn, implies that the angular frequency is

$$\omega = kv = 1100 \text{ rad/s} = 1.1 \times 10^3 \text{ rad/s}.$$

(d) The figure depicts two traveling waves (both going in the $-x$ direction) of equal amplitude y_m. The amplitude of their resultant wave, as shown in the figure, is $y'_m = 4.00$ mm. Eq. 16-52 applies:

$$y'_m = 2 y_m \cos(\tfrac{1}{2}\phi_2) \quad \Rightarrow \quad \phi_2 = 2\cos^{-1}(2.00/9.00) = 2.69 \text{ rad}.$$

(e) In making the plus-or-minus sign choice in $y = y_m \sin(kx \pm \omega t + \phi)$, we recall the discussion in section 16-5, where it shown that sinusoidal waves traveling in the $-x$ direction are of the form $y = y_m \sin(kx + \omega t + \phi)$. Here, ϕ should be thought of as the phase *difference* between the two waves (that is, $\phi_1 = 0$ for wave 1 and $\phi_2 = 2.69$ rad for wave 2).

In summary, the waves have the forms (with SI units understood):

$$y_1 = (0.00900)\sin(16x + 1100 t) \quad \text{and} \quad y_2 = (0.00900)\sin(16x + 1100 t + 2.7).$$

34. (a) We use Eq. 16-26 and Eq. 16-33 with $\mu = 0.00200$ kg/m and $y_m = 0.00300$ m. These give $v = \sqrt{\tau/\mu} = 775$ m/s and

$$P_{avg} = \tfrac{1}{2} \mu v \omega^2 y_m^2 = 10 \text{ W}.$$

(b) In this situation, the waves are two separate string (no superposition occurs). The answer is clearly twice that of part (a); $P = 20$ W.

(c) Now they are on the same string. If they are interfering constructively (as in Fig. 16-16(a)) then the amplitude y_m is doubled which means its square y_m^2 increases by a factor of 4. Thus, the answer now is four times that of part (a); $P = 40$ W.

(d) Eq. 16-52 indicates in this case that the amplitude (for their superposition) is $2\,y_m\cos(0.2\pi) = 1.618$ times the original amplitude y_m. Squared, this results in an increase in the power by a factor of 2.618. Thus, $P = 26$ W in this case.

(e) Now the situation depicted in Fig. 16-16(b) applies, so $P = 0$.

35. The phasor diagram is shown below: y_{1m} and y_{2m} represent the original waves and y_m represents the resultant wave. The phasors corresponding to the two constituent waves make an angle of 90° with each other, so the triangle is a right triangle. The Pythagorean theorem gives

$$y_m^2 = y_{1m}^2 + y_{2m}^2 = (3.0\,\text{cm})^2 + (4.0\,\text{cm})^2 = (25\,\text{cm})^2.$$

Thus $y_m = 5.0$ cm.

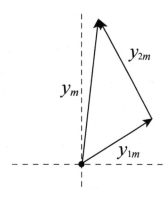

36. (a) As shown in Figure 16-16(b) in the textbook, the least-amplitude resultant wave is obtained when the phase difference is π rad.

(b) In this case, the amplitude is $(8.0 \text{ mm} - 5.0 \text{ mm}) = 3.0$ mm.

(c) As shown in Figure 16-16(a) in the textbook, the greatest-amplitude resultant wave is obtained when the phase difference is 0 rad.

(d) In the part (c) situation, the amplitude is (8.0 mm + 5.0 mm) = 13 mm.

(e) Using phasor terminology, the angle "between them" in this case is $\pi/2$ rad (90°), so the Pythagorean theorem applies:

$$\sqrt{(8.0 \text{ mm})^2 + (5.0 \text{ mm})^2} = 9.4 \text{ mm}.$$

37. The phasor diagram is shown on the right. We use the cosine theorem:

$$y_m^2 = y_{m1}^2 + y_{m2}^2 - 2y_{m1}y_{m2}\cos\theta = y_{m1}^2 + y_{m2}^2 + 2y_{m1}y_{m2}\cos\phi.$$

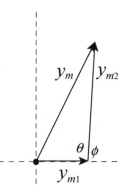

We solve for $\cos\phi$:

$$\cos\phi = \frac{y_m^2 - y_{m1}^2 - y_{m2}^2}{2y_{m1}y_{m2}} = \frac{(9.0\text{ mm})^2 - (5.0\text{ mm})^2 - (7.0\text{ mm})^2}{2(5.0\text{ mm})(7.0\text{ mm})} = 0.10.$$

The phase constant is therefore $\phi = 84°$.

38. We see that y_1 and y_3 cancel (they are 180°) out of phase, and y_2 cancels with y_4 because their phase difference is also equal to π rad (180°). There is no resultant wave in this case.

39. (a) Using the phasor technique, we think of these as two "vectors" (the first of "length" 4.6 mm and the second of "length" 5.60 mm) separated by an angle of $\phi = 0.8\pi$ radians (or 144°). Standard techniques for adding vectors then lead to a resultant vector of length 3.29 mm.

(b) The angle (relative to the first vector) is equal to 88.8° (or 1.55 rad).

(c) Clearly, it should in "in phase" with the result we just calculated, so its phase angle relative to the first phasor should be also 88.8° (or 1.55 rad).

40. (a) The wave speed is given by

$$v = \sqrt{\frac{\tau}{\mu}} = \sqrt{\frac{7.00\text{ N}}{2.00 \times 10^{-3}\text{ kg}/1.25\text{ m}}} = 66.1 \text{ m/s}.$$

(b) The wavelength of the wave with the lowest resonant frequency f_1 is $\lambda_1 = 2L$, where $L = 125$ cm. Thus,

$$f_1 = \frac{v}{\lambda_1} = \frac{66.1 \text{ m/s}}{2(1.25 \text{ m})} = 26.4 \text{ Hz}.$$

41. Possible wavelengths are given by $\lambda = 2L/n$, where L is the length of the wire and n is an integer. The corresponding frequencies are given by $f = v/\lambda = nv/2L$, where v is the wave speed. The wave speed is given by $v = \sqrt{\tau/\mu} = \sqrt{\tau L/M}$, where τ is the tension in the wire, μ is the linear mass density of the wire, and M is the mass of the wire. $\mu = M/L$ was used to obtain the last form. Thus

$$f_n = \frac{n}{2L}\sqrt{\frac{\tau L}{M}} = \frac{n}{2}\sqrt{\frac{\tau}{LM}} = \frac{n}{2}\sqrt{\frac{250\ \text{N}}{(10.0\ \text{m})(0.100\ \text{kg})}} = n\ (7.91\ \text{Hz}).$$

(a) The lowest frequency is $f_1 = 7.91$ Hz.

(b) The second lowest frequency is $f_2 = 2(7.91\ \text{Hz}) = 15.8$ Hz.

(c) The third lowest frequency is $f_3 = 3(7.91\ \text{Hz}) = 23.7$ Hz.

42. The nth resonant frequency of string A is

$$f_{n,A} = \frac{v_A}{2l_A}n = \frac{n}{2L}\sqrt{\frac{\tau}{\mu}},$$

while for string B it is

$$f_{n,B} = \frac{v_B}{2l_B}n = \frac{n}{8L}\sqrt{\frac{\tau}{\mu}} = \frac{1}{4}f_{n,A}.$$

(a) Thus, we see $f_{1,A} = f_{4,B}$. That is, the fourth harmonic of B matches the frequency of A's first harmonic.

(b) Similarly, we find $f_{2,A} = f_{8,B}$.

(c) No harmonic of B would match $f_{3,A} = \frac{3v_A}{2l_A} = \frac{3}{2L}\sqrt{\frac{\tau}{\mu}}$,

43. (a) The wave speed is given by $v = \sqrt{\tau/\mu}$, where τ is the tension in the string and μ is the linear mass density of the string. Since the mass density is the mass per unit length, $\mu = M/L$, where M is the mass of the string and L is its length. Thus

$$v = \sqrt{\frac{\tau L}{M}} = \sqrt{\frac{(96.0\ \text{N})(8.40\ \text{m})}{0.120\ \text{kg}}} = 82.0\ \text{m/s}.$$

(b) The longest possible wavelength λ for a standing wave is related to the length of the string by $L = \lambda/2$, so $\lambda = 2L = 2(8.40\ \text{m}) = 16.8$ m.

(c) The frequency is $f = v/\lambda = (82.0 \text{ m/s})/(16.8 \text{ m}) = 4.88$ Hz.

44. The string is flat each time the particle passes through its equilibrium position. A particle may travel up to its positive amplitude point and back to equilibrium during this time. This describes *half* of one complete cycle, so we conclude $T = 2(0.50 \text{ s}) = 1.0$ s. Thus, $f = 1/T = 1.0$ Hz, and the wavelength is

$$\lambda = \frac{v}{f} = \frac{10 \text{ cm/s}}{1.0 \text{ Hz}} = 10 \text{ cm}.$$

45. (a) Eq. 16–26 gives the speed of the wave:

$$v = \sqrt{\frac{\tau}{\mu}} = \sqrt{\frac{150 \text{ N}}{7.20 \times 10^{-3} \text{ kg/m}}} = 144.34 \text{ m/s} \approx 1.44 \times 10^2 \text{ m/s}.$$

(b) From the figure, we find the wavelength of the standing wave to be

$$\lambda = (2/3)(90.0 \text{ cm}) = 60.0 \text{ cm}.$$

(c) The frequency is

$$f = \frac{v}{\lambda} = \frac{1.44 \times 10^2 \text{ m/s}}{0.600 \text{ m}} = 241 \text{ Hz}.$$

46. Use Eq. 16–66 (for the resonant frequencies) and Eq. 16–26 ($v = \sqrt{\tau/\mu}$) to find f_n:

$$f_n = \frac{nv}{2L} = \frac{n}{2L}\sqrt{\frac{\tau}{\mu}}$$

which gives $f_3 = (3/2L)\sqrt{\tau_i/\mu}$.

(a) When $\tau_f = 4\tau_i$, we get the new frequency

$$f_3' = \frac{3}{2L}\sqrt{\frac{\tau_f}{\mu}} = 2f_3.$$

(b) And we get the new wavelength $\lambda_3' = \dfrac{v'}{f_3'} = \dfrac{2L}{3} = \lambda_3$.

47. (a) The resonant wavelengths are given by $\lambda = 2L/n$, where L is the length of the string and n is an integer, and the resonant frequencies are given by $f = v/\lambda = nv/2L$, where v is the wave speed. Suppose the lower frequency is associated with the integer n. Then, since there are no resonant frequencies between, the higher frequency is associated

with $n + 1$. That is, $f_1 = nv/2L$ is the lower frequency and $f_2 = (n + 1)v/2L$ is the higher. The ratio of the frequencies is

$$\frac{f_2}{f_1} = \frac{n+1}{n}.$$

The solution for n is

$$n = \frac{f_1}{f_2 - f_1} = \frac{315 \text{ Hz}}{420 \text{ Hz} - 315 \text{ Hz}} = 3.$$

The lowest possible resonant frequency is $f = v/2L = f_1/n = (315 \text{ Hz})/3 = 105 \text{ Hz}$.

(b) The longest possible wavelength is $\lambda = 2L$. If f is the lowest possible frequency then

$$v = \lambda f = 2Lf = 2(0.75 \text{ m})(105 \text{ Hz}) = 158 \text{ m/s}.$$

48. Using Eq. 16-26, we find the wave speed to be

$$v = \sqrt{\frac{\tau}{\mu}} = \sqrt{\frac{65.2 \times 10^6 \text{ N}}{3.35 \text{ kg/m}}} = 4412 \text{ m/s}.$$

The corresponding resonant frequencies are

$$f_n = \frac{nv}{2L} = \frac{n}{2L}\sqrt{\frac{\tau}{\mu}}, \quad n = 1, 2, 3, \ldots$$

(a) The wavelength of the wave with the lowest (fundamental) resonant frequency f_1 is $\lambda_1 = 2L$, where $L = 347$ m. Thus,

$$f_1 = \frac{v}{\lambda_1} = \frac{4412 \text{ m/s}}{2(347 \text{ m})} = 6.36 \text{ Hz}.$$

(b) The frequency difference between successive modes is

$$\Delta f = f_n - f_{n-1} = \frac{v}{2L} = \frac{4412 \text{ m/s}}{2(347 \text{ m})} = 6.36 \text{ Hz}.$$

49. The harmonics are integer multiples of the fundamental, which implies that the difference between any successive pair of the harmonic frequencies is equal to the fundamental frequency. Thus, $f_1 = (390 \text{ Hz} - 325 \text{ Hz}) = 65$ Hz. This further implies that the next higher resonance above 195 Hz should be $(195 \text{ Hz} + 65 \text{ Hz}) = 260$ Hz.

50. Since the rope is fixed at both ends, then the phrase "second-harmonic standing wave pattern" describes the oscillation shown in Figure 16–23(b), where (see Eq. 16–65)

$$\lambda = L \quad \text{and} \quad f = \frac{v}{L}.$$

(a) Comparing the given function with Eq. 16-60, we obtain $k = \pi/2$ and $\omega = 12\pi$ rad/s. Since $k = 2\pi/\lambda$ then

$$\frac{2\pi}{\lambda} = \frac{\pi}{2} \Rightarrow \lambda = 4.0\,\text{m} \Rightarrow L = 4.0\,\text{m}.$$

(b) Since $\omega = 2\pi f$ then $2\pi f = 12\pi$ rad/s, which yields

$$f = 6.0\,\text{Hz} \Rightarrow v = f\lambda = 24\,\text{m/s}.$$

(c) Using Eq. 16–26, we have

$$v = \sqrt{\frac{\tau}{\mu}} \Rightarrow 24\,\text{m/s} = \sqrt{\frac{200\,\text{N}}{m/(4.0\,\text{m})}}$$

which leads to $m = 1.4$ kg.

(d) With

$$f = \frac{3v}{2L} = \frac{3(24\,\text{m/s})}{2(4.0\,\text{m})} = 9.0\,\text{Hz}$$

The period is $T = 1/f = 0.11$ s.

51. (a) The amplitude of each of the traveling waves is half the maximum displacement of the string when the standing wave is present, or 0.25 cm.

(b) Each traveling wave has an angular frequency of $\omega = 40\pi$ rad/s and an angular wave number of $k = \pi/3$ cm^{-1}. The wave speed is

$$v = \omega/k = (40\pi\,\text{rad/s})/(\pi/3\,\text{cm}^{-1}) = 1.2\times 10^2\,\text{cm/s}.$$

(c) The distance between nodes is half a wavelength: $d = \lambda/2 = \pi/k = \pi/(\pi/3\,\text{cm}^{-1}) = 3.0$ cm. Here $2\pi/k$ was substituted for λ.

(d) The string speed is given by $u(x, t) = \partial y/\partial t = -\omega y_m \sin(kx)\sin(\omega t)$. For the given coordinate and time,

$$u = -(40\pi\,\text{rad/s})(0.50\,\text{cm})\sin\left[\left(\frac{\pi}{3}\,\text{cm}^{-1}\right)(1.5\,\text{cm})\right]\sin\left[(40\pi\,\text{s}^{-1})\left(\frac{9}{8}\,\text{s}\right)\right] = 0.$$

52. The nodes are located from vanishing of the spatial factor $\sin 5\pi x = 0$ for which the solutions are

$$5\pi x = 0, \pi, 2\pi, 3\pi, \ldots \quad \Rightarrow \quad x = 0, \frac{1}{5}, \frac{2}{5}, \frac{3}{5}, \ldots$$

(a) The smallest value of x which corresponds to a node is $x = 0$.

(b) The second smallest value of x which corresponds to a node is $x = 0.20$ m.

(c) The third smallest value of x which corresponds to a node is $x = 0.40$ m.

(d) Every point (except at a node) is in simple harmonic motion of frequency $f = \omega/2\pi = 40\pi/2\pi = 20$ Hz. Therefore, the period of oscillation is $T = 1/f = 0.050$ s.

(e) Comparing the given function with Eq. 16–58 through Eq. 16–60, we obtain

$$y_1 = 0.020\sin(5\pi x - 40\pi t) \quad \text{and} \quad y_2 = 0.020\sin(5\pi x + 40\pi t)$$

for the two traveling waves. Thus, we infer from these that the speed is $v = \omega/k = 40\pi/5\pi = 8.0$ m/s.

(f) And we see the amplitude is $y_m = 0.020$ m.

(g) The derivative of the given function with respect to time is

$$u = \frac{\partial y}{\partial t} = -(0.040)(40\pi)\sin(5\pi x)\sin(40\pi t)$$

which vanishes (for all x) at times such as $\sin(40\pi t) = 0$. Thus,

$$40\pi t = 0, \pi, 2\pi, 3\pi, \ldots \quad \Rightarrow \quad t = 0, \frac{1}{40}, \frac{2}{40}, \frac{3}{40}, \ldots$$

Thus, the first time in which all points on the string have zero transverse velocity is when $t = 0$ s.

(h) The second time in which all points on the string have zero transverse velocity is when $t = 1/40$ s $= 0.025$ s.

(i) The third time in which all points on the string have zero transverse velocity is when $t = 2/40$ s $= 0.050$ s.

53. (a) The waves have the same amplitude, the same angular frequency, and the same angular wave number, but they travel in opposite directions. We take them to be

$$y_1 = y_m \sin(kx - \omega t), \quad y_2 = y_m \sin(kx + \omega t).$$

The amplitude y_m is half the maximum displacement of the standing wave, or 5.0×10^{-3} m.

(b) Since the standing wave has three loops, the string is three half-wavelengths long: $L = 3\lambda/2$, or $\lambda = 2L/3$. With $L = 3.0$ m, $\lambda = 2.0$ m. The angular wave number is

$$k = 2\pi/\lambda = 2\pi/(2.0 \text{ m}) = 3.1 \text{ m}^{-1}.$$

(c) If v is the wave speed, then the frequency is

$$f = \frac{v}{\lambda} = \frac{3v}{2L} = \frac{3(100 \text{ m/s})}{2(3.0 \text{ m})} = 50 \text{ Hz}.$$

The angular frequency is the same as that of the standing wave, or

$$\omega = 2\pi f = 2\pi(50 \text{ Hz}) = 314 \text{ rad/s}.$$

(d) The two waves are

$$y_1 = (5.0 \times 10^{-3} \text{ m}) \sin\left[(3.14 \text{ m}^{-1})x - (314 \text{ s}^{-1})t\right]$$

and

$$y_2 = (5.0 \times 10^{-3} \text{ m}) \sin\left[(3.14 \text{ m}^{-1})x + (314 \text{ s}^{-1})t\right].$$

Thus, if one of the waves has the form $y(x,t) = y_m \sin(kx + \omega t)$, then the other wave must have the form $y'(x,t) = y_m \sin(kx - \omega t)$. The sign in front of ω for $y'(x,t)$ is minus.

54. From the $x = 0$ plot (and the requirement of an anti-node at $x = 0$), we infer a standing wave function of the form $y(x,t) = -(0.04)\cos(kx)\sin(\omega t)$, where $\omega = 2\pi/T = \pi$ rad/s, with length in meters and time in seconds. The parameter k is determined by the existence of the node at $x = 0.10$ (presumably the *first* node that one encounters as one moves from the origin in the positive x direction). This implies $k(0.10) = \pi/2$ so that $k = 5\pi$ rad/m.

(a) With the parameters determined as discussed above and $t = 0.50$ s, we find

$$y(0.20 \text{ m}, 0.50 \text{ s}) = -0.04\cos(kx)\sin(\omega t) = 0.040 \text{ m}.$$

(b) The above equation yields $y(0.30 \text{ m}, 0.50 \text{ s}) = -0.04\cos(kx)\sin(\omega t) = 0$.

(c) We take the derivative with respect to time and obtain, at $t = 0.50$ s and $x = 0.20$ m,

$$u = \frac{dy}{dt} = -0.04\omega \cos(kx)\cos(\omega t) = 0.$$

d) The above equation yields $u = -0.13$ m/s at $t = 1.0$ s.

(e) The sketch of this function at $t = 0.50$ s for $0 \leq x \leq 0.40$ m is shown below:

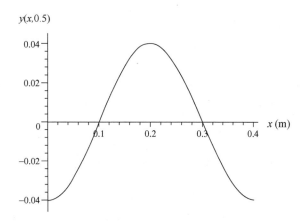

55. (a) The angular frequency is $\omega = 8.00\pi/2 = 4.00\pi$ rad/s, so the frequency is

$$f = \omega/2\pi = (4.00\pi \text{ rad/s})/2\pi = 2.00 \text{ Hz}.$$

(b) The angular wave number is $k = 2.00\pi/2 = 1.00\pi$ m^{-1}, so the wavelength is

$$\lambda = 2\pi/k = 2\pi/(1.00\pi \text{ m}^{-1}) = 2.00 \text{ m}.$$

(c) The wave speed is

$$v = \lambda f = (2.00 \text{ m})(2.00 \text{ Hz}) = 4.00 \text{ m/s}.$$

(d) We need to add two cosine functions. First convert them to sine functions using $\cos \alpha = \sin(\alpha + \pi/2)$, then apply

$$\cos\alpha + \cos\beta = \sin\left(\alpha + \frac{\pi}{2}\right) + \sin\left(\beta + \frac{\pi}{2}\right) = 2\sin\left(\frac{\alpha+\beta+\pi}{2}\right)\cos\left(\frac{\alpha+\beta}{2}\right)$$

$$= 2\cos\left(\frac{\alpha+\beta}{2}\right)\cos\left(\frac{\alpha-\beta}{2}\right)$$

Letting $\alpha = kx$ and $\beta = \omega t$, we find

$$y_m \cos(kx + \omega t) + y_m \cos(kx - \omega t) = 2y_m \cos(kx)\cos(\omega t).$$

Nodes occur where $\cos(kx) = 0$ or $kx = n\pi + \pi/2$, where n is an integer (including zero). Since $k = 1.0\pi$ m^{-1}, this means $x = \left(n + \frac{1}{2}\right)(1.00\,\text{m})$. Thus, the smallest value of x which corresponds to a node is $x = 0.500$ m ($n=0$).

(e) The second smallest value of x which corresponds to a node is $x = 1.50$ m ($n=1$).

(f) The third smallest value of x which corresponds to a node is $x = 2.50$ m ($n=2$).

(g) The displacement is a maximum where $\cos(kx) = \pm 1$. This means $kx = n\pi$, where n is an integer. Thus, $x = n(1.00\text{ m})$. The smallest value of x which corresponds to an anti-node (maximum) is $x = 0$ ($n=0$).

(h) The second smallest value of x which corresponds to an anti-node (maximum) is $x = 1.00$ m ($n=1$).

(i) The third smallest value of x which corresponds to an anti-node (maximum) is $x = 2.00$ m ($n=2$).

56. Reference to point A as an anti-node suggests that this is a standing wave pattern and thus that the waves are traveling in opposite directions. Thus, we expect one of them to be of the form $y = y_m \sin(kx + \omega t)$ and the other to be of the form $y = y_m \sin(kx - \omega t)$.

(a) Using Eq. 16-60, we conclude that $y_m = \frac{1}{2}(9.0\text{ mm}) = 4.5$ mm, due to the fact that the amplitude of the standing wave is $\frac{1}{2}(1.80\text{ cm}) = 0.90$ cm $= 9.0$ mm.

(b) Since one full cycle of the wave (one wavelength) is 40 cm, $k = 2\pi/\lambda \approx 16$ m^{-1}.

(c) The problem tells us that the time of half a full period of motion is 6.0 ms, so $T = 12$ ms and Eq. 16-5 gives $\omega = 5.2 \times 10^2$ rad/s.

(d) The two waves are therefore

$$y_1(x, t) = (4.5\text{ mm}) \sin[(16\text{ m}^{-1})x + (520\text{ s}^{-1})t]$$

and

$$y_2(x, t) = (4.5\text{ mm}) \sin[(16\text{ m}^{-1})x - (520\text{ s}^{-1})t].$$

If one wave has the form $y(x,t) = y_m \sin(kx + \omega t)$ as in y_1, then the other wave must be of the form $y'(x,t) = y_m \sin(kx - \omega t)$ as in y_2. Therefore, the sign in front of ω is minus.

57. Recalling the discussion in section 16-12, we observe that this problem presents us with a standing wave condition with amplitude 12 cm. The angular wave number and frequency are noted by comparing the given waves with the form $y = y_m \sin(kx \pm \omega t)$.

The anti-node moves through 12 cm in simple harmonic motion, just as a mass on a vertical spring would move from its upper turning point to its lower turning point – which occurs during a half-period. Since the period T is related to the angular frequency by Eq. 15-5, we have

$$T = \frac{2\pi}{\omega} = \frac{2\pi}{4.00\,\pi} = 0.500 \text{ s}.$$

Thus, in a time of $t = \frac{1}{2}T = 0.250$ s, the wave moves a distance $\Delta x = vt$ where the speed of the wave is $v = \frac{\omega}{k} = 1.00$ m/s. Therefore, $\Delta x = (1.00 \text{ m/s})(0.250 \text{ s}) = 0.250$ m.

58. With the string fixed on both ends, using Eq. 16-66 and Eq. 16-26, the resonant frequencies can be written as

$$f = \frac{nv}{2L} = \frac{n}{2L}\sqrt{\frac{\tau}{\mu}} = \frac{n}{2L}\sqrt{\frac{mg}{\mu}}, \quad n = 1, 2, 3, \ldots$$

(a) The mass that allows the oscillator to set up the 4th harmonic ($n = 4$) on the string is

$$m = \left.\frac{4L^2 f^2 \mu}{n^2 g}\right|_{n=4} = \frac{4(1.20 \text{ m})^2 (120 \text{ Hz})^2 (0.00160 \text{ kg/m})}{(4)^2 (9.80 \text{ m/s}^2)} = 0.846 \text{ kg}$$

(b) If the mass of the block is $m = 1.00$ kg, the corresponding n is

$$n = \sqrt{\frac{4L^2 f^2 \mu}{g}} = \sqrt{\frac{4(1.20 \text{ m})^2 (120 \text{ Hz})^2 (0.00160 \text{ kg/m})}{9.80 \text{ m/s}^2}} = 3.68$$

which is not an integer. Therefore, the mass cannot set up a standing wave on the string.

59. (a) The frequency of the wave is the same for both sections of the wire. The wave speed and wavelength, however, are both different in different sections. Suppose there are n_1 loops in the aluminum section of the wire. Then,

$$L_1 = n_1 \lambda_1 / 2 = n_1 v_1 / 2f,$$

where λ_1 is the wavelength and v_1 is the wave speed in that section. In this consideration, we have substituted $\lambda_1 = v_1/f$, where f is the frequency. Thus $f = n_1 v_1 / 2L_1$. A similar expression holds for the steel section: $f = n_2 v_2 / 2L_2$. Since the frequency is the same for the two sections, $n_1 v_1 / L_1 = n_2 v_2 / L_2$. Now the wave speed in the aluminum section is given by $v_1 = \sqrt{\tau/\mu_1}$, where μ_1 is the linear mass density of the aluminum wire. The mass of aluminum in the wire is given by $m_1 = \rho_1 A L_1$, where ρ_1 is the mass density (mass per unit volume) for aluminum and A is the cross-sectional area of the wire. Thus

$$\mu_1 = \rho_1 A L_1 / L_1 = \rho_1 A$$

and $v_1 = \sqrt{\tau/\rho_1 A}$. A similar expression holds for the wave speed in the steel section: $v_2 = \sqrt{\tau/\rho_2 A}$. We note that the cross-sectional area and the tension are the same for the two sections. The equality of the frequencies for the two sections now leads to $n_1 / L_1 \sqrt{\rho_1} = n_2 / L_2 \sqrt{\rho_2}$, where A has been canceled from both sides. The ratio of the integers is

$$\frac{n_2}{n_1} = \frac{L_2 \sqrt{\rho_2}}{L_1 \sqrt{\rho_1}} = \frac{(0.866\,\text{m}) \sqrt{7.80 \times 10^3\,\text{kg/m}^3}}{(0.600\,\text{m}) \sqrt{2.60 \times 10^3\,\text{kg/m}^3}} = 2.50.$$

The smallest integers that have this ratio are $n_1 = 2$ and $n_2 = 5$. The frequency is

$$f = n_1 v_1 / 2L_1 = (n_1 / 2L_1) \sqrt{\tau / \rho_1 A}.$$

The tension is provided by the hanging block and is $\tau = mg$, where m is the mass of the block. Thus,

$$f = \frac{n_1}{2L_1}\sqrt{\frac{mg}{\rho_1 A}} = \frac{2}{2(0.600\,\text{m})}\sqrt{\frac{(10.0\,\text{kg})(9.80\,\text{m/s}^2)}{(2.60 \times 10^3\,\text{kg/m}^3)(1.00 \times 10^{-6}\,\text{m}^2)}} = 324\,\text{Hz}.$$

(b) The standing wave pattern has two loops in the aluminum section and five loops in the steel section, or seven loops in all. There are eight nodes, counting the end points.

60. With the string fixed on both ends, using Eq. 16-66 and Eq. 16-26, the resonant frequencies can be written as

$$f = \frac{nv}{2L} = \frac{n}{2L}\sqrt{\frac{\tau}{\mu}} = \frac{n}{2L}\sqrt{\frac{mg}{\mu}}, \quad n = 1, 2, 3, \ldots$$

The mass that allows the oscillator to set up the nth harmonic on the string is

$$m = \frac{4L^2 f^2 \mu}{n^2 g}.$$

Thus, we see that the block mass is inversely proportional to the harmonic number squared. Thus, if the 447 gram block corresponds to harmonic number n then

$$\frac{447}{286.1} = \frac{(n+1)^2}{n^2} = \frac{n^2 + 2n + 1}{n^2} = 1 + \frac{2n+1}{n^2}.$$

Therefore, $\frac{447}{286.1} - 1 = 0.5624$ must equal an odd integer $(2n + 1)$ divided by a squared integer (n^2). That is, multiplying 0.5624 by a square (such as 1, 4, 9, 16, etc) should give us a number very close (within experimental uncertainty) to an odd number (1, 3, 5, …). Trying this out in succession (starting with multiplication by 1, then by 4, …), we find that multiplication by 16 gives a value very close to 9; we conclude $n = 4$ (so $n^2 = 16$ and $2n + 1 = 9$). Plugging $m = 0.447$ kg, $n = 4$, and the other values given in the problem, we find

$$\mu = 0.000845 \text{ kg/m} = 0.845 \text{ g/m}.$$

61. (a) The phasor diagram is shown here: y_1, y_2, and y_3 represent the original waves and y_m represents the resultant wave.

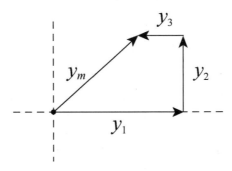

The horizontal component of the resultant is $y_{mh} = y_1 - y_3 = y_1 - y_1/3 = 2y_1/3$. The vertical component is $y_{mv} = y_2 = y_1/2$. The amplitude of the resultant is

$$y_m = \sqrt{y_{mh}^2 + y_{mv}^2} = \sqrt{\left(\frac{2y_1}{3}\right)^2 + \left(\frac{y_1}{2}\right)^2} = \frac{5}{6}y_1 = 0.83 y_1.$$

(b) The phase constant for the resultant is

$$\phi = \tan^{-1}\left(\frac{y_{mv}}{y_{mh}}\right) = \tan^{-1}\left(\frac{y_1/2}{2y_1/3}\right) = \tan^{-1}\left(\frac{3}{4}\right) = 0.644 \text{ rad} = 37°.$$

(c) The resultant wave is

$$y = \frac{5}{6}y_1 \sin(kx - \omega t + 0.644 \text{ rad}).$$

The graph below shows the wave at time $t = 0$. As time goes on it moves to the right with speed $v = \omega/k$.

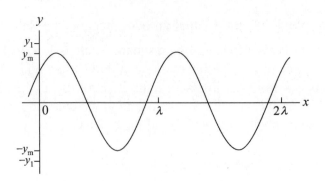

62. Setting $x = 0$ in $y = y_m \sin(kx - \omega t + \phi)$ gives $y = y_m \sin(-\omega t + \phi)$ as the function being plotted in the graph. We note that it has a positive "slope" (referring to its t-derivative) at $t = 0$:

$$\frac{dy}{dt} = \frac{d\, y_m \sin(-\omega t + \phi)}{dt} = -y_m \omega \cos(-\omega t + \phi) > 0 \text{ at } t = 0.$$

This implies that $-\cos(\phi) > 0$ and consequently that ϕ is in either the second or third quadrant. The graph shows (at $t = 0$) $y = 2.00$ mm, and (at some later t) $y_m = 6.00$ mm. Therefore,

$$y = y_m \sin(-\omega t + \phi)\big|_{t=0} \Rightarrow \phi = \sin^{-1}(\tfrac{1}{3}) = 0.34 \text{ rad or } 2.8 \text{ rad}$$

(bear in mind that $\sin(\theta) = \sin(\pi - \theta)$), and we must choose $\phi = 2.8$ rad because this is about 161° and is in second quadrant. Of course, this answer added to $2n\pi$ is still a valid answer (where n is any integer), so that, for example, $\phi = 2.8 - 2\pi = -3.48$ rad is also an acceptable result.

63. We compare the resultant wave given with the standard expression (Eq. 16–52) to obtain $k = 20\,\text{m}^{-1} = 2\pi/\lambda$, $2y_m \cos(\tfrac{1}{2}\phi) = 3.0$ mm, and $\tfrac{1}{2}\phi = 0.820$ rad.

(a) Therefore, $\lambda = 2\pi/k = 0.31$ m.

(b) The phase difference is $\phi = 1.64$ rad.

(c) And the amplitude is $y_m = 2.2$ mm.

64. Setting $x = 0$ in $a_y = -\omega^2 y$ (see the solution to part (b) of Sample Problem 16-2) where $y = y_m \sin(kx - \omega t + \phi)$ gives $a_y = -\omega^2 y_m \sin(-\omega t + \phi)$ as the function being plotted in the graph. We note that it has a negative "slope" (referring to its t-derivative) at $t = 0$:

$$\frac{d\,a_y}{dt} = \frac{d(-\omega^2 y_m \sin(-\omega t + \phi))}{dt} = y_m\omega^3 \cos(-\omega t + \phi) < 0 \text{ at } t = 0.$$

This implies that $\cos\phi < 0$ and consequently that ϕ is in either the second or third quadrant. The graph shows (at t = 0) $a_y = -100$ m/s², and (at another t) $a_{max} = 400$ m/s². Therefore,

$$a_y = -a_{max}\sin(-\omega t + \phi)\big|_{t=0} \Rightarrow \phi = \sin^{-1}(\tfrac{1}{4}) = 0.25 \text{ rad or } 2.9 \text{ rad}$$

(bear in mind that $\sin\theta = \sin(\pi - \theta)$), and we must choose $\phi = 2.9$ rad because this is about 166° and is in the second quadrant. Of course, this answer added to $2n\pi$ is still a valid answer (where n is any integer), so that, for example, $\phi = 2.9 - 2\pi = -3.4$ rad is also an acceptable result.

65. We note that

$$dy/dt = -\omega\cos(kx - \omega t + \phi),$$

which we will refer to as $u(x,t)$. so that the ratio of the function $y(x,t)$ divided by $u(x,t)$ is $-\tan(kx - \omega t + \phi)/\omega$. With the given information (for $x = 0$ and $t = 0$) then we can take the inverse tangent of this ratio to solve for the phase constant:

$$\phi = \tan^{-1}\left(\frac{-\omega\, y(0,0)}{u(0,0)}\right) = \tan^{-1}\left(\frac{-(440)(0.0045)}{-0.75}\right) = 1.2 \text{ rad}.$$

66. (a) Recalling the discussion in §16-5, we see that the speed of the wave given by a function with argument $x - 5.0t$ (where x is in centimeters and t is in seconds) must be 5.0 cm/s.

(b) In part (c), we show several "snapshots" of the wave: the one on the left is as shown in Figure 16–48 (at $t = 0$), the middle one is at $t = 1.0$ s, and the rightmost one is at $t = 2.0$ s. It is clear that the wave is traveling to the right (the +x direction).

(c) The third picture in the sequence below shows the pulse at 2.0 s. The horizontal scale (and, presumably, the vertical one also) is in centimeters.

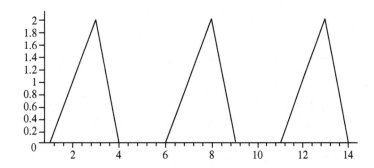

(d) The leading edge of the pulse reaches $x = 10$ cm at $t = (10 - 4.0)/5 = 1.2$ s. The particle (say, of the string that carries the pulse) at that location reaches a maximum displacement $h = 2$ cm at $t = (10 - 3.0)/5 = 1.4$ s. Finally, the trailing edge of the pulse departs from $x = 10$ cm at $t = (10 - 1.0)/5 = 1.8$ s. Thus, we find for $h(t)$ at $x = 10$ cm (with the horizontal axis, t, in seconds):

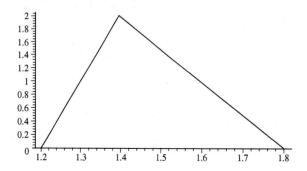

67. (a) The displacement of the string is assumed to have the form $y(x, t) = y_m \sin(kx - \omega t)$. The velocity of a point on the string is

$$u(x, t) = \partial y/\partial t = -\omega y_m \cos(kx - \omega t)$$

and its maximum value is $u_m = \omega y_m$. For this wave the frequency is $f = 120$ Hz and the angular frequency is $\omega = 2\pi f = 2\pi(120 \text{ Hz}) = 754$ rad/s. Since the bar moves through a distance of 1.00 cm, the amplitude is half of that, or $y_m = 5.00 \times 10^{-3}$ m. The maximum speed is

$$u_m = (754 \text{ rad/s})(5.00 \times 10^{-3} \text{ m}) = 3.77 \text{ m/s}.$$

(b) Consider the string at coordinate x and at time t and suppose it makes the angle θ with the x axis. The tension is along the string and makes the same angle with the x axis. Its transverse component is $\tau_{trans} = \tau \sin \theta$. Now θ is given by $\tan \theta = \partial y/\partial x = k y_m \cos(kx - \omega t)$ and its maximum value is given by $\tan \theta_m = k y_m$. We must calculate the angular wave number k. It is given by $k = \omega/v$, where v is the wave speed. The wave speed is given by $v = \sqrt{\tau/\mu}$, where τ is the tension in the rope and μ is the linear mass density of the rope. Using the data given,

$$v = \sqrt{\frac{90.0 \text{ N}}{0.120 \text{ kg/m}}} = 27.4 \text{ m/s}$$

and

$$k = \frac{754 \text{ rad/s}}{27.4 \text{ m/s}} = 27.5 \text{ m}^{-1}.$$

Thus,

$$\tan \theta_m = (27.5 \text{ m}^{-1})(5.00 \times 10^{-3} \text{ m}) = 0.138$$

and $\theta = 7.83°$. The maximum value of the transverse component of the tension in the string is

$$\tau_{trans} = (90.0 \text{ N}) \sin 7.83° = 12.3 \text{ N}.$$

We note that $\sin \theta$ is nearly the same as $\tan \theta$ because θ is small. We can approximate the maximum value of the transverse component of the tension by $\tau k y_m$.

(c) We consider the string at x. The transverse component of the tension pulling on it due to the string to the left is $-\tau(\partial y/\partial x) = -\tau k y_m \cos(kx - \omega t)$ and it reaches its maximum value when $\cos(kx - \omega t) = -1$. The wave speed is

$$u = \partial y/\partial t = -\omega y_m \cos(kx - \omega t)$$

and it also reaches its maximum value when $\cos(kx - \omega t) = -1$. The two quantities reach their maximum values at the same value of the phase. When $\cos(kx - \omega t) = -1$ the value of $\sin(kx - \omega t)$ is zero and the displacement of the string is $y = 0$.

(d) When the string at any point moves through a small displacement Δy, the tension does work $\Delta W = \tau_{trans} \Delta y$. The rate at which it does work is

$$P = \frac{\Delta W}{\Delta t} = \tau_{trans} \frac{\Delta y}{\Delta t} = \tau_{trans} u.$$

P has its maximum value when the transverse component τ_{trans} of the tension and the string speed u have their maximum values. Hence the maximum power is $(12.3 \text{ N})(3.77 \text{ m/s}) = 46.4 \text{ W}$.

(e) As shown above $y = 0$ when the transverse component of the tension and the string speed have their maximum values.

(f) The power transferred is zero when the transverse component of the tension and the string speed are zero.

(g) $P = 0$ when $\cos(kx - \omega t) = 0$ and $\sin(kx - \omega t) = \pm 1$ at that time. The string displacement is $y = \pm y_m = \pm 0.50$ cm.

68. We use Eq. 16-52 in interpreting the figure.

(a) Since $y' = 6.0$ mm when $\phi = 0$, then Eq. 16-52 can be used to determine $y_m = 3.0$ mm.

(b) We note that $y' = 0$ when the shift distance is 10 cm; this occurs because $\cos(\phi/2) = 0$ there $\Rightarrow \phi = \pi$ rad or ½ cycle. Since a full cycle corresponds to a distance of one full wavelength, this ½ cycle shift corresponds to a distance of $\lambda/2$. Therefore, $\lambda = 20$ cm $\Rightarrow k = 2\pi/\lambda = 31$ m^{-1}.

(c) Since $f = 120$ Hz, $\omega = 2\pi f = 754$ rad/s $\approx 7.5\times 10^2$ rad/s.

(d) The sign in front of ω is minus since the waves are traveling in the $+x$ direction.

The results may be summarized as $y = (3.0 \text{ mm}) \sin[(31.4 \text{ m}^{-1})x - (754 \text{ s}^{-1})t]]$ (this applies to each wave when they are in phase).

69. (a) We take the form of the displacement to be $y(x, t) = y_m \sin(kx - \omega t)$. The speed of a point on the cord is

$$u(x, t) = \partial y/\partial t = -\omega y_m \cos(kx - \omega t),$$

and its maximum value is $u_m = \omega y_m$. The wave speed, on the other hand, is given by $v = \lambda/T = \omega/k$. The ratio is

$$\frac{u_m}{v} = \frac{\omega y_m}{\omega/k} = k y_m = \frac{2\pi y_m}{\lambda}.$$

(b) The ratio of the speeds depends only on the ratio of the amplitude to the wavelength. Different waves on different cords have the same ratio of speeds if they have the same amplitude and wavelength, regardless of the wave speeds, linear densities of the cords, and the tensions in the cords.

70. We write the expression for the displacement in the form $y(x, t) = y_m \sin(kx - \omega t)$.

(a) The amplitude is $y_m = 2.0$ cm $= 0.020$ m, as given in the problem.

(b) The angular wave number k is $k = 2\pi/\lambda = 2\pi/(0.10 \text{ m}) = 63 \text{ m}^{-1}$

(c) The angular frequency is $\omega = 2\pi f = 2\pi(400 \text{ Hz}) = 2510$ rad/s $= 2.5\times 10^3$ rad/s.

(d) A minus sign is used before the ωt term in the argument of the sine function because the wave is traveling in the positive x direction.

Using the results above, the wave may be written as

$$y(x,t) = (2.00 \text{ cm}) \sin\left(\left(62.8 \text{ m}^{-1}\right)x - \left(2510 \text{ s}^{-1}\right)t\right).$$

(e) The (transverse) speed of a point on the cord is given by taking the derivative of y:

$$u(x,t) = \frac{\partial y}{\partial t} = -\omega y_m \cos(kx - \omega t)$$

which leads to a maximum speed of $u_m = \omega y_m = (2510 \text{ rad/s})(0.020 \text{ m}) = 50$ m/s.

(f) The speed of the wave is

$$v = \frac{\lambda}{T} = \frac{\omega}{k} = \frac{2510\,\text{rad/s}}{62.8\,\text{rad/m}} = 40\,\text{m/s}.$$

71. (a) The amplitude is $y_m = 1.00$ cm $= 0.0100$ m, as given in the problem.

(b) Since the frequency is $f = 550$ Hz, the angular frequency is $\omega = 2\pi f = 3.46\times10^3$ rad/s.

(c) The angular wave number is $k = \omega/v = (3.46\times10^3\text{ rad/s})/(330\text{ m/s}) = 10.5$ rad/m.

(d) Since the wave is traveling in the $-x$ direction, the sign in front of ω is plus and the argument of the trig function is $kx + \omega t$.

The results may be summarized as

$$y(x,t) = y_m \sin(kx + \omega t) = y_m \sin\left[2\pi f\left(\frac{x}{v} + t\right)\right]$$

$$= (0.010\,\text{m})\sin\left[2\pi(550\,\text{Hz})\left(\frac{x}{330\,\text{m/s}} + t\right)\right]$$

$$= (0.010\,\text{m})\sin[(10.5\text{ rad/s})\,x + (3.46\times10^3\text{ rad/s})t].$$

72. We orient one phasor along the x axis with length 3.0 mm and angle 0 and the other at 70° (in the first quadrant) with length 5.0 mm. Adding the components, we obtain

$$(3.0\text{ mm}) + (5.0\text{ mm})\cos(70°) = 4.71\text{ mm}\quad\text{along }x\text{ axis}$$
$$(5.0\text{ mm})\sin(70°) = 4.70\text{ mm}\quad\text{along }y\text{ axis}.$$

(a) Thus, amplitude of the resultant wave is $\sqrt{(4.71\text{ mm})^2 + (4.70\text{ mm})^2} = 6.7$ mm.

(b) And the angle (phase constant) is $\tan^{-1}(4.70/4.71) = 45°$.

73. (a) Using $v = f\lambda$, we obtain

$$f = \frac{240\,\text{m/s}}{3.2\,\text{m}} = 75\text{ Hz}.$$

(b) Since frequency is the reciprocal of the period, we find

$$T = \frac{1}{f} = \frac{1}{75\,\text{Hz}} = 0.0133\,\text{s} \approx 13\,\text{ms}.$$

74. By Eq. 16–66, the higher frequencies are integer multiples of the lowest (the fundamental).

(a) The frequency of the second harmonic is $f_2 = 2(440) = 880$ Hz.

(b) The frequency of the third harmonic is and $f_3 = 3(440) = 1320$ Hz.

75. We make use of Eq. 16–65 with $L = 120$ cm.

(a) The longest wavelength for waves traveling on the string if standing waves are to be set up is $\lambda_1 = 2L/1 = 240$ cm.

(b) The second longest wavelength for waves traveling on the string if standing waves are to be set up is $\lambda_2 = 2L/2 = 120$ cm.

(c) The third longest wavelength for waves traveling on the string if standing waves are to be set up is $\lambda_3 = 2L/3 = 80.0$ cm.

The three standing waves are shown below:

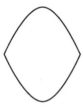 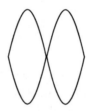 

76. (a) At $x = 2.3$ m and $t = 0.16$ s the displacement is

$$y(x,t) = 0.15\sin[(0.79)(2.3) - 13(0.16)]\,\text{m} = -0.039\,\text{m}.$$

(b) We choose $y_m = 0.15$ m, so that there would be nodes (where the wave amplitude is zero) in the string as a result.

(c) The second wave must be traveling with the same speed and frequency. This implies $k = 0.79$ m^{-1},

(d) and $\omega = 13$ rad/s.

(e) The wave must be traveling in $-x$ direction, implying a plus sign in front of ω.

Thus, its general form is $y'(x,t) = (0.15\,\text{m})\sin(0.79x + 13t)$.

(f) The displacement of the standing wave at $x = 2.3$ m and $t = 0.16$ s is

$$y(x,t) = -0.039\,\text{m} + (0.15\,\text{m})\sin[(0.79)(2.3) + 13(0.16)] = -0.14\,\text{m}.$$

77. (a) The wave speed is

$$v = \sqrt{\frac{\tau}{\mu}} = \sqrt{\frac{120\,\text{N}}{8.70 \times 10^{-3}\,\text{kg}/1.50\,\text{m}}} = 144\,\text{m/s}.$$

(b) For the one-loop standing wave we have $\lambda_1 = 2L = 2(1.50\,\text{m}) = 3.00\,\text{m}$.

(c) For the two-loop standing wave $\lambda_2 = L = 1.50\,\text{m}$.

(d) The frequency for the one-loop wave is $f_1 = v/\lambda_1 = (144\,\text{m/s})/(3.00\,\text{m}) = 48.0\,\text{Hz}$.

(e) The frequency for the two-loop wave is $f_2 = v/\lambda_2 = (144\,\text{m/s})/(1.50\,\text{m}) = 96.0\,\text{Hz}$.

78. We use $P = \frac{1}{2}\mu v \omega^2 y_m^2 \propto v f^2 \propto \sqrt{\tau} f^2$.

(a) If the tension is quadrupled, then $P_2 = P_1 \sqrt{\frac{\tau_2}{\tau_1}} = P_1 \sqrt{\frac{4\tau_1}{\tau_1}} = 2P_1$.

(b) If the frequency is halved, then $P_2 = P_1 \left(\frac{f_2}{f_1}\right)^2 = P_1 \left(\frac{f_1/2}{f_1}\right)^2 = \frac{1}{4}P_1$.

79. We use Eq. 16-2, Eq. 16-5, Eq. 16-9, Eq. 16-13, and take the derivative to obtain the transverse speed u.

(a) The amplitude is $y_m = 2.0\,\text{mm}$.

(b) Since $\omega = 600\,\text{rad/s}$, the frequency is found to be $f = 600/2\pi \approx 95\,\text{Hz}$.

(c) Since $k = 20\,\text{rad/m}$, the velocity of the wave is $v = \omega/k = 600/20 = 30\,\text{m/s}$ in the $+x$ direction.

(d) The wavelength is $\lambda = 2\pi/k \approx 0.31\,\text{m}$, or 31 cm.

(e) We obtain

$$u = \frac{dy}{dt} = -\omega y_m \cos(kx - \omega t) \Rightarrow u_m = \omega y_m$$

so that the maximum transverse speed is $u_m = (600)(2.0) = 1200\,\text{mm/s}$, or 1.2 m/s.

80. (a) Since the string has four loops its length must be two wavelengths. That is, $\lambda = L/2$, where λ is the wavelength and L is the length of the string. The wavelength is related to the frequency f and wave speed v by $\lambda = v/f$, so $L/2 = v/f$ and

$$L = 2v/f = 2(400 \text{ m/s})/(600 \text{ Hz}) = 1.3 \text{ m}.$$

(b) We write the expression for the string displacement in the form $y = y_m \sin(kx)\cos(\omega t)$, where y_m is the maximum displacement, k is the angular wave number, and ω is the angular frequency. The angular wave number is

$$k = 2\pi/\lambda = 2\pi f/v = 2\pi(600 \text{ Hz})/(400 \text{ m/s}) = 9.4 \text{m}^{-1}$$

and the angular frequency is

$$\omega = 2\pi f = 2\pi(600 \text{ Hz}) = 3800 \text{ rad/s}.$$

With $y_m = 2.0$ mm, the displacement is given by

$$y(x,t) = (2.0 \text{ mm})\sin[(9.4 \text{m}^{-1})x]\cos[(3800 \text{ s}^{-1})t].$$

81. To oscillate in four loops means $n = 4$ in Eq. 16-65 (treating both ends of the string as effectively "fixed"). Thus, $\lambda = 2(0.90 \text{ m})/4 = 0.45$ m. Therefore, the speed of the wave is $v = f\lambda = 27$ m/s. The mass-per-unit-length is

$$\mu = m/L = (0.044 \text{ kg})/(0.90 \text{ m}) = 0.049 \text{ kg/m}.$$

Thus, using Eq. 16-26, we obtain the tension:

$$\tau = v^2 \mu = (27 \text{ m/s})^2 (0.049 \text{ kg/m}) = 36 \text{ N}.$$

82. (a) This distance is determined by the longitudinal speed:

$$d_\ell = v_\ell t = (2000 \text{ m/s})(40 \times 10^{-6} \text{ s}) = 8.0 \times 10^{-2} \text{ m}.$$

(b) Assuming the acceleration is constant (justified by the near-straightness of the curve $a = 300/40 \times 10^{-6}$) we find the stopping distance d:

$$v^2 = v_o^2 + 2ad \Rightarrow d = \frac{(300)^2 (40 \times 10^{-6})}{2(300)}$$

which gives $d = 6.0 \times 10^{-3}$ m. This and the radius r form the legs of a right triangle (where r is opposite from $\theta = 60°$). Therefore,

$$\tan 60° = \frac{r}{d} \Rightarrow r = d\tan 60° = 1.0\times 10^{-2}\,\text{m}.$$

83. (a) Let the cross-sectional area of the wire be A and the density of steel be ρ. The tensile stress is given by τ/A where τ is the tension in the wire. Also, $\mu = \rho A$. Thus,

$$v_{max} = \sqrt{\frac{\tau_{max}}{\mu}} = \sqrt{\frac{\tau_{max}/A}{\rho}} = \sqrt{\frac{7.00\times 10^8\,\text{N/m}^2}{7800\,\text{kg/m}^3}} = 3.00\times 10^2\,\text{m/s}$$

(b) The result does not depend on the diameter of the wire.

84. (a) Let the displacements of the wave at (y,t) be $z(y,t)$. Then

$$z(y,t) = z_m \sin(ky - \omega t),$$

where $z_m = 3.0$ mm, $k = 60$ cm^{-1}, and $\omega = 2\pi/T = 2\pi/0.20$ s $= 10\pi$ s^{-1}. Thus

$$z(y,t) = (3.0\,\text{mm})\sin\left[(60\,\text{cm}^{-1})y - (10\pi\,\text{s}^{-1})t\right].$$

(b) The maximum transverse speed is $u_m = \omega z_m = (2\pi/0.20\,\text{s})(3.0\,\text{mm}) = 94$ mm/s.

85. (a) With length in centimeters and time in seconds, we have

$$u = \frac{dy}{dt} = -60\pi \cos\left(\frac{\pi x}{8} - 4\pi t\right).$$

Thus, when $x = 6$ and $t = \frac{1}{4}$, we obtain

$$u = -60\pi \cos\frac{-\pi}{4} = \frac{-60\pi}{\sqrt{2}} = -133$$

so that the *speed* there is 1.33 m/s.

(b) The numerical coefficient of the cosine in the expression for u is -60π. Thus, the maximum *speed* is 1.88 m/s.

(c) Taking another derivative,

$$a = \frac{du}{dt} = -240\pi^2 \sin\left(\frac{\pi x}{8} - 4\pi t\right)$$

so that when $x = 6$ and $t = \frac{1}{4}$ we obtain $a = -240\pi^2 \sin(-\pi/4)$ which yields $a = 16.7$ m/s^2.

(d) The numerical coefficient of the sine in the expression for a is $-240\pi^2$. Thus, the maximum acceleration is 23.7 m/s^2.

86. Repeating the steps of Eq. 16-47 → Eq. 16-53, but applying

$$\cos\alpha + \cos\beta = 2\cos\left(\frac{\alpha+\beta}{2}\right)\cos\left(\frac{\alpha-\beta}{2}\right)$$

(see Appendix E) instead of Eq. 16-50, we obtain $y' = [0.10\cos\pi x]\cos 4\pi t$, with SI units understood.

(a) For non-negative x, the smallest value to produce $\cos\pi x = 0$ is $x = 1/2$, so the answer is $x = 0.50$ m.

(b) Taking the derivative,

$$u' = \frac{dy'}{dt} = [0.10\cos\pi x](-4\pi\sin 4\pi t)$$

We observe that the last factor is zero when $t = 0, \frac{1}{4}, \frac{1}{2}, \frac{3}{4}, \ldots$. Thus, the value of the first time the particle at $x=0$ has zero velocity is $t = 0$.

(c) Using the result obtained in (b), the second time where the velocity at $x=0$ vanishes would be $t = 0.25$ s,

(d) and the third time is $t = 0.50$ s.

87. (a) From the frequency information, we find $\omega = 2\pi f = 10\pi$ rad/s. A point on the rope undergoing simple harmonic motion (discussed in Chapter 15) has maximum speed as it passes through its "middle" point, which is equal to $y_m\omega$. Thus,

$$5.0 \text{ m/s} = y_m\omega \implies y_m = 0.16 \text{ m}.$$

(b) Because of the oscillation being in the *fundamental* mode (as illustrated in Fig. 16-23(a) in the textbook), we have $\lambda = 2L = 4.0$ m. Therefore, the speed of waves along the rope is $v = f\lambda = 20$ m/s. Then, with $\mu = m/L = 0.60$ kg/m, Eq. 16-26 leads to

$$v = \sqrt{\frac{\tau}{\mu}} \implies \tau = \mu v^2 = 240 \text{ N} \approx 2.4\times 10^2 \text{ N}.$$

(c) We note that for the fundamental, $k = 2\pi/\lambda = \pi/L$, and we observe that the anti-node having zero displacement at $t = 0$ suggests the use of sine instead of cosine for the simple harmonic motion factor. Now, *if* the fundamental mode is the only one present (so the amplitude calculated in part (a) is indeed the amplitude of the fundamental wave pattern) then we have

$$y = (0.16 \text{ m}) \sin\left(\frac{\pi x}{2}\right) \sin(10\pi t) = (0.16 \text{ m}) \sin[(1.57 \text{ m}^{-1})x]\sin[(31.4 \text{ rad/s})t]$$

88. (a) The frequency is $f = 1/T = 1/4$ Hz, so $v = f\lambda = 5.0$ cm/s.

(b) We refer to the graph to see that the maximum transverse speed (which we will refer to as u_m) is 5.0 cm/s. Recalling from Ch. 11 the simple harmonic motion relation $u_m = y_m \omega = y_m 2\pi f$, we have

$$5.0 = y_m \left(2\pi \frac{1}{4}\right) \Rightarrow y_m = 3.2 \text{ cm.}$$

(c) As already noted, $f = 0.25$ Hz.

(d) Since $k = 2\pi/\lambda$, we have $k = 10\pi$ rad/m. There must be a sign difference between the t and x terms in the argument in order for the wave to travel to the right. The figure shows that at $x = 0$, the transverse velocity function is $0.050 \sin \pi t / 2$. Therefore, the function $u(x,t)$ is

$$u(x,t) = 0.050 \sin\left(\frac{\pi}{2}t - 10\pi x\right)$$

with lengths in meters and time in seconds. Integrating this with respect to time yields

$$y(x,t) = -\frac{2(0.050)}{\pi} \cos\left(\frac{\pi}{2}t - 10\pi x\right) + C$$

where C is an integration constant (which we will assume to be zero). The sketch of this function at $t = 2.0$ s for $0 \leq x \leq 0.20$ m is shown below.

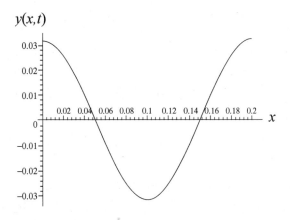

89. (a) The wave speed is

$$v = \sqrt{\frac{F}{\mu}} = \sqrt{\frac{k\Delta\ell}{m/(\ell+\Delta\ell)}} = \sqrt{\frac{k\Delta\ell(\ell+\Delta\ell)}{m}}.$$

(b) The time required is

$$t = \frac{2\pi(\ell+\Delta\ell)}{v} = \frac{2\pi(\ell+\Delta\ell)}{\sqrt{k\Delta\ell(\ell+\Delta\ell)/m}} = 2\pi\sqrt{\frac{m}{k}}\sqrt{1+\frac{\ell}{\Delta\ell}}.$$

Thus if $\ell/\Delta\ell \gg 1$, then $t \propto \sqrt{\ell/\Delta\ell} \propto 1/\sqrt{\Delta\ell}$; and if $\ell/\Delta\ell \ll 1$, then $t \approx 2\pi\sqrt{m/k} = $ const.

90. (a) The wave number for each wave is $k = 25.1/\text{m}$, which means $\lambda = 2\pi/k = 250.3$ mm. The angular frequency is $\omega = 440/\text{s}$; therefore, the period is $T = 2\pi/\omega = 14.3$ ms. We plot the superposition of the two waves $y = y_1 + y_2$ over the time interval $0 \le t \le 15$ ms. The first two graphs below show the oscillatory behavior at $x = 0$ (the graph on the left) and at $x = \lambda/8 \approx 31$ mm. The time unit is understood to be the millisecond and vertical axis (y) is in millimeters.

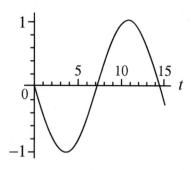
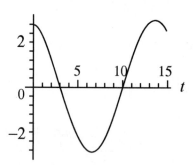

The following three graphs show the oscillation at $x = \lambda/4 = 62.6$ mm ≈ 63 mm (graph on the left), at $x = 3\lambda/8 \approx 94$ mm (middle graph), and at $x = \lambda/2 \approx 125$ mm.

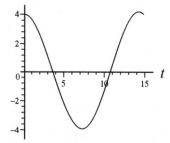
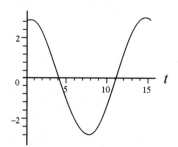
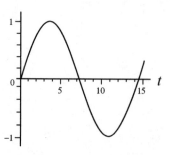

(b) We can think of wave y_1 as being made of two smaller waves going in the same direction, a wave y_{1a} of amplitude 1.50 mm (the same as y_2) and a wave y_{1b} of amplitude 1.00 mm. It is made clear in §16-12 that two equal-magnitude oppositely-moving waves form a standing wave pattern. Thus, waves y_{1a} and y_2 form a standing wave, which leaves

y_{1b} as the remaining traveling wave. Since the argument of y_{1b} involves the subtraction $kx - \omega t$, then y_{1b} travels in the $+x$ direction.

(c) If y_2 (which travels in the $-x$ direction, which for simplicity will be called "leftward") had the larger amplitude, then the system would consist of a standing wave plus a leftward moving wave. A simple way to obtain such a situation would be to interchange the amplitudes of the given waves.

(d) Examining carefully the vertical axes, the graphs above certainly suggest that the largest amplitude of oscillation is $y_{max} = 4.0$ mm and occurs at $x = \lambda/4 = 62.6$ mm.

(e) The smallest amplitude of oscillation is $y_{min} = 1.0$ mm and occurs at $x = 0$ and at $x = \lambda/2 = 125$ mm.

(f) The largest amplitude can be related to the amplitudes of y_1 and y_2 in a simple way: $y_{max} = y_{1m} + y_{2m}$, where $y_{1m} = 2.5$ mm and $y_{2m} = 1.5$ mm are the amplitudes of the original traveling waves.

(g) The smallest amplitudes is $y_{min} = y_{1m} - y_{2m}$, where $y_{1m} = 2.5$ mm and $y_{2m} = 1.5$ mm are the amplitudes of the original traveling waves.

91. Using Eq. 16-50, we have

$$y' = \left[0.60\cos\frac{\pi}{6}\right]\sin\left(5\pi x - 200\pi t + \frac{\pi}{6}\right)$$

with length in meters and time in seconds (see Eq. 16-55 for comparison).

(a) The amplitude is seen to be

$$0.60\cos\frac{\pi}{6} = 0.3\sqrt{3} = 0.52\,\text{m}.$$

(b) Since $k = 5\pi$ and $\omega = 200\pi$, then (using Eq. 16-12) $v = \frac{\omega}{k} = 40\,\text{m/s}$.

(c) $k = 2\pi/\lambda$ leads to $\lambda = 0.40$ m.

92. (a) For visible light

$$f_{min} = \frac{c}{\lambda_{max}} = \frac{3.0\times 10^8\,\text{m/s}}{700\times 10^{-9}\,\text{m}} = 4.3\times 10^{14}\,\text{Hz}$$

and

$$f_{max} = \frac{c}{\lambda_{min}} = \frac{3.0 \times 10^8 \text{ m/s}}{400 \times 10^{-9} \text{ m}} = 7.5 \times 10^{14} \text{ Hz}.$$

(b) For radio waves

$$\lambda_{min} = \frac{c}{\lambda_{max}} = \frac{3.0 \times 10^8 \text{ m/s}}{300 \times 10^6 \text{ Hz}} = 1.0 \text{ m}$$

and

$$\lambda_{max} = \frac{c}{\lambda_{min}} = \frac{3.0 \times 10^8 \text{ m/s}}{1.5 \times 10^6 \text{ Hz}} = 2.0 \times 10^2 \text{ m}.$$

(c) For X rays

$$f_{min} = \frac{c}{\lambda_{max}} = \frac{3.0 \times 10^8 \text{ m/s}}{5.0 \times 10^{-9} \text{ m}} = 6.0 \times 10^{16} \text{ Hz}$$

and

$$f_{max} = \frac{c}{\lambda_{min}} = \frac{3.0 \times 10^8 \text{ m/s}}{1.0 \times 10^{-11} \text{ m}} = 3.0 \times 10^{19} \text{ Hz}.$$

93. (a) Centimeters are to be understood as the length unit and seconds as the time unit. Making sure our (graphing) calculator is in radians mode, we find

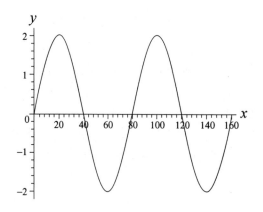

(b) The previous graph is at $t = 0$, and this next one is at $t = 0.050$ s.

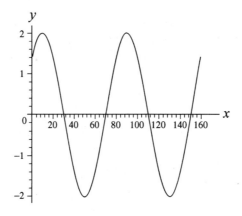

And the final one, shown below, is at $t = 0.010$ s.

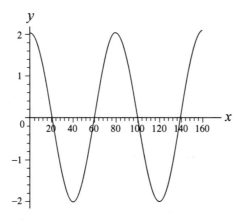

(c) The wave can be written as $y(x,t) = y_m \sin(kx + \omega t)$, where $v = \omega/k$ is the speed of propagation. From the problem statement, we see that $\omega = 2\pi/0.40 = 5\pi$ rad/s and $k = 2\pi/80 = \pi/40$ rad/cm. This yields $v = 2.0 \times 10^2$ cm/s $= 2.0$ m/s

(d) These graphs (as well as the discussion in the textbook) make it clear that the wave is traveling in the $-x$ direction.

Chapter 17

1. (a) When the speed is constant, we have $v = d/t$ where $v = 343$ m/s is assumed. Therefore, with $t = 15/2$ s being the time for sound to travel to the far wall we obtain $d = (343$ m/s$) \times (15/2$ s$)$ which yields a distance of 2.6 km.

(b) Just as the $\frac{1}{2}$ factor in part (a) was $1/(n + 1)$ for $n = 1$ reflection, so also can we write

$$d = (343 \text{ m/s}) \left(\frac{15 \text{ s}}{n+1} \right) \quad \Rightarrow \quad n = \frac{(343)(15)}{d} - 1$$

for multiple reflections (with d in meters). For $d = 25.7$ m, we find $n = 199 \approx 2.0 \times 10^2$.

2. The time it takes for a soldier in the rear end of the column to switch from the left to the right foot to stride forward is $t = 1$ min$/120 = 1/120$ min $= 0.50$ s. This is also the time for the sound of the music to reach from the musicians (who are in the front) to the rear end of the column. Thus the length of the column is

$$l = vt = (343 \text{ m/s})(0.50 \text{ s}) = 1.7 \times 10^2 \text{m}.$$

3. (a) The time for the sound to travel from the kicker to a spectator is given by d/v, where d is the distance and v is the speed of sound. The time for light to travel the same distance is given by d/c, where c is the speed of light. The delay between seeing and hearing the kick is $\Delta t = (d/v) - (d/c)$. The speed of light is so much greater than the speed of sound that the delay can be approximated by $\Delta t = d/v$. This means $d = v \, \Delta t$. The distance from the kicker to spectator A is

$$d_A = v \, \Delta t_A = (343 \text{ m/s})(0.23 \text{ s}) = 79 \text{ m}.$$

(b) The distance from the kicker to spectator B is $d_B = v \, \Delta t_B = (343$ m/s$)(0.12$ s$) = 41$ m.

(c) Lines from the kicker to each spectator and from one spectator to the other form a right triangle with the line joining the spectators as the hypotenuse, so the distance between the spectators is

$$D = \sqrt{d_A^2 + d_B^2} = \sqrt{(79 \text{ m})^2 + (41 \text{ m})^2} = 89 \text{ m}.$$

4. The density of oxygen gas is

$$\rho = \frac{0.0320\,\text{kg}}{0.0224\,\text{m}^3} = 1.43\,\text{kg/m}^3.$$

From $v = \sqrt{B/\rho}$ we find

$$B = v^2\rho = (317\,\text{m/s})^2\,(1.43\,\text{kg/m}^3) = 1.44\times 10^5\,\text{Pa}.$$

5. Let t_f be the time for the stone to fall to the water and t_s be the time for the sound of the splash to travel from the water to the top of the well. Then, the total time elapsed from dropping the stone to hearing the splash is $t = t_f + t_s$. If d is the depth of the well, then the kinematics of free fall gives

$$d = \frac{1}{2}gt_f^2 \;\Rightarrow\; t_f = \sqrt{2d/g}.$$

The sound travels at a constant speed v_s, so $d = v_s t_s$, or $t_s = d/v_s$. Thus the total time is $t = \sqrt{2d/g} + d/v_s$. This equation is to be solved for d. Rewrite it as $\sqrt{2d/g} = t - d/v_s$ and square both sides to obtain

$$2d/g = t^2 - 2(t/v_s)d + (1 + v_s^2)d^2.$$

Now multiply by gv_s^2 and rearrange to get

$$gd^2 - 2v_s(gt + v_s)d + gv_s^2 t^2 = 0.$$

This is a quadratic equation for d. Its solutions are

$$d = \frac{2v_s(gt+v_s) \pm \sqrt{4v_s^2(gt+v_s)^2 - 4g^2 v_s^2 t^2}}{2g}.$$

The physical solution must yield $d = 0$ for $t = 0$, so we take the solution with the negative sign in front of the square root. Once values are substituted the result $d = 40.7$ m is obtained.

6. Using Eqs. 16-13 and 17-3, the speed of sound can be expressed as

$$v = \lambda f = \sqrt{\frac{B}{\rho}},$$

where $B = -(dp/dV)/V$. Since V, λ and ρ are not changed appreciably, the frequency ratio becomes

$$\frac{f_s}{f_i} = \frac{v_s}{v_i} = \sqrt{\frac{B_s}{B_i}} = \sqrt{\frac{(dp/dV)_s}{(dp/dV)_i}}.$$

Thus, we have

$$\frac{(dV/dp)_s}{(dV/dp)_i} = \frac{B_i}{B_s} = \left(\frac{f_i}{f_s}\right)^2 = \left(\frac{1}{0.333}\right)^2 = 9.00.$$

7. If d is the distance from the location of the earthquake to the seismograph and v_s is the speed of the S waves then the time for these waves to reach the seismograph is $t_s = d/v_s$. Similarly, the time for P waves to reach the seismograph is $t_p = d/v_p$. The time delay is

$$\Delta t = (d/v_s) - (d/v_p) = d(v_p - v_s)/v_s v_p,$$

so

$$d = \frac{v_s v_p \Delta t}{(v_p - v_s)} = \frac{(4.5 \text{ km/s})(8.0 \text{ km/s})(3.0 \text{ min})(60 \text{ s/min})}{8.0 \text{ km/s} - 4.5 \text{ km/s}} = 1.9 \times 10^3 \text{ km}.$$

We note that values for the speeds were substituted as given, in km/s, but that the value for the time delay was converted from minutes to seconds.

8. Let ℓ be the length of the rod. Then the time of travel for sound in air (speed v_s) will be $t_s = \ell/v_s$. And the time of travel for compressional waves in the rod (speed v_r) will be $t_r = \ell/v_r$. In these terms, the problem tells us that

$$t_s - t_r = 0.12 \text{ s} = \ell\left(\frac{1}{v_s} - \frac{1}{v_r}\right).$$

Thus, with $v_s = 343$ m/s and $v_r = 15 v_s = 5145$ m/s, we find $\ell = 44$ m.

9. (a) Using $\lambda = v/f$, where v is the speed of sound in air and f is the frequency, we find

$$\lambda = \frac{343 \text{ m/s}}{4.50 \times 10^6 \text{ Hz}} = 7.62 \times 10^{-5} \text{ m}.$$

(b) Now, $\lambda = v/f$, where v is the speed of sound in tissue. The frequency is the same for air and tissue. Thus

$$\lambda = (1500 \text{ m/s})/(4.50 \times 10^6 \text{ Hz}) = 3.33 \times 10^{-4} \text{ m}.$$

10. (a) The amplitude of a sinusoidal wave is the numerical coefficient of the sine (or cosine) function: $p_m = 1.50$ Pa.

(b) We identify $k = 0.9\pi$ and $\omega = 315\pi$ (in SI units), which leads to $f = \omega/2\pi = 158$ Hz.

(c) We also obtain $\lambda = 2\pi/k = 2.22$ m.

(d) The speed of the wave is $v = \omega/k = 350$ m/s.

11. Without loss of generality we take $x = 0$, and let $t = 0$ be when $s = 0$. This means the phase is $\phi = -\pi/2$ and the function is $s = (6.0 \text{ nm})\sin(\omega t)$ at $x = 0$. Noting that $\omega = 3000$ rad/s, we note that at $t = \sin^{-1}(1/3)/\omega = 0.1133$ ms the displacement is $s = +2.0$ nm. Doubling that time (so that we consider the excursion from -2.0 nm to $+2.0$ nm) we conclude that the time required is $2(0.1133 \text{ ms}) = 0.23$ ms.

12. The key idea here is that the time delay Δt is due to the distance d that each wavefront must travel to reach your left ear (L) after it reaches your right ear (R).

(a) From the figure, we find $\Delta t = \dfrac{d}{v} = \dfrac{D\sin\theta}{v}$.

(b) Since the speed of sound in water is now v_w, with $\theta = 90°$, we have

$$\Delta t_w = \frac{D\sin 90°}{v_w} = \frac{D}{v_w}.$$

(c) The apparent angle can be found by substituting D/v_w for Δt:

$$\Delta t = \frac{D\sin\theta}{v} = \frac{D}{v_w}.$$

Solving for θ with $v_w = 1482$ m/s (see Table 17-1), we obtain

$$\theta = \sin^{-1}\left(\frac{v}{v_w}\right) = \sin^{-1}\left(\frac{343 \text{ m/s}}{1482 \text{ m/s}}\right) = \sin^{-1}(0.231) = 13°$$

13. (a) Consider a string of pulses returning to the stage. A pulse which came back just before the previous one has traveled an extra distance of $2w$, taking an extra amount of time $\Delta t = 2w/v$. The frequency of the pulse is therefore

$$f = \frac{1}{\Delta t} = \frac{v}{2w} = \frac{343 \text{ m/s}}{2(0.75 \text{ m})} = 2.3\times 10^2 \text{ Hz}.$$

(b) Since $f \propto 1/w$, the frequency would be higher if w were smaller.

14. (a) The period is $T = 2.0$ ms (or 0.0020 s) and the amplitude is $\Delta p_m = 8.0$ mPa (which is equivalent to 0.0080 N/m²). From Eq. 17-15 we get

$$s_m = \frac{\Delta p_m}{v\rho\omega} = \frac{\Delta p_m}{v\rho(2\pi/T)} = 6.1 \times 10^{-9} \text{ m}.$$

where $\rho = 1.21$ kg/m³ and $v = 343$ m/s.

(b) The angular wave number is $k = \omega/v = 2\pi/vT = 9.2$ rad/m.

(c) The angular frequency is $\omega = 2\pi/T = 3142$ rad/s $\approx 3.1 \times 10^3$ rad/s.

The results may be summarized as $s(x, t) = (6.1 \text{ nm}) \cos[(9.2 \text{ m}^{-1})x - (3.1 \times 10^3 \text{ s}^{-1})t]$.

(d) Using similar reasoning, but with the new values for density ($\rho' = 1.35$ kg/m³) and speed ($v' = 320$ m/s), we obtain

$$s_m = \frac{\Delta p_m}{v'\rho'\omega} = \frac{\Delta p_m}{v'\rho'(2\pi/T)} = 5.9 \times 10^{-9} \text{ m}.$$

(e) The angular wave number is $k = \omega/v' = 2\pi/v'T = 9.8$ rad/m.

(f) The angular frequency is $\omega = 2\pi/T = 3142$ rad/s $\approx 3.1 \times 10^3$ rad/s.

The new displacement function is $s(x, t) = (5.9 \text{ nm}) \cos[(9.8 \text{ m}^{-1})x - (3.1 \times 10^3 \text{ s}^{-1})t]$.

15. The problem says "At one instant.." and we choose that instant (without loss of generality) to be $t = 0$. Thus, the displacement of "air molecule A" at that instant is

$$s_A = +s_m = s_m \cos(kx_A - \omega t + \phi)|_{t=0} = s_m \cos(kx_A + \phi),$$

where $x_A = 2.00$ m. Regarding "air molecule B" we have

$$s_B = +\frac{1}{3}s_m = s_m \cos(kx_B - \omega t + \phi)|_{t=0} = s_m \cos(kx_B + \phi).$$

These statements lead to the following conditions:

$$kx_A + \phi = 0$$
$$kx_B + \phi = \cos^{-1}(1/3) = 1.231$$

where $x_B = 2.07$ m. Subtracting these equations leads to

$$k(x_B - x_A) = 1.231 \quad \Rightarrow \quad k = 17.6 \text{ rad/m}.$$

Using the fact that $k = 2\pi/\lambda$ we find $\lambda = 0.357$ m, which means

$$f = v/\lambda = 343/0.357 = 960 \text{ Hz}.$$

Another way to complete this problem (once k is found) is to use $kv = \omega$ and then the fact that $\omega = 2\pi f$.

16. Let the separation between the point and the two sources (labeled 1 and 2) be x_1 and x_2, respectively. Then the phase difference is

$$\Delta\phi = \phi_1 - \phi_2 = 2\pi\left(\frac{x_1}{\lambda} + ft\right) - 2\pi\left(\frac{x_2}{\lambda} + ft\right) = \frac{2\pi(x_1 - x_2)}{\lambda} = \frac{2\pi(4.40\,\text{m} - 4.00\,\text{m})}{(330\,\text{m/s})/540\,\text{Hz}} = 4.12\,\text{rad}.$$

17. (a) The problem is asking at how many angles will there be "loud" resultant waves, and at how many will there be "quiet" ones? We note that at all points (at large distance from the origin) along the x axis there will be quiet ones; one way to see this is to note that the path-length difference (for the waves traveling from their respective sources) divided by wavelength gives the (dimensionless) value 3.5, implying a half-wavelength (180°) phase difference (destructive interference) between the waves. To distinguish the destructive interference along the $+x$ axis from the destructive interference along the $-x$ axis, we label one with +3.5 and the other –3.5. This labeling is useful in that it suggests that the complete enumeration of the quiet directions in the upper-half plane (including the x axis) is: –3.5, –2.5, –1.5, –0.5, +0.5, +1.5, +2.5, +3.5. Similarly, the complete enumeration of the loud directions in the upper-half plane is: –3, –2, –1, 0, +1, +2, +3. Counting also the "other" –3, –2, –1, 0, +1, +2, +3 values for the *lower*-half plane, then we conclude there are a total of $7 + 7 = 14$ "loud" directions.

(b) The discussion about the "quiet" directions was started in part (a). The number of values in the list: –3.5, –2.5, –1.5, –0.5, +0.5, +1.5, +2.5, +3.5 along with –2.5, –1.5, –0.5, +0.5, +1.5, +2.5 (for the lower-half plane) is 14. There are 14 "quiet" directions.

18. At the location of the detector, the phase difference between the wave which traveled straight down the tube and the other one which took the semi-circular detour is

$$\Delta\phi = k\Delta d = \frac{2\pi}{\lambda}(\pi r - 2r).$$

For $r = r_{\min}$ we have $\Delta\phi = \pi$, which is the smallest phase difference for a destructive interference to occur. Thus,

$$r_{\min} = \frac{\lambda}{2(\pi - 2)} = \frac{40.0\,\text{cm}}{2(\pi - 2)} = 17.5\,\text{cm}.$$

19. Let L_1 be the distance from the closer speaker to the listener. The distance from the other speaker to the listener is $L_2 = \sqrt{L_1^2 + d^2}$, where d is the distance between the speakers. The phase difference at the listener is $\phi = 2\pi(L_2 - L_1)/\lambda$, where λ is the wavelength.

For a minimum in intensity at the listener, $\phi = (2n + 1)\pi$, where n is an integer. Thus,

$$\lambda = 2(L_2 - L_1)/(2n + 1).$$

The frequency is

$$f = \frac{v}{\lambda} = \frac{(2n+1)v}{2\left(\sqrt{L_1^2 + d^2} - L_1\right)} = \frac{(2n+1)(343\,\text{m/s})}{2\left(\sqrt{(3.75\,\text{m})^2 + (2.00\,\text{m})^2} - 3.75\,\text{m}\right)} = (2n+1)(343\,\text{Hz}).$$

Now $20{,}000/343 = 58.3$, so $2n + 1$ must range from 0 to 57 for the frequency to be in the audible range. This means n ranges from 0 to 28.

(a) The lowest frequency that gives minimum signal is ($n = 0$) $f_{\text{min},1} = 343$ Hz.

(b) The second lowest frequency is ($n = 1$) $f_{\text{min},2} = [2(1)+1]343\,\text{Hz} = 1029\,\text{Hz} = 3f_{\text{min},1}$. Thus, the factor is 3.

(c) The third lowest frequency is ($n=2$) $f_{\text{min},3} = [2(2)+1]343\,\text{Hz} = 1715\,\text{Hz} = 5f_{\text{min},1}$. Thus, the factor is 5.

For a maximum in intensity at the listener, $\phi = 2n\pi$, where n is any positive integer. Thus $\lambda = (1/n)\left(\sqrt{L_1^2 + d^2} - L_1\right)$ and

$$f = \frac{v}{\lambda} = \frac{nv}{\sqrt{L_1^2 + d^2} - L_1} = \frac{n(343\,\text{m/s})}{\sqrt{(3.75\,\text{m})^2 + (2.00\,\text{m})^2} - 3.75\,\text{m}} = n(686\,\text{Hz}).$$

Since $20{,}000/686 = 29.2$, n must be in the range from 1 to 29 for the frequency to be audible.

(d) The lowest frequency that gives maximum signal is ($n = 1$) $f_{\text{max},1} = 686$ Hz.

(e) The second lowest frequency is ($n = 2$) $f_{\text{max},2} = 2(686\,\text{Hz}) = 1372\,\text{Hz} = 2f_{\text{max},1}$. Thus, the factor is 2.

(f) The third lowest frequency is ($n = 3$) $f_{\text{max},3} = 3(686\,\text{Hz}) = 2058\,\text{Hz} = 3f_{\text{max},1}$. Thus, the factor is 3.

20. (a) To be out of phase (and thus result in destructive interference if they superpose) means their path difference must be $\lambda/2$ (or $3\lambda/2$ or $5\lambda/2$ or ...). Here we see their path difference is L, so we must have (in the least possibility) $L = \lambda/2$, or $q = L/\lambda = 0.5$.

(b) As noted above, the next possibility is $L = 3\lambda/2$, or $q = L/\lambda = 1.5$.

21. Building on the theory developed in §17 – 5, we set $\Delta L/\lambda = n - 1/2$, $n = 1, 2, ...$ in order to have destructive interference. Since $v = f\lambda$, we can write this in terms of frequency:

$$f_{min,n} = \frac{(2n-1)v}{2\Delta L} = (n-1/2)(286 \text{ Hz})$$

where we have used $v = 343$ m/s (note the remarks made in the textbook at the beginning of the exercises and problems section) and $\Delta L = (19.5 - 18.3)$ m $= 1.2$ m.

(a) The lowest frequency that gives destructive interference is ($n = 1$)

$$f_{min,1} = (1-1/2)(286 \text{ Hz}) = 143 \text{ Hz}.$$

(b) The second lowest frequency that gives destructive interference is ($n = 2$)

$$f_{min,2} = (2-1/2)(286 \text{ Hz}) = 429 \text{ Hz} = 3(143 \text{ Hz}) = 3f_{min,1}.$$

So the factor is 3.

(c) The third lowest frequency that gives destructive interference is ($n = 3$)

$$f_{min,3} = (3-1/2)(286 \text{ Hz}) = 715 \text{ Hz} = 5(143 \text{ Hz}) = 5f_{min,1}.$$

So the factor is 5.

Now we set $\Delta L/\lambda = \frac{1}{2}$ (even numbers) — which can be written more simply as "(all integers $n = 1, 2, ...$)" — in order to establish constructive interference. Thus,

$$f_{max,n} = \frac{nv}{\Delta L} = n(286 \text{ Hz}).$$

(d) The lowest frequency that gives constructive interference is ($n = 1$) $f_{max,1} = (286 \text{ Hz})$.

(e) The second lowest frequency that gives constructive interference is ($n = 2$)

$$f_{\max,2} = 2(286 \text{ Hz}) = 572 \text{ Hz} = 2f_{\max,1}.$$

Thus, the factor is 2.

(f) The third lowest frequency that gives constructive interference is ($n = 3$)

$$f_{\max,3} = 3(286 \text{ Hz}) = 858 \text{ Hz} = 3f_{\max,1}.$$

Thus, the factor is 3.

22. (a) The problem indicates that we should ignore the decrease in sound amplitude which means that all waves passing through point P have equal amplitude. Their superposition at P if $d = \lambda/4$ results in a net effect of zero there since there are four sources (so the first and third are $\lambda/2$ apart and thus interfere destructively; similarly for the second and fourth sources).

(b) Their superposition at P if $d = \lambda/2$ also results in a net effect of zero there since there are an even number of sources (so the first and second being $\lambda/2$ apart will interfere destructively; similarly for the waves from the third and fourth sources).

(c) If $d = \lambda$ then the waves from the first and second sources will arrive at P in phase; similar observations apply to the second and third, and to the third and fourth sources. Thus, four waves interfere constructively there with net amplitude equal to $4s_m$.

23. (a) If point P is infinitely far away, then the small distance d between the two sources is of no consequence (they seem effectively to be the same distance away from P). Thus, there is no perceived phase difference.

(b) Since the sources oscillate in phase, then the situation described in part (a) produces fully constructive interference.

(c) For finite values of x, the difference in source positions becomes significant. The path lengths for waves to travel from S_1 and S_2 become now different. We interpret the question as asking for the behavior of the absolute value of the phase difference $|\Delta\phi|$, in which case any change from zero (the answer for part (a)) is certainly an increase.

The path length difference for waves traveling from S_1 and S_2 is

$$\Delta \ell = \sqrt{d^2 + x^2} - x \quad \text{for } x > 0.$$

The phase difference in "cycles" (in absolute value) is therefore

$$|\Delta\phi| = \frac{\Delta\ell}{\lambda} = \frac{\sqrt{d^2 + x^2} - x}{\lambda}.$$

Thus, in terms of λ, the phase difference is identical to the path length difference: $|\Delta\phi| = \Delta\ell > 0$. Consider $\Delta\ell = \lambda/2$. Then $\sqrt{d^2 + x^2} = x + \lambda/2$. Squaring both sides, rearranging, and solving, we find

$$x = \frac{d^2}{\lambda} - \frac{\lambda}{4}.$$

In general, if $\Delta\ell = \xi\lambda$ for some multiplier $\xi > 0$, we find

$$x = \frac{d^2}{2\xi\lambda} - \frac{1}{2}\xi\lambda = \frac{64.0}{\xi} - \xi$$

where we have used $d = 16.0$ m and $\lambda = 2.00$ m.

(d) For $\Delta\ell = 0.50\lambda$, or $\xi = 0.50$, we have $x = (64.0/0.50 - 0.50)$ m $= 127.5$ m ≈ 128 m.

(e) For $\Delta\ell = 1.00\lambda$, or $\xi = 1.00$, we have $x = (64.0/1.00 - 1.00)$ m $= 63.0$ m.

(f) For $\Delta\ell = 1.50\lambda$, or $\xi = 1.50$, we have $x = (64.0/1.50 - 1.50)$ m $= 41.2$ m.

Note that since whole cycle phase differences are equivalent (as far as the wave superposition goes) to zero phase difference, then the $\xi = 1, 2$ cases give constructive interference. A shift of a half-cycle brings "troughs" of one wave in superposition with "crests" of the other, thereby canceling the waves; therefore, the $\xi = \frac{1}{2}, \frac{3}{2}, \frac{5}{2}$ cases produce destructive interference.

24. (a) Since intensity is power divided by area, and for an isotropic source the area may be written $A = 4\pi r^2$ (the area of a sphere), then we have

$$I = \frac{P}{A} = \frac{1.0\,\text{W}}{4\pi(1.0\,\text{m})^2} = 0.080\,\text{W/m}^2.$$

(b) This calculation may be done exactly as shown in part (a) (but with $r = 2.5$ m instead of $r = 1.0$ m), or it may be done by setting up a ratio. We illustrate the latter approach. Thus,

$$\frac{I'}{I} = \frac{P/4\pi(r')^2}{P/4\pi r^2} = \left(\frac{r}{r'}\right)^2$$

leads to $I' = (0.080\,\text{W/m}^2)(1.0/2.5)^2 = 0.013\,\text{W/m}^2$.

25. The intensity is the rate of energy flow per unit area perpendicular to the flow. The rate at which energy flow across every sphere centered at the source is the same, regardless of the sphere radius, and is the same as the power output of the source. If P is

the power output and I is the intensity a distance r from the source, then $P = IA = 4\pi r^2 I$, where $A (= 4\pi r^2)$ is the surface area of a sphere of radius r. Thus

$$P = 4\pi(2.50 \text{ m})^2 (1.91 \times 10^{-4} \text{ W/m}^2) = 1.50 \times 10^{-2} \text{ W}.$$

26. Sample Problem 17-5 shows that a decibel difference $\Delta\beta$ is directly related to an intensity ratio (which we write as $\mathcal{R} = I'/I$). Thus,

$$\Delta\beta = 10\log(\mathcal{R}) \Rightarrow \mathcal{R} = 10^{\Delta\beta/10} = 10^{0.1} = 1.26.$$

27. The intensity is given by $I = \frac{1}{2}\rho v \omega^2 s_m^2$, where ρ is the density of air, v is the speed of sound in air, ω is the angular frequency, and s_m is the displacement amplitude for the sound wave. Replace ω with $2\pi f$ and solve for s_m:

$$s_m = \sqrt{\frac{I}{2\pi^2 \rho v f^2}} = \sqrt{\frac{1.00 \times 10^{-6} \text{ W/m}^2}{2\pi^2 (1.21 \text{ kg/m}^3)(343 \text{ m/s})(300 \text{ Hz})^2}} = 3.68 \times 10^{-8} \text{ m}.$$

28. (a) The intensity is given by $I = P/4\pi r^2$ when the source is "point-like." Therefore, at $r = 3.00$ m,

$$I = \frac{1.00 \times 10^{-6} \text{ W}}{4\pi(3.00 \text{ m})^2} = 8.84 \times 10^{-9} \text{ W/m}^2.$$

(b) The sound level there is

$$\beta = 10 \log\left(\frac{8.84 \times 10^{-9} \text{ W/m}^2}{1.00 \times 10^{-12} \text{ W/m}^2}\right) = 39.5 \text{ dB}.$$

29. (a) Let I_1 be the original intensity and I_2 be the final intensity. The original sound level is $\beta_1 = (10 \text{ dB}) \log(I_1/I_0)$ and the final sound level is $\beta_2 = (10 \text{ dB}) \log(I_2/I_0)$, where I_0 is the reference intensity. Since $\beta_2 = \beta_1 + 30$ dB which yields

$$(10 \text{ dB}) \log(I_2/I_0) = (10 \text{ dB}) \log(I_1/I_0) + 30 \text{ dB},$$

or

$$(10 \text{ dB}) \log(I_2/I_0) - (10 \text{ dB}) \log(I_1/I_0) = 30 \text{ dB}.$$

Divide by 10 dB and use $\log(I_2/I_0) - \log(I_1/I_0) = \log(I_2/I_1)$ to obtain $\log(I_2/I_1) = 3$. Now use each side as an exponent of 10 and recognize that $10^{\log(I_2/I_1)} = I_2/I_1$. The result is $I_2/I_1 = 10^3$. The intensity is increased by a factor of 1.0×10^3.

(b) The pressure amplitude is proportional to the square root of the intensity so it is increased by a factor of $\sqrt{1000} \approx 32$.

30. (a) Eq. 17-29 gives the relation between sound level β and intensity I, namely

$$I = I_0 10^{(\beta/10\text{dB})} = (10^{-12}\,\text{W/m}^2)10^{(\beta/10\text{dB})} = 10^{-12+(\beta/10\text{dB})}\,\text{W/m}^2$$

Thus we find that for a $\beta = 70$ dB level we have a high intensity value of $I_{\text{high}} = 10\ \mu\text{W/m}^2$.

(b) Similarly, for $\beta = 50$ dB level we have a low intensity value of $I_{\text{low}} = 0.10\ \mu\text{W/m}^2$.

(c) Eq. 17-27 gives the relation between the displacement amplitude and I. Using the values for density and wave speed, we find $s_m = 70$ nm for the high intensity case.

(d) Similarly, for the low intensity case we have $s_m = 7.0$ nm.

We note that although the intensities differed by a factor of 100, the amplitudes differed by only a factor of 10.

31. We use $\beta = 10\log(I/I_0)$ with $I_0 = 1 \times 10^{-12}$ W/m^2 and Eq. 17–27 with $\omega = 2\pi f = 2\pi(260\text{ Hz})$, $v = 343$ m/s and $\rho = 1.21$ kg/m^3.

$$I = I_0\left(10^{8.5}\right) = \frac{1}{2}\rho v(2\pi f)^2 s_m^2 \quad \Rightarrow \quad s_m = 7.6\times 10^{-7}\,\text{m} = 0.76\ \mu\text{m}.$$

32. (a) Since $\omega = 2\pi f$, Eq. 17-15 leads to

$$\Delta p_m = v\rho(2\pi f)s_m \quad \Rightarrow \quad s_m = \frac{1.13\times 10^{-3}\,\text{Pa}}{2\pi(1665\,\text{Hz})(343\,\text{m/s})(1.21\,\text{kg/m}^3)}$$

which yields $s_m = 0.26$ nm. The nano prefix represents 10^{-9}. We use the speed of sound and air density values given at the beginning of the exercises and problems section in the textbook.

(b) We can plug into Eq. 17–27 or into its equivalent form, rewritten in terms of the pressure amplitude:

$$I = \frac{1}{2}\frac{(\Delta p_m)^2}{\rho v} = \frac{1}{2}\frac{(1.13\times 10^{-3}\,\text{Pa})^2}{(1.21\,\text{kg/m}^3)(343\,\text{m/s})} = 1.5\ \text{nW/m}^2.$$

33. We use $\beta = 10\log(I/I_0)$ with $I_0 = 1 \times 10^{-12}$ W/m^2 and $I = P/4\pi r^2$ (an assumption we are asked to make in the problem). We estimate $r \approx 0.3$ m (distance from knuckle to ear) and find

$$P \approx 4\pi(0.3\,\text{m})^2(1\times 10^{-12}\,\text{W/m}^2)10^{6.2} = 2\times 10^{-6}\,\text{W} = 2\ \mu\text{W}.$$

34. The difference in sound level is given by Eq. 17-37:

$$\Delta\beta = \beta_f - \beta_i = (10 \text{ db})\log\left(\frac{I_f}{I_i}\right).$$

Thus, if $\Delta\beta = 5.0$ db, then $\log(I_f/I_i) = 1/2$, which implies that $I_f = \sqrt{10}I_i$. On the other hand, the intensity at a distance r from the source is $I = \frac{P}{4\pi r^2}$, where P is the power of the source. A fixed P implies that $I_i r_i^2 = I_f r_f^2$. Thus, with $r_i = 1.2$ m, we obtain

$$r_f = \left(\frac{I_i}{I_f}\right)^{1/2} r_i = \left(\frac{1}{10}\right)^{1/4}(1.2 \text{ m}) = 0.67 \text{ m}.$$

35. (a) The intensity is

$$I = \frac{P}{4\pi r^2} = \frac{30.0 \text{ W}}{(4\pi)(200 \text{ m})^2} = 5.97\times 10^{-5} \text{ W/m}^2.$$

(b) Let A (= 0.750 cm^2) be the cross-sectional area of the microphone. Then the power intercepted by the microphone is

$$P' = IA = 0 = (6.0\times 10^{-5} \text{ W/m}^2)(0.750 \text{ cm}^2)(10^{-4} \text{ m}^2/\text{cm}^2) = 4.48\times 10^{-9} \text{ W}.$$

36. Combining Eqs.17-28 and 17-29 we have $\beta = 10 \log\left(\frac{P}{I_o 4\pi r^2}\right)$. Taking differences (for sounds A and B) we find

$$\Delta\beta = 10 \log\left(\frac{P_A}{I_o 4\pi r^2}\right) - 10 \log\left(\frac{P_B}{I_o 4\pi r^2}\right) = 10 \log\left(\frac{P_A}{P_B}\right)$$

using well-known properties of logarithms. Thus, we see that $\Delta\beta$ is independent of r and can be evaluated anywhere.

(a) We can solve the above relation (once we know $\Delta\beta = 5.0$) for the ratio of powers; we find $P_A/P_B \approx 3.2$.

(b) At $r = 1000$ m it is easily seen (in the graph) that $\Delta\beta = 5.0$ dB. This is the same $\Delta\beta$ we expect to find, then, at $r = 10$ m.

37. (a) As discussed on page 408, the average potential energy transport rate is the same as that of the kinetic energy. This implies that the (average) rate for the total energy is

$$\left(\frac{dE}{dt}\right)_{avg} = 2\left(\frac{dK}{dt}\right)_{avg} = 2\left(\tfrac{1}{4}\rho A v \omega^2 s_m^2\right)$$

using Eq. 17-44. In this equation, we substitute $\rho = 1.21$ kg/m^3, $A = \pi r^2 = \pi(0.020\text{ m})^2$, $v = 343$ m/s, $\omega = 3000$ rad/s, $s_m = 12 \times 10^{-9}$ m, and obtain the answer 3.4×10^{-10} W.

(b) The second string is in a separate tube, so there is no question about the waves superposing. The total rate of energy, then, is just the addition of the two: $2(3.4 \times 10^{-10}$ W$) = 6.8 \times 10^{-10}$ W.

(c) Now we *do* have superposition, with $\phi = 0$, so the resultant amplitude is twice that of the individual wave which leads to the energy transport rate being four times that of part (a). We obtain $4(3.4 \times 10^{-10}$ W$) = 1.4 \times 10^{-9}$ W.

(d) In this case $\phi = 0.4\pi$, which means (using Eq. 17-39)

$$s_m' = 2 s_m \cos(\phi/2) = 1.618 s_m.$$

This means the energy transport rate is $(1.618)^2 = 2.618$ times that of part (a). We obtain $2.618(3.4 \times 10^{-10}$ W$) = 8.8 \times 10^{-10}$ W.

(e) The situation is as shown in Fig. 17-14(b). The answer is zero.

38. (a) Using Eq. 17–39 with $v = 343$ m/s and $n = 1$, we find $f = nv/2L = 86$ Hz for the fundamental frequency in a nasal passage of length $L = 2.0$ m (subject to various assumptions about the nature of the passage as a "bent tube open at both ends").

(b) The sound would be perceptible as *sound* (as opposed to just a general vibration) of very low frequency.

(c) Smaller L implies larger f by the formula cited above. Thus, the female's sound is of higher pitch (frequency).

39. (a) From Eq. 17–53, we have

$$f = \frac{nv}{2L} = \frac{(1)(250\text{ m/s})}{2(0.150\text{ m})} = 833\text{ Hz}.$$

(b) The frequency of the wave on the string is the same as the frequency of the sound wave it produces during its vibration. Consequently, the wavelength in air is

$$\lambda = \frac{v_{\text{sound}}}{f} = \frac{348\text{ m/s}}{833\text{ Hz}} = 0.418\text{ m}.$$

40. The distance between nodes referred to in the problem means that $\lambda/2 = 3.8$ cm, or $\lambda = 0.076$ m. Therefore, the frequency is

$$f = v/\lambda = (1500 \text{ m/s})/(0.076 \text{ m}) \approx 20 \times 10^3 \text{ Hz}.$$

41. (a) We note that 1.2 = 6/5. This suggests that both even and odd harmonics are present, which means the pipe is open at both ends (see Eq. 17-39).

(b) Here we observe 1.4 = 7/5. This suggests that only odd harmonics are present, which means the pipe is open at only one end (see Eq. 17-41).

42. At the beginning of the exercises and problems section in the textbook, we are told to assume $v_{\text{sound}} = 343$ m/s unless told otherwise. The second harmonic of pipe A is found from Eq. 17–39 with $n = 2$ and $L = L_A$, and the third harmonic of pipe B is found from Eq. 17–41 with $n = 3$ and $L = L_B$. Since these frequencies are equal, we have

$$\frac{2v_{\text{sound}}}{2L_A} = \frac{3v_{\text{sound}}}{4L_B} \Rightarrow L_B = \frac{3}{4}L_A.$$

(a) Since the fundamental frequency for pipe A is 300 Hz, we immediately know that the second harmonic has $f = 2(300 \text{ Hz}) = 600$ Hz. Using this, Eq. 17–39 gives

$$L_A = (2)(343 \text{ m/s})/2(600 \text{ s}^{-1}) = 0.572 \text{ m}.$$

(b) The length of pipe B is $L_B = \frac{3}{4}L_A = 0.429$ m.

43. (a) When the string (fixed at both ends) is vibrating at its lowest resonant frequency, exactly one-half of a wavelength fits between the ends. Thus, $\lambda = 2L$. We obtain

$$v = f\lambda = 2Lf = 2(0.220 \text{ m})(920 \text{ Hz}) = 405 \text{ m/s}.$$

(b) The wave speed is given by $v = \sqrt{\tau/\mu}$, where τ is the tension in the string and μ is the linear mass density of the string. If M is the mass of the (uniform) string, then $\mu = M/L$. Thus,

$$\tau = \mu v^2 = (M/L)v^2 = [(800 \times 10^{-6} \text{ kg})/(0.220 \text{ m})](405 \text{ m/s})^2 = 596 \text{ N}.$$

(c) The wavelength is $\lambda = 2L = 2(0.220 \text{ m}) = 0.440$ m.

(d) The frequency of the sound wave in air is the same as the frequency of oscillation of the string. The wavelength is different because the wave speed is different. If v_a is the speed of sound in air the wavelength in air is

$$\lambda_a = v_a/f = (343 \text{ m/s})/(920 \text{ Hz}) = 0.373 \text{ m}.$$

44. The frequency is $f = 686$ Hz and the speed of sound is $v_{sound} = 343$ m/s. If L is the length of the air-column, then using Eq. 17–41, the water height is (in unit of meters)

$$h = 1.00 - L = 1.00 - \frac{nv}{4f} = 1.00 - \frac{n(343)}{4(686)} = (1.00 - 0.125n) \text{ m}$$

where $n = 1, 3, 5, \ldots$ with only one end closed.

(a) There are 4 values of n ($n = 1, 3, 5, 7$) which satisfies $h > 0$.

(b) The smallest water height for resonance to occur corresponds to $n = 7$ with $h = 0.125$ m.

(c) The second smallest water height corresponds to $n = 5$ with $h = 0.375$ m.

45. (a) Since the pipe is open at both ends there are displacement antinodes at both ends and an integer number of half-wavelengths fit into the length of the pipe. If L is the pipe length and λ is the wavelength then $\lambda = 2L/n$, where n is an integer. If v is the speed of sound then the resonant frequencies are given by $f = v/\lambda = nv/2L$. Now $L = 0.457$ m, so

$$f = n(344 \text{ m/s})/2(0.457 \text{ m}) = 376.4n \text{ Hz}.$$

To find the resonant frequencies that lie between 1000 Hz and 2000 Hz, first set $f = 1000$ Hz and solve for n, then set $f = 2000$ Hz and again solve for n. The results are 2.66 and 5.32, which imply that $n = 3, 4$, and 5 are the appropriate values of n. Thus, there are 3 frequencies.

(b) The lowest frequency at which resonance occurs is ($n = 3$) $f = 3(376.4 \text{ Hz}) = 1129$ Hz.

(c) The second lowest frequency at which resonance occurs is ($n = 4$)

$$f = 4(376.4 \text{ Hz}) = 1506 \text{ Hz}.$$

46. (a) Since the difference between consecutive harmonics is equal to the fundamental frequency (see section 17-6) then $f_1 = (390 - 325)$ Hz $= 65$ Hz. The next harmonic after 195 Hz is therefore $(195 + 65)$ Hz $= 260$ Hz.

(b) Since $f_n = nf_1$ then $n = 260/65 = 4$.

(c) Only *odd* harmonics are present in tube B so the difference between consecutive harmonics is equal to *twice* the fundamental frequency in this case (consider taking differences of Eq. 17-41 for various values of n). Therefore,

$$f_1 = \tfrac{1}{2}(1320 - 1080) \text{ Hz} = 120 \text{ Hz}.$$

The next harmonic after 600 Hz is consequently [600 + 2(120)] Hz = 840 Hz.

(d) Since $f_n = nf_1$ (for n odd) then $n = 840/120 = 7$.

47. The string is fixed at both ends so the resonant wavelengths are given by $\lambda = 2L/n$, where L is the length of the string and n is an integer. The resonant frequencies are given by $f = v/\lambda = nv/2L$, where v is the wave speed on the string. Now $v = \sqrt{\tau/\mu}$, where τ is the tension in the string and μ is the linear mass density of the string. Thus $f = (n/2L)\sqrt{\tau/\mu}$. Suppose the lower frequency is associated with $n = n_1$ and the higher frequency is associated with $n = n_1 + 1$. There are no resonant frequencies between so you know that the integers associated with the given frequencies differ by 1. Thus $f_1 = (n_1/2L)\sqrt{\tau/\mu}$ and

$$f_2 = \frac{n_1+1}{2L}\sqrt{\frac{\tau}{\mu}} = \frac{n_1}{2L}\sqrt{\frac{\tau}{\mu}} + \frac{1}{2L}\sqrt{\frac{\tau}{\mu}} = f_1 + \frac{1}{2L}\sqrt{\frac{\tau}{\mu}}.$$

This means $f_2 - f_1 = (1/2L)\sqrt{\tau/\mu}$ and

$$\tau = 4L^2\mu(f_2 - f_1)^2 = 4(0.300\,\text{m})^2(0.650\times10^{-3}\,\text{kg/m})(1320\,\text{Hz} - 880\,\text{Hz})^2 = 45.3\,\text{N}.$$

48. (a) Using Eq. 17–39 with $n = 1$ (for the fundamental mode of vibration) and 343 m/s for the speed of sound, we obtain

$$f = \frac{(1)v_{\text{sound}}}{4L_{\text{tube}}} = \frac{343\,\text{m/s}}{4(1.20\,\text{m})} = 71.5\,\text{Hz}.$$

(b) For the wire (using Eq. 17–53) we have

$$f' = \frac{nv_{\text{wire}}}{2L_{\text{wire}}} = \frac{1}{2L_{\text{wire}}}\sqrt{\frac{\tau}{\mu}}$$

where $\mu = m_{\text{wire}}/L_{\text{wire}}$. Recognizing that $f = f'$ (both the wire and the air in the tube vibrate at the same frequency), we solve this for the tension τ.

$$\tau = (2L_{\text{wire}}\,f)^2\left(\frac{m_{\text{wire}}}{L_{\text{wire}}}\right) = 4f^2 m_{\text{wire}} L_{\text{wire}} = 4(71.5\,\text{Hz})^2(9.60\times10^{-3}\,\text{kg})(0.330\,\text{m}) = 64.8\,\text{N}.$$

49. The top of the water is a displacement node and the top of the well is a displacement anti-node. At the lowest resonant frequency exactly one-fourth of a wavelength fits into the depth of the well. If d is the depth and λ is the wavelength then $\lambda = 4d$. The frequency is $f = v/\lambda = v/4d$, where v is the speed of sound. The speed of sound is given by

$v = \sqrt{B/\rho}$, where B is the bulk modulus and ρ is the density of air in the well. Thus $f = (1/4d)\sqrt{B/\rho}$ and

$$d = \frac{1}{4f}\sqrt{\frac{B}{\rho}} = \frac{1}{4(7.00\,\text{Hz})}\sqrt{\frac{1.33\times 10^5\,\text{Pa}}{1.10\,\text{kg/m}^3}} = 12.4\,\text{m}.$$

50. We observe that "third lowest ... frequency" corresponds to harmonic number $n_A = 3$ for pipe A which is open at both ends. Also, "second lowest ... frequency" corresponds to harmonic number $n_B = 3$ for pipe B which is closed at one end.

(a) Since the frequency of B matches the frequency of A, using Eqs. 17-39 and 17-41, we have

$$f_A = f_B \quad \Rightarrow \quad \frac{3v}{2L_A} = \frac{3v}{4L_B}$$

which implies $L_B = L_A/2 = (1.20\,\text{m})/2 = 0.60\,\text{m}$. Using Eq. 17-40, the corresponding wavelength is

$$\lambda = \frac{4L_B}{3} = \frac{4(0.60\,\text{m})}{3} = 0.80\,\text{m}.$$

The change from node to anti-node requires a distance of $\lambda/4$ so that every increment of 0.20 m along the x axis involves a switch between node and anti-node. Since the closed end is a node, the next node appears at $x = 0.40$ m So there are 2 nodes. The situation corresponds to that illustrated in Fig. 17-15(b) with $n = 3$.

(b) The smallest value of x where a node is present is $x = 0$.

(c) The second smallest value of x where a node is present is $x = 0.40$m.

(d) Using $v = 343$ m/s, we find $f_3 = v/\lambda = 429$ Hz. Now, we find the fundamental resonant frequency by dividing by the harmonic number, $f_1 = f_3/3 = 143$ Hz.

51. Let the period be T. Then the beat frequency is $1/T - 440\,\text{Hz} = 4.00\,\text{beats/s}$. Therefore, $T = 2.25 \times 10^{-3}$ s. The string that is "too tightly stretched" has the higher tension and thus the higher (fundamental) frequency.

52. Since the beat frequency equals the difference between the frequencies of the two tuning forks, the frequency of the first fork is either 381 Hz or 387 Hz. When mass is added to this fork its frequency decreases (recall, for example, that the frequency of a mass-spring oscillator is proportional to $1/\sqrt{m}$). Since the beat frequency also decreases the frequency of the first fork must be greater than the frequency of the second. It must be 387 Hz.

53. Each wire is vibrating in its fundamental mode so the wavelength is twice the length of the wire ($\lambda = 2L$) and the frequency is

$$f = v/\lambda = (1/2L)\sqrt{\tau/\mu},$$

where $v = \sqrt{\tau/\mu}$ is the wave speed for the wire, τ is the tension in the wire, and μ is the linear mass density of the wire. Suppose the tension in one wire is τ and the oscillation frequency of that wire is f_1. The tension in the other wire is $\tau + \Delta\tau$ and its frequency is f_2. You want to calculate $\Delta\tau/\tau$ for $f_1 = 600$ Hz and $f_2 = 606$ Hz. Now, $f_1 = (1/2L)\sqrt{\tau/\mu}$ and $f_2 = (1/2L)\sqrt{(\tau + \Delta\tau)/\mu}$, so

$$f_2/f_1 = \sqrt{(\tau + \Delta\tau)/\tau} = \sqrt{1 + (\Delta\tau/\tau)}.$$

This leads to $\Delta\tau/\tau = (f_2/f_1)^2 - 1 = [(606\,\text{Hz})/(600\,\text{Hz})]^2 - 1 = 0.020$.

54. (a) The number of different ways of picking up a pair of tuning forks out of a set of five is $5!/(2!3!) = 10$. For each of the pairs selected, there will be one beat frequency. If these frequencies are all different from each other, we get the maximum possible number of 10.

(b) First, we note that the minimum number occurs when the frequencies of these forks, labeled 1 through 5, increase in equal increments: $f_n = f_1 + n\Delta f$, where $n = 2, 3, 4, 5$. Now, there are only 4 different beat frequencies: $f_{beat} = n\Delta f$, where $n = 1, 2, 3, 4$.

55. In the general Doppler shift equation, the trooper's speed is the source speed and the speeder's speed is the detector's speed. The Doppler effect formula, Eq. 17–47, and its accompanying rule for choosing ± signs, are discussed in §17-10. Using that notation, we have $v = 343$ m/s,

$$v_D = v_S = 160 \text{ km/h} = (160000 \text{ m})/(3600 \text{ s}) = 44.4 \text{ m/s},$$

and $f = 500$ Hz. Thus,

$$f' = (500 \text{ Hz})\left(\frac{343 \text{ m/s} - 44.4 \text{ m/s}}{343 \text{ m/s} - 44.4 \text{ m/s}}\right) = 500 \text{ Hz} \Rightarrow \Delta f = 0.$$

56. The Doppler effect formula, Eq. 17–47, and its accompanying rule for choosing ± signs, are discussed in §17-10. Using that notation, we have $v = 343$ m/s, $v_D = 2.44$ m/s, $f' = 1590$ Hz and $f = 1600$ Hz. Thus,

$$f' = f\left(\frac{v + v_D}{v + v_S}\right) \Rightarrow v_S = \frac{f}{f'}(v + v_D) - v = 4.61 \text{ m/s}.$$

57. We use $v_S = r\omega$ (with $r = 0.600$ m and $\omega = 15.0$ rad/s) for the linear speed during circular motion, and Eq. 17–47 for the Doppler effect (where $f = 540$ Hz, and $v = 343$ m/s for the speed of sound).

(a) The lowest frequency is
$$f' = f\left(\frac{v+0}{v+v_S}\right) = 526 \text{ Hz}.$$

(b) The highest frequency is
$$f' = f\left(\frac{v+0}{v-v_S}\right) = 555 \text{ Hz}.$$

58. We are combining two effects: the reception of a moving object (the truck of speed $u = 45.0$ m/s) of waves emitted by a stationary object (the motion detector), and the subsequent emission of those waves by the moving object (the truck) which are picked up by the stationary detector. This could be figured in two steps, but is more compactly computed in one step as shown here:

$$f_{\text{final}} = f_{\text{initial}}\left(\frac{v+u}{v-u}\right) = (0.150 \text{ MHz})\left(\frac{343 \text{ m/s} + 45 \text{ m/s}}{343 \text{ m/s} - 45 \text{ m/s}}\right) = 0.195 \text{ MHz}.$$

59. In this case, the intruder is moving *away* from the source with a speed u satisfying $u/v \ll 1$. The Doppler shift (with $u = -0.950$ m/s) leads to

$$f_{\text{beat}} = |f_r - f_s| \approx \frac{2|u|}{v} f_s = \frac{2(0.95 \text{ m/s})(28.0 \text{ kHz})}{343 \text{ m/s}} = 155 \text{ Hz}.$$

60. We use Eq. 17–47 with $f = 1200$ Hz and $v = 329$ m/s.

(a) In this case, $v_D = 65.8$ m/s and $v_S = 29.9$ m/s, and we choose signs so that f' is larger than f:
$$f' = f\left(\frac{329 \text{ m/s} + 65.8 \text{ m/s}}{329 \text{ m/s} - 29.9 \text{ m/s}}\right) = 1.58 \times 10^3 \text{ Hz}.$$

(b) The wavelength is $\lambda = v/f' = 0.208$ m.

(c) The wave (of frequency f') "emitted" by the moving reflector (now treated as a "source," so $v_S = 65.8$ m/s) is returned to the detector (now treated as a detector, so $v_D = 29.9$ m/s) and registered as a new frequency f'':

$$f'' = f'\left(\frac{329 \text{ m/s} + 29.9 \text{ m/s}}{329 \text{ m/s} - 65.8 \text{ m/s}}\right) = 2.16 \times 10^3 \text{ Hz}.$$

(d) This has wavelength $v/f'' = 0.152$ m.

61. We denote the speed of the French submarine by u_1 and that of the U.S. sub by u_2.

(a) The frequency as detected by the U.S. sub is

$$f_1' = f_1\left(\frac{v+u_2}{v-u_1}\right) = (1.000\times 10^3 \text{ Hz})\left(\frac{5470 \text{ km/h} + 70.00 \text{ km/h}}{5470 \text{ km/h} - 50.00 \text{ km/h}}\right) = 1.022 \times 10^3 \text{ Hz}.$$

(b) If the French sub were stationary, the frequency of the reflected wave would be $f_r = f_1(v+u_2)/(v-u_2)$. Since the French sub is moving towards the reflected signal with speed u_1, then

$$f_r' = f_r\left(\frac{v+u_1}{v}\right) = f_1\frac{(v+u_1)(v+u_2)}{v(v-u_2)} = \frac{(1.000\times 10^3 \text{ Hz})(5470+50.00)(5470+70.00)}{(5470)(5470-70.00)}$$

$$= 1.045\times 10^3 \text{ Hz}.$$

62. When the detector is stationary (with respect to the air) then Eq. 17-47 gives

$$f' = \frac{f}{1-v_s/v}$$

where v_s is the speed of the source (assumed to be approaching the detector in the way we've written it, above). The difference between the approach and the recession is

$$f' - f'' = f\left(\frac{1}{1-v_s/v} - \frac{1}{1+v_s/v}\right) = f\left(\frac{2v_s/v}{1-(v_s/v)^2}\right)$$

which, after setting $(f'-f'')/f = 1/2$, leads to an equation which can be solved for the ratio v_s/v. The result is $\sqrt{5} - 2 = 0.236$. Thus, $v_s/v = 0.236$.

63. As a result of the Doppler effect, the frequency of the reflected sound as heard by the bat is

$$f_r = f'\left(\frac{v+u_{bat}}{v-u_{bat}}\right) = (3.9\times 10^4 \text{ Hz})\left(\frac{v+v/40}{v-v/40}\right) = 4.1\times 10^4 \text{ Hz}.$$

64. The "third harmonic" refers to a resonant frequency $f_3 = 3f_1$, where f_1 is the fundamental lowest resonant frequency. When the source is stationary, with respect to the air, then Eq. 17-47 gives

$$f' = f\left(1-\frac{v_d}{v}\right)$$

where v_d is the speed of the detector (assumed to be moving away from the source, in the way we've written it, above). The problem, then, wants us to find v_d such that $f' = f_1$ when the emitted frequency is $f = f_3$. That is, we require $1 - v_d/v = 1/3$. Clearly, the solution to this is $v_d/v = 2/3$ (independent of length and whether one or both ends are open [the latter point being due to the fact that the odd harmonics occur in both systems]). Thus,

(a) For tube 1, $v_d = 2v/3$.

(b) For tube 2, $v_d = 2v/3$.

(c) For tube 3, $v_d = 2v/3$.

(d) For tube 4, $v_d = 2v/3$.

65. (a) The expression for the Doppler shifted frequency is

$$f' = f \frac{v \pm v_D}{v \mp v_S},$$

where f is the unshifted frequency, v is the speed of sound, v_D is the speed of the detector (the uncle), and v_S is the speed of the source (the locomotive). All speeds are relative to the air. The uncle is at rest with respect to the air, so $v_D = 0$. The speed of the source is $v_S = 10$ m/s. Since the locomotive is moving away from the uncle the frequency decreases and we use the plus sign in the denominator. Thus

$$f' = f \frac{v}{v + v_S} = (500.0 \text{ Hz}) \left(\frac{343 \text{ m/s}}{343 \text{ m/s} + 10.00 \text{ m/s}} \right) = 485.8 \text{ Hz}.$$

(b) The girl is now the detector. Relative to the air she is moving with speed $v_D = 10.00$ m/s toward the source. This tends to increase the frequency and we use the plus sign in the numerator. The source is moving at $v_S = 10.00$ m/s away from the girl. This tends to decrease the frequency and we use the plus sign in the denominator. Thus $(v + v_D) = (v + v_S)$ and $f' = f = 500.0$ Hz.

(c) Relative to the air the locomotive is moving at $v_S = 20.00$ m/s away from the uncle. Use the plus sign in the denominator. Relative to the air the uncle is moving at $v_D = 10.00$ m/s toward the locomotive. Use the plus sign in the numerator. Thus

$$f' = f \frac{v + v_D}{v + v_S} = (500.0 \text{ Hz}) \left(\frac{343 \text{ m/s} + 10.00 \text{ m/s}}{343 \text{ m/s} + 20.00 \text{ m/s}} \right) = 486.2 \text{ Hz}.$$

(d) Relative to the air the locomotive is moving at $v_S = 20.00$ m/s away from the girl and the girl is moving at $v_D = 20.00$ m/s toward the locomotive. Use the plus signs in both the numerator and the denominator. Thus $(v + v_D) = (v + v_S)$ and $f' = f = 500.0$ Hz.

66. We use Eq. 17–47 with $f = 500$ Hz and $v = 343$ m/s. We choose signs to produce $f' > f$.

(a) The frequency heard in still air is

$$f' = (500 \text{ Hz}) \left(\frac{343 \text{ m/s} + 30.5 \text{ m/s}}{343 \text{ m/s} - 30.5 \text{ m/s}} \right) = 598 \text{ Hz}.$$

(b) In a frame of reference where the air seems still, the velocity of the detector is $30.5 - 30.5 = 0$, and that of the source is $2(30.5)$. Therefore,

$$f' = (500 \text{ Hz}) \left(\frac{343 \text{ m/s} + 0}{343 \text{ m/s} - 2(30.5 \text{ m/s})} \right) = 608 \text{ Hz}.$$

(c) We again pick a frame of reference where the air seems still. Now, the velocity of the source is $30.5 - 30.5 = 0$, and that of the detector is $2(30.5)$. Consequently,

$$f' = (500 \text{ Hz}) \left(\frac{343 \text{ m/s} + 2(30.5 \text{ m/s})}{343 \text{ m/s} - 0} \right) = 589 \text{ Hz}.$$

67. The Doppler shift formula, Eq. 17–47, is valid only when both u_S and u_D are measured with respect to a stationary medium (i.e., no wind). To modify this formula in the presence of a wind, we switch to a new reference frame in which there is no wind.

(a) When the wind is blowing from the source to the observer with a speed w, we have $u'_S = u'_D = w$ in the new reference frame that moves together with the wind. Since the observer is now approaching the source while the source is backing off from the observer, we have, in the new reference frame,

$$f' = f \left(\frac{v + u'_D}{v + u'_S} \right) = f \left(\frac{v + w}{v + w} \right) = 2.0 \times 10^3 \text{ Hz}.$$

In other words, there is no Doppler shift.

(b) In this case, all we need to do is to reverse the signs in front of both u'_D and u'_S. The result is that there is still no Doppler shift:

$$f' = f \left(\frac{v - u'_D}{v - u'_S} \right) = f \left(\frac{v - w}{v - w} \right) = 2.0 \times 10^3 \text{ Hz}.$$

In general, there will always be no Doppler shift as long as there is no relative motion between the observer and the source, regardless of whether a wind is present or not.

68. We note that 1350 km/h is $v_S = 375$ m/s. Then, with $\theta = 60°$, Eq. 17-57 gives $v = 3.3 \times 10^2$ m/s.

69. (a) The half angle θ of the Mach cone is given by $\sin\theta = v/v_S$, where v is the speed of sound and v_S is the speed of the plane. Since $v_S = 1.5v$, $\sin\theta = v/1.5v = 1/1.5$. This means $\theta = 42°$.

(b) Let h be the altitude of the plane and suppose the Mach cone intersects Earth's surface a distance d behind the plane. The situation is shown on the diagram below, with P indicating the plane and O indicating the observer. The cone angle is related to h and d by $\tan\theta = h/d$, so $d = h/\tan\theta$. The shock wave reaches O in the time the plane takes to fly the distance d:

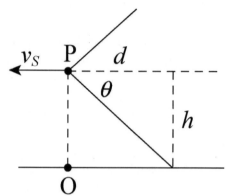

$$t = \frac{d}{v} = \frac{h}{v\tan\theta} = \frac{5000 \text{ m}}{1.5(331 \text{ m/s})\tan 42°} = 11 \text{ s}.$$

70. The altitude H and the horizontal distance x for the legs of a right triangle, so we have

$$H = x\tan\theta = v_p t \tan\theta = 1.25vt\sin\theta$$

where v is the speed of sound, v_p is the speed of the plane and

$$\theta = \sin^{-1}\left(\frac{v}{v_p}\right) = \sin^{-1}\left(\frac{v}{1.25v}\right) = 53.1°.$$

Thus the altitude is

$$H = x\tan\theta = (1.25)(330 \text{ m/s})(60 \text{ s})(\tan 53.1°) = 3.30 \times 10^4 \text{ m}.$$

71. (a) Incorporating a term ($\lambda/2$) to account for the phase shift upon reflection, then the path difference for the waves (when they come back together) is

$$\sqrt{L^2 + (2d)^2} - L + \lambda/2 = \Delta(\text{path}).$$

Setting this equal to the condition needed to destructive interference ($\lambda/2$, $3\lambda/2$, $5\lambda/2$ ...) leads to $d = 0, 2.10$ m, ... Since the problem explicitly excludes the $d = 0$ possibility, then our answer is $d = 2.10$ m.

(b) Setting this equal to the condition needed to constructive interference ($\lambda, 2\lambda, 3\lambda \ldots$) leads to $d = 1.47$ m, … Our answer is $d = 1.47$ m.

72. When the source is stationary (with respect to the air) then Eq. 17-47 gives

$$f' = f\left(1 - \frac{v_d}{v}\right),$$

where v_d is the speed of the detector (assumed to be moving away from the source, in the way we've written it, above). The difference between the approach and the recession is

$$f'' - f' = f\left[\left(1 + \frac{v_d}{v}\right) - \left(1 - \frac{v_d}{v}\right)\right] = f\left(2\frac{v_d}{v}\right)$$

which, after setting $(f'' - f')/f = 1/2$, leads to an equation which can be solved for the ratio v_d/v. The result is 1/4. Thus, $v_d/v = 0.250$.

73. (a) Adapting Eq. 17-39 to the notation of this chapter, we have

$$s_m' = 2 s_m \cos(\phi/2) = 2(12 \text{ nm}) \cos(\pi/6) = 20.78 \text{ nm}.$$

Thus, the amplitude of the resultant wave is roughly 21 nm.

(b) The wavelength ($\lambda = 35$ cm) does not change as a result of the superposition.

(c) Recalling Eq. 17-47 (and the accompanying discussion) from the previous chapter, we conclude that the standing wave amplitude is 2(12 nm) = 24 nm when they are traveling in opposite directions.

(d) Again, the wavelength ($\lambda = 35$ cm) does not change as a result of the superposition.

74. (a) The separation distance between points A and B is one-quarter of a wavelength; therefore, $\lambda = 4(0.15 \text{ m}) = 0.60$ m. The frequency, then, is

$$f = v/\lambda = (343 \text{ m/s})/(0.60 \text{ m}) = 572 \text{ Hz}.$$

(b) The separation distance between points C and D is one-half of a wavelength; therefore, $\lambda = 2(0.15 \text{ m}) = 0.30$ m. The frequency, then, is

$$f = v/\lambda = (343 \text{ m/s})/(0.30 \text{ m}) = 1144 \text{ Hz (or approximately 1.14 kHz)}.$$

75. Any phase changes associated with the reflections themselves are rendered inconsequential by the fact that there are an even number of reflections. The additional path length traveled by wave A consists of the vertical legs in the zig-zag path: $2L$. To be

(minimally) out of phase means, therefore, that $2L = \lambda/2$ (corresponding to a half-cycle, or 180°, phase difference). Thus, $L = \lambda/4$, or $L/\lambda = 1/4 = 0.25$.

76. Since they are approaching each other, the sound produced (of emitted frequency f) by the flatcar-trumpet received by an observer on the ground will be of higher pitch f'. In these terms, we are told $f' - f = 4.0$ Hz, and consequently that $f'/f = 444/440 = 1.0091$. With v_S designating the speed of the flatcar and $v = 343$ m/s being the speed of sound, the Doppler equation leads to

$$\frac{f'}{f} = \frac{v+0}{v-v_S} \Rightarrow v_S = (343 \text{ m/s})\frac{1.0091-1}{1.0091} = 3.1 \text{ m/s}.$$

77. The siren is between you and the cliff, moving away from you and towards the cliff. Both "detectors" (you and the cliff) are stationary, so $v_D = 0$ in Eq. 17–47 (and see the discussion in the textbook immediately after that equation regarding the selection of ± signs). The source is the siren with $v_S = 10$ m/s. The problem asks us to use $v = 330$ m/s for the speed of sound.

(a) With $f = 1000$ Hz, the frequency f_y you hear becomes

$$f_y = f\left(\frac{v+0}{v+v_S}\right) = 970.6 \text{ Hz} \approx 9.7 \times 10^2 \text{ Hz}.$$

(b) The frequency heard by an observer at the cliff (and thus the frequency of the sound reflected by the cliff, ultimately reaching your ears at some distance from the cliff) is

$$f_c = f\left(\frac{v+0}{v-v_S}\right) = 1031.3 \text{ Hz} \approx 1.0 \times 10^3 \text{ Hz}.$$

(c) The beat frequency is $f_c - f_y = 60$ beats/s (which, due to specific features of the human ear, is too large to be perceptible).

78. Let r stand for the ratio of the source speed to the speed of sound. Then, Eq. 17-55 (plus the fact that frequency is inversely proportional to wavelength) leads to

$$2\left(\frac{1}{1+r}\right) = \frac{1}{1-r}.$$

Solving, we find $r = 1/3$. Thus, $v_s/v = 0.33$.

79. The source being isotropic means $A_{\text{sphere}} = 4\pi r^2$ is used in the intensity definition $I = P/A$, which further implies

$$\frac{I_2}{I_1} = \frac{P/4\pi r_2^2}{P/4\pi r_1^2} = \left(\frac{r_1}{r_2}\right)^2.$$

(a) With $I_1 = 9.60 \times 10^{-4}$ W/m², $r_1 = 6.10$ m, and $r_2 = 30.0$ m, we find

$$I_2 = (9.60 \times 10^{-4} \text{ W/m}^2)(6.10/30.0)^2 = 3.97 \times 10^{-5} \text{ W/m}^2.$$

(b) Using Eq. 17–27 with $I_1 = 9.60 \times 10^{-4}$ W/m², $\omega = 2\pi(2000 \text{ Hz})$, $v = 343$ m/s and $\rho = 1.21$ kg/m³, we obtain

$$s_m = \sqrt{\frac{2I}{\rho v \omega^2}} = 1.71 \times 10^{-7} \text{ m}.$$

(c) Eq. 17-15 gives the pressure amplitude:

$$\Delta p_m = \rho v \omega s_m = 0.893 \text{ Pa}.$$

80. When $\phi = 0$ it is clear that the superposition wave has amplitude $2\Delta p_m$. For the other cases, it is useful to write

$$\Delta p_1 + \Delta p_2 = \Delta p_m \left(\sin(\omega t) + \sin(\omega t - \phi)\right) = \left(2\Delta p_m \cos\frac{\phi}{2}\right)\sin\left(\omega t - \frac{\phi}{2}\right).$$

The factor in front of the sine function gives the amplitude Δp_r. Thus, $\Delta p_r / \Delta p_m = 2\cos(\phi/2)$.

(a) When $\phi = 0$, $\Delta p_r / \Delta p_m = 2\cos(0) = 2.00$.

(b) When $\phi = \pi/2$, $\Delta p_r / \Delta p_m = 2\cos(\pi/4) = \sqrt{2} = 1.41$.

(c) When $\phi = \pi/3$, $\Delta p_r / \Delta p_m = 2\cos(\pi/6) = \sqrt{3} = 1.73$.

(d) When $\phi = \pi/4$, $\Delta p_r / \Delta p_m = 2\cos(\pi/8) = 1.85$.

81. (a) With $r = 10$ m in Eq. 17–28, we have

$$I = \frac{P}{4\pi r^2} \Rightarrow P = 10 \text{ W}.$$

(b) Using that value of P in Eq. 17–28 with a new value for r, we obtain

$$I = \frac{P}{4\pi(5.0)^2} = 0.032 \frac{\text{W}}{\text{m}^2}.$$

Alternatively, a ratio $I'/I = (r/r')^2$ could have been used.

(c) Using Eq. 17–29 with $I = 0.0080$ W/m^2, we have

$$\beta = 10\log\frac{I}{I_0} = 99\,\text{dB}$$

where $I_0 = 1.0 \times 10^{-12}$ W/m^2.

82. We use $v = \sqrt{B/\rho}$ to find the bulk modulus B:

$$B = v^2\rho = (5.4\times 10^3\,\text{m/s})^2(2.7\times 10^3\,\text{kg/m}^3) = 7.9\times 10^{10}\,\text{Pa}.$$

83. Let the frequencies of sound heard by the person from the left and right forks be f_l and f_r, respectively.

(a) If the speeds of both forks are u, then $f_{l,r} = fv/(v \pm u)$ and

$$f_{\text{beat}} = |f_r - f_l| = fv\left(\frac{1}{v-u} - \frac{1}{v+u}\right) = \frac{2fuv}{v^2 - u^2} = \frac{2(440\,\text{Hz})(3.00\,\text{m/s})(343\,\text{m/s})}{(343\,\text{m/s})^2 - (3.00\,\text{m/s})^2}$$
$$= 7.70\,\text{Hz}.$$

(b) If the speed of the listener is u, then $f_{l,r} = f(v \pm u)/v$ and

$$f_{\text{beat}} = |f_l - f_r| = 2f\left(\frac{u}{v}\right) = 2(440\,\text{Hz})\left(\frac{3.00\,\text{m/s}}{343\,\text{m/s}}\right) = 7.70\,\text{Hz}.$$

84. The rule: if you divide the time (in seconds) by 3, then you get (approximately) the straight-line distance d. We note that the speed of sound we are to use is given at the beginning of the problem section in the textbook, and that the speed of light is very much larger than the speed of sound. The proof of our rule is as follows:

$$t = t_{\text{sound}} - t_{\text{light}} \approx t_{\text{sound}} = \frac{d}{v_{\text{sound}}} = \frac{d}{343\,\text{m/s}} = \frac{d}{0.343\,\text{km/s}}.$$

Cross-multiplying yields (approximately) $(0.3\,\text{km/s})t = d$ which (since $1/3 \approx 0.3$) demonstrates why the rule works fairly well.

85. (a) The intensity is given by $I = \frac{1}{2}\rho v\omega^2 s_m^2$, where ρ is the density of the medium, v is the speed of sound, ω is the angular frequency, and s_m is the displacement amplitude. The displacement and pressure amplitudes are related by $\Delta p_m = \rho v\omega s_m$, so $s_m = \Delta p_m/\rho v\omega$ and I

$= (\Delta p_m)^2/2\rho v$. For waves of the same frequency the ratio of the intensity for propagation in water to the intensity for propagation in air is

$$\frac{I_w}{I_a} = \left(\frac{\Delta p_{mw}}{\Delta p_{ma}}\right)^2 \frac{\rho_a v_a}{\rho_w v_w},$$

where the subscript a denotes air and the subscript w denotes water. Since $I_a = I_w$,

$$\frac{\Delta p_{mw}}{\Delta p_{ma}} = \sqrt{\frac{\rho_w v_w}{\rho_a v_a}} = \sqrt{\frac{(0.998 \times 10^3 \text{ kg/m}^3)(1482 \text{ m/s})}{(1.21 \text{ kg/m}^3)(343 \text{ m/s})}} = 59.7.$$

The speeds of sound are given in Table 17-1 and the densities are given in Table 15-1.

(b) Now, $\Delta p_{mw} = \Delta p_{ma}$, so

$$\frac{I_w}{I_a} = \frac{\rho_a v_a}{\rho_w v_w} = \frac{(1.21 \text{ kg/m}^3)(343 \text{ m/s})}{(0.998 \times 10^3 \text{ kg/m}^3)(1482 \text{ m/s})} = 2.81 \times 10^{-4}.$$

86. We use $\Delta\beta_{12} = \beta_1 - \beta_2 = (10 \text{ dB}) \log(I_1/I_2)$.

(a) Since $\Delta\beta_{12} = (10 \text{ dB}) \log(I_1/I_2) = 37$ dB, we get

$$I_1/I_2 = 10^{37 \text{ dB}/10 \text{ dB}} = 10^{3.7} = 5.0 \times 10^3.$$

(b) Since $\Delta p_m \propto s_m \propto \sqrt{I}$, we have

$$\Delta p_{m1}/\Delta p_{m2} = \sqrt{I_1/I_2} = \sqrt{5.0 \times 10^3} = 71.$$

(c) The displacement amplitude ratio is $s_{m1}/s_{m2} = \sqrt{I_1/I_2} = 71$.

87. (a) When the right side of the instrument is pulled out a distance d the path length for sound waves increases by $2d$. Since the interference pattern changes from a minimum to the next maximum, this distance must be half a wavelength of the sound. So $2d = \lambda/2$, where λ is the wavelength. Thus $\lambda = 4d$ and, if v is the speed of sound, the frequency is

$$f = v/\lambda = v/4d = (343 \text{ m/s})/4(0.0165 \text{ m}) = 5.2 \times 10^3 \text{ Hz}.$$

(b) The displacement amplitude is proportional to the square root of the intensity (see Eq. 17–27). Write $\sqrt{I} = C s_m$, where I is the intensity, s_m is the displacement amplitude, and C is a constant of proportionality. At the minimum, interference is destructive and the displacement amplitude is the difference in the amplitudes of the individual waves: $s_m =$

$s_{SAD} - s_{SBD}$, where the subscripts indicate the paths of the waves. At the maximum, the waves interfere constructively and the displacement amplitude is the sum of the amplitudes of the individual waves: $s_m = s_{SAD} + s_{SBD}$. Solve

$$\sqrt{100} = C(s_{SAD} - s_{SBD}) \quad \text{and} \quad \sqrt{900} = C(s_{SAD} - s_{SBD})$$

for s_{SAD} and s_{SBD}. Adding the equations give

$$s_{SAD} = (\sqrt{100} + \sqrt{900})/2C = 20/C,$$

while subtracting them yields

$$s_{SBD} = (\sqrt{900} - \sqrt{100})/2C = 10/C.$$

Thus, the ratio of the amplitudes is $s_{SAD}/s_{SBD} = 2$.

(c) Any energy losses, such as might be caused by frictional forces of the walls on the air in the tubes, result in a decrease in the displacement amplitude. Those losses are greater on path B since it is longer than path A.

88. The angle is $\sin^{-1}(v/v_s) = \sin^{-1}(343/685) = 30°$.

89. The round-trip time is $t = 2L/v$ where we estimate from the chart that the time between clicks is 3 ms. Thus, with $v = 1372$ m/s, we find $L = \frac{1}{2}vt = 2.1$ m.

90. The wave is written as $s(x,t) = s_m \cos(kx \pm \omega t)$.

(a) The amplitude s_m is equal to the maximum displacement: $s_m = 0.30$ cm.

(b) Since $\lambda = 24$ cm, the angular wave number is $k = 2\pi/\lambda = 0.26$ cm^{-1}.

(c) The angular frequency is $\omega = 2\pi f = 2\pi(25 \text{ Hz}) = 1.6 \times 10^2$ rad/s.

(d) The speed of the wave is $v = \lambda f = (24 \text{ cm})(25 \text{ Hz}) = 6.0 \times 10^2$ cm/s.

(e) Since the direction of propagation is $-x$, the sign is plus, i.e., $s(x,t) = s_m \cos(kx + \omega t)$.

91. The source being a "point source" means $A_{\text{sphere}} = 4\pi r^2$ is used in the intensity definition $I = P/A$, which further implies

$$\frac{I_2}{I_1} = \frac{P/4\pi r_2^2}{P/4\pi r_1^2} = \left(\frac{r_1}{r_2}\right)^2.$$

From the discussion in §17-5, we know that the intensity ratio between "barely audible" and the "painful threshold" is $10^{-12} = I_2/I_1$. Thus, with $r_2 = 10000$ m, we find

$$r_1 = r_2\sqrt{10^{-12}} = 0.01\,\text{m} = 1\,\text{cm}.$$

92. (a) The time it takes for sound to travel in air is $t_a = L/v$, while it takes $t_m = L/v_m$ for the sound to travel in the metal. Thus,

$$\Delta t = t_a - t_m = \frac{L}{v} - \frac{L}{v_m} = \frac{L(v_m - v)}{v_m v}.$$

(b) Using the values indicated (see Table 17-1), we obtain

$$L = \frac{\Delta t}{1/v - 1/v_m} = \frac{1.00\,\text{s}}{1/(343\,\text{m/s}) - 1/(5941\,\text{m/s})} = 364\,\text{m}.$$

93. (a) We observe that "third lowest ... frequency" corresponds to harmonic number $n = 5$ for such a system. Using Eq. 17–41, we have

$$f = \frac{nv}{4L} \Rightarrow 750\,\text{Hz} = \frac{5v}{4(0.60\,\text{m})}$$

so that $v = 3.6\times 10^2$ m/s.

(b) As noted, $n = 5$; therefore, $f_1 = 750/5 = 150$ Hz.

94. We note that waves 1 and 3 differ in phase by π radians (so they cancel upon superposition). Waves 2 and 4 also differ in phase by π radians (and also cancel upon superposition). Consequently, there is no resultant wave.

95. Since they oscillate out of phase, then their waves will cancel (producing a node) at a point exactly midway between them (the midpoint of the system, where we choose $x = 0$). We note that Figure 17-14, and the $n = 3$ case of Figure 17-15(a) have this property (of a node at the midpoint). The distance Δx between nodes is $\lambda/2$, where $\lambda = v/f$ and $f = 300$ Hz and $v = 343$ m/s. Thus, $\Delta x = v/2f = 0.572$ m.

Therefore, nodes are found at the following positions:

$$x = n\Delta x = n(0.572\,\text{m}),\ n = 0, \pm 1, \pm 2, \ldots$$

(a) The shortest distance from the midpoint where nodes are found is $\Delta x = 0$.

(b) The second shortest distance from the midpoint where nodes are found is $\Delta x = 0.572$ m.

(c) The third shortest distance from the midpoint where nodes are found is $2\Delta x = 1.14$ m.

96. (a) With $f = 686$ Hz and $v = 343$ m/s, then the "separation between adjacent wavefronts" is $\lambda = v/f = 0.50$ m.

(b) This is one of the effects which are part of the Doppler phenomena. Here, the wavelength shift (relative to its "true" value in part (a)) equals the source speed v_s (with appropriate $\pm$ sign) relative to the speed of sound v:

$$\frac{\Delta \lambda}{\lambda} = \pm \frac{v_s}{v}.$$

In front of the source, the shift in wavelength is $-(0.50 \text{ m})(110 \text{ m/s})/(343 \text{ m/s}) = -0.16$ m, and the wavefront separation is 0.50 m $- 0.16$ m $= 0.34$ m.

(c) Behind the source, the shift in wavelength is $+(0.50 \text{ m})(110 \text{ m/s})/(343 \text{ m/s}) = +0.16$ m, and the wavefront separation is 0.50 m $+ 0.16$ m $= 0.66$ m.

97. We use $I \propto r^{-2}$ appropriate for an isotropic source. We have

$$\frac{I_{r=d}}{I_{r=D-d}} = \frac{(D-d)^2}{D^2} = \frac{1}{2},$$

where $d = 50.0$ m. We solve for

$$D: D = \sqrt{2}d/(\sqrt{2}-1) = \sqrt{2}(50.0 \text{ m})/(\sqrt{2}-1) = 171 \text{ m}.$$

98. (a) Using $m = 7.3 \times 10^7$ kg, the initial gravitational potential energy is $U = mgy = 3.9 \times 10^{11}$ J, where $h = 550$ m. Assuming this converts primarily into kinetic energy during the fall, then $K = 3.9 \times 10^{11}$ J just before impact with the ground. Using instead the mass estimate $m = 1.7 \times 10^8$ kg, we arrive at $K = 9.2 \times 10^{11}$ J.

(b) The process of converting this kinetic energy into other forms of energy (during the impact with the ground) is assumed to take $\Delta t = 0.50$ s (and in the average sense, we take the "power" P to be wave-energy/Δt). With 20% of the energy going into creating a seismic wave, the intensity of the body wave is estimated to be

$$I = \frac{P}{A_{\text{hemisphere}}} = \frac{(0.20)K/\Delta t}{\frac{1}{2}(4\pi r^2)} = 0.63 \text{ W/m}^2$$

using $r = 200 \times 10^3$ m and the smaller value for K from part (a). Using instead the larger estimate for K, we obtain $I = 1.5$ W/m^2.

(c) The surface area of a cylinder of "height" d is $2\pi rd$, so the intensity of the surface wave is

$$I = \frac{P}{A_{cylinder}} = \frac{(0.20)K/\Delta t}{(2\pi rd)} = 25\times 10^3 \text{ W/m}^2$$

using $d = 5.0$ m, $r = 200 \times 10^3$ m and the smaller value for K from part (a). Using instead the larger estimate for K, we obtain $I = 58$ kW/m^2.

(d) Although several factors are involved in determining which seismic waves are most likely to be detected, we observe that on the basis of the above findings we should expect the more intense waves (the surface waves) to be more readily detected.

99. (a) The period is the reciprocal of the frequency:

$$T = 1/f = 1/(90 \text{ Hz}) = 1.1 \times 10^{-2} \text{ s}.$$

(b) Using $v = 343$ m/s, we find $\lambda = v/f = 3.8$ m.

100. (a) The problem asks for the source frequency f. We use Eq. 17–47 with great care (regarding its ± sign conventions).

$$f' = f\left(\frac{340 \text{ m/s} - 16 \text{ m/s}}{340 \text{ m/s} - 40 \text{ m/s}}\right)$$

Therefore, with $f' = 950$ Hz, we obtain $f = 880$ Hz.

(b) We now have

$$f' = f\left(\frac{340 \text{ m/s} + 16 \text{ m/s}}{340 \text{ m/s} + 40 \text{ m/s}}\right)$$

so that with $f = 880$ Hz, we find $f' = 824$ Hz.

101. (a) The blood is moving towards the right (towards the detector), because the Doppler shift in frequency is an *increase*: $\Delta f > 0$.

(b) The reception of the ultrasound by the blood and the subsequent remitting of the signal by the blood back toward the detector is a two-step process which may be compactly written as

$$f + \Delta f = f\left(\frac{v + v_x}{v - v_x}\right)$$

where $v_x = v_{blood} \cos\theta$. If we write the ratio of frequencies as $R = (f + \Delta f)/f$, then the solution of the above equation for the speed of the blood is

$$v_{blood} = \frac{(R-1)v}{(R+1)\cos\theta} = 0.90 \, \text{m/s}$$

where $v = 1540$ m/s, $\theta = 20°$, and $R = 1 + 5495/5 \times 10^6$.

(c) We interpret the question as asking how Δf (still taken to be positive, since the detector is in the "forward" direction) changes as the detection angle θ changes. Since larger θ means smaller horizontal component of velocity v_x then we expect Δf to decrease towards zero as θ is increased towards 90°.

102. Pipe A (which can only support odd harmonics – see Eq. 17-41) has length L_A. Pipe B (which supports both odd and even harmonics [any value of n] – see Eq. 17-39) has length $L_B = 4L_A$. Taking ratios of these equations leads to the condition:

$$\left(\frac{n}{2}\right)_B = (n_{odd})_A .$$

Solving for n_B we have $n_B = 2n_{odd}$.

(a) Thus, the smallest value of n_B at which a harmonic frequency of B matches that of A is $n_B = 2(1) = 2$.

(b) The second smallest value of n_B at which a harmonic frequency of B matches that of A is $n_B = 2(3) = 6$.

(c) The third smallest value of n_B at which a harmonic frequency of B matches that of A is $n_B = 2(5) = 10$.

103. The points and the least-squares fit is shown in the graph that follows.

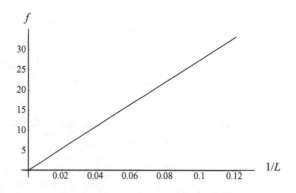

The graph has frequency in Hertz along the vertical axis and $1/L$ in inverse meters along the horizontal axis. The function found by the least squares fit procedure is $f = 276(1/L) + 0.037$. We shall assume that this fits either the model of an open organ pipe (mathematically similar to a string fixed at both ends) or that of a pipe closed at one end.

(a) In a tube with two open ends, $f = v/2L$. If the least-squares slope of 276 fits the first model, then a value of

$$v = 2(276 \text{ m/s}) = 553 \text{ m/s} \approx 5.5 \times 10^2 \text{ m/s}$$

is implied.

(b) In a tube with only one open end, $f = v/4L$, and we find $v = 4(276 \text{ m/s}) = 1106 \text{ m/s} \approx 1.1 \times 10^3$ m/s which is more "in the ballpark" of the 1400 m/s value cited in the problem.

(c) This suggests that the acoustic resonance involved in this situation is more closely related to the $n = 1$ case of Figure 17-15(b) than to Figure 17-14.

104. (a) Since the source is moving toward the wall, the frequency of the sound as received at the wall is

$$f' = f\left(\frac{v}{v - v_S}\right) = (440 \text{ Hz})\left(\frac{343 \text{ m/s}}{343 \text{ m/s} - 20.0 \text{ m/s}}\right) = 467 \text{ Hz}.$$

(b) Since the person is moving with a speed u toward the reflected sound with frequency f', the frequency registered at the source is

$$f_r = f'\left(\frac{v + u}{v}\right) = (467 \text{ Hz})\left(\frac{343 \text{ m/s} + 20.0 \text{ m/s}}{343 \text{ m/s}}\right) = 494 \text{ Hz}.$$

105. Using Eq. 17-47 with great care (regarding its $\pm$ sign conventions), we have

$$f' = (440 \text{ Hz})\left(\frac{340 \text{ m/s} - 80.0 \text{ m/s}}{340 \text{ m/s} - 54.0 \text{ m/s}}\right) = 400 \text{ Hz}.$$

106. (a) Let P be the power output of the source. This is the rate at which energy crosses the surface of any sphere centered at the source and is therefore equal to the product of the intensity I at the sphere surface and the area of the sphere. For a sphere of radius r, $P = 4\pi r^2 I$ and $I = P/4\pi r^2$. The intensity is proportional to the square of the displacement amplitude s_m. If we write $I = Cs_m^2$, where C is a constant of proportionality, then $Cs_m^2 = P/4\pi r^2$. Thus,

$$s_m = \sqrt{P/4\pi r^2 C} = \left(\sqrt{P/4\pi C}\right)(1/r).$$

The displacement amplitude is proportional to the reciprocal of the distance from the source. We take the wave to be sinusoidal. It travels radially outward from the source, with points on a sphere of radius r in phase. If ω is the angular frequency and k is the angular wave number then the time dependence is $\sin(kr - \omega t)$. Letting $b = \sqrt{P/4\pi C}$, the displacement wave is then given by

$$s(r,t) = \sqrt{\frac{P}{4\pi C}}\frac{1}{r}\sin(kr - \omega t) = \frac{b}{r}\sin(kr - \omega t).$$

(b) Since s and r both have dimensions of length and the trigonometric function is dimensionless, the dimensions of b must be length squared.

107. (a) The problem is asking at how many angles will there be "loud" resultant waves, and at how many will there be "quiet" ones? We consider the resultant wave (at large distance from the origin) along the $+x$ axis; we note that the path-length difference (for the waves traveling from their respective sources) divided by wavelength gives the (dimensionless) value $n = 3.2$, implying a sort of intermediate condition between constructive interference (which would follow if, say, $n = 3$) and destructive interference (such as the $n = 3.5$ situation found in the solution to the previous problem) between the waves. To distinguish this resultant along the $+x$ axis from the similar one along the $-x$ axis, we label one with $n = +3.2$ and the other $n = -3.2$. This labeling facilitates the complete enumeration of the loud directions in the upper-half plane: $n = -3, -2, -1, 0, +1, +2, +3$. Counting also the "other" $-3, -2, -1, 0, +1, +2, +3$ values for the *lower*-half plane, then we conclude there are a total of $7 + 7 = 14$ "loud" directions.

(b) The labeling also helps us enumerate the quiet directions. In the upper-half plane we find: $n = -2.5, -1.5, -0.5, +0.5, +1.5, +2.5$. This is duplicated in the lower half plane, so the total number of quiet directions is $6 + 6 = 12$.

108. The source being isotropic means $A_{\text{sphere}} = 4\pi r^2$ is used in the intensity definition $I = P/A$. Since intensity is proportional to the square of the amplitude (see Eq. 17–27), this further implies

$$\frac{I_2}{I_1} = \left(\frac{s_{m2}}{s_{m1}}\right)^2 = \frac{P/4\pi r_2^2}{P/4\pi r_1^2} = \left(\frac{r_1}{r_2}\right)^2$$

or $s_{m2}/s_{m1} = r_1/r_2$.

(a) $I = P/4\pi r^2 = (10\text{ W})/4\pi(3.0\text{ m})^2 = 0.088\text{ W/m}^2$.

(b) Using the notation A instead of s_m for the amplitude, we find

$$\frac{A_4}{A_3} = \frac{3.0\text{ m}}{4.0\text{ m}} = 0.75.$$

109. (a) In regions where the speed is constant, it is equal to distance divided by time. Thus, we conclude that the time difference is

$$\Delta t = \left(\frac{L-d}{V} + \frac{d}{V - \Delta V} \right) - \frac{L}{V}$$

where the first term is the travel time through bone and rock and the last term is the expected travel time purely through rock. Solving for d and simplifying, we obtain

$$d = \Delta t \frac{V(V - \Delta V)}{\Delta V} \approx \Delta t \frac{V^2}{\Delta V}.$$

(b) If we estimate $d \approx 10$ cm (as the lower limit of a range that goes up to a diameter of 20 cm), then the above expression (with the numerical values given in the problem) leads to $\Delta t = 0.8$ μs (as the lower limit of a range that goes up to a time difference of 1.6 μs).

110. (a) We expect the center of the star to be a displacement node. The star has spherical symmetry and the waves are spherical. If matter at the center moved it would move equally in all directions and this is not possible.

(b) We assume the oscillation is at the lowest resonance frequency. Then, exactly one-fourth of a wavelength fits the star radius. If λ is the wavelength and R is the star radius then $\lambda = 4R$. The frequency is $f = v/\lambda = v/4R$, where v is the speed of sound in the star. The period is $T = 1/f = 4R/v$.

(c) The speed of sound is $v = \sqrt{B/\rho}$, where B is the bulk modulus and ρ is the density of stellar material. The radius is $R = 9.0 \times 10^{-3} R_s$, where R_s is the radius of the Sun (6.96 $\times 10^8$ m). Thus

$$T = 4R\sqrt{\frac{\rho}{B}} = 4(9.0 \times 10^{-3})(6.96 \times 10^8 \text{ m})\sqrt{\frac{1.0 \times 10^{10} \text{ kg/m}^3}{1.33 \times 10^{22} \text{ Pa}}} = 22 \text{ s}.$$

111. We find the difference in the two applications of the Doppler formula:

$$f_2 - f_1 = 37 \text{ Hz} = f\left(\frac{340 \text{ m/s} + 25 \text{ m/s}}{340 \text{ m/s} - 15 \text{ m/s}} - \frac{340 \text{ m/s}}{340 \text{ m/s} - 15 \text{ m/s}} \right) = f\left(\frac{25 \text{ m/s}}{340 \text{ m/s} - 15 \text{ m/s}} \right)$$

which leads to $f = 4.8 \times 10^2$ Hz.

112. (a) We proceed by dividing the (velocity) equation involving the new (fundamental) frequency f' by the equation when the frequency f is 440 Hz to obtain

$$\frac{f'\lambda}{f\lambda} = \sqrt{\frac{\tau'/\mu}{\tau/\mu}} \quad \Rightarrow \quad \frac{f'}{f} = \sqrt{\frac{\tau'}{\tau}}$$

where we are making an assumption that the mass-per-unit-length of the string does not change significantly. Thus, with $\tau' = 1.2\tau$, we have $f'/440 = \sqrt{1.2}$, which gives $f' = 482\,\text{Hz}$.

(b) In this case, neither tension nor mass-per-unit-length change, so the wave speed v is unchanged. Hence, using Eq. 17–38 with $n = 1$,

$$f'\lambda' = f\lambda \quad \Rightarrow \quad f'(2L') = f(2L)$$

Since $L' = \tfrac{2}{3}L$, we obtain $f' = \tfrac{3}{2}(440) = 660\,\text{Hz}$.

Chapter 18

1. Let T_L be the temperature and p_L be the pressure in the left-hand thermometer. Similarly, let T_R be the temperature and p_R be the pressure in the right-hand thermometer. According to the problem statement, the pressure is the same in the two thermometers when they are both at the triple point of water. We take this pressure to be p_3. Writing Eq. 18-5 for each thermometer,

$$T_L = (273.16\,\text{K})\left(\frac{p_L}{p_3}\right) \quad \text{and} \quad T_R = (273.16\,\text{K})\left(\frac{p_R}{p_3}\right),$$

we subtract the second equation from the first to obtain

$$T_L - T_R = (273.16\,\text{K})\left(\frac{p_L - p_R}{p_3}\right).$$

First, we take $T_L = 373.125$ K (the boiling point of water) and $T_R = 273.16$ K (the triple point of water). Then, $p_L - p_R = 120$ torr. We solve

$$373.125\,\text{K} - 273.16\,\text{K} = (273.16\,\text{K})\left(\frac{120\,\text{torr}}{p_3}\right)$$

for p_3. The result is $p_3 = 328$ torr. Now, we let $T_L = 273.16$ K (the triple point of water) and T_R be the unknown temperature. The pressure difference is $p_L - p_R = 90.0$ torr. Solving the equation

$$273.16\,\text{K} - T_R = (273.16\,\text{K})\left(\frac{90.0\,\text{torr}}{328\,\text{torr}}\right)$$

for the unknown temperature, we obtain $T_R = 348$ K.

2. We take p_3 to be 80 kPa for both thermometers. According to Fig. 18-6, the nitrogen thermometer gives 373.35 K for the boiling point of water. Use Eq. 18-5 to compute the pressure:

$$p_N = \frac{T}{273.16\,\text{K}}\,p_3 = \left(\frac{373.35\,\text{K}}{273.16\,\text{K}}\right)(80\,\text{kPa}) = 109.343\,\text{kPa}.$$

The hydrogen thermometer gives 373.16 K for the boiling point of water and

$$p_H = \left(\frac{373.16\,\text{K}}{273.16\,\text{K}}\right)(80\,\text{kPa}) = 109.287\,\text{kPa}.$$

(a) The difference is $p_N - p_H = 0.056$ kPa ≈ 0.06 kPa.

(b) The pressure in the nitrogen thermometer is higher than the pressure in the hydrogen thermometer.

3. From Eq. 18-6, we see that the limiting value of the pressure ratio is the same as the absolute temperature ratio: $(373.15\,\text{K})/(273.16\,\text{K}) = 1.366$.

4. (a) Let the reading on the Celsius scale be x and the reading on the Fahrenheit scale be y. Then $y = \frac{9}{5}x + 32$. For $x = -71°C$, this gives $y = -96°F$.

(b) The relationship between y and x may be inverted to yield $x = \frac{5}{9}(y - 32)$. Thus, for $y = 134$ we find $x \approx 56.7$ on the Celsius scale.

5. (a) Let the reading on the Celsius scale be x and the reading on the Fahrenheit scale be y. Then $y = \frac{9}{5}x + 32$. If we require $y = 2x$, then we have

$$2x = \frac{9}{5}x + 32 \quad \Rightarrow \quad x = (5)(32) = 160°C$$

which yields $y = 2x = 320°F$.

(b) In this case, we require $y = \frac{1}{2}x$ and find

$$\frac{1}{2}x = \frac{9}{5}x + 32 \quad \Rightarrow \quad x = -\frac{(10)(32)}{13} \approx -24.6°C$$

which yields $y = x/2 = -12.3°F$.

6. We assume scales X and Y are linearly related in the sense that reading x is related to reading y by a linear relationship $y = mx + b$. We determine the constants m and b by solving the simultaneous equations:

$$-70.00 = m(-125.0) + b$$
$$-30.00 = m(375.0) + b$$

which yield the solutions $m = 40.00/500.0 = 8.000 \times 10^{-2}$ and $b = -60.00$. With these values, we find x for $y = 50.00$:

$$x = \frac{y-b}{m} = \frac{50.00 + 60.00}{0.08000} = 1375°X.$$

7. We assume scale X is a linear scale in the sense that if its reading is x then it is related to a reading y on the Kelvin scale by a linear relationship $y = mx + b$. We determine the constants m and b by solving the simultaneous equations:

$$373.15 = m(-53.5) + b$$
$$273.15 = m(-170) + b$$

which yield the solutions $m = 100/(170 - 53.5) = 0.858$ and $b = 419$. With these values, we find x for $y = 340$:

$$x = \frac{y-b}{m} = \frac{340 - 419}{0.858} = -92.1°X.$$

8. The change in length for the aluminum pole is

$$\Delta\ell = \ell_0 \alpha_{Al} \Delta T = (33\,\text{m})(23 \times 10^{-6}/\text{C}°)(15\,°\text{C}) = 0.011\,\text{m}.$$

9. Since a volume is the product of three lengths, the change in volume due to a temperature change ΔT is given by $\Delta V = 3\alpha V \Delta T$, where V is the original volume and α is the coefficient of linear expansion. See Eq. 18-11. Since $V = (4\pi/3)R^3$, where R is the original radius of the sphere, then

$$\Delta V = 3\alpha \left(\frac{4\pi}{3} R^3\right) \Delta T = (23 \times 10^{-6}/\text{C}°)(4\pi)(10\,\text{cm})^3 (100\,°\text{C}) = 29\,\text{cm}^3.$$

The value for the coefficient of linear expansion is found in Table 18-2.

10. (a) The coefficient of linear expansion α for the alloy is

$$\alpha = \frac{\Delta L}{L\Delta T} = \frac{10.015\,\text{cm} - 10.000\,\text{cm}}{(10.01\,\text{cm})(100°\text{C} - 20.000°\text{C})} = 1.88 \times 10^{-5}/\text{C}°.$$

Thus, from 100°C to 0°C we have

$$\Delta L = L\alpha\Delta T = (10.015\,\text{cm})(1.88 \times 10^{-5}/\text{C}°)(0°\text{C} - 100°\text{C}) = -1.88 \times 10^{-2}\,\text{cm}.$$

The length at 0°C is therefore $L' = L + \Delta L = (10.015\,\text{cm} - 0.0188\,\text{cm}) = 9.996\,\text{cm}$.

(b) Let the temperature be T_x. Then from 20°C to T_x we have

$$\Delta L = 10.009 \text{ cm} - 10.000 \text{ cm} = \alpha L \Delta T = (1.88 \times 10^{-5}/\text{C}°)(10.000 \text{ cm}) \Delta T,$$

giving $\Delta T = 48$ °C. Thus, $T_x = (20°\text{C} + 48 °\text{C}) = 68°\text{C}$.

11. The new diameter is

$$D = D_0(1 + \alpha_{Al}\Delta T) = (2.725 \text{ cm})[1 + (23 \times 10^{-6}/\text{C}°)(100.0°\text{C} - 0.000°\text{C})] = 2.731 \text{ cm}.$$

12. The increase in the surface area of the brass cube (which has six faces), which had side length is L at $20°$, is

$$\Delta A = 6(L + \Delta L)^2 - 6L^2 \approx 12L\Delta L = 12\alpha_b L^2 \Delta T = 12 \ (19 \times 10^{-6}/\text{C}°) \ (30 \text{ cm})^2 (75°\text{C} - 20°\text{C})$$
$$= 11 \text{ cm}^2.$$

13. The volume at 30°C is given by

$$V' = V(1 + \beta\Delta T) = V(1 + 3\alpha\Delta T) = (50.00 \text{ cm}^3)[1 + 3(29.00 \times 10^{-6}/\text{C}°) \ (30.00°\text{C} - 60.00°\text{C})]$$
$$= 49.87 \text{ cm}^3$$

where we have used $\beta = 3\alpha$.

14. (a) We use $\rho = m/V$ and

$$\Delta\rho = \Delta(m/V) = m\Delta(1/V) \approx -m\Delta V/V^2 = -\rho(\Delta V/V) = -3\rho(\Delta L/L).$$

The percent change in density is

$$\frac{\Delta\rho}{\rho} = -3\frac{\Delta L}{L} = -3(0.23\%) = -0.69\%.$$

(b) Since $\alpha = \Delta L/(L\Delta T) = (0.23 \times 10^{-2})/(100°\text{C} - 0.0°\text{C}) = 23 \times 10^{-6}/\text{C}°$, the metal is aluminum (using Table 18-2).

15. If V_c is the original volume of the cup, α_a is the coefficient of linear expansion of aluminum, and ΔT is the temperature increase, then the change in the volume of the cup is $\Delta V_c = 3\alpha_a V_c \Delta T$. See Eq. 18-11. If β is the coefficient of volume expansion for glycerin then the change in the volume of glycerin is $\Delta V_g = \beta V_c \Delta T$. Note that the original volume of glycerin is the same as the original volume of the cup. The volume of glycerin that spills is

$$\Delta V_g - \Delta V_c = (\beta - 3\alpha_a)V_c\Delta T = \left[(5.1 \times 10^{-4}/\text{C}°) - 3(23 \times 10^{-6}/\text{C}°)\right](100 \text{ cm}^3)(6.0°\text{C})$$
$$= 0.26 \text{ cm}^3.$$

16. The change in length for the section of the steel ruler between its 20.05 cm mark and 20.11 cm mark is

$$\Delta L_s = L_s \alpha_s \Delta T = (20.11\,\text{cm})(11 \times 10^{-6}/\text{C}°)(270°\text{C} - 20°\text{C}) = 0.055\,\text{cm}.$$

Thus, the actual change in length for the rod is

$$\Delta L = (20.11\text{ cm} - 20.05\text{ cm}) + 0.055\text{ cm} = 0.115\text{ cm}.$$

The coefficient of thermal expansion for the material of which the rod is made is then

$$\alpha = \frac{\Delta L}{\Delta T} = \frac{0.115\text{ cm}}{270°\text{C} - 20°\text{C}} = 23 \times 10^{-6}/\text{C}°.$$

17. After the change in temperature the diameter of the steel rod is $D_s = D_{s0} + \alpha_s D_{s0}\,\Delta T$ and the diameter of the brass ring is $D_b = D_{b0} + \alpha_b D_{b0}\,\Delta T$, where D_{s0} and D_{b0} are the original diameters, α_s and α_b are the coefficients of linear expansion, and ΔT is the change in temperature. The rod just fits through the ring if $D_s = D_b$. This means

$$D_{s0} + \alpha_s D_{s0}\,\Delta T = D_{b0} + \alpha_b D_{b0}\,\Delta T.$$

Therefore,

$$\Delta T = \frac{D_{s0} - D_{b0}}{\alpha_b D_{b0} - \alpha_s D_{s0}} = \frac{3.000\,\text{cm} - 2.992\,\text{cm}}{(19.00 \times 10^{-6}/\text{C}°)(2.992\,\text{cm}) - (11.00 \times 10^{-6}/\text{C}°)(3.000\,\text{cm})}$$
$$= 335.0\,°\text{C}.$$

The temperature is $T = (25.00°\text{C} + 335.0\,°\text{C}) = 360.0°\text{C}$.

18. (a) Since $A = \pi D^2/4$, we have the differential $dA = 2(\pi D/4)dD$. Dividing the latter relation by the former, we obtain $dA/A = 2\,dD/D$. In terms of Δ's, this reads

$$\frac{\Delta A}{A} = 2\frac{\Delta D}{D} \quad \text{for} \quad \frac{\Delta D}{D} \ll 1.$$

We can think of the factor of 2 as being due to the fact that area is a two-dimensional quantity. Therefore, the area increases by $2(0.18\%) = 0.36\%$.

(b) Assuming that all dimensions are allowed to freely expand, then the thickness increases by 0.18%.

(c) The volume (a three-dimensional quantity) increases by $3(0.18\%) = 0.54\%$.

(d) The mass does not change.

(e) The coefficient of linear expansion is

$$\alpha = \frac{\Delta D}{D \Delta T} = \frac{0.18 \times 10^{-2}}{100°C} = 1.8 \times 10^{-5} /C°.$$

19. The initial volume V_0 of the liquid is $h_0 A_0$ where A_0 is the initial cross-section area and $h_0 = 0.64$ m. Its final volume is $V = hA$ where $h - h_0$ is what we wish to compute. Now, the area expands according to how the glass expands, which we analyze as follows: Using $A = \pi r^2$, we obtain

$$dA = 2\pi r\, dr = 2\pi r \left(r \alpha\, dT\right) = 2\alpha(\pi r^2) dT = 2\alpha A\, dT.$$

Therefore, the height is

$$h = \frac{V}{A} = \frac{V_0\left(1 + \beta_{\text{liquid}} \Delta T\right)}{A_0\left(1 + 2\alpha_{\text{glass}} \Delta T\right)}.$$

Thus, with $V_0/A_0 = h_0$ we obtain

$$h - h_0 = h_0 \left(\frac{1 + \beta_{\text{liquid}} \Delta T}{1 + 2\alpha_{\text{glass}} \Delta T} - 1\right) = (0.64)\left(\frac{1 + \left(4 \times 10^{-5}\right)(10°)}{1 + 2\left(1 \times 10^{-5}\right)(10°)}\right) = 1.3 \times 10^{-4} \text{ m}.$$

20. We divide Eq. 18-9 by the time increment Δt and equate it to the (constant) speed $v = 100 \times 10^{-9}$ m/s.

$$v = \alpha L_0 \frac{\Delta T}{\Delta t}$$

where $L_0 = 0.0200$ m and $\alpha = 23 \times 10^{-6}/C°$. Thus, we obtain

$$\frac{\Delta T}{\Delta t} = 0.217 \frac{C°}{s} = 0.217 \frac{K}{s}.$$

21. Consider half the bar. Its original length is $\ell_0 = L_0/2$ and its length after the temperature increase is $\ell = \ell_0 + \alpha \ell_0 \Delta T$. The old position of the half-bar, its new position, and the distance x that one end is displaced form a right triangle, with a hypotenuse of length ℓ, one side of length ℓ_0, and the other side of length x. The Pythagorean theorem yields

$$x^2 = \ell^2 - \ell_0^2 = \ell_0^2(1 + \alpha \Delta T)^2 - \ell_0^2.$$

Since the change in length is small we may approximate $(1 + \alpha \Delta T)^2$ by $1 + 2\alpha \Delta T$, where the small term $(\alpha \Delta T)^2$ was neglected. Then,

$$x^2 = \ell_0^2 + 2\ell_0^2 \alpha \Delta T - \ell_0^2 = 2\ell_0^2 \alpha \Delta T$$

and

$$x = \ell_0\sqrt{2\alpha \Delta T} = \frac{3.77\,\text{m}}{2}\sqrt{2(25\times 10^{-6}/\text{C}°)(32°\text{C})} = 7.5\times 10^{-2}\,\text{m}.$$

22. (a) The specific heat is given by $c = Q/m(T_f - T_i)$, where Q is the heat added, m is the mass of the sample, T_i is the initial temperature, and T_f is the final temperature. Thus, recalling that a change in Celsius degrees is equal to the corresponding change on the Kelvin scale,

$$c = \frac{314\,\text{J}}{(30.0\times 10^{-3}\,\text{kg})(45.0°\text{C} - 25.0°\text{C})} = 523\,\text{J/kg}\cdot\text{K}.$$

(b) The molar specific heat is given by

$$c_m = \frac{Q}{N(T_f - T_i)} = \frac{314\,\text{J}}{(0.600\,\text{mol})(45.0°\text{C} - 25.0°\text{C})} = 26.2\,\text{J/mol}\cdot\text{K}.$$

(c) If N is the number of moles of the substance and M is the mass per mole, then $m = NM$, so

$$N = \frac{m}{M} = \frac{30.0\times 10^{-3}\,\text{kg}}{50\times 10^{-3}\,\text{kg/mol}} = 0.600\,\text{mol}.$$

23. We use $Q = cm\Delta T$. The textbook notes that a nutritionist's "Calorie" is equivalent to 1000 cal. The mass m of the water that must be consumed is

$$m = \frac{Q}{c\Delta T} = \frac{3500\times 10^3\,\text{cal}}{(1\,\text{g/cal}\cdot\text{C}°)(37.0°\text{C} - 0.0°\text{C})} = 94.6\times 10^4\,\text{g},$$

which is equivalent to 9.46×10^4 g/(1000 g/liter) = 94.6 liters of water. This is certainly too much to drink in a single day!

24. The amount of water m that is frozen is

$$m = \frac{Q}{L_F} = \frac{50.2\,\text{kJ}}{333\,\text{kJ/kg}} = 0.151\,\text{kg} = 151\,\text{g}.$$

Therefore the amount of water which remains unfrozen is 260 g − 151 g = 109 g.

25. The melting point of silver is 1235 K, so the temperature of the silver must first be raised from 15.0° C (= 288 K) to 1235 K. This requires heat

$$Q = cm(T_f - T_i) = (236 \text{ J/kg} \cdot \text{K})(0.130 \text{ kg})(1235°\text{C} - 288°\text{C}) = 2.91 \times 10^4 \text{ J}.$$

Now the silver at its melting point must be melted. If L_F is the heat of fusion for silver this requires

$$Q = mL_F = (0.130 \text{ kg})(105 \times 10^3 \text{ J/kg}) = 1.36 \times 10^4 \text{ J}.$$

The total heat required is (2.91×10^4 J + 1.36×10^4 J) = 4.27×10^4 J.

26. (a) The water (of mass m) releases energy in two steps, first by lowering its temperature from 20°C to 0°C, and then by freezing into ice. Thus the total energy transferred from the water to the surroundings is

$$Q = c_w m \Delta T + L_F m = (4190 \text{ J/kg} \cdot \text{K})(125 \text{ kg})(20°\text{C}) + (333 \text{ kJ/kg})(125 \text{ kg}) = 5.2 \times 10^7 \text{ J}.$$

(b) Before all the water freezes, the lowest temperature possible is 0°C, below which the water must have already turned into ice.

27. The mass $m = 0.100$ kg of water, with specific heat $c = 4190$ J/kg·K, is raised from an initial temperature $T_i = 23$°C to its boiling point $T_f = 100$°C. The heat input is given by $Q = cm(T_f - T_i)$. This must be the power output of the heater P multiplied by the time t; $Q = Pt$. Thus,

$$t = \frac{Q}{P} = \frac{cm(T_f - T_i)}{P} = \frac{(4190 \text{ J/kg} \cdot \text{K})(0.100 \text{ kg})(100°\text{C} - 23°\text{C})}{200 \text{ J/s}} = 160 \text{ s}.$$

28. The work the man has to do to climb to the top of Mt. Everest is given by

$$W = mgy = (73.0 \text{ kg})(9.80 \text{ m/s}^2)(8840 \text{ m}) = 6.32 \times 10^6 \text{ J}.$$

Thus, the amount of butter needed is

$$m = \frac{(6.32 \times 10^6 \text{ J})\left(\frac{1.00 \text{ cal}}{4.186 \text{ J}}\right)}{6000 \text{ cal/g}} \approx 250 \text{ g}.$$

29. Let the mass of the steam be m_s and that of the ice be m_i. Then

$$L_F m_c + c_w m_c (T_f - 0.0°\text{C}) = m_s L_s + m_s c_w (100°\text{C} - T_f),$$

where $T_f = 50$°C is the final temperature. We solve for m_s:

$$m_s = \frac{L_F m_c + c_w m_c (T_f - 0.0°C)}{L_s + c_w (100°C - T_f)} = \frac{(79.7\,\text{cal}/\text{g})(150\,\text{g}) + (1\,\text{cal}/\text{g}\cdot°C)(150\,\text{g})(50°C - 0.0°C)}{539\,\text{cal}/\text{g} + (1\,\text{cal}/\text{g}\cdot C°)(100°C - 50°C)}$$
$$= 33\,\text{g}.$$

30. (a) Using Eq. 18-17, the heat transferred to the water is

$$Q_w = c_w m_w \Delta T + L_V m_s = (1\,\text{cal}/\text{g}\cdot C°)(220\,\text{g})(100°C - 20.0°C) + (539\,\text{cal}/\text{g})(5.00\,\text{g})$$
$$= 20.3\,\text{kcal}.$$

(b) The heat transferred to the bowl is

$$Q_b = c_b m_b \Delta T = (0.0923\,\text{cal}/\text{g}\cdot C°)(150\,\text{g})(100°C - 20.0°C) = 1.11\,\text{kcal}.$$

(c) If the original temperature of the cylinder be T_i, then $Q_w + Q_b = c_c m_c (T_i - T_f)$, which leads to

$$T_i = \frac{Q_w + Q_b}{c_c m_c} + T_f = \frac{20.3\,\text{kcal} + 1.11\,\text{kcal}}{(0.0923\,\text{cal}/\text{g}\cdot C°)(300\,\text{g})} + 100°C = 873°C.$$

31. We note from Eq. 18-12 that 1 Btu = 252 cal. The heat relates to the power, and to the temperature change, through $Q = Pt = cm\Delta T$. Therefore, the time t required is

$$t = \frac{cm\Delta T}{P} = \frac{(1000\,\text{cal}/\text{kg}\cdot C°)(40\,\text{gal})(1000\,\text{kg}/264\,\text{gal})(100°F - 70°F)(5°C/9°F)}{(2.0\times 10^5\,\text{Btu}/\text{h})(252.0\,\text{cal}/\text{Btu})(1\,\text{h}/60\,\text{min})}$$
$$= 3.0\,\text{min}.$$

The metric version proceeds similarly:

$$t = \frac{c\rho V \Delta T}{P} = \frac{(4190\,\text{J/kg}\cdot C°)(1000\,\text{kg/m}^3)(150\,\text{L})(1\,\text{m}^3/1000\,\text{L})(38°C - 21°C)}{(59000\,\text{J/s})(60\,\text{s}/1\,\text{min})}$$
$$= 3.0\,\text{min}.$$

32. We note that the heat capacity of sample B is given by the reciprocal of the slope of the line in Figure 18-32(b) (compare with Eq. 18-14). Since the reciprocal of that slope is 16/4 = 4 kJ/kg·C°, then c_B = 4000 J/kg·C° = 4000 J/kg·K (since a change in Celsius is equivalent to a change in Kelvins). Now, following the same procedure as shown in Sample Problem 18-4, we find

$$c_A m_A (T_f - T_A) + c_B m_B (T_f - T_B) = 0$$

$$c_A (5.0\,\text{kg})(40°C - 100°C) + (4000\,\text{J/kg}\cdot C°)(1.5\,\text{kg})(40°C - 20°C) = 0$$

which leads to $c_A = 4.0\times 10^2$ J/kg·K.

33. The power consumed by the system is

$$P = \left(\frac{1}{20\%}\right)\frac{cm\Delta T}{t} = \left(\frac{1}{20\%}\right)\frac{(4.18\,\text{J/g}\cdot°\text{C})(200\times10^3\,\text{cm}^3)(1\,\text{g/cm}^3)(40°\text{C}-20°\text{C})}{(1.0\,\text{h})(3600\,\text{s/h})}$$
$$= 2.3\times10^4\,\text{W}.$$

The area needed is then $A = \dfrac{2.3\times10^4\,\text{W}}{700\,\text{W/m}^2} = 33\,\text{m}^2$.

34. While the sample is in its liquid phase, its temperature change (in absolute values) is $|\Delta T| = 30\,°\text{C}$. Thus, with $m = 0.40$ kg, the absolute value of Eq. 18-14 leads to

$$|Q| = cm|\Delta T| = (3000\,\text{J/kg}\cdot°\text{C})(0.40\,\text{kg})(30\,°\text{C}) = 36000\,\text{J}.$$

The rate (which is constant) is

$$P = |Q|/t = (36000\,\text{J})/(40\,\text{min}) = 900\,\text{J/min},$$

which is equivalent to 15 Watts.

(a) During the next 30 minutes, a phase change occurs which is described by Eq. 18-16:

$$|Q| = Pt = (900\,\text{J/min})(30\,\text{min}) = 27000\,\text{J} = Lm.$$

Thus, with $m = 0.40$ kg, we find $L = 67500$ J/kg ≈ 68 kJ/kg.

(b) During the final 20 minutes, the sample is solid and undergoes a temperature change (in absolute values) of $|\Delta T| = 20\,\text{C}°$. Now, the absolute value of Eq. 18-14 leads to

$$c = \frac{|Q|}{m|\Delta T|} = \frac{Pt}{m|\Delta T|} = \frac{(900)(20)}{(0.40)(20)} = 2250\,\frac{\text{J}}{\text{kg}\cdot\text{C}°} \approx 2.3\,\frac{\text{kJ}}{\text{kg}\cdot\text{C}°}.$$

35. We denote the ice with subscript I and the coffee with c, respectively. Let the final temperature be T_f. The heat absorbed by the ice is

$$Q_I = \lambda_F m_I + m_I c_w (T_f - 0°\text{C}),$$

and the heat given away by the coffee is $|Q_c| = m_w c_w (T_I - T_f)$. Setting $Q_I = |Q_c|$, we solve for T_f:

$$T_f = \frac{m_w c_w T_I - \lambda_F m_I}{(m_I + m_c)c_w} = \frac{(130\,\text{g})(4190\,\text{J/kg}\cdot\text{C}°)(80.0°\text{C}) - (333\times10^3\,\text{J/g})(12.0\,\text{g})}{(12.0\,\text{g}+130\,\text{g})(4190\,\text{J/kg}\cdot\text{C}°)}$$
$$= 66.5°\text{C}.$$

Note that we work in Celsius temperature, which poses no difficulty for the J/kg·K values of specific heat capacity (see Table 18-3) since a change of Kelvin temperature is numerically equal to the corresponding change on the Celsius scale. Therefore, the temperature of the coffee will cool by $|\Delta T| = 80.0°C - 66.5°C = 13.5 C°$.

36. (a) Eq. 18-14 (in absolute value) gives

$$|Q| = (4190 \text{ J/ kg} \cdot °C)(0.530 \text{ kg})(40 \text{ °C}) = 88828 \text{ J}.$$

Since $\frac{dQ}{dt}$ is assumed constant (we will call it P) then we have

$$P = \frac{88828 \text{ J}}{40 \text{ min}} = \frac{88828 \text{ J}}{2400 \text{ s}} = 37 \text{ W}.$$

(b) During that same time (used in part (a)) the ice warms by 20 C°. Using Table 18-3 and Eq. 18-14 again we have

$$m_{\text{ice}} = \frac{Q}{c_{\text{ice}} \Delta T} = \frac{88828}{(2220)(20°)} = 2.0 \text{ kg}.$$

(c) To find the ice produced (by freezing the water that has already reached 0°C – so we concerned with the 40 min $< t <$ 60 min time span), we use Table 18-4 and Eq. 18-16:

$$m_{\text{water becoming ice}} = \frac{Q_{20 \text{ min}}}{L_F} = \frac{44414}{333000} = 0.13 \text{ kg}.$$

37. To accomplish the phase change at 78°C,

$$Q = L_V m = (879 \text{ kJ/kg})(0.510 \text{ kg}) = 448.29 \text{ kJ}$$

must be removed. To cool the liquid to –114°C,

$$Q = cm|\Delta T| = (2.43 \text{ kJ/ kg} \cdot \text{K})(0.510 \text{ kg})(192 \text{ K}) = 237.95 \text{ kJ},$$

must be removed. Finally, to accomplish the phase change at –114°C,

$$Q = L_F m = (109 \text{ kJ/kg})(0.510 \text{ kg}) = 55.59 \text{ kJ}$$

must be removed. The grand total of heat removed is therefore (448.29 + 237.95 + 55.59) kJ = 742 kJ.

38. The heat needed is found by integrating the heat capacity:

$$Q = \int_{T_i}^{T_f} cm \, dT = m \int_{T_i}^{T_f} c \, dT = (2.09) \int_{5.0°C}^{15.0°C} (0.20 + 0.14T + 0.023T^2) \, dT$$

$$= (2.0)(0.20T + 0.070T^2 + 0.00767T^3)\Big|_{5.0}^{15.0} \text{ (cal)}$$

$$= 82 \text{ cal.}$$

39. We compute with Celsius temperature, which poses no difficulty for the J/kg·K values of specific heat capacity (see Table 18-3) since a change of Kelvin temperature is numerically equal to the corresponding change on the Celsius scale. If the equilibrium temperature is T_f then the energy absorbed as heat by the ice is

$$Q_I = L_F m_I + c_w m_I (T_f - 0°C),$$

while the energy transferred as heat from the water is $Q_w = c_w m_w (T_f - T_i)$. The system is insulated, so $Q_w + Q_I = 0$, and we solve for T_f:

$$T_f = \frac{c_w m_w T_i - L_F m_I}{(m_I + m_c) c_w}.$$

(a) Now $T_i = 90°C$ so

$$T_f = \frac{(4190 \text{ J/kg} \cdot \text{C°})(0.500 \text{ kg})(90°C) - (333 \times 10^3 \text{ J/kg})(0.500 \text{ kg})}{(0.500 \text{ kg} + 0.500 \text{ kg})(4190 \text{ J/kg} \cdot \text{C°})} = 5.3°C.$$

(b) Since no ice has remained at $T_f = 5.3°C$, we have $m_f = 0$.

(c) If we were to use the formula above with $T_i = 70°C$, we would get $T_f < 0$, which is impossible. In fact, not all the ice has melted in this case and the equilibrium temperature is $T_f = 0°C$.

(d) The amount of ice that melts is given by

$$m_I' = \frac{c_w m_w (T_i - 0°C)}{L_F} = \frac{(4190 \text{ J/kg} \cdot \text{C°})(0.500 \text{ kg})(70 \text{ C°})}{333 \times 10^3 \text{ J/kg}} = 0.440 \text{ kg.}$$

Therefore, the amount of (solid) ice remaining is $m_f = m_I - m_I' = 500$ g $- 440$ g $= 60.0$ g, and (as mentioned) we have $T_f = 0°C$ (because the system is an ice-water mixture in thermal equilibrium).

40. (a) Using Eq. 18-32, we find the rate of energy conducted upward to be

$$P_{\text{cond}} = \frac{Q}{t} = kA \frac{T_H - T_C}{L} = (0.400 \text{ W/m} \cdot °C) A \frac{5.0 \text{ °C}}{0.12 \text{ m}} = (16.7 A) \text{ W.}$$

Recall that a change in Celsius temperature is numerically equivalent to a change on the Kelvin scale.

(b) The heat of fusion in this process is $Q = L_F m$, where $L_F = 3.33 \times 10^5$ J/kg. Differentiating the expression with respect to t and equating the result with P_{cond}, we have

$$P_{cond} = \frac{dQ}{dt} = L_F \frac{dm}{dt}.$$

Thus, the rate of mass converted from liquid to ice is

$$\frac{dm}{dt} = \frac{P_{cond}}{L_F} = \frac{16.7A \text{ W}}{3.33 \times 10^5 \text{ J/kg}} = (5.02 \times 10^{-5} A) \text{ kg/s}.$$

(c) Since $m = \rho V = \rho A h$, differentiating both sides of the expression gives

$$\frac{dm}{dt} = \frac{d}{dt}(\rho A h) = \rho A \frac{dh}{dt}.$$

Thus, the rate of change of the icicle length is

$$\frac{dh}{dt} = \frac{1}{\rho A} \frac{dm}{dt} = \frac{5.02 \times 10^{-5} \text{ kg/m}^2 \cdot \text{s}}{1000 \text{ kg/m}^3} = 5.02 \times 10^{-8} \text{ m/s}$$

41. (a) We work in Celsius temperature, which poses no difficulty for the J/kg·K values of specific heat capacity (see Table 18-3) since a change of Kelvin temperature is numerically equal to the corresponding change on the Celsius scale. There are three possibilities:

• None of the ice melts and the water-ice system reaches thermal equilibrium at a temperature that is at or below the melting point of ice.

• The system reaches thermal equilibrium at the melting point of ice, with some of the ice melted.

• All of the ice melts and the system reaches thermal equilibrium at a temperature at or above the melting point of ice.

First, suppose that no ice melts. The temperature of the water decreases from $T_{Wi} = 25°C$ to some final temperature T_f and the temperature of the ice increases from $T_{Ii} = -15°C$ to T_f. If m_W is the mass of the water and c_W is its specific heat then the water rejects heat

$$|Q| = c_W m_W (T_{Wi} - T_f).$$

If m_I is the mass of the ice and c_I is its specific heat then the ice absorbs heat

$$Q = c_I m_I (T_f - T_{Ii}).$$

Since no energy is lost to the environment, these two heats (in absolute value) must be the same. Consequently,

$$c_W m_W (T_{Wi} - T_f) = c_I m_I (T_f - T_{Ii}).$$

The solution for the equilibrium temperature is

$$T_f = \frac{c_W m_W T_{Wi} + c_I m_I T_{Ii}}{c_W m_W + c_I m_I}$$

$$= \frac{(4190\,\text{J/kg}\cdot\text{K})(0.200\,\text{kg})(25°\text{C}) + (2220\,\text{J/kg}\cdot\text{K})(0.100\,\text{kg})(-15°\text{C})}{(4190\,\text{J/kg}\cdot\text{K})(0.200\,\text{kg}) + (2220\,\text{J/kg}\cdot\text{K})(0.100\,\text{kg})}$$

$$= 16.6°\text{C}.$$

This is above the melting point of ice, which invalidates our assumption that no ice has melted. That is, the calculation just completed does not take into account the melting of the ice and is in error. Consequently, we start with a new assumption: that the water and ice reach thermal equilibrium at $T_f = 0°\text{C}$, with mass m ($< m_I$) of the ice melted. The magnitude of the heat rejected by the water is

$$|Q| = c_W m_W T_{Wi},$$

and the heat absorbed by the ice is

$$Q = c_I m_I (0 - T_{Ii}) + m L_F,$$

where L_F is the heat of fusion for water. The first term is the energy required to warm all the ice from its initial temperature to 0°C and the second term is the energy required to melt mass m of the ice. The two heats are equal, so

$$c_W m_W T_{Wi} = -c_I m_I T_{Ii} + m L_F.$$

This equation can be solved for the mass m of ice melted:

$$m = \frac{c_W m_W T_{Wi} + c_I m_I T_{Ii}}{L_F}$$

$$= \frac{(4190\,\text{J/kg}\cdot\text{K})(0.200\,\text{kg})(25°\text{C}) + (2220\,\text{J/kg}\cdot\text{K})(0.100\,\text{kg})(-15°\text{C})}{333 \times 10^3\,\text{J/kg}}$$

$$= 5.3 \times 10^{-2}\,\text{kg} = 53\,\text{g}.$$

Since the total mass of ice present initially was 100 g, there *is* enough ice to bring the water temperature down to 0°C. This is then the solution: the ice and water reach thermal equilibrium at a temperature of 0°C with 53 g of ice melted.

(b) Now there is less than 53 g of ice present initially. All the ice melts and the final temperature is above the melting point of ice. The heat rejected by the water is

$$|Q| = c_W m_W (T_{Wi} - T_f)$$

and the heat absorbed by the ice and the water it becomes when it melts is

$$Q = c_I m_I (0 - T_{Ii}) + c_W m_I (T_f - 0) + m_I L_F.$$

The first term is the energy required to raise the temperature of the ice to 0°C, the second term is the energy required to raise the temperature of the melted ice from 0°C to T_f, and the third term is the energy required to melt all the ice. Since the two heats are equal,

$$c_W m_W (T_{Wi} - T_f) = c_I m_I (-T_{Ii}) + c_W m_I T_f + m_I L_F.$$

The solution for T_f is

$$T_f = \frac{c_W m_W T_{Wi} + c_I m_I T_{Ii} - m_I L_F}{c_W (m_W + m_I)}.$$

Inserting the given values, we obtain $T_f = 2.5°C$.

42. If the ring diameter at 0.000°C is D_{r0} then its diameter when the ring and sphere are in thermal equilibrium is

$$D_r = D_{r0} (1 + \alpha_c T_f),$$

where T_f is the final temperature and α_c is the coefficient of linear expansion for copper. Similarly, if the sphere diameter at T_i (= 100.0°C) is D_{s0} then its diameter at the final temperature is

$$D_s = D_{s0} [1 + \alpha_a (T_f - T_i)],$$

where α_a is the coefficient of linear expansion for aluminum. At equilibrium the two diameters are equal, so

$$D_{r0}(1 + \alpha_c T_f) = D_{s0}[1 + \alpha_a (T_f - T_i)].$$

The solution for the final temperature is

$$T_f = \frac{D_{r0} - D_{s0} + D_{s0}\alpha_a T_i}{D_{s0}\alpha_a - D_{r0}\alpha_c}$$

$$= \frac{2.54000\,\text{cm} - 2.54508\,\text{cm} + (2.54508\,\text{cm})(23\times 10^{-6}/\text{C}°)(100.0°\text{C})}{(2.54508\,\text{cm})(23\times 10^{-6}/\text{C}°) - (2.54000\,\text{cm})(17\times 10^{-6}/\text{C}°)}$$

$$= 50.38°\text{C}.$$

The expansion coefficients are from Table 18-2 of the text. Since the initial temperature of the ring is 0°C, the heat it absorbs is $Q = c_c m_r T_f$, where c_c is the specific heat of copper and m_r is the mass of the ring. The heat rejected up by the sphere is

$$|Q| = c_a m_s (T_i - T_f)$$

where c_a is the specific heat of aluminum and m_s is the mass of the sphere. Since these two heats are equal,

$$c_c m_r T_f = c_a m_s (T_i - T_f),$$

we use specific heat capacities from the textbook to obtain

$$m_s = \frac{c_c m_r T_f}{c_a (T_i - T_f)} = \frac{(386\,\text{J/kg}\cdot\text{K})(0.0200\,\text{kg})(50.38°\text{C})}{(900\,\text{J/kg}\cdot\text{K})(100°\text{C} - 50.38°\text{C})} = 8.71\times 10^{-3}\,\text{kg}.$$

43. Over a cycle, the internal energy is the same at the beginning and end, so the heat Q absorbed equals the work done: $Q = W$. Over the portion of the cycle from A to B the pressure p is a linear function of the volume V and we may write

$$p = \frac{10}{3}\,\text{Pa} + \left(\frac{20}{3}\,\text{Pa/m}^3\right) V,$$

where the coefficients were chosen so that $p = 10$ Pa when $V = 1.0$ m³ and $p = 30$ Pa when $V = 4.0$ m³. The work done by the gas during this portion of the cycle is

$$W_{AB} = \int_1^4 p\,dV = \int_1^4 \left(\frac{10}{3} + \frac{20}{3}V\right) dV = \left(\frac{10}{3}V + \frac{10}{3}V^2\right)\Bigg|_1^4$$

$$= \left(\frac{40}{3} + \frac{160}{3} - \frac{10}{3} - \frac{10}{3}\right) \text{J} = 60\,\text{J}.$$

The BC portion of the cycle is at constant pressure and the work done by the gas is

$$W_{BC} = p\Delta V = (30\,\text{Pa})(1.0\,\text{m}^3 - 4.0\,\text{m}^3) = -90\,\text{J}.$$

The CA portion of the cycle is at constant volume, so no work is done. The total work done by the gas is

$$W = W_{AB} + W_{BC} + W_{CA} = 60 \text{ J} - 90 \text{ J} + 0 = -30 \text{ J}$$

and the total heat absorbed is $Q = W = -30$ J. This means the gas loses 30 J of energy in the form of heat.

44. (a) Since work is done *on* the system (perhaps to compress it) we write $W = -200$ J.

(b) Since heat leaves the system, we have $Q = -70.0$ cal $= -293$ J.

(c) The change in internal energy is $\Delta E_{int} = Q - W = -293 \text{ J} - (-200 \text{ J}) = -93$ J.

45. (a) One part of path A represents a constant pressure process. The volume changes from 1.0 m³ to 4.0 m³ while the pressure remains at 40 Pa. The work done is

$$W_A = p\Delta V = (40 \text{ Pa})(4.0 \text{ m}^3 - 1.0 \text{ m}^3) = 1.2 \times 10^2 \text{ J}.$$

(b) The other part of the path represents a constant volume process. No work is done during this process. The total work done over the entire path is 120 J. To find the work done over path B we need to know the pressure as a function of volume. Then, we can evaluate the integral $W = \int p\, dV$. According to the graph, the pressure is a linear function of the volume, so we may write $p = a + bV$, where a and b are constants. In order for the pressure to be 40 Pa when the volume is 1.0 m³ and 10 Pa when the volume is 4.00 m³ the values of the constants must be $a = 50$ Pa and $b = -10$ Pa/m³. Thus,

$$p = 50 \text{ Pa} - (10 \text{ Pa/m}^3)V$$

and

$$W_B = \int_1^4 p\, dV = \int_1^4 (50 - 10V)\, dV = (50V - 5V^2)\Big|_1^4 = 200 \text{ J} - 50 \text{ J} - 80 \text{ J} + 5.0 \text{ J} = 75 \text{ J}.$$

(c) One part of path C represents a constant pressure process in which the volume changes from 1.0 m³ to 4.0 m³ while p remains at 10 Pa. The work done is

$$W_C = p\Delta V = (10 \text{ Pa})(4.0 \text{ m}^3 - 1.0 \text{ m}^3) = 30 \text{ J}.$$

The other part of the process is at constant volume and no work is done. The total work is 30 J. We note that the work is different for different paths.

46. During process $A \to B$, the system is expanding, doing work on its environment, so $W > 0$, and since $\Delta E_{int} > 0$ is given then $Q = W + \Delta E_{int}$ must also be positive.

(a) $Q > 0$.

(b) $W > 0$.

During process $B \to C$, the system is neither expanding nor contracting. Thus,

(c) $W = 0$.

(d) The sign of ΔE_{int} must be the same (by the first law of thermodynamics) as that of Q which is given as positive. Thus, $\Delta E_{int} > 0$.

During process $C \to A$, the system is contracting. The environment is doing work on the system, which implies $W < 0$. Also, $\Delta E_{int} < 0$ because $\Sigma \Delta E_{int} = 0$ (for the whole cycle) and the other values of ΔE_{int} (for the other processes) were positive. Therefore, $Q = W + \Delta E_{int}$ must also be negative.

(e) $Q < 0$.

(f) $W < 0$.

(g) $\Delta E_{int} < 0$.

(h) The area of a triangle is $\frac{1}{2}$ (base)(height). Applying this to the figure, we find $|W_{net}| = \frac{1}{2}(2.0\,\text{m}^3)(20\,\text{Pa}) = 20\,\text{J}$. Since process $C \to A$ involves larger negative work (it occurs at higher average pressure) than the positive work done during process $A \to B$, then the net work done during the cycle must be negative. The answer is therefore $W_{net} = -20$ J.

47. We note that there is no work done in the process going from d to a, so $Q_{da} = \Delta E_{int\,da} = 80$ J. Also, since the total change in internal energy around the cycle is zero, then

$$\Delta E_{int\,ac} + \Delta E_{int\,cd} + \Delta E_{int\,da} = 0$$

$$-200\,\text{J} + \Delta E_{int\,cd} + 80\,\text{J} = 0$$

which yields $\Delta E_{int\,cd} = 120$ J. Thus, applying the first law of thermodynamics to the c to d process gives the work done as

$$W_{cd} = Q_{cd} - \Delta E_{int\,cd} = 180\,\text{J} - 120\,\text{J} = 60\,\text{J}.$$

48. (a) We note that process a to b is an expansion, so $W > 0$ for it. Thus, $W_{ab} = +5.0$ J. We are told that the change in internal energy during that process is $+3.0$ J, so application of the first law of thermodynamics for that process immediately yields $Q_{ab} = +8.0$ J.

(b) The net work $(+1.2\,\text{J})$ is the same as the net heat $(Q_{ab} + Q_{bc} + Q_{ca})$, and we are told that $Q_{ca} = +2.5$ J. Thus we readily find $Q_{bc} = (1.2 - 8.0 - 2.5)\,\text{J} = -9.3$ J.

49. (a) The change in internal energy ΔE_{int} is the same for path *iaf* and path *ibf*. According to the first law of thermodynamics, $\Delta E_{int} = Q - W$, where Q is the heat absorbed and W is the work done by the system. Along *iaf*

$$\Delta E_{int} = Q - W = 50 \text{ cal} - 20 \text{ cal} = 30 \text{ cal}.$$

Along *ibf*,

$$W = Q - \Delta E_{int} = 36 \text{ cal} - 30 \text{ cal} = 6.0 \text{ cal}.$$

(b) Since the curved path is traversed from *f* to *i* the change in internal energy is –30 cal and $Q = \Delta E_{int} + W = -30 \text{ cal} - 13 \text{ cal} = -43 \text{ cal}$.

(c) Let $\Delta E_{int} = E_{int, f} - E_{int, i}$. Then, $E_{int, f} = \Delta E_{int} + E_{int, i} = 30 \text{ cal} + 10 \text{ cal} = 40 \text{ cal}$.

(d) The work W_{bf} for the path *bf* is zero, so $Q_{bf} = E_{int, f} - E_{int, b} = 40 \text{ cal} - 22 \text{ cal} = 18 \text{ cal}$.

(e) For the path *ibf*, $Q = 36$ cal so $Q_{ib} = Q - Q_{bf} = 36 \text{ cal} - 18 \text{ cal} = 18 \text{ cal}$.

50. Since the process is a complete cycle (beginning and ending in the same thermodynamic state) the change in the internal energy is zero and the heat absorbed by the gas is equal to the work done by the gas: $Q = W$. In terms of the contributions of the individual parts of the cycle $Q_{AB} + Q_{BC} + Q_{CA} = W$ and

$$Q_{CA} = W - Q_{AB} - Q_{BC} = +15.0 \text{ J} - 20.0 \text{ J} - 0 = -5.0 \text{ J}.$$

This means 5.0 J of energy leaves the gas in the form of heat.

51. The rate of heat flow is given by

$$P_{cond} = kA \frac{T_H - T_C}{L},$$

where k is the thermal conductivity of copper (401 W/m·K), A is the cross-sectional area (in a plane perpendicular to the flow), L is the distance along the direction of flow between the points where the temperature is T_H and T_C. Thus,

$$P_{cond} = \frac{(401 \text{ W/m} \cdot \text{K})(90.0 \times 10^{-4} \text{ m}^2)(125°\text{C} - 10.0°\text{C})}{0.250 \text{ m}} = 1.66 \times 10^3 \text{ J/s}.$$

The thermal conductivity is found in Table 18-6 of the text. Recall that a change in Kelvin temperature is numerically equivalent to a change on the Celsius scale.

52. (a) We estimate the surface area of the average human body to be about 2 m² and the skin temperature to be about 300 K (somewhat less than the internal temperature of 310 K). Then from Eq. 18-37

$$P_r = \sigma\varepsilon A T^4 \approx (5.67\times 10^{-8}\,\text{W/m}^2\cdot\text{K}^4)(0.9)(2.0\,\text{m}^2)(300\,\text{K})^4 = 8\times 10^2\,\text{W}.$$

(b) The energy lost is given by

$$\Delta E = P_r \Delta t = (8\times 10^2\,\text{W})(30\,\text{s}) = 2\times 10^4\,\text{J}.$$

53. (a) Recalling that a change in Kelvin temperature is numerically equivalent to a change on the Celsius scale, we find that the rate of heat conduction is

$$P_{\text{cond}} = \frac{kA(T_H - T_C)}{L} = \frac{(401\,\text{W/m}\cdot\text{K})(4.8\times 10^{-4}\,\text{m}^2)(100\,^\circ\text{C})}{1.2\,\text{m}} = 16\,\text{J/s}.$$

(b) Using Table 18-4, the rate at which ice melts is

$$\left|\frac{dm}{dt}\right| = \frac{P_{\text{cond}}}{L_F} = \frac{16\,\text{J/s}}{333\,\text{J/g}} = 0.048\,\text{g/s}.$$

54. We refer to the polyurethane foam with subscript p and silver with subscript s. We use Eq. 18–32 to find $L = kR$.

(a) From Table 18-6 we find $k_p = 0.024$ W/m·K so

$$L_p = k_p R_p$$
$$= (0.024\,\text{W/m}\cdot\text{K})(30\,\text{ft}^2\cdot\text{F}^\circ\cdot\text{h/Btu})(1\,\text{m}/3.281\,\text{ft})^2(5\text{C}^\circ/9\text{F}^\circ)(3600\,\text{s/h})(1\,\text{Btu}/1055\,\text{J})$$
$$= 0.13\,\text{m}.$$

(b) For silver $k_s = 428$ W/m·K, so

$$L_s = k_s R_s = \left(\frac{k_s R_s}{k_p R_p}\right) L_p = \left[\frac{428(30)}{0.024(30)}\right](0.13\,\text{m}) = 2.3\times 10^3\,\text{m}.$$

55. We use Eqs. 18-38 through 18-40. Note that the surface area of the sphere is given by $A = 4\pi r^2$, where $r = 0.500$ m is the radius.

(a) The temperature of the sphere is $T = (273.15 + 27.00)\,\text{K} = 300.15\,\text{K}$. Thus

$$P_r = \sigma\varepsilon A T^4 = (5.67\times 10^{-8}\,\text{W/m}^2\cdot\text{K}^4)(0.850)(4\pi)(0.500\,\text{m})^2(300.15\,\text{K})^4$$
$$= 1.23\times 10^3\,\text{W}.$$

(b) Now, $T_{\text{env}} = 273.15 + 77.00 = 350.15$ K so

$$P_a = \sigma\varepsilon A T_{env}^4 = (5.67\times10^{-8}\text{ W/m}^2\cdot\text{K}^4)(0.850)(4\pi)(0.500\text{ m})^2(350.15\text{ K})^4 = 2.28\times10^3\text{ W}.$$

(c) From Eq. 18-40, we have

$$P_n = P_a - P_r = 2.28\times10^3\text{ W} - 1.23\times10^3\text{ W} = 1.05\times10^3\text{ W}.$$

56. (a) The surface area of the cylinder is given by

$$A_1 = 2\pi r_1^2 + 2\pi r_1 h_1 = 2\pi(2.5\times10^{-2}\text{ m})^2 + 2\pi(2.5\times10^{-2}\text{ m})(5.0\times10^{-2}\text{ m}) = 1.18\times10^{-2}\text{ m}^2,$$

its temperature is $T_1 = 273 + 30 = 303$ K, and the temperature of the environment is $T_{env} = 273 + 50 = 323$ K. From Eq. 18-39 we have

$$P_1 = \sigma\varepsilon A_1(T_{env}^4 - T^4) = (0.85)(1.18\times10^{-2}\text{ m}^2)((323\text{ K})^4 - (303\text{ K})^4) = 1.4\text{ W}.$$

(b) Let the new height of the cylinder be h_2. Since the volume V of the cylinder is fixed, we must have $V = \pi r_1^2 h_1 = \pi r_2^2 h_2$. We solve for h_2:

$$h_2 = \left(\frac{r_1}{r_2}\right)^2 h_1 = \left(\frac{2.5\text{ cm}}{0.50\text{ cm}}\right)^2 (5.0\text{ cm}) = 125\text{ cm} = 1.25\text{ m}.$$

The corresponding new surface area A_2 of the cylinder is

$$A_2 = 2\pi r_2^2 + 2\pi r_2 h_2 = 2\pi(0.50\times10^{-2}\text{ m})^2 + 2\pi(0.50\times10^{-2}\text{ m})(1.25\text{ m}) = 3.94\times10^{-2}\text{ m}^2.$$

Consequently,

$$\frac{P_2}{P_1} = \frac{A_2}{A_1} = \frac{3.94\times10^{-2}\text{ m}^2}{1.18\times10^{-2}\text{ m}^2} = 3.3.$$

57. We use $P_{cond} = kA\Delta T/L \propto A/L$. Comparing cases (a) and (b) in Figure 18–44, we have

$$P_{cond\,b} = \left(\frac{A_b L_a}{A_a L_b}\right) P_{cond\,a} = 4 P_{cond\,a}.$$

Consequently, it would take 2.0 min/4 = 0.50 min for the same amount of heat to be conducted through the rods welded as shown in Fig. 18-44(b).

58. (a) As in Sample Problem 18-6, we take the rate of conductive heat transfer through each layer to be the same. Thus, the rate of heat transfer across the entire wall P_w is equal

to the rate across layer 2 (P_2). Using Eq. 18-37 and canceling out the common factor of area A, we obtain

$$\frac{T_H - T_c}{(L_1/k_1 + L_2/k_2 + L_3/k_3)} = \frac{\Delta T_2}{(L_2/k_2)} \Rightarrow \frac{45\ C°}{(1 + 7/9 + 35/80)} = \frac{\Delta T_2}{(7/9)}$$

which leads to $\Delta T_2 = 15.8\ °C$.

(b) We expect (and this is supported by the result in the next part) that greater conductivity should mean a larger rate of conductive heat transfer.

(c) Repeating the calculation above with the new value for k_2, we have

$$\frac{45\ C°}{(1 + 7/11 + 35/80)} = \frac{\Delta T_2}{(7/11)}$$

which leads to $\Delta T_2 = 13.8\ °C$. This is less than our part (a) result which implies that the temperature gradients across layers 1 and 3 (the ones where the parameters did not change) are greater than in part (a); those larger temperature gradients lead to larger conductive heat currents (which is basically a statement of "Ohm's law as applied to heat conduction").

59. (a) We use

$$P_{cond} = kA \frac{T_H - T_C}{L}$$

with the conductivity of glass given in Table 18-6 as 1.0 W/m·K. We choose to use the Celsius scale for the temperature: a temperature difference of

$$T_H - T_C = 72°F - (-20°F) = 92\ °F$$

is equivalent to $\frac{5}{9}(92) = 51.1 C°$. This, in turn, is equal to 51.1 K since a change in Kelvin temperature is entirely equivalent to a Celsius change. Thus,

$$\frac{P_{cond}}{A} = k \frac{T_H - T_C}{L} = (1.0\ W/m \cdot K)\left(\frac{51.1°C}{3.0 \times 10^{-3} m}\right) = 1.7 \times 10^4\ W/m^2.$$

(b) The energy now passes in succession through 3 layers, one of air and two of glass. The heat transfer rate P is the same in each layer and is given by

$$P_{cond} = \frac{A(T_H - T_C)}{\Sigma L/k}$$

where the sum in the denominator is over the layers. If L_g is the thickness of a glass layer, L_a is the thickness of the air layer, k_g is the thermal conductivity of glass, and k_a is the thermal conductivity of air, then the denominator is

$$\sum \frac{L}{k} = \frac{2L_g}{k_g} + \frac{L_a}{k_a} = \frac{2L_g k_a + L_a k_g}{k_a k_g}.$$

Therefore, the heat conducted per unit area occurs at the following rate:

$$\frac{P_{cond}}{A} = \frac{(T_H - T_C)k_a k_g}{2L_g k_a + L_a k_g} = \frac{(51.1°C)(0.026\,\text{W/m}\cdot\text{K})(1.0\,\text{W/m}\cdot\text{K})}{2(3.0\times 10^{-3}\,\text{m})(0.026\,\text{W/m}\cdot\text{K})+(0.075\,\text{m})(1.0\,\text{W/m}\cdot\text{K})}$$
$$= 18\,\text{W/m}^2.$$

60. The surface area of the ball is $A = 4\pi R^2 = 4\pi(0.020\,\text{m})^2 = 5.03\times 10^{-3}\,\text{m}^2$. Using Eq. 18-37 with $T_i = 35+273 = 308\,\text{K}$ and $T_f = 47+273 = 320\,\text{K}$, the power required to maintain the temperature is

$$P_r = \sigma \varepsilon A(T_f^4 - T_i^4) \approx (5.67\times 10^{-8}\,\text{W/m}^2\cdot\text{K}^4)(0.80)(5.03\times 10^{-3}\,\text{m}^2)\left[(320\,\text{K})^4 - (308\,\text{K})^4\right]$$
$$= 0.34\,\text{W}.$$

Thus, the heat each bee must produce during the 20-minutes interval is

$$\frac{Q}{N} = \frac{P_r t}{N} = \frac{(0.34\,\text{W})(20\,\text{min})(60\,\text{s/min})}{500} = 0.81\,\text{J}.$$

61. We divide both sides of Eq. 18-32 by area A, which gives us the (uniform) rate of heat conduction per unit area:

$$\frac{P_{cond}}{A} = k_1 \frac{T_H - T_1}{L_1} = k_4 \frac{T - T_C}{L_4}$$

where $T_H = 30°C$, $T_1 = 25°C$ and $T_C = -10°C$. We solve for the unknown T.

$$T = T_C + \frac{k_1 L_4}{k_4 L_1}(T_H - T_1) = -4.2°C.$$

62. (a) For each individual penguin, the surface area that radiates is the sum of the top surface area and the sides:

$$A_r = a + 2\pi rh = a + 2\pi\sqrt{\frac{a}{\pi}}h = a + 2h\sqrt{\pi a},$$

where we have used $r = \sqrt{a/\pi}$ (from $a = \pi r^2$) for the radius of the cylinder. For the huddled cylinder, the radius is $r' = \sqrt{Na/\pi}$ (since $Na = \pi r'^2$), and the total surface area is

$$A_h = Na + 2\pi r'h = Na + 2\pi\sqrt{\frac{Na}{\pi}}h = Na + 2h\sqrt{N\pi a}.$$

Since the power radiated is proportional to the surface area, we have

$$\frac{P_h}{NP_r} = \frac{A_h}{NA_r} = \frac{Na + 2h\sqrt{N\pi a}}{N(a + 2h\sqrt{\pi a})} = \frac{1 + 2h\sqrt{\pi/Na}}{1 + 2h\sqrt{\pi/a}}.$$

With $N = 1000$, $a = 0.34$ m^2 and $h = 1.1$ m, the ratio is

$$\frac{P_h}{NP_r} = \frac{1 + 2h\sqrt{\pi/Na}}{1 + 2h\sqrt{\pi/a}} = \frac{1 + 2(1.1\text{ m})\sqrt{\pi/(1000 \cdot 0.34\text{ m}^2)}}{1 + 2(1.1\text{ m})\sqrt{\pi/(0.34\text{ m}^2)}} = 0.16.$$

(b) The total radiation loss is reduced by $1.00 - 0.16 = 0.84$, or 84%.

63. We assume (although this should be viewed as a "controversial" assumption) that the top surface of the ice is at $T_C = -5.0°$C. Less controversial are the assumptions that the bottom of the body of water is at $T_H = 4.0°$C and the interface between the ice and the water is at $T_X = 0.0°$C. The primary mechanism for the heat transfer through the total distance $L = 1.4$ m is assumed to be conduction, and we use Eq. 18-34:

$$\frac{k_{\text{water}}A(T_H - T_X)}{L - L_{\text{ice}}} = \frac{k_{\text{ice}}A(T_X - T_C)}{L_{\text{ice}}} \Rightarrow \frac{(0.12)A(4.0° - 0.0°)}{1.4 - L_{\text{ice}}} = \frac{(0.40)A(0.0° + 5.0°)}{L_{\text{ice}}}.$$

We cancel the area A and solve for thickness of the ice layer: $L_{\text{ice}} = 1.1$ m.

64. (a) Using Eq. 18-32, the rate of energy flow through the surface is

$$P_{\text{cond}} = \frac{kA(T_s - T_w)}{L} = (0.026\text{ W/m} \cdot \text{K})(4.00 \times 10^{-6}\text{ m}^2)\frac{300°\text{C} - 100°\text{C}}{1.0 \times 10^{-4}\text{ m}} = 0.208\text{W} \approx 0.21\text{ W}.$$

(Recall that a change in Celsius temperature is numerically equivalent to a change on the Kelvin scale.)

(b) With $P_{\text{cond}}t = L_V m = L_V(\rho V) = L_V(\rho Ah)$, the drop will last a duration of

$$t = \frac{L_V \rho Ah}{P_{\text{cond}}} = \frac{(2.256 \times 10^6\text{ J/kg})(1000\text{ kg/m}^3)(4.00 \times 10^{-6}\text{ m}^2)(1.50 \times 10^{-3}\text{ m})}{0.208\text{W}} = 65\text{ s}.$$

65. Let h be the thickness of the slab and A be its area. Then, the rate of heat flow through the slab is

$$P_{cond} = \frac{kA(T_H - T_C)}{h}$$

where k is the thermal conductivity of ice, T_H is the temperature of the water (0°C), and T_C is the temperature of the air above the ice (–10°C). The heat leaving the water freezes it, the heat required to freeze mass m of water being $Q = L_F m$, where L_F is the heat of fusion for water. Differentiate with respect to time and recognize that $dQ/dt = P_{cond}$ to obtain

$$P_{cond} = L_F \frac{dm}{dt}.$$

Now, the mass of the ice is given by $m = \rho A h$, where ρ is the density of ice and h is the thickness of the ice slab, so $dm/dt = \rho A(dh/dt)$ and

$$P_{cond} = L_F \rho A \frac{dh}{dt}.$$

We equate the two expressions for P_{cond} and solve for dh/dt:

$$\frac{dh}{dt} = \frac{k(T_H - T_C)}{L_F \rho h}.$$

Since 1 cal = 4.186 J and 1 cm = 1×10^{-2} m, the thermal conductivity of ice has the SI value

$$k = (0.0040 \text{ cal/s·cm·K}) (4.186 \text{ J/cal})/(1 \times 10^{-2} \text{ m/cm}) = 1.674 \text{ W/m·K}.$$

The density of ice is $\rho = 0.92$ g/cm^3 = 0.92×10^3 kg/m^3. Thus,

$$\frac{dh}{dt} = \frac{(1.674 \text{ W/m·K})(0°C + 10°C)}{(333 \times 10^3 \text{ J/kg})(0.92 \times 10^3 \text{ kg/m}^3)(0.050 \text{ m})} = 1.1 \times 10^{-6} \text{ m/s} = 0.40 \text{ cm/h}.$$

66. The condition that the energy lost by the beverage can due to evaporation equals the energy gained via radiation exchange implies

$$L_V \frac{dm}{dt} = P_{rad} = \sigma \varepsilon A (T_{env}^4 - T^4).$$

The total area of the top and side surfaces of the can is

$$A = \pi r^2 + 2\pi r h = \pi (0.022 \text{ m})^2 + 2\pi (0.022 \text{ m})(0.10 \text{ m}) = 1.53 \times 10^{-2} \text{ m}^2.$$

With $T_{env} = 32°C = 305$ K, $T = 15°C = 288$ K and $\varepsilon = 1$, the rate of water mass loss is

$$\frac{dm}{dt} = \frac{\sigma \varepsilon A}{L_V}(T_{env}^4 - T^4) = \frac{(5.67 \times 10^{-8} \text{ W/m}^2 \cdot \text{K}^4)(1.0)(1.53 \times 10^{-2} \text{ m}^2)}{2.256 \times 10^6 \text{ J/kg}}\left[(305 \text{ K})^4 - (288 \text{ K})^4\right]$$

$= 6.82 \times 10^{-7}$ kg/s ≈ 0.68 mg/s.

67. We denote the total mass M and the melted mass m. The problem tells us that Work/$M = p/\rho$, and that all the work is assumed to contribute to the phase change $Q = Lm$ where $L = 150 \times 10^3$ J/kg. Thus,

$$\frac{p}{\rho}M = Lm \Rightarrow m = \frac{5.5 \times 10^6}{1200}\frac{M}{150 \times 10^3}$$

which yields $m = 0.0306M$. Dividing this by 0.30 M (the mass of the fats, which we are told is equal to 30% of the total mass), leads to a percentage 0.0306/0.30 = 10%.

68. As is shown in the textbook for Sample Problem 18-4, we can express the final temperature in the following way:

$$T_f = \frac{m_A c_A T_A + m_B c_B T_B}{m_A c_A + m_B c_B} = \frac{c_A T_A + c_B T_B}{c_A + c_B}$$

where the last equality is made possible by the fact that $m_A = m_B$. Thus, in a graph of T_f versus T_A, the "slope" must be $c_A /(c_A + c_B)$, and the "y intercept" is $c_B /(c_A + c_B)T_B$. From the observation that the "slope" is equal to 2/5 we can determine, then, not only the ratio of the heat capacities but also the coefficient of T_B in the "y intercept"; that is,

$$c_B /(c_A + c_B)T_B = (1 - \text{"slope"})T_B.$$

(a) We observe that the "y intercept" is 150 K, so

$$T_B = 150/(1 - \text{"slope"}) = 150/(3/5)$$

which yields $T_B = 2.5 \times 10^2$ K.

(b) As noted already, $c_A /(c_A + c_B) = \frac{2}{5}$, so $5c_A = 2c_A + 2c_B$, which leads to $c_B/c_A = \frac{3}{2} = 1.5$.

69. The graph shows that the absolute value of the temperature change is $|\Delta T| = 25$ °C. Since a Watt is a Joule per second, we reason that the energy removed is

$$|Q| = (2.81 \text{ J/s})(20 \text{ min})(60 \text{ s/min}) = 3372 \text{ J}.$$

Thus, with $m = 0.30$ kg, the absolute value of Eq. 18-14 leads to

$$c = \frac{|Q|}{m\,|\Delta T|} = 4.5\times 10^2 \text{ J/kg·K}.$$

70. Let $m_w = 14$ kg, $m_c = 3.6$ kg, $m_m = 1.8$ kg, $T_{i1} = 180°C$, $T_{i2} = 16.0°C$, and $T_f = 18.0°C$. The specific heat c_m of the metal then satisfies

$$(m_w c_w + m_c c_m)(T_f - T_{i2}) + m_m c_m (T_f - T_{i1}) = 0$$

which we solve for c_m:

$$c_m = \frac{m_w c_w (T_{i2} - T_f)}{m_c (T_f - T_{i2}) + m_m (T_f - T_{i1})} = \frac{(14\text{kg})(4.18\text{kJ/kg·K})(16.0°C - 18.0°C)}{(3.6\text{kg})(18.0°C - 16.0°C) + (1.8\text{kg})(18.0°C - 180°C)}$$
$$= 0.41 \text{ kJ/kg·C°} = 0.41 \text{ kJ/kg·K}.$$

71. Its initial volume is $5^3 = 125$ cm^3, and using Table 18-2, Eq. 18-10 and Eq. 18-11, we find

$$\Delta V = (125\text{ m}^3)(3\times 23\times 10^{-6}/\text{C°})(50.0\text{ C°}) = 0.432 \text{ cm}^3.$$

72. (a) We denote $T_H = 100°C$, $T_C = 0°C$, the temperature of the copper-aluminum junction by T_1. and that of the aluminum-brass junction by T_2. Then,

$$P_{\text{cond}} = \frac{k_c A}{L}(T_H - T_1) = \frac{k_a A}{L}(T_1 - T_2) = \frac{k_b A}{L}(T_2 - T_c).$$

We solve for T_1 and T_2 to obtain

$$T_1 = T_H + \frac{T_C - T_H}{1 + k_c(k_a + k_b)/k_a k_b} = 100°C + \frac{0.00°C - 100°C}{1 + 401(235 + 109)/[(235)(109)]} = 84.3°C$$

(b) and

$$T_2 = T_c + \frac{T_H - T_C}{1 + k_b(k_c + k_a)/k_c k_a} = 0.00°C + \frac{100°C - 0.00°C}{1 + 109(235 + 401)/[(235)(401)]}$$
$$= 57.6°C.$$

73. The work (the "area under the curve") for process 1 is $4p_i V_i$, so that

$$U_b - U_a = Q_1 - W_1 = 6p_i V_i$$

by the First Law of Thermodynamics.

(a) Path 2 involves more work than path 1 (note the triangle in the figure of area $\frac{1}{2}(4V_i)(p_i/2) = p_iV_i$). With $W_2 = 4p_iV_i + p_iV_i = 5p_iV_i$, we obtain

$$Q_2 = W_2 + U_b - U_a = 5p_iV_i + 6p_iV_i = 11p_iV_i.$$

(b) Path 3 starts at a and ends at b so that $\Delta U = U_b - U_a = 6p_iV_i$.

74. We use $P_{cond} = kA(T_H - T_C)/L$. The temperature T_H at a depth of 35.0 km is

$$T_H = \frac{P_{cond}L}{kA} + T_C = \frac{(54.0\times10^{-3}\text{ W/m}^2)(35.0\times10^3\text{ m})}{2.50\text{ W/m}\cdot\text{K}} + 10.0°\text{C} = 766°\text{C}.$$

75. The volume of the disk (thought of as a short cylinder) is $\pi r^2 L$ where $L = 0.50$ cm is its thickness and $r = 8.0$ cm is its radius. Eq. 18-10, Eq. 18-11 and Table 18-2 (which gives $\alpha = 3.2\times10^{-6}/\text{C}°$) lead to

$$\Delta V = (\pi r^2 L)(3\alpha)(60°\text{C} - 10°\text{C}) = 4.83\times10^{-2}\text{ cm}^3.$$

76. We use $Q = cm\Delta T$ and $m = \rho V$. The volume of water needed is

$$V = \frac{m}{\rho} = \frac{Q}{\rho C \Delta T} = \frac{(1.00\times10^6\text{ kcal/day})(5\text{ days})}{(1.00\times10^3\text{ kg/m}^3)(1.00\text{ kcal/kg})(50.0°\text{C} - 22.0°\text{C})} = 35.7\text{ m}^3.$$

77. We have $W = \int p\, dV$ (Eq. 18-24). Therefore,

$$W = a\int V^2 dV = \frac{a}{3}(V_f^3 - V_i^3) = 23\text{ J}.$$

78. (a) The rate of heat flow is

$$P_{cond} = \frac{kA(T_H - T_C)}{L} = \frac{(0.040\text{ W/m}\cdot\text{K})(1.8\text{ m}^2)(33°\text{C} - 1.0°\text{C})}{1.0\times10^{-2}\text{ m}} = 2.3\times10^2\text{ J/s}.$$

(b) The new rate of heat flow is

$$P'_{cond} = \frac{k'P_{cond}}{k} = \frac{(0.60\text{ W/m}\cdot\text{K})(230\text{ J/s})}{0.040\text{ W/m}\cdot\text{K}} = 3.5\times10^3\text{ J/s},$$

which is about 15 times as fast as the original heat flow.

79. We note that there is no work done in process $c \to b$, since there is no change of volume. We also note that the *magnitude* of work done in process $b \to c$ is given, but not

its sign (which we identify as negative as a result of the discussion in §18-8). The total (or *net*) heat transfer is $Q_{net} = [(-40) + (-130) + (+400)]$ J = 230 J. By the First Law of Thermodynamics (or, equivalently, conservation of energy), we have

$$Q_{net} = W_{net}$$
$$230\,\text{J} = W_{a \to c} + W_{c \to b} + W_{b \to a}$$
$$= W_{a \to c} + 0 + (-80\,\text{J})$$

Therefore, $W_{a \to c} = 3.1 \times 10^2$ J.

80. If the window is L_1 high and L_2 wide at the lower temperature and $L_1 + \Delta L_1$ high and $L_2 + \Delta L_2$ wide at the higher temperature then its area changes from $A_1 = L_1 L_2$ to

$$A_2 = (L_1 + \Delta L_1)(L_2 + \Delta L_2) \approx L_1 L_2 + L_1 \Delta L_2 + L_2 \Delta L_1$$

where the term $\Delta L_1 \Delta L_2$ has been omitted because it is much smaller than the other terms, if the changes in the lengths are small. Consequently, the change in area is

$$\Delta A = A_2 - A_1 = L_1 \Delta L_2 + L_2 \Delta L_1.$$

If ΔT is the change in temperature then $\Delta L_1 = \alpha L_1 \Delta T$ and $\Delta L_2 = \alpha L_2 \Delta T$, where α is the coefficient of linear expansion. Thus

$$\Delta A = \alpha(L_1 L_2 + L_1 L_2)\,\Delta T = 2\alpha L_1 L_2 \Delta T$$
$$= 2(9 \times 10^{-6}/\text{C}°)(30\,\text{cm})(20\,\text{cm})(30°\text{C})$$
$$= 0.32\,\text{cm}^2.$$

81. Following the method of Sample Problem 18-4 (particularly its third Key Idea), we have

$$(900\,\tfrac{\text{J}}{\text{kg}\cdot\text{C}°})(2.50\,\text{kg})(T_f - 92.0°\text{C}) + (4190\,\tfrac{\text{J}}{\text{kg}\cdot\text{C}°})(8.00\,\text{kg})(T_f - 5.0°\text{C}) = 0$$

where Table 18-3 has been used. Thus we find $T_f = 10.5°$C.

82. We use $Q = -\lambda_F m_{ice} = W + \Delta E_{int}$. In this case $\Delta E_{int} = 0$. Since $\Delta T = 0$ for the ideal gas, then the work done on the gas is

$$W' = -W = \lambda_F m_i = (333\,\text{J/g})(100\,\text{g}) = 33.3\,\text{kJ}.$$

83. This is similar to Sample Problem 18-3. An important difference with part (b) of that sample problem is that, in this case, the final state of the H$_2$O is *all liquid* at $T_f > 0$. As discussed in part (a) of that sample problem, there are three steps to the total process:

$$Q = m\,[c_{ice}(0\ C° - (-150\ C°)) + L_F + c_{liquid}(T_f - 0\ C°)]$$

Thus,

$$T_f = \frac{Q/m - (c_{ice}(150°) + L_F)}{c_{liquid}} = 79.5°C.$$

84. We take absolute values of Eq. 18-9 and Eq. 12-25:

$$|\Delta L| = L\alpha|\Delta T| \quad \text{and} \quad \left|\frac{F}{A}\right| = E\left|\frac{\Delta L}{L}\right|.$$

The ultimate strength for steel is $(F/A)_{rupture} = S_u = 400 \times 10^6$ N/m^2 from Table 12-1. Combining the above equations (eliminating the ratio $\Delta L/L$), we find the rod will rupture if the temperature change exceeds

$$|\Delta T| = \frac{S_u}{E\alpha} = \frac{400 \times 10^6\ \text{N/m}^2}{(200 \times 10^9\ \text{N/m}^2)(11 \times 10^{-6}/C°)} = 182°C.$$

Since we are dealing with a temperature decrease, then, the temperature at which the rod will rupture is $T = 25.0°C - 182°C = -157°C$.

85. The problem asks for 0.5% of E, where $E = Pt$ with $t = 120$ s and P given by Eq. 18-38. Therefore, with $A = 4\pi r^2 = 5.0 \times 10^{-3}$ m^2, we obtain

$$(0.005)Pt = (0.005)\sigma\varepsilon AT^4 t = 8.6\ \text{J}.$$

86. From the law of cosines, with $\phi = 59.95°$, we have

$$L_{Invar}^2 = L_{alum}^2 + L_{steel}^2 - 2L_{alum}L_{steel}\cos\phi$$

Plugging in $L = L_0(1 + \alpha\Delta T)$, dividing by L_0 (which is the same for all sides) and ignoring terms of order $(\Delta T)^2$ or higher, we obtain

$$1 + 2\alpha_{Invar}\Delta T = 2 + 2(\alpha_{alum} + \alpha_{steel})\Delta T - 2(1 + (\alpha_{alum} + \alpha_{steel})\Delta T)\cos\phi.$$

This is rearranged to yield

$$\Delta T = \frac{\cos\phi - \tfrac{1}{2}}{(\alpha_{alum} + \alpha_{steel})(1 - \cos\phi) - \alpha_{Invar}} = \approx 46°C,$$

so that the final temperature is $T = 20.0° + \Delta T = 66°$ C. Essentially the same argument, but arguably more elegant, can be made in terms of the differential of the above cosine law expression.

87. We assume the ice is at 0°C to being with, so that the only heat needed for melting is that described by Eq. 18-16 (which requires information from Table 18-4). Thus,

$$Q = Lm = (333 \text{ J/g})(1.00 \text{ g}) = 333 \text{ J}.$$

88. Let the initial water temperature be T_{wi} and the initial thermometer temperature be T_{ti}. Then, the heat absorbed by the thermometer is equal (in magnitude) to the heat lost by the water:

$$c_t m_t \left(T_f - T_{ti} \right) = c_w m_w \left(T_{wi} - T_f \right).$$

We solve for the initial temperature of the water:

$$T_{wi} = \frac{c_t m_t \left(T_f - T_{ti} \right)}{c_w m_w} + T_f = \frac{(0.0550 \text{ kg})(0.837 \text{ kJ/kg} \cdot \text{K})(44.4 - 15.0) \text{ K}}{(4.18 \text{ kJ/kg} \cdot \text{C}°)(0.300 \text{ kg})} + 44.4°\text{C}$$

$$= 45.5°\text{C}.$$

89. For a cylinder of height h, the surface area is $A_c = 2\pi rh$, and the area of a sphere is $A_o = 4\pi R^2$. The net radiative heat transfer is given by Eq. 18-40.

(a) We estimate the surface area A of the body as that of a cylinder of height 1.8 m and radius $r = 0.15$ m plus that of a sphere of radius $R = 0.10$ m. Thus, we have $A \approx A_c + A_o = 1.8$ m². The emissivity $\varepsilon = 0.80$ is given in the problem, and the Stefan-Boltzmann constant is found in §18-11: $\sigma = 5.67 \times 10^{-8}$ W/m²·K⁴. The "environment" temperature is $T_{env} = 303$ K, and the skin temperature is $T = \tfrac{5}{9}(102 - 32) + 273 = 312$ K. Therefore,

$$P_{net} = \sigma \varepsilon A \left(T_{env}^4 - T^4 \right) = -86 \text{ W}.$$

The corresponding sign convention is discussed in the textbook immediately after Eq. 18-40. We conclude that heat is being lost by the body at a rate of roughly 90 W.

(b) Half the body surface area is roughly $A = 1.8/2 = 0.9$ m². Now, with $T_{env} = 248$ K, we find

$$|P_{net}| = |\sigma \varepsilon A \left(T_{env}^4 - T^4 \right)| \approx 2.3 \times 10^2 \text{ W}.$$

(c) Finally, with $T_{env} = 193$ K (and still with $A = 0.9$ m²) we obtain $|P_{net}| = 3.3 \times 10^2$ W.

90. One method is to simply compute the change in length in each edge ($x_0 = 0.200$ m and $y_0 = 0.300$ m) from Eq. 18-9 ($\Delta x = 3.6 \times 10^{-5}$ m and $\Delta y = 5.4 \times 10^{-5}$ m) and then compute the area change:

$$A - A_0 = (x_0 + \Delta x)(y_0 + \Delta y) - x_0 y_0 = 2.16 \times 10^{-5} \text{ m}^2.$$

Another (though related) method uses $\Delta A = 2\alpha A_0 \Delta T$ (valid for $\Delta A/A \ll 1$) which can be derived by taking the differential of $A = xy$ and replacing d's with Δ's.

91. (a) Let the number of weight lift repetitions be N. Then $Nmgh = Q$, or (using Eq. 18-12 and the discussion preceding it)

$$N = \frac{Q}{mgh} = \frac{(3500\,\text{Cal})(4186\,\text{J/Cal})}{(80.0\,\text{kg})(9.80\,\text{m/s}^2)(1.00\,\text{m})} \approx 1.87 \times 10^4.$$

(b) The time required is

$$t = (18700)(2.00\,\text{s})\left(\frac{1.00\,\text{h}}{3600\,\text{s}}\right) = 10.4\,\text{h}.$$

92. The heat needed is

$$Q = (10\%)mL_F = \left(\frac{1}{10}\right)(200,000\,\text{metric tons})(1000\,\text{kg/metric ton})(333\,\text{kJ/kg})$$
$$= 6.7 \times 10^{12}\,\text{J}.$$

93. The net work may be computed as a sum of works (for the individual processes involved) or as the "area" (with appropriate ± sign) inside the figure (representing the cycle). In this solution, we take the former approach (sum over the processes) and will need the following fact related to processes represented in pV diagrams:

$$\text{for straight line} \quad \text{Work} = \frac{p_i + p_f}{2}\Delta V$$

which is easily verified using the definition Eq. 18-25. The cycle represented by the "triangle" BC consists of three processes:

- "tilted" straight line from (1.0 m³, 40 Pa) to (4.0 m³, 10 Pa), with

$$\text{Work} = \frac{40\,\text{Pa} + 10\,\text{Pa}}{2}(4.0\,\text{m}^3 - 1.0\,\text{m}^3) = 75\,\text{J}$$

- horizontal line from (4.0 m³, 10 Pa) to (1.0 m³, 10 Pa), with

$$\text{Work} = (10\,\text{Pa})(1.0\,\text{m}^3 - 4.0\,\text{m}^3) = -30\,\text{J}$$

- vertical line from (1.0 m³, 10 Pa) to (1.0 m³, 40 Pa), with

$$\text{Work} = \frac{10\,\text{Pa} + 40\,\text{Pa}}{2}\left(1.0\,\text{m}^3 - 1.0\,\text{m}^3\right) = 0$$

(a) and (b) Thus, the total work during the *BC* cycle is (75 − 30) J = 45 J. During the *BA* cycle, the "tilted" part is the same as before, and the main difference is that the horizontal portion is at higher pressure, with Work = (40 Pa)(−3.0 m³) = −120 J. Therefore, the total work during the *BA* cycle is (75 − 120) J = − 45 J.

94. For isotropic materials, the coefficient of linear expansion α is related to that for volume expansion by $\alpha = \tfrac{1}{3}\beta$ (Eq. 18-11). The radius of Earth may be found in the Appendix. With these assumptions, the radius of the Earth should have increased by approximately

$$\Delta R_E = R_E \alpha \Delta T = \left(6.4\times 10^3\,\text{km}\right)\left(\frac{1}{3}\right)\left(3.0\times 10^{-5}/\text{K}\right)(3000\,\text{K} - 300\,\text{K}) = 1.7\times 10^2\,\text{km}.$$

95. (a) Regarding part (a), it is important to recognize that the problem is asking for the total work done during the two-step "path": $a \to b$ followed by $b \to c$. During the latter part of this "path" there is no volume change and consequently no work done. Thus, the answer to part (b) is also the answer to part (a). Since ΔU for process $c \to a$ is −160 J, then $U_c - U_a = 160$ J. Therefore, using the First Law of Thermodynamics, we have

$$\begin{aligned}160 &= U_c - U_b + U_b - U_a \\ &= Q_{b\to c} - W_{b\to c} + Q_{a\to b} - W_{a\to b} \\ &= 40 - 0 + 200 - W_{a\to b}\end{aligned}$$

Therefore, $W_{a\to b\to c} = W_{a\to b} = 80$ J.

(b) $W_{a\to b} = 80$ J.

96. Since the combination "$p_1 V_1$" appears frequently in this derivation we denote it as "x." Thus for process 1, the heat transferred is $Q_1 = 5x = \Delta E_{\text{int 1}} + W_1$, and for path 2 (which consists of two steps, one at constant volume followed by an expansion accompanied by a linear pressure decrease) it is $Q_2 = 5.5x = \Delta E_{\text{int 2}} + W_2$. If we subtract these two expressions and make use of the fact that internal energy is state function (and thus has the same value for path 1 as for path 2) then we have

$$5.5x - 5x = W_2 - W_1 = \text{"area" inside the triangle} = \frac{1}{2}(2V_1)(p_2 - p_1).$$

Thus, dividing both sides by $x\,(= p_1 V_1)$, we find

$$0.5 = \frac{p_2}{p_1} - 1$$

which leads immediately to the result: $p_2/p_1 = 1.5$.

97. The cube has six faces, each of which has an area of $(6.0 \times 10^{-6} \text{ m})^2$. Using Kelvin temperatures and Eq. 18-40, we obtain

$$P_{net} = \sigma \varepsilon A (T_{env}^4 - T^4)$$
$$= \left(5.67 \times 10^{-8} \frac{\text{W}}{\text{m}^2 \cdot \text{K}^4}\right)(0.75)\left(2.16 \times 10^{-10} \text{ m}^2\right)\left((123.15 \text{ K})^4 - (173.15 \text{ K})^4\right)$$
$$= -6.1 \times 10^{-9} \text{ W}.$$

98. We denote the density of the liquid as ρ, the rate of liquid flowing in the calorimeter as μ, the specific heat of the liquid as c, the rate of heat flow as P, and the temperature change as ΔT. Consider a time duration dt, during this time interval, the amount of liquid being heated is $dm = \mu \rho dt$. The energy required for the heating is

$$dQ = Pdt = c(dm)\Delta T = c\mu\rho\Delta T dt.$$

Thus,

$$c = \frac{P}{\rho\mu\Delta T} = \frac{250 \text{ W}}{(8.0 \times 10^{-6} \text{ m}^3/\text{s})(0.85 \times 10^3 \text{ kg/m}^3)(15°\text{C})}$$
$$= 2.5 \times 10^3 \text{ J/kg} \cdot \text{C}° = 2.5 \times 10^3 \text{ J/kg} \cdot \text{K}.$$

99. Consider the object of mass m_1 falling through a distance h. The loss of its mechanical energy is $\Delta E = m_1 gh$. This amount of energy is then used to heat up the temperature of water of mass m_2: $\Delta E = m_1 gh = Q = m_2 c \Delta T$. Thus, the maximum possible rise in water temperature is

$$\Delta T = \frac{m_1 gh}{m_2 c} = \frac{(6.00 \text{ kg})(9.8 \text{ m/s}^2)(50.0 \text{ m})}{(0.600 \text{ kg})(4190 \text{ J/kg} \cdot \text{C}°)} = 1.17°\text{C}.$$

Chapter 19

1. (a) Eq. 19-3 yields $n = M_{sam}/M = 2.5/197 = 0.0127$ mol.

(b) The number of atoms is found from Eq. 19-2:

$$N = nN_A = (0.0127)(6.02 \times 10^{23}) = 7.64 \times 10^{21}.$$

2. Each atom has a mass of $m = M/N_A$, where M is the molar mass and N_A is the Avogadro constant. The molar mass of arsenic is 74.9 g/mol or 74.9×10^{-3} kg/mol. Therefore, 7.50×10^{24} arsenic atoms have a total mass of

$$(7.50 \times 10^{24})(74.9 \times 10^{-3} \text{ kg/mol})/(6.02 \times 10^{23} \text{ mol}^{-1}) = 0.933 \text{ kg}.$$

3. With $V = 1.0 \times 10^{-6}$ m^3, $p = 1.01 \times 10^{-13}$ Pa, and $T = 293$ K, the ideal gas law gives

$$n = \frac{pV}{RT} = \frac{(1.01 \times 10^{-13} \text{ Pa})(1.0 \times 10^{-6} \text{ m}^3)}{(8.31 \text{ J/mol} \cdot \text{K})(293 \text{ K})} = 4.1 \times 10^{-23} \text{ mole}.$$

Consequently, Eq. 19-2 yields $N = nN_A = 25$ molecules. We can express this as a ratio (with V now written as 1 cm^3) $N/V = 25$ molecules/cm^3.

4. (a) We solve the ideal gas law $pV = nRT$ for n:

$$n = \frac{pV}{RT} = \frac{(100 \text{ Pa})(1.0 \times 10^{-6} \text{ m}^3)}{(8.31 \text{ J/mol} \cdot \text{K})(220 \text{ K})} = 5.47 \times 10^{-8} \text{ mol}.$$

(b) Using Eq. 19-2, the number of molecules N is

$$N = nN_A = (5.47 \times 10^{-6} \text{ mol})(6.02 \times 10^{23} \text{ mol}^{-1}) = 3.29 \times 10^{16} \text{ molecules}.$$

5. Since (standard) air pressure is 101 kPa, then the initial (absolute) pressure of the air is $p_i = 266$ kPa. Setting up the gas law in ratio form (where $n_i = n_f$ and thus cancels out — see Sample Problem 19-1), we have

$$\frac{p_f V_f}{p_i V_i} = \frac{T_f}{T_i}$$

which yields

$$p_f = p_i\left(\frac{V_i}{V_f}\right)\left(\frac{T_f}{T_i}\right) = (266\,\text{kPa})\left(\frac{1.64\times 10^{-2}\,\text{m}^3}{1.67\times 10^{-2}\,\text{m}^3}\right)\left(\frac{300\,\text{K}}{273\,\text{K}}\right) = 287\,\text{kPa}.$$

Expressed as a gauge pressure, we subtract 101 kPa and obtain 186 kPa.

6. (a) With $T = 283$ K, we obtain

$$n = \frac{pV}{RT} = \frac{(100\times 10^3\,\text{Pa})(2.50\,\text{m}^3)}{(8.31\,\text{J/mol}\cdot\text{K})(283\,\text{K})} = 106\,\text{mol}.$$

(b) We can use the answer to part (a) with the new values of pressure and temperature, and solve the ideal gas law for the new volume, or we could set up the gas law in ratio form as in Sample Problem 19-1 (where $n_i = n_f$ and thus cancels out):

$$\frac{p_f V_f}{p_i V_i} = \frac{T_f}{T_i}$$

which yields a final volume of

$$V_f = V_i\left(\frac{p_i}{p_f}\right)\left(\frac{T_f}{T_i}\right) = (2.50\,\text{m}^3)\left(\frac{100\,\text{kPa}}{300\,\text{kPa}}\right)\left(\frac{303\,\text{K}}{283\,\text{K}}\right) = 0.892\,\text{m}^3.$$

7. (a) In solving $pV = nRT$ for n, we first convert the temperature to the Kelvin scale: $T = (40.0 + 273.15)\,\text{K} = 313.15\,\text{K}$. And we convert the volume to SI units: $1000\,\text{cm}^3 = 1000\times 10^{-6}\,\text{m}^3$. Now, according to the ideal gas law,

$$n = \frac{pV}{RT} = \frac{(1.01\times 10^5\,\text{Pa})(1000\times 10^{-6}\,\text{m}^3)}{(8.31\,\text{J/mol}\cdot\text{K})(313.15\,\text{K})} = 3.88\times 10^{-2}\,\text{mol}.$$

(b) The ideal gas law $pV = nRT$ leads to

$$T = \frac{pV}{nR} = \frac{(1.06\times 10^5\,\text{Pa})(1500\times 10^{-6}\,\text{m}^3)}{(3.88\times 10^{-2}\,\text{mol})(8.31\,\text{J/mol}\cdot\text{K})} = 493\,\text{K}.$$

We note that the final temperature may be expressed in degrees Celsius as 220°C.

8. The pressure p_1 due to the first gas is $p_1 = n_1 RT/V$, and the pressure p_2 due to the second gas is $p_2 = n_2 RT/V$. So the total pressure on the container wall is

$$p = p_1 + p_2 = \frac{n_1 RT}{V} + \frac{n_2 RT}{V} = (n_1 + n_2)\frac{RT}{V}.$$

The fraction of P due to the second gas is then

$$\frac{p_2}{p} = \frac{n_2 RT/V}{(n_1+n_2)(RT/V)} = \frac{n_2}{n_1+n_2} = \frac{0.5}{2+0.5} = 0.2.$$

9. (a) Eq. 19-45 (which gives 0) implies $Q = W$. Then Eq. 19-14, with $T = (273 + 30.0)$K leads to gives $Q = -3.14 \times 10^3$ J, or $|Q| = 3.14 \times 10^3$ J.

(b) That negative sign in the result of part (a) implies the transfer of heat is *from* the gas.

10. The initial and final temperatures are $T_i = 5.00°C = 278$ K and $T_f = 75.0°C = 348$ K, respectively. Using ideal-gas law with $V_i = V_f$, we find the final pressure to be

$$\frac{p_f V_f}{p_i V_i} = \frac{T_f}{T_i} \Rightarrow p_f = \frac{T_f}{T_i} p_i = \left(\frac{348\,\text{K}}{278\,\text{K}}\right)(1.00\,\text{atm}) = 1.25\,\text{atm}.$$

11. Using Eq. 19-14, we note that since it is an isothermal process (involving an ideal gas) then $Q = W = nRT \ln(V_f/V_i)$ applies at any point on the graph. An easy one to read is $Q = 1000$ J and $V_f = 0.30$ m^3, and we can also infer from the graph that $V_i = 0.20$ m^3. We are told that $n = 0.825$ mol, so the above relation immediately yields $T = 360$ K.

12. Since the pressure is constant the work is given by $W = p(V_2 - V_1)$. The initial volume is $V_1 = (AT_1 - BT_1^2)/p$, where $T_1=315$ K is the initial temperature, $A = 24.9$ J/K and $B=0.00662$ J/K^2. The final volume is $V_2 = (AT_2 - BT_2^2)/p$, where $T_2=315$ K. Thus

$$W = A(T_2 - T_1) - B(T_2^2 - T_1^2)$$
$$= (24.9\,\text{J/K})(325\,\text{K} - 315\,\text{K}) - (0.00662\,\text{J/K}^2)[(325\,\text{K})^2 - (315\,\text{K})^2] = 207\,\text{J}.$$

13. Suppose the gas expands from volume V_i to volume V_f during the isothermal portion of the process. The work it does is

$$W = \int_{V_i}^{V_f} p\,dV = nRT \int_{V_i}^{V_f} \frac{dV}{V} = nRT \ln \frac{V_f}{V_i},$$

where the ideal gas law $pV = nRT$ was used to replace p with nRT/V. Now $V_i = nRT/p_i$ and $V_f = nRT/p_f$, so $V_f/V_i = p_i/p_f$. Also replace nRT with $p_i V_i$ to obtain

$$W = p_i V_i \ln \frac{p_i}{p_f}.$$

Since the initial gauge pressure is 1.03×10^5 Pa,

$$p_i = 1.03 \times 10^5 \text{ Pa} + 1.013 \times 10^5 \text{ Pa} = 2.04 \times 10^5 \text{ Pa}.$$

The final pressure is atmospheric pressure: $p_f = 1.013 \times 10^5$ Pa. Thus

$$W = (2.04 \times 10^5 \text{ Pa})(0.14 \text{ m}^3) \ln\left(\frac{2.04 \times 10^5 \text{ Pa}}{1.013 \times 10^5 \text{ Pa}}\right) = 2.00 \times 10^4 \text{ J}.$$

During the constant pressure portion of the process the work done by the gas is $W = p_f(V_i - V_f)$. The gas starts in a state with pressure p_f, so this is the pressure throughout this portion of the process. We also note that the volume decreases from V_f to V_i. Now $V_f = p_i V_i / p_f$, so

$$W = p_f\left(V_i - \frac{p_i V_i}{p_f}\right) = (p_f - p_i)V_i = (1.013 \times 10^5 \text{ Pa} - 2.04 \times 10^5 \text{ Pa})(0.14 \text{ m}^3)$$
$$= -1.44 \times 10^4 \text{ J}.$$

The total work done by the gas over the entire process is

$$W = 2.00 \times 10^4 \text{ J} - 1.44 \times 10^4 \text{ J} = 5.60 \times 10^3 \text{ J}.$$

14. (a) At the surface, the air volume is

$$V_1 = Ah = \pi(1.00 \text{ m})^2 (4.00 \text{ m}) = 12.57 \text{ m}^3 \approx 12.6 \text{ m}^3.$$

(b) The temperature and pressure of the air inside the submarine at the surface are $T_1 = 20°C = 293$ K and $p_1 = p_0 = 1.00$ atm. On the other hand, at depth $h = 80$ m, we have $T_2 = -30°C = 243$ K and

$$p_2 = p_0 + \rho g h = 1.00 \text{ atm} + (1024 \text{ kg/m}^3)(9.80 \text{ m/s}^2)(80.0 \text{ m})\frac{1.00 \text{ atm}}{1.01 \times 10^5 \text{ Pa}}$$
$$= 1.00 \text{ atm} + 7.95 \text{ atm} = 8.95 \text{ atm}.$$

Therefore, using ideal-gas law, $pV = NkT$, the air volume at this depth would be

$$\frac{p_1 V_1}{p_2 V_2} = \frac{T_1}{T_2} \Rightarrow V_2 = \left(\frac{p_1}{p_2}\right)\left(\frac{T_2}{T_1}\right) V_1 = \left(\frac{1.00 \text{ atm}}{8.95 \text{ atm}}\right)\left(\frac{243 \text{ K}}{293 \text{ K}}\right)(12.57 \text{ m}^3) = 1.16 \text{ m}^3.$$

(c) The decrease in volume is $\Delta V = V_1 - V_2 = 11.44 \text{ m}^3$. Using Eq. 19-5, the amount of air this volume corresponds to is

$$n = \frac{p\Delta V}{RT} = \frac{(8.95 \text{ atm})(1.01\times 10^5 \text{ Pa/atm})(11.44 \text{ m}^3)}{(8.31 \text{ J/mol}\cdot\text{K})(243\text{ K})} = 5.10\times 10^3 \text{ mol}.$$

Thus, in order for the submarine to maintain the original air volume in the chamber, 5.10×10^3 mol of air must be released.

15. (a) At point a, we know enough information to compute n:

$$n = \frac{pV}{RT} = \frac{(2500 \text{ Pa})(1.0 \text{ m}^3)}{(8.31 \text{ J/mol}\cdot\text{K})(200\text{ K})} = 1.5 \text{ mol}.$$

(b) We can use the answer to part (a) with the new values of pressure and volume, and solve the ideal gas law for the new temperature, or we could set up the gas law as in Sample Problem 19-1 in terms of ratios (note: $n_a = n_b$ and cancels out):

$$\frac{p_b V_b}{p_a V_a} = \frac{T_b}{T_a} \Rightarrow T_b = (200\text{ K})\left(\frac{7.5 \text{ kPa}}{2.5 \text{ kPa}}\right)\left(\frac{3.0 \text{ m}^3}{1.0 \text{ m}^3}\right)$$

which yields an absolute temperature at b of $T_b = 1.8\times 10^3$ K.

(c) As in the previous part, we choose to approach this using the gas law in ratio form (see Sample Problem 19-1):

$$\frac{p_c V_c}{p_a V_a} = \frac{T_c}{T_a} \Rightarrow T_c = (200\text{ K})\left(\frac{2.5 \text{ kPa}}{2.5 \text{ kPa}}\right)\left(\frac{3.0 \text{ m}^3}{1.0 \text{ m}^3}\right)$$

which yields an absolute temperature at c of $T_c = 6.0\times 10^2$ K.

(d) The net energy added to the gas (as heat) is equal to the net work that is done as it progresses through the cycle (represented as a right triangle in the pV diagram shown in Fig. 19-21). This, in turn, is related to $\pm$ "area" inside that triangle (with area = $\frac{1}{2}$(base)(height)), where we choose the plus sign because the volume change at the largest pressure is an *increase*. Thus,

$$Q_{\text{net}} = W_{\text{net}} = \frac{1}{2}(2.0 \text{ m}^3)(5.0\times 10^3 \text{ Pa}) = 5.0\times 10^3 \text{ J}.$$

16. We assume that the pressure of the air in the bubble is essentially the same as the pressure in the surrounding water. If d is the depth of the lake and ρ is the density of water, then the pressure at the bottom of the lake is $p_1 = p_0 + \rho g d$, where p_0 is atmospheric pressure. Since $p_1 V_1 = nRT_1$, the number of moles of gas in the bubble is

$$n = p_1V_1/RT_1 = (p_0 + \rho g d)V_1/RT_1,$$

where V_1 is the volume of the bubble at the bottom of the lake and T_1 is the temperature there. At the surface of the lake the pressure is p_0 and the volume of the bubble is $V_2 = nRT_2/p_0$. We substitute for n to obtain

$$V_2 = \frac{T_2}{T_1}\frac{p_0 + \rho g d}{p_0}V_1$$

$$= \left(\frac{293\,\text{K}}{277\,\text{K}}\right)\left(\frac{1.013\times 10^5\,\text{Pa} + (0.998\times 10^3\,\text{kg/m}^3)(9.8\,\text{m/s}^2)(40\,\text{m})}{1.013\times 10^5\,\text{Pa}}\right)(20\,\text{cm}^3)$$

$$= 1.0\times 10^2\,\text{cm}^3.$$

17. When the valve is closed the number of moles of the gas in container A is $n_A = p_AV_A/RT_A$ and that in container B is $n_B = 4p_BV_A/RT_B$. The total number of moles in both containers is then

$$n = n_A + n_B = \frac{V_A}{R}\left(\frac{p_A}{T_A} + \frac{4p_B}{T_B}\right) = \text{const.}$$

After the valve is opened the pressure in container A is $p'_A = Rn'_AT_A/V_A$ and that in container B is $p'_B = Rn'_BT_B/4V_A$. Equating p'_A and p'_B, we obtain $Rn'_AT_A/V_A = Rn'_BT_B/4V_A$, or $n'_B = (4T_A/T_B)n'_A$. Thus,

$$n = n'_A + n'_B = n'_A\left(1 + \frac{4T_A}{T_B}\right) = n_A + n_B = \frac{V_A}{R}\left(\frac{p_A}{T_A} + \frac{4p_B}{T_B}\right).$$

We solve the above equation for n'_A:

$$n'_A = \frac{V}{R}\frac{(p_A/T_A + 4p_B/T_B)}{(1 + 4T_A/T_B)}.$$

Substituting this expression for n'_A into $p'V_A = n'_ART_A$, we obtain the final pressure:

$$p' = \frac{n'_ART_A}{V_A} = \frac{p_A + 4p_BT_A/T_B}{1 + 4T_A/T_B} = 2.0\times 10^5\,\text{Pa}.$$

18. Appendix F gives $M = 4.00\times 10^{-3}$ kg/mol (Table 19-1 gives this to fewer significant figures). Using Eq. 19-22, we obtain

$$v_{\text{rms}} = \sqrt{\frac{3RT}{M}} = \sqrt{\frac{3(8.31\,\text{J/mol}\cdot\text{K})(1000\,\text{K})}{4.00\times 10^{-3}\,\text{kg/mol}}} = 2.50\times 10^3\,\text{m/s}.$$

19. According to kinetic theory, the rms speed is

$$v_{rms} = \sqrt{\frac{3RT}{M}}$$

where T is the temperature and M is the molar mass. See Eq. 19-34. According to Table 19-1, the molar mass of molecular hydrogen is 2.02 g/mol = 2.02 × 10^{-3} kg/mol, so

$$v_{rms} = \sqrt{\frac{3\,(8.31\,\text{J/mol}\cdot\text{K})(2.7\,\text{K})}{2.02\times 10^{-3}\,\text{kg/mol}}} = 1.8\times 10^2 \text{ m/s}.$$

20. The molar mass of argon is 39.95 g/mol. Eq. 19–22 gives

$$v_{rms} = \sqrt{\frac{3RT}{M}} = \sqrt{\frac{3(8.31\,\text{J/mol}\cdot\text{K})(313\,\text{K})}{39.95\times 10^{-3}\,\text{kg/mol}}} = 442 \text{ m/s}.$$

21. Table 19-1 gives $M = 28.0$ g/mol for nitrogen. This value can be used in Eq. 19-22 with T in Kelvins to obtain the results. A variation on this approach is to set up ratios, using the fact that Table 19-1 also gives the rms speed for nitrogen gas at 300 K (the value is 517 m/s). Here we illustrate the latter approach, using v for v_{rms}:

$$\frac{v_2}{v_1} = \frac{\sqrt{3RT_2/M}}{\sqrt{3RT_1/M}} = \sqrt{\frac{T_2}{T_1}}.$$

(a) With $T_2 = (20.0 + 273.15)$ K ≈ 293 K, we obtain

$$v_2 = (517\,\text{m/s})\sqrt{\frac{293\,\text{K}}{300\,\text{K}}} = 511\,\text{m/s}.$$

(b) In this case, we set $v_3 = \tfrac{1}{2}v_2$ and solve $v_3/v_2 = \sqrt{T_3/T_2}$ for T_3:

$$T_3 = T_2\left(\frac{v_3}{v_2}\right)^2 = (293\,\text{K})\left(\frac{1}{2}\right)^2 = 73.0\,\text{K}$$

which we write as 73.0 – 273 = – 200°C.

(c) Now we have $v_4 = 2v_2$ and obtain

$$T_4 = T_2\left(\frac{V_4}{V_2}\right)^2 = (293\,\text{K})(4) = 1.17 \times 10^3\,\text{K}$$

which is equivalent to 899°C.

22. First we rewrite Eq. 19-22 using Eq. 19-4 and Eq. 19-7:

$$v_{rms} = \sqrt{\frac{3RT}{M}} = \sqrt{\frac{3(kN_A)T}{(mN_A)}} = \sqrt{\frac{3kT}{M}}.$$

The mass of the electron is given in the problem, and $k = 1.38 \times 10^{-23}$ J/K is given in the textbook. With $T = 2.00 \times 10^6$ K, the above expression gives $v_{rms} = 9.53 \times 10^6$ m/s. The pressure value given in the problem is not used in the solution.

23. In the reflection process, only the normal component of the momentum changes, so for one molecule the change in momentum is $2mv\cos\theta$, where m is the mass of the molecule, v is its speed, and θ is the angle between its velocity and the normal to the wall. If N molecules collide with the wall, then the change in their total momentum is $2Nmv\cos\theta$, and if the total time taken for the collisions is Δt, then the average rate of change of the total momentum is $2(N/\Delta t)mv\cos\theta$. This is the average force exerted by the N molecules on the wall, and the pressure is the average force per unit area:

$$p = \frac{2}{A}\left(\frac{N}{\Delta t}\right)mv\cos\theta$$

$$= \left(\frac{2}{2.0 \times 10^{-4}\,\text{m}^2}\right)(1.0 \times 10^{23}\,\text{s}^{-1})(3.3 \times 10^{-27}\,\text{kg})(1.0 \times 10^3\,\text{m/s})\cos 55°$$

$$= 1.9 \times 10^3\,\text{Pa}.$$

We note that the value given for the mass was converted to kg and the value given for the area was converted to m².

24. We can express the ideal gas law in terms of density using $n = M_{sam}/M$:

$$pV = \frac{M_{sam}RT}{M} \Rightarrow \rho = \frac{pM}{RT}.$$

We can also use this to write the rms speed formula in terms of density:

$$v_{rms} = \sqrt{\frac{3RT}{M}} = \sqrt{\frac{3(pM/\rho)}{M}} = \sqrt{\frac{3p}{\rho}}.$$

(a) We convert to SI units: $\rho = 1.24 \times 10^{-2}$ kg/m^3 and $p = 1.01 \times 10^3$ Pa. The rms speed is $\sqrt{3(1010)/0.0124} = 494$ m/s.

(b) We find M from $\rho = pM/RT$ with $T = 273$ K.

$$M = \frac{\rho RT}{p} = \frac{(0.0124 \text{ kg/m}^3)(8.31 \text{ J/mol} \cdot \text{K})(273 \text{ K})}{1.01 \times 10^3 \text{ Pa}} = 0.0279 \text{ kg/mol} = 27.9 \text{ g/mol}.$$

(c) From Table 19.1, we identify the gas to be N_2.

25. (a) Eq. 19-24 gives $K_{avg} = \frac{3}{2}(1.38 \times 10^{-23} \text{ J/K})(273 \text{ K}) = 5.65 \times 10^{-21}$ J.

(b) For $T = 373$ K, the average translational kinetic energy is $K_{avg} = 7.72 \times 10^{-21}$ J.

(c) The unit mole may be thought of as a (large) collection: 6.02×10^{23} molecules of ideal gas, in this case. Each molecule has energy specified in part (a), so the large collection has a total kinetic energy equal to

$$K_{mole} = N_A K_{avg} = (6.02 \times 10^{23})(5.65 \times 10^{-21} \text{ J}) = 3.40 \times 10^3 \text{ J}.$$

(d) Similarly, the result from part (b) leads to

$$K_{mole} = (6.02 \times 10^{23})(7.72 \times 10^{-21} \text{ J}) = 4.65 \times 10^3 \text{ J}.$$

26. The average translational kinetic energy is given by $K_{avg} = \frac{3}{2}kT$, where k is the Boltzmann constant (1.38×10^{-23} J/K) and T is the temperature on the Kelvin scale. Thus

$$K_{avg} = \frac{3}{2}(1.38 \times 10^{-23} \text{ J/K})(1600 \text{ K}) = 3.31 \times 10^{-20} \text{ J}.$$

27. (a) We use $\varepsilon = L_V/N$, where L_V is the heat of vaporization and N is the number of molecules per gram. The molar mass of atomic hydrogen is 1 g/mol and the molar mass of atomic oxygen is 16 g/mol so the molar mass of H_2O is (1.0 + 1.0 + 16) = 18 g/mol. There are $N_A = 6.02 \times 10^{23}$ molecules in a mole so the number of molecules in a gram of water is $(6.02 \times 10^{23} \text{ mol}^{-1})/(18 \text{ g/mol}) = 3.34 \times 10^{22}$ molecules/g. Thus

$$\varepsilon = (539 \text{ cal/g})/(3.34 \times 10^{22}/\text{g}) = 1.61 \times 10^{-20} \text{ cal} = 6.76 \times 10^{-20} \text{ J}.$$

(b) The average translational kinetic energy is

$$K_{\text{avg}} = \frac{3}{2}kT = \frac{3}{2}(1.38 \times 10^{-23} \text{ J/K})[(32.0+273.15)\text{K}] = 6.32 \times 10^{-21} \text{ J}.$$

The ratio $\varepsilon/K_{\text{avg}}$ is $(6.76 \times 10^{-20} \text{ J})/(6.32 \times 10^{-21} \text{ J}) = 10.7$.

28. We solve Eq. 19-25 for d:

$$d = \sqrt{\frac{1}{\lambda \pi \sqrt{2}\,(N/V)}} = \sqrt{\frac{1}{(0.80 \times 10^5 \text{ cm})\,\pi\sqrt{2}\,(2.7 \times 10^{19}/\text{cm}^3)}}$$

which yields $d = 3.2 \times 10^{-8}$ cm, or 0.32 nm.

29. (a) According to Eq. 19-25, the mean free path for molecules in a gas is given by

$$\lambda = \frac{1}{\sqrt{2}\pi d^2 N/V},$$

where d is the diameter of a molecule and N is the number of molecules in volume V. Substitute $d = 2.0 \times 10^{-10}$ m and $N/V = 1 \times 10^6$ molecules/m³ to obtain

$$\lambda = \frac{1}{\sqrt{2}\pi(2.0 \times 10^{-10} \text{ m})^2\,(1 \times 10^6 \text{ m}^{-3})} = 6 \times 10^{12} \text{ m}.$$

(b) At this altitude most of the gas particles are in orbit around Earth and do not suffer randomizing collisions. The mean free path has little physical significance.

30. Using $v = f\lambda$ with $v = 331$ m/s (see Table 17-1) with Eq. 19-2 and Eq. 19-25 leads to

$$f = \frac{v}{\left(\dfrac{1}{\sqrt{2}\pi d^2(N/V)}\right)} = (331\,\text{m/s})\pi\sqrt{2}\,(3.0 \times 10^{-10}\text{ m})^2\left(\frac{nN_A}{V}\right)$$

$$= \left(8.0 \times 10^7\,\frac{\text{m}^3}{\text{s}\cdot\text{mol}}\right)\left(\frac{n}{V}\right) = \left(8.0 \times 10^7\,\frac{\text{m}^3}{\text{s}\cdot\text{mol}}\right)\left(\frac{1.01 \times 10^5 \text{ Pa}}{(8.31 \text{ J/mol}\cdot\text{K})\,(273.15\,\text{K})}\right)$$

$$= 3.5 \times 10^9 \text{ Hz}.$$

where we have used the ideal gas law and substituted $n/V = p/RT$. If we instead use $v = 343$ m/s (the "default value" for speed of sound in air, used repeatedly in Ch. 17), then the answer is 3.7×10^9 Hz.

31. (a) We use the ideal gas law $pV = nRT = NkT$, where p is the pressure, V is the volume, T is the temperature, n is the number of moles, and N is the number of molecules.

The substitutions $N = nN_A$ and $k = R/N_A$ were made. Since 1 cm of mercury = 1333 Pa, the pressure is $p = (10^{-7})(1333 \text{ Pa}) = 1.333 \times 10^{-4}$ Pa. Thus,

$$\frac{N}{V} = \frac{p}{kT} = \frac{1.333 \times 10^{-4} \text{ Pa}}{(1.38 \times 10^{-23} \text{ J/K})(295 \text{ K})} = 3.27 \times 10^{16} \text{ molecules/m}^3 = 3.27 \times 10^{10} \text{ molecules/cm}^3.$$

(b) The molecular diameter is $d = 2.00 \times 10^{-10}$ m, so, according to Eq. 19-25, the mean free path is

$$\lambda = \frac{1}{\sqrt{2}\pi d^2 N/V} = \frac{1}{\sqrt{2}\pi (2.00 \times 10^{-10} \text{ m})^2 (3.27 \times 10^{16} \text{ m}^{-3})} = 172 \text{ m}.$$

32. (a) We set up a ratio using Eq. 19-25:

$$\frac{\lambda_{Ar}}{\lambda_{N_2}} = \frac{1/\left(\pi\sqrt{2}d_{Ar}^2(N/V)\right)}{1/\left(\pi\sqrt{2}d_{N_2}^2(N/V)\right)} = \left(\frac{d_{N_2}}{d_{Ar}}\right)^2.$$

Therefore, we obtain

$$\frac{d_{Ar}}{d_{N_2}} = \sqrt{\frac{\lambda_{N_2}}{\lambda_{Ar}}} = \sqrt{\frac{27.5 \times 10^{-6} \text{ cm}}{9.9 \times 10^{-6} \text{ cm}}} = 1.7.$$

(b) Using Eq. 19-2 and the ideal gas law, we substitute $N/V = N_A n/V = N_A p/RT$ into Eq. 19-25 and find

$$\lambda = \frac{RT}{\pi\sqrt{2}d^2 pN_A}.$$

Comparing (for the same species of molecule) at two different pressures and temperatures, this leads to

$$\frac{\lambda_2}{\lambda_1} = \left(\frac{T_2}{T_1}\right)\left(\frac{p_1}{p_2}\right).$$

With $\lambda_1 = 9.9 \times 10^{-6}$ cm, $T_1 = 293$ K (the same as T_2 in this part), $p_1 = 750$ torr and $p_2 = 150$ torr, we find $\lambda_2 = 5.0 \times 10^{-5}$ cm.

(c) The ratio set up in part (b), using the same values for quantities with subscript 1, leads to $\lambda_2 = 7.9 \times 10^{-6}$ cm for $T_2 = 233$ K and $p_2 = 750$ torr.

33. (a) The average speed is

$$v_{avg} = \frac{1}{N}\sum_{i=1}^{N} v_i = \frac{1}{10}[4(200 \text{ m/s}) + 2(500 \text{ m/s}) + 4(600 \text{ m/s})] = 420 \text{ m/s}.$$

(b) The rms speed is

$$v_{rms} = \sqrt{\frac{1}{N}\sum_{i=1}^{N} v_i^2} = \sqrt{\frac{1}{10}[4(200 \text{ m/s})^2 + 2(500 \text{ m/s})^2 + 4(600 \text{ m/s})^2]} = 458 \text{ m/s}$$

(c) Yes, $v_{rms} > v_{avg}$.

34. (a) The average speed is

$$v_{avg} = \frac{\sum n_i v_i}{\sum n_i} = \frac{[2(1.0) + 4(2.0) + 6(3.0) + 8(4.0) + 2(5.0)] \text{ cm/s}}{2+4+6+8+2} = 3.2 \text{ cm/s}.$$

(b) From $v_{rms} = \sqrt{\sum n_i v_i^2 / \sum n_i}$ we get

$$v_{rms} = \sqrt{\frac{2(1.0)^2 + 4(2.0)^2 + 6(3.0)^2 + 8(4.0)^2 + 2(5.0)^2}{2+4+6+8+2}} \text{ cm/s} = 3.4 \text{ cm/s}.$$

(c) There are eight particles at $v = 4.0$ cm/s, more than the number of particles at any other single speed. So 4.0 cm/s is the most probable speed.

35. (a) The average speed is $\bar{v} = \dfrac{\sum v}{N}$, where the sum is over the speeds of the particles and N is the number of particles. Thus

$$\bar{v} = \frac{(2.0 + 3.0 + 4.0 + 5.0 + 6.0 + 7.0 + 8.0 + 9.0 + 10.0 + 11.0) \text{ km/s}}{10} = 6.5 \text{ km/s}.$$

(b) The rms speed is given by $v_{rms} = \sqrt{\dfrac{\sum v^2}{N}}$. Now

$$\sum v^2 = [(2.0)^2 + (3.0)^2 + (4.0)^2 + (5.0)^2 + (6.0)^2 + (7.0)^2 + (8.0)^2 + (9.0)^2 + (10.0)^2 + (11.0)^2] \text{ km}^2/\text{s}^2 = 505 \text{ km}^2/\text{s}^2$$

so

$$v_{rms} = \sqrt{\frac{505 \text{ km}^2/\text{s}^2}{10}} = 7.1 \text{ km/s}.$$

36. (a) From the graph we see that $v_p = 400$ m/s. Using the fact that $M = 28$ g/mol = 0.028 kg/mol for nitrogen (N_2) gas, Eq. 19-35 can then be used to determine the absolute temperature. We obtain $T = \frac{1}{2} M v_p^2 / R = 2.7 \times 10^2$ K.

(b) Comparing with Eq. 19-34, we conclude $v_{rms} = \sqrt{3/2} \, v_p = 4.9 \times 10^2$ m/s.

37. The rms speed of molecules in a gas is given by $v_{rms} = \sqrt{3RT/M}$, where T is the temperature and M is the molar mass of the gas. See Eq. 19-34. The speed required for escape from Earth's gravitational pull is $v = \sqrt{2gr_e}$, where g is the acceleration due to gravity at Earth's surface and r_e (= 6.37×10^6 m) is the radius of Earth. To derive this expression, take the zero of gravitational potential energy to be at infinity. Then, the gravitational potential energy of a particle with mass m at Earth's surface is

$$U = -GMm/r_e^2 = -mgr_e,$$

where $g = GM/r_e^2$ was used. If v is the speed of the particle, then its total energy is $E = -mgr_e + \frac{1}{2} mv^2$. If the particle is just able to travel far away, its kinetic energy must tend toward zero as its distance from Earth becomes large without bound. This means $E = 0$ and $v = \sqrt{2gr_e}$. We equate the expressions for the speeds to obtain $\sqrt{3RT/M} = \sqrt{2gr_e}$. The solution for T is $T = 2gr_e M/3R$.

(a) The molar mass of hydrogen is 2.02×10^{-3} kg/mol, so for that gas

$$T = \frac{2(9.8 \, \text{m/s}^2)(6.37 \times 10^6 \, \text{m})(2.02 \times 10^{-3} \, \text{kg/mol})}{3(8.31 \, \text{J/mol} \cdot \text{K})} = 1.0 \times 10^4 \text{ K}.$$

(b) The molar mass of oxygen is 32.0×10^{-3} kg/mol, so for that gas

$$T = \frac{2(9.8 \, \text{m/s}^2)(6.37 \times 10^6 \, \text{m})(32.0 \times 10^{-3} \, \text{kg/mol})}{3(8.31 \, \text{J/mol} \cdot \text{K})} = 1.6 \times 10^5 \text{ K}.$$

(c) Now, $T = 2g_m r_m M / 3R$, where $r_m = 1.74 \times 10^6$ m is the radius of the Moon and $g_m = 0.16g$ is the acceleration due to gravity at the Moon's surface. For hydrogen, the temperature is

$$T = \frac{2(0.16)(9.8 \, \text{m/s}^2)(1.74 \times 10^6 \, \text{m})(2.02 \times 10^{-3} \, \text{kg/mol})}{3(8.31 \, \text{J/mol} \cdot \text{K})} = 4.4 \times 10^2 \text{ K}.$$

(d) For oxygen, the temperature is

$$T = \frac{2(0.16)(9.8\,\text{m/s}^2)(1.74\times10^6\,\text{m})(32.0\times10^{-3}\,\text{kg/mol})}{3(8.31\,\text{J/mol}\cdot\text{K})} = 7.0\times10^3\,\text{K}.$$

(e) The temperature high in Earth's atmosphere is great enough for a significant number of hydrogen atoms in the tail of the Maxwellian distribution to escape. As a result the atmosphere is depleted of hydrogen.

(f) On the other hand, very few oxygen atoms escape. So there should be much oxygen high in Earth's upper atmosphere.

38. We divide Eq. 19-31 by Eq. 19-22:

$$\frac{v_{\text{avg2}}}{v_{\text{rms1}}} = \frac{\sqrt{8RT/\pi M_2}}{\sqrt{3RT/M_1}} = \sqrt{\frac{8M_1}{3\pi M_2}}$$

which, for $v_{\text{avg2}} = 2v_{\text{rms1}}$, leads to

$$\frac{m_1}{m_2} = \frac{M_1}{M_2} = \frac{3\pi}{8}\left(\frac{v_{\text{avg2}}}{v_{\text{rms1}}}\right)^2 = \frac{3\pi}{2} = 4.7.$$

39. (a) The root-mean-square speed is given by $v_{\text{rms}} = \sqrt{3RT/M}$. See Eq. 19-34. The molar mass of hydrogen is 2.02×10^{-3} kg/mol, so

$$v_{\text{rms}} = \sqrt{\frac{3(8.31\,\text{J/mol}\cdot\text{K})(4000\,\text{K})}{2.02\times10^{-3}\,\text{kg/mol}}} = 7.0\times10^3\,\text{m/s}.$$

(b) When the surfaces of the spheres that represent an H_2 molecule and an Ar atom are touching, the distance between their centers is the sum of their radii:

$$d = r_1 + r_2 = 0.5\times10^{-8}\,\text{cm} + 1.5\times10^{-8}\,\text{cm} = 2.0\times10^{-8}\,\text{cm}.$$

(c) The argon atoms are essentially at rest so in time t the hydrogen atom collides with all the argon atoms in a cylinder of radius d and length vt, where v is its speed. That is, the number of collisions is $\pi d^2 vtN/V$, where, N/V is the concentration of argon atoms. The number of collisions per unit time is

$$\frac{\pi d^2 vN}{V} = \pi(2.0\times10^{-10}\,\text{m})^2(7.0\times10^3\,\text{m/s})(4.0\times10^{25}\,\text{m}^{-3}) = 3.5\times10^{10}\,\text{collisions/s}.$$

40. We divide Eq. 19-35 by Eq. 19-22:

$$\frac{v_P}{v_{rms}} = \frac{\sqrt{2RT_2/M}}{\sqrt{3RT_1/M}} = \sqrt{\frac{2T_2}{3T_1}}$$

which, for $v_P = v_{rms}$, leads to

$$\frac{T_2}{T_1} = \frac{3}{2}\left(\frac{v_P}{v_{rms}}\right)^2 = \frac{3}{2}.$$

41. (a) The distribution function gives the fraction of particles with speeds between v and $v + dv$, so its integral over all speeds is unity: $\int P(v)\,dv = 1$. Evaluate the integral by calculating the area under the curve in Fig. 19-24. The area of the triangular portion is half the product of the base and altitude, or $\frac{1}{2}av_0$. The area of the rectangular portion is the product of the sides, or av_0. Thus,

$$\int P(v)dv = \frac{1}{2}av_0 + av_0 = \frac{3}{2}av_0,$$

so $\frac{3}{2}av_0 = 1$ and $av_0 = 2/3 = 0.67$.

(b) The average speed is given by $v_{avg} = \int vP(v)\,dv$. For the triangular portion of the distribution $P(v) = av/v_0$, and the contribution of this portion is

$$\frac{a}{v_0}\int_0^{v_0} v^2 dv = \frac{a}{3v_0}v_0^3 = \frac{av_0^2}{3} = \frac{2}{9}v_0,$$

where $2/3v_0$ was substituted for a. $P(v) = a$ in the rectangular portion, and the contribution of this portion is

$$a\int_{v_0}^{2v_0} v\,dv = \frac{a}{2}\left(4v_0^2 - v_0^2\right) = \frac{3a}{2}v_0^2 = v_0.$$

Therefore,

$$v_{avg} = \frac{2}{9}v_0 + v_0 = 1.22v_0 \quad \Rightarrow \quad \frac{v_{avg}}{v_0} = 1.22.$$

(c) The mean-square speed is given by $v_{rms}^2 = \int v^2 P(v)\,dv$. The contribution of the triangular section is

$$\frac{a}{v_0}\int_0^{v_0} v^3 dv = \frac{a}{4v_0}v_0^4 = \frac{1}{6}v_0^2.$$

The contribution of the rectangular portion is

$$a\int_{v_0}^{2v_0} v^2 dv = \frac{a}{3}\left(8v_0^3 - v_0^3\right) = \frac{7a}{3}v_0^3 = \frac{14}{9}v_0^2.$$

Thus,

$$v_{rms} = \sqrt{\frac{1}{6}v_0^2 + \frac{14}{9}v_0^2} = 1.31 v_0 \quad \Rightarrow \quad \frac{v_{rms}}{v_0} = 1.31.$$

(d) The number of particles with speeds between $1.5v_0$ and $2v_0$ is given by $N\int_{1.5v_0}^{2v_0} P(v)dv$. The integral is easy to evaluate since $P(v) = a$ throughout the range of integration. Thus the number of particles with speeds in the given range is

$$Na(2.0v_0 - 1.5v_0) = 0.5N\, av_0 = N/3,$$

where $2/3v_0$ was substituted for a. In other words, the fraction of particles in this range is 1/3 or 0.33.

42. The internal energy is

$$E_{int} = \frac{3}{2}nRT = \frac{3}{2}(1.0\,\text{mol})(8.31\,\text{J/mol}\cdot\text{K})(273\,\text{K}) = 3.4\times 10^3\,\text{J}.$$

43. (a) The work is zero in this process since volume is kept fixed.

(b) Since $C_V = \frac{3}{2}R$ for an ideal monatomic gas, then Eq. 19-39 gives $Q = +374$ J.

(c) $\Delta E_{int} = Q - W = +374$ J.

(d) Two moles are equivalent to $N = 12 \times 10^{23}$ particles. Dividing the result of part (c) by N gives the average translational kinetic energy change per atom: 3.11×10^{-22} J.

44. (a) Since the process is a constant-pressure expansion,

$$W = p\Delta V = nR\Delta T = (2.02\,\text{mol})(8.31\,\text{J/mol}\cdot\text{K})(15\,\text{K}) = 249\,\text{J}.$$

(b) Now, $C_p = \frac{5}{2}R$ in this case, so $Q = nC_p\Delta T = +623$ J by Eq. 19-46.

(c) The change in the internal energy is $\Delta E_{int} = Q - W = +374$ J.

(d) The change in the average kinetic energy per atom is

$$\Delta K_{avg} = \Delta E_{int}/N = +3.11 \times 10^{-22}\,\text{J}.$$

45. When the temperature changes by ΔT the internal energy of the first gas changes by $n_1 C_1 \Delta T$, the internal energy of the second gas changes by $n_2 C_2 \Delta T$, and the internal energy of the third gas changes by $n_3 C_3 \Delta T$. The change in the internal energy of the composite gas is

$$\Delta E_{int} = (n_1\, C_1 + n_2\, C_2 + n_3\, C_3)\, \Delta T.$$

This must be $(n_1 + n_2 + n_3)\, C_V\, \Delta T$, where C_V is the molar specific heat of the mixture. Thus,

$$C_V = \frac{n_1 C_1 + n_2 C_2 + n_3 C_3}{n_1 + n_2 + n_3}.$$

With n_1=2.40 mol, C_{V1}=12.0 J/mol·K for gas 1, n_2=1.50 mol, C_{V2}=12.8 J/mol·K for gas 2, and n_3=3.20 mol, C_{V3}=20.0 J/mol·K for gas 3, we obtain C_V=15.8 J/mol·K for the mixture.

46. Two formulas (other than the first law of thermodynamics) will be of use to us. It is straightforward to show, from Eq. 19-11, that for any process that is depicted as a *straight line* on the pV diagram — the work is

$$W_{straight} = \left(\frac{p_i + p_f}{2}\right) \Delta V$$

which includes, as special cases, $W = p\Delta V$ for constant-pressure processes and $W = 0$ for constant-volume processes. Further, Eq. 19-44 with Eq. 19-51 gives

$$E_{int} = n\left(\frac{f}{2}\right)RT = \left(\frac{f}{2}\right)pV$$

where we have used the ideal gas law in the last step. We emphasize that, in order to obtain work and energy in Joules, pressure should be in Pascals (N / m^2) and volume should be in cubic meters. The degrees of freedom for a diatomic gas is $f = 5$.

(a) The internal energy change is

$$E_{int\,c} - E_{int\,a} = \frac{5}{2}(p_c V_c - p_a V_a) = \frac{5}{2}\left((2.0\times10^3\,\text{Pa})(4.0\,\text{m}^3) - (5.0\times10^3\,\text{Pa})(2.0\,\text{m}^3)\right)$$
$$= -5.0\times10^3\ \text{J}.$$

(b) The work done during the process represented by the diagonal path is

$$W_{diag} = \left(\frac{p_a + p_c}{2}\right)(V_c - V_a) = (3.5\times10^3\,\text{Pa})(2.0\,\text{m}^3)$$

which yields $W_{diag} = 7.0\times10^3$ J. Consequently, the first law of thermodynamics gives

$$Q_{diag} = \Delta E_{int} + W_{diag} = (-5.0\times10^3 + 7.0\times10^3)\text{ J} = 2.0\times10^3 \text{ J}.$$

(c) The fact that ΔE_{int} only depends on the initial and final states, and not on the details of the "path" between them, means we can write $\Delta E_{int} = E_{int\,c} - E_{int\,a} = -5.0\times10^3$ J for the indirect path, too. In this case, the work done consists of that done during the constant pressure part (the horizontal line in the graph) plus that done during the constant volume part (the vertical line):

$$W_{indirect} = (5.0\times10^3 \text{ Pa})(2.0\,\text{m}^3) + 0 = 1.0\times10^4 \text{ J}.$$

Now, the first law of thermodynamics leads to

$$Q_{indirect} = \Delta E_{int} + W_{indirect} = (-5.0\times10^3 + 1.0\times10^4)\text{ J} = 5.0\times10^3 \text{ J}.$$

47. Argon is a monatomic gas, so $f = 3$ in Eq. 19-51, which provides

$$C_V = \frac{3}{2}R = \frac{3}{2}(8.31 \text{ J/mol·K})\left(\frac{1 \text{ cal}}{4.186 \text{ J}}\right) = 2.98 \frac{\text{cal}}{\text{mol·C}°}$$

where we have converted Joules to calories, and taken advantage of the fact that a Celsius degree is equivalent to a unit change on the Kelvin scale. Since (for a given substance) M is effectively a conversion factor between grams and moles, we see that c_V (see units specified in the problem statement) is related to C_V by $C_V = c_V\,M$ where $M = mN_A$, and m is the mass of a single atom (see Eq. 19-4).

(a) From the above discussion, we obtain

$$m = \frac{M}{N_A} = \frac{C_V/c_V}{N_A} = \frac{2.98/0.075}{6.02\times10^{23}} = 6.6\times10^{-23} \text{ g}.$$

(b) The molar mass is found to be $M = C_V/c_V = 2.98/0.075 = 39.7$ g/mol which should be rounded to 40 g/mol since the given value of c_V is specified to only two significant figures.

48. (a) According to the first law of thermodynamics $Q = \Delta E_{int} + W$. When the pressure is a constant $W = p\,\Delta V$. So

$$\Delta E_{int} = Q - p\Delta V = 20.9 \text{ J} - (1.01\times10^5 \text{ Pa})(100 \text{ cm}^3 - 50 \text{ cm}^3)\left(\frac{1\times10^{-6} \text{ m}^3}{1 \text{ cm}^3}\right) = 15.9 \text{ J}.$$

(b) The molar specific heat at constant pressure is

$$C_p = \frac{Q}{n\Delta T} = \frac{Q}{n(p\Delta V/nR)} = \frac{R}{p}\frac{Q}{\Delta V} = \frac{(8.31\text{ J/mol}\cdot\text{K})(20.9\text{ J})}{(1.01\times 10^5\text{ Pa})(50\times 10^{-6}\text{ m}^3)} = 34.4\text{ J/mol}\cdot\text{K}.$$

(c) Using Eq. 19-49, $C_V = C_p - R = 26.1$ J/mol·K.

49. (a) From Table 19-3, $C_V = \tfrac{5}{2}R$ and $C_p = \tfrac{7}{2}R$. Thus, Eq. 19-46 yields

$$Q = nC_p\Delta T = (3.00)\left(\frac{7}{2}(8.31)\right)(40.0) = 3.49\times 10^3\text{ J}.$$

(b) Eq. 19-45 leads to

$$\Delta E_{\text{int}} = nC_V\Delta T = (3.00)\left(\frac{5}{2}(8.31)\right)(40.0) = 2.49\times 10^3\text{ J}.$$

(c) From either $W = Q - \Delta E_{\text{int}}$ or $W = p\Delta T = nR\Delta T$, we find $W = 997$ J.

(d) Eq. 19-24 is written in more convenient form (for this problem) in Eq. 19-38. Thus, the increase in kinetic energy is

$$\Delta K_{\text{trans}} = \Delta(NK_{\text{avg}}) = n\left(\frac{3}{2}R\right)\Delta T \approx 1.49\times 10^3\text{ J}.$$

Since $\Delta E_{\text{int}} = \Delta K_{\text{trans}} + \Delta K_{\text{rot}}$, the increase in rotational kinetic energy is

$$\Delta K_{\text{rot}} = \Delta E_{\text{int}} - \Delta K_{\text{trans}} = 2.49\times 10^3\text{ J} - 1.49\times 10^3\text{ J} = 1.00\times 10^3\text{ J}.$$

Note that had there been no rotation, all the energy would have gone into the translational kinetic energy.

50. Referring to Table 19-3, Eq. 19-45 and Eq. 19-46, we have

$$\Delta E_{\text{int}} = nC_V\Delta T = \frac{5}{2}nR\Delta T$$

$$Q = nC_p\Delta T = \frac{7}{2}nR\Delta T.$$

Dividing the equations, we obtain

$$\frac{\Delta E_{\text{int}}}{Q} = \frac{5}{7}.$$

Thus, the given value $Q = 70$ J leads to $\Delta E_{\text{int}} = 50$ J.

51. The fact that they rotate but do not oscillate means that the value of f given in Table 19-3 is relevant. Thus, Eq. 19-46 leads to

$$Q = nC_p \Delta T = n\left(\frac{7}{2}R\right)(T_f - T_i) = nRT_i\left(\frac{7}{2}\right)\left(\frac{T_f}{T_i} - 1\right)$$

where $T_i = 273$ K and $n = 1.0$ mol. The ratio of absolute temperatures is found from the gas law in ratio form (see Sample Problem 19-1). With $p_f = p_i$ we have

$$\frac{T_f}{T_i} = \frac{V_f}{V_i} = 2.$$

Therefore, the energy added as heat is

$$Q = (1.0\,\text{mol})(8.31\,\text{J/mol}\cdot\text{K})(273\,\text{K})\left(\frac{7}{2}\right)(2-1) \approx 8.0\times 10^3\,\text{J}.$$

52. (a) Using $M = 32.0$ g/mol from Table 19-1 and Eq. 19-3, we obtain

$$n = \frac{M_{\text{sam}}}{M} = \frac{12.0\,\text{g}}{32.0\,\text{g/mol}} = 0.375\,\text{mol}.$$

(b) This is a constant pressure process with a diatomic gas, so we use Eq. 19-46 and Table 19-3. We note that a change of Kelvin temperature is numerically the same as a change of Celsius degrees.

$$Q = nC_p \Delta T = n\left(\frac{7}{2}R\right)\Delta T = (0.375\,\text{mol})\left(\frac{7}{2}\right)(8.31\,\text{J/mol}\cdot\text{K})(100\,\text{K}) = 1.09\times 10^3\,\text{J}.$$

(c) We could compute a value of ΔE_{int} from Eq. 19-45 and divide by the result from part (b), or perform this manipulation algebraically to show the generality of this answer (that is, many factors will be seen to cancel). We illustrate the latter approach:

$$\frac{\Delta E_{\text{int}}}{Q} = \frac{n\left(\frac{5}{2}R\right)\Delta T}{n\left(\frac{7}{2}R\right)\Delta T} = \frac{5}{7} \approx 0.714.$$

53. (a) Since the process is at constant pressure, energy transferred as heat to the gas is given by $Q = nC_p\,\Delta T$, where n is the number of moles in the gas, C_p is the molar specific heat at constant pressure, and ΔT is the increase in temperature. For a diatomic ideal gas $C_p = \frac{7}{2}R$. Thus,

$$Q = \frac{7}{2} nR\Delta T = \frac{7}{2}(4.00\,\text{mol})(8.31\,\text{J/mol}\cdot\text{K})(60.0\,\text{K}) = 6.98\times 10^3\,\text{J}.$$

(b) The change in the internal energy is given by $\Delta E_{\text{int}} = nC_V\Delta T$, where C_V is the specific heat at constant volume. For a diatomic ideal gas $C_V = \frac{5}{2}R$, so

$$\Delta E_{\text{int}} = \frac{5}{2}nR\Delta T = \frac{5}{2}(4.00\,\text{mol})(8.31\,\text{J/mol.K})(60.0\,\text{K}) = 4.99\times 10^3\,\text{J}.$$

(c) According to the first law of thermodynamics, $\Delta E_{\text{int}} = Q - W$, so

$$W = Q - \Delta E_{\text{int}} = 6.98\times 10^3\,\text{J} - 4.99\times 10^3\,\text{J} = 1.99\times 10^3\,\text{J}.$$

(d) The change in the total translational kinetic energy is

$$\Delta K = \frac{3}{2}nR\Delta T = \frac{3}{2}(4.00\,\text{mol})(8.31\,\text{J/mol}\cdot\text{K})(60.0\,\text{K}) = 2.99\times 10^3\,\text{J}.$$

54. (a) We use Eq. 19-54 with $V_f/V_i = \frac{1}{2}$ for the gas (assumed to obey the ideal gas law).

$$p_i V_i^\gamma = p_f V_f^\gamma \Rightarrow \frac{p_f}{p_i} = \left(\frac{V_i}{V_f}\right)^\gamma = (2.00)^{1.3}$$

which yields $p_f = (2.46)(1.0\,\text{atm}) = 2.46\,\text{atm}$.

(b) Similarly, Eq. 19-56 leads to

$$T_f = T_i\left(\frac{V_i}{V_f}\right)^{\gamma-1} = (273\,\text{K})(1.23) = 336\,\text{K}.$$

(c) We use the gas law in ratio form (see Sample Problem 19-1) and note that when $p_1 = p_2$ then the ratio of volumes is equal to the ratio of (absolute) temperatures. Consequently, with the subscript 1 referring to the situation (of small volume, high pressure, and high temperature) the system is in at the end of part (a), we obtain

$$\frac{V_2}{V_1} = \frac{T_2}{T_1} = \frac{273\,\text{K}}{336\,\text{K}} = 0.813.$$

The volume V_1 is half the original volume of one liter, so

$$V_2 = 0.813(0.500\,\text{L}) = 0.406\,\text{L}.$$

55. (a) Let p_i, V_i, and T_i represent the pressure, volume, and temperature of the initial state of the gas. Let p_f, V_f, and T_f represent the pressure, volume, and temperature of the final state. Since the process is adiabatic $p_i V_i^\gamma = p_f V_f^\gamma$, so

$$p_f = \left(\frac{V_i}{V_f}\right)^\gamma p_i = \left(\frac{4.3\,\text{L}}{0.76\,\text{L}}\right)^{1.4} (1.2\,\text{atm}) = 13.6\,\text{atm} \approx 14\,\text{atm}.$$

We note that since V_i and V_f have the same units, their units cancel and p_f has the same units as p_i.

(b) The gas obeys the ideal gas law $pV = nRT$, so $p_i V_i / p_f V_f = T_i / T_f$ and

$$T_f = \frac{p_f V_f}{p_i V_i} T_i = \left[\frac{(13.6\,\text{atm})(0.76\,\text{L})}{(1.2\,\text{atm})(4.3\,\text{L})}\right](310\,\text{K}) = 6.2 \times 10^2\,\text{K}.$$

56. The fact that they rotate but do not oscillate means that the value of f given in Table 19-3 is relevant. In §19-11, it is noted that $\gamma = C_p/C_V$ so that we find $\gamma = 7/5$ in this case. In the state described in the problem, the volume is

$$V = \frac{nRT}{p} = \frac{(2.0\,\text{mol})(8.31\,\text{J/mol}\cdot\text{K})(300\,\text{K})}{1.01 \times 10^5\,\text{N/m}^2} = 0.049\,\text{m}^3.$$

Consequently,

$$pV^\gamma = (1.01 \times 10^5\,\text{N/m}^2)(0.049\,\text{m}^3)^{1.4} = 1.5 \times 10^3\,\text{N}\cdot\text{m}^{2.2}.$$

57. Since ΔE_{int} does not depend on the type of process,

$$(\Delta E_{\text{int}})_{\text{path 2}} = (\Delta E_{\text{int}})_{\text{path 1}}.$$

Also, since (for an ideal gas) it only depends on the temperature variable (so $\Delta E_{\text{int}} = 0$ for isotherms), then

$$(\Delta E_{\text{int}})_{\text{path 1}} = \sum (\Delta E_{\text{int}})_{\text{adiabat}}.$$

Finally, since $Q = 0$ for adiabatic processes, then (for path 1)

$$(\Delta E_{\text{int}})_{\text{adiabatic expansion}} = -W = -40\,\text{J}$$

$$(\Delta E_{\text{int}})_{\text{adiabatic compression}} = -W = -(-25)\,\text{J} = 25\,\text{J}.$$

Therefore, $(\Delta E_{\text{int}})_{\text{path 2}} = -40\,\text{J} + 25\,\text{J} = -15\,\text{J}$.

58. Let p_1, V_1 and T_1 represent the pressure, volume, and temperature of the air at $y_1 = 4267$ m. Similarly, let p, V and T be the pressure, volume, and temperature of the air at $y = 1567$ m. Since the process is adiabatic $p_1 V_1^\gamma = pV^\gamma$. Combining with ideal-gas law, $pV = NkT$, we obtain

$$pV^\gamma = p(T/p)^\gamma = p^{1-\gamma} T^\gamma = \text{constant} \quad \Rightarrow \quad p^{1-\gamma} T^\gamma = p_1^{1-\gamma} T_1^\gamma$$

With $p = p_0 e^{-ay}$ and $\gamma = 4/3$ (which gives $(1-\gamma)/\gamma = -1/4$), the temperature at the end of the decent is

$$T = \left(\frac{p_1}{p}\right)^{\frac{1-\gamma}{\gamma}} T_1 = \left(\frac{p_0 e^{-ay_1}}{p_0 e^{-ay}}\right)^{\frac{1-\gamma}{\gamma}} T_1 = e^{-a(y-y_1)/4} T_1 = e^{-(1.16\times 10^{-4}/\text{m})(1567\,\text{m} - 4267\,\text{m})/4} (268\text{ K})$$
$$= (1.08)(268\text{ K}) = 290\text{ K} = 17°\text{C}$$

59. The aim of this problem is to emphasize what it means for the internal energy to be a state function. Since path 1 and path 2 start and stop at the same places, then the internal energy change along path 1 is equal to that along path 2. Now, during isothermal processes (involving an ideal gas) the internal energy change is zero, so the only step in path 1 that we need to examine is step 2. Eq. 19-28 then immediately yields –20 J as the answer for the internal energy change.

60. Let p_i, V_i, and T_i represent the pressure, volume, and temperature of the initial state of the gas, and let p_f, V_f, and T_f be the pressure, volume, and temperature of the final state. Since the process is adiabatic $p_i V_i^\gamma = p_f V_f^\gamma$. Combining with ideal-gas law, $pV = NkT$, we obtain

$$p_i V_i^\gamma = p_i (T_i/p_i)^\gamma = p_i^{1-\gamma} T_i^\gamma = \text{constant} \quad \Rightarrow \quad p_i^{1-\gamma} T_i^\gamma = p_f^{1-\gamma} T_f^\gamma$$

With $\gamma = 4/3$ which gives $(1-\gamma)/\gamma = -1/4$, the temperature at the end of the adiabatic expansion is

$$T_f = \left(\frac{p_i}{p_f}\right)^{\frac{1-\gamma}{\gamma}} T_i = \left(\frac{5.00\text{ atm}}{1.00\text{ atm}}\right)^{-1/4} (278\text{ K}) = 186\text{ K} = -87°\text{C}.$$

61. (a) Eq. 19-54, $p_i V_i^\gamma = p_f V_f^\gamma$, leads to

$$p_f = p_i \left(\frac{V_i}{V_f}\right)^\gamma \quad \Rightarrow \quad 4.00\text{ atm} = (1.00\text{ atm})\left(\frac{200\text{ L}}{74.3\text{ L}}\right)^\gamma$$

which can be solved to yield

$$\gamma = \frac{\ln(p_f/p_i)}{\ln(V_i/V_f)} = \frac{\ln(4.00\,\text{atm}/1.00\,\text{atm})}{\ln(200\,\text{L}/74.3\,\text{L})} = 1.4 = \frac{7}{5}.$$

This implies that the gas is diatomic (see Table 19-3).

(b) One can now use either Eq. 19-56 (as illustrated in part (a) of Sample Problem 19-9) or use the ideal gas law itself. Here we illustrate the latter approach:

$$\frac{P_f V_f}{P_i V_i} = \frac{nRT_f}{nRT_i} \quad \Rightarrow \quad T_f = 446\,\text{K}.$$

(c) Again using the ideal gas law: $n = P_i V_i/RT_i = 8.10$ moles. The same result would, of course, follow from $n = P_f V_f/RT_f$.

62. Using Eq. 19-53 in Eq. 18-25 gives

$$W = p_i V_i^\gamma \int_{V_i}^{V_f} V^{-\gamma} dV = p_i V_i^\gamma \frac{V_f^{1-\gamma} - V_i^{1-\gamma}}{1-\gamma}.$$

Using Eq. 19-54 we can write this as

$$W = p_i V_i \frac{1-(p_f/p_i)^{1-1/\gamma}}{1-\gamma}$$

In this problem, $\gamma = 7/5$ (see Table 19-3) and $P_f/P_i = 2$. Converting the initial pressure to Pascals we find $P_i V_i = 24240$ J. Plugging in, then, we obtain $W = -1.33 \times 10^4$ J.

63. In the following $C_V = \frac{3}{2}R$ is the molar specific heat at constant volume, $C_p = \frac{5}{2}R$ is the molar specific heat at constant pressure, ΔT is the temperature change, and n is the number of moles.

The process $1 \to 2$ takes place at constant volume.

(a) The heat added is

$$Q = nC_V \Delta T = \frac{3}{2} nR\Delta T = \frac{3}{2}(1.00\,\text{mol})(8.31\,\text{J/mol}\cdot\text{K})(600\,\text{K} - 300\,\text{K}) = 3.74 \times 10^3\,\text{J}.$$

(b) Since the process takes place at constant volume the work W done by the gas is zero, and the first law of thermodynamics tells us that the change in the internal energy is

$$\Delta E_{int} = Q = 3.74 \times 10^3 \text{ J}.$$

(c) The work W done by the gas is zero.

The process $2 \rightarrow 3$ is adiabatic.

(d) The heat added is zero.

(e) The change in the internal energy is

$$\Delta E_{int} = nC_V \Delta T = \frac{3}{2}nR\Delta T = \frac{3}{2}(1.00\text{ mol})(8.31\text{ J/mol}\cdot\text{K})(455\text{ K} - 600\text{ K}) = -1.81 \times 10^3 \text{ J}.$$

(f) According to the first law of thermodynamics the work done by the gas is

$$W = Q - \Delta E_{int} = +1.81 \times 10^3 \text{ J}.$$

The process $3 \rightarrow 1$ takes place at constant pressure.

(g) The heat added is

$$Q = nC_p \Delta T = \frac{5}{2}nR\Delta T = \frac{5}{2}(1.00\text{ mol})(8.31\text{ J/mol}\cdot\text{K})(300\text{ K} - 455\text{ K}) = -3.22 \times 10^3 \text{ J}.$$

(h) The change in the internal energy is

$$\Delta E_{int} = nC_V \Delta T = \frac{3}{2}nR\Delta T = \frac{3}{2}(1.00\text{ mol})(8.31\text{ J/mol}\cdot\text{K})(300\text{ K} - 455\text{ K}) = -1.93 \times 10^3 \text{ J}.$$

(i) According to the first law of thermodynamics the work done by the gas is

$$W = Q - \Delta E_{int} = -3.22 \times 10^3 \text{ J} + 1.93 \times 10^3 \text{ J} = -1.29 \times 10^3 \text{ J}.$$

(j) For the entire process the heat added is

$$Q = 3.74 \times 10^3 \text{ J} + 0 - 3.22 \times 10^3 \text{ J} = 520 \text{ J}.$$

(k) The change in the internal energy is

$$\Delta E_{int} = 3.74 \times 10^3 \text{ J} - 1.81 \times 10^3 \text{ J} - 1.93 \times 10^3 \text{ J} = 0.$$

(l) The work done by the gas is

$$W = 0 + 1.81 \times 10^3 \text{ J} - 1.29 \times 10^3 \text{ J} = 520 \text{ J}.$$

(m) We first find the initial volume. Use the ideal gas law $p_1 V_1 = nRT_1$ to obtain

$$V_1 = \frac{nRT_1}{p_1} = \frac{(1.00 \text{ mol})(8.31 \text{ J/mol} \cdot \text{K})(300 \text{ K})}{(1.013 \times 10^5 \text{ Pa})} = 2.46 \times 10^{-2} \text{ m}^3.$$

(n) Since $1 \to 2$ is a constant volume process $V_2 = V_1 = 2.46 \times 10^{-2} \text{ m}^3$. The pressure for state 2 is

$$p_2 = \frac{nRT_2}{V_2} = \frac{(1.00 \text{ mol})(8.31 \text{ J/mol} \cdot \text{K})(600 \text{ K})}{2.46 \times 10^{-2} \text{ m}^3} = 2.02 \times 10^5 \text{ Pa}.$$

This is approximately equal to 2.00 atm.

(o) $3 \to 1$ is a constant pressure process. The volume for state 3 is

$$V_3 = \frac{nRT_3}{p_3} = \frac{(1.00 \text{ mol})(8.31 \text{ J/mol} \cdot \text{K})(455 \text{ K})}{1.013 \times 10^5 \text{ Pa}} = 3.73 \times 10^{-2} \text{ m}^3.$$

(p) The pressure for state 3 is the same as the pressure for state 1: $p_3 = p_1 = 1.013 \times 10^5$ Pa (1.00 atm)

64. Using the ideal gas law, one mole occupies a volume equal to

$$V = \frac{nRT}{p} = \frac{(1)(8.31)(50.0)}{1.00 \times 10^{-8}} = 4.16 \times 10^{10} \text{ m}^3.$$

Therefore, the number of molecules per unit volume is

$$\frac{N}{V} = \frac{nN_A}{V} = \frac{(1)(6.02 \times 10^{23})}{4.16 \times 10^{10}} = 1.45 \times 10^{13} \frac{\text{molecules}}{\text{m}^3}.$$

Using $d = 20.0 \times 10^{-9}$ m, Eq. 19-25 yields

$$\lambda = \frac{1}{\sqrt{2}\pi d^2 \left(\frac{N}{V}\right)} = 38.8 \text{ m}.$$

65. We note that $\Delta K = n\left(\frac{3}{2}R\right)\Delta T$ according to the discussion in §19-5 and §19-9. Also, $\Delta E_{\text{int}} = nC_V \Delta T$ can be used for each of these processes (since we are told this is an ideal gas). Finally, we note that Eq. 19-49 leads to $C_p = C_V + R \approx 8.0$ cal/mol·K after we

convert Joules to calories in the ideal gas constant value (Eq. 19-6): $R \approx 2.0$ cal/mol·K. The first law of thermodynamics $Q = \Delta E_{int} + W$ applies to each process.

- Constant volume process with $\Delta T = 50$ K and $n = 3.0$ mol.

(a) Since the change in the internal energy is $\Delta E_{int} = (3.0)(6.00)(50) = 900$ cal, and the work done by the gas is $W = 0$ for constant volume processes, the first law gives $Q = 900 + 0 = 900$ cal.

(b) As shown in part (a), $W = 0$.

(c) The change in the internal energy is, from part (a), $\Delta E_{int} = (3.0)(6.00)(50) = 900$ cal.

(d) The change in the total translational kinetic energy is

$$\Delta K = (3.0)\left(\tfrac{3}{2}(2.0)\right)(50) = 450 \text{ cal}.$$

- Constant pressure process with $\Delta T = 50$ K and $n = 3.0$ mol.

(e) $W = p\Delta V$ for constant pressure processes, so (using the ideal gas law)

$$W = nR\Delta T = (3.0)(2.0)(50) = 300 \text{ cal}.$$

The first law gives $Q = (900 + 300)$ cal $= 1200$ cal.

(f) From (e), we have $W = 300$ cal.

(g) The change in the internal energy is $\Delta E_{int} = (3.0)(6.00)(50) = 900$ cal.

(h) The change in the translational kinetic energy is $\Delta K = (3.0)\left(\tfrac{3}{2}(2.0)\right)(50) = 450$ cal.

- Adiabiatic process with $\Delta T = 50$ K and $n = 3.0$ mol.

(i) $Q = 0$ by definition of "adiabatic."

(j) The first law leads to $W = Q - E_{int} = 0 - 900$ cal $= -900$ cal.

(k) The change in the internal energy is $\Delta E_{int} = (3.0)(6.00)(50) = 900$ cal.

(l) As in part (d) and (h), $\Delta K = (3.0)\left(\tfrac{3}{2}(2.0)\right)(50) = 450$ cal.

66. The ratio is

$$\frac{mgh}{mv_{rms}^2/2} = \frac{2gh}{v_{rms}^2} = \frac{2Mgh}{3RT}$$

where we have used Eq. 19-22 in that last step. With $T = 273$ K, $h = 0.10$ m and $M = 32$ g/mol $= 0.032$ kg/mol, we find the ratio equals 9.2×10^{-6}.

67. In this solution we will use non-standard notation: writing ρ for *weight*-density (instead of mass-density), where ρ_c refers to the cool air and ρ_h refers to the hot air. Then the condition required by the problem is

$$F_{net} = F_{buoyant} - \text{hot-air-weight} - \text{balloon-weight}$$

$$2.67 \times 10^3 \text{ N} = \rho_c V - \rho_h V - 2.45 \times 10^3 \text{ N}$$

where $V = 2.18 \times 10^3$ m^3 and $\rho_c = 11.9$ N/m^3. This condition leads to $\rho_h = 9.55$ N/m^3. Using the ideal gas law to write ρ_h as PMg/RT where $P = 101000$ Pascals and $M = 0.028$ kg/m^3 (as suggested in the problem), we conclude that the temperature of the enclosed air should be 349 K.

68. (a) In the free expansion from state 0 to state 1 we have $Q = W = 0$, so $\Delta E_{int} = 0$, which means that the temperature of the ideal gas has to remain unchanged. Thus the final pressure is

$$p_1 = \frac{p_0 V_0}{V_1} = \frac{p_0 V_0}{3.00 V_0} = \frac{1}{3.00} p_0 \Rightarrow \frac{p_1}{p_0} = \frac{1}{3.00} = 0.333.$$

(b) For the adiabatic process from state 1 to 2 we have $p_1 V_1^\gamma = p_2 V_2^\gamma$, i.e.,

$$\frac{1}{3.00} p_0 (3.00 V_0)^\gamma = (3.00)^{\frac{1}{3}} p_0 V_0^\gamma$$

which gives $\gamma = 4/3$. The gas is therefore polyatomic.

(c) From $T = pV/nR$ we get

$$\frac{\bar{K}_2}{\bar{K}_1} = \frac{T_2}{T_1} = \frac{p_2}{p_1} = (3.00)^{1/3} = 1.44.$$

69. (a) By Eq. 19-28, $W = -374$ J (since the process is an adiabatic compression).

(b) $Q = 0$ since the process is adiabatic.

(c) By first law of thermodynamics, the change in internal energy is $\Delta E_{int} = Q - W = +374$ J.

(d) The change in the average kinetic energy per atom is

$$\Delta K_{avg} = \Delta E_{int}/N = +3.11 \times 10^{-22} \text{ J}.$$

70. (a) With work being given by

$$W = p\Delta V = (250)(-0.60) \text{ J} = -150 \text{ J},$$

and the heat transfer given as –210 J, then the change in internal energy is found from the first law of thermodynamics to be $[-210 - (-150)]$ J = –60 J.

(b) Since the pressures (and also the number of moles) don't change in this process, then the volume is simply proportional to the (absolute) temperature. Thus, the final temperature is ¼ of the initial temperature. The answer is 90 K.

71. This is very similar to Sample Problem 19-4 (and we use similar notation here) except for the use of Eq. 19-31 for v_{avg} (whereas in that Sample Problem, its value was just assumed). Thus,

$$f = \frac{\text{speed}}{\text{distance}} = \frac{v_{avg}}{\lambda} = \frac{p d^2}{k}\left(\frac{16\pi R}{MT}\right).$$

Therefore, with $p = 2.02 \times 10^3$ Pa, $d = 290 \times 10^{-12}$ m and $M = 0.032$ kg/mol (see Table 19-1), we obtain $f = 7.03 \times 10^9 \text{ s}^{-1}$.

72. Eq. 19-25 gives the mean free path:

$$\lambda = \frac{1}{\sqrt{2}\, d^2\, \pi\, \varepsilon_o\, (N/V)} = \frac{n R T}{\sqrt{2}\, d^2\, \pi\, \varepsilon_o\, P N}$$

where we have used the ideal gas law in that last step. Thus, the change in the mean free path is

$$\Delta\lambda = \frac{n R \Delta T}{\sqrt{2}\, d^2\, \pi\, \varepsilon_o\, P N} = \frac{R Q}{\sqrt{2}\, d^2\, \pi\, \varepsilon_o\, P N\, C_p}$$

where we have used Eq. 19-46. The constant pressure molar heat capacity is $(7/2)R$ in this situation, so (with $N = 9 \times 10^{23}$ and $d = 250 \times 10^{-12}$ m) we find

$$\Delta\lambda = 1.52 \times 10^{-9} \text{ m} = 1.52 \text{ nm}.$$

73. (a) The volume has increased by a factor of 3, so the pressure must decrease accordingly (since the temperature does not change in this process). Thus, the final pressure is one-third of the original 6.00 atm. The answer is 2.00 atm.

(b) We note that Eq. 19-14 can be written as $P_i V_i \ln(V_f/V_i)$. Converting "atm" to "Pa" (a Pascal is equivalent to a N/m^2) we obtain $W = 333$ J.

(c) The gas is monatomic so $\gamma = 5/3$. Eq. 19-54 then yields $P_f = 0.961$ atm.

(d) Using Eq. 19-53 in Eq. 18-25 gives

$$W = p_i V_i^\gamma \int_{V_i}^{V_f} V^{-\gamma} dV = p_i V_i^\gamma \frac{V_f^{1-\gamma} - V_i^{1-\gamma}}{1-\gamma} = \frac{p_f V_f - p_i V_i}{1-\gamma}$$

where in the last step Eq. 19-54 has been used. Converting "atm" to "Pa", we obtain $W = 236$ J.

74. (a) With $P_1 = (20.0)(1.01 \times 10^5$ Pa$)$ and $V_1 = 0.0015$ m^3, the ideal gas law gives

$$P_1 V_1 = nRT_1 \quad \Rightarrow \quad T_1 = 121.54 \text{ K} \approx 122 \text{ K}.$$

(b) From the information in the problem, we deduce that $T_2 = 3T_1 = 365$ K.

(c) We also deduce that $T_3 = T_1$ which means $\Delta T = 0$ for this process. Since this involves an ideal gas, this implies the change in internal energy is zero here.

75. (a) We use $p_i V_i^\gamma = p_f V_f^\gamma$ to compute γ.

$$\gamma = \frac{\ln(p_i/p_f)}{\ln(V_f/V_i)} = \frac{\ln(1.0 \text{ atm}/1.0 \times 10^5 \text{ atm})}{\ln(1.0 \times 10^3 \text{ L}/1.0 \times 10^6 \text{ L})} = \frac{5}{3}.$$

Therefore the gas is monatomic.

(b) Using the gas law in ratio form (see Sample Problem 19-1), the final temperature is

$$T_f = T_i \frac{p_f V_f}{p_i V_i} = (273 \text{ K}) \frac{(1.0 \times 10^5 \text{ atm})(1.0 \times 10^3 \text{ L})}{(1.0 \text{ atm})(1.0 \times 10^6 \text{ L})} = 2.7 \times 10^4 \text{ K}.$$

(c) The number of moles of gas present is

$$n = \frac{p_i V_i}{RT_i} = \frac{(1.01 \times 10^5 \text{ Pa})(1.0 \times 10^3 \text{ cm}^3)}{(8.31 \text{ J/mol} \cdot \text{K})(273 \text{ K})} = 4.5 \times 10^4 \text{ mol}.$$

(d) The total translational energy per mole before the compression is

$$K_i = \frac{3}{2} RT_i = \frac{3}{2}(8.31 \text{ J/mol} \cdot \text{K})(273 \text{ K}) = 3.4 \times 10^3 \text{ J}.$$

(e) After the compression,

$$K_f = \frac{3}{2}RT_f = \frac{3}{2}(8.31 \text{ J/mol·K})(2.7\times10^4 \text{ K}) = 3.4\times10^5 \text{ J}.$$

(f) Since $v_{rms}^2 \propto T$, we have

$$\frac{v_{rms,i}^2}{v_{rms,f}^2} = \frac{T_i}{T_f} = \frac{273 \text{ K}}{2.7\times10^4 \text{ K}} = 0.010.$$

76. We label the various states of the ideal gas as follows: it starts expanding adiabatically from state 1 until it reaches state 2, with $V_2 = 4$ m^3; then continues on to state 3 isothermally, with $V_3 = 10$ m^3; and eventually getting compressed adiabatically to reach state 4, the final state. For the adiabatic process $1 \to 2$ $p_1V_1^\gamma = p_2V_2^\gamma$, for the isothermal process $2 \to 3$ $p_2V_2 = p_3V_3$, and finally for the adiabatic process $3 \to 4$ $p_3V_3^\gamma = p_4V_4^\gamma$. These equations yield

$$p_4 = p_3\left(\frac{V_3}{V_4}\right)^\gamma = p_2\left(\frac{V_2}{V_3}\right)\left(\frac{V_3}{V_4}\right)^\gamma = p_1\left(\frac{V_1}{V_2}\right)^\gamma\left(\frac{V_2}{V_3}\right)\left(\frac{V_3}{V_4}\right)^\gamma.$$

We substitute this expression for p_4 into the equation $p_1V_1 = p_4V_4$ (since $T_1 = T_4$) to obtain $V_1V_3 = V_2V_4$. Solving for V_4 we obtain

$$V_4 = \frac{V_1V_3}{V_2} = \frac{(2.0 \text{ m}^3)(10 \text{ m}^3)}{4.0 \text{ m}^3} = 5.0 \text{ m}^3.$$

77. (a) The final pressure is

$$p_f = \frac{p_iV_i}{V_f} = \frac{(32 \text{ atm})(1.0 \text{ L})}{4.0 \text{ L}} = 8.0 \text{ atm},$$

(b) For the isothermal process the final temperature of the gas is $T_f = T_i = 300$ K.

(c) The work done is

$$W = nRT_i \ln\left(\frac{V_f}{V_i}\right) = p_iV_i \ln\left(\frac{V_f}{V_i}\right) = (32 \text{ atm})(1.01\times10^5 \text{ Pa/atm})(1.0\times10^{-3} \text{ m}^3)\ln\left(\frac{4.0 \text{ L}}{1.0 \text{ L}}\right)$$
$$= 4.4\times10^3 \text{ J}.$$

For the adiabatic process $p_iV_i^\gamma = p_fV_f^\gamma$. Thus,

(d) The final pressure is

$$p_f = p_i \left(\frac{V_i}{V_f}\right)^\gamma = (32\,\text{atm})\left(\frac{1.0\,\text{L}}{4.0\,\text{L}}\right)^{5/3} = 3.2\,\text{atm}.$$

(e) The final temperature is

$$T_f = \frac{p_f V_f T_i}{p_i V_i} = \frac{(3.2\,\text{atm})(4.0\,\text{L})(300\,\text{K})}{(32\,\text{atm})(1.0\,\text{L})} = 120\,\text{K}.$$

(f) The work done is

$$W = Q - \Delta E_{\text{int}} = -\Delta E_{\text{int}} = -\frac{3}{2} nR\Delta T = -\frac{3}{2}\left(p_f V_f - p_i V_i\right)$$

$$= -\frac{3}{2}\left[(3.2\,\text{atm})(4.0\,\text{L}) - (32\,\text{atm})(1.0\,\text{L})\right](1.01\times 10^5\,\text{Pa/atm})(10^{-3}\,\text{m}^3/\text{L})$$

$$= 2.9\times 10^3\,\text{J}.$$

(g) If the gas is diatomic, then $\gamma = 1.4$, and the final pressure is

$$p_f = p_i \left(\frac{V_i}{V_f}\right)^\gamma = (32\,\text{atm})\left(\frac{1.0\,\text{L}}{4.0\,\text{L}}\right)^{1.4} = 4.6\,\text{atm}.$$

(h) The final temperature is

$$T_f = \frac{p_f V_f T_i}{p_i V_i} = \frac{(4.6\,\text{atm})(4.0\,\text{L})(300\,\text{K})}{(32\,\text{atm})(1.0\,\text{L})} = 170\,\text{K}.$$

(i) The work done is

$$W = Q - \Delta E_{\text{int}} = -\frac{5}{2} nR\Delta T = -\frac{5}{2}\left(p_f V_f - p_i V_i\right)$$

$$= -\frac{5}{2}\left[(4.6\,\text{atm})(4.0\,\text{L}) - (32\,\text{atm})(1.0\,\text{L})\right](1.01\times 10^5\,\text{Pa/atm})(10^{-3}\,\text{m}^3/\text{L})$$

$$= 3.4\times 10^3\,\text{J}.$$

78. We write $T = 273$ K and use Eq. 19-14:

$$W = (1.00\,\text{mol})(8.31\,\text{J/mol}\cdot\text{K})(273\,\text{K})\ln\left(\frac{16.8}{22.4}\right)$$

which yields $W = -653$ J. Recalling the sign conventions for work stated in Chapter 18, this means an external agent does 653 J of work *on* the ideal gas during this process.

79. (a) We use $pV = nRT$. The volume of the tank is

$$V = \frac{nRT}{p} = \frac{\left(\frac{300\text{g}}{17\text{ g/mol}}\right)(8.31\text{ J/mol}\cdot\text{K})(350\text{K})}{1.35\times 10^6\text{ Pa}} = 3.8\times 10^{-2}\text{ m}^3 = 38\text{ L}.$$

(b) The number of moles of the remaining gas is

$$n' = \frac{p'V}{RT'} = \frac{(8.7\times 10^5\text{ Pa})(3.8\times 10^{-2}\text{ m}^3)}{(8.31\text{ J/mol}\cdot\text{K})(293\text{K})} = 13.5\text{ mol}.$$

The mass of the gas that leaked out is then $\Delta m = 300\text{ g} - (13.5\text{ mol})(17\text{ g/mol}) = 71$ g.

80. We solve

$$\sqrt{\frac{3RT}{M_\text{helium}}} = \sqrt{\frac{3R(293\text{K})}{M_\text{hydrogen}}}$$

for T. With the molar masses found in Table 19-1, we obtain

$$T = (293\text{K})\left(\frac{4.0}{2.02}\right) = 580\text{ K}$$

which is equivalent to 307°C.

81. It is recommended to look over §19-7 before doing this problem.

(a) We normalize the distribution function as follows:

$$\int_0^{v_o} P(v)\,dv = 1 \Rightarrow C = \frac{3}{v_o^3}.$$

(b) The average speed is

$$\int_0^{v_o} vP(v)\,dv = \int_0^{v_o} v\left(\frac{3v^2}{v_o^3}\right) dv = \frac{3}{4}v_o.$$

(c) The rms speed is the square root of

$$\int_0^{v_o} v^2 P(v)\,dv = \int_0^{v_o} v^2\left(\frac{3v^2}{v_o^3}\right) dv = \frac{3}{5}v_o^2.$$

Therefore, $v_{rms} = \sqrt{3/5} v_\circ \approx 0.775 v_\circ$.

82. To model the "uniform rates" described in the problem statement, we have expressed the volume and the temperature functions as follows:

$$V = V_i + \left(\frac{V_f - V_i}{\tau_f}\right) t \qquad \text{and} \qquad T = T_i + \left(\frac{T_f - T_i}{\tau_f}\right) t$$

where $V_i = 0.616$ m^3, $V_f = 0.308$ m^3, $\tau_f = 7200$ s, $T_i = 300$ K and $T_f = 723$ K.

(a) We can take the derivative of V with respect to t and use that to evaluate the cumulative work done (from $t = 0$ until $t = \tau$):

$$W = \int p\, dV = \int \left(\frac{nRT}{V}\right)\left(\frac{dV}{dt}\right) dt = 12.2\, \tau + 238113\, \ln(14400 - \tau) - 2.28 \times 10^6$$

with SI units understood. With $\tau = \tau_f$ our result is $W = -77169$ J ≈ -77.2 kJ, or $|W| \approx 77.2$ kJ.

The graph of cumulative work is shown below. The graph for work done is purely negative because the gas is being compressed (work is being done *on* the gas).

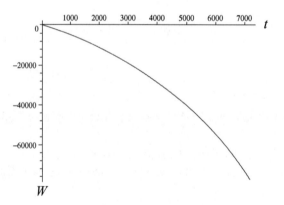

(b) With $C_V = \frac{3}{2} R$ (since it's a monatomic ideal gas) then the (infinitesimal) change in internal energy is $nC_V\, dT = \frac{3}{2} nR \left(\frac{dT}{dt}\right) dt$ which involves taking the derivative of the temperature expression listed above. Integrating this and adding this to the work done gives the cumulative heat absorbed (from $t = 0$ until $t = \tau$):

$$Q = \int \left(\frac{nRT}{V}\right)\left(\frac{dV}{dt}\right) + \frac{3}{2} nR\left(\frac{dT}{dt}\right) dt = 30.5\, \tau + 238113\, \ln(14400 - \tau) - 2.28 \times 10^6$$

with SI units understood. With $\tau = \tau_f$ our result is $Q_{total} = 54649$ J $\approx 5.46 \times 10^4$ J.

The graph cumulative heat is shown below. We see that $Q > 0$ since the gas is absorbing heat.

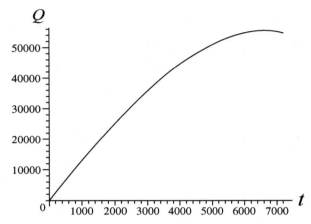

(c) Defining $C = \dfrac{Q_{\text{total}}}{n(T_f - T_i)}$ we obtain $C = 5.17$ J/mol·K. We note that this is considerably smaller than the constant-volume molar heat C_V.

We are now asked to consider this to be a two-step process (time dependence is no longer an issue) where the first step is isothermal and the second step occurs at constant volume (the ending values of pressure, volume and temperature being the same as before).

(d) Eq. 19-14 readily yields $W = -43222$ J $\approx -4.32 \times 10^4$ J (or $|W| \approx 4.32 \times 10^4$ J), where it is important to keep in mind that no work is done in a process where the volume is held constant.

(e) In step 1 the heat is equal to the work (since the internal energy does not change during an isothermal ideal gas process), and step 2 the heat is given by Eq. 19-39. The total heat is therefore $88595 \approx 8.86 \times 10^4$ J.

(f) Defining a molar heat capacity in the same manner as we did in part (c), we now arrive at $C = 8.38$ J/ mol·K.

83. (a) The temperature is $10.0°C \rightarrow T = 283$ K. Then, with $n = 3.50$ mol and $V_f/V_0 = 3/4$, we use Eq. 19-14:

$$W = nRT \ln\left(\frac{V_f}{V_0}\right) = -2.37\,\text{kJ}.$$

(b) The internal energy change ΔE_{int} vanishes (for an ideal gas) when $\Delta T = 0$ so that the First Law of Thermodynamics leads to $Q = W = -2.37$ kJ. The negative value implies that the heat transfer is from the sample to its environment.

84. (a) Since $n/V = p/RT$, the number of molecules per unit volume is

$$\frac{N}{V} = \frac{nN_A}{V} = N_A\left(\frac{p}{RT}\right)(6.02\times10^{23})\frac{1.01\times10^5\,\text{Pa}}{(8.31\frac{\text{J}}{\text{mol·K}})(293\,\text{K})} = 2.5\times10^{25}\frac{\text{molecules}}{\text{m}^3}.$$

(b) Three-fourths of the 2.5×10^{25} value found in part (a) are nitrogen molecules with $M = 28.0$ g/mol (using Table 19-1), and one-fourth of that value are oxygen molecules with $M = 32.0$ g/mol. Consequently, we generalize the $M_{\text{sam}} = NM/N_A$ expression for these two species of molecules and write

$$\frac{3}{4}(2.5\times10^{25})\frac{28.0}{6.02\times10^{23}} + \frac{1}{4}(2.5\times10^{25})\frac{32.0}{6.02\times10^{23}} = 1.2\times10^3\,\text{g}.$$

85. For convenience, the "int" subscript for the internal energy will be omitted in this solution. Recalling Eq. 19-28, we note that $\sum_{\text{cycle}} E = 0$, which gives

$$\Delta E_{A\to B} + \Delta E_{B\to C} + \Delta E_{C\to D} + \Delta E_{D\to E} + \Delta E_{E\to A} = 0.$$

Since a gas is involved (assumed to be ideal), then the internal energy does not change when the temperature does not change, so

$$\Delta E_{A\to B} = \Delta E_{D\to E} = 0.$$

Now, with $\Delta E_{E\to A} = 8.0$ J given in the problem statement, we have

$$\Delta E_{B\to C} + \Delta E_{C\to D} + 8.0\,\text{J} = 0.$$

In an adiabatic process, $\Delta E = -W$, which leads to $-5.0\,\text{J} + \Delta E_{C\to D} + 8.0\,\text{J} = 0$, and we obtain $\Delta E_{C\to D} = -3.0$ J.

86. (a) The work done in a constant-pressure process is $W = p\Delta V$. Therefore,

$$W = (25\,\text{N/m}^2)(1.8\,\text{m}^3 - 3.0\,\text{m}^3) = -30\,\text{J}.$$

The sign conventions discussed in the textbook for Q indicate that we should write -75 J for the energy which leaves the system in the form of heat. Therefore, the first law of thermodynamics leads to

$$\Delta E_{\text{int}} = Q - W = (-75\,\text{J}) - (-30\,\text{J}) = -45\,\text{J}.$$

(b) Since the pressure is constant (and the number of moles is presumed constant), the ideal gas law in ratio form (see Sample Problem 19-1) leads to

$$T_2 = T_1 \left(\frac{V_2}{V_1}\right) = (300\,\text{K}) \left(\frac{1.8\,\text{m}^3}{3.0\,\text{m}^3}\right) = 1.8 \times 10^2\,\text{K}.$$

It should be noted that this is consistent with the gas being monatomic (that is, if one assumes $C_V = \frac{3}{2}R$ and uses Eq. 19-45, one arrives at this same value for the final temperature).

87. (a) The p-V diagram is shown below. Note that o obtain the above graph, we have chosen $n = 0.37$ moles for concreteness, in which case the horizontal axis (which we note starts not at zero but at 1) is to be interpreted in units of cubic centimeters, and the vertical axis (the absolute pressure) is in kilopascals. However, the constant volume temp-increase process described in the third step (see problem statement) is difficult to see in this graph since it coincides with the pressure axis.

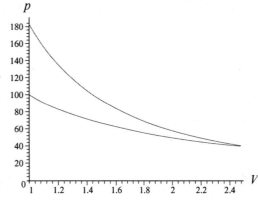

(b) We note that the change in internal energy is zero for an ideal gas isothermal process, so (since the net change in the internal energy must be zero for the entire cycle) the increase in internal energy in step 3 must equal (in magnitude) its decease in step 1. By Eq. 19-28, we see this number must be 125 J.

(c) As implied by Eq. 19-29, this is equivalent to heat being added *to the gas*.

88. (a) The ideal gas law leads to

$$V = \frac{nRT}{p} = \frac{(1.00\,\text{mol})(8.31\,\text{J/mol}\cdot\text{K})(273\,\text{K})}{1.01 \times 10^5\,\text{Pa}}$$

which yields $V = 0.0225\,\text{m}^3 = 22.5$ L. If we use the standard pressure value given in Appendix D, 1 atm $= 1.013 \times 10^5$ Pa, then our answer rounds more properly to 22.4 L.

(b) From Eq. 19-2, we have $N = 6.02 \times 10^{23}$ molecules in the volume found in part (a) (which may be expressed as $V = 2.24 \times 10^4\,\text{cm}^3$), so that

$$\frac{N}{V} = \frac{6.02 \times 10^{23}}{2.24 \times 10^4\,\text{cm}^3} = 2.69 \times 10^{19}\,\text{molecules/cm}^3.$$

Chapter 20

1. An isothermal process is one in which $T_i = T_f$ which implies $\ln(T_f/T_i) = 0$. Therefore, with $V_f/V_i = 2.00$, Eq. 20-4 leads to

$$\Delta S = nR \ln\left(\frac{V_f}{V_i}\right) = (2.50 \text{ mol})(8.31 \text{ J/mol}\cdot\text{K})\ln(2.00) = 14.4 \text{ J/K}.$$

2. From Eq. 20-2, we obtain

$$Q = T\Delta S = (405 \text{ K})(46.0 \text{ J/K}) = 1.86\times 10^4 \text{ J}.$$

3. We use the following relation derived in Sample Problem 20-2:

$$\Delta S = mc \ln\left(\frac{T_f}{T_i}\right).$$

(a) The energy absorbed as heat is given by Eq. 19-14. Using Table 19-3, we find

$$Q = cm\Delta T = \left(386 \frac{\text{J}}{\text{kg}\cdot\text{K}}\right)(2.00 \text{ kg})(75 \text{ K}) = 5.79\times 10^4 \text{ J}$$

where we have used the fact that a change in Kelvin temperature is equivalent to a change in Celsius degrees.

(b) With $T_f = 373.15$ K and $T_i = 298.15$ K, we obtain

$$\Delta S = (2.00 \text{ kg})\left(386 \frac{\text{J}}{\text{kg}\cdot\text{K}}\right) \ln\left(\frac{373.15}{298.15}\right) = 173 \text{ J/K}.$$

4. (a) This may be considered a reversible process (as well as isothermal), so we use $\Delta S = Q/T$ where $Q = Lm$ with $L = 333$ J/g from Table 19-4. Consequently,

$$\Delta S = \frac{(333 \text{ J/g})(12.0 \text{ g})}{273 \text{ K}} = 14.6 \text{ J/K}.$$

(b) The situation is similar to that described in part (a), except with $L = 2256$ J/g, $m = 5.00$ g, and $T = 373$ K. We therefore find $\Delta S = 30.2$ J/K.

5. (a) Since the gas is ideal, its pressure p is given in terms of the number of moles n, the volume V, and the temperature T by $p = nRT/V$. The work done by the gas during the isothermal expansion is

$$W = \int_{V_1}^{V_2} p\,dV = nRT \int_{V_1}^{V_2} \frac{dV}{V} = nRT \ln \frac{V_2}{V_1}.$$

We substitute $V_2 = 2.00 V_1$ to obtain

$$W = nRT \ln 2.00 = (4.00 \text{ mol})(8.31 \text{ J/mol}\cdot\text{K})(400 \text{ K}) \ln 2.00 = 9.22 \times 10^3 \text{ J}.$$

(b) Since the expansion is isothermal, the change in entropy is given by

$$\Delta S = \int (1/T)\,dQ = Q/T,$$

where Q is the heat absorbed. According to the first law of thermodynamics, $\Delta E_{int} = Q - W$. Now the internal energy of an ideal gas depends only on the temperature and not on the pressure and volume. Since the expansion is isothermal, $\Delta E_{int} = 0$ and $Q = W$. Thus,

$$\Delta S = \frac{W}{T} = \frac{9.22 \times 10^3 \text{ J}}{400 \text{ K}} = 23.1 \text{ J/K}.$$

(c) $\Delta S = 0$ for all reversible adiabatic processes.

6. An isothermal process is one in which $T_i = T_f$ which implies $\ln(T_f/T_i) = 0$. Therefore, Eq. 20-4 leads to

$$\Delta S = nR \ln\left(\frac{V_f}{V_i}\right) \Rightarrow n = \frac{22.0}{(8.31)\ln(3.4/1.3)} = 2.75 \text{ mol}.$$

7. (a) The energy that leaves the aluminum as heat has magnitude $Q = m_a c_a (T_{ai} - T_f)$, where m_a is the mass of the aluminum, c_a is the specific heat of aluminum, T_{ai} is the initial temperature of the aluminum, and T_f is the final temperature of the aluminum-water system. The energy that enters the water as heat has magnitude $Q = m_w c_w (T_f - T_{wi})$, where m_w is the mass of the water, c_w is the specific heat of water, and T_{wi} is the initial temperature of the water. The two energies are the same in magnitude since no energy is lost. Thus,

$$m_a c_a (T_{ai} - T_f) = m_w c_w (T_f - T_{wi}) \Rightarrow T_f = \frac{m_a c_a T_{ai} + m_w c_w T_{wi}}{m_a c_a + m_w c_w}.$$

The specific heat of aluminum is 900 J/kg·K and the specific heat of water is 4190 J/kg·K. Thus,

$$T_f = \frac{(0.200 \text{ kg})(900 \text{ J/kg} \cdot \text{K})(100°C) + (0.0500 \text{ kg})(4190 \text{ J/kg} \cdot \text{K})(20°C)}{(0.200 \text{ kg})(900 \text{ J/kg} \cdot \text{K}) + (0.0500 \text{ kg})(4190 \text{ J/kg} \cdot \text{K})}$$

$$= 57.0°C = 330 \text{ K}.$$

(b) Now temperatures must be given in Kelvins: $T_{ai} = 393$ K, $T_{wi} = 293$ K, and $T_f = 330$ K. For the aluminum, $dQ = m_a c_a dT$ and the change in entropy is

$$\Delta S_a = \int \frac{dQ}{T} = m_a c_a \int_{T_{ai}}^{T_f} \frac{dT}{T} = m_a c_a \ln \frac{T_f}{T_{ai}} = (0.200 \text{ kg})(900 \text{ J/kg} \cdot \text{K}) \ln \left(\frac{330 \text{ K}}{373 \text{ K}} \right)$$

$$= -22.1 \text{ J/K}.$$

(c) The entropy change for the water is

$$\Delta S_w = \int \frac{dQ}{T} = m_w c_w \int_{T_{wi}}^{T_f} \frac{dT}{T} = m_w c_w \ln \frac{T_f}{T_{wi}} = (0.0500 \text{ kg})(4190 \text{ J/kg.K}) \ln \left(\frac{330 \text{ K}}{293 \text{ K}} \right)$$

$$= +24.9 \text{ J/K}.$$

(d) The change in the total entropy of the aluminum-water system is

$$\Delta S = \Delta S_a + \Delta S_w = -22.1 \text{ J/K} + 24.9 \text{ J/K} = +2.8 \text{ J/K}.$$

8. We follow the method shown in Sample Problem 20-2. Since

$$\Delta S = mc \int_{T_i}^{T_f} \frac{dT}{T} = mc \ln(T_f / T_i),$$

then with $\Delta S = 50$ J/K, $T_f = 380$ K, $T_i = 280$ K and $m = 0.364$ kg, we obtain $c = 4.5 \times 10^2$ J/kg·K.

9. This problem is similar to Sample Problem 20-2. The only difference is that we need to find the mass m of each of the blocks. Since the two blocks are identical the final temperature T_f is the average of the initial temperatures:

$$T_f = \frac{1}{2}(T_i + T_f') = \frac{1}{2}(305.5 \text{ K} + 294.5 \text{ K}) = 300.0 \text{ K}.$$

Thus from $Q = mc\Delta T$ we find the mass m:

$$m = \frac{Q}{c\Delta T} = \frac{215 \text{ J}}{(386 \text{ J/kg} \cdot \text{K})(300.0 \text{ K} - 294.5 \text{ K})} = 0.101 \text{ kg}.$$

(a) The change in entropy for block L is

$$\Delta S_L = mc \ln\left(\frac{T_f}{T_{iL}}\right) = (0.101 \text{ kg})(386 \text{ J/kg} \cdot \text{K}) \ln\left(\frac{300.0 \text{ K}}{305.5 \text{ K}}\right) = -0.710 \text{ J/K}.$$

(b) Since the temperature of the reservoir is virtually the same as that of the block, which gives up the same amount of heat as the reservoir absorbs, the change in entropy $\Delta S'_L$ of the reservoir connected to the left block is the opposite of that of the left block: $\Delta S'_L = -\Delta S_L = +0.710$ J/K.

(c) The entropy change for block R is

$$\Delta S_R = mc \ln\left(\frac{T_f}{T_{iR}}\right) = (0.101 \text{ kg})(386 \text{ J/kg} \cdot \text{K}) \ln\left(\frac{300.0 \text{ K}}{294.5 \text{ K}}\right) = +0.723 \text{ J/K}.$$

(d) Similar to the case in part (b) above, the change in entropy $\Delta S'_R$ of the reservoir connected to the right block is given by $\Delta S'_R = -\Delta S_R = -0.723$ J/K.

(e) The change in entropy for the two-block system is

$$\Delta S_L + \Delta S_R = -0.710 \text{ J/K} + 0.723 \text{ J/K} = +0.013 \text{ J/K}.$$

(f) The entropy change for the entire system is given by

$$\Delta S = \Delta S_L + \Delta S'_L + \Delta S_R + \Delta S'_R = \Delta S_L - \Delta S_L + \Delta S_R - \Delta S_R = 0,$$

which is expected of a reversible process.

10. We concentrate on the first term of Eq. 20-4 (the second term is zero because the final and initial temperatures are the same, and because $\ln(1) = 0$). Thus, the entropy change is

$$\Delta S = nR \ln(V_f/V_i) \, .$$

Noting that $\Delta S = 0$ at $V_f = 0.40$ m^3, we are able to deduce that $V_i = 0.40$ m^3. We now examine the point in the graph where $\Delta S = 32$ J/K and $V_f = 1.2$ m^3; the above expression can now be used to solve for the number of moles. We obtain $n = 3.5$ mol.

11. (a) We refer to the copper block as block 1 and the lead block as block 2. The equilibrium temperature T_f satisfies

$$m_1 c_1 (T_f - T_{i,1}) + m_2 c_2 (T_f - T_{i2}) = 0,$$

which we solve for T_f:

$$T_f = \frac{m_1 c_1 T_{i,1} + m_2 c_2 T_{i,2}}{m_1 c_1 + m_2 c_2} = \frac{(50.0 \text{ g})(386 \text{ J/kg}\cdot\text{K})(400 \text{ K}) + (100 \text{ g})(128 \text{ J/kg}\cdot\text{K})(200 \text{ K})}{(50.0 \text{ g})(386 \text{ J/kg}\cdot\text{K}) + (100 \text{ g})(128 \text{ J/kg}\cdot\text{K})}$$
$$= 320 \text{ K}.$$

(b) Since the two-block system in thermally insulated from the environment, the change in internal energy of the system is zero.

(c) The change in entropy is

$$\Delta S = \Delta S_1 + \Delta S_2 = m_1 c_1 \ln\left(\frac{T_f}{T_{i,1}}\right) + m_2 c_2 \ln\left(\frac{T_f}{T_{i,2}}\right)$$
$$= (50.0 \text{ g})(386 \text{ J/kg}\cdot\text{K})\ln\left(\frac{320 \text{ K}}{400 \text{ K}}\right) + (100 \text{ g})(128 \text{ J/kg}\cdot\text{K})\ln\left(\frac{320 \text{ K}}{200 \text{ K}}\right)$$
$$= +1.72 \text{ J/K}.$$

12. We use Eq. 20-1:

$$\Delta S = \int \frac{nC_V \, dT}{T} = nA \int_{5.00}^{10.0} T^2 \, dT = \frac{nA}{3}\left[(10.0)^3 - (5.00)^3\right] = 0.0368 \text{ J/K}.$$

13. The connection between molar heat capacity and the degrees of freedom of a diatomic gas is given by setting $f = 5$ in Eq. 19-51. Thus, $C_V = 5R/2$, $C_p = 7R/2$, and $\gamma = 7/5$. In addition to various equations from Chapter 19, we also make use of Eq. 20-4 of this chapter. We note that we are asked to use the ideal gas constant as R and not plug in its numerical value. We also recall that isothermal means constant-temperature, so $T_2 = T_1$ for the $1 \rightarrow 2$ process. The statement (at the end of the problem) regarding "per mole" may be taken to mean that n may be set identically equal to 1 wherever it appears.

(a) The gas law in ratio form (see Sample Problem 19-1) is used to obtain

$$p_2 = p_1\left(\frac{V_1}{V_2}\right) = \frac{p_1}{3} \quad \Rightarrow \quad \frac{p_2}{p_1} = \frac{1}{3} = 0.333.$$

(b) The adiabatic relations Eq. 19-54 and Eq. 19-56 lead to

$$p_3 = p_1\left(\frac{V_1}{V_3}\right)^\gamma = \frac{p_1}{3^{1.4}} \quad \Rightarrow \quad \frac{p_3}{p_1} = \frac{1}{3^{1.4}} = 0.215.$$

(c) Similarly, we find

$$T_3 = T_1\left(\frac{V_1}{V_3}\right)^{\gamma-1} = \frac{T_1}{3^{0.4}} \quad \Rightarrow \quad \frac{T_3}{T_1} = \frac{1}{3^{0.4}} = 0.644.$$

• process 1 → 2

(d) The work is given by Eq. 19-14:

$$W = nRT_1 \ln (V_2/V_1) = RT_1 \ln 3 = 1.10 RT_1.$$

Thus, $W/nRT_1 = \ln 3 = 1.10$.

(e) The internal energy change is $\Delta E_{int} = 0$ since this is an ideal gas process without a temperature change (see Eq. 19-45). Thus, the energy absorbed as heat is given by the first law of thermodynamics: $Q = \Delta E_{int} + W \approx 1.10 RT_1$, or $Q/nRT_1 = \ln 3 = 1.10$.

(f) $\Delta E_{int} = 0$ or $\Delta E_{int}/nRT_1 = 0$

(g) The entropy change is $\Delta S = Q/T_1 = 1.10 R$, or $\Delta S/R = 1.10$.

• process 2 → 3

(h) The work is zero since there is no volume change. Therefore, $W/nRT_1 = 0$

(i) The internal energy change is

$$\Delta E_{int} = nC_V (T_3 - T_2) = (1)\left(\frac{5}{2}R\right)\left(\frac{T_1}{3^{0.4}} - T_1\right) \approx -0.889 RT_1 \Rightarrow \frac{\Delta E_{int}}{nRT_1} \approx -0.889.$$

This ratio (−0.889) is also the value for Q/nRT_1 (by either the first law of thermodynamics or by the definition of C_V).

(j) $\Delta E_{int}/nRT_1 = -0.889$.

(k) For the entropy change, we obtain

$$\frac{\Delta S}{R} = n \ln\left(\frac{V_3}{V_1}\right) + n\frac{C_V}{R} \ln\left(\frac{T_3}{T_1}\right) = (1)\ \ln(1) + (1)\left(\frac{5}{2}\right) \ln\left(\frac{T_1/3^{0.4}}{T_1}\right) = 0 + \frac{5}{2}\ln(3^{-0.4}) \approx -1.10 \ .$$

• process 3 → 1

(l) By definition, $Q = 0$ in an adiabatic process, which also implies an absence of entropy change (taking this to be a reversible process). The internal change must be the negative of the value obtained for it in the previous process (since all the internal energy changes must add up to zero, for an entire cycle, and its change is zero for process 1 → 2), so $\Delta E_{int} = +0.889 RT_1$. By the first law of thermodynamics, then,

$$W = Q - \Delta E_{int} = -0.889 RT_1,$$

or $W/nRT_1 = -0.889$.

(m) $Q = 0$ in an adiabatic process.

(n) $\Delta E_{int}/nRT_1 = +0.889$.

(o) $\Delta S/nR = 0$.

14. (a) It is possible to motivate, starting from Eq. 20-3, the notion that heat may be found from the integral (or "area under the curve") of a curve in a TS diagram, such as this one. Either from calculus, or from geometry (area of a trapezoid), it is straightforward to find the result for a "straight-line" path in the TS diagram:

$$Q_{straight} = \left(\frac{T_i + T_f}{2}\right)\Delta S$$

which could, in fact, be *directly* motivated from Eq. 20-3 (but it is important to bear in mind that this is rigorously true only for a process which forms a straight line in a graph that plots T versus S). This leads to

$$Q = (300\text{ K})(15\text{ J/K}) = 4.5 \times 10^3 \text{ J}$$

for the energy absorbed as heat by the gas.

(b) Using Table 19-3 and Eq. 19-45, we find

$$\Delta E_{int} = n\left(\frac{3}{2}R\right)\Delta T = (2.0\text{ mol})(8.31\text{ J/mol}\cdot\text{K})(200\text{ K} - 400\text{ K}) = -5.0\times 10^3 \text{ J}.$$

(c) By the first law of thermodynamics,

$$W = Q - \Delta E_{int} = 4.5\text{ kJ} - (-5.0\text{ kJ}) = 9.5\text{ kJ}.$$

15. The ice warms to 0°C, then melts, and the resulting water warms to the temperature of the lake water, which is 15°C. As the ice warms, the energy it receives as heat when the temperature changes by dT is $dQ = mc_I dT$, where m is the mass of the ice and c_I is the specific heat of ice. If T_i (= 263 K) is the initial temperature and T_f (= 273 K) is the final temperature, then the change in its entropy is

$$\Delta S = \int\frac{dQ}{T} = mc_I \int_{T_i}^{T_f}\frac{dT}{T} = mc_I \ln\frac{T_f}{T_i} = (0.010\text{ kg})(2220\text{ J/kg}\cdot\text{K})\ln\left(\frac{273\text{ K}}{263\text{ K}}\right) = 0.828\text{ J/K}.$$

Melting is an isothermal process. The energy leaving the ice as heat is mL_F, where L_F is the heat of fusion for ice. Thus,

$$\Delta S = Q/T = mL_F/T = (0.010 \text{ kg})(333 \times 10^3 \text{ J/kg})/(273 \text{ K}) = 12.20 \text{ J/K}.$$

For the warming of the water from the melted ice, the change in entropy is

$$\Delta S = mc_w \ln \frac{T_f}{T_i},$$

where c_w is the specific heat of water (4190 J/kg · K). Thus,

$$\Delta S = (0.010 \text{ kg})(4190 \text{ J/kg} \cdot \text{K}) \ln \left(\frac{288 \text{ K}}{273 \text{ K}} \right) = 2.24 \text{ J/K}.$$

The total change in entropy for the ice and the water it becomes is

$$\Delta S = 0.828 \text{ J/K} + 12.20 \text{ J/K} + 2.24 \text{ J/K} = 15.27 \text{ J/K}.$$

Since the temperature of the lake does not change significantly when the ice melts, the change in its entropy is $\Delta S = Q/T$, where Q is the energy it receives as heat (the negative of the energy it supplies the ice) and T is its temperature. When the ice warms to 0°C,

$$Q = -mc_I(T_f - T_i) = -(0.010 \text{ kg})(2220 \text{ J/kg} \cdot \text{K})(10 \text{ K}) = -222 \text{ J}.$$

When the ice melts,

$$Q = -mL_F = -(0.010 \text{ kg})(333 \times 10^3 \text{ J/kg}) = -3.33 \times 10^3 \text{ J}.$$

When the water from the ice warms,

$$Q = -mc_w(T_f - T_i) = -(0.010 \text{ kg})(4190 \text{ J/kg} \cdot \text{K})(15 \text{ K}) = -629 \text{ J}.$$

The total energy leaving the lake water is

$$Q = -222 \text{ J} - 3.33 \times 10^3 \text{ J} - 6.29 \times 10^2 \text{ J} = -4.18 \times 10^3 \text{ J}.$$

The change in entropy is

$$\Delta S = -\frac{4.18 \times 10^3 \text{ J}}{288 \text{ K}} = -14.51 \text{ J/K}.$$

The change in the entropy of the ice-lake system is $\Delta S = (15.27 - 14.51) \text{ J/K} = 0.76 \text{ J/K}$.

16. (a) Work is done only for the *ab* portion of the process. This portion is at constant pressure, so the work done by the gas is

$$W = \int_{V_0}^{4V_0} p_0 \, dV = p_0(4.00V_0 - 1.00V_0) = 3.00 p_0 V_0 \Rightarrow \frac{W}{p_0 V} = 3.00$$

(b) We use the first law: $\Delta E_{int} = Q - W$. Since the process is at constant volume, the work done by the gas is zero and $E_{int} = Q$. The energy Q absorbed by the gas as heat is $Q = nC_V \Delta T$, where C_V is the molar specific heat at constant volume and ΔT is the change in temperature. Since the gas is a monatomic ideal gas, $C_V = 3R/2$. Use the ideal gas law to find that the initial temperature is

$$T_b = \frac{p_b V_b}{nR} = \frac{4 p_0 V_0}{nR}$$

and that the final temperature is

$$T_c = \frac{p_c V_c}{nR} = \frac{(2p_0)(4V_0)}{nR} = \frac{8 p_0 V_0}{nR}.$$

Thus,

$$Q = \frac{3}{2} nR \left(\frac{8 p_0 V_0}{nR} - \frac{4 p_0 V_0}{nR} \right) = 6.00 p_0 V_0.$$

The change in the internal energy is $\Delta E_{int} = 6 p_0 V_0$ or $\Delta E_{int}/p_0 V_0 = 6.00$. Since $n = 1$ mol, this can also be written $Q = 6.00 RT_0$.

(c) For a complete cycle, $\Delta E_{int} = 0$

(d) Since the process is at constant volume, use $dQ = nC_V dT$ to obtain

$$\Delta S = \int \frac{dQ}{T} = nC_V \int_{T_b}^{T_c} \frac{dT}{T} = nC_V \ln \frac{T_c}{T_b}.$$

Substituting $C_V = \frac{3}{2}R$ and using the ideal gas law, we write

$$\frac{T_c}{T_b} = \frac{p_c V_c}{p_b V_b} = \frac{(2p_0)(4V_0)}{p_0(4V_0)} = 2.$$

Thus, $\Delta S = \frac{3}{2} nR \ln 2$. Since $n = 1$, this is $\Delta S = \frac{3}{2} R \ln 2 = 8.64$ J/K..

(e) For a complete cycle, $\Delta E_{int} = 0$ and $\Delta S = 0$.

17. (a) The final mass of ice is (1773 g + 227 g)/2 = 1000 g. This means 773 g of water froze. Energy in the form of heat left the system in the amount mL_F, where m is the mass of the water that froze and L_F is the heat of fusion of water. The process is isothermal, so the change in entropy is

$$\Delta S = Q/T = -mL_F/T = -(0.773 \text{ kg})(333 \times 10^3 \text{ J/kg})/(273 \text{ K}) = -943 \text{ J/K}.$$

(b) Now, 773 g of ice is melted. The change in entropy is

$$\Delta S = \frac{Q}{T} = \frac{mL_F}{T} = +943 \text{ J/K}.$$

(c) Yes, they are consistent with the second law of thermodynamics. Over the entire cycle, the change in entropy of the water-ice system is zero even though part of the cycle is irreversible. However, the system is not closed. To consider a closed system, we must include whatever exchanges energy with the ice and water. Suppose it is a constant-temperature heat reservoir during the freezing portion of the cycle and a Bunsen burner during the melting portion. During freezing the entropy of the reservoir increases by 943 J/K. As far as the reservoir-water-ice system is concerned, the process is adiabatic and reversible, so its total entropy does not change. The melting process is irreversible, so the total entropy of the burner-water-ice system increases. The entropy of the burner either increases or else decreases by less than 943 J/K.

18. In coming to equilibrium, the heat lost by the 100 cm³ of liquid water (of mass m_w = 100 g and specific heat capacity c_w = 4190 J/kg·K) is absorbed by the ice (of mass m_i which melts and reaches $T_f > 0$ °C). We begin by finding the equilibrium temperature:

$$\sum Q = 0$$

$$Q_{\text{warm water cools}} + Q_{\text{ice warms to 0°}} + Q_{\text{ice melts}} + Q_{\text{melted ice warms}} = 0$$

$$c_w m_w (T_f - 20°) + c_i m_i (0° - (-10°)) + L_F m_i + c_w m_i (T_f - 0°) = 0$$

which yields, after using L_F = 333000 J/kg and values cited in the problem, T_f = 12.24 ° which is equivalent to T_f = 285.39 K. Sample Problem 19-2 shows that

$$\Delta S_{\text{temp change}} = mc \ln\left(\frac{T_2}{T_1}\right)$$

for processes where $\Delta T = T_2 - T_1$, and Eq. 20-2 gives

$$\Delta S_{\text{melt}} = \frac{L_F m}{T_o}$$

for the phase change experienced by the ice (with $T_o = 273.15$ K). The total entropy change is (with T in Kelvins)

$$\Delta S_{system} = m_w c_w \ln\left(\frac{285.39}{293.15}\right) + m_i c_i \ln\left(\frac{273.15}{263.15}\right) + m_i c_w \ln\left(\frac{285.39}{273.15}\right) + \frac{L_F m_i}{273.15}$$
$$= (-11.24 + 0.66 + 1.47 + 9.75) \text{J/K} = 0.64 \text{ J/K}.$$

19. We consider a three-step reversible process as follows: the supercooled water drop (of mass m) starts at state 1 ($T_1 = 268$ K), moves on to state 2 (still in liquid form but at $T_2 = 273$ K), freezes to state 3 ($T_3 = T_2$), and then cools down to state 4 (in solid form, with $T_4 = T_1$). The change in entropy for each of the stages is given as follows:

$$\Delta S_{12} = mc_w \ln(T_2/T_1),$$

$$\Delta S_{23} = -mL_F/T_2,$$

$$\Delta S_{34} = mc_I \ln(T_4/T_3) = mc_I \ln(T_1/T_2) = -mc_I \ln(T_2/T_1).$$

Thus the net entropy change for the water drop is

$$\Delta S = \Delta S_{12} + \Delta S_{23} + \Delta S_{34} = m(c_w - c_I) \ln\left(\frac{T_2}{T_1}\right) - \frac{mL_F}{T_2}$$
$$= (1.00 \text{ g})(4.19 \text{ J/g·K} - 2.22 \text{ J/g·K}) \ln\left(\frac{273 \text{ K}}{268 \text{ K}}\right) - \frac{(1.00 \text{ g})(333 \text{ J/g})}{273 \text{ K}}$$
$$= -1.18 \text{ J/K}.$$

20. (a) We denote the mass of the ice (which turns to water and warms to T_f) as m and the mass of original-water (which cools from 80° down to T_f) as m'. From $\Sigma Q = 0$ we have

$$L_F m + cm(T_f - 0°) + cm'(T_f - 80°) = 0.$$

Since $L_F = 333 \times 10^3$ J/kg, $c = 4190$ J/(kg·C°), $m' = 0.13$ kg and $m = 0.012$ kg, we find $T_f = 66.5°$C, which is equivalent to 339.67 K.

(b) Using Eq. 20-2, the process of ice at 0° C turning to water at 0° C involves an entropy change of

$$\frac{Q}{T} = \frac{L_F m}{273.15 \text{ K}} = 14.6 \text{ J/K}.$$

(c) Using Eq. 20-1, the process of $m = 0.012$ kg of water warming from 0° C to 66.5° C involves an entropy change of

$$\int_{273.15}^{339.67} \frac{cm \, dT}{T} = cm \ln\left(\frac{339.67}{273.15}\right) = 11.0 \text{ J/K}.$$

(d) Similarly, the cooling of the original-water involves an entropy change of

$$\int_{353.15}^{339.67} \frac{cm' \, dT}{T} = cm' \ln\left(\frac{339.67}{353.15}\right) = -21.2 \text{ J/K}$$

(e) The net entropy change in this calorimetry experiment is found by summing the previous results; we find (by using more precise values than those shown above) ΔS_{net} = 4.39 J/K.

21. We note that the connection between molar heat capacity and the degrees of freedom of a monatomic gas is given by setting $f = 3$ in Eq. 19-51. Thus, $C_V = 3R/2$, $C_p = 5R/2$, and $\gamma = 5/3$.

(a) Since this is an ideal gas, Eq. 19-45 holds, which implies $\Delta E_{int} = 0$ for this process. Eq. 19-14 also applies, so that by the first law of thermodynamics,

$$Q = 0 + W = nRT_1 \ln V_2/V_1 = p_1 V_1 \ln 2 \rightarrow Q/p_1 V_1 = \ln 2 = 0.693.$$

(b) The gas law in ratio form (see Sample Problem 19-1) implies that the pressure decreased by a factor of 2 during the isothermal expansion process to $V_2 = 2.00 V_1$, so that it needs to increase by a factor of 4 in this step in order to reach a final pressure of $p_2 = 2.00 p_1$. That same ratio form now applied to this constant-volume process, yielding $4.00 = T_2 T_1$ which is used in the following:

$$Q = nC_V \Delta T = n\left(\frac{3}{2}R\right)(T_2 - T_1) = \frac{3}{2} nRT_1 \left(\frac{T_2}{T_1} - 1\right) = \frac{3}{2} p_1 V_1 (4-1) = \frac{9}{2} p_1 V_1$$

or $Q/p_1 V_1 = 9/2 = 4.50$.

(c) The work done during the isothermal expansion process may be obtained by using Eq. 19-14:

$$W = nRT_1 \ln V_2/V_1 = p_1 V_1 \ln 2.00 \rightarrow W/p_1 V_1 = \ln 2 = 0.693.$$

(d) In step 2 where the volume is kept constant, $W = 0$.

(e) The change in internal energy can be calculated by combining the above results and applying the first law of thermodynamics:

$$\Delta E_{int} = Q_{total} - W_{total} = \left(p_1 V_1 \ln 2 + \frac{9}{2} p_1 V_1\right) - (p_1 V_1 \ln 2 + 0) = \frac{9}{2} p_1 V_1$$

or $\Delta E_{int}/p_1 V_1 = 9/2 = 4.50$.

(f) The change in entropy may be computed by using Eq. 20-4:

$$\Delta S = R \ln\left(\frac{2.00 V_1}{V_1}\right) + C_V \ln\left(\frac{4.00 T_1}{T_1}\right) = R \ln 2.00 + \left(\frac{3}{2} R\right) \ln (2.00)^2$$
$$= R \ln 2.00 + 3R \ln 2.00 = 4R \ln 2.00 = 23.0 \text{ J/K}.$$

The second approach consists of an isothermal (constant T) process in which the volume halves, followed by an isobaric (constant p) process.

(g) Here the gas law applied to the first (isothermal) step leads to a volume half as big as the original. Since $\ln(1/2.00) = -\ln 2.00$, the reasoning used above leads to

$$Q = -p_1 V_1 \ln 2.00 \Rightarrow Q/p_1 V_1 = -\ln 2.00 = -0.693.$$

(h) To obtain a final volume twice as big as the original, in this step we need to increase the volume by a factor of 4.00. Now, the gas law applied to this isobaric portion leads to a temperature ratio $T_2/T_1 = 4.00$. Thus,

$$Q = C_p \Delta T = \frac{5}{2} R(T_2 - T_1) = \frac{5}{2} RT_1 \left(\frac{T_2}{T_1} - 1\right) = \frac{5}{2} p_1 V_1 (4 - 1) = \frac{15}{2} p_1 V_1$$

or $Q/p_1 V_1 = 15/2 = 7.50$.

(i) During the isothermal compression process, Eq. 19-14 gives

$$W = nRT_1 \ln V_2/V_1 = p_1 V_1 \ln(-1/2.00) = -p_1 V_1 \ln 2.00 \Rightarrow W/p_1 V_1 = -\ln 2 = -0.693.$$

(j) The initial value of the volume, for this part of the process, is $V_i = V_1/2$, and the final volume is $V_f = 2V_1$. The pressure maintained during this process is $p' = 2.00 p_1$. The work is given by Eq. 19-16:

$$W = p' \Delta V = p'(V_f - V_i) = (2.00 p_1)\left(2.00 V_1 - \frac{1}{2} V_1\right) = 3.00 p_1 V_1 \Rightarrow W/p_1 V_1 = 3.00.$$

(k) Using the first law of thermodynamics, the change in internal energy is

$$\Delta E_{int} = Q_{total} - W_{total} = \left(\frac{15}{2} p_1 V_1 - p_1 V_1 \ln 2.00\right) - (3 p_1 V_1 - p_1 V_1 \ln 2.00) = \frac{9}{2} p_1 V_1$$

or $\Delta E_{int}/p_1 V_1 = 9/2 = 4.50$. The result is the same as that obtained in part (e).

(l) Similarly, $\Delta S = 4R \ln 2.00 = 23.0$ J/K. the same as that obtained in part (f).

22. (a) The final pressure is

$$p_f = (5.00 \text{ kPa})e^{(V_i-V_f)/a} = (5.00 \text{ kPa})e^{(1.00 \text{ m}^3 - 2.00 \text{ m}^3)/1.00 \text{ m}^3} = 1.84 \text{ kPa}.$$

(b) We use the ratio form of the gas law (see Sample Problem 19-1) to find the final temperature of the gas:

$$T_f = T_i \left(\frac{p_f V_f}{p_i V_i}\right) = (600 \text{ K})\frac{(1.84 \text{ kPa})(2.00 \text{ m}^3)}{(5.00 \text{ kPa})(1.00 \text{ m}^3)} = 441 \text{ K}.$$

For later purposes, we note that this result can be written "exactly" as $T_f = T_i (2e^{-1})$. In our solution, we are avoiding using the "one mole" datum since it is not clear how precise it is.

(c) The work done by the gas is

$$W = \int_{V_i}^{V_f} p \, dV = \int_{V_i}^{V_f} (5.00 \text{ kPa}) e^{(V_i-V)/a} dV = (5.00 \text{ kPa}) e^{V_i/a} \cdot \left[-ae^{-V/a}\right]_{V_i}^{V_f}$$

$$= (5.00 \text{ kPa})e^{1.00}(1.00 \text{ m}^3)\left(e^{-1.00} - e^{-2.00}\right)$$

$$= 3.16 \text{ kJ}.$$

(d) Consideration of a two-stage process, as suggested in the hint, brings us simply to Eq. 20-4. Consequently, with $C_V = \frac{3}{2}R$ (see Eq. 19-43), we find

$$\Delta S = nR \ln\left(\frac{V_f}{V_i}\right) + n\left(\frac{3}{2}R\right)\ln\left(\frac{T_f}{T_i}\right) = nR\left(\ln 2 + \frac{3}{2}\ln(2e^{-1})\right) = \frac{p_i V_i}{T_i}\left(\ln 2 + \frac{3}{2}\ln 2 + \frac{3}{2}\ln e^{-1}\right)$$

$$= \frac{(5000 \text{ Pa})(1.00 \text{ m}^3)}{600 \text{ K}}\left(\frac{5}{2}\ln 2 - \frac{3}{2}\right)$$

$$= 1.94 \text{ J/K}.$$

23. We solve (b) first.

(b) For a Carnot engine, the efficiency is related to the reservoir temperatures by Eq. 20-13. Therefore,

$$T_H = \frac{T_H - T_L}{\varepsilon} = \frac{75 \text{ K}}{0.22} = 341 \text{ K}$$

which is equivalent to 68°C.

(a) The temperature of the cold reservoir is $T_L = T_H - 75 = 341 \text{ K} - 75 \text{ K} = 266 \text{ K}$.

24. Eq. 20-13 leads to

$$\varepsilon = 1 - \frac{T_L}{T_H} = 1 - \frac{373 \text{ K}}{7 \times 10^8 \text{ K}} = 0.9999995$$

quoting more figures than are significant. As a percentage, this is $\varepsilon = 99.99995\%$.

25. (a) The efficiency is

$$\varepsilon = \frac{T_H - T_L}{T_H} = \frac{(235-115)\text{K}}{(235+273)\text{K}} = 0.236 = 23.6\%.$$

We note that a temperature difference has the same value on the Kelvin and Celsius scales. Since the temperatures in the equation must be in Kelvins, the temperature in the denominator is converted to the Kelvin scale.

(b) Since the efficiency is given by $\varepsilon = |W|/|Q_H|$, the work done is given by

$$|W| = \varepsilon |Q_H| = 0.236(6.30 \times 10^4 \text{ J}) = 1.49 \times 10^4 \text{ J}.$$

26. The answers to this exercise do not depend on the engine being of the Carnot design. Any heat engine that intakes energy as heat (from, say, consuming fuel) equal to $|Q_H|$ = 52 kJ and exhausts (or discards) energy as heat equal to $|Q_L|$ = 36 kJ will have these values of efficiency ε and net work W.

(a) Eq. 20-12 gives

$$\varepsilon = 1 - \left|\frac{Q_L}{Q_H}\right| = 0.31 = 31\%.$$

(b) Eq. 20-8 gives

$$W = |Q_H| - |Q_L| = 16 \text{ kJ}.$$

27. With T_L = 290 k, we find

$$\varepsilon = 1 - \frac{T_L}{T_H} \Rightarrow T_H = \frac{T_L}{1-\varepsilon} = \frac{290 \text{ K}}{1-0.40}$$

which yields the (initial) temperature of the high-temperature reservoir: T_H = 483 K. If we replace ε = 0.40 in the above calculation with ε = 0.50, we obtain a (final) high temperature equal to T_H' = 580 K. The difference is

$$T_H' - T_H = 580 \text{ K} - 483 \text{ K} = 97 \text{ K}.$$

28. (a) Eq. 20-13 leads to

$$\varepsilon = 1 - \frac{T_L}{T_H} = 1 - \frac{333\text{ K}}{373\text{ K}} = 0.107.$$

We recall that a Watt is Joule-per-second. Thus, the (net) work done by the cycle per unit time is the given value 500 J/s. Therefore, by Eq. 20-11, we obtain the heat input per unit time:

$$\varepsilon = \frac{W}{|Q_H|} \Rightarrow \frac{0.500 \text{ kJ/s}}{0.107} = 4.67 \text{ kJ/s}.$$

(b) Considering Eq. 20-8 on a per unit time basis, we find (4.67 − 0.500) kJ/s = 4.17 kJ/s for the rate of heat exhaust.

29. (a) Energy is added as heat during the portion of the process from a to b. This portion occurs at constant volume (V_b), so $Q_{in} = nC_V \Delta T$. The gas is a monatomic ideal gas, so $C_V = 3R/2$ and the ideal gas law gives

$$\Delta T = (1/nR)(p_b V_b - p_a V_a) = (1/nR)(p_b - p_a) V_b.$$

Thus, $Q_{in} = \frac{3}{2}(p_b - p_a)V_b$. V_b and p_b are given. We need to find p_a. Now p_a is the same as p_c and points c and b are connected by an adiabatic process. Thus, $p_c V_c^\gamma = p_b V_b^\gamma$ and

$$p_a = p_c = \left(\frac{V_b}{V_c}\right)^\gamma p_b = \left(\frac{1}{8.00}\right)^{5/3} (1.013 \times 10^6 \text{ Pa}) = 3.167 \times 10^4 \text{ Pa}.$$

The energy added as heat is

$$Q_{in} = \frac{3}{2}(1.013 \times 10^6 \text{ Pa} - 3.167 \times 10^4 \text{ Pa})(1.00 \times 10^{-3} \text{ m}^3) = 1.47 \times 10^3 \text{ J}.$$

(b) Energy leaves the gas as heat during the portion of the process from c to a. This is a constant pressure process, so

$$Q_{out} = nC_p \Delta T = \frac{5}{2}(p_a V_a - p_c V_c) = \frac{5}{2} p_a (V_a - V_c)$$

$$= \frac{5}{2}(3.167 \times 10^4 \text{ Pa})(-7.00)(1.00 \times 10^{-3} \text{ m}^3) = -5.54 \times 10^2 \text{ J},$$

or $|Q_{out}| = 5.54 \times 10^2$ J. The substitutions $V_a - V_c = V_a - 8.00\, V_a = -7.00\, V_a$ and $C_p = \frac{5}{2}R$ were made.

(c) For a complete cycle, the change in the internal energy is zero and

$$W = Q = 1.47 \times 10^3 \text{ J} - 5.54 \times 10^2 \text{ J} = 9.18 \times 10^2 \text{ J}.$$

(d) The efficiency is

$$\varepsilon = W/Q_{in} = (9.18 \times 10^2 \text{ J})/(1.47 \times 10^3 \text{ J}) = 0.624 = 62.4\%.$$

30. From Fig. 20-28, we see $Q_H = 4000$ J at $T_H = 325$ K. Combining Eq. 20-11 with Eq. 20-13, we have

$$\frac{W}{Q_H} = 1 - \frac{T_C}{T_H} \quad \Rightarrow \quad W = 923 \text{ J}.$$

Now, for $T'_H = 550$ K, we have

$$\frac{W}{Q'_H} = 1 - \frac{T_C}{T'_H} \quad \Rightarrow \quad Q'_H = 1692 \text{ J} \approx 1.7 \text{ kJ}$$

31. (a) The net work done is the rectangular "area" enclosed in the pV diagram:

$$W = (V - V_0)(p - p_0) = (2V_0 - V_0)(2p_0 - p_0) = V_0 p_0.$$

Inserting the values stated in the problem, we obtain $W = 2.27$ kJ.

(b) We compute the energy added as heat during the "heat-intake" portions of the cycle using Eq. 19-39, Eq. 19-43, and Eq. 19-46:

$$Q_{abc} = nC_V(T_b - T_a) + nC_p(T_c - T_b) = n\left(\frac{3}{2}R\right)T_a\left(\frac{T_b}{T_a} - 1\right) + n\left(\frac{5}{2}R\right)T_a\left(\frac{T_c}{T_a} - \frac{T_b}{T_a}\right)$$

$$= nRT_a\left(\frac{3}{2}\left(\frac{T_b}{T_a} - 1\right) + \frac{5}{2}\left(\frac{T_c}{T_a} - \frac{T_b}{T_a}\right)\right) = p_0V_0\left(\frac{3}{2}(2-1) + \frac{5}{2}(4-2)\right)$$

$$= \frac{13}{2}p_0V_0$$

where, to obtain the last line, the gas law in ratio form has been used (see Sample Problem 19-1). Therefore, since $W = p_0V_0$, we have $Q_{abc} = 13W/2 = 14.8$ kJ.

(c) The efficiency is given by Eq. 20-11:

$$\varepsilon = \frac{W}{|Q_H|} = \frac{2}{13} = 0.154 = 15.4\%.$$

(d) A Carnot engine operating between T_c and T_a has efficiency equal to

$$\varepsilon = 1 - \frac{T_a}{T_c} = 1 - \frac{1}{4} = 0.750 = 75.0\%$$

where the gas law in ratio form has been used.

(e) This is greater than our result in part (c), as expected from the second law of thermodynamics.

32. (a) Using Eq. 19-54 for process $D \to A$ gives

$$p_D V_D^{\gamma} = p_A V_A^{\gamma} \quad \Rightarrow \quad \frac{p_0}{32}(8V_0)^{\gamma} = p_0 V_0^{\gamma}$$

which leads to $8^{\gamma} = 32 \Rightarrow \gamma = 5/3$. The result (see §19-9 and §19-11) implies the gas is monatomic.

(b) The input heat is that absorbed during process $A \to B$:

$$Q_H = nC_p \Delta T = n\left(\frac{5}{2}R\right)T_A\left(\frac{T_B}{T_A}-1\right) = nRT_A\left(\frac{5}{2}\right)(2-1) = p_0 V_0 \left(\frac{5}{2}\right)$$

and the exhaust heat is that liberated during process $C \to D$:

$$Q_L = nC_p \Delta T = n\left(\frac{5}{2}R\right)T_D\left(1-\frac{T_L}{T_D}\right) = nRT_D\left(\frac{5}{2}\right)(1-2) = -\frac{1}{4}p_0 V_0\left(\frac{5}{2}\right)$$

where in the last step we have used the fact that $T_D = \frac{1}{4}T_A$ (from the gas law in ratio form — see Sample Problem 19-1). Therefore, Eq. 20-12 leads to

$$\varepsilon = 1 - \left|\frac{Q_L}{Q_H}\right| = 1 - \frac{1}{4} = 0.75 = 75\%.$$

33. (a) We use $\varepsilon = |W/Q_H|$. The heat absorbed is $|Q_H| = \frac{|W|}{\varepsilon} = \frac{8.2\,\text{kJ}}{0.25} = 33\,\text{kJ}$.

(b) The heat exhausted is then $|Q_L| = |Q_H| - |W| = 33\,\text{kJ} - 8.2\,\text{kJ} = 25\,\text{kJ}$.

(c) Now we have $|Q_H| = \frac{|W|}{\varepsilon} = \frac{8.2\,\text{kJ}}{0.31} = 26\,\text{kJ}$.

(d) Similarly, $|Q_C| = |Q_H| - |W| = 26\,\text{kJ} - 8.2\,\text{kJ} = 18\,\text{kJ}$.

861

34. All terms are assumed to be positive. The total work done by the two-stage system is $W_1 + W_2$. The heat-intake (from, say, consuming fuel) of the system is Q_1 so we have (by Eq. 20-11 and Eq. 20-8)

$$\varepsilon = \frac{W_1 + W_2}{Q_1} = \frac{(Q_1 - Q_2) + (Q_2 - Q_3)}{Q_1} = 1 - \frac{Q_3}{Q_1}.$$

Now, Eq. 20-10 leads to

$$\frac{Q_1}{T_1} = \frac{Q_2}{T_2} = \frac{Q_3}{T_3}$$

where we assume Q_2 is absorbed by the second stage at temperature T_2. This implies the efficiency can be written

$$\varepsilon = 1 - \frac{T_3}{T_1} = \frac{T_1 - T_3}{T_1}.$$

35. (a) The pressure at 2 is $p_2 = 3.00 p_1$, as given in the problem statement. The volume is $V_2 = V_1 = nRT_1/p_1$. The temperature is

$$T_2 = \frac{p_2 V_2}{nR} = \frac{3.00 p_1 V_1}{nR} = 3.00 T_1 \Rightarrow \frac{T_2}{T_1} = 3.00.$$

(b) The process $2 \to 3$ is adiabatic, so $T_2 V_2^{\gamma-1} = T_3 V_3^{\gamma-1}$. Using the result from part (a), $V_3 = 4.00 V_1$, $V_2 = V_1$ and $\gamma = 1.30$, we obtain

$$\frac{T_3}{T_1} = \frac{T_3}{T_2/3.00} = 3.00 \left(\frac{V_2}{V_3}\right)^{\gamma-1} = 3.00 \left(\frac{1}{4.00}\right)^{0.30} = 1.98.$$

(c) The process $4 \to 1$ is adiabatic, so $T_4 V_4^{\gamma-1} = T_1 V_1^{\gamma-1}$. Since $V_4 = 4.00 V_1$, we have

$$\frac{T_4}{T_1} = \left(\frac{V_1}{V_4}\right)^{\gamma-1} = \left(\frac{1}{4.00}\right)^{0.30} = 0.660.$$

(d) The process $2 \to 3$ is adiabatic, so $p_2 V_2^\gamma = p_3 V_3^\gamma$ or $p_3 = (V_2/V_3)^\gamma p_2$. Substituting $V_3 = 4.00 V_1$, $V_2 = V_1$, $p_2 = 3.00 p_1$ and $\gamma = 1.30$, we obtain

$$\frac{p_3}{p_1} = \frac{3.00}{(4.00)^{1.30}} = 0.495.$$

(e) The process $4 \to 1$ is adiabatic, so $p_4 V_4^\gamma = p_1 V_1^\gamma$ and

$$\frac{p_4}{p_1} = \left(\frac{V_1}{V_4}\right)^\gamma = \frac{1}{(4.00)^{1.30}} = 0.165,$$

where we have used $V_4 = 4.00 V_1$.

(f) The efficiency of the cycle is $\varepsilon = W/Q_{12}$, where W is the total work done by the gas during the cycle and Q_{12} is the energy added as heat during the $1 \to 2$ portion of the cycle, the only portion in which energy is added as heat. The work done during the portion of the cycle from 2 to 3 is $W_{23} = \int p \, dV$. Substitute $p = p_2 V_2^\gamma / V^\gamma$ to obtain

$$W_{23} = p_2 V_2^\gamma \int_{V_2}^{V_3} V^{-\gamma} dV = \left(\frac{p_2 V_2^\gamma}{\gamma - 1}\right)\left(V_2^{1-\gamma} - V_3^{1-\gamma}\right).$$

Substitute $V_2 = V_1$, $V_3 = 4.00 V_1$, and $p_3 = 3.00 p_1$ to obtain

$$W_{23} = \left(\frac{3 p_1 V_1}{1 - \gamma}\right)\left(1 - \frac{1}{4^{\gamma-1}}\right) = \left(\frac{3 n R T_1}{\gamma - 1}\right)\left(1 - \frac{1}{4^{\gamma-1}}\right).$$

Similarly, the work done during the portion of the cycle from 4 to 1 is

$$W_{41} = \left(\frac{p_1 V_1^\gamma}{\gamma - 1}\right)\left(V_4^{1-\gamma} - V_1^{1-\gamma}\right) = -\left(\frac{p_1 V_1}{\gamma - 1}\right)\left(1 - \frac{1}{4^{\gamma-1}}\right) = -\left(\frac{n R T_1}{\gamma - 1}\right)\left(1 - \frac{1}{4^{\gamma-1}}\right).$$

No work is done during the $1 \to 2$ and $3 \to 4$ portions, so the total work done by the gas during the cycle is

$$W = W_{23} + W_{41} = \left(\frac{2 n R T_1}{\gamma - 1}\right)\left(1 - \frac{1}{4^{\gamma-1}}\right).$$

The energy added as heat is

$$Q_{12} = n C_V (T_2 - T_1) = n C_V (3 T_1 - T_1) = 2 n C_V T_1,$$

where C_V is the molar specific heat at constant volume. Now

$$\gamma = C_p / C_V = (C_V + R)/C_V = 1 + (R/C_V),$$

so $C_V = R/(\gamma - 1)$. Here C_p is the molar specific heat at constant pressure, which for an ideal gas is $C_p = C_V + R$. Thus, $Q_{12} = 2 n R T_1/(\gamma - 1)$. The efficiency is

$$\varepsilon = \frac{2nRT_1}{\gamma-1}\left(1-\frac{1}{4^{\gamma-1}}\right)\frac{\gamma-1}{2nRT_1} = 1-\frac{1}{4^{\gamma-1}}.$$

With $\gamma = 1.30$, the efficiency is $\varepsilon = 0.340$ or 34.0%.

36. Eq. 20-10 still holds (particularly due to its use of absolute values), and energy conservation implies $|W| + Q_L = Q_H$. Therefore, with $T_L = 268.15$ K and $T_H = 290.15$ K, we find

$$|Q_H| = |Q_L|\left(\frac{T_H}{T_L}\right) = (|Q_H| - |W|)\left(\frac{290.15}{268.15}\right)$$

which (with $|W| = 1.0$ J) leads to $|Q_H| = |W|\left(\dfrac{1}{1-268.15/290.15}\right) = 13$ J.

37. A Carnot refrigerator working between a hot reservoir at temperature T_H and a cold reservoir at temperature T_L has a coefficient of performance K that is given by

$$K = \frac{T_L}{T_H - T_L}.$$

For the refrigerator of this problem, $T_H = 96°$ F $= 309$ K and $T_L = 70°$ F $= 294$ K, so

$$K = (294 \text{ K})/(309 \text{ K} - 294 \text{ K}) = 19.6.$$

The coefficient of performance is the energy Q_L drawn from the cold reservoir as heat divided by the work done: $K = |Q_L|/|W|$. Thus,

$$|Q_L| = K|W| = (19.6)(1.0 \text{ J}) = 20 \text{ J}.$$

38. (a) Eq. 20-15 provides

$$K_C = \frac{|Q_L|}{|Q_H|-|Q_L|} \Rightarrow |Q_H| = |Q_L|\left(\frac{1+K_C}{K_C}\right)$$

which yields $|Q_H| = 49$ kJ when $K_C = 5.7$ and $|Q_L| = 42$ kJ.

(b) From §20-5 we obtain

$$|W| = |Q_H| - |Q_L| = 49.4 \text{ kJ} - 42.0 \text{ kJ} = 7.4 \text{ kJ}$$

if we take the initial 42 kJ datum to be accurate to three figures. The given temperatures are not used in the calculation; in fact, it is possible that the given room temperature value is not meant to be the high temperature for the (reversed) Carnot cycle — since it does not lead to the given K_C using Eq. 20-16.

39. The coefficient of performance for a refrigerator is given by $K = |Q_L|/|W|$, where Q_L is the energy absorbed from the cold reservoir as heat and W is the work done during the refrigeration cycle, a negative value. The first law of thermodynamics yields $Q_H + Q_L - W = 0$ for an integer number of cycles. Here Q_H is the energy ejected to the hot reservoir as heat. Thus, $Q_L = W - Q_H$. Q_H is negative and greater in magnitude than W, so $|Q_L| = |Q_H| - |W|$. Thus,

$$K = \frac{|Q_H| - |W|}{|W|}.$$

The solution for $|W|$ is $|W| = |Q_H|/(K + 1)$. In one hour,

$$|W| = \frac{7.54 \text{ MJ}}{3.8 + 1} = 1.57 \text{ MJ}.$$

The rate at which work is done is $(1.57 \times 10^6 \text{ J})/(3600 \text{ s}) = 440 \text{ W}$.

40. (a) Using Eq. 20-14 and Eq. 20-16, we obtain

$$|W| = \frac{|Q_L|}{K_C} = (1.0 \text{ J}) \left(\frac{300 \text{ K} - 280 \text{ K}}{280 \text{ K}} \right) = 0.071 \text{ J}.$$

(b) A similar calculation (being sure to use absolute temperature) leads to 0.50 J in this case.

(c) With $T_L = 100$ K, we obtain $|W| = 2.0$ J.

(d) Finally, with the low temperature reservoir at 50 K, an amount of work equal to $|W| = 5.0$ J is required.

41. The efficiency of the engine is defined by $\varepsilon = W/Q_1$ and is shown in the text to be

$$\varepsilon = \frac{T_1 - T_2}{T_1} \quad \Rightarrow \quad \frac{W}{Q_1} = \frac{T_1 - T_2}{T_1}.$$

The coefficient of performance of the refrigerator is defined by $K = Q_4/W$ and is shown in the text to be

$$K = \frac{T_4}{T_3 - T_4} \quad \Rightarrow \quad \frac{Q_4}{W} = \frac{T_4}{T_3 - T_4}.$$

Now $Q_4 = Q_3 - W$, so

$$(Q_3 - W)/W = T_4/(T_3 - T_4).$$

The work done by the engine is used to drive the refrigerator, so W is the same for the two. Solve the engine equation for W and substitute the resulting expression into the refrigerator equation. The engine equation yields $W = (T_1 - T_2)Q_1/T_1$ and the substitution yields

$$\frac{T_4}{T_3 - T_4} = \frac{Q_3}{W} - 1 = \frac{Q_3 T_1}{Q_1(T_1 - T_2)} - 1.$$

Solving for Q_3/Q_1, we obtain

$$\frac{Q_3}{Q_1} = \left(\frac{T_4}{T_3 - T_4} + 1\right)\left(\frac{T_1 - T_2}{T_1}\right) = \left(\frac{T_3}{T_3 - T_4}\right)\left(\frac{T_1 - T_2}{T_1}\right) = \frac{1 - (T_2/T_1)}{1 - (T_4/T_3)}.$$

With $T_1 = 400$ K, $T_2 = 150$ K, $T_3 = 325$ K, and $T_4 = 225$ K, the ratio becomes $Q_3/Q_1 = 2.03$.

42. (a) Eq. 20-13 gives the Carnot efficiency as $1 - T_L/T_H$. This gives 0.222 in this case. Using this value with Eq. 20-11 leads to

$$W = (0.222)(750 \text{ J}) = 167 \text{ J}.$$

(b) Now, Eq. 20-16 gives $K_C = 3.5$. Then, Eq. 20-14 yields $|W| = 1200/3.5 = 343$ J.

43. We are told $K = 0.27 K_C$ where

$$K_C = \frac{T_L}{T_H - T_L} = \frac{294 \text{ K}}{307 \text{ K} - 294 \text{ K}} = 23$$

where the Fahrenheit temperatures have been converted to Kelvins. Expressed on a per unit time basis, Eq. 20-14 leads to

$$\frac{|W|}{t} = \frac{|Q_L|/t}{K} = \frac{4000 \text{ Btu/h}}{(0.27)(23)} = 643 \text{ Btu/h}.$$

Appendix D indicates 1 But/h = 0.0003929 hp, so our result may be expressed as $|W|/t = 0.25$ hp.

44. The work done by the motor in $t = 10.0$ min is $|W| = Pt = (200 \text{ W})(10.0 \text{ min})(60 \text{ s/min}) = 1.20 \times 10^5$ J. The heat extracted is then

$$|Q_L| = K|W| = \frac{T_L|W|}{T_H - T_L} = \frac{(270 \text{ K})(1.20 \times 10^5 \text{ J})}{300 \text{ K} - 270 \text{ K}} = 1.08 \times 10^6 \text{ J}.$$

45. We need nine labels:

Label	Number of molecules on side 1	Number of molecules on side 2
I	8	0
II	7	1
III	6	2
IV	5	3
V	4	4
VI	3	5
VII	2	6
VIII	1	7
IX	0	8

The multiplicity W is computing using Eq. 20-20. For example, the multiplicity for label IV is

$$W = \frac{8!}{(5!)(3!)} = \frac{40320}{(120)(6)} = 56$$

and the corresponding entropy is (using Eq. 20-21)

$$S = k \ln W = \left(1.38 \times 10^{-23} \text{ J/K}\right) \ln(56) = 5.6 \times 10^{-23} \text{ J/K}.$$

In this way, we generate the following table:

Label	W	S
I	1	0
II	8	2.9×10^{-23} J/K
III	28	4.6×10^{-23} J/K
IV	56	5.6×10^{-23} J/K
V	70	5.9×10^{-23} J/K
VI	56	5.6×10^{-23} J/K
VII	28	4.6×10^{-23} J/K
VIII	8	2.9×10^{-23} J/K
IX	1	0

46. (a) We denote the configuration with n heads out of N trials as $(n; N)$. We use Eq. 20-20:

$$W(25;50) = \frac{50!}{(25!)(50-25)!} = 1.26 \times 10^{14}.$$

(b) There are 2 possible choices for each molecule: it can either be in side 1 or in side 2 of the box. If there are a total of N independent molecules, the total number of available states of the N-particle system is

$$N_{total} = 2 \times 2 \times 2 \times \cdots \times 2 = 2^N.$$

With $N = 50$, we obtain $N_{total} = 2^{50} = 1.13 \times 10^{15}$.

(c) The percentage of time in question is equal to the probability for the system to be in the central configuration:

$$p(25;50) = \frac{W(25;50)}{2^{50}} = \frac{1.26 \times 10^{14}}{1.13 \times 10^{15}} = 11.1\%.$$

With $N = 100$, we obtain

(d) $W(N/2, N) = N!/[(N/2)!]^2 = 1.01 \times 10^{29}$,

(e) $N_{total} = 2^N = 1.27 \times 10^{30}$,

(f) and $p(N/2;N) = W(N/2, N)/N_{total} = 8.0\%$.

Similarly, for $N = 200$, we obtain

(g) $W(N/2, N) = 9.25 \times 10^{58}$,

(h) $N_{total} = 1.61 \times 10^{60}$,

(i) and $p(N/2; N) = 5.7\%$.

(j) As N increases the number of available microscopic states increase as 2^N, so there are more states to be occupied, leaving the probability less for the system to remain in its central configuration. Thus, the time spent in there decreases with an increase in N.

47. (a) Suppose there are n_L molecules in the left third of the box, n_C molecules in the center third, and n_R molecules in the right third. There are $N!$ arrangements of the N molecules, but $n_L!$ are simply rearrangements of the n_L molecules in the right third, $n_C!$ are rearrangements of the n_C molecules in the center third, and $n_R!$ are rearrangements of the n_R molecules in the right third. These rearrangements do not produce a new configuration. Thus, the multiplicity is

$$W = \frac{N!}{n_L! n_C! n_R!}.$$

(b) If half the molecules are in the right half of the box and the other half are in the left half of the box, then the multiplicity is

$$W_B = \frac{N!}{(N/2)!(N/2)!}.$$

If one-third of the molecules are in each third of the box, then the multiplicity is

$$W_A = \frac{N!}{(N/3)!(N/3)!(N/3)!}.$$

The ratio is

$$\frac{W_A}{W_B} = \frac{(N/2)!(N/2)!}{(N/3)!(N/3)!(N/3)!}.$$

(c) For $N = 100$,

$$\frac{W_A}{W_B} = \frac{50!\,50!}{33!\,33!\,34!} = 4.2 \times 10^{16}.$$

48. Using Hooke's law, we find the spring constant to be

$$k = \frac{F_s}{x_s} = \frac{1.50 \text{ N}}{0.0350 \text{ m}} = 42.86 \text{ N/m}.$$

To find the rate of change of entropy with a small additional stretch, we use Eq. 20-7 (see also Sample Problem 20-3) and obtain

$$\left|\frac{dS}{dx}\right| = \frac{k|x|}{T} = \frac{(42.86 \text{ N/m})(0.0170 \text{ m})}{275 \text{ K}} = 2.65 \times 10^{-3} \text{ J/K} \cdot \text{m}.$$

49. Using Eq. 19-34 and Eq. 19-35, we arrive at

$$\Delta v = (\sqrt{3} - \sqrt{2})\sqrt{RT/M}$$

(a) We find, with $M = 28$ g/mol $= 0.028$ kg/mol (see Table 19-1), $\Delta v_i = 87$ m/s at 250 K,

(b) and $\Delta v_f = 122 \approx 1.2 \times 10^2$ m/s at 500 K.

(c) The expression above for Δv implies

$$T = \frac{M}{R(\sqrt{3} - \sqrt{2})^2}(\Delta v)^2$$

which we can plug into Eq. 20-4 to yield

$$\Delta S = nR \ln(V_f/V_i) + nC_V \ln(T_f/T_i) = 0 + nC_V \ln[(\Delta v_f)^2/(\Delta v_i)^2] = 2nC_V \ln(\Delta v_f/\Delta v_i).$$

Using Table 19-3 to get $C_V = 5R/2$ (see also Table 19-2) we then find, for $n = 1.5$ mol, $\Delta S = 22$ J/K.

50. The net work is figured from the (positive) isothermal expansion (Eq. 19-14) and the (negative) constant-pressure compression (Eq. 19-48). Thus,

$$W_{net} = nRT_H \ln(V_{max}/V_{min}) + nR(T_L - T_H)$$

where $n = 3.4$, $T_H = 500$ K, $T_L = 200$ K and $V_{max}/V_{min} = 5/2$ (same as the ratio T_H/T_L). Therefore, $W_{net} = 4468$ J. Now, we identify the "input heat" as that transferred in steps 1 and 2:

$$Q_{in} = Q_1 + Q_2 = nC_V(T_H - T_L) + nRT_H \ln(V_{max}/V_{min})$$

where $C_V = 5R/2$ (see Table 19-3). Consequently, $Q_{in} = 34135$ J. Dividing these results gives the efficiency: $W_{net}/Q_{in} = 0.131$ (or about 13.1%).

51. (a) If T_H is the temperature of the high-temperature reservoir and T_L is the temperature of the low-temperature reservoir, then the maximum efficiency of the engine is

$$\varepsilon = \frac{T_H - T_L}{T_H} = \frac{(800+40) \text{ K}}{(800+273) \text{ K}} = 0.78 \text{ or } 78\%.$$

(b) The efficiency is defined by $\varepsilon = |W|/|Q_H|$, where W is the work done by the engine and Q_H is the heat input. W is positive. Over a complete cycle, $Q_H = W + |Q_L|$, where Q_L is the heat output, so $\varepsilon = W/(W + |Q_L|)$ and $|Q_L| = W[(1/\varepsilon) - 1]$. Now $\varepsilon = (T_H - T_L)/T_H$, where T_H is the temperature of the high-temperature heat reservoir and T_L is the temperature of the low-temperature reservoir. Thus,

$$\frac{1}{\varepsilon} - 1 = \frac{T_L}{T_H - T_L} \text{ and } |Q_L| = \frac{WT_L}{T_H - T_L}.$$

The heat output is used to melt ice at temperature $T_i = -40°C$. The ice must be brought to $0°C$, then melted, so

$$|Q_L| = mc(T_f - T_i) + mL_F,$$

where m is the mass of ice melted, T_f is the melting temperature ($0°C$), c is the specific heat of ice, and L_F is the heat of fusion of ice. Thus,

$$WT_L/(T_H - T_L) = mc(T_f - T_i) + mL_F.$$

We differentiate with respect to time and replace dW/dt with P, the power output of the engine, and obtain

$$PT_L/(T_H - T_L) = (dm/dt)[c(T_f - T_i) + L_F].$$

Therefore,

$$\frac{dm}{dt} = \left(\frac{PT_L}{T_H - T_L}\right)\left(\frac{1}{c(T_f - T_i) + L_F}\right).$$

Now, $P = 100 \times 10^6$ W, $T_L = 0 + 273 = 273$ K, $T_H = 800 + 273 = 1073$ K, $T_i = -40 + 273 = 233$ K, $T_f = 0 + 273 = 273$ K, $c = 2220$ J/kg·K, and $L_F = 333 \times 10^3$ J/kg, so

$$\frac{dm}{dt} = \left[\frac{(100\times 10^6 \text{ J/s})(273 \text{ K})}{1073 \text{ K} - 273 \text{ K}}\right]\left[\frac{1}{(2220 \text{ J/kg}\cdot\text{K})(273 \text{ K} - 233 \text{ K}) + 333\times 10^3 \text{ J/kg}}\right]$$

$$= 82 \text{ kg/s}.$$

We note that the engine is now operated between 0°C and 800°C.

52. (a) Combining Eq. 20-11 with Eq. 20-13, we obtain

$$|W| = |Q_H|\left(1 - \frac{T_L}{T_H}\right) = (500 \text{ J})\left(1 - \frac{260 \text{ K}}{320 \text{ K}}\right) = 93.8 \text{ J}.$$

(b) Combining Eq. 20-14 with Eq. 20-16, we find

$$|W| = \frac{|Q_L|}{\left(\frac{T_L}{T_H - T_L}\right)} = \frac{1000 \text{ J}}{\left(\frac{260 \text{ K}}{320 \text{ K} - 260 \text{ K}}\right)} = 231 \text{ J}.$$

53. (a) Starting from $\sum Q = 0$ (for calorimetry problems) we can derive (when no phase changes are involved)

$$T_f = \frac{c_1 m_1 T_1 + c_2 m_2 T_2}{c_1 m_1 + c_2 m_2} = 40.9°C,$$

which is equivalent to 314 K.

(b) From Eq. 20-1, we have

$$\Delta S_{copper} = \int_{353}^{314} \frac{cm\, dT}{T} = (386)(0.600)\ln\left(\frac{314}{353}\right) = -27.1 \text{ J/K}.$$

(c) For water, the change in entropy is

$$\Delta S_{water} = \int_{283}^{314} \frac{cm\, dT}{T} = (4190)(0.0700)\ln\left(\frac{314}{283}\right) = 30.5 \text{ J/K}.$$

(d) The net result for the system is (30.5 − 27.1) J/K = 3.4 J/K. (Note: these calculations are fairly sensitive to round-off errors. To arrive at this final answer, the value 273.15 was used to convert to Kelvins, and all intermediate steps were retained to full calculator accuracy.)

54. For an isothermal ideal gas process, we have $Q = W = nRT \ln(V_f/V_i)$. Thus,

$$\Delta S = Q/T = W/T = nR \ln(V_f/V_i)$$

(a) $V_f/V_i = (0.800)/(0.200) = 4.00$, $\Delta S = (0.55)(8.31)\ln(4.00) = 6.34$ J/K.

(b) $V_f/V_i = (0.800)/(0.200) = 4.00$, $\Delta S = (0.55)(8.31)\ln(4.00) = 6.34$ J/K.

(c) $V_f/V_i = (1.20)/(0.300) = 4.00$, $\Delta S = (0.55)(8.31)\ln(4.00) = 6.34$ J/K.

(d) $V_f/V_i = (1.20)/(0.300) = 4.00$, $\Delta S = (0.55)(8.31)\ln(4.00) = 6.34$ J/K.

55. Except for the phase change (which just uses Eq. 20-2), this has some similarities with Sample Problem 20-2. Using constants available in the Chapter 19 tables, we compute

$$\Delta S = m[c_{\text{ice}} \ln(273/253) + \frac{L_f}{273} + c_{\text{water}} \ln(313/273)] = 1.18 \times 10^3 \text{ J/K}.$$

56. Eq. 20-4 yields

$$\Delta S = nR \ln(V_f/V_i) + nC_V \ln(T_f/T_i) = 0 + nC_V \ln(425/380)$$

where $n = 3.20$ and $C_V = \frac{3}{2}R$ (Eq. 19-43). This gives 4.46 J/K.

57. (a) It is a reversible set of processes returning the system to its initial state; clearly, $\Delta S_{\text{net}} = 0$.

(b) Process 1 is adiabatic and reversible (as opposed to, say, a free expansion) so that Eq. 20-1 applies with $dQ = 0$ and yields $\Delta S_1 = 0$.

(c) Since the working substance is an ideal gas, then an isothermal process implies $Q = W$, which further implies (regarding Eq. 20-1) $dQ = p\, dV$. Therefore,

$$\int \frac{dQ}{T} = \int \frac{p\, dV}{\left(\frac{pV}{nR}\right)} = nR \int \frac{dV}{V}$$

which leads to $\Delta S_3 = nR \ln(1/2) = -23.0 \text{ J/K}$.

(d) By part (a), $\Delta S_1 + \Delta S_2 + \Delta S_3 = 0$. Then, part (b) implies $\Delta S_2 = -\Delta S_3$. Therefore, $\Delta S_2 = 23.0$ J/K.

58. (a) The most obvious input-heat step is the constant-volume process. Since the gas is monatomic, we know from Chapter 19 that $C_V = \frac{3}{2}R$. Therefore,

$$Q_V = nC_V \Delta T = (1.0 \text{ mol})\left(\frac{3}{2}\right)\left(8.31 \frac{\text{J}}{\text{mol} \cdot \text{K}}\right)(600 \text{ K} - 300 \text{ K}) = 3740 \text{ J}.$$

Since the heat transfer during the isothermal step is positive, we may consider it also to be an input-heat step. The isothermal Q is equal to the isothermal work (calculated in the next part) because $\Delta E_{int} = 0$ for an ideal gas isothermal process (see Eq. 19-45). Borrowing from the part (b) computation, we have

$$Q_{\text{isotherm}} = nRT_H \ln 2 = (1 \text{ mol})\left(8.31 \frac{\text{J}}{\text{mol} \cdot \text{K}}\right)(600 \text{ K}) \ln 2 = 3456 \text{ J}.$$

Therefore, $Q_H = Q_V + Q_{\text{isotherm}} = 7.2 \times 10^3$ J.

(b) We consider the sum of works done during the processes (noting that no work is done during the constant-volume step). Using Eq. 19-14 and Eq. 19-16, we have

$$W = nRT_H \ln\left(\frac{V_{\max}}{V_{\min}}\right) + p_{\min}(V_{\min} - V_{\max})$$

where (by the gas law in ratio form, as illustrated in Sample Problem 19-1) the volume ratio is

$$\frac{V_{\max}}{V_{\min}} = \frac{T_H}{T_L} = \frac{600 \text{ K}}{300 \text{ K}} = 2.$$

Thus, the net work is

$$W = nRT_H \ln 2 + p_{\min} V_{\min}\left(1 - \frac{V_{\max}}{V_{\min}}\right) = nRT_H \ln 2 + nRT_L(1 - 2) = nR(T_H \ln 2 - T_L)$$

$$= (1 \text{ mol})\left(8.31 \frac{\text{J}}{\text{mol} \cdot \text{K}}\right)((600 \text{ K}) \ln 2 - (300 \text{ K}))$$

$$= 9.6 \times 10^2 \text{ J}.$$

(c) Eq. 20-11 gives

$$\varepsilon = \frac{W}{Q_H} = 0.134 \approx 13\%.$$

59. (a) Processes 1 and 2 both require the input of heat, which is denoted Q_H. Noting that rotational degrees of freedom are not involved, then, from the discussion in Chapter 19, $C_V = 3R/2$, $C_p = 5R/2$, and $\gamma = 5/3$. We further note that since the working substance is an ideal gas, process 2 (being isothermal) implies $Q_2 = W_2$. Finally, we note that the volume ratio in process 2 is simply 8/3. Therefore,

$$Q_H = Q_1 + Q_2 = nC_V(T' - T) + nRT'\ln\frac{8}{3}$$

which yields (for $T = 300$ K and $T' = 800$ K) the result $Q_H = 25.5 \times 10^3$ J.

(b) The net work is the net heat ($Q_1 + Q_2 + Q_3$). We find Q_3 from $nC_p(T - T') = -20.8 \times 10^3$ J. Thus, $W = 4.73 \times 10^3$ J.

(c) Using Eq. 20-11, we find that the efficiency is

$$\varepsilon = \frac{|W|}{|Q_H|} = \frac{4.73 \times 10^3}{25.5 \times 10^3} = 0.185 \text{ or } 18.5\%.$$

60. (a) Starting from $\sum Q = 0$ (for calorimetry problems) we can derive (when no phase changes are involved)

$$T_f = \frac{c_1 m_1 T_1 + c_2 m_2 T_2}{c_1 m_1 + c_2 m_2} = -44.2°C,$$

which is equivalent to 229 K.

(b) From Eq. 20-1, we have

$$\Delta S_{\text{tungsten}} = \int_{303}^{229} \frac{cm\,dT}{T} = (134)(0.045)\ln\left(\frac{229}{303}\right) = -1.69 \text{ J/K}.$$

(c) Also,

$$\Delta S_{\text{silver}} = \int_{153}^{229} \frac{cm\,dT}{T} = (236)(0.0250)\ln\left(\frac{229}{153}\right) = 2.38 \text{ J/K}.$$

(d) The net result for the system is $(2.38 - 1.69)$ J/K $= 0.69$ J/K. (Note: these calculations are fairly sensitive to round-off errors. To arrive at this final answer, the value 273.15 was used to convert to Kelvins, and all intermediate steps were retained to full calculator accuracy.)

61. From the formula for heat conduction, Eq. 19-32, using Table 19-6, we have

$$H = kA\frac{T_H - T_C}{L} = (401)\,(\pi(0.02)^2)\,270/1.50$$

which yields $H = 90.7$ J/s. Using Eq. 20-2, this is associated with an entropy rate-of-decrease of the high temperature reservoir (at 573 K) equal to

$$\Delta S/t = -90.7/573 = -0.158 \text{ (J/K)/s}.$$

And it is associated with an entropy rate-of-increase of the low temperature reservoir (at 303 K) equal to

$$\Delta S/t = +90.7/303 = 0.299 \text{ (J/K)/s}.$$

The net result is $(0.299 - 0.158)$ (J/K)/s = 0.141 (J/K)/s.

62. (a) Eq. 20-14 gives $K = 560/150 = 3.73$.

(b) Energy conservation requires the exhaust heat to be $560 + 150 = 710$ J.

63. (a) Eq. 20-15 can be written as $|Q_H| = |Q_L|(1 + 1/K_C) = (35)(1 + \frac{1}{4.6}) = 42.6$ kJ.

(b) Similarly, Eq. 20-14 leads to $|W| = |Q_L|/K = 35/4.6 = 7.61$ kJ.

64. (a) A good way to (mathematically) think of this is: consider the terms when you expand

$$(1 + x)^4 = 1 + 4x + 6x^2 + 4x^3 + x^4.$$

The coefficients correspond to the multiplicities. Thus, the smallest coefficient is 1.

(b) The largest coefficient is 6.

(c) Since the logarithm of 1 is zero, then Eq. 20-21 gives $S = 0$ for the least case.

(d) $S = k \ln(6) = 2.47 \times 10^{-23}$ J/K.

65. (a) Eq. 20-2 gives the entropy change for each reservoir (each of which, by definition, is able to maintain constant temperature conditions within itself). The net entropy change is therefore

$$\Delta S = \frac{+|Q|}{273 + 24} + \frac{-|Q|}{273 + 130} = 4.45 \text{ J/K}$$

where we set $|Q| = 5030$ J.

(b) We have assumed that the conductive heat flow in the rod is "steady-state"; that is, the situation described by the problem has existed and will exist for "long times." Thus there are no entropy change terms included in the calculation for elements of the rod itself.

66. Eq. 20-10 gives

$$\left|\frac{Q_{to}}{Q_{from}}\right| = \frac{T_{to}}{T_{from}} = \frac{300\,\text{K}}{4.0\,\text{K}} = 75.$$

67. We adapt the discussion of §20-7 to 3 and 5 particles (as opposed to the 6 particle situation treated in that section).

(a) The least multiplicity configuration is when all the particles are in the same half of the box. In this case, using Eq. 20-20, we have

$$W = \frac{3!}{3!0!} = 1.$$

(b) Similarly for box B, $W = 5!/(5!0!) = 1$ in the "least" case.

(c) The most likely configuration in the 3 particle case is to have 2 on one side and 1 on the other. Thus,

$$W = \frac{3!}{2!1!} = 3.$$

(d) The most likely configuration in the 5 particle case is to have 3 on one side and 2 on the other. Thus,

$$W = \frac{5!}{3!2!} = 10.$$

(e) We use Eq. 20-21 with our result in part (c) to obtain

$$S = k \ln W = (1.38 \times 10^{-23}) \ln 3 = 1.5 \times 10^{-23} \text{ J/K}.$$

(f) Similarly for the 5 particle case (using the result from part (d)), we find

$$S = k \ln 10 = 3.2 \times 10^{-23} \text{ J/K}.$$

68. A metric ton is 1000 kg, so that the heat generated by burning 380 metric tons during one hour is $(380000 \text{ kg})(28 \text{ MJ/kg}) = 10.6 \times 10^6$ MJ. The work done in one hour is

$$W = (750 \text{ MJ/s})(3600 \text{ s}) = 2.7 \times 10^6 \text{ MJ}$$

where we use the fact that a Watt is a Joule-per-second. By Eq. 20-11, the efficiency is

$$\varepsilon = \frac{2.7 \times 10^6 \text{ MJ}}{10.6 \times 10^6 \text{ MJ}} = 0.253 = 25\%.$$

69. Since the volume of the monatomic ideal gas is kept constant it does not do any work in the heating process. Therefore the heat Q it absorbs is equal to the change in its inertial energy: $dQ = dE_{int} = \frac{3}{2} nR\, dT$. Thus,

$$\Delta S = \int \frac{dQ}{T} = \int_{T_i}^{T_f} \frac{(3nR/2)\,dT}{T} = \frac{3}{2} nR \ln\left(\frac{T_f}{T_i}\right) = \frac{3}{2}(1.00 \text{ mol})\left(8.31 \frac{\text{J}}{\text{mol}\cdot\text{K}}\right) \ln\left(\frac{400\text{ K}}{300\text{ K}}\right)$$
$$= 3.59 \text{ J/K}.$$

70. With the pressure kept constant,

$$dQ = nC_p\, dT = n(C_V + R)\, dT = \left(\frac{3}{2} nR + nR\right) dT = \frac{5}{2} nR\, dT,$$

so we need to replace the factor 3/2 in the last problem by 5/2. The rest is the same. Thus the answer now is

$$\Delta S = \frac{5}{2} nR \ln\left(\frac{T_f}{T_i}\right) = \frac{5}{2}(1.00 \text{ mol})\left(8.31 \frac{\text{J}}{\text{mol}\cdot\text{K}}\right) \ln\left(\frac{400\text{ K}}{300\text{ K}}\right) = 5.98 \text{ J/K}.$$

71. The change in entropy in transferring a certain amount of heat Q from a heat reservoir at T_1 to another one at T_2 is $\Delta S = \Delta S_1 + \Delta S_2 = Q(1/T_2 - 1/T_1)$.

(a) $\Delta S = (260 \text{ J})(1/100\text{ K} - 1/400\text{ K}) = 1.95 \text{ J/K}$.

(b) $\Delta S = (260 \text{ J})(1/200\text{ K} - 1/400\text{ K}) = 0.650 \text{ J/K}$.

(c) $\Delta S = (260 \text{ J})(1/300\text{ K} - 1/400\text{ K}) = 0.217 \text{ J/K}$.

(d) $\Delta S = (260 \text{ J})(1/360\text{ K} - 1/400\text{ K}) = 0.072 \text{ J/K}$.

(e) We see that as the temperature difference between the two reservoirs decreases, so does the change in entropy.

72. The Carnot efficiency (Eq. 20-13) depends linearly on T_L so that we can take a derivative

$$\varepsilon = 1 - \frac{T_L}{T_H} \Rightarrow \frac{d\varepsilon}{dT_L} = -\frac{1}{T_H}$$

and quickly get to the result. With $d\varepsilon \to \Delta\varepsilon = 0.100$ and $T_H = 400$ K, we find $dT_L \to \Delta T_L = -40$ K.

73. (a) We use Eq. 20-16. For configuration A

$$W_A = \frac{N!}{(N/2)!(N/2)!} = \frac{50!}{(25!)(25!)} = 1.26 \times 10^{14}.$$

(b) For configuration B

$$W_B = \frac{N!}{(0.6N)!(0.4N)!} = \frac{50!}{[0.6(50)]![0.4(50)]!} = 4.71 \times 10^{13}.$$

(c) Since all microstates are equally probable,

$$f = \frac{W_B}{W_A} = \frac{1265}{3393} \approx 0.37.$$

We use these formulas for $N = 100$. The results are

(d) $W_A = \dfrac{N!}{(N/2)!(N/2)!} = \dfrac{100!}{(50!)(50!)} = 1.01 \times 10^{29}.$

(e) $W_B = \dfrac{N!}{(0.6N)!(0.4N)!} = \dfrac{100!}{[0.6(100)]![0.4(100)]!} = 1.37 \times 10^{28}.$

(f) and f $W_B/W_A \approx 0.14$.

Similarly, using the same formulas for $N = 200$, we obtain

(g) $W_A = 9.05 \times 10^{58}$,

(h) $W_B = 1.64 \times 10^{57}$,

(i) and $f = 0.018$.

(j) We see from the calculation above that f decreases as N increases, as expected.

74. (a) From Eq. 20-1, we infer $Q = \int T\, dS$, which corresponds to the "area under the curve" in a T-S diagram. Thus, since the area of a rectangle is (height)×(width), we have $Q_{1 \to 2} = (350)(2.00) = 700\text{J}$.

(b) With no "area under the curve" for process $2 \to 3$, we conclude $Q_{2 \to 3} = 0$.

(c) For the cycle, the (net) heat should be the "area inside the figure," so using the fact that the area of a triangle is ½ (base) × (height), we find

$$Q_{net} = \frac{1}{2}(2.00)(50) = 50 \text{ J} .$$

(d) Since we are dealing with an ideal gas (so that $\Delta E_{int} = 0$ in an isothermal process), then

$$W_{1 \to 2} = Q_{1 \to 2} = 700 \text{ J} .$$

(e) Using Eq. 19-14 for the isothermal work, we have

$$W_{1 \to 2} = nRT \ln \frac{V_2}{V_1} .$$

where $T = 350$ K. Thus, if $V_1 = 0.200$ m³, then we obtain

$$V_2 = V_1 \exp(W/nRT) = (0.200) e^{0.12} = 0.226 \text{ m}^3 .$$

(f) Process $2 \to 3$ is adiabatic; Eq. 19-56 applies with $\gamma = 5/3$ (since only translational degrees of freedom are relevant, here).

$$T_2 V_2^{\gamma - 1} = T_3 V_3^{\gamma - 1}$$

This yields $V_3 = 0.284$ m³.

(g) As remarked in part (d), $\Delta E_{int} = 0$ for process $1 \to 2$.

(h) We find the change in internal energy from Eq. 19-45 (with $C_V = \frac{3}{2} R$):

$$\Delta E_{int} = nC_V (T_3 - T_2) = -1.25 \times 10^3 \text{ J} .$$

(i) Clearly, the net change of internal energy for the entire cycle is zero. This feature of a closed cycle is as true for a T-S diagram as for a p-V diagram.

(j) For the adiabatic ($2 \to 3$) process, we have $W = -\Delta E_{int}$. Therefore, $W = 1.25 \times 10^3$ J. Its positive value indicates an expansion.

75. Since the inventor's claim implies that less heat (typically from burning fuel) is needed to operate his engine than, say, a Carnot engine (for the same magnitude of net work), then $Q_H' < Q_H$ (See Fig. 20-35(a)) which implies that the Carnot (ideal refrigerator) unit is delivering more heat to the high temperature reservoir than engine X draws from it. This (using also energy conservation) immediately implies Fig. 20-35(b) which violates the second law.